T0188951

Wireless Communication Electronics

Robert Sobot

Wireless Communication Electronics

Introduction to RF Circuits and Design Techniques

Second Edition

Robert Sobot
École Nationale Supérieure de l'Électronique
et de ses Applications
Cergy-Pontoise, France

ISBN 978-3-030-48632-7 ISBN 978-3-030-48630-3 (eBook)
https://doi.org/10.1007/978-3-030-48630-3

This Springer imprint is published by the registered company Springer Nature Switzerland AG
The registered company address is: Gewerbestrasse 11, 6330 Cham, Switzerland

To Allen

Preface

Preface to the Second Edition

This textbook is the updated, significantly modified, and reorganized version of the first edition. Some material was removed, some added, all in the attempt to produce a more focused book with the explanations better suited for the students who are entering this topic.

In this edition's structure, chapter after chapter, the course lectures, examples, case studies, and problems follow the design of AM RF receiver circuit that my Canadian students were studying and designing as the laboratory requirement of the course. Thus, these textbook example circuits have been demonstrated in the practical part of the course. For pedagogical reasons, the complexity and challenges of the receiver's subcircuits are adjusted so that they are appropriate for senior undergraduate students who are entering RF circuit design for the first time. The assumed prerequisites for this course are the first course in circuit theory and analogue electronics.

The increased volume of this textbook prohibits me to give more space to "drilling" problems that are required for practicing the presented ideas and concepts. Instead, in the second edition of my "By Example" exercise book, I give a number of solved tutorial type exercises and problems that are needed to further develop the design skills required for this subject.

Cergy-Pontoise, Île-de-France
March 8, 2019

Robert Sobot

Preface to the First Edition

This textbook originated in my lecture notes for the "Communication Electronics I" undergraduate course that I have offered over the last 6 years to the students at The University of Western Ontario in London, Ontario, Canada. The book covers the transitional area between low frequency and high frequency wireless circuits. Specifically, it introduces the fundamental physical principles related to the operation of a typical wireless radio communication system.

By no means have I attempted to touch upon all the possible topics related to wireless transmission systems. Most modern textbooks cover a large number of topics with a relatively low level of details, which are usually left as an "exercise to the reader". In this textbook I have chosen to discuss the subject in more depth, and thus provide detailed mathematical derivations, applied approximations, and analogies. The chosen topics are, in my experience, suitable for a one semester, 4 h per week, senior undergraduate engineering course. My intent was to tell a logical story that flows smoothly from one chapter to the next, hoping that the reader will find it easy to follow.

My main inspiration in writing this book came from my students, who at the beginning of the semester would always ask: "What do I need to study for this course?" Having a choice between writing a textbook that covers many topics at a high level or the one that covers fewer fundamental principles but in more detail, I chose the latter. All of the material in this book is considered the basic knowledge that is expected to have been acquired by aspiring engineers entering the field of wireless communication electronics.

Therefore, the intended audience for this book are, primarily, senior undergraduate engineering students preparing for their carriers in communication electronics. At the same time, my hope is that graduate engineering students will find this book a useful reference for some of the topics that have been only touched upon in the previous stages of their education or are explained from a different point of view. Finally, the practicing junior RF engineers may find this book a handy source for the quick answers that are routinely omitted from most textbooks.

London, ON, Canada
August 12, 2011

Robert Sobot

Acknowledgements

I would like to acknowledge all those wonderful books that I used as the source of my knowledge and to say thank you to their authors for providing me with the insights that otherwise I would not have been able to acquire. Under their influence, I was able to expand my own horizons, which is what acquiring of the knowledge is all about. Hence I do want to acknowledge their contributions that are clearly visible throughout this book, which are now being passed on to my readers.

In professional life, one learns both from written sources and from experience. The experience comes from the interaction with people that we meet and projects that we work on. I am grateful to my former colleagues who I was fortunate to have as my technical mentors on really inspirational projects, first at the Institute of Microelectronic Technologies and Single Crystals, University of Belgrade, former Yugoslavia, then at PMC-Sierra Burnaby, BC, Canada, where I gained most of my experiences of the real engineering world.

I would like to acknowledge the contributions of my colleagues at Western University in Canada, specifically of Professor John MacDougall, who initialized and restructured the course into the form of "design and build" and of Professors Alan Webster, Zine Eddine Abid, and Serguei Primak who taught the course at various times.

I would like to thank all of my former and current students who relentlessly keep asking "Why?" and "How did you get this?" I hope that the material compiled in this book contains answers to at least some of those questions and that it will encourage them to keep asking questions with unconstrained curiosity about all the phenomena that surround us.

Sincere gratitude goes to my publisher and editors for their support and making this book possible.

Most of all, I want to thank my son for patiently growing up along with this book, for hanging around my desk, asking questions, and for making me laugh.

Contents

Abbreviations

AC	Alternating current
A/D	Analogue to digital
ADC	Analogue to digital converter
AF	Audio frequency
AFC	Automatic frequency control
AM	Amplitude modulation
BJT	Bipolar junction transistor
BW	Bandwidth
CMOS	Complementary metal-oxide semiconductor
CRTC	Canadian Radio-Television and Telecommunications Commission
CW	Continuous wave
D/A	Digital to analogue
DAC	Digital to analogue converter
dB	Decibel
dBm	Decibel with respect to 1 mW
DC	Direct current
DR	Dynamic range
ELF	Extremely low frequency
EM	Electromagnetic
eV	Electron volts
FCC	Federal Communications Commission
FET	Field-effect transistor
FFT	Fast Fourier transform
FM	Frequency modulation
GHz	Gigahertz
HF	High frequency
Hz	Hertz
IC	Integrated circuit
$\Im(z)$	Imaginary part of a complex number z
IF	Intermediate frequency
I/O	Input–output
IR	Infrared
JFET	Junction field-effect transistor
KVL	Kirchhoff's voltage law
KCL	Kirchhoff's current law
LC	Inductive–capacitive
LF	Low frequency

LNA	Low-noise amplifier
LO	Local oscillator
MOS	Metal-oxide semiconductor
NF	Noise figure
PCB	Printed circuit board
PLL	Phase-locked loop
PM	Phase modulation
pp	Peak-to-peak
ppm	Parts per million
pwl	Piecewise linear
Q	Quality factor
$\Re(z)$	Real part of a complex number z
RADAR	RAdio Detection and Ranging
RF	Radio frequency
RFC	RF choke
RMS	Root mean square
SAW	Surface acoustic wave
SHF	Super high frequency
SINAD	Signal-to-noise plus distortion
S/N	Signal to noise
SNR	Signal-to-noise ratio
SPICE	Simulation program with integrated circuit emphasis
TC	Temperature coefficient
THD	Total harmonic distortion
UHF	Ultra high frequency
UV	Ultraviolet
V_T	Thermal voltage, $V_T = kT/q$
V_t	PN junction threshold voltage
VCO	Voltage-controlled oscillator
V/F	Voltage to frequency
VHF	Very high frequency
V/I	Voltage/current
VLF	Very low frequency
VSWR	Voltage standing wave ratio

Part I

Basic Concepts and Definitions

Introduction

1

Wireless transmission of information over vast distances is one of the finest examples of Clarke's third law, which states that "any sufficiently advanced technology is indistinguishable from magic". Even though a radio represents one of the most ingenious achievements of humankind and is now taken for granted; for the majority of the modern human population (including some of its highly educated members), this phenomenon still appears to be magical. This chapter introduces fundamental concepts and definitions in physics, mathematics and engineering with the intention of preparing you, the reader, for the material that follows; in return, that material is expected to reduce, if not completely remove, the "magic" part of the subject.

1.1 Concept of Energy and Information

All modern engineering disciplines are derived from the fundamental physical (and somewhat philosophical) concepts, most importantly energy, matter, space, and time. Thus, while developing engineering theories and practical design techniques these concepts are accepted as already defined.

Although the word *energy*, which derives from Greek $\epsilon\nu\epsilon\rho\gamma\iota\alpha$ (energeia), was used by Aristotle way back in the fourth century BC, it is still one of the most ambiguous concepts in science. Twenty-four centuries later, this topic was addressed by R. Feynman in his famous *Lectures on Physics* where he said:

> There is a fact, or if you wish, a *law*, governing all natural phenomena that are known to date. There is no known exception to this law—it is exact so far as we know. The law is called the *conservation of energy*. It states that there is a certain quantity, which we call energy, that does not change in manifold changes which nature undergoes. That is a most abstract idea, because it is a mathematical principle; it says that there is a numerical quantity which does not change when something happens. It is not a description of a mechanism, or anything concrete; it is just a strange fact that we can calculate some number and when we finish watching nature go through her tricks and calculate the number again, it is the same.

The concept of energy was famously united with the concept of *matter* by Einstein through his $E = mc^2$ mass–energy equivalence equation.[1] The arena needed to describe the interactions of energy and matter is then set by introducing a medium called *space*. In order to keep these interactions

[1] Strict relativistic version of this equation is $E_r = \sqrt{(m_0c^2)^2 + (pc)^2}$. This equation reduces to $E = mc^2$ when the momentum term (pc) is zero. For photons where $m_0 = 0$, the equation reduces to $E_r = pc$.

R. Sobot, *Wireless Communication Electronics*,
https://doi.org/10.1007/978-3-030-48630-3_1

"catalogued", i.e. to be able to tell them apart, the last fundamental concept, the concept of *time*, had to be introduced. With this set of fundamental physical concepts: energy, space, and time, science is able to develop detailed models that can correctly describe present state and predict the future behaviour of many of the phenomena in this world.

For the purpose of our discussion, we may accept a rather vague definition of energy as "the ability to do work", while the work itself is defined in terms of both time and space. Hence, the process of transmitting (i.e. doing the work of carrying) a bit of information is equivalent to the process of moving a packet of energy from point A to point B in space and time, which brings us back to the main topic of this book. We refer to these streams of energy as "messages" or "signals" originating at the *transmitting* side and terminating at the *receiving* side, with variations that are observed in *time*. Note that this broad definition does not favour any particular physical form of the signal, it does not matter whether these signals take the form of smoke clouds rising in the sky, a message in a bottle, sound caused by a distant thunderstorm, digital bits of data travelling from one computer to another through the network, or the light arriving to Earth from a star faraway. As long as the message has any meaning to the receiver, we say that signal transmission has taken place.

Example 1: Mass–Energy Equivalence Equation
An average sized snowflake consists of approximately $n = 6.68559 \times 10^{19}$ molecules. Assuming the complete matter of the snowflake is converted into energy, estimate for how long a laptop computer whose average power consumption is $P = 25\text{W}$ could be powered?

Solution 1:
Einstein discovered the mass–energy equivalence concept that is formulated as

$$E = \sqrt{(mc^2)^2 + (pc)^2} \tag{1.1}$$

where m is the mass of object whose equivalent energy is calculated, c is the velocity of light in vacuum, and p is the momentum of the mass m in motion. However, if the mass is not moving, then $p = 0$ and (1.1) reduces into its simple form

$$E = mc^2 \tag{1.2}$$

A snowflake consists of water molecules (H_2O), thus, first we find its total mass (m_S) by adding masses of all n individual molecules that constitute the snowflake. Each water molecule consists of two hydrogen atoms (atomic weight $H = 1.00794\,\text{g/mol}$) and one atom of oxygen (atomic weight $O = 15.9994\,\text{g/mol}$). Thus, a single water molecule has atomic weight of:

$$\text{H}_2\text{O} = 2 \times H + O = 18.01528\,\text{g/mol} \tag{1.3}$$

or, if the mass is expressed in g, then

$$\text{H}_2\text{O} = \frac{\text{H}_2\text{O}}{N_A} = \frac{18.01528\,\text{g/mol}}{6.0221415 \times 10^{23}\,\text{g/mol}} = 2.99151 \times 10^{-23}\,\text{g} \tag{1.4}$$

then the complete snowflake has mass m_S of

(continued)

$$m_S = n \times H_2O = 6.68559 \times 10^{19} \times 2.99151 \times 10^{-23}\,\text{g}$$
$$= 2\,\text{mg} = 2 \times 10^{-6}\,\text{kg} \tag{1.5}$$

Velocity of the snowflake is negligible relative to the speed of light, thus we calculate the equivalent energy as:

$$E = m\,c^2 = 2 \times 10^{-6}\,\text{kg} \times 2.99792458 \times 10^{8}\,\text{m/s} = 1.79751036 \times 10^{11}\,\text{J} \tag{1.6}$$

Power is defined as the energy transfer rate $E = P/t$, therefore in order to provide the average power of $P = 25\text{W}$ the total energy E must be distributed over the following time

$$t = \frac{E}{P} = \frac{1.79751036 \times 10^{11}\,\text{J}}{25\,\text{W}} = 7.190 \times 10^{9}\,\text{s} \approx 228\,\text{years} \tag{1.7}$$

We conclude that if we were able to completely convert, for instance, only 2mg of any matter (including a snowflake) into energy we would be able to provide power to our hand-held electronic equipment for many years. However, until we develop such a source of energy we must obey the limits imposed by the capacity of our modern batteries and design our circuits accordingly.

1.2 Wireless Transmission of Signals

Strictly speaking, wireless transmission of signals, i.e. transmission of signals between two points in space without any visible physical connection between them, has been available to us since the dawn of humanity. Most of us communicate with other people using our voice without additional special equipment. Our vocal cords and hearing system create a wonderful wireless communication system; engineering efforts merely represent attempts to increase its range.

In the most general sense, a transmission (communication) system consists of: (a) a transmitter; (b) a transmitting medium; and (c) a receiver (see Fig. 1.1), existing for the sole purpose of moving a message between the transmitter and the receiver. That is to say, the vocal cord–ear system is called a "transceiver" because it is capable of both receiving and transmitting a signal, in this case transmitted in the form of sound, while the air between the transmitter and the receiver serves as the transmitting medium. To complete the system, there must be mutual agreement put in place regarding

Fig. 1.1 A wireless system consisting of a transmitter (vocal cords), transmitting media (air, in this case), and the receiver (hearing system)

the *message encoding*, in this case the choice of spoken language that both receiving and transmitting side understand.

Our bodies are also capable of receiving signals encoded in the form of light, by means of our visual cortex. In this case, only the receiving channel is available to us; for a message encoded in light, the human body is only a receiver—it cannot produce "light rays". Strictly speaking our body emits the infrared radiation (IR), however, it is not really an encoded message—it merely reveals the existence of the transmitting source.

Humans have always needed to extend the distance over which messages can travel, which has resulted in the development of various communication systems. For example, carrier pigeons, writing systems, telegraph, radio, television, satellite systems, and cellphones all serve the same purpose of extending the distance that a message created by a person can travel in time and space. As an example, the message contained in this book will be received by readers who are widely spread in both time and space.

1.2.1 A Short History of Wireless Technology

In modern terminology, it is assumed that the term "wireless communication" refers to an electronic system for transmitting messages that consists of an electronic transmitter, an electronic receiver, and radio waves. While most of us have a vague idea what a radio wave is, it is not that simple to describe it in plain words. For the time being, we accept that the term "wave" symbolizes the *flow of energy*.

In the nineteenth century, interest in the phenomenon of electricity, magnetism, and light was at its height. A number of scientists worked on related problems and a long series of studies culminated in Maxwell's equations of the electromagnetic (EM) field, first published in 1865, which describe electricity, magnetism, and light in one uniform system. Consequently, all major laws in electrical engineering can be derived from his equations. In the May 24, 1940 issue of Science, Einstein said:

> The precise formulation of the time–space laws was the work of Maxwell. Imagine his feelings when the differential equations he had formulated proved to him that electromagnetic fields spread in the form of polarized waves, and at the speed of light! To few men in the world has such an experience been vouchsafed ... it took physicists some decades to grasp the full significance of Maxwell's discovery, so bold was the leap that his genius forced upon the conceptions of his fellow-workers.

It goes without saying that studying Maxwell's equations and their derivatives is of the highest importance for electrical engineers.

In 1887, Hertz ventured to prove the theory of electromagnetism experimentally, eventually performing his famous "spark experiment" that proved the existence of radio waves, as predicted by Maxwell. His simple experimental setup consisted of a coil and two copper plates with spherical probes connected to a battery. Each time it was turned on and off, this structure would create a spark jumping across the small gap between the spherical probes. A short distance away, there was another copper ring with a short gap between two small spherical terminals. Each time the spark was created in the main apparatus, Hertz noticed a spark in the other copper ring. Wireless transmission was born. As often happens, Hertz did not realize the full practical implications of his discovery; he stated:[2]

> It's of no use whatsoever[...] this is just an experiment that proves Maestro Maxwell was right – we just have these mysterious electromagnetic waves that we cannot see with the naked eye. But they are there.

The same year, Tesla, who for most of his life was obsessed with the wireless transfer of energy, was granted a patent on a rotating magnetic field, originally conceived in 1882. By 1891, Tesla

[2] *Annalen der Physik*, vol. 270, no. 7, p. 551–569, May, 1888.

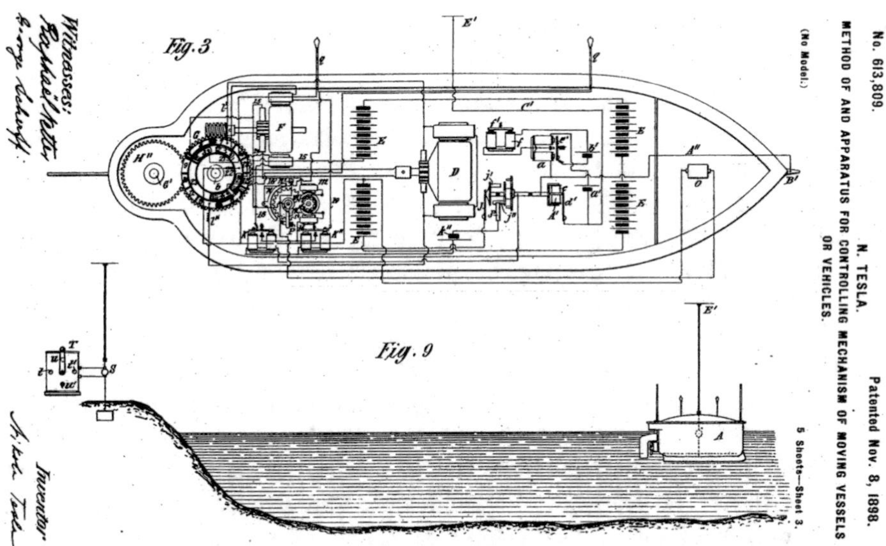

Fig. 1.2 Patent of Nikola Tesla's remotely controlled boat using radio waves, first demonstrated in 1897 in the Hudson river, New York

had invented the "Tesla Coil", a type of resonant circuit that can produce high-voltage (HV), high-frequency (HF) alternating currents (AC), which he proposed could be used for "telecommunication of information.[3]" In 1897, Tesla demonstrated the first radio communication system, which he used to control a model boat with his wireless transmitter and receiver (an inductively coupled system),[4] which started the era of practical wireless communications (see Fig. 1.2). On March 20, 1900 Tesla was issued a patent on the radio transmission of electrical energy.[5]

If Tesla is considered the father of practical wireless communications, Marconi should be considered the father of commercial radio communications. In 1901, he demonstrated the first wireless communication system for transmitting Morse-coded messages across the Atlantic. His demonstrations set in motion the wide use of radio for wireless communications, especially with ships (the Titanic disaster also helped the cause). He established the first transatlantic radio service and built the first commercial stations for the British short wave service. It is also recorded in history that Tesla was not pleased with the attention Marconi was getting while using Tesla's patented technology. Nevertheless, it took until 1943 for the US Supreme Court to invalidate Marconi's patents in favour of Tesla, stating:[6]

> The Tesla patent No. 645,576, applied for 2 September 1897 and allowed 20 March 1900, disclosed a four-circuit system, having two circuits each at transmitter and receiver, and recommended that all four circuits be tuned to the same frequency. [...He] recognized that his apparatus could, without change, be used for wireless communication, which is dependent upon the transmission of electrical energy. [...]
>
> Marconi's reputation as the man who first achieved successful radio transmission rests on his original patent, which became reissue No. 11,913, and which is not here [320 U.S. 1, 38] in question. That reputation, however well-deserved, does not entitle him to a patent for every later improvement which he claims in the radio field.

[3]Indeed, as late as the 1920s, Tesla coils were used in commercial radio transmitters.

[4]US Patent 613,809, November 8, 1898.

[5]US Patent 645,576, Applied for on September 3, 1897.

[6]U.S. Supreme Court, "Marconi Wireless Telegraph co. of America v. United States". 320 U.S. 1. Nos. 369, 373, April 9–12, 1943.

> Patent cases, like others, must be decided not by weighing the reputations of the litigants, but by careful study of the merits of their respective contentions and proofs.

Feud stories like this one repeatedly happen throughout history; bitter rivalry and disputes over inventions are not exceptions, rather they are the rule. In another example, even though Bell was the first to receive a patent for the invention of the telephone in 1876, several other scientists demonstrated working prototypes as early as 1857 when Meucci developed a voice communication apparatus but, apparently, did not have enough money for the full patent fee. He was granted a *caveat* (i.e. a provisional patent) in 1871, which expired in 1874, leaving an opening for Bell's patent.[7]

The most basic wireless data transmission is possible simply by repeating Hertz's experiment many times, i.e. switching on and off the transmitting coil. Morse was the first to formalize a "time sharing" scheme for encoding a message, known as "Morse code". Transmitting Morse code wirelessly requires only a simple tuned circuit being constantly turned on and off. However, it was not possible to transmit voice messages until 1904 when Fleming invented the thermionic valve (i.e. vacuum tube). This thermal device (which functions as a diode) was the key element needed for radio communication systems. Two years later, the addition of a third terminal was a natural development leading to the invention of a triode (a vacuum tube that functions as an amplifying element) (see Fig. 1.3). Again, Fleming argued bitterly with De Forest about ownership of these ideas. At the same time, Armstrong (still an undergraduate student) used the triode to create a "regenerative circuit" topology and patented it in 1914.[8] It should be remembered that virtually all modern radio equipment, including the radio receiver topology studied in this book, trace their history back to this "heterodyne" topology (later expanded into "superheterodyne").

Although use of the term "radio" may imply the exclusion of television, that is not the case. The television should be looked at as no more than a sophisticated radio. To be historically correct, it has to be stated that television was invented by the many scientists and engineers who made incremental contributions while radio and television systems were being developed in parallel. It is worth mentioning that the first patent for an electro–mechanical television system was granted to Nipkow, a university student, back in 1884.[9] In 1925, Baird demonstrated a system that paved the way to the first practical use of television in 1929, when regular television broadcasts started in Germany.

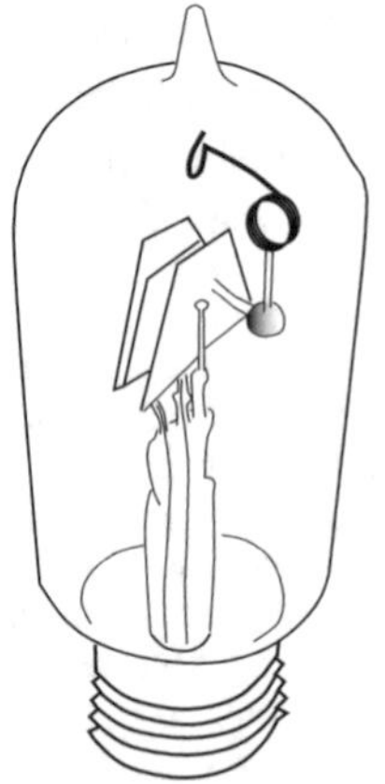

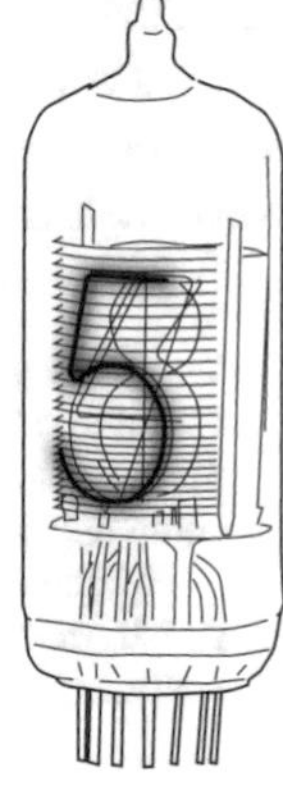

Fig. 1.3 The first electronic valve (*left*), designed by Fleming in 1904, and an alphanumeric valve used in electronic equipment to display numbers and letters (*right*) before modern semiconductor displays were invented

[7] https://en.wikipedia.org/wiki/Antonio_Meucci, archived December 19, 2019.

[8] US Patent 1,113,149, October 6, 1914.

[9] https://en.wikipedia.org/wiki/Paul_Gottlieb_Nipkow, archived December 19, 2019.

After the groundbreaking work on radio transmission in the early twentieth century, it is safe to say that there have been no new fundamental advances ever since. Incremental advances can be credited only to engineering ingenuity and technological improvements, most notably the invention of the transistor in 1948 (the three scientists who invented it received a Nobel Prize but never talked to each other again) and the integrated circuit (IC) in 1958 (Kilby, a newly employed engineer at Texas Instruments who did not yet have the right to a vacation, spent his summer working on making this concept practical—it earned him a Nobel Prize and a place in history[10]).

To conclude this short historical review, the importance of radio development is such that several engineers and scientists who have made major contributions also earned a Nobel Prize. They have also served as inspiration for the generations of engineers who have followed in their footsteps.

1.2.2 Nature of Waves

As hinted at in the previous sections, our understanding of real nature of "waves" is more intuitive than exact. Dropping a rock into a pond creates circular ripples that expand both in space and time (which we can visualize, as in Fig. 1.4); a "wave" of spectators can travel around a packed stadium at a soccer game (when each spectator stands up and sits down at the right moment). These familiar examples of waves are perceived by our vision.

We are also accustomed to talking about sound waves because we can detect them with our hearing system. It is a bit more complicated to envision sound waves in our mind because we would need to "see" air pressure regions that change from low pressure to high pressure and back. The situation becomes even more difficult if we try to envision light waves. Attempts to explain the nature of light waves led scientists into the development of theories of relativity and quantum mechanics, and touched the deepest philosophical questions of human existence.

At the fundamental level, we can easily accept that water in a pond carries the water waves; we can also accept that sound waves are carried by air; the natural question is to wonder what carries light waves. After all, light waves come from outer space. Is the space empty? What are the waves? When passing through airport security, how does the machine knows whether we are carrying metallic objects without touching us? How does MRI equipment, which always stays outside the body, make detailed pictures of the inside?

To answer these kind of questions and create a meaningful model that correctly describes the observed reality, Faraday introduced the concept of a *field*. This abstract concept, expanded by Maxwell and many others, underlines many of the little mysteries we encounter in everyday life. Although it has proved very useful, being a very abstract concept the concept of field still does not answer the fundamental question of *what* waves are. Nevertheless, it does help us visualize

Fig. 1.4 Water ripples created in a pond by a small rock. Although the water particles move vertically, the wave expands horizontally while carrying away the kinetic energy of the falling rock

[10]https://en.wikipedia.org/wiki/Jack_Kilby, archived December 19, 2019.

something that otherwise would have been beyond the reach of our senses. For example, the existence of magnetic "field lines" is easily demonstrated by the elementary school experiment when iron filings were sprinkled on paper above a bar magnet (as in Fig. 1.5 (left)). It should be noted, however, that the field lines *do not really exist*. Instead, each point of 3D space surrounding the magnet could be associated with a *field vector* that quantifies the filed's strength and direction at the given point (as in Fig. 1.5 (right)).

Visualizing the magnetic field certainly helps us to imagine other fields, especially the EM field that was introduced by Maxwell to explain the wave nature of the light. According to Maxwell's equations, a spatially varying electric field generates a time-varying magnetic field and vice versa. Starting, for instance, with a time-varying electric field, magnetic and electric fields are successively generated indefinitely. As an oscillating electric field generates an oscillating magnetic field, the magnetic field in turn generates an oscillating electric field, and so on. These time-varying fields together form an EM wave that propagates in space (see Fig. 1.6). A less obvious observation is that once the EM wave is established, its source can be removed without further influencing the already existing wave. In free space, an EM wave propagating in the z direction is described as:

$$E_x = E_{0x} \sin(\omega t - \beta z) \tag{1.8}$$

$$H_y = H_{0y} \sin(\omega t - \beta z) \tag{1.9}$$

where E_x is x-directed electrical field vector, E_{0x} is its maximum amplitude in [V/m], H_y is y-directed magnetic field vector (which is orthogonal to both the electrical filed vector **x** and the wave propagation vector **z**), H_{0y} is its maximum amplitude in [A/m], ω is radial frequency in [rad/s], and β is the *propagation constant* defined as

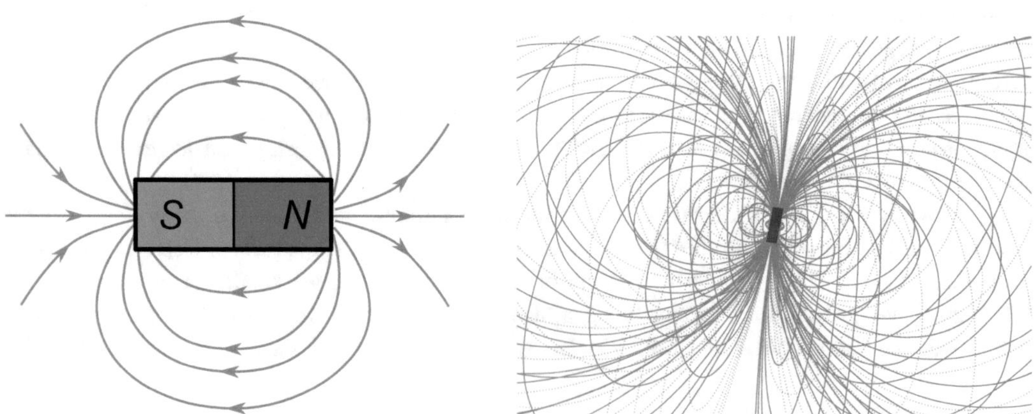

Fig. 1.5 The direction of magnetic field lines represented by the alignment of iron filings sprinkled on paper placed above a bar magnet. The mutual attraction of opposite poles of the iron filings results in the formation of elongated clusters of filings along "field lines" (left), a 3D vector field representation (right)

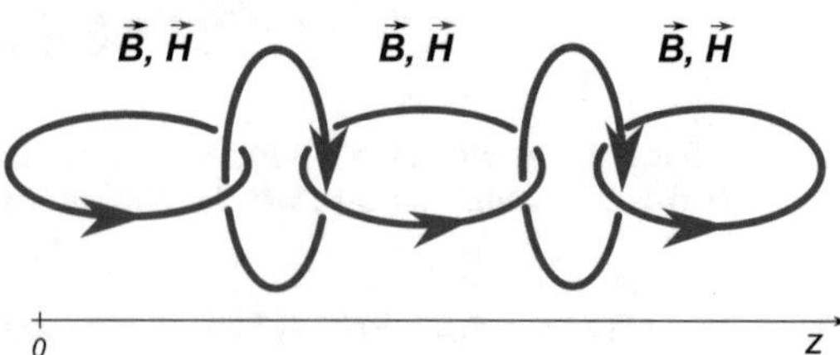

Fig. 1.6 An electromagnetic wave may be imagined as a self-propagating transverse oscillating wave of electric and magnetic fields

$$\beta = \frac{2\pi}{\lambda} \tag{1.10}$$

where λ is the wavelength, which is defined in Sect. 1.3.3. Expanding on the propagation constant we define *phase velocity* v_p as

$$v_p = \frac{\omega}{\beta} \tag{1.11}$$

which gives us information about how fast the wave phase propagates in space. A way to visualize phase velocity is by focusing on one single point on the wave (for example, on the crest) and follow it in space.

Motion of electric charges is, by definition, *electric current*. Electric current creates moving magnetic field, which in return creates moving electric filed. Once the process is started, the initial source of this moving EM filed (i.e. the moving electric charge) can be removed; the EM field keeps moving in space in self-perpetual motion. By experiment[11] the EM wave speed, i.e. its phase velocity, c_0 was found to be the same as the speed of light in a vacuum:

$$c_0 = \frac{1}{\sqrt{\mu_0\,\epsilon_0 0}} = 299\,792\,458\ \text{m/s} \approx 3 \times 10^8\ \text{m/s} \tag{1.12}$$

where μ_0 is the magnetic constant (or, vacuum permeability) and ϵ_0 is the electric constant (or, vacuum permittivity). Maxwell concluded that EM waves (i.e. radio waves) and light are fundamentally the same thing, hence Maxwell's equations deal with this moving EM wave and the relationship between electric and magnetic fields. It was also established that a given light frequency f_0 stays constant in various transmission media (e.g. vacuum, air, water...), which means that speed and wavelength are reduced relative to their values in vacuum, where the reduction factor n is known as the "refractive index" or "index of refraction"

$$n = \frac{c}{v} = \frac{\lambda_0 \cancel{f_0}}{\lambda_1 \cancel{f_0}} \tag{1.13}$$

where c is the speed of light in a vacuum, v is its phase velocity in the given media, λ_0 is its wavelength in a vacuum, and λ_1 is its wavelength in the media.

Going back to the ripples in the pond (Fig. 1.4) and dropping a cork into the water, we can see that the cork moves only in the *vertical* direction, indicating that the water particles do not move *away* from the centre of the ripples. That is to say, it is not the particles of matter that propagate through space in the z direction but the wave carrying the energy of the disturbed particles while they vibrate around their nominal positions (in the x or y directions) in synchronicity with their neighbours. These repetitive "up" and "down" vibrations are usually referred to as "oscillations".

Using experimental methods, wave propagation speeds for sound and light waves through various materials were established. For instance, it was established that a sound wave travels at a speed of 343m/s through dry air at 20°C, or approximately one kilometre in three seconds. Similarly, the speed of a light wave in a vacuum was established as $c = 299\,792\,458$ m/s, which is often rounded to 300 000 km/s. As an illustration, at this speed, it takes sunlight 8min and 19s to reach Earth.

[11] ϵ_0 was originally measured based on Coulomb's law in vacuum.

Example 2: Measuring the Speed of Light

(1) By timing the eclipses of the Jupiter moon Io, in 1676 Rømer estimated that light would take about 22min to travel a distance equal to the diameter of Earth's orbit around the Sun, which was known to be approximately $d = 2.98 \times 10^{11}$m. What was his estimated speed of light?

(2) In the series of experiments, in 1922–1924 Michelson established that speed of light is $c = 299\,796$ km/s by using a sophisticated setup of mirrors placed at 35km distance from each other. How much time it takes light to travel that distance?

Solution 2: The velocity is the ratio of distance and time, thus

(1)

$$v = \frac{d}{t} = \frac{2.98 \times 10^{11}\text{m}}{22 \times 60\text{s}} = 2.26 \times 10^{8}\,\text{m/s}$$

(2)

$$t = \frac{d}{v} = \frac{35 \times 10^{3}\text{m}}{299\,796 \times 10^{3}\,\text{m/s}} = 117\,\mu\text{s}$$

Example 3: Speeds of Light and Sound

Estimate the distance of lightning if approximately nine seconds pass between the time the lightning flash is registered and the time the thunder is heard.

Solution 3: The speed of sound is negligible relative to the speed of light; thus, in 9s the sound travels approximately 3km. Therefore, we can ignore the delay of light travelling 3km (which is about 10μs) and just estimate that the lightning happened at approximately 3km away.

1.2.2.1 Maxwell Equations Approximation

Existence of the interleaved self-perpetuating magnetic and electric fields (Fig. 1.6) is fundamental to the propagation of EM waves and, therefore, to wireless communication systems. Their relationship is described by the set of Maxwell's equations. Because there is a big gap in the required theoretical background, complexity and design methods between linear low-frequency circuits and, for instance, the millimeter wave RF circuits (see Fig. 1.7), being an introductory level RF circuits book, focus is only on relatively low-frequency non-linear RF circuits. By focusing on low-frequency RF circuits, we are able to use most of the methods acquired from previous linear LF electronics courses and to apply approximate methods derived from Maxwell's equations under condition of low frequencies. After the low-frequency approximation is applied, i.e. $d \ll \lambda$ (that is, the transmission distance is much shorter than the signal wavelength), Kirchhoff's current law (KCL) and Kirchhoff's voltage law (KVL) equations are used instead of the full set of Maxwell's equations (see Fig. 1.8). Thus for detailed study of exact Maxwell's equations the reader is directed to another courses in electromagnetic.

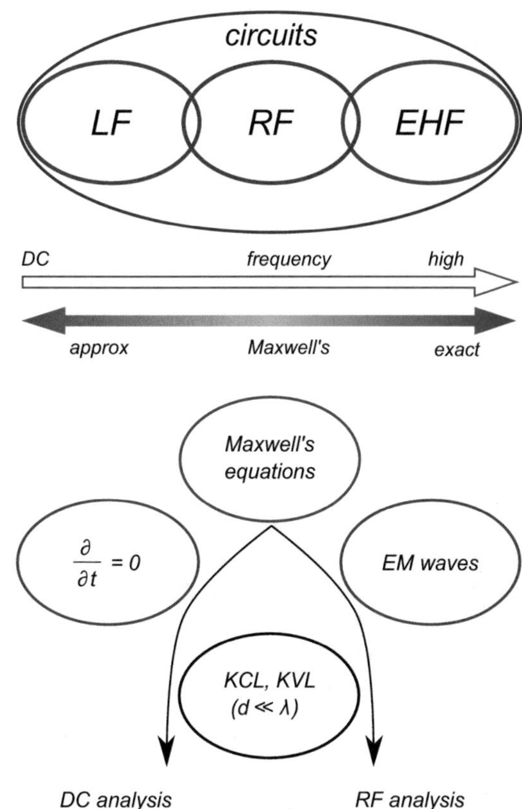

Fig. 1.7 The relationship between EHF, RF, and LF design methods with respect to exact and approximated Maxwell's equations

Fig. 1.8 The role of Maxwell's equations relative to electrical circuit analysis

1.3 Waveform Definitions

Following the qualitative introduction of waves in the previous section, we now introduce a set of specific characteristics to help us quantify general wave function properties. We focus on a single sinusoidal functions because, in accordance with the Fourier transform, any general complicated wave can be represented mathematically as the sum of one or more sinusoidal functions. A vertical cross-section of water ripples, an instantaneous picture of a piano string producing a single note, and the time domain plot of a voltage signal recorded at the terminals of an electrical resonator all resemble the familiar shape of the sinusoid. In analogy with the sounds of a single note, these single sinusoidal functions are referred to as "single tones" or simply just "tones" (even though we cannot really hear them in their original form).

1.3.1 Amplitude

Exploiting further the analogy of a sound wave created by a piano string playing a single note (for example, A), the wave amplitude is manifested by the volume of the tone. The harder the string is struck, the more violently it vibrates (i.e. the greater the displacement) or, equivalently, the greater is the amplitude of the sound wave. Figure 1.9 shows the amplitude change in time for two independent sinusoidal waves (A_1 and A_2).

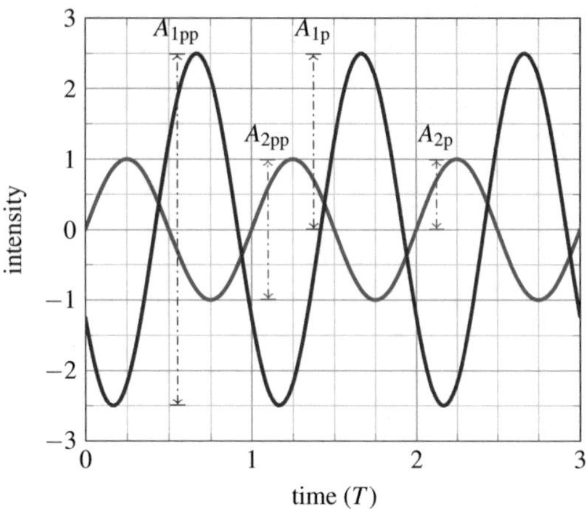

Fig. 1.9 A loud sound is symbolized by a large amplitude (A_1), while a relatively weak sound is quantified by a small amplitude (A_2)

Wave amplitude is quantified in two ways. It can be measured in the positive direction from the zero point (that is, the average value) to the wave's maximum on the vertical scale (displacement), for example, amplitudes A_1 and A_2 in Fig. 1.9. It can also be measured by the distance between the wave's extreme vertical points, for example, amplitudes A_{1pp} and A_{2pp} in Fig. 1.9, where the index "pp" is pronounced "peak-to-peak". By definition, the numerical value of the peak-to-peak (pp) amplitude A_{1pp} (or A_{2pp}) is double that of the single-sided (peak) value A_1 (or A_2).

1.3.2 Frequency

Various notes played on a piano, for example `A` and `B`, are perceived by our brains as different pitches. This quality of a sound wave is directly related to the amount of time required by the wave to complete one full pattern, that is, to complete one "period" (measured in seconds). In other words, this is the time required for the string to complete one full movement up, down, and back again along the displacement axis in Fig. 1.10. This particular time is marked as T_1 for the A_1 waveform and as T_2 for the A_2 waveform.

A period T is measured between two adjacent extreme amplitude points or at any other two points on adjacent slopes of the same kind (i.e. either two up-slopes or two down-slopes) that have the same displacement value. A shorter period T implies a greater number of patterns being repeated in a given time—the waveform has *higher frequency*. Frequency is measured in hertz (abbreviated as Hz), where "1Hz" means that one full wave cycle took a second to complete; in other words, the associated period $T = 1$s. In Fig. 1.10, the waveform with T_2 has a frequency four times higher than the waveform with T_1. For example, the middle `C` tone played on a piano has a frequency of 261Hz. The full frequency range[12] of piano tones is 27–3,516Hz. Young people with normal hearing can perceive tones in the range of approximately 20Hz–20kHz. Similarly, human eyes distinguish the various frequencies of light and our brain perceives them as various colours. The visible frequency band for most people is approximately $(400–790) \times 10^{12}$ Hz (i.e. 400–790 THz). This huge bandwidth represents almost unlimited resource for signal transmission. As defined, the period and frequency of a waveform are inversely proportional:

[12]In radio terminology, "range" means distance that the waveform can travel. A "range of frequencies" is referred to as a *"band" or "bandwidth"*.

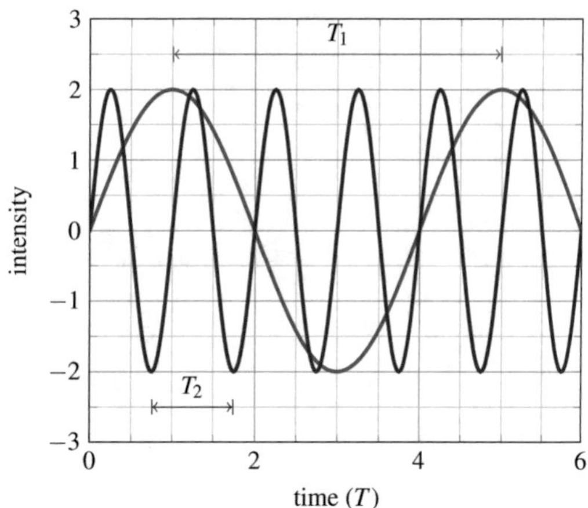

Fig. 1.10 Lower frequency tone (red) has longer period T_1 relative to the higher frequency tone (blue). In this example the A_2 waveform has four times higher frequency than the A_1 waveform

$$f \stackrel{\text{def}}{=} \frac{1}{T} \qquad \text{[Hz]} \tag{1.14}$$

where f is the frequency in [Hz] and T is period in seconds. A practical representation of sinusoidal waveforms is based on a mathematical model known as a *rotating phasor* (see Fig. 1.18). In a geometrical sense, the time to accomplish one pattern is directly mapped onto the time required to accomplish one full rotation around the circle. The usefulness of the model comes from equivalency between one full movement along the `displacement` axis and one full circle rotation of the phasor, which is expressed in angle units as 2π, i.e.

$$\omega \stackrel{\text{def}}{=} 2\pi f = \frac{2\pi}{T} \qquad \left[\frac{\text{rad}}{\text{s}}\right] \tag{1.15}$$

where ω is called the "radian frequency", associated with one full rotation of 2π, thus the distance travelled in one second.[13] The use of radian frequency leads into common analytical form of a sinusoidal function $x(t)$ as

$$x(t) = A_m \sin(\omega t + \phi) \tag{1.16}$$

where A_m is the maximal amplitude, ϕ is the "initial phase" introduced in Sect. 1.3.5. Also, it is common engineering practice to use the term "single tone" (or just "tone") while referring to a wave that, mathematically speaking, consists of a single sinusoidal waveform (1.16) even in cases when the wave is not a sound wave. The term "wave" refers to the conceptual phenomenon; "waveform" refers to a graphical representation of a wave. These terms are often used interchangeably.

[13]Circumference of a unity radius circle is $2\pi \times 1 = 2\pi$.

Example 4: Waveform Definitions
For a voltage disturbance wave travelling at the speed of light and described as

$$v_1(t) = \sin(20\pi \times 10^6\, t)$$

find: (1) its maximum amplitude; (2) its frequency; (3) its period; (4) its phase at time $t = 0$s.

Solution 4: By inspection of the given wave $v_1(t)$ equation and its comparison with the general analytical form of a sine wave given by (1.16), we write

$$v_1(t) = 1\text{V} \times \sin(2\pi \times 10 \times 10^6\, t + 0)$$

thus,

1. being a voltage waveform its maximum value is $A_m = 1$V,
2. by definition (1.15), radial frequency is $\omega = 2\pi\, f = 2\pi \times 10 \times 10^6$Hz, therefore $f = 10 \times 10^6\text{Hz} = 10$MHz,
3. by definition (1.14), period is $T = 1/f = 1/10 \times 10^6\text{Hz} = 100$ns,
4. and, it can be written that $v_1(t) = \sin(\omega\, t + 0)$, thus the initial phase (i.e. at $t = 0$) is $\phi = 0°$

1.3.3 Wavelength

We note that Fig. 1.4 shows a wave frozen in time: after the water wavefront has travelled outwards in space from the point where the rock hit the water to its last position. Again, (ideally) the water particles have only vertical movement, i.e. it is only the displacement (energy) that travels horizontally. From Fig. 1.4, it is possible to measure the horizontal distance between any two wave peaks in space. This *spatial* dimension is denoted as wavelength λ. If, instead of a single frame, the full movie was available to us, then it would be possible to measure the same event in the time domain, namely, the period T for any given particle of water to complete the full up, down, and back again vertical swing. In addition, we realize that the period T is the same time taken by the wavefront to travel distance λ. Figure 1.10 shows the vertical displacement of a *single* wave particle in time and Fig. 1.11 shows the vertical displacement of *all* wave particles in space (measured horizontally from the wave starting point to its end).

As in any other case of linear motion in classical physics, knowing two of the three parameters (i.e. the distance travelled, the time taken for the trip, and the average speed) enables calculation of the third parameter.

Therefore, wavelength is expressed by the equation

$$\lambda = \nu\, T = \frac{\nu\ [\text{m}/\cancel{\text{s}}]}{f\ [1/\cancel{\text{s}}]} \qquad [\text{m}] \tag{1.17}$$

where λ is the wavelength (i.e. horizontal distance travelled by the disturbance while completing one full cycle of the vertical disturbance) in metres, T is the time in seconds needed by the waveform to travel horizontal distance λ while completing one full vertical cycle, and ν is the wave propagation

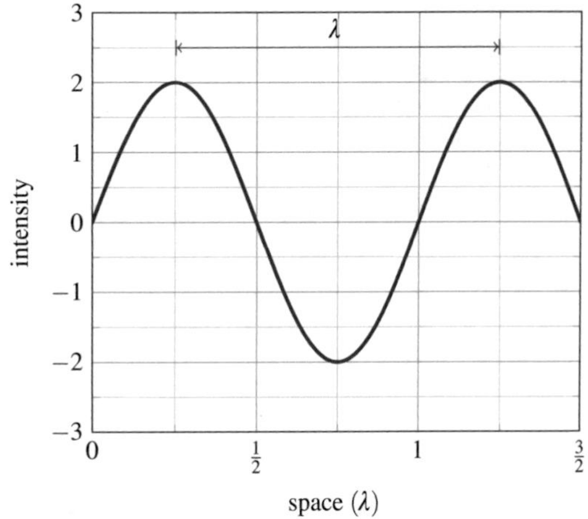

Fig. 1.11 By measuring the distance in space between the peaks (or any other pairs of equivalent points, as shown) the spatial dimension, wavelength λ, is established

speed[14] in metres per second (denoted by c in the special case of the speed of light). To summarize, it should be noted that it is the frequency of the wave that determines the pitch (or color). The wavelength is a secondary phenomena depending on the speed of the wave in the given transmission medium.

Example 5: Wavelength Calculations
Calculate wavelengths of EM waves at the following frequencies: $f_1 = 3\text{kHz}$ (i.e. in the audio range), $f_2 = 3\text{MHz}$ (i.e. frequency of a simple LC oscillator), and $f_3 = 3\text{GHz}$ (i.e. close to the operational frequency of cellphones).

Solution 5: EM waves have phase velocity of $c_0 \approx 3 \times 10^8\,[\text{m/s}]$, hence, after substituting (1.11) into (1.10) we write

$$\lambda = \frac{2\pi}{\beta} = \frac{2\pi\ v_p}{\omega} = \frac{v_p}{f} \tag{1.18}$$

which results in

$$\lambda_1 \approx \frac{3 \times 10^8\ \text{m/s}}{3\text{kHz}} = 100 \times 10^3\text{m};\quad \lambda_2 \approx \frac{3 \times 10^8\ \text{m/s}}{3\text{MHz}} = 100\text{m};$$

$$\lambda_3 \approx \frac{3 \times 10^8\ \text{m/s}}{3\text{GHz}} = 100 \times 10^{-3}\text{m} = 100\text{mm}$$

1.3.4 Wave Envelope

Figure 1.12 shows an important case of a waveform found in the radio frequency (RF) communication systems. This waveform consists of a high-frequency tone (blue) whose amplitude is changing in

[14]The correct term should be velocity, but most books use (wrongly) the speed instead.

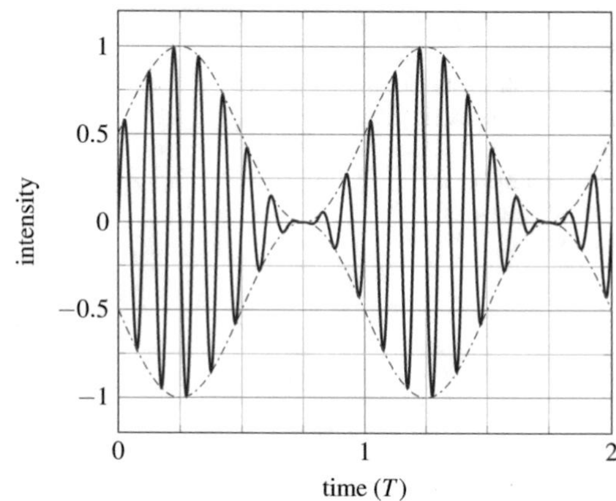

Fig. 1.12 A high-frequency waveform (solid line) and its low-frequency embedded envelope (dash-dot line)

accordance with a low-frequency sinusoidal function (red). The low-frequency waveform (not necessarily sinusoidal) that is "riding" on the high-frequency peaks is very important in communications; it is referred to as the "envelope" of the high-frequency tone and the high-frequency waveform is referred to as the "carrier". In reality, the envelope waveform is the transmitted message that must be recovered by the receiver, it is not readily available as Fig. 1.12 may imply. Instead, it is the carrier's amplitude that holds the message.

Theoretical and practical techniques for imprinting an arbitrary envelope over a carrier (that originally had a constant amplitude) and for extracting the envelope and discarding the high-frequency carrier are the main subjects not only of this book, but also of the RF circuit design field in general. The process of "imprinting" the envelope signal into the carrier is referred to as "modulation" and the process of envelope extraction is known as "demodulation".

Example 6: AM Modulated Waveform

Referring to Fig. 1.12, estimate frequencies of carrier f_C and modulating f_M waveforms if horizontal time axes is shown: (1) in units of [ms]; (2) in units of [µs]; (3) in units of [ns];

Solution 6: Assuming given time unit scaling,

1. Carrier waveform, in blue, takes $t_C = 6\text{ms}$ to finish 12 full periods,

$$t_C = 12 \times T \Rightarrow 6\text{ms} = \frac{12}{f_C} \Rightarrow f_C = \frac{12}{6\text{ms}} = 2\text{kHz}$$

At the same time, modulating waveform (i.e. red envelope) takes $t_M = 4\text{ms}$ to finish one full period,

$$t_M = 1 \times T \Rightarrow 4\text{ms} = \frac{1}{f_C} \Rightarrow f_M = \frac{1}{4\text{ms}} = 250\text{Hz}$$

2. Using the same approach, after replacing the time units, we find

$$f_C = 2\text{MHz}; \quad f_M = \frac{1}{4\mu\text{s}} = 250\text{kHz} \quad (t_C = 6\mu\text{s}; \; t_M = 4\mu\text{s})$$

$$f_C = 2\text{GHz}; \quad f_M = \frac{1}{4\text{ns}} = 250\text{MHz} \quad (t_C = 6\text{ns}; \; t_M = 4\text{ns})$$

1.3.5 Phase, Phase Difference, and Signal Velocity

A stand-alone single-tone wave is fully described by its amplitude, frequency (or, equivalently, its period), and *phase*. The concept of a phase is derived from the rotating phasor model and it assumes sine as the default waveform function because at time $t = 0$ its phase is zero. Consequently, one period T in the time domain is mapped onto an angle of a circle, i.e. $T = 2\pi$ radians (or 360°).

$$T \text{ [s]} \stackrel{\text{def}}{=} \frac{1}{f} \text{ [Hz]} \stackrel{\text{def}}{=} 2\pi \text{ [rad]} \stackrel{\text{def}}{=} 360^\circ \tag{1.19}$$

Note that numerical value for time T (measured in seconds, thus an absolute unit) is scaled to a number 2π (with no unit, thus a relative unit); these two measuring units are used interchangeably.

Phase measured at a specific point is referred to as "instantaneous phase"; the phase measured at a point in time $t = 0$ is the known as the "initial phase". Since the initial value of a sine function (A function in Fig. 1.13) at $t = 0$ (i.e. $\sin(\omega \times 0 + \phi) = 0$) is zero, its initial angle (or phase) is zero. At the same time, cosine waveform (B function in Fig. 1.13) has the initial value equal to one, thus its initial phase must be $\phi = \pi/2$ (or 90°) because $\sin(\omega \times 0 + \pi/2) = \sin \pi/2 = \cos 0 = 1$.

Relation of *two* co-existing single-tone waves can be described in terms of their "phase difference". This term refers to *difference* between instantaneous phases of two waveforms. With two sinusoidal waves, once the amplitudes are normalized and compared, it is important to compare their frequencies. If the frequencies are not the same, then there is not much left to say, but to note existence of two relatively independent waves whose phase difference is not defined.

However, if two waves do have the same frequency (and not necessarily the same amplitude) and one of them is declared the "reference", then it makes sense to ask question which wave "arrives first", that is to say, what is their phase difference? Under the condition of the frequency equality, the phase difference is either *constant*, Fig. 1.13, or, by definition, is not defined. In the given example at $t = 1$, Fig. 1.13, amplitude of A is zero (point ($\mathscr{A}$), i.e. $\phi = 0$) while at the same time amplitude B equals one (point ($\mathscr{B}$), i.e. $\phi = \pi/2$) the phase difference between A and B at any given moment is therefore $\Delta\phi = \pi/2$. In this special case, where the peaks of one wave coincide with zero-crossing points of the second, the phase difference is one-quarter of the cycle, i.e. 90°, and the two waves are said to be "in quadrature". Note that the quadrature signals are very important and widely used in radio

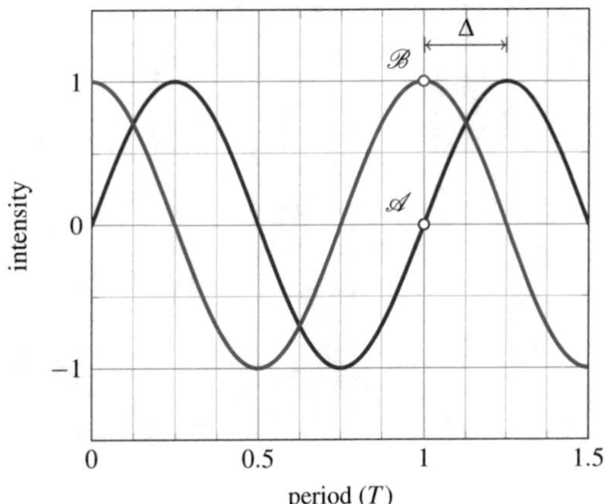

Fig. 1.13 Two single-tone waveforms with normalized amplitudes and the same frequency that have phase difference $\Delta = \pi/2$

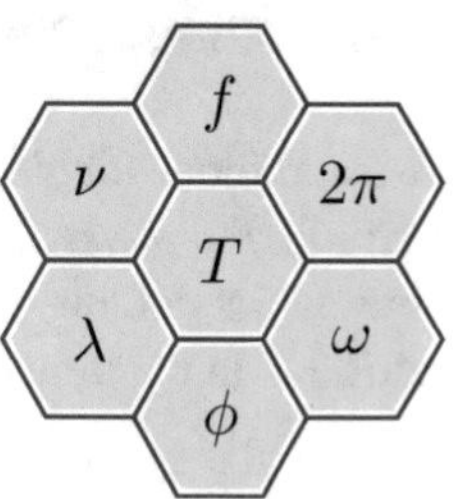

Fig. 1.14 Interrelated variables needed to fully describe a sine waveform

communication systems. Summary plot of relevant variables that are needed to fully characterize a sine waveform is shown in Fig. 1.14.

When phase difference does exist, it is said that one wave either "leads" or "lags" the second one by Δ seconds (or, equivalently, by Y degrees), again, this measurement is *relative*.

Example 7: Phase Difference

Calculate differences in the times of arrival Δt (which is absolute measure) for EM wave pairs with phase difference of $\Delta\phi = {}^{\pi}/_{2}$ (which is relative measure) at each of the following frequencies: $f_1 = 1\text{kHz}$, $f_2 = 1\text{MHz}$, $f_3 = 1\text{GHz}$.

Solution 7: Conversion of the given frequencies into their equivalent periods is as follows:

$$T_1 = \frac{1}{f_1} = \frac{1}{1\text{kHz}} = 1\text{ms}\,; \quad T_2 = \frac{1}{f_2} = \frac{1}{1\text{MHz}} = 1\,\mu\text{s}\,; \quad T_3 = \frac{1}{f_3} = \frac{1}{1\text{GHz}} = 1\text{ns}$$

then, by knowing that period $T \overset{\text{def}}{=} 2\pi$ angle (i.e. one full cycle), we conclude that ${}^{\pi}/_{2}$ is equivalent to ${}^{T}/_{4}$ of their respective waveforms. Therefore, we write

$$\Delta t_1 = \frac{T_1}{4} = \frac{1\text{ms}}{4} = 250\mu\text{s}\,; \quad \Delta t_2 = \frac{T_2}{4} = \frac{1\mu\text{s}}{4} = 250\text{ns}\,; \quad \Delta t_3 = \frac{T_3}{4} = \frac{1\text{ns}}{4} = 250\text{ps}$$

which illustrates how the phase difference translates into time of arrival differences.

Example 8: Waveform Definitions

For a voltage disturbance wave given in Example 4 travelling at the speed of light

$$v_1(t) = \sin(20\pi \times 10^6\, t)$$

find its wavelength. Assuming second wave $v_2(t)$ with the same maximum amplitude and with the phase difference of $\Delta\phi = +45°$, find its amplitude at time $t = 0$ and space distance between one of its peaks and the first following peak of $v_1(t)$.

Solution 8: By inspection of wave v_1 equation, we write:

1. by definition (1.17), wavelength is $\lambda = c\,T = 3 \times 10^8\ {}^{\text{m}}/_{\text{s}} \times 100 \times 10^{-9}\text{s} \approx 30\text{m}$

(continued)

2. the second wave is leading with phase difference of $\Delta\phi = 45° = \pi/4 = 2\pi/8 = T/8$; its amplitude at $t = 0$s is therefore $v_2 = 1\text{V}\ \sin(\omega \times 0 + \pi/4) = 1\text{V}\ \sin(\pi/4) = 1/\sqrt{2}\,\text{V} \approx 0.707\,\text{V}$
3. the phase difference is $T/8$, therefore the distance in space has to be $\lambda/8 = 30\text{m}/8 = 3.75\text{m}$

Example 9: Index of Refraction
If light crossing boundary from an optical fibre to air changes its wavelength from $\lambda_1 = 452\text{nm}$ to $\lambda_0 = 633\text{nm}$, calculate: (1) index of refraction of fibre; (2) speed of light in fibre; (3) frequency of light in fibre; and (4) frequency of light in air.

Solution 9:

(1) From definition of refraction index (1.13), (the refraction index of dry air is as same as for vacuum) it follows that

$$n = \frac{\lambda_0}{\lambda_1} = \frac{633\text{nm}}{452\text{nm}} = 1.400$$

(2) From definition (1.13) it follows that

$$v = \frac{c}{n} = \frac{299\,792\,458\ \text{m/s}}{1.4} = 214\,137\,470\ \text{m/s}$$

(3) The frequency in the fibre can be found from its speed and wavelength:

$$\nu = \frac{v}{\lambda_1} = \frac{214\,137\,470\ \text{m/s}}{452\text{nm}} = 4.74 \times 10^{14}\text{Hz}$$

(4) The frequency in the air can be found from its speed and wavelength:

$$\nu = \frac{c}{\lambda_0} = \frac{299\,792\,458\ \text{m/s}}{633\text{nm}} = 4.74 \times 10^{14}\text{Hz}$$

1.3.6 Average of Sine Function

For a given periodic function $f(x)$, mathematical definition of its average value $\langle f(x)\rangle$ is given by the following integral:

$$\langle f(x)\rangle \stackrel{\text{def}}{=} \frac{1}{T}\int_0^T f(x)\,dx \tag{1.20}$$

whose geometrical interpretation is that $\langle f(x)\rangle$ is height of a rectangle whose surface area is equal to the area of the surface under the function (assuming the same x-interval). Thus, value of $\langle f(x)\rangle$ is a *constant*, in other words it is at "DC" level.

An important case to engineering is the average value of a sine function $f(t) = A\sin\omega t$, that is found by definite integral (i.e. surface of the area) over one period $T = 2\pi$

$$\langle f(x)\rangle \stackrel{\text{def}}{=} \frac{1}{T}\int_0^T f(x)\,dx = \frac{1}{2\pi}\int_0^{2\pi} A\sin\omega t\,dx = \frac{A}{2\pi}\,(-\cos\omega t)\Big|_0^{2\pi} = \frac{A}{2\pi}\,(-1+1) = 0 \tag{1.21}$$

That is to say that, as per (1.21), average value of a sine waveform is *zero* over integer number of cycles and regardless of the initial phase. A geometrical interpretation is that each period consists of one negative and one positive half-cycle, both having the same area. Since the cosine and sine functions are related, $\cos\omega = \sin(\omega - \pi/2)$, for the purposes of this discussion, it does not matter whether the sine or the cosine function is used in the analysis. This result has significant consequence on the way we design circuits and interface between its subsequent stages.

A periodic function $A(t)$ that fluctuates around an average value other than zero may be thought of as being composed of a constant DC component I_{CM} and a time-varying sinusoidal component AC added together (see Fig. 1.15), i.e.

$$A(t) = A_{\text{DC}} + A_{\text{AC}} = I_{\text{CM}} + I_m \sin\omega t \tag{1.22}$$

where I_{CM} is the constant value and I_m is the maximum sine amplitude. Obviously, if we want to keep the intensity of sinusoidal function positive at all times, then $I_{\text{CM}} > I_m$ (often $I_{\text{CM}} \gg I_m$).

1.3.6.1 Average of Sine Product

A very important case in engineering is the product of two sine waves. Let us consider two sine wave functions, with frequencies ω_1 and ω_2 and an initial phase difference θ at $t = 0$,

$$A = a\sin(\omega_1 t) \tag{1.23}$$

$$B = b\sin(\omega_2 t - \theta) \tag{1.24}$$

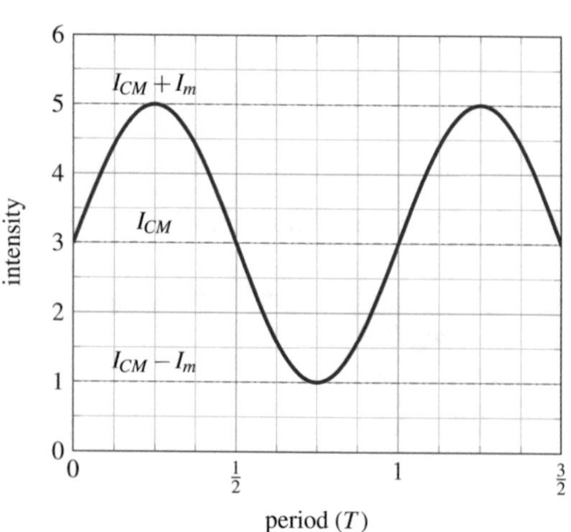

Fig. 1.15 A sinusoidal signal $A(t)$ whose DC component, i.e. average level, is $I_{\text{CM}} = 3$ and AC amplitude is $I_m = 2$

so that their product $x = A\,B$ is simply written as[15]

$$
\begin{aligned}
x &= ab\sin(\omega_1 t)\,\sin(\omega_2 t - \theta)\\
&= \frac{a\,b}{2}\{\cos[(\omega_1 - \omega_2)\,t + \theta] - \cos[(\omega_1 + \omega_2)\,t - \theta]\}\\
&= \frac{a\,b}{2}(x_1 - x_2)
\end{aligned}
\tag{1.25}
$$

and the average value $\langle x \rangle$ is then calculated as the sum of averages of the two terms x_1 and x_2. When $\omega_1 \neq \omega_2$, average of the first term $\langle x_1 \rangle$ is

$$\langle x_1 \rangle = \langle \cos[(\omega_1 - \omega_2)\,t + \theta] \rangle = 0 \tag{1.26}$$

for an integer number of cycles $n\,T$. Note, from (1.14), that the first term has a period of $T = 1/(f_1 - f_2)$. Following the same argument, the same results are obtained for the second term,

$$\langle x_2 \rangle = \langle \cos[(\omega_1 + \omega_1)\,t - \theta] \rangle = 0 \tag{1.27}$$

which is to say that, for the case of $\omega_1 \neq \omega_1$, average value over integer number of cycles of the product of two sine waves is also zero.

However, for case of equal frequencies $\omega_1 = \omega_2 = \omega$, (1.25) becomes

$$x = \frac{ab}{2}\cos\theta - \frac{ab}{2}\cos(2\omega\,t - \theta) \tag{1.28}$$

where the average of the second term $\cos(2\omega\,t - \theta)$ is zero, which leads to

$$\langle x \rangle = \frac{ab}{2}\cos\theta \tag{1.29}$$

In this case the average value depends upon the phase difference (the two frequencies are identical) and can, therefore, be adjusted to zero or anywhere between $(\pm ab/2)$. This observation is very important for RF design, because operation of RF circuits for wireless communication is based on perfect frequency relationships among multiple sinusoidal signals.

Example 10: Average of a Waveform
Starting from definition (1.20), derive the average value (i.e. CM as illustrated in Fig. 1.15) of the following waveform:

$$v(t) = 1\text{V} + \sin(\omega\,t) \tag{1.30}$$

where $\omega = 2\pi f$.

(continued)

[15] Use the trigonometric identity $\sin(\alpha)\sin(\beta) = 1/2[\cos(\alpha - \beta) - \cos(\alpha + \beta)]$.

Solution 10: By definition (1.20), a sine waveform is periodic with $T = 2\pi$, while the given $v(t)$ is simply sum of a sine and 1V_{DC} level, thus by definition (1.20)

$$\begin{aligned}
\langle v(t)\rangle &= \frac{1}{T}\int_0^T v(t)\,dt = \frac{1}{2\pi}\int_0^{2\pi}[1\text{V} + \sin(\omega t)]\,dt \\
&= \frac{1}{2\pi}\int_0^{2\pi} dt + \frac{1}{2\pi}\int_0^{2\pi}\sin(\omega t)\,dt \\
&= \frac{1}{2\pi}\,t\big|_0^{2\pi} + \frac{1}{2\pi}\frac{1}{\omega}\int_0^{2\pi}\sin(x)\,dx \to 0 \\
&\therefore \\
\langle v(t)\rangle &= 1\text{V}
\end{aligned}$$

which illustrates the definition of common mode (CM), or average value of a sine waveform. Also, we note that this value is found simply by inspection of (1.30).

1.3.7 The "High-Frequency" Concept

The term "high frequency" (HF) is often used, and it is valid to ask how the term *high frequency* is actually defined. Is there any particular number, for example 1kHz or 1GHz, that is accepted as "high frequency" or is there something else that is important to notice? In order to answer this question, let us take a look at a simple, one-dimensional wave of an electric field travelling the z direction along the conductive wire's length, whose length is d, as

$$E_x = E_{0x}\cos(\omega t - kz) \tag{1.31}$$

where E_x is the electric wave field component along the x coordinate, E_{0x} is its maximum amplitude, ω is the angular frequency, $k = |\mathbf{k}| = {}^{\omega}/_{c} = {}^{2\pi}/_{\lambda}$ is the wave vector value, z is the space coordinate showing the direction of wave propagation (which is perpendicular to the electric field vector), and the initial phase ϕ is assumed to be zero. The electric wave travels inside a "long" conductive wire aligned along the z coordinate, where the wave equation (1.31) explicitly shows the time t and space z arguments of the electric field. We note that, in this case, the term "long" implies that the wire length d is measured in units of the wavelength λ. In other words, this is a *relative* measurement where the wire length d is measured by the number of wavelengths λ. For example, Fig. 1.16 shows the wire length to be $d \approx 2.25\lambda$. Therefore, for a given physical wire length d, whether the wire is quantified as "long" or "short" depends strictly on the signal frequency, i.e. its wavelength. Hence, a "short" wire is one where the wire length d is comparable to or shorter than the signal wavelength, i.e. $d \approx \lambda$ or $d \ll \lambda$, while a "long" wire implies that the wire length d is much longer than the wavelength, i.e. $d \gg \lambda$, regardless of whether the signal frequency is 60Hz, 1kHz, 1GHz or any other number. The engineering rule of thumb is to estimate the wire length as "very short" if $d \leqslant {}^{\lambda}/_{10}$; where the wide grey area between "short" and "long" is assumed.

As a thought experiment, let us imagine that the time for this wave field has stopped (except for the little ant), so that the ant can observe and closely examine a "single frame" of this movie, i.e. $t = const$, while walking along a long conductive wire (see Fig. 1.16). Because this is a long

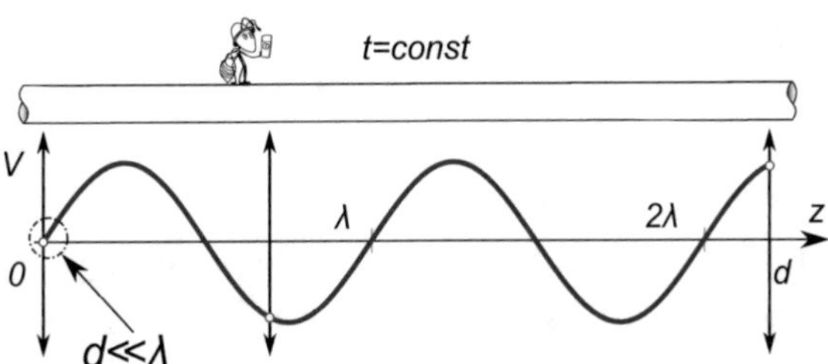

Fig. 1.16 A unidirectional wavefront inside a conductive wire, with a single time frame shown in space. The measured voltage amplitude along the wire drastically depends upon the current location in space along the z axis

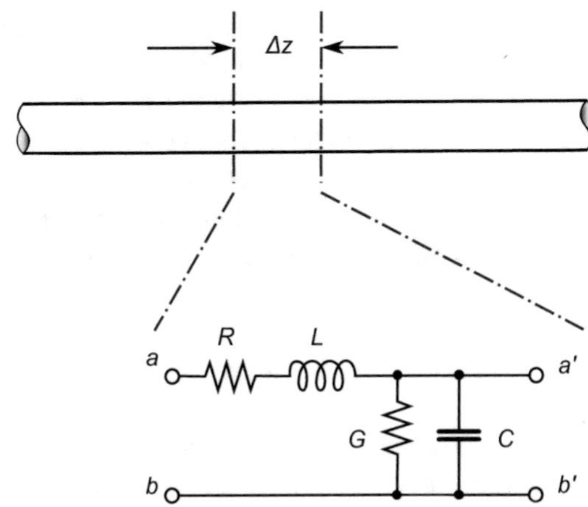

Fig. 1.17 Long conductor (relative to the λ) is divided into infinitesimally short sections $\Delta z \ll \lambda$, where each section is then modelled using distributed circuit elements R, L, C, and G

wire, the waveform goes through more than a full cycle in space, which is to say that the measured potential along the wire varies between its minimum and maximum amplitude values in accordance with (1.31). A direct consequence of this situation is that the wire is *not equipotential.* For example, compare the current that would flow into a branch connected at a point corresponding to $z = \lambda/4$ against the currents flowing into branches connected at points where $z = 2\lambda$ or $z = 3\lambda/4$.

That consequence arises because KVL derived from Maxwell's equations assumes that the wire length is $d = 0$ (or, equivalently, $\lambda = \infty$), which in reality is not the case, not even for a DC signal (due to the thermal losses). The spatial behaviour of the voltage (and its corresponding current) must be taken into account in cases when the signal wavelength is comparable to the conductor length (i.e. in the case of a "high-frequency signal") and Kirchhoff's circuit laws cannot be applied in their approximated form. The discovery of this wavelength to wire length relationship led to the development of a mathematical model known as the transmission line model.

In order to circumvent the above problem, a long conductor carrying a high-frequency signal is split into a number of short-length sections Δz (mathematically $\Delta z \to 0$), which is to say that KVL is valid when applied to each section Δz separately (see Fig. 1.17). The physical properties of the wire section are then modelled using distributed electrical parameters R, L, C, and G, where the corresponding electrical units are expressed in terms of the unit length, i.e. Ω/m, H/m, F/m, and S/m, respectively. Analysis of each section is therefore reduced to the analysis of a traditional circuit with lumped RLC parameters.

Electric circuit representation of the line sections is a very useful modelling tool because it:

1. Is a very intuitive model that is consistent with the two-port network methodology
2. Permits analysis using KVL and KCL

It has the following limitations:

1. It is a one-dimensional model that does not include leaking fields and interference with other components
2. Material nonlinearities are mostly ignored

In conclusion, KVL and KCL models are definitely applicable at DC and for "low-frequency" signals. For example, a 60Hz signal ($\lambda \approx 5\,000$km) can be analysed using Kirchhoff's laws if the signal is measured on a small PCB (with a wire length of, say, $d = 10$cm). However, if the 60Hz signal is carried across a continent, i.e. $\lambda \sim d$, then a more accurate transmission line model must be used. Similarly, a 1GHz signal ($\lambda \approx 300$mm) must be treated with the transmission line model if used on a 10cm long PCB but the KVL model would result in a "close-enough" solution if the 1GHz signal is carried by a 100µm long wire inside an IC. Finally, an analysis of antennas and EM wave propagation through space must include full Maxwell's equations.

A solid understanding of these two extreme approximations, i.e. low frequency (LF) and high frequency (HF), is important for mastering RF circuit design. In this introductory book, however, we employ only low-frequency, quasi-static RF circuit design techniques for purposes of mastering the basic RF design principles without too much emphasis on specific properties of high and ultra-high-frequency systems, which are the subject of more advanced RF courses.

1.4 Electronic Signals

In electronic communication systems, the useful information, i.e. the signal, is embedded and carried in the form of voltage or current, or both. Time domain variations of either of these two variables are then modelled using appropriate mathematical functions. For example, digital information is transmitted by switching between two fixed voltage levels, which is modelled by using the pulse function. In mathematically simpler case of analog wireless radio communications the transmitted signal is embedded into a sinusoidal "carrier" function.

1.4.1 DC and AC Signals

A *signal* is defined as any time-varying event being observed. In electronic communications, signals are processed in form of either *current* or *voltage*; signal transmission can be either wired or wireless.

Two general categories of electronic signals are DC signals that have constant amplitude in time (for example, a battery voltage) and AC signals that have varying amplitude in time (for example, voltage amplitude measured at the wall power outlet). Further, an AC signal can be either periodic or aperiodic. Examples of periodic AC signal shapes are sinusoidal, square, and saw waveforms, i.e. signals consisting of fixed, time-repetitive patterns. An example of an aperiodic electronic AC signal waveform is thermal noise. Being constants, DC signals therefore have a simpler mathematical representation and treatment relative to periodic AC signals. On the other hand, aperiodic, or random, signals are significantly more complicated than periodic signals and they are treated using mathematical tools from statistical analysis.

In this section, we review terminology related only to the most important form of AC signals, sinusoidal signals. Without being concerned about the nature of the signal, how it was generated, or what physical quantity it represents, a general sine-wave function is represented by

$$a(t) = A_p \sin(\omega t + \phi) \tag{1.32}$$

where

$a(t)$ is the instantaneous value of time-varying quantity (voltage, current, power…)
A_p is the maximum or peak amplitude
ω is the angular frequency (related to frequency as $\omega = 2\pi f$)
ϕ is the initial phase (often assumed zero)
t is the time variable

Figure 1.18 shows two common representation of AC signal (1.32), namely a phasor (or rotating vector) and its equivalent time domain graph. As an example, two sinusoidal waveforms are shown, labelled $i(t)$ and $v(t)$. The instantaneous angular positions of their respective phasors show that their phase difference is $\varphi = {}^{\pi}/_{2}$, and that $v(t)$ is at the angular position $\alpha = {}^{\pi}/_{6}$ while angular position of $i(t)$ is at $\alpha + \varphi = {}^{2\pi}/_{3}$ (we say that $i(t)$ "leads" $v(t)$). By following the horizontal guidelines, we find the same angles on the time axis. In the time domain it is visible that all subsequent time domain points $(v, v', v'', \ldots)$ and $(i, i', i'', \ldots)$ separated by $T = 2\pi$ are also mapped back to the same phasor vector. In addition, assuming EM wave propagating in vacuum (i.e. with the velocity c), we can find the instantaneous position of these points in space, relative to the wavelength λ. To help the conversion between period, wavelength, and phase, the horizontal axis is labelled in all three units.

1.4.2 Single-Ended and Differential Signals

Typical signals, such as the sinusoids in Fig. 1.18, are also known as *single-ended* signals because they consists of only one waveform that is referenced to the local zero level (a.k.a. "ground"). In this section we introduce an important signal form, known as a *differential signal*, which is created by using two single-ended sinusoidal waveforms in the following relationship. They have

1. equal amplitudes,

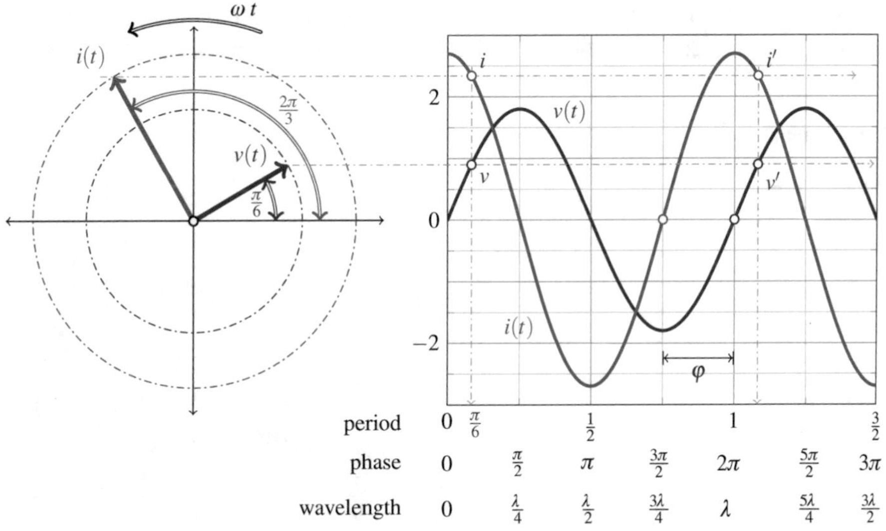

Fig. 1.18 Phasor representations of two signals; i.e. rotating vectors (left), and their equivalent time domain sinusoidal function (right) as measured in three units: period, phase, and wavelength

2. equal frequencies,
3. equal common mode
4. *opposite* phases, i.e. the phase difference is π.

Let us consider two sinusoidal signals v_1 and v_2 as

$$v_1 = V_{\text{CM}} + V_m \sin\omega t \tag{1.33}$$

$$v_2 = V_{\text{CM}} - V_m \sin\omega t \tag{1.34}$$

where (1.33) and (1.34) formalize the required relationship between the two waveforms, shown in Fig. 1.19. If these two signals are added, then the sum is, obviously, $v_1 + v_2 = 2\,V_{\text{CM}}$, which is a DC signal and the v_1 and v_2 waveforms are lost. However, if they are *subtracted*, then we write

$$v_{\text{diff}}(t) = v_1 - v_2 = \cancel{V_{\text{CM}}} + V_m \sin\omega t - (\cancel{V_{\text{CM}}} - V_m \sin\omega t) = 2\,V_m \sin\omega t \tag{1.35}$$

which is a very interesting result because this time DC level is cancelled while the original waveform[16]is preserved, amplified by a factor of two and shifted down to the zero common mode value. This amplification is achieved simply by the addition of two signals (one of which was negative). We note that the gain factor of two is significant, especially when we have weak signals to start with.

Let us further assume that two conductive wires carrying the v_1 and v_2 signals are physically located close to each other. With that assumption, any interference noise signal $n(t)$ is equally added to both v_1 and v_2, i.e. the two noisy signals are

$$v_1 = n(t) + V_{\text{CM}} + V_m \sin\omega t \tag{1.36}$$

$$v_2 = n(t) + V_{\text{CM}} - V_m \sin\omega t \tag{1.37}$$

which, after subtraction results again in (1.35). In other words, the common interfering signal is removed from the differential signal. These two properties of differential signals, namely the gain and the immunity to common noise, are beneficial and important enough that most modern, high-performance signal-processing circuits are designed to process differential signals. However, for the

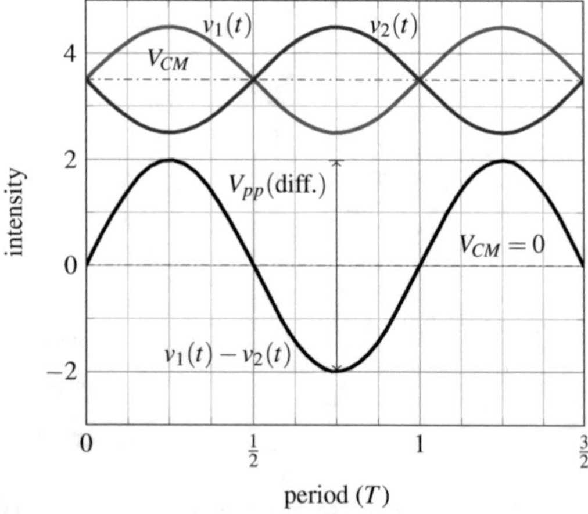

Fig. 1.19 Differential signal, $v_1 - v_2$, is constructed using two single-ended signals, v_1 and v_2

[16]Except for the phase difference, the two initial waveforms are identical.

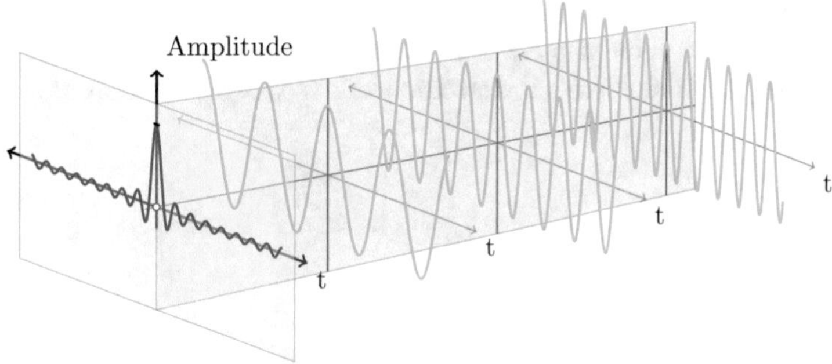

Fig. 1.20 Constructive addition of sine creates Dirac function if infinite series is used. In this plot, the sum (in the front) is created by summing the first ten sine, only the first three are shown

sake of simplicity and accepted educational methodology, all circuits in this textbook are assumed to be single-ended, leaving differential architectures for more advanced microelectronics courses.

1.4.3 Constructive and Destructive Signal Interactions

The phase difference between two periodic signals is very important from the perspective of their sum because in a circuit network signals are constantly added, either through KCL or KVL. Result of the addition, as already illustrated in the previous sections, can be drastically different depending on the phase difference. This signal addition can be either intentional (as in signal processing) or unintentional (as in cross-talk interference). We keep in mind that the concept of signal addition applies to all signals, not only to the single tones. For example, one harmonic within the complicated signal spectrum can be removed from the spectrum with destructive addition of the external tone, i.e. the one with the same frequency and amplitude but opposite phase.

To illustrate an important theoretical and practical case of constructive and destructive signal interaction, Fig. 1.20 shows how addition of multiple sine waveforms results in a waveform that is non-zero inside a very localized interval, while its amplitude is close to zero elsewhere. The addition that includes an infinite number of sine waveforms would create Dirac's delta function.

In summary, both constructive and destructive signal additions are used intentionally in signal processing as, for example, in high-end noise-cancellation audio headphones.

1.5 Signal Metrics

Periodic signals are arguably the most important category of signals in the theory of RF communication systems. Because of the phase differences, in RF communication circuits at any given moment in time either voltage or current may be equal to zero. In that case, regardless which one of the two equals zero, the signal power also equals to zero. In other words, at the moment there is no useful signal transmission. Therefore, it is more important to follow RF signal *power* levels (often expressed in [dB]) instead of the instantaneous values of individual voltages and currents.

1.5.1 Power

By definition, the electric potential at a point r in a static electric field $\mathbf{E}$ is given by the line integral

$$\Delta V_{\mathbf{E}} = -\int_0^L \mathbf{E} \cdot d\mathbf{l} \tag{1.38}$$

where L is an arbitrary path connecting the point in infinity (i.e. with zero potential) to point r, and $d\mathbf{l}$ is the unity path element. Note the dot product of the two vectors in the integral. In physical terms, (1.38) represents the electric work W (scalar variable) of the electric field along the integral path,

$$W = q \int_0^l \mathbf{E} \cdot d\mathbf{l} = q\, \Delta V_{\mathbf{E}} \tag{1.39}$$

$$\therefore$$

$$V = \frac{dW}{dq} \text{ [V]} \tag{1.40}$$

where voltage V is measured in volts [V]. Therefore, we say that work represents energy that is needed to move a charged particle q over a certain distance.

Strictly, an electric current I is defined either as the rate of change of charge in time or, equivalently, the current density within the total conducting surface

$$I \stackrel{\text{def}}{=} \frac{dQ}{dt} = \int_S \mathbf{J} \cdot d\mathbf{s} \text{ [A]} \tag{1.41}$$

where current I is measured in amperes [A], Q is the total amount of charge through the cross-sectional area S, dt is the differential unit of time, $\mathbf{J}$ is the current density vector, and $\mathbf{s}$ is the vector of the conducting surface element oriented in space.

From the engineering perspective, it is important to establish not only the amount of energy needed to perform a work, but also the rate of energy exchange, i.e. the rate of either generation or absorption of energy. That brings us to the concept of *power* P (a scalar variable), which quantifies how fast, for a given amount of energy, the work is finished. Or, in a strictly mathematical sense, after substituting (1.40) and (1.41), the electrical power P is

$$P \stackrel{\text{def}}{=} \frac{dW}{dt} = \frac{dW}{dQ}\frac{dQ}{dt} = V\, I \text{ [W]} \tag{1.42}$$

where power P is measured in watts [W]. As a side note, all definitions introduced in this section assumed either static or quasi-static (i.e. steady state) electric field.

1.5.2 Root Mean Square (RMS)

One possible view of a resistor is that it is a device that converts electrical energy into heat energy, which is then dissipated either intentionally (as in a stove heater, for example) or as wasted energy (as in a bulb, for example). Hence, it is important to know how much power is dissipated by the resistor in case of both DC and AC over an integer number of cycles. To do so, let us first consider the simple problem of calculating electrical power P dissipated by an ideal resistor R while conducting direct (i.e. constant in time) current I. After Ohm's law, electric power (1.42) may be rewritten as

$$P = V\, I = I^2\, R = \frac{V^2}{R} \tag{1.43}$$

which, for a given resistance R, is dependent upon the current's (or the voltage's) squared value.

On the other hand, the average value of $i(t) = I_m \sin \omega t$ is zero, however, the average dissipated power is not zero. Therefore, to find the answer for the case of periodic AC current (e.g. $i(t) = I_m \sin \omega t$), the calculation of the constant current term I^2 in (1.43) has to be replaced with the average value of time-varying quadratic current, i.e. $\langle i(t)^2 \rangle$, which, by definition, represents a "quadratic mean" or *root mean square* (RMS) of the current. Hence, calculation of the equivalent dissipated power for a sine waveform is as follows:[17]

$$\begin{aligned}
\langle P(t) \rangle &= \langle i(t)^2 \rangle\, R \overset{\text{def}}{=} I_{\text{rms}}^2\, R \\
&= \left[\sqrt{\frac{1}{T} \int_0^T |i(t)|^2\, dt} \right]^2 R = \left[\sqrt{\frac{1}{T} \int_0^T (I_m \sin \omega t)^2\, dt} \right]^2 R \\
&= \left[\sqrt{\frac{I_m^2}{T} \int_0^T \sin^2 \omega t\, dt} \right]^2 R = \left[\sqrt{\frac{I_m^2}{T} \int_0^T \frac{1 - \cos(2\omega t)}{2} dt} \right]^2 R \\
&= \left[\sqrt{\frac{I_m^2}{T} \left[\frac{t}{2} - \frac{\sin(2\omega t)}{4\omega} \right]_0^T} \right]^2 R = \left[\sqrt{\frac{I_m^2}{\cancel{T}} \left[\frac{\cancel{T}}{2} - \cancel{\frac{1}{4\omega} \sin(2\omega t) \Big|_0^{T}}^{\,0} \right]} \right]^2 R \\
&= \left[\frac{I_m}{\sqrt{2}} \right]^2 R
\end{aligned} \tag{1.44}$$

That is to say, equivalent effective direct current (DC) of a sinusoidal alternating current is the AC peak divided by the square root of two.[18]

If several sinusoidal functions with various frequencies are added together, for example

$$i = a \sin(\omega_1 t + \alpha) + b \sin(\omega_2 t + \beta) + c \sin(\omega_3 t + \gamma) + \cdots \tag{1.45}$$

then, the RMS value the sum (1.45) must be squared, however, in this case all inter-products between terms at different frequencies may be ignored because the average values of those products are zero, which leads to

$$I_{\text{rms}} = \sqrt{\frac{a^2}{2} + \frac{b^2}{2} + \frac{c^2}{2} + \cdots} \tag{1.46}$$

This result shows that, when calculating the power of a multi-tone signal, each tone's power can be calculated separately, which is the property exploited in Fourier's analysis.

[17]Use the trigonometric identity $\sin^2 \alpha = 1/2(1 - \cos(2\alpha))$.

[18]In the case of a square wave, $I_{\text{rms}} = I_m$, while for a sawtooth wave, $I_{\text{rms}} = I_m/\sqrt{3}$.

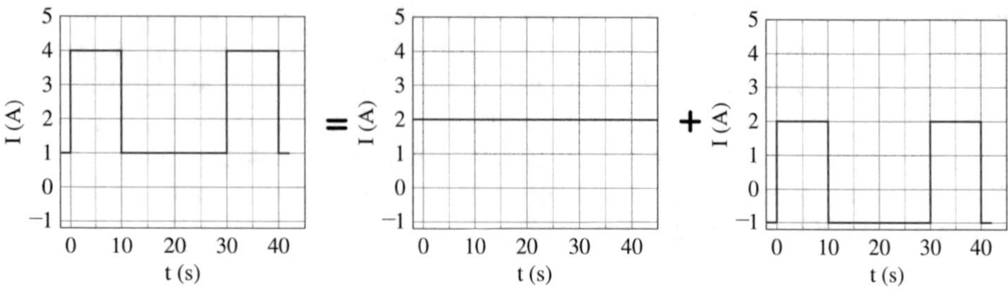

Fig. 1.21 Square waveform used in Example 11

Example 11: Waveform Definitions, RMS
Calculate common mode level I_{CM}, AC amplitude I_m, RMS value of the AC component, and RMS value of square signal in Fig. 1.21 (left), value of current I is in [A] and time in [ms]

Solution 11: The integration of a quasi-linear function is reduced to a simple addition over the period T. By inspection, the function period is $T = 30$ms; thus we write,

1. The common mode, Fig. 1.21 (middle), i.e. DC level

$$\langle I \rangle = \frac{4\text{A} \times 10\text{ms} + 1\text{A} \times 20\text{ms}}{30\text{ms}} = 2\text{A} \tag{1.47}$$

2. The AC component is found by knowing that a waveform is effectively the sum of its DC and AC components. Therefore, the AC waveform must have $I_{AC} = 2A$ during the first 10ms (so that the sum of its DC and AC components is $2\text{A(AC)} + 2\text{A(DC)} = 4\text{A(total)}$, as found above) and $I_{AC} = -1$A from 10ms to 30ms (following the same reasoning), see Fig. 1.21 (right).
3. The RMS value can be calculated, by definition, first for the AC component as

$$I_{\text{rms}}(\text{AC}) = \sqrt{\frac{(2\text{A})^2 \times 10\cancel{\text{ms}} + (-1\text{A})^2 \times 20\cancel{\text{ms}}}{30\cancel{\text{ms}}}} = 1.414\text{A} \tag{1.48}$$

4. RMS of the complete square waveform is calculated as

$$I_{\text{rms}} = \sqrt{\frac{(4\text{A})^2 \times 10\cancel{\text{ms}} + (1\text{A})^2 \times 20\cancel{\text{ms}}}{30\cancel{\text{ms}}}} = 2.45\text{A} \tag{1.49}$$

or, alternatively, the total RMS value could be calculated as the sum of the RMS squares of the DC and AC components, as

$$I_{\text{rms}} = \sqrt{I_{\text{DC}}^2 + I_{\text{rms}}^2(\text{AC})} = \sqrt{2^2 + 1.414^2}\text{A} = 2.45\text{A} \tag{1.50}$$

which gives the same result because the RMS value of a DC level is a constant number.

1.5.3 AC Signal Power

So far, we have introduced AC through a pure resistive network. In general, we need to expand our analysis to include inductive and capacitive elements as well. Being energy storage components, these *reactive elements* may cause reversal of energy flow (i.e. power flow) within the network. Simply put, once charged they can serve as the temporary source of energy if the external voltage level becomes lower than their internal voltage. Consequently, it is common in the engineering community to define three "types" of power: real power P (i.e. power delivered to a pure resistive network); reactive power Q (i.e. power delivered to reactive components L and C); and complex power S (i.e. power delivered to a general RLC network); where the modulus of complex power $|S|$ is referred to as apparent power. At any given moment, the instantaneous power delivered to any circuit element or network is given by product $p(t) = v(t)\, i(t)$, where $p(t)$ is the instantaneous power, $v(t)$ is the instantaneous voltage, and $i(t)$ is the instantaneous current.

However, in the case of alternating currents and voltages, there is a very important consequence to notice, which we show here. Let us assume that instantaneous values of current and voltage in one branch of a circuit are given as follows:

$$i(t) = I_p \sin \omega t \tag{1.51}$$

$$v(t) = V_p \sin(\omega t + \phi) \tag{1.52}$$

In other words, there is a phase difference of ϕ between the current and voltage of that particular branch. Then, the instantaneous power is calculated as

$$p(t) = v(t)\, i(t) = V_p I_p \sin(\omega t)\, \sin(\omega t + \phi) \tag{1.53}$$

Equation (1.53) suggests that at some instances in time the power is positive and at other instances the power is negative (see Fig. 1.22). In order to correctly interpret the above statement, by convention the sign of power indicates the direction of energy flow, where "positive power" indicates that external world is supplying power to the circuit, while "negative power" indicates that the circuit is delivering power to the world. This is possible only if some devices capable of storing energy are present in the circuit, i.e. inductors or capacitors.

By definition (1.20), we find the average power of this circuit branch as

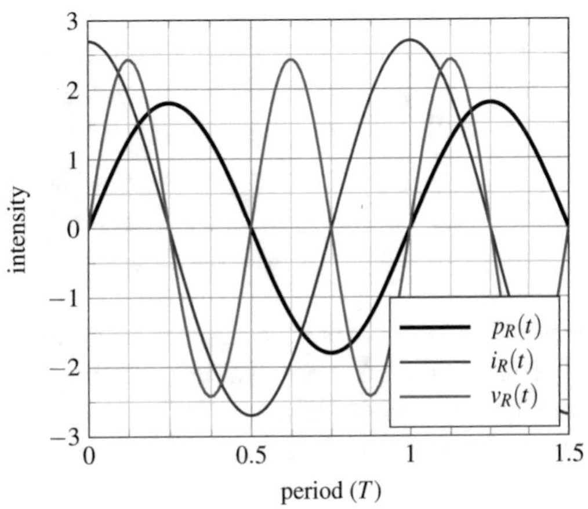

Fig. 1.22 Instantaneous voltage, current, and power in an AC circuit branch with phase difference $\phi = \pi/2$

$$\begin{aligned}\langle P\rangle &= \frac{1}{T}\int_0^T p(t)\,dt = \frac{1}{T}\int_0^T v(t)\,i(t)\,dt = \frac{V_p I_p}{T}\int_0^T \sin(\omega\,t)\sin(\omega\,t+\phi)\,dt\\ &= \frac{V_p I_p}{T}\int_0^T \frac{1}{2}\left[\cos(\cancel{\omega t}-\cancel{\omega t}-\phi)-\cos(\omega\,t+\omega\,t+\phi)\right]\,dt\\ &= \frac{V_p I_p}{2T}\int_0^T \left[\cos\phi-\cos(2\omega\,t+\phi)\right]\,dt = \frac{V_p I_p}{2\cancel{T}}\left[\cancel{T}\cos\phi-\cancelto{0}{\int_0^T \cos(2\omega t+\phi)\,dt}\right]\end{aligned}$$

therefore,

$$\langle P\rangle = \frac{V_p\,I_p}{2}\cos\phi \tag{1.54}$$

We note that AC power depends upon the cosine of the phase difference between the corresponding current and voltage. Consequently, in special cases when the phase difference $\phi = \pm 90°$ (i.e. in a purely reactive circuit), the AC power factor $\cos\phi$ is zero. When the power factor $\cos\phi = 1$ (i.e. in a purely resistive circuit) the power is at maximum. Therefore, a power factor less than one always indicates the presence of reactive (i.e. L and C) components in the circuit.

1.5.4 The Decibel Scale

In wireless communication systems, it is common to have RF transmitter delivering signals at power levels of the order of watts, kilowatts or even megawatts. As a comparison, the signal power level at the receiving antenna can be only a few picowatts. That is, the power ratio of the transmitted and received signals may be as large as 1 000 000 000 000 000:1. Clearly, using absolute numbers is not the most convenient way of presenting RF signal relations. By definition, the dB is a logarithmic unit of measurement that expresses the magnitude of a physical quantity (usually power) relative to a specified or implied reference level. Its logarithmic nature allows very large and very small ratios to be represented by a convenient number. Being a simple ratio of two quantities, the dB is a *dimensionless* unit. The Bel scale is defined as the logarithm base 10 of the power ratio. One Bel is a factor of 10, two Bels are a factor of 100, and so on. It is common, however, to use a more practical dB unit, so that 10dB is a power ratio of 10, 20dB is a ratio of 100, and so on, see Table 1.1. It is useful to remember that 3dB is a power ratio of ≈ 2, 6dB is a power ratio of ≈ 4, and so on. Every doubling of power in absolute numbers is equivalent to adding 3dB while every division by factor two is equivalent to adding -3dB. Thus, power ratio (i.e. power gain G) is expressed in dB as

$$G_{\text{dB}} \stackrel{\text{def}}{=} 10\log\frac{P_2}{P_1} \tag{1.55}$$

where P_1 and P_2 are the two signal powers being compared, for example, the input and output powers of an amplifier. It is convenient to know that when G_{dB} is a positive number, it indicates that $P_2 > P_1$ (often referred to as "gain"), while a negative G_{dB} number indicates that $P_1 > P_2$ (often referred to as "loss"). Naturally, 0dB means that $P_2 = P_1$. If we want to express a voltage (or current) ratio (i.e.

Table 1.1 Decibel scale for power ratios of ten (left) and two (right)

Ratio	Calculation	[dB]	Ratio	Calculation	[dB]
⋮	⋮	⋮	⋮	⋮	⋮
1/1000	10 log(1/1000) = 10 × (−3)	−30	1/8	10 log(1/8) = 10 × (−0.9)	−9
1/100	10 log(1/100) = 10 × (−2)	−20	1/4	10 log(1/4) = 10 × (−0.6)	−6
1/10	10 log(1/10) = 10 × (−1)	−10	1/2	10 log(1/2) = 10 × (−0.3)	−3
1	10 log(1/1) = 10 × (0)	0	1	10 log(1/1) = 10 × (0)	0
10	10 log(10) = 10 × (1)	10	2	10 log(2) = 10 × (0.3)	3
100	10 log(100) = 10 × (2)	20	4	10 log(4) = 10 × (0.6)	6
1000	10 log(1000) = 10 × (3)	30	8	10 log(8) = 10 × (0.9)	9
⋮	⋮	⋮	⋮	⋮	⋮

Table 1.2 Decibel scale for voltage or current ratios of ten (left) and two (right)

Ratio	Calculation	[dB]	Ratio	Calculation	[dB]
⋮	⋮	⋮	⋮	⋮	⋮
1/1000	20 log(1/1000) = 20 × (−3)	−60	1/8	20 log(1/8) = 20 × (−0.9)	−18
1/100	20 log(1/100) = 20 × (−2)	−40	1/4	20 log(1/4) = 20 × (−0.6)	−12
1/10	20 log(1/10) = 20 × (−1)	−20	1/2	20 log(1/2) = 20 × (−0.3)	−6
1	20 log(1/1) = 20 × (0)	0	1	20 log(1/1) = 20 × (0)	0
10	20 log(10) = 20 × (1)	20	2	20 log(2) = 20 × (0.3)	6
100	20 log(100) = 20 × (2)	40	4	20 log(4) = 20 × (0.6)	12
1000	20 log(1000) = 20 × (3)	60	8	20 log(8) = 20 × (0.9)	18
⋮	⋮	⋮	⋮	⋮	⋮

a voltage or current gain A) of two signals v_2 and v_1 in the dB scale, and assuming that both signals are measured across the same impedance Z, then the gain is expressed in dB as

$$A_{\text{dB}} \stackrel{\text{def}}{=} 10\log\frac{P_2}{P_1} = 10\log\frac{v_2^2/\cancel{Z}}{v_1^2/\cancel{Z}} = 10\log\left(\frac{v_2}{v_1}\right)^2 = 20\log\frac{v_2}{v_1} \tag{1.56}$$

which is to say that a voltage (or current) ratio of 10 equals 20dB gain, a ratio of 0.1 equals −20dB gain, a ratio of 100 is equal to 40dB gain, etc., see Table 1.2. It is handy to practice mental conversion between ratios and dB units by taking the number of zeros in the ratio exponent and multiplying it by 10 for power or by 20 for voltage or current; the final number is then in dB units.

Because dB numbers are dimensionless they do not say anything about the absolute power levels being compared. Hence, given a specific gain, we can only conclude whether there was power amplification or power loss. From such a statement, however, we cannot conclude neither what kind of gain it is (i.e. power, voltage, or current) nor which two absolute signal values are being compared. Therefore, for low-power applications, the standard reference value for power specification is defined in the form of the dBm scale, which is set to compare a given power level relative to the absolute power level of $P_1 = 1\text{mW}$. After substituting the 1mW level in (1.55), we write

$$G_{\text{dBm}} \stackrel{\text{def}}{=} 10\log\frac{P_2}{1\text{mW}} \tag{1.57}$$

indicating that 1mW of power is equivalent to 0dBm. Similarly, if an amplifier delivers 10mW of power, it is usually expressed as 10dBm gain, 100mW as 20dBm, etc. Note that, due to the same scale, in power calculations the units of dB and dBm are added to or subtracted from each other, i.e. they are interchangeable as long as we keep the 1mW absolute reference in mind. In order to facilitate RF circuits analysis, a brief summary is given in Tables 1.1 and 1.2. Note that every increase by factor of ten (i.e. "decade") power increases by 10dB while each doubling of power adds 3dB. Furthermore, division by the same factor only changes the sign of the result. Similarly, for voltages or currents, each decade adds 20dB and each doubling of voltage or current amplitude adds 6dB. Again, division by the same factors only changes the sign of the result. A special and very important case

$$20\ \log(\sqrt{2}) \approx 3\,\text{dB} \tag{1.58}$$

is known as the "3dB point" as it is used for definition of a filter's bandwidth. In the case of complex transfer function, the "$\sqrt{2}$" term is its module when both real and imaginary parts are equal.

Example 12: Units of dB/dBm
A cell phone transmits $P_1 = +30\text{dBm}$ of signal power from its antenna. At the receiving side the signal power is $P_2 = 5\text{pW}$. Calculate the propagation loss in the transmitting medium.

Solution 12: We convert the received power into dBm units as

$$P_2 = 10\log\frac{P_2}{1\text{mW}} = 10\log\frac{5\text{pW}}{1\text{mW}} = -83\text{dBm}$$

Therefore the total signal power loss is the difference between the power levels at the end $P_2 = -83\text{dBm}$ and at the beginning, i.e. $P_1 = +30\text{dBm}$, which is to say

$$A = P_2 - P_1 = -83\text{dBm} - 30\text{dBm} = -113\text{dBm}$$

1.6 Complex Transfer Functions

As a consequence of impedance being dependent upon frequency, it is unavoidable that circuit transfer functions $H(j\omega)$ are inherently complex,

$$H(j\omega) = \Re(H(j\omega)) + j\Im(H(j\omega)) \tag{1.59}$$

thus they must be analysed in the frequency domain. One of the practical methods to show this frequency dependence is to use Bode plots where convenience of logarithmic scale becomes evident. In the following sections we review five basic "building blocks" (i.e. basic mathematical forms) of Bode plots. More complicated transfer complex functions can be factored into some combinations of these basic forms.

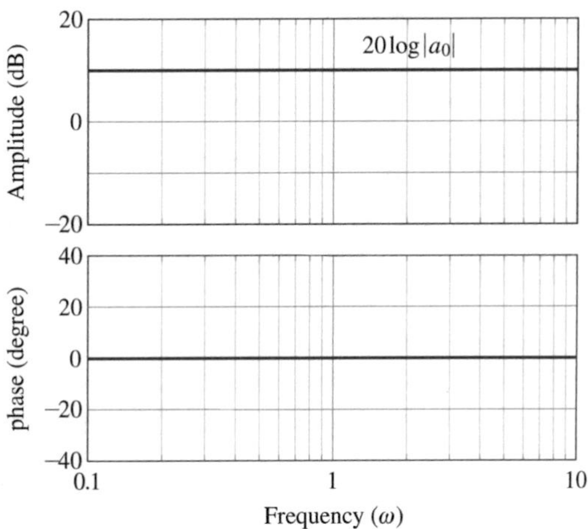

Fig. 1.23 Bode plot of a constant function, normalized to $\omega = 1$

1.6.1 First Order Function, Form 1

Simplest form of transfer function $H(j\omega)$ is a *constant*, which is trivial case of complex transfer function when $\Im\,(H(j\omega)) = 0$, that is

$$H_1(j\omega) = a_0 \qquad (a_0 \in \Re) \tag{1.60}$$

which, in complex plane[19] is analysed as two functions, one for its amplitude $|H_1(j\omega)|$ and one for its phase $\angle H_1(j\omega)$, i.e.

$$|H_1(j\omega)| = \sqrt{\Re(H_1(j\omega))^2 + \Im(H_1(j\omega))^2} = \sqrt{a_0^2 + 0^2} = |a_0| \tag{1.61}$$

$$\angle H_1(j\omega) = \arctan\frac{\Im(H_1(j\omega))}{\Re(H_1(j\omega))} = \arctan\frac{0}{a_0} = 0^\circ \tag{1.62}$$

Being a simple real number that does not depend on frequency, function (1.61) represents DC gain of a transfer function. The corresponding Bode plot of any transfer function $H(j\omega)$, in effect, consists of two frequency dependent plots: one for the gain function, in this case (1.61), that is found as the module of $|H_1(j\omega)|$ in its logarithmic form, and one for the phase function (1.62), i.e.

$$20\log\left(|H_1(j\omega)|\right) = 20\log\left(|a_0|\right) \tag{1.63}$$

$$\angle H_1(j\omega) = 0^\circ \tag{1.64}$$

Direct calculations of (1.63) and (1.64) are given in the form of Bode plot, Fig. 1.23, where in this case the two functions are trivial, thus their exact Bode plot is identical to its linear form.

[19]See Appendix C for complex algebra reminder.

1.6.2 First Order Function, Form 2

Complex function that is strictly imaginary (i.e. $\Re(H_2(j\omega)) = 0$) is written in the form

$$H_2(j\omega) = j\frac{\omega}{\omega_0} \qquad (j^2 = -1,\ \omega_0 = \texttt{const.}) \tag{1.65}$$

where ω_0 is a given frequency. Form (1.65) is convenient to normalize the transfer function so that when $\omega = \omega_0$ then amplitude $|H_2(\omega_0)| = 1$ (i.e. it equals 0dB).

After rewriting function (1.65) in its logarithmic form

$$20\log\left(H_2(j\omega)\right) = 20\log\left(j\frac{\omega}{\omega_0}\right) \quad [\text{dB}] \tag{1.66}$$

amplitude and phase of complex function $H_2(j\omega)$ are calculated as follows.

1. **amplitude of** $H_2(j\omega)$: module of (1.66) is calculated as

$$\begin{aligned} 20\log|H_2(j\omega)| &\stackrel{\text{def}}{=} 20\log\left[\sqrt{\Re(H_2(j\omega))^2 + \Im(H_2(j\omega))^2}\right] \\ &= 20\log\left[\sqrt{0 + \left(\frac{\omega}{\omega_0}\right)^2}\,\right] = 20\log\left|\frac{\omega}{\omega_0}\right| \quad [\text{dB}] \end{aligned} \tag{1.67}$$

$$\therefore$$

$$20\log|H_2(\omega_0)| = 20\log(1) = 0\text{dB}$$

Form (1.67) shows that in log scale $|H_2(j\omega)|$ is a linear function of $\omega > 0$ and its zero is found at $\omega = \omega_0$. Also, $|H_2(j\omega)| > 0$, thus on logarithmic scale its vertical asymptote is at $\omega = 0$. Consequently, this linear function (1.67) is shown over an interval including at least two decades centred at $\omega = \omega_0$, that is from $(\omega_1 = 0.1\ \omega_0)$ to $(\omega_2 = 10\ \omega_0)$, but obviously never at $\omega = 0$. Being a linear function of the form (1.67), $|H_2(j\omega)|$ has +20dB slope for voltage (current) signals (which illustrates the "voltage gain changes 20dB per decade" phrase), see Fig. 1.24 (top).

2. **phase of** $H_2(j\omega)$: this function is strictly imaginary, i.e. its form is $z = 0 + j\,\Im$ thus, by definition,[20] its phase ϕ is constant $\phi = +\pi/2$, because

$$\angle H_2(j\omega) \stackrel{\text{def}}{=} \arctan\frac{\Im(H_2(j\omega))}{\Re(H_2(j\omega))} = \arctan\frac{\frac{\omega}{\omega_0}}{0} = \arctan(+\infty) = +\frac{\pi}{2} \tag{1.68}$$

Therefore, the gain and phase equations of $H_2(j\omega)$ are summarized as follows:

$$20\log|H_2(j\omega)| = 20\log\left|\frac{\omega}{\omega_0}\right| \tag{1.69}$$

$$\angle H_2(j\omega) = +\frac{\pi}{2} = +90^\circ \tag{1.70}$$

[20] See Appendix C for complex algebra reminder.

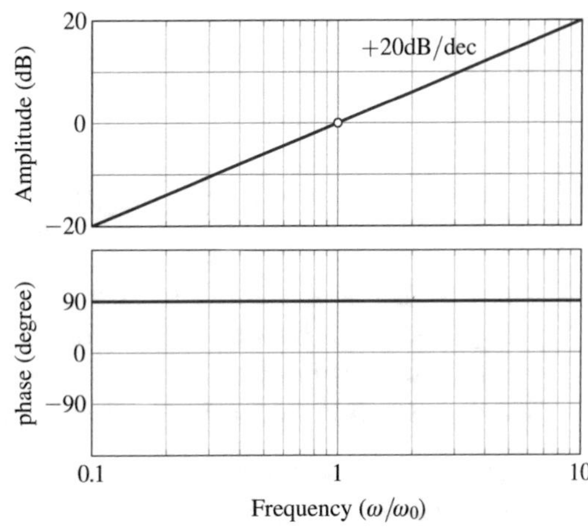

Fig. 1.24 Bode plot of a first order transfer function (normalized to ω/ω_0) whose amplitude slope is +20dB/decade, and the phase is constant $+\pi/2$

Direct calculations of (1.69) and (1.70) are given in the form of Bode plot, Fig. 1.24, where in this case the two functions are trivial, thus their exact Bode plot is identical to its linear form.

1.6.3 First Order Function, Form 3

First order complex function $H_3(j\omega)$ that takes general form $z = 1 + jb$, that is to say $\Re(z) = 1$, so that its normalized form is written as

$$H_3(j\omega) = 1 + j\frac{\omega}{\omega_0} \tag{1.71}$$

which is a very convenient form because the fact that $\Re(H_3(j\omega)) = 1$ simplifies its analysis, as we see in the following sections. Therefore, its logarithmic form is written as

$$20\log H_3(j\omega) = 20\log\left(1 + j\frac{\omega}{\omega_0}\right) \tag{1.72}$$

1. **amplitude of** $H_3(j\omega)$: Exact amplitude is written by inspection of (1.72) as

$$20\log|H_3(j\omega)| \overset{\text{def}}{=} 20\log\left(\sqrt{\Re(H_3(j\omega))^2 + \Im(H_3(j\omega))^2}\right)$$

$$= 20\log\sqrt{1^2 + \left(\frac{\omega}{\omega_0}\right)^2} \tag{1.73}$$

$$\therefore$$

$$20\log|H_3(\omega_0)| = 20\log\sqrt{1^2 + \left(\frac{\cancel{\omega_0}}{\cancel{\omega_0}}\right)^2} = 20\log\sqrt{2} = +3\text{dB} \tag{1.74}$$

2. **phase of** $H_3(j\omega)$: it is convenient to write the exact expression for phase as,

$$\angle H_3(j\omega) \overset{\text{def}}{=} \arctan \frac{\Im(H_3(j\omega))}{\Re(H_3(j\omega))} = \arctan \frac{\frac{\omega}{\omega_0}}{1} = \arctan \frac{\omega}{\omega_0} \quad (1.75)$$

$$\therefore$$

$$\angle H_3(\omega_0) = \arctan \left(\frac{\cancel{\omega_0}}{\cancel{\omega_0}} \right) = \arctan(1) = +\frac{\pi}{4} = +45^\circ \quad (1.76)$$

which is also general result for any complex function at the point when $\Re(H(j\omega))$ equals $\Im(H(j\omega))$. This is the well known property of a right triangle whose catheti have equal lengths (as justified by Pythagoras' theorem). Both real and imaginary parts of this function are positive, thus the phase is in the first quadrant and arctan function gives angles between 0° and 90°.

Being non-trivial functions, (1.73) and (1.75), Bode plot is created by *piecewise linear approximations*, as follows:

1. *HF region* ($\omega \gg \omega_0$): in the high-frequency range relative to ω_0 the amplitude function (1.73) degenerates into a linear function because if ($\omega \gg \omega_0$), then ($\omega/\omega_0 \approx \omega$) as well as ($\omega/\omega_0 \gg 1$). With these approximations we conclude that ($1^2 + (\omega/\omega_0)^2 \approx \omega^2$), thus

$$20 \log \left[\lim_{\omega \gg \omega_0} |H_3(j\omega)| \right] = 20 \log \left[\lim_{\omega \gg \omega_0} \sqrt{1^2 + \left(\frac{\omega}{\omega_0}\right)^2} \right] \quad (1.77)$$

$$\approx 20 \log |\omega| \quad (\omega \geq \omega_0) \quad [\text{dB}] \quad (1.78)$$

At the same time, the phase function (1.75) follows the limit of arctan as

$$\lim_{\omega \to \infty} \arctan \frac{\omega}{\omega_0} = \arctan(\infty) = +\frac{\pi}{2} = +90^\circ \quad (1.79)$$

To help visualize behaviour of arctan function and its limits, we draw the associated right angle triangle, where the signs of horizontal and vertical catheti are specifically shown so that there is not ambiguity[21] about the quadrant where corresponding angle is found. In respect to notification in (1.79) and Fig. 1.25, as well as definition $\tan\alpha \overset{\text{def}}{=} y/x$, in this case we write that the vertical catheti in Fig. 1.25 equal to ω and the horizontal catheti equal to ω_0. As $\omega \to \infty$, i.e. vertical catheti become longer relative to ω_0, it force $\alpha \to 90^\circ$. Similarly, as $\omega \to 0$, i.e. vertical catheti become shorter relative to ω_0, it force $\alpha \to 0^\circ$.

2. *Crossover point* ($\omega = \omega_0$): at frequency ω_0, in respect to (1.73) and (1.75), the amplitude and phase functions have the following values:

$$20 \log |H_3(\omega_0)| = 20 \log \sqrt{2} = +3\text{dB} \quad (1.80)$$

$$\angle H_3(\omega_0) = \arctan(1) = +\frac{\pi}{4} = +45^\circ \quad (1.81)$$

We note that in HF region we approximated the gain function with zero when ($\omega = \omega_0$), therefore there is +3dB discrepancy at this particular point.

3. *LF region* ($\omega \ll \omega_0$): in this region amplitude function (1.73) degenerates into a linear function because if ($\omega \ll \omega_0$), then ($\omega/\omega_0 \approx 0$) as well as ($1^2 + (\omega/\omega_0)^2 \approx 1^2$), thus

[21] See discussion about differences between $\arctan(x)$ vs. $\arctan 2(x)$ functions in Sect. 1.6.5.

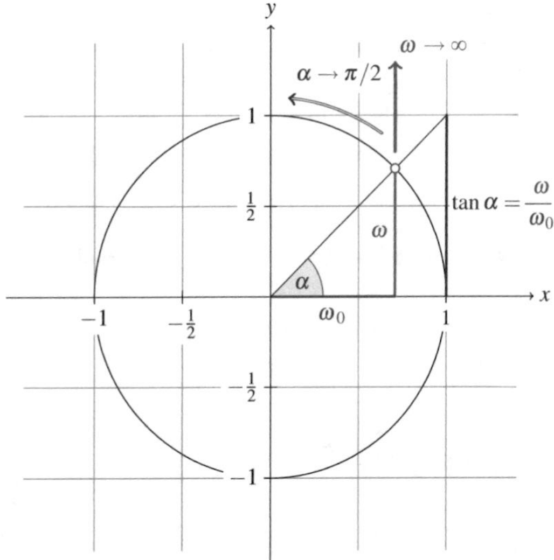

Fig. 1.25 Geometrical interpretation of arctan limit in the first quadrant

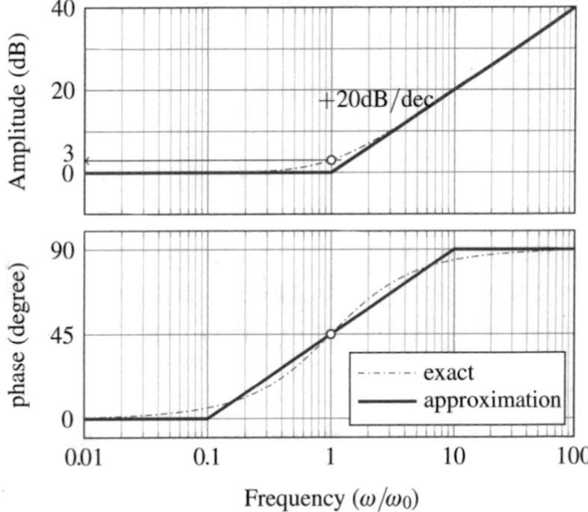

Fig. 1.26 Bode plot of a first order transfer function (normalized to ω/ω_0) whose amplitude slope is +20dB/decade, and the phase is between 0 and $+\pi/2$

$$20\log \lim_{\omega\to 0} |H_3(j\omega)| = 20\log\left[\lim_{\omega\to 0}\sqrt{1^2 + \left(\frac{\omega}{\omega_0}\right)^2}\right] = 20\log(1) = 0\text{dB} \tag{1.82}$$

At the same time, the phase function (1.75) follows the limit of arctan as

$$\lim_{\omega\to 0} \arctan\frac{\omega}{\omega_0} = \arctan(0) = 0° \tag{1.83}$$

The above piecewise linear approximations of amplitude and phase functions are shown in Fig. 1.26 as solid lines. It shows that, from DC until ω_0, amplitude of this function is approximated with a zero, then it continues with linear section whose gain +20dB. Similarly, the single point $\angle H_3(j\omega_0) = +45°$ as well as the two extreme limiting phase values determine the approximated phase function. That is to say, starting from DC the phase function is approximated with a constant value (here, 0°) until one decade before 3dB point, i.e. from DC to $0.1\omega_0$ point. High-frequency approximation starts one decade after 3dB point, i.e. from $10\omega_0$ point to infinity. In between these two constant values there is linear phase approximation (here, between 0° and 90°) spreading over two decades from $0.1\omega_0$ to $10\omega_0$ and crossing the +45° level at 3dB point.

While doing fast hand analysis it is sufficient to quickly draw the piecewise linear approximation version of this transfer function, where ω_0 is given. We keep in mind the 3dB difference between the exact and approximated values of the amplitude function at ω_0.

1.6.4 First Order Function, Form 4

First order complex function $H_4(j\omega)$ with general form that is inverse of the one in Sect. 1.6.3 is written in the following forms:

$$H_4(j\omega) = \frac{1}{1 + j\frac{\omega}{\omega_0}} \tag{1.84}$$

thus, the complex form of (1.84) is derived as follows:

$$H_4(j\omega) = \frac{1}{1 + j\frac{\omega}{\omega_0}}\,\frac{1 - j\frac{\omega}{\omega_0}}{1 - j\frac{\omega}{\omega_0}} = \frac{1}{\sqrt{1 + (\omega/\omega_0)^2}} - j\frac{\omega/\omega_0}{\sqrt{1 + (\omega/\omega_0)^2}} \tag{1.85}$$

$$= \Re(H_4(j\omega)) - j\,\Im(H_4(j\omega)) \tag{1.86}$$

where positive real part and negative imaginary part of a complex function imply that its phase must be in the fourth quadrant of the complex plane,[22] see Fig. 1.27. Logarithmic form of this function is written with the help of log identities[23]

$$\begin{aligned} 20\log H_4(j\omega) &= 20\log\left(\frac{1}{1 + j\frac{\omega}{\omega_0}}\right) \\ &= \cancelto{0}{20\log(1)} - 20\log\left(1 + j\frac{\omega}{\omega_0}\right) \\ &= 0 - 20\log(H_3(j\omega)) = -20\log(H_3(j\omega)) \end{aligned} \tag{1.87}$$

which is convenient because we already analysed Bode plot of $H_3(j\omega)$ in Sect. 1.6.3. Knowing that the "−" sign in front of a function simply mirrors the function relative to its horizontal axis and having already analysed $H_3(j\omega)$ function, it is straightforward to draw Bode plot of $H_4(j\omega)$. Nevertheless, for the sake of completeness, we write

1. **amplitude of** $H_4(j\omega)$: the exact logarithmic form of the amplitude function is written by inspection of (1.87) as

$$20\log|H_4(j\omega)| = -20\log\left(\sqrt{\Re(H_4(j\omega))^2 + \Im(H_4(j\omega))^2}\right)$$

[22] See Appendix C for complex algebra reminder.

[23] $\log(a/b) = \log a - \log b$.

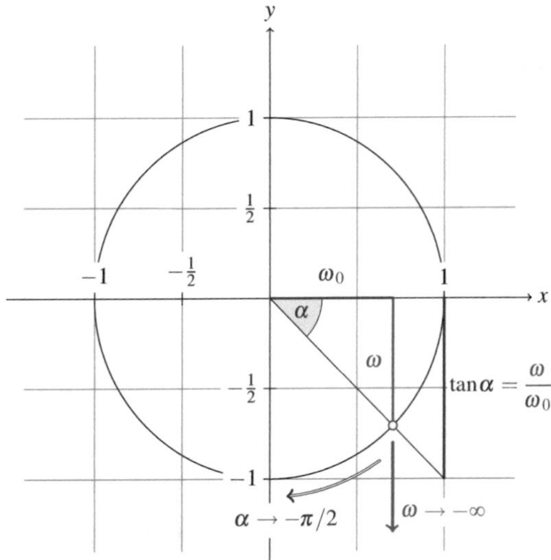

Fig. 1.27 Geometrical interpretation of arctan limit in the fourth quadrant

$$= -20 \log \left(\sqrt{1^2 + \left(\frac{\omega}{\omega_0} \right)^2} \right) \tag{1.88}$$

$\therefore$

$$20 \log |H_4(\omega_0)| = -20 \log \sqrt{2} = -3\text{dB} \tag{1.89}$$

2. **phase of** $H_4(j\omega)$: it is convenient to write the exact expression for phase as,

$$\angle H_4(j\omega) = \arctan \frac{\Im(H_4(j\omega))}{\Re(H_4(j\omega))} = \arctan \frac{-\frac{\omega/\omega_0}{\sqrt{1+(\omega/\omega_0)^2}}}{\frac{1}{\sqrt{1+(\omega/\omega_0)^2}}} = -\arctan \frac{\omega}{\omega_0} \tag{1.90}$$

$\therefore$

$$\angle H_4(\omega_0) = -\arctan(1) = -\frac{\pi}{4} = -45^\circ \tag{1.91}$$

which is also the well known result when $\Re(H)$ equals $\Im(H)$. As a consequence of positive real part and negative imaginary part, the phase is found in the fourth quadrant as -45°, which is an important detail specific for this function, see Fig. 1.27.

Piecewise linear approximations of (1.88) and (1.90) are derived as follows:

1. *HF region* ($\omega \gg \omega_0$): in the high-frequency range relative to the given ω_0 the amplitude function (1.73) degenerates into a linear function because if ($\omega \gg \omega_0$), then ($\omega/\omega_0 \approx \omega$) as well as ($\omega/\omega_0 \gg 1$). With these approximations we conclude that ($1^2 + (\omega/\omega_0)^2 \approx \omega^2$), thus

$$20 \log \left[\lim_{\omega \gg \omega_0} |H_4(j\omega)| \right] = -20 \log \left[\lim_{\omega \gg \omega_0} \sqrt{1^2 + \left(\frac{\omega}{\omega_0} \right)^2} \right]$$

$$= -20 \log |\omega| \quad (\omega > 0) \quad [\text{dB}] \tag{1.92}$$

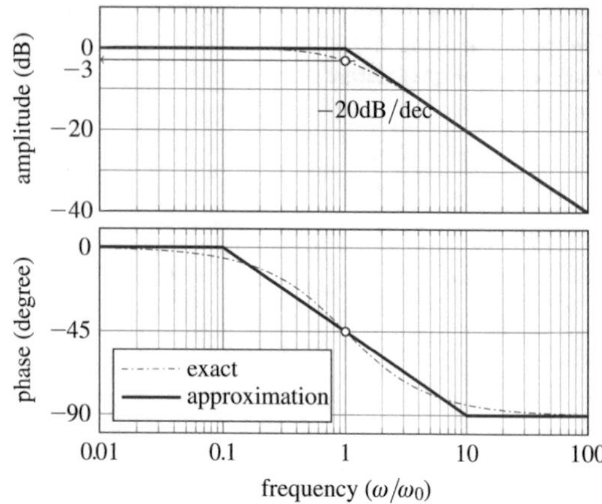

Fig. 1.28 Bode plot of a first order transfer function whose amplitude slope is −20dB/decade, and the phase is between 0 and $-\pi/2$

and its phase function follows the limit of $\arctan(\alpha)$ in the forth quadrant as

$$-\lim_{\omega\to\infty} \arctan\frac{\omega}{\omega_0} = -\arctan(\infty) = -\frac{\pi}{2} = -90^\circ \tag{1.93}$$

2. *Crossover point* ($\omega = \omega_0$): at the specific given frequency ω_0 the amplitude and phase functions have the following values:

$$20\log|H_4(\omega_0)| = -20\log\sqrt{2} = -3\text{dB} \tag{1.94}$$

$$\angle H_4(\omega_0) = -\arctan(1) = -\frac{\pi}{4} = -45^\circ \tag{1.95}$$

3. *LF region* ($\omega \ll \omega_0$): for low frequencies relative to ω_0 amplitude function (1.88) degenerates into a linear function because if ($\omega \ll \omega_0$), then ($\omega/\omega_0 \approx 0$) as well as ($1^2 + (\omega/\omega_0)^2 \approx 1^2$), thus

$$20\log\lim_{\omega\to 0}|H_4(j\omega)| = -20\log\left[\lim_{\omega\to 0}\sqrt{1^2+\left(\frac{\omega}{\omega_0}\right)^2}\right] = -20\log(1) = 0\text{dB} \tag{1.96}$$

At the same time, the phase function (1.90) follows the limit of arctan as

$$-\lim_{\omega\to 0}\arctan\frac{\omega}{\omega_0} = -\arctan(0) = 0^\circ \tag{1.97}$$

Gain and phase piecewise linear approximations are summarized in Fig. 1.28. From DC until ω_0, amplitude function is approximated with a horizontal linear section at level zero, then it continues with another linear section with −20dB slope. We note that $\angle H_4(j\omega_0) = -45^\circ$. That is to say, starting from DC the phase function is approximated as 0° until one decade before 3dB point, i.e. in DC to $0.1\omega_0$ range. High-frequency approximation starts from $10\omega_0$ relative to 3dB point until infinity. In the range from $0.1\omega_0$ to $10\omega_0$ the phase function is approximated as a linear function from 0° and −90°, therefore it must cross the −45° point at −3dB frequency.

Example 13: Bode Plot: First Order Function
Show Bode plot of the following transfer function:

$$H(\omega) = \frac{100}{j\omega + 10} \tag{1.98}$$

Solution 13: This is a first order function, it is necessary to rewrite it in some of its basic forms.

$$H(\omega) = \frac{100}{j\omega + 10} = 100\frac{1}{j\omega + 10} = \frac{100}{10}\frac{1}{1 + j\dfrac{\omega}{10}} = 10\frac{1}{1 + j\dfrac{\omega}{10}} \tag{1.99}$$

Therefore, by inspection of (1.99) we find that $\omega_0 = 10$, and logarithmic form of gain function (1.99) is

$$20 \log H(\omega) = 20 \log\left(10\frac{1}{1 + j\dfrac{\omega}{10}}\right) = \underbrace{\underbrace{20 \log 10}_{\text{Sect. 1.6.1}}}_{①} + \cancelto{0}{20 \log(1)} \underbrace{\underbrace{-20 \log\left(1 + j\frac{\omega}{10}\right)}_{\text{Sect. 1.6.4},\omega_0=10}}_{④}$$

The first term is a constant term whose value is 20dB (form 1, Sect. 1.6.1), the second term equals zero, and the third term is form that is introduced as form 4 in Sect. 1.6.4.

In addition, its corresponding phase function is

$$\angle H(\omega) = \underbrace{\underbrace{0^\circ}_{\text{Sect. 1.6.1}}}_{①} \underbrace{\underbrace{- \arctan\frac{\omega}{10}}_{\text{Sect. 1.6.4},\omega_0=10}}_{④}$$

Gain and phase Bode plots are created as simple superposition of the linear sections in the gain and phase graphs, Fig. 1.29.

1.6.5 Second-Order Transfer Function

A general form of a second-order transfer function is in the form

$$H_5(j\omega) = \frac{\omega_0^2}{s^2 + 2\xi\omega_0 s + \omega_0^2} \tag{1.100}$$

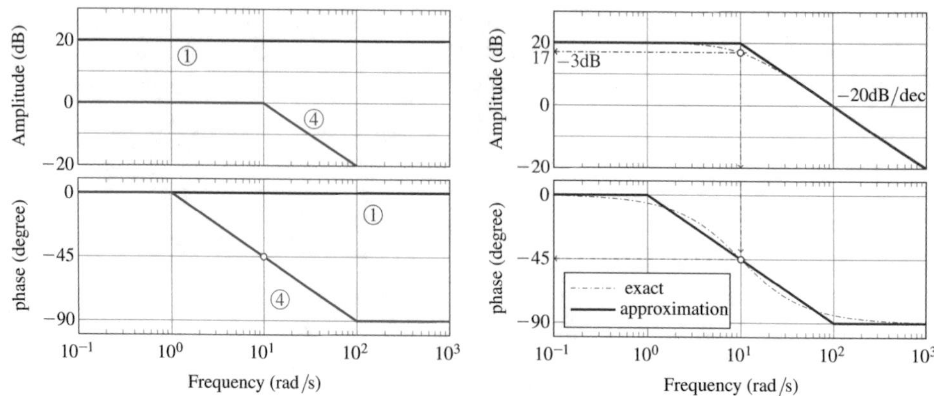

Fig. 1.29 Linear approximation of two gain and phase (left), and their respective sums along with the exact solutions (right)

where the alternative forms are derived as

$$H_5(j\omega) = \frac{1}{\left(\dfrac{s}{\omega_0}\right)^2 + \dfrac{2\xi}{\omega_0}s + 1} = \frac{\omega_0^2}{(\omega_0^2 - \omega^2) + j\,2\xi\omega_0\,\omega} \tag{1.101}$$

$$= \frac{\omega_0^2\,(\omega_0^2 - \omega^2)}{\sqrt{(\omega_0^2 - \omega^2)^2 + (2\xi\omega_0\,\omega)^2}} - j\,\frac{\omega_0^2\,(2\xi\omega_0\,\omega)}{\sqrt{(\omega_0^2 - \omega^2)^2 + (2\xi\omega_0\,\omega)^2}} \tag{1.102}$$

$$= \Re(H_5(j\omega)) - j\,\Im(H_5(j\omega)) \tag{1.103}$$

where $s = j\omega$ is complex variable. Quadratic polynomial in the denominator must have two zeros (which are therefore poles of transfer function $H_5(j\omega)$) that are calculated either by the polynomial factorization methods or by using the well known formula for zeros of quadratic function

$$s_{1,2} = -\xi\omega_0 \pm \omega_0\sqrt{\xi^2 - 1} \tag{1.104}$$

When $\xi > 1$ poles of the quadratic function are real and can be factorized into the first order forms. However, the case that is of special interest is when $0 < \xi < 1$, i.e. when poles of quadratic function are complex-conjugate pair. In that case we write logarithmic form of (1.101) as

$$20\log H_5(j\omega) = 20\log\frac{\omega_0^2}{(\omega_0^2 - \omega^2) + j\,(2\xi\omega_0\,\omega)} \tag{1.105}$$

$$= 20\log\left(\omega_0^2\right) - 20\log\left[(\omega_0^2 - \omega^2) + j\,(2\xi\omega_0\,\omega)\right] \tag{1.106}$$

1. **amplitude of** $H_5(j\omega)$: the exact equation for amplitude is written by inspection of (1.106)

$$20\log|H_5(j\omega)| = 20\log\left(\omega_0^2\right) - 20\log\sqrt{(\omega_0^2 - \omega^2)^2 + (2\xi\omega_0\,\omega)^2} \tag{1.107}$$

$$\therefore$$

$$20\log|H_5(\omega_0)| = 20\log\left(\omega_0^2\right) - 20\log\left(2\,\xi\,\omega_0^2\right)$$

$$= 20 \log \frac{\omega_0^2}{2\xi\, \omega_0^2} = 20 \log \frac{1}{2\xi} \tag{1.108}$$

which is to say that exact value of $|H_5(\omega_0)|$ depends on parameter ($\xi > 0$), for the moment we note that if $\xi = 1/2$, then $|H_5(\omega_0)| = 0\text{dB}$.

2. **phase of** $H_5(j\omega)$: we write the exact expression for phase where ($\omega, \omega_0, \xi > 0$) of (1.102), as

$$\angle H_5(j\omega) = \arctan \frac{\Im(H_5(j\omega))}{\Re(H_5(j\omega))} = \arctan \frac{-2\xi\omega_0\,\omega}{\omega_0^2 - \omega^2} = \arctan \frac{-2\xi \dfrac{\omega}{\omega_0}}{1 - \left(\dfrac{\omega}{\omega_0}\right)^2} \tag{1.109}$$

Function (1.109) is ambiguous relative to a simple $\arctan(x)$ function in the first quadrant. Let us assume $\xi = 1/2$ and $x = \omega/\omega_0$ so that we can write (1.109) as

$$\angle H_5(x) = \arctan \frac{-x}{1 - x^2} \tag{1.110}$$

where if $\omega > \omega_0$, then $x > 1$ forcing $1-x^2 < 0$, which puts the phase into the third quadrant (both numerator and denominator are negative). Similarly, if $\omega < \omega_0$, then $x < 1$ forcing $1 - x^2 > 0$, which puts the phase into the forth quadrant (numerator is negative and denominator is positive). When $\omega \to \omega_0$, i.e. $x \to 1$ so that denominator in (1.110) tends to zero and forces the argument of arctan function in (1.110) to infinity. Thus, we have to pay attention how to calculate the limiting value

$$\lim_{x\to 1} \left(\arctan \frac{-x}{1 - x^2} \right) \tag{1.111}$$

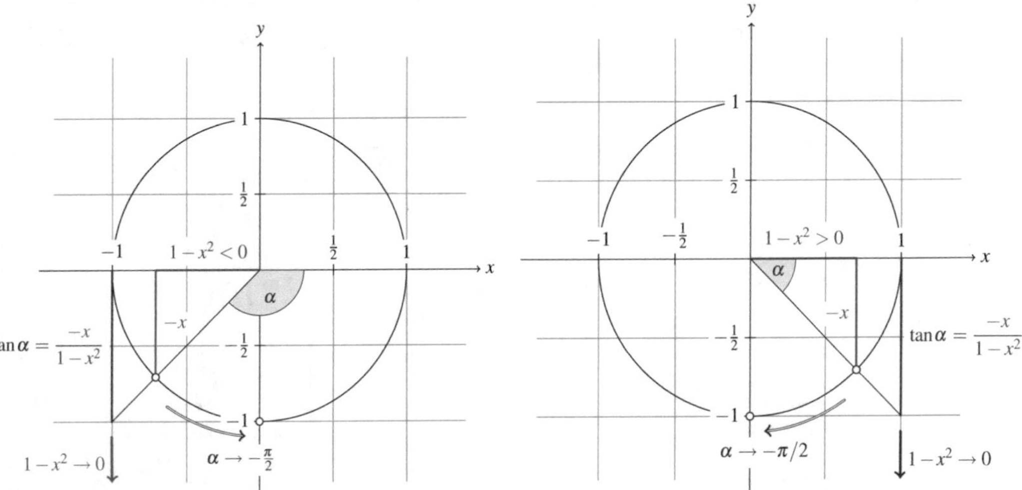

Fig. 1.30 Geometrical interpretation of $\tan(\alpha)$ limit of (1.110) in the third and fourth quadrants

which is ambiguous for numerical programs lacking `atan2(y,x)` mathematical function. We use the graphical representation of $\tan(\alpha)$ function and its limit when $\omega \to \omega_0$, Fig. 1.30, in other words $1 - x^2 \to 0$, to reach conclusion that

$$\angle H_5(\omega_0) = -\frac{\pi}{2} = -90^\circ \tag{1.112}$$

3. **damping factor** ξ of $H_5(j\omega)$: both amplitude and phase depend on ξ, (1.108) and (1.109), so it is important to examine this dependence. Calculation of amplitude $|H(j\omega)|$ according to (1.108) shows that for small values of ξ the amplitude can cause large "peaking" close to ω_0, while larger values of ξ keep the amplitude below its asymptotic level, see Fig. 1.31. Position of the peaking frequency and range of parameter ξ is found by searching maximum of (1.107), i.e.

$$\frac{d}{d\omega} 20 \log |H_5(j\omega)| = 0$$

$$\therefore$$

$$\frac{d}{d\omega}\left[20 \log\left(\omega_0^2\right) - 20 \log \sqrt{(\omega_0^2 - \omega^2)^2 + (2\xi\omega_0\,\omega)^2}\right] = 0$$

$$\therefore$$

$$-\frac{40\omega\,(\omega^2 + 2\omega_0^2\,\xi^2 - \omega_0^2)}{\ln(10)\left[(\omega_0^2 - \omega^2)^2 + 4\omega_0^2\,\xi^2\omega^2\right]} = 0$$

which is to say that maximum of $|H(j\omega)|$ occurs at frequency ω_m when

$$\omega_m^2 + 2\omega_0^2\,\xi^2 - \omega_0^2 = 0 \quad \therefore \quad \omega_m = \omega_0\sqrt{1 - 2\xi^2} \tag{1.113}$$

leading into conclusion that, for the maximum to occur, ξ is limited to interval $0 \le \xi \le 1/\sqrt{2}$ (so that the argument of the square root function is positive). Numerical calculations of phase function (1.109) are shown in Fig. 1.31 (right).

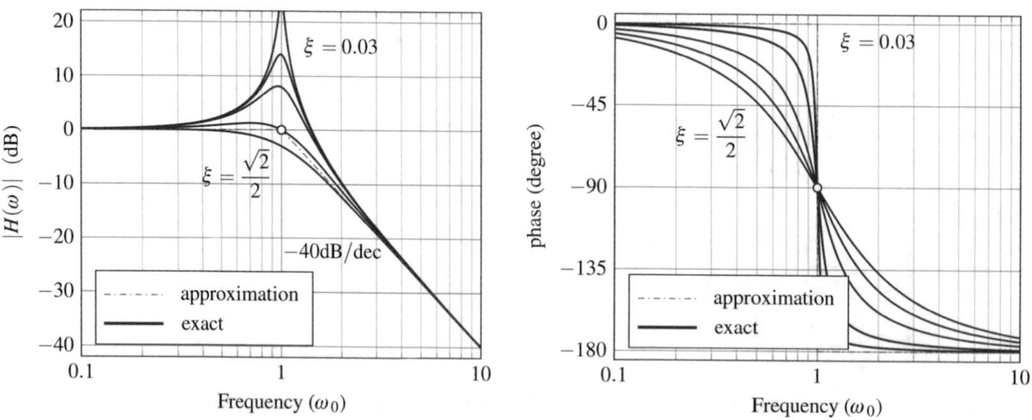

Fig. 1.31 Calculated values (left) and Bode plot of phase function of second-order function given in Sect. 1.6.5

4. **piecewise linear approximation**: hand-drawn Bode plot is realized by using *piecewise linear approximation* of (1.107) and (1.109) as the following limits and approximations show

   a. *HF region* ($\omega \gg \omega_0$): in the high-frequency range relative to the given ω_0 (and knowing that $0 \leq \xi \leq 1/\sqrt{2}$) the amplitude function (1.107) degenerates into a linear function as

$$
\begin{aligned}
20\log\left[\lim_{\omega\gg\omega_0}|H_5(j\omega)|\right] &= \lim_{\omega\gg\omega_0}\left[20\log\omega_0^2 - 20\log\sqrt{\left[(\omega_0^2-\omega^2)^2+(2\xi\omega_0\,\omega)^2\right]}\right]\\
&= 20\log\omega_0^2 - 20\log\sqrt{\lim_{\omega\gg\omega_0}\left[(\omega_0^2-\omega^2)^2+(2\xi\omega_0\,\omega)^2\right]}\\
&= 20\log\omega_0^2 - 20\log\sqrt{\lim_{\omega\gg\omega_0}\left[\omega^4+(2\xi\omega_0)^2\,\omega^2\right]}\\
&= 20\log\omega_0^2 - 20\log\omega^2\\
&= 40\log\omega_0 - 40\log\omega = 40\log\frac{\omega_0}{\omega}
\end{aligned}
\tag{1.114}
$$

   which is to say that linear approximation of (1.107) in log–log scale consists of constant function "$40\log\omega_0$" and linear function "$-40\log\omega$" whose slope is −40dB/dec, see (1.114) and Fig. 1.31 (left). Additionally, to resolve the phase function we must find two limits of (1.110): $\omega \to \infty$ and $\omega \to 0$, which is equivalent to

$$
\lim_{x\to\infty}\tan f(x) = \lim_{x\to\infty}\tan\left(\frac{-x}{1-x^2}\right) = \lim_{x\to\infty}\left(\frac{-x}{-x^2}\right) = \lim_{x\to\infty}\tan\left(\frac{1}{x}\right) \to 0 \tag{1.115}
$$

   This limit is shown in Fig. 1.32 (left), leading into the conclusion that

$$
\lim_{\omega\to\infty}\arctan\frac{-2\xi\dfrac{\omega}{\omega_0}}{1-\left(\dfrac{\omega}{\omega_0}\right)^2} = -\pi = -180^\circ \tag{1.116}
$$

   because in this case the argument of arctan function is in the third quadrant, not in the fourth.

   b. *Crossover point* ($\omega = \omega_0$): at 3dB point we have already found that the amplitude function depends on ξ and that phase function crosses −90° value,

$$
20\log|H_5(\omega_0)| = -20\log(2\xi)\ \ [\text{dB}] \tag{1.117}
$$

$$
\angle H_5(\omega_0) = -\frac{\pi}{2} = -90^\circ \tag{1.118}
$$

   c. *LF region*($\omega \ll \omega_0$): limit $\omega \to 0$ is much easier to find, a simple algebra gives

$$
\begin{aligned}
20\log\lim_{\omega\to0}|H_5(j\omega)| &= \lim_{\omega\to0}\left[20\log\left(\omega_0^2\right) - 20\log\sqrt{\left[(\omega_0^2-\omega^2)^2+(2\xi\omega_0\,\omega)^2\right]}\right]\\
&= 20\log\left(\omega_0^2\right) - \lim_{\omega\to0}\left[20\log\sqrt{\omega_0^4}\right] = 20\log\left(\omega_0^2\right) - 20\log\left(\omega_0^2\right)\\
&= 0\text{dB}
\end{aligned}
\tag{1.119}
$$

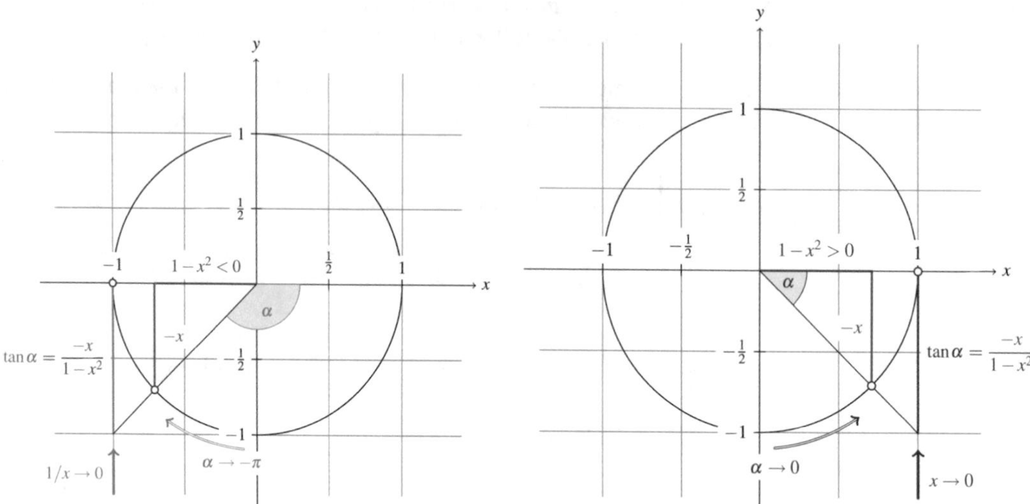

Fig. 1.32 Geometrical interpretation of arctan limit in the third and fourth quadrants

Similarly, the phase function limit shown in Fig. 1.32 (right) is also found analytically as

$$\lim_{\omega\to 0} \arctan \frac{2\xi \frac{\omega}{\omega_0} \to 0}{1-\left(\frac{\omega}{\omega_0}\right)^2 \to 0} = \arctan(0) = 0^\circ \tag{1.120}$$

The above piecewise linear approximations of amplitude and phase functions are summarized in Fig. 1.31. The graphs shows that, from DC until ω_0, amplitude of this function is approximated as zero, then it continues with another linear section with -40dB slope. The peaking point is noted as per (1.108). On the other hand, at ω_0 the phase function experiences rapid switch from zero to -180° values, where the $\angle H_5(j\omega_0) = -90^\circ$ is common crossover point for all values of ξ.

1.6.6 Gain and Phase Margins

Additional information that is inherently found in Bode plots are known as "gain margin" (GM) and "phase margin" (PM) values. Knowing these two numbers is important to quantify *stability* of a system being evaluated. That is it say, if open-loop transfer function is to be used in a closed-loop configuration, these two margins must be calculated.

1. **Gain margin**: we note frequency ω_M, which is found when phase equals -180° (a.k.a. "phase crossover frequency"), Fig. 1.33 (bottom). This point leads into the gain margin point, Fig. 1.33 (top). Then, the gain margin is measured on the amplitude graph as

$$GM = 0 - |H(j\omega_M)| \quad [\text{dB}] \tag{1.121}$$

2. **Phase margin**: we note frequency ω_0 when amplitude equals 0dB (a.k.a. "gain crossover frequency"), Fig. 1.33 (top). This frequency point leads into the phase margin definition that is found in the phase plot, Fig. 1.33 (bottom). The phase margin is measured on the phase graph as

$$PM = \phi(\omega_0) - (-180^\circ) \tag{1.122}$$

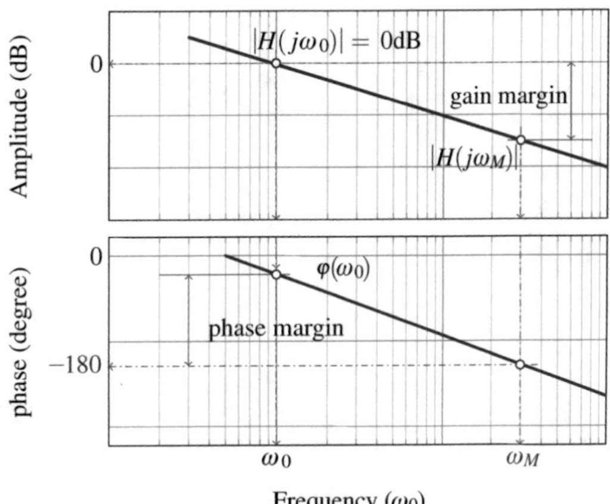

Fig. 1.33 Bode plot illustrating definitions of gain and phase margins

In summary, one possible strategy to create Bode plot is to:

1. factorize given transfer function into the basic forms studied in this section,
2. transform the factorized transfer function into sum of terms by exploiting properties of a logarithmic function
3. apply linear approximations to each of the factorized terms and find limits at LF, HF range and the crossover frequencies ω_0 ("±3dB" points)
4. draw piecewise linear graphs by performing a simple addition within each frequency region

By doing so, it is possible to and derive Bode plots by hands, which is subsequently confirmed by numerical methods.

Example 14: Bode Plot: Second-Order Function, Real Poles
Show Bode plot of the following transfer function:

$$H(\omega) = 100\,\frac{s+1}{s^2+110\,s+1000} \tag{1.123}$$

where $s = j\omega$.

Solution 14: This is second-order function, thus, in order to find out if roots of its denominator are reel or complex it is necessary to factorize its denominator.

$$\begin{aligned}
H(j\omega) &= 100\,\frac{s+1}{s^2+110\,s+1000} = 100\,\frac{s+1}{s^2+10\,s+100\,s+1000}\\
&= 100\,\frac{s+1}{s\,(s+10)+100\,(s+10)} = 100\,\frac{s+1}{(s+10)(s+100)}\\
&= \cancel{100}\,\frac{1+\dfrac{s}{1}}{10\left(1+\dfrac{s}{10}\right)\cancel{100}\left(1+\dfrac{s}{100}\right)} = \frac{1}{10}\,\frac{1+j\,\dfrac{\omega}{1}}{\left(1+j\,\dfrac{\omega}{10}\right)\left(1+j\,\dfrac{\omega}{100}\right)}
\end{aligned}$$

(continued)

This factorized form is suitable for conversion into sum of the basic form simply by rewriting it in its logarithmic form,

$$
\begin{aligned}
20\log H(j\omega) &= 20\log\left[\frac{1}{10}\,\frac{1+j\dfrac{\omega}{1}}{\left(1+j\dfrac{\omega}{10}\right)\left(1+j\dfrac{\omega}{100}\right)}\right]\\
&= \underbrace{\underbrace{20\log\frac{1}{10}}_{\text{Sect. 1.6.1}}}_{\textcircled{1}}
\underbrace{\underbrace{+20\log\left(1+j\frac{\omega}{1}\right)}_{\text{Sect. 1.6.3},\omega_0=1}}_{\textcircled{2}}
\underbrace{\underbrace{-20\log\left(1+j\frac{\omega}{10}\right)}_{\text{Sect. 1.6.4},\omega_0=10}}_{\textcircled{3}}
\underbrace{\underbrace{-20\log\left(1+j\frac{\omega}{100}\right)}_{\text{Sect. 1.6.4},\omega_0=100}}_{\textcircled{4}}
\end{aligned}
$$

Obviously, this example is a second-order function whose poles are real, thus it is possible to decompose it into the basic first order functions. Each summing term in the logarithmic form is in effect one of the already studied basic forms, as annotated.

Similarly, the phase plot is created as the sum of the corresponding linear terms,

$$
\angle H(\omega) = \underbrace{\underbrace{0^\circ}_{\text{Sect. 1.6.1}}}_{\textcircled{1}}
\underbrace{\underbrace{+\arctan\frac{\omega}{1}}_{\text{Sect. 1.6.2},\omega_0=1}}_{\textcircled{2}}
\underbrace{\underbrace{-\arctan\frac{\omega}{10}}_{\text{Sect. 1.6.3},\omega_0=10}}_{\textcircled{3}}
\underbrace{\underbrace{-\arctan\frac{\omega}{100}}_{\text{Sect. 1.6.3},\omega_0=100}}_{\textcircled{4}}
$$

Once the gain and phase logarithmic forms are factorized, first we plot graphs all four terms (both amplitude and phase), then we create plots of the total sums, Fig. 1.34.

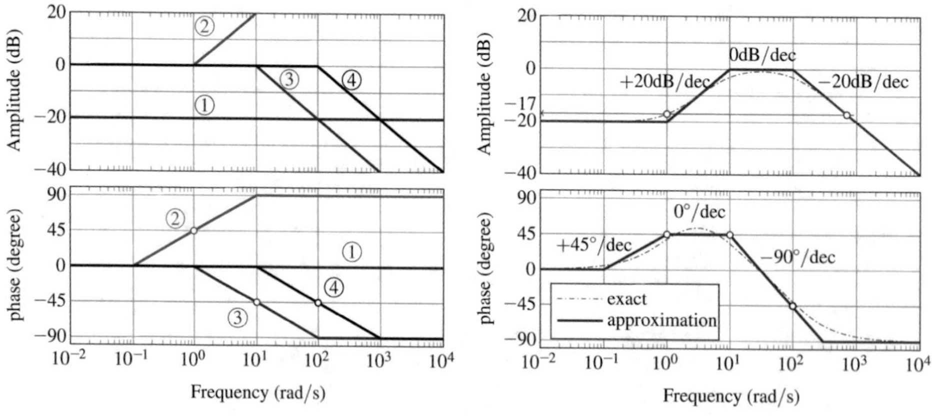

Fig. 1.34 Linear approximations (left), with their sums and with the exact solutions (right)

Example 15: Bode Plot: Second-Order Function, Complex Poles
Show Bode plot of the following transfer function, a case of second-order function with complex roots (where $s = j\omega$):

$$H(s) = 10^2 \, \frac{s+10}{s^2+40\,s+10^4} \tag{1.124}$$

Solution 15: Transfer function (1.124) is a second-order function with complex poles. It can be rewritten as

$$H(s) = 100\,\frac{(s+10)}{s^2+40\,s+10^4} = \cancel{100}\,\frac{\cancel{10}}{10^{\cancel{4}}}\,\frac{1+\dfrac{s}{10}}{\left(\dfrac{s}{10^2}\right)^2+\left(\dfrac{40}{10^4}\right)s+1} \tag{1.125}$$

Discriminant of the quadratic equation is

$$\Delta = 40^2 - 4\times 1\times 10^4 = -38.4\times 10^3 < 0$$

therefore, indeed this second-order function has complex poles. In accordance to (1.101) by inspection of (1.125) we write

$$\omega_0^2 = 10^4 \quad \therefore \quad \omega_0 = 100$$

$$2\xi\,\omega_0 = 40 \quad \therefore \quad \xi = \frac{40}{2\times 100} = 0.2$$

from (1.108) and (1.113) we find $H(\omega_0) = 8\text{dB}$ at $\omega_m = 95.9\text{rad/s}$. Logarithm of (1.125) is

$$\begin{aligned} 20\log H(j\omega) &= 20\log\left[\frac{1}{10}\left(1+\frac{s}{10}\right)\frac{10^4}{s^2+40\,s+10^4}\right] \\ &= \underbrace{\underbrace{-\,20\log 10}_{\text{Sect. 1.6.1}}}_{①} + \underbrace{\underbrace{20\log\left(1+j\,\frac{\omega}{10}\right)}_{\text{Sect. 1.6.3},\omega_0=10}}_{②} + \underbrace{\underbrace{20\log\left(\frac{10^4}{s^2+40\,s+10^4}\right)}_{\text{Sect. 1.6.5},\omega_0=100,\xi=0.2}}_{③} \end{aligned} \tag{1.126}$$

Similarly, the phase plot is created as the sum of the corresponding linear terms (Fig. 1.35),

$$\angle H(\omega) = \underbrace{\underbrace{0^\circ}_{\text{Sect. 1.6.1}}}_{①} + \underbrace{\underbrace{\arctan\frac{\omega}{10}}_{\text{Sect. 1.6.3},\omega_0=10}}_{②} - \underbrace{\underbrace{\arctan\frac{-0.4\,(\omega/100)}{1-(\omega/100)^2}}_{\text{Sect. 1.6.5},\omega_0=100,\xi=0.2}}_{③}$$

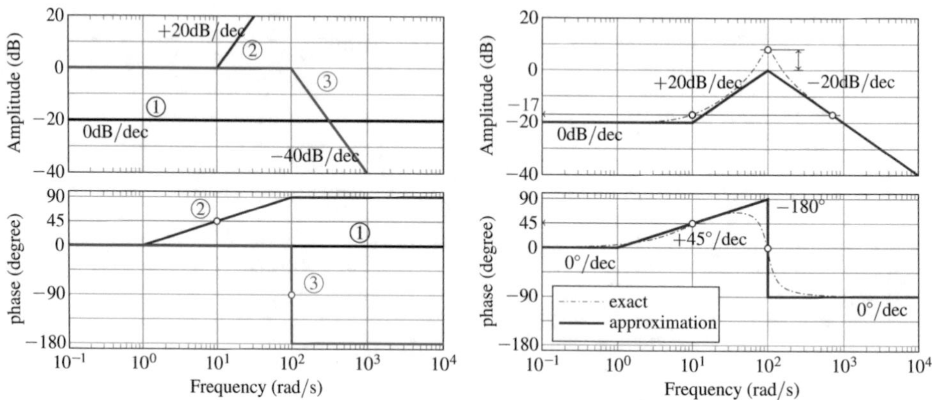

Fig. 1.35 Linear approximations (left), with their sums and the exact solutions (right)

1.7 Summary

In this chapter, we have surveyed the fundamental principles required to understand the transmission of information over long distances using waves. The philosophy of building communication systems is driven by the main constraint that, at the end, the information must be detectable by the human senses. For example, our hearing is capable of distinguishing sound waves in the range of 20–20 000Hz, thus this frequency range is almost always built in some of the communication circuits. Unlike sound, light and other EM radiation does not need matter to travel through and it travels at a speed of close to 300 000 000m/s. Thus, it is natural to use it as our main means of transporting messages.

The conversion of the original sound wave to a radio wave at a particular frequency, and back, is done by electronic circuitry designed specifically to implement the mathematical operations of upconversion and downconversion. Details of these two primary steps in wireless communications were worked out using Maxwell's equations and basic trigonometry. Since the communication system is based on an application of electricity, the rest of this book is devoted to a detailed study of the general principles of electricity and time-varying electrical signals, the mathematical principles behind radio, and the design of practical electronic circuits that are synthesized to implement the required mathematical equations.

Problems

1.1 Calculate period, wavelength, the propagation constant, and phase velocity an electromagnetic wave in free space for the following frequencies: $f_1 = 10\text{MHz}$, $f_2 = 100\text{MHz}$, $f_3 = 10\text{GHz}$.

1.2 Derive the average and rms values of a sine, square, triangle, and sawtooth waveforms; assume their amplitudes to be $V_p = \pm 1\text{V}$.

1.3 Instantaneous voltage of a waveform is described as $v(t) = V_m \cos(\omega t + \phi)$ where $\omega = (2\pi \times 10^3)\,\text{rad/s}$ and $\phi = \pi/4$. Calculate its frequency f as well as its instantaneous phase $\varphi(t)$ at $t = 125\mu\text{s}$.

1.4 Two sine waveforms with the same frequency and amplitude of $V_{pp} = 2$V travel along a conductive wire, however, their phase difference of $\phi = \pi/2$. At the end of the wire, there is a $R = 1\text{k}\Omega$ loading resistor connected to the ground. At the moment t_0 of the arrival and assuming that one of the two waveforms is at its maximum amplitude, calculate current $i(t_0)$ generated in the loading resistor.

In addition, express the phase difference Δphi of these two waveforms as the fraction of their period, then calculate their time Δt and spatial Δx differences at any given point along the wire. In addition, give the answers for each of the three following frequencies: (a) $f_1 = 10$kHz, (b) $f_2 = 10$MHz, and (c) $f_3 = 10$GHz.

1.5 Instantaneous voltage of a waveform is described as

$$v(t) = \cos(2\pi \times 1 \times 10^3\, t) + \frac{1}{2}\cos(2\pi \times 2 \times 10^3\, t) + \frac{1}{3}\cos(2\pi \times 3 \times 10^3\, t)$$

Using any plotting software, on the same graph, plot $v(t)$ together with its three single-tone terms over time window of at least two periods of the slowest tone. Comment on the generated waveform.

1.6 A sinusoidal wave is defined as $v(t) = 10\text{V} \sin(100\, t + 45°)$. Determine its: (a) amplitude; (b) $v_{\mathtt{rms}}$ value; (c) frequency; (d) period; (e) phase at time $t = 1$s; and (f) equivalent cosine form function.

1.7 An arbitrary waveform $v(t)$ consists of $DC = 1$V, the fundamental tone $v_0 = 2 \sin \omega t$, and the second harmonic $v_2 = {}^3/_2 \sin 2\,\omega t$. Draw a frequency domain sketch (approximately to scale) of this $v(t)$ waveform.

1.8 How much energy is used by a 60W amplifier turned on over $t = 8$h? Compare with a 1000W amplifier turned on over $t = 60$s.

1.9 The voltage and current values of a 50Hz sine waveforms are given as: $v(t) = 220 \sin(\omega\, t + {}^{\pi}/_3)$V and $i(t) = 10 \sin(\omega\, t - {}^{2\pi}/_3)$A. Calculate the values of the instantaneous power $p(t)$ and the average power $\langle P \rangle$ generated by this voltage and current.

1.10 Calculate the total power gain of the two amplifiers in Fig. 1.36. Comment on the calculations when using the two different gain units shown in the illustration.

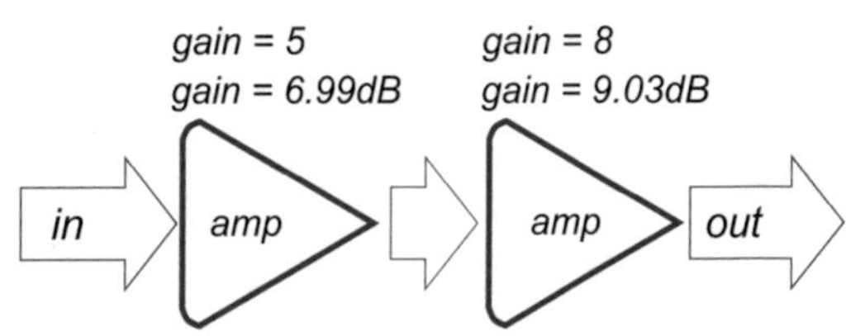

Fig. 1.36 Illustration for the Problem 1.10

Basic Behavioural and Device Models

2

At the system level, analysis and design of electronic circuits is based on the set of fundamental building blocks represented by their behavioural models. In the initial phase of the design it is important to validate the intended functionality of the overall circuit at the level of mathematics, without regard for the implementation details. Therefore, the knowledge of functionality of fundamental devices, namely a switch, voltage and current sources, and RLC devices as well as their respective impedances, is the prerequisite for the next phase in the design process. In this chapter we review these fundamental devices and their transfer functions.

2.1 "Black Box" Technique

The "black box" approach is based on the idea that in order to create systems' behavioural model it is not necessary to know its exact internal structure, instead it is sufficient to know its stimulus/response (i.e. its I/O) transfer function. In the case of electronic circuits, the transfer function is found in terms of the input/output voltage/current characteristics both in time and frequency domains where the transfer function may be either linear or nonlinear.

Therefore, in total, there are four possible I/O voltage/current transfer functions:

1. $(V_{in} \rightarrow V_{out})$: this relationship defines system's "voltage gain" transfer function as

$$v_{out} = A_v \times v_{in} \quad \therefore \quad A_v \stackrel{\text{def}}{=} \frac{v_{out}}{v_{in}} \quad \left[\frac{\text{V}}{\text{V}}\right] \tag{2.1}$$

2. $(I_{in} \rightarrow I_{out})$: this relationship defines system's "current gain" transfer function as

$$i_{out} = A_i \times i_{in} \quad \therefore \quad A_i \stackrel{\text{def}}{=} \frac{i_{out}}{i_{in}} \quad \left[\frac{\text{A}}{\text{A}}\right] \tag{2.2}$$

3. $(I_{in} \rightarrow V_{out})$: this relationship defines system's "resistance" transfer function as

$$v_{out} = R \times i_{in} \quad \therefore \quad R \stackrel{\text{def}}{=} \frac{v_{out}}{i_{in}} \quad \left[\frac{\text{V}}{\text{A}} \stackrel{\text{def}}{=} \Omega\right] \tag{2.3}$$

R. Sobot, *Wireless Communication Electronics*,
https://doi.org/10.1007/978-3-030-48630-3_2

4. $(V_{in} \rightarrow I_{out})$: this relationship defines system's "g_m" transfer function as

$$i_{out} = g_m \times v_{in} \quad \therefore \quad g_m \overset{\text{def}}{=} \frac{1}{R} = \frac{i_{out}}{v_{in}} \quad \left[\frac{\text{A}}{\text{V}} \overset{\text{def}}{=} \text{S}\right] \tag{2.4}$$

In order to create complete model of a system, all four of the above transfer functions are equally important. We determine these I/O functions first by theoretical analysis then by the experiment. Therefore, the final model of a device is often created by a mixture of theoretical and experimental parameters. In general, to determine a system's transfer functions, first we apply an input stimulus to one of the available terminals, then we collect the output data at the remaining terminals. We need to systematically choose one terminal as "input" and other as "output", characterize its I/O transfer function, then proceed to choose another pair of terminals until all combinations are exhausted.

In the following sections we look at some of the most typical experimental "signatures" of black boxes and the associated models of ideal elements.

2.2 Two-terminal Models

The two-terminal model is very useful because the I/O function simply shows "gain" in terms of I/O variables, namely voltage and current, where this I/O function may be either linear or nonlinear. In addition, the I/O functions are studied both in time and frequency domains.

2.2.1 Ideal Resistor

The data-fitting method is applied to experimental data, see Fig. 2.1 (left), to produce the analytical model in the form of a linear function, Fig. 2.1 (right)

$$V_{out} = a_1 \, I_{in} + a_0 \tag{2.5}$$

where a_1 and a_0 coefficients are deduced from the linear properties of the experimental graph.

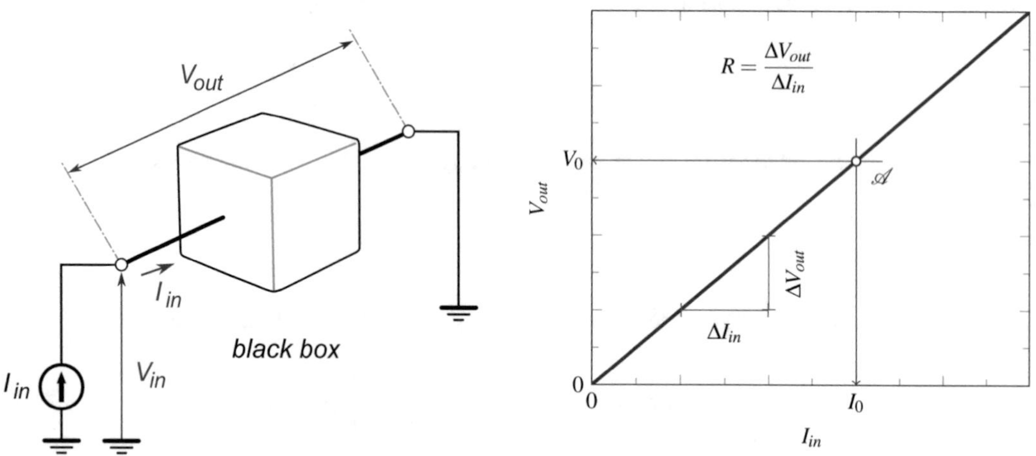

Fig. 2.1 Concept of "black box" and its, for example, ideal linear V/I transfer function

Example 16: Black Box Method
If the transfer function in Fig. 2.1 (right) is drawn using 10 µA and 1 V per square, then what basic electrical component may be hidden in this black box?

Solution 16: In mathematics, the horizontal axis is dedicated to the input variable, while the vertical axis shows the actual value of function, i.e. it shows the output variable. Transfer function of this black box is linear, therefore either ratio or derivative of voltage/current gives

$$R = \frac{\Delta V_{out}}{\Delta I_{in}} = \frac{V_{out}}{I_{in}} = \frac{1\,\text{V}}{10\,\mu\text{A}} = 100\,\text{k}\Omega$$

which is to say that the black box may contain a whole another universe, but for purposes of this setup it may be used as a simple linear 100 kΩ resistor.

2.2.2 Ideal Switch

Let us assume an experimental setup as in Fig. 2.2 (left) where, because one of the terminals serves as the ground reference, we write $V_{out} = V_{in}$. Furthermore, the experiment produces either vertical or horizontal linear response, as in Fig. 2.2 (right), which makes us to conclude that there is an ideal switch in the black box. To reach this conclusion, we apply the following reasoning.

By definition, resistance is the ratio of the current to the potential difference at an arbitrary "input" terminal of the black box. After applying current/voltage test stimulus, we find two distinct I/O relations: (a) I_{in} current is zero (regardless of the voltage V_{in}) or (b) voltage V_{in} is zero (regardless of the I_{in} current entering the box). Thus, these two characteristics can be described as the input resistance perceived between the input terminal of the black box and the ground. In the form of Ohm's law, we describe these two states as

$$R_{ON} = \frac{V_{out}}{I_{in}} = \frac{\Delta V_{out}}{\Delta I_{in}} = \frac{0}{I_{in}} = 0\ [\Omega] \tag{2.6}$$

$$R_{OFF} = \frac{V_{out}}{I_{in}} = \frac{\Delta V_{out}}{\Delta I_{in}} = \frac{V_{out}}{0} = \infty\ [\Omega] \tag{2.7}$$

For example, state indicated as R_{OFF} shows that current through the box is zero even if $V_{in} = \pm\infty$. What is more, $I_{in} = const. \Rightarrow \Delta I_{in} = 0$, which is dynamic resistance, i.e. resistance to *change* of current. The equivalent description is to say that the black box contains an *infinite* resistance. Similarly, the second state associated with a switch is when the voltage across the two terminal equals zero, i.e. $V_{out} = 0$, even if $I_{in} = \pm\infty$. What is more, because $V_{in} = const. \Rightarrow \Delta V_{in} = 0$ its dynamic resistance also equals zero. In other words, we say that in this state, i.e. $R_{SW(ON)}$, the black box contains zero resistance, or simply a "short connection".

Both static and dynamic resistances imply that there is a switch in the box. Due to the existence of two distinct states, mathematical model of a two-state switch may be formalized as

$$R(\text{switch}) = \begin{cases} 0, & \text{for } V_{out} = 0 \\ \infty, & \text{for } I_{in} = 0 \end{cases}$$

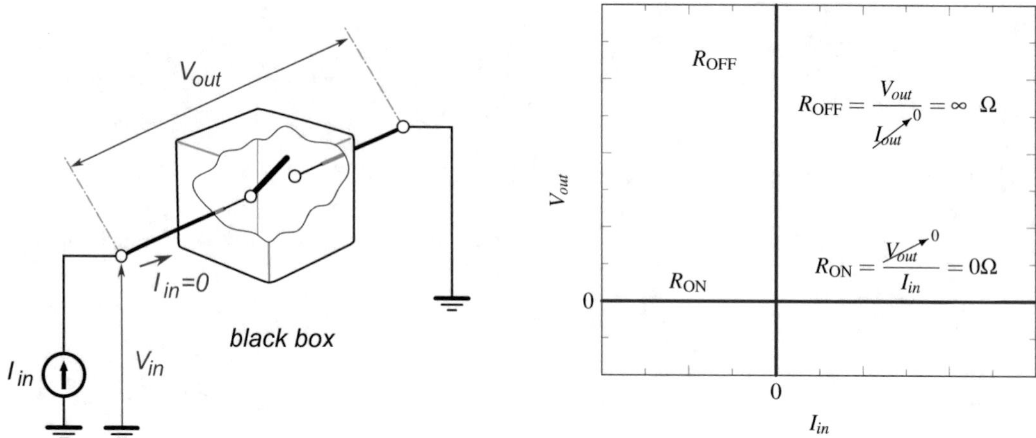

Fig. 2.2 Experimental setup (left) and two linear input–output transfer functions (right) associated with a switch

In practice, as we find in the following chapters, in its idealized form a diode serves as an electronic switch that either blocks or permits the flow of current.

2.2.3 Ideal Voltage Source

By definition, ideal voltage source V_0 is an element capable of holding constant voltage across its terminals regardless of I_{in} current level, Fig. 2.3. We check both static (when V_0 is constant but I_{in} extremely large) and dynamic (i.e. derivative of the output voltage) response to calculate resistance associated with the ideal voltage source

$$R_{V_0} = \frac{\Delta V_{out}}{\Delta I_{in}} = \frac{0}{\Delta I_{in}} = 0\ [\Omega] \quad \text{or, in the extreme static case} \tag{2.8}$$

$$R_{V_0} = \frac{V_0}{I_{in}} = \frac{V_0}{\infty} = 0\ [\Omega] \tag{2.9}$$

reveals its important property: both static and dynamic resistance of a voltage source is zero. What is more, at the same time and as opposed to the closed switch, there is a source of "internal energy" that is capable of holding a constant (i.e. $\Delta V_0 = 0$) non-zero V_0 level even if $I_{in} = \pm\infty$. In a special case, however, when $V_0 = 0\,\text{V}$ the internal voltage source is indistinguishable from a closed switch (or, a short connection).

2.2.4 Ideal Current Source

Experimental setup in Fig. 2.4, where we can write $V_{out} = V_{in}$ and $I_{out} = I_{in}$, reveals another type of energy source in the black box. This time, we find that $I_{out} = const. \Rightarrow \Delta I_{out} = 0$ even if $V_{in} = \pm\infty$. Its static (when $I_{out} = I_0 = \text{const.}$ and $V_{out} \to \infty$) and dynamic (i.e. $\Delta I_{out} = 0$) resistances are therefore,

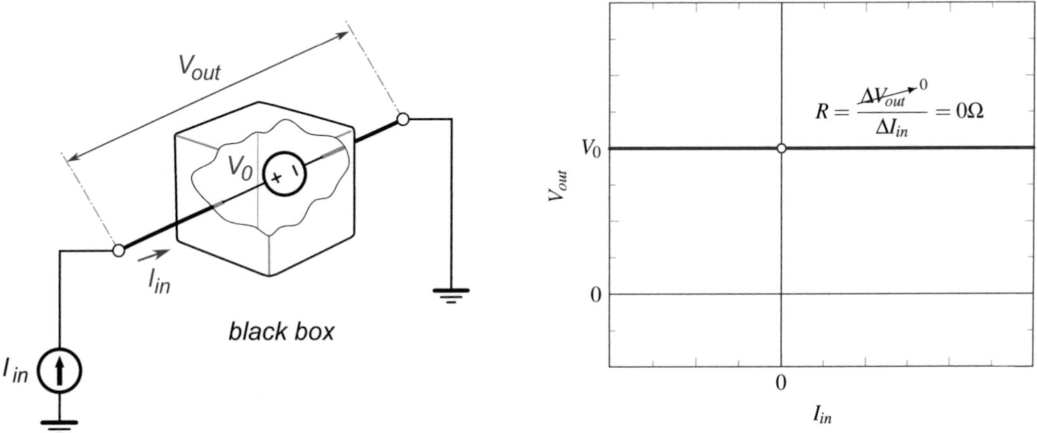

Fig. 2.3 Experimental setup (left) and a linear input–output transfer function (right) of an ideal voltage source

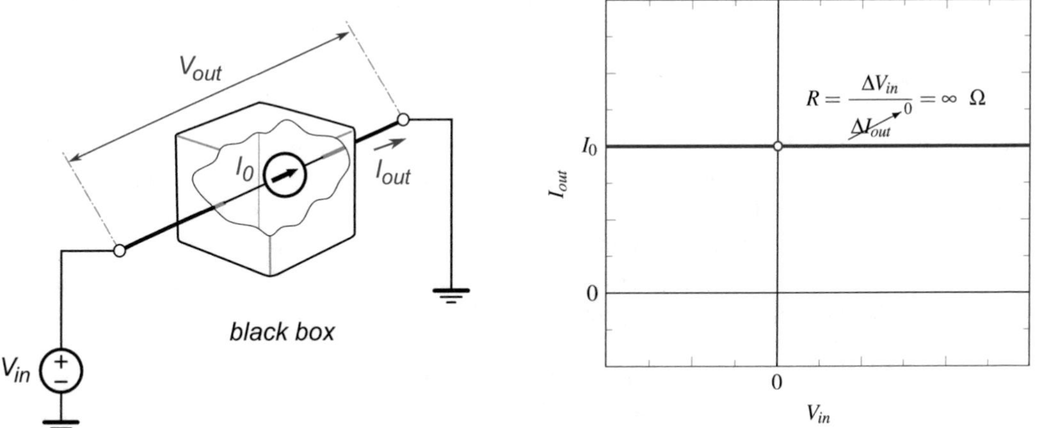

Fig. 2.4 Experimental setup (left) and a linear input–output transfer function (right) associated with a current source

$$R_{I_0} = \frac{V_{out}}{I_0} = \frac{\infty}{I_0} = \infty\ [\Omega] \quad \text{and} \tag{2.10}$$

$$R_{I_0} = \frac{\Delta V_{out}}{\Delta I_{in}} = \frac{\Delta V_{out}}{0} = \infty\ [\Omega] \tag{2.11}$$

We note difference with the open switch case (whose resistance can also be infinite) that this time the current level is not zero, thus indicating the internal source of energy, which is not property of a switch. However, in the special case of $I_0 = 0$ it is not possible to make that distinction. In conclusion, existence of an infinite resistance at the box terminals along with a non-zero constant current indicates the existence of a current source.

As a general note, all of these three ideal elements are capable of handling "infinite" power levels, as implied by $P = V\,I$ when either of the two variables is allowed to take infinite value, which is not possible in the case of real components. Nevertheless, these ideal models are extensively used in circuit analysis because they represent "envelope" case, i.e. extreme limits. In addition, the "envelope" analysis simplifies the circuit complexity and enables hand analysis of circuits.

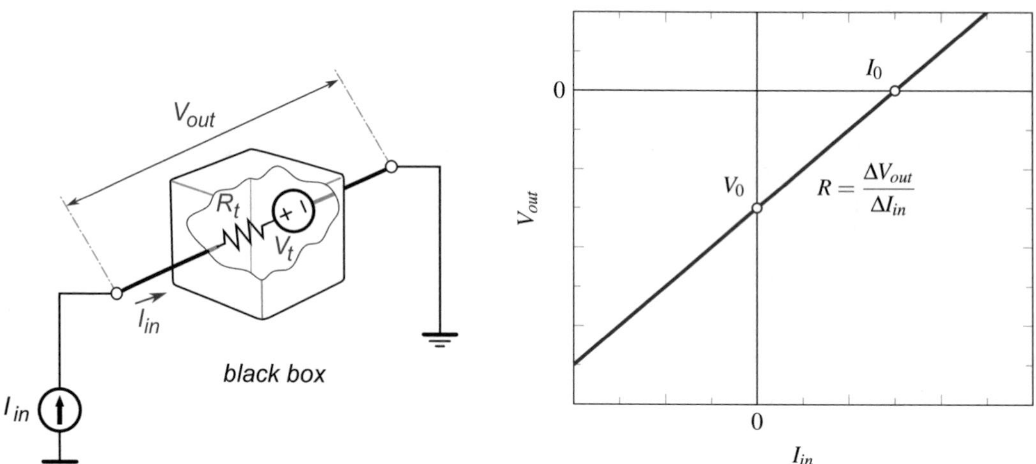

Fig. 2.5 Experimental setup (left) and linear input–output transfer function (right) of a Thévenin/Norton generator

2.2.5 Voltage/Current Generators

Transfer function that does not cross the (0, 0) graph origin Fig. 2.5 (right) indicates that the black box contains not only a resistor R_t (as exposed by its "signature" in the form of a non-vertical non-horizontal linear function) but also the internal energy source, either Thévenin V_t or Norton type. The internal resistance is deduced again by the slope of its linear transfer function, while the generator type is decided based on the convenience and equivalence between the two models. Therefore, it is useful to realize that Thévenin/Norton generator can be derived as a linear superposition of resistor and ideal voltage/current generator.

Aside from the already introduced ideal voltage/current sources, there is a special type of ideal voltage or current generators that are extensively used to model active circuits. This group of generators is known as a *voltage/current controlled* devices, Fig. 2.6. There is the total of four dependent courses, two controlled by voltage and two controlled by current: voltage controlled voltage source (VCVS), voltage controlled current source (VCCS), current controlled voltage source (CCVS), and current controlled current source (CCCS). We note that the controlling voltage or current references, i_{IN} and v_{IN}, are found elsewhere in the given circuit. Physicality of the input/output multiplying constant (i.e. gain) depends on physicality of the input output variables, thus we define four types of gains: voltage to voltage (A_v), voltage to current (g_m), current to voltage (R), and current to current (A_i). In order to visually distinguish this group of generators from the independent generators, most textbooks use diamond shaped symbols as opposed to the traditional circle shapes, Fig. 2.6.

2.2.6 Summary of Ideal Element Models

As shown in the previous section, the black box model is very practical for understanding fundamental functionality of any system. Again, thinking in terms of the internal resistance while "looking into" the box terminals greatly simplifies the overall analysis. As a summary, in Fig. 2.7 these principal models are shown next to each other.

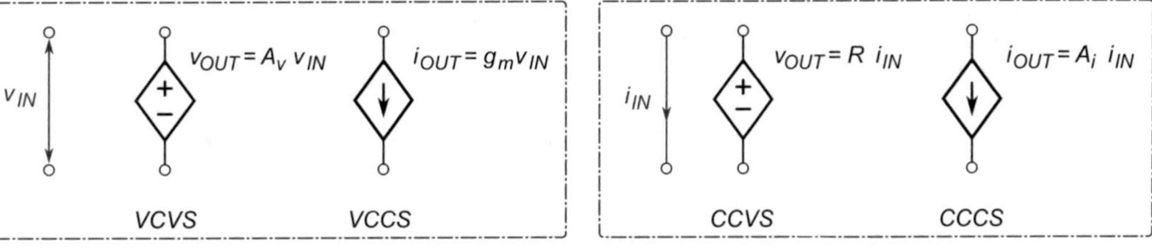

Fig. 2.6 Symbols for voltage controlled voltage source (VCVS), voltage controlled current source (VCCS), current controlled voltage source (CCVS), and current controlled current source (CCCS)

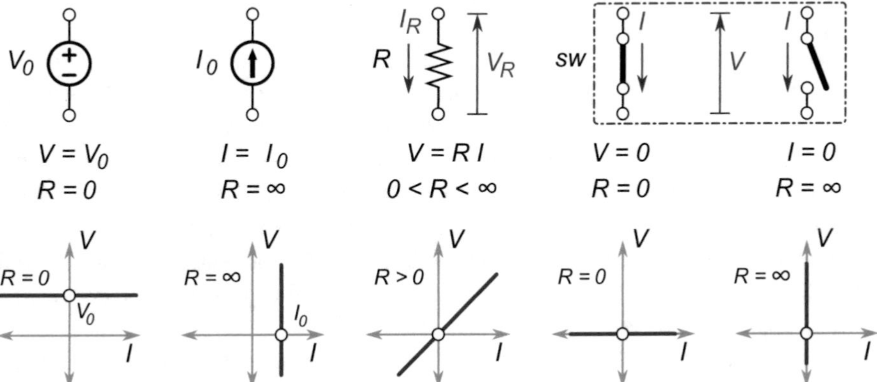

Fig. 2.7 Summary of ideal elements and their transfer characteristics

2.3 Impedance

From the perspective of an engineer working on any of the topics derived from general electronics (e.g. RF electronics, microelectronics, etc.), an "impedance" is one of the key variables that controls development and functionality of any circuit.

By definition, impedance Z is seen as *gain* (i.e. multiplication) factor between the two fundamental variables in electronics: current (I) and voltage (V), Fig. 2.8, as formalized by Ohm's law:

$$V = Z\, I \tag{2.12}$$

It is an accepted convention that an "impedance" is dependent on frequency, as opposed to "resistance" that is frequency independent, both of which are defined by (2.12). Thus, each of the three basic circuits elements (i.e. resistor R, capacitor C, and inductor L) used in circuit theory, as well as any general black box, is characterized by its current to voltage gain characteristics, i.e. its *impedance*. Therefore, in a broad sense, all impedances are grouped in two categories, either dependent upon frequency ("complex") or not ("real"). It is an accepted convention to reserve the term *resistance* R for *real* resistive, i.e. frequency independent components, and to use the term *reactance* for the equivalent resistance of an inductor X_L or a capacitor X_C at a given frequency ω.

By definition, a "reactance" is described only by the *imaginary* term "$j\Im$" (including the $j = \sqrt{-1}$ part, which in complex algebra notation takes care of the phase) of a complex number $Z = \Re + j\Im$. By the same convention, for example, a serial combination of a resistance and either capacitance or inductance is referred to as *impedance* Z. Two important parameters of any impedance are its absolute

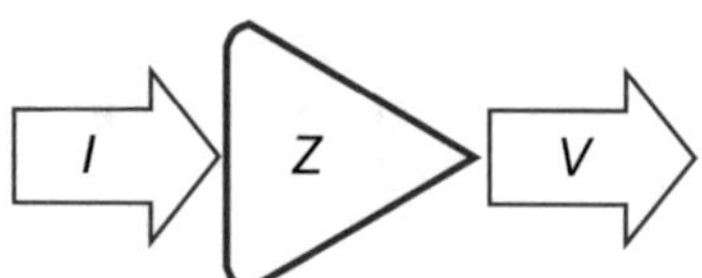

Fig. 2.8 Impedance's functional model seen as a current to voltage amplifier with the gain Z

value $|Z(\omega)|$ and argument $\phi(\omega)$ (also referred to as a *phase*), where radian frequency $\omega = 2\pi f$ in the units of [rad/Hz]. In the complex plane, the real and imaginary axes are set at $\pi/2$ angle relative to each other, thus the absolute value and argument of a complex number are calculated using the Pythagorean theorem and trigonometric identities.[1] For the time being, we note that both L/R ratio and RC product have dimensions of *time*.

With these definitions in mind, it should be noted that a resistor may be seen as an ideal, real, and *fixed* impedance (thus, whose $\Im(Z_R) = 0$, i.e. its phase $\phi(Z_R) = 0$) with the "R" value being the only variable to set. On the other hand, a capacitor and an inductor may be seen as ideal, complex, and *programable* impedances (thus, whose phase is found by $\phi(Z_{L,C}) = \arctan[\Im(Z_{L,C})/\Re(Z_{L,C})]$. That is to say, there are *two* variables available to control their overall impedances, one being "L" or "C" value, and the other being the working frequency ω. Consequently, the same complex component presents different impedance to each of the signal's frequency components. For example, in the audio range of frequencies between 20 Hz–20 kHz the same $C = 1\,\text{nF}$ capacitor is perceived as resistance anywhere from approximately 8 MΩ–8 kΩ.

In the following sections, we use "black box" method to evaluate impedance characteristics of (R,L,C) electronic elements in respect of DC, AC, and transient (i.e. a rapid step change) stimulus.

2.3.1 Linear Resistor

Without looking inside a black box, a linear resistor is recognized by its linear "signature" in the voltage–current graph that is generated by stimulating the input terminal by current $I_{in} = I_R$ and reading voltage $V_{out} = V_R$ across its terminals, Fig. 2.9. We note that, by definition, its transfer function must cross (0, 0) point simply because if there is no current flowing through a resistor (i.e. $I_R = 0$) there cannot be voltage generated at its terminals because ($V_R = I_R \times R$). Among a few transfer functions that are shown, R_1–R_5, we note two cases that are different, R_1 (horizontal) and R_5 (vertical), which we recognize as open/closed switch (i.e. zero resistance short connection or infinite resistance, as discussed in the previous sections). Liner, non-zero, non-infinite resistors have transfer functions similar to R_2–R_4.

2.3.1.1 DC State of a Circuit with Resistor

Linear resistor, together with voltage/current sources and switches, is an element whose impedance Z_R is frequency independent, i.e.

$$Z_R = R \neq f(\omega) \Rightarrow R \in \mathbb{R}, \angle Z_R = 0° \tag{2.13}$$

As a side note, a real resistor is the only *heat generating* element. An ideal resistor with its resistance R is a linear device whose voltage V_R and current I_R at its terminals are proportional to each other in accordance with Ohm's law (2.12) as

[1] See definitions of module and phase of a complex number in Appendix C.

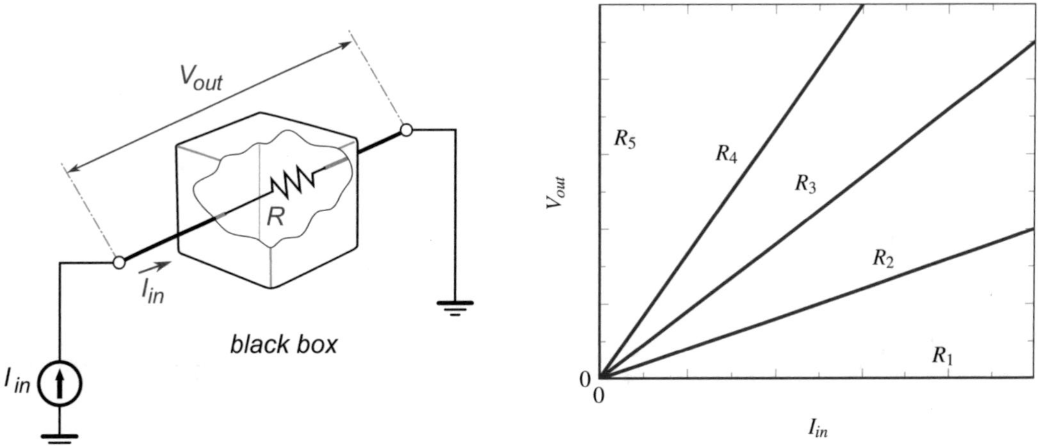

Fig. 2.9 Experimental setup (left) and linear input–output transfer functions (right) associated with a linear resistor

$$V_R = Z_R\, I_R = R\, I_R \tag{2.14}$$

where resistance $Z_R = R$ is the proportionality constant, Fig. 2.9. In other words, the main function of a linear resistor is to generate a voltage difference at its terminals that is proportional to the current flowing through. An *ideal* resistive element is capable of absorbing an *infinite* amount of power (i.e. its internal current and/or voltage can be raised to infinity). It is important to note that (2.14) does not include frequency ω variable, which means that ideal resistor is a frequency independent component.

Practical resistors are created using a slab of a material, for example, metallic or semiconductor, whose geometrical properties control the overall resistance to DC current, Fig. 2.10 (left), as

$$R_{DC} = \rho_{cond}\,\frac{l}{S} = \frac{1}{\sigma_{cond}}\,\frac{l}{S} \tag{2.15}$$

where R_{DC} is the DC wire resistance, l is the wire length, $\rho = 1/\sigma$ is the metal resistivity constant, σ is the metal conductivity constant, S is the wire's cross-sectional area ($S = \pi a^2$ in the case of a round wire whose radius is a). Consequently, exceeding a certain level of internal power results in overheating and permanent destruction of a real resistor.

Static relationship between the input current I_R and the output voltage V_R is obviously a linear function, Fig. 2.10 (right), thus "linear resistor". When calculated at any given point ($\mathscr{A}$) of the linear function the voltage to current ratio gives the same "gain" (i.e. slope or derivative, which is inherent mathematical property of a linear function),

$$R = \frac{V_0}{I_0} = \frac{dV}{dI} \tag{2.16}$$

Geometrical interpretation is that R as calculated by (2.16) is simply a ratio of the vertical (i.e. the output variable V_0) to horizontal catheti (i.e. the input variable I_0) of right-angled triangle, which in the case of a linear function is also the first derivative, Fig. 2.10 (right). Since this form of graph

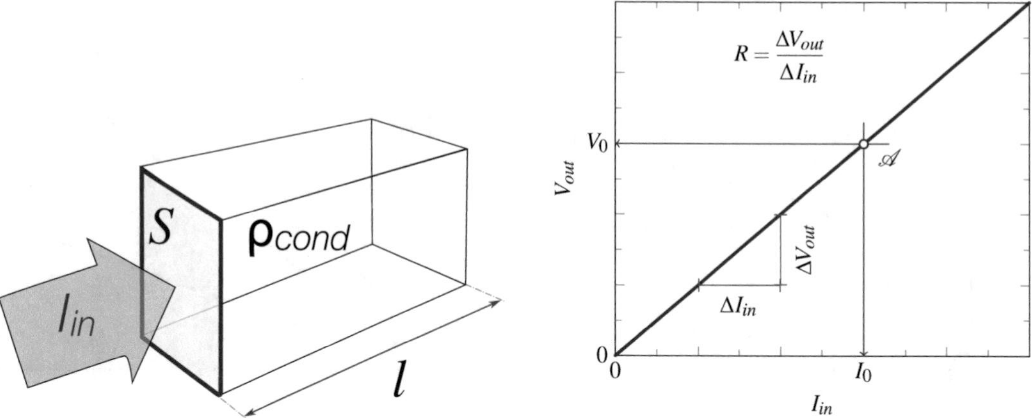

Fig. 2.10 A slab of a material (left) whose geometry defines the overall resistance (right)

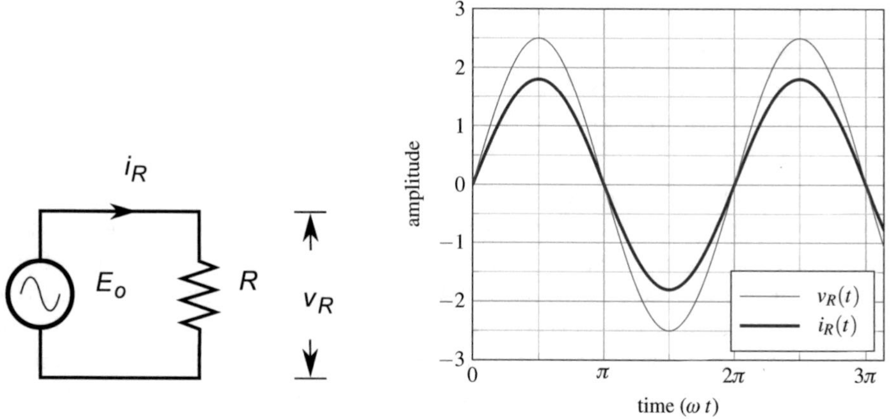

Fig. 2.11 Schematic diagram of a circuit with purely resistive load and sine voltage generator, (left), and the corresponding voltage–current time domain plot (right)

assumes I_R to be the input variable and V_R the output variable, thus their ratio (output variable divided by the input variable) must have units of resistance.

It is equally convenient to show voltage–current relationship by using V_R as the input variable and I_R as the output variable, Fig. 2.10 (right), which is expressed by this variant of Ohm's law

$$\frac{1}{R} = \frac{I_0}{V_0} \stackrel{\text{def}}{=} g_m \quad \left[\frac{1}{\Omega} \stackrel{\text{def}}{=} \text{S}\right] \tag{2.17}$$

where the inverse resistance has the unit of "Siemens" [S]. In this case, instead of using term "voltage to current gain" we say "g_m gain" (or, simply "g_m").

2.3.1.2 AC Steady State of a Circuit with Resistor

As opposed to DC static relationship where the input output variables are not functions of time, dynamic (or, AC) voltage–current relationship is studied by using a sinusoidal stimulus. Parallel connection of a sine voltage generator and purely resistive load is shown in Fig. 2.11 (left).

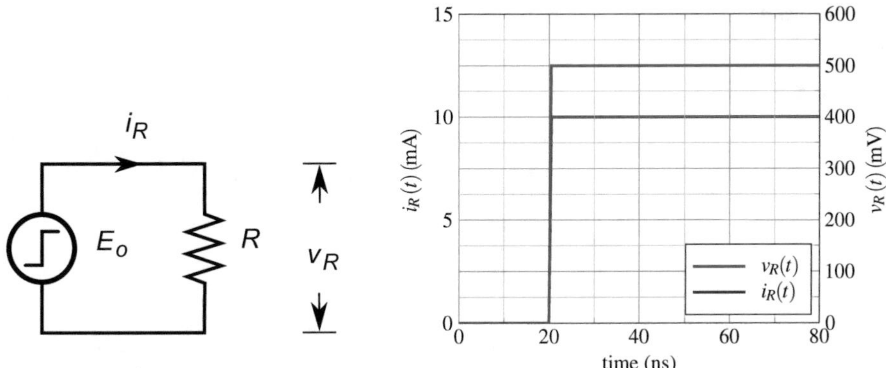

Fig. 2.12 Schematic diagram of a circuit with purely resistive load and step voltage generator, (left), and the corresponding voltage–current time domain plot (right)

At any given point in time, obviously, ratio of the output and input sine amplitudes must give the same value of R because it still obeys the general (2.12) relationship. This time, however, there is another information to observe. Phase difference (i.e. time delay) between voltage and current is zero, Fig. 2.11, which is as expected for a real function.

2.3.1.3 Transient Resistive Current

In addition to static (i.e. DC) and steady state (AC) responses, it is important to characterize response of resistor to a very rapid "step function" stimulus. The goal is to find out if there is a limit to how fast V_R is generated after the I_R stimulus.

Since (2.14) does not contain time variable (i.e. ω), conclusion is that in the case of ideal resistor there should not be any time delay between output voltages, regardless of how fast the input current changes. Numerical simulations dully confirm that step input current generates immediately step output voltage, Fig. 2.12 (right).

2.3.1.4 Realistic Resistor

The electrical behaviour of a real resistor is much more complicated than the basic ideal model (2.14) suggests. The main reason is that a real resistor is build using several materials, each of them with a different conductivity constant σ, a different temperature coefficient (TC), and so on. In reality, even a simple wire on its own, which is commonly used to make the lead terminals for through-hole resistors, turns into a very complicated device once AC starts to flow through. In addition, there is a capacitive effect between the wire turns, between the resistor's body and the environment, and between the two wire terminals.

A numerical analysis (see Fig. 2.13) of a resistor model whose DC resistance value was designed to be R shows that there are four distinctly different frequency regions where the resistor's behaviour is drastically different. Although the numbers shown in Fig. 2.13 are specific only for this particular example, the curve shape is similar for other examples.

- DC to 7 MHz: in this frequency region, the resistance is dominant and constant. Note that the resistor's value does not change significantly with the frequency increase, i.e. $|Z| \approx R \neq f(\omega)$.
- 20 MHz–1 GHz: in this frequency region, the capacitive behaviour is dominant, which is illustrated by the linear (in log scale) drop in the impedance amplitude. It is consistent with the capacitive impedance behaviour (capacitive impedance is inversely proportional to the frequency).

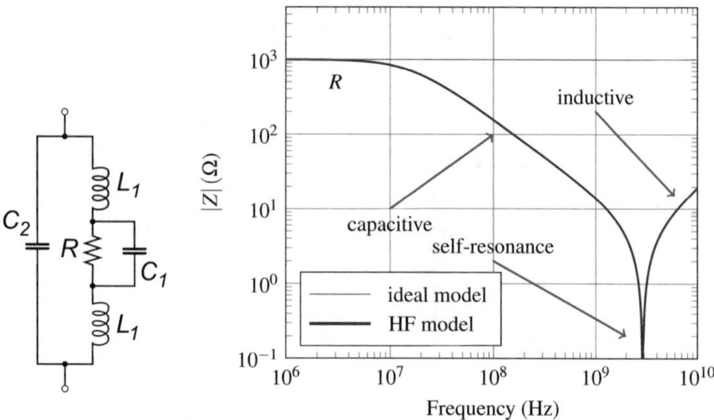

Fig. 2.13 HF model (left) and frequency characteristics (right) of a typical wire-wound $R = 1\,\text{k}\Omega$ resistor

- close to 3 GHz: A narrow sharp region is a very important property of any physical object. For the time being, let us only remember the frequency of the minimal point as the "self-resonating" frequency (see Chap. 10).
- Above 3 GHz: At very high frequencies, inductive behaviour is most prominent, which is characterized by the linear increase in the impedance amplitude as the frequency increases, which is the typical behaviour of an ideal inductor.

This example illustrates the complexity of the behaviour associated with real components due to frequency dependence that is caused by their internal parasitics, which directly limits the useful operating range of frequencies of real components.

2.3.2 Nonlinear Resistor

Definition of a linear resistance in its basic form, (2.14), applies only to ideal *linear* resistor components whose voltage vs. current derivative is constant at any point, Fig. 2.14 (left); thus, direct application of (2.14) is sufficient. That is to say, when resistance is calculated as simple division of constant voltage V_0 and constant current I_0 it is referred to as "DC resistance". For that reason, resistance calculation at any point $(\mathscr{A}, \mathscr{B}, \mathscr{C})$ in Fig. 2.14 (left)

$$R_{DC} = \frac{V_0}{I_0} = \frac{V_A}{I_A} = \frac{V_B}{I_B} = \frac{V_C}{I_C} = \left.\frac{dV_0}{dI_0}\right|_{(\mathscr{A},\mathscr{B},\mathscr{C})} \geq 0 \tag{2.18}$$

gives the same and positive R_{DC} value, regardless whether a simple ratio or derivative is used to calculate the associated resistances. However, resistance is inherently *nonlinear* in terms of the voltage–current relationship at the resistor's terminals, Fig. 2.14 (right). Therefore, *change* in voltage that corresponds to *change* in current gives a different value of point-by-point resistance.

Consequently, there are two ways to calculate resistance at any given voltage–current point: (a) "static" or "DC" resistance that is simply V/I ratio and (b) "small change" or "AC" resistance found as derivatives at the given points $(\mathscr{A}, \mathscr{B}, \mathscr{C})$, also referred to as "biasing points". For nonlinear functions these two calculations produce very different resistances, which is illustrated as follows.

One of the three possible forms of Ohm's law, which describes the relationship of time-varying voltage and current, is used to derive "AC resistance" as

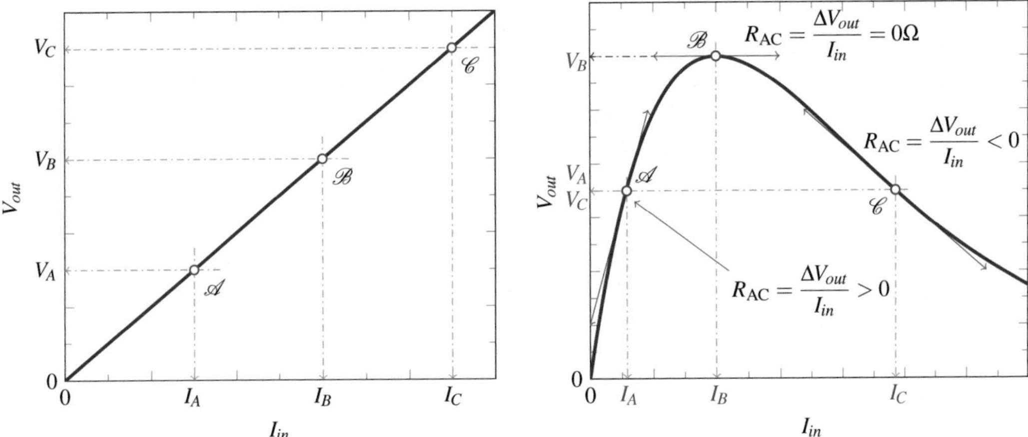

Fig. 2.14 Voltage–current characteristics of a linear element (left), and a nonlinear element (right)

$$v(t) = i(t)\,R \Rightarrow R = \frac{v(t)}{i(t)} \tag{2.19}$$

$$\therefore$$

$$R_{AC} \overset{\text{def}}{=} \lim_{\Delta\to 0} \frac{\Delta v}{\Delta i} = \left.\frac{dv}{di}\right|_{(V_0, I_0)} \tag{2.20}$$

at any moment in time, which should be interpreted as being valid only at one particular voltage–current biasing point (V_0, I_0). Let us take a look at three possible types of biasing points, Fig. 2.14 (right), using relative units as marked on the graph.

1. *Point* $\mathscr{A}$: DC and AC resistances are calculated as

$$R_{DC}|_{\mathscr{A}} = \frac{V_A}{I_A} \approx \frac{3.5}{0.5} \approx 7 > 0\,; \qquad R_{AC}|_{\mathscr{A}} = \left.\frac{dV_0}{dI_0}\right|_{\mathscr{A}} \approx \left.\frac{\Delta V_0}{\Delta I_0}\right|_{\mathscr{A}} \approx \frac{4}{1} \approx 4 > 0 \tag{2.21}$$

   where, in the first approximation, derivative is found graphically by following the tangent line. Two results R_{DC} and R_{AC} may or may not be too much different, nevertheless both are certainly positive values.

2. *Point* $\mathscr{B}$: similarly, DC and AC resistances are calculated as

$$R_{DC}|_{\mathscr{B}} = \frac{V_B}{I_B} = \frac{6}{2} = 3 > 0\,; \qquad R_{AC}|_{\mathscr{B}} = \left.\frac{dV_0}{dI_0}\right|_{\mathscr{B}} = \left.\frac{0}{dI_0}\right|_{\mathscr{B}} = 0 \tag{2.22}$$

   where, in the first approximation, the $R_{DC} > 0$ while at the same time $R_{AC} = 0$, because at the point $\mathscr{B}$ voltage is constant for a relatively small current change (the tangent is horizontal). That is, at this biasing point $\mathscr{B}$ a small AC input signal would not perceive resistance at all.

3. *Point* $\mathscr{C}$: at the biasing point $\mathscr{C}$, the two resistances are calculated as

$$R_{DC}|_{\mathscr{C}} = \frac{V_C}{I_C} \approx \frac{3.5}{5} = 0.7 > 0\,; \qquad R_{AC}|_{\mathscr{C}} = \left.\frac{dV_0}{dI_0}\right|_{\mathscr{C}} \approx \left.\frac{\Delta V_0}{\Delta I_0}\right|_{\mathscr{C}} \approx \frac{3}{-3} \approx -1 < 0 \tag{2.23}$$

We note that, for nonlinear functions, small signal (i.e. AC) resistance may be even negative. Interpretation is that (as opposed to linear resistance where *increase* in current always results in *increase* of voltage at the resistor terminals) negative resistance is a device or a system where *increase* in current results in *decrease* of voltage at its terminals. What is more, it is possible to have two different DC resistances with two different currents but *same* voltages (for example, at biasing points $\mathscr{A}$ and $\mathscr{C}$).

In the above discussion, derivatives (i.e. AC values) are approximated by simply looking at the slope of tangent at the respective biasing points. We accept terminology that the word "derivative" is interpreted as "change of the output variable caused by *small change* of the input variable". Stress is on the "small change" part of the phrase.

In summary, for nonlinear devices there are not one but two valid results for resistance at the same (V_0, I_0) biasing point,

$$R_{DC} = \frac{V_0}{I_0}$$
$$R_{AC} = \left.\frac{dv}{di}\right|_{(V_0,I_0)} \tag{2.24}$$

hence the need to specify the corresponding AC and DC resistances of nonlinear devices, such as diodes and transistors.

2.3.2.1 Power Dissipated in a Resistive Load

In this section, we review power dissipated in a resistive load in terms of its voltage–current–power relationship, Fig. 2.15. Because an AC signal, mathematically described by sinusoidal function, constantly changes in time, it is convenient to use its RMS value to quantify the energy transfer between the source and the load. We already know (Sect. 1.5.1) that electrical power is the product of voltage and current. In the same section, (1.54) showed that the phase difference between voltage and current waveforms is an important factor. By inspection of (2.14), we note that resistance by itself does not have a frequency-dependent component, hence its voltage v_R and current i_R measured at the terminals must be in phase, as shown in Fig. 2.15 (right). It is important to observe that:[2]

- Because the voltage and the current through a resistor are in phase, i.e. for half a cycle both are positive and for half a cycle both are negative, the power is always positive (it always flows out of the generator into the resistor and dissipates in heat).
- The power cycle is half the signal cycle.
- Even though average values for voltage and current are zero, the average power is halfway between its minimum and peak values, i.e. positive.

The effective DC voltage E is what is commonly referred to as the "RMS" value of the AC peak voltage V_m. For sine waveform, the same $1/\sqrt{2}$ factor applies to the current peak value, which leads to a relation for the RMS power as

$$P_{\text{rms}} = \frac{1}{\sqrt{2}} V_m \times \frac{1}{\sqrt{2}} I_m = \frac{1}{2} V_m\, I_m \tag{2.25}$$

To conclude, because the square of the average does not always equal to the average of the squares, the average and RMS values are not always equal.

[2]Reminder: $a\sin(x) \times b\sin x = \frac{ab}{2}(1 - \cos(2x))$.

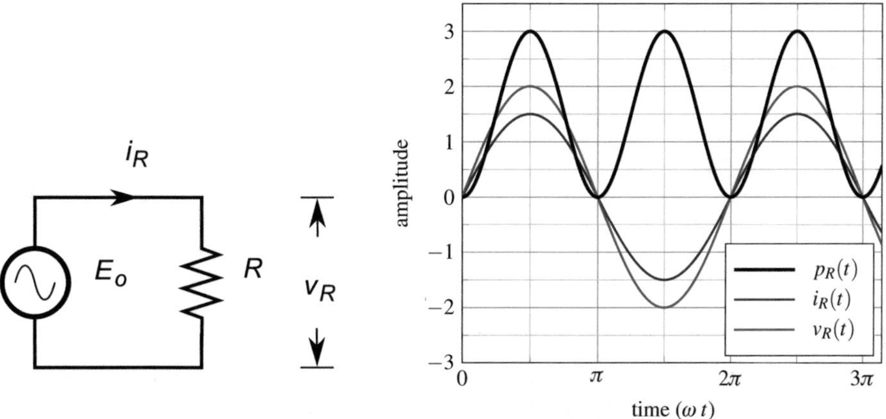

Fig. 2.15 Schematic diagram of a circuit with purely restive load and AC voltage generator, (left), and the corresponding voltage–current–power time domain plot (right). In this particular example, $E_0(t) = 2\sin(\omega t) = v_R$ [V], and $i_R = 1.5\sin(\omega t)$ [mA]

Example 17: AC Power Dissipated by a Resistor
Referring to Fig. 2.15, assume the following data: $v(t) = 2\sin(\omega t)$ [V], and $i(t) = 1.5\sin(\omega t)$ [mA]. Calculate: (a) average power dissipated in the resistor; (b) its resistance; (c) equivalent DC voltage that would generate the same power level as the average power calculated in (a).

Solution 17:

- Peak power is calculated as $P_m = V_m I_m = 2\,\text{V} \times 1.5\,\text{mA} = 3\,\text{mW}$, so the average power is calculated as $(3\,\text{mW} + 0\,\text{mW})/2 = 1.5\,\text{mW}$.
- At any moment, the resistor value is $R = v(t)/i(t) = 2\,\text{V}/1.5\,\text{mA} = 1.333\,\text{k}\Omega$.
- The equivalent DC voltage V that is needed to generate a power level equal to the average power is $V = \sqrt{P\,R} = \sqrt{1.5\,\text{mW} \times 1.333\,\text{k}\Omega} = 1.414\,\text{V}$.

2.3.3 Capacitor

As a physical component, most commonly used capacitor shape is a parallel plate capacitor, which is made of two thin metal sheets with thin insulating layer sandwiched in between, Fig. 2.16.

Similarly to (2.15), relationship between geometrical properties of a parallel plate capacitor and its capacitance is given as

$$C = \epsilon\,\frac{S}{d} = \epsilon_0 \epsilon_r\,\frac{S}{d} \tag{2.26}$$

where C is capacitance that is directly proportional to the overlapping surface area S of the two conducting plates, to the permittivity of the insulating layer $\epsilon = \epsilon_0\epsilon_r$, and inversely proportional to the separation distance d between the plates, Fig. 2.16. Relative permittivity of the insulating material is ϵ_r, ϵ_0 is the vacuum permittivity. Even though (2.26) applies only to a plate capacitor, capacitance of many other shapes can be approximated reasonably well with the same formula.

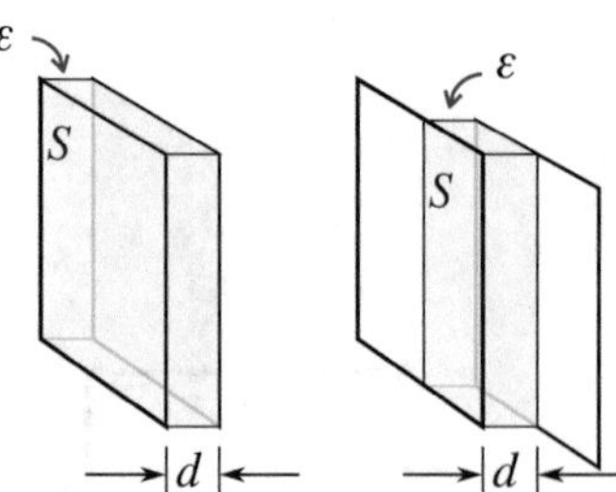

Fig. 2.16 Parallel plate capacitor C_1, (left), and partially overlapping plates creating C_2 (right), therefore $C_2 < C_1$

2.3.3.1 Capacitance

Primary property of a capacitor is that it can store a charge on its plate and serve as a temporary energy source (as same as a voltage battery). That is why a capacitor is also referred to as an "energy storage device", where the energy is stored in its internal electric field. We define *capacitance* C as the proportionality (i.e., gain) constant that defines the total number of electrons Q that are required to create $V = 1\,\text{V}$ potential *difference* between the *two objects*, i.e.

$$Q \overset{\text{def}}{=} C \times V \tag{2.27}$$

where the charge Q and potential V hold as long as the two objects are physically separated by a distance d, in other words, there is no DC path between the two plates.

2.3.3.2 Capacitive Reactance

Recalling that electric current $i(t)$ is defined as change of charge relative to time, i.e. mathematically, it is equivalent to first derivative of (2.27), i.e.

$$i(t) \overset{\text{def}}{=} \frac{d}{dt}Q = C\,\frac{dv(t)}{dt} \tag{2.28}$$

Therefore, general definition of capacitance C is that it is the proportionality constant between *instantaneous current* $i(t)$ and *change of voltage in time* $dv(t)/dt$. In other words, change in *time* is inherent property of capacitance.

1. **DC case**: if there is *no voltage change*, i.e. $v_C(t) = \texttt{const}$ at the capacitor terminals, then the term $dv/dt = 0$ in (2.28) becomes zero, as a consequence the capacitive current $i_C = C \times 0 = 0$ must be zero. We, therefore, conclude that, once the charging process is over and the terminal voltage is stable, a capacitor does not let DC current through (caused by the constant voltage), which is as same as saying that capacitive DC resistance is *infinite*,

$$Z_C(DC) \doteq \frac{v_C}{i_C} = \frac{v_C}{0} = \infty \tag{2.29}$$

2. **AC case**: when capacitive voltage changes periodically in time by following sinusoidal function is known as *steady state*. Steady state signals change by rate of radian frequency (1.15), $\omega = 2\pi/T$, therefore the maximum capacitor voltage V_m changes with the same rate. Mathematically, we define a periodic voltage as $v_C(t) = V_m \cos\omega t$ and we are able to rewrite (2.28) as

$$i_C(\omega t) = C\frac{d}{dt}v_C(\omega t) = C\,\frac{d}{dt}V_m \cos\omega t = \omega C\;V_m(-\sin\omega t)$$

$$= \omega C\, V_m \cos\left(\omega t + \frac{\pi}{2}\right)$$

$$= \omega C \underbrace{v_C(\omega t + \pi/2)}_{v_C(t) \text{ delayed by } \pi/2} \tag{2.30}$$

We recall from complex algebra that a real function delayed by $\pi/2$ is as same as if the function is rotated by $\pi/2$ to j direction in the complex plane, thus we write

$$I_C = \omega C\ j\, V_C$$

.

$$Z_C(j\omega) \stackrel{\text{def}}{=} \frac{V_C}{I_C} = \frac{1}{j\omega C} = -j\frac{1}{\omega C} \;\; [\Omega] \tag{2.31}$$

where the capital letters where used to denote steady state variables (i.e. complex "phasors"). Equation (2.31) shows that value of capacitive reactance $Z_C(j\omega)$ is *inversely* proportional to frequency, starting with infinite resistance at DC and tending to zero as frequency increases. In addition, due to $-j$ factor, (2.31) therefore shows $-\pi/2$ phase difference between voltage and current components.

The initial assumption of infinite resistance associated with the insulating layer is an abstraction that, of course, cannot be achieved in reality. Good insulators, however, block DC current reasonably well, which makes the capacitor leakage current negligible. As a result, a good charged up capacitor can hold its charge over very long period of time when the leakage current is close to zero.

Example 18: Capacitor Impedance $Z_C(j\omega)$
To illustrate capacitor impedance Z_C relative to frequency, calculate impedance of a C = 159 pF capacitor at the following frequencies: 1 Hz, 100 Hz, 10 kHz, 1 MHz, 100 MHz, and 1 GHz.

Solution 18: Direct implementation of (2.31) results in $|Z_C| = 1/2\pi f C$, therefore we write frequency to impedance ($f \rightarrow Z_C$) pairs of this capacitor as: (1 Hz → 1 GΩ), (100 Hz → 10 MΩ), (10 kHz → 100 kΩ), (1 MHz → 1 kΩ), (100 MHz → 10 Ω), and (1 GHz → 1 Ω).

2.3.3.3 HF Capacitor Model

Because the real dielectric materials are lossy, which is to say that there is a small current flow under all conditions, i.e. the real capacitor insulator is leaky, the finite resistance of the insulating material is calculated in the same way as any other resistive material using (2.15). This parasitic resistance is perceived as being in parallel with the desired capacitance (effectively it provides DC path between the capacitor terminals).

The corresponding electric equivalent circuit for a real capacitor therefore includes the desired capacitance C, parasitic series resistance R_S and inductance L of lead wires, and dielectric loss resistance R_C, Fig. 2.17 (left). The overall impedance of real capacitor Z_{Cr} is then calculated as

$$Z_{Cr} = (R_S + j\omega L) + \frac{1}{G_C + j\omega C} \tag{2.32}$$

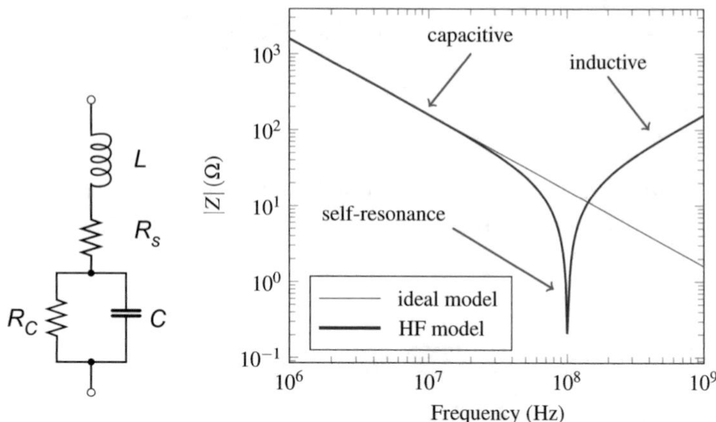

Fig. 2.17 Equivalent electrical circuit model of a high-frequency plate capacitor (left), frequency domain plot (right)

Similar to real resistors, numerical analysis of a realistic capacitor model in Fig. 2.17 (left) shows that there are two distinctly different frequency regions where the capacitor's behaviour is drastically different. For purposes of comparison, an impedance amplitude of an ideal capacitor is shown in the same plot. Although the numbers used to create Fig. 2.17 (right) are specific only for this particular numerical example, the curve shape is similar for other examples as well.

1. *Below* 30 MHz: inside this frequency region the intended function, i.e. the capacitance, closely follows the one for ideal capacitor, which is to say that parasitic components are negligible.
2. 30 MHz–200 MHz: Again, resonant behaviour of real capacitor is clearly visible with self-resonant frequency at approximately 100 MHz.
3. *Above* 100 MHz: in this frequency region inductive parasitics are dominant turning this capacitor into an inductor, and the desired capacitive function is completely suppressed.

This example illustrates complexity of behaviour associated with real capacitive components in frequency domain due to their internal parasitics, which limits the useful frequency range of operation.

2.3.3.4 AC Steady State of a Circuit with Capacitor

Parallel connection of a sine generator and purely capacitive load is shown in Fig. 2.18 (left). In this section we take a closer look at the important characteristics of this class of circuits in terms of the AC voltage–current relationship.

Although it may look trivial, (2.28) is very important for understanding the voltage–current relationship in Fig. 2.18, because it states that the AC current through a capacitor depends on the *rate* of voltage change, i.e. on its first derivative with respect to time. At the beginning of the voltage waveform in Fig. 2.18, i.e. at $t = 0$, the capacitor is discharged and, according to (2.28), the voltage v_C must also be zero. However, at that moment the rate of voltage change is *the highest*, which means that, according to (2.28), the corresponding current i_C must be at its maximum value.

Moving along the voltage waveform, e.g. at the point $t = \pi/2$ its value is at the maximum (i.e. its derivative is zero), therefore, the voltage rate of change is zero. Consequently, the corresponding current i_C also must be zero, which is exactly what the plot in Fig. 2.18 (right) shows. Once this analysis is done for all points in time, we reach the conclusion that the current waveform also takes sinusoidal shape, however it is *out of phase*, i.e. 90° *ahead* of the voltage waveform. This "ahead" wording is often a source of confusion because it is valid to ask question "How the current could possibly know its value a quarter of the cycle ahead in time?" No it does not know. Instead to the actual voltage value that is quarter of the cycle ahead in time, at any given point in time the current

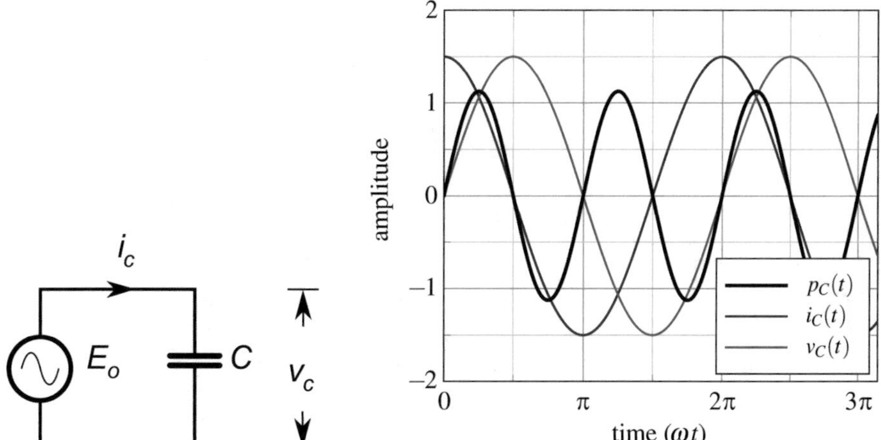

Fig. 2.18 Schematic diagram of a circuit with purely capacitive load and AC voltage generator, (left), and an example of the corresponding voltage–current–power time domain plot (right)

value is proportional to the instantaneous *rate of change* of the corresponding voltage value. This voltage–current relationship is commonly described with phrase "current leads the voltage by 90°".

For the specific numerical example in Fig. 2.18 important points to observe are:

(a) because the voltage and current through a capacitor are out of phase, the power changes from its most positive value, through the zero, to the most negative value, and back. Its waveform also follows a sinusoidal shape. That means, for half of its cycle the power flows *out of the generator* into the capacitor where it is stored in the form of electrical field. For the other half the power flows *out of the capacitor* back into the voltage generator;
(b) the power cycle is half the signal cycle;
(c) the average power is zero, i.e. energy keeps bouncing back and forth between the source and the capacitor. In short, in an ideal capacitor there is no thermal power dissipation.

2.3.3.5 Transient Capacitive Current

Given the initial condition that capacitor C is not charged at the beginning, the abrupt voltage change is introduced by the pulse function, Fig. 2.19. The time domain, non-steady-state behaviour of the capacitor is analysed as follows.

At any given moment in time, the source voltage E_0 is split between voltages across the resistor v_R and capacitor v_C. At the beginning the capacitor is not charged, which is to say that $v_C = 0$ (both plates at the same potential). At the moment $t = t_0$ when the voltage E_0 level becomes abruptly high, the source voltage is distributed only over the resistor (because the initial voltage $v_C(t_0) = 0$) and the current abruptly jumps to the $i(0) = E_0/R$ level. However, as soon as the current starts to flow, the charges are lending on the capacitor plate, which is equivalent of saying that the capacitor's voltage v_C starts to rise at very high rate (limited only with the initial current $i(0)$). As a consequence, less voltage is left across the resistor, which by itself further lowers the current, while the rate of the current change is being constantly reduced. Theoretically, this process that is very common in nature keeps going on forever, thus, it is known either as *exponential decay* or *natural growth.*

Mathematically, the exponential decay process is modelled by a first order differential equation. To help us visualize a pulse shape we note that the two constant voltage pulse levels (i.e. the "low" and the "high") by themselves are considered DC voltages while they last. Another way to say it is

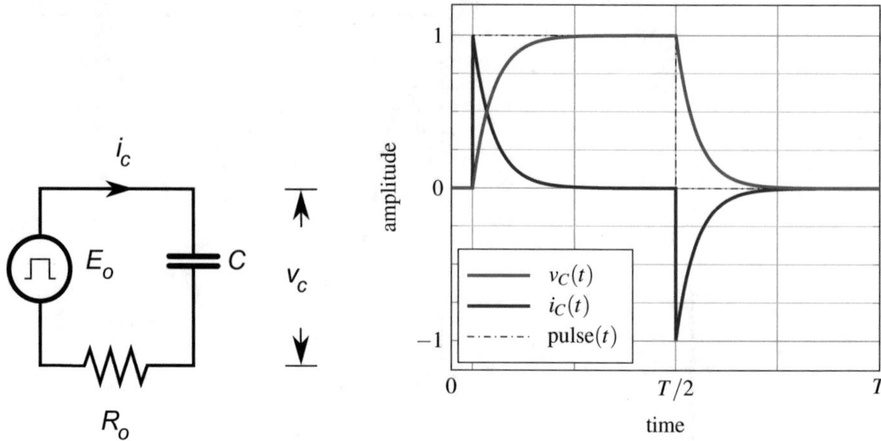

Fig. 2.19 Schematic diagram of a circuit with a capacitor, limiting resistor R_0, and `pulse` voltage generator E_0, (left), and the corresponding voltage–current time domain plot (right)

that a pulse shape could have been created with a DC voltage source and an ON/OFF switch. Hence, KVL equation and its derivative are

$$E_0 = v_R(t) + v_C(t) = i(t)R + \frac{q(t)}{C} \tag{2.33}$$

therefore, derivative of both sides is

$$0 = R\frac{di(t)}{dt} + \frac{1}{C}\frac{dq(t)}{dt} \quad \therefore \quad 0 = \frac{di(t)}{dt} + \frac{1}{RC}\,i(t) \tag{2.34}$$

Solution to the first order differential equation (2.34), with the initial condition $i(t_0) = E_0/R$, is

$$i(t) = i_C(t) = \frac{E_0}{R}\,e^{-t/\tau_0} \tag{2.35}$$

where

$$\tau_0 = RC \tag{2.36}$$

is the *time constant* of the system. After substituting (2.35) into (2.33) we find voltage across the capacitor as

$$E_0 = i(t)\,R + v_C(t) \quad \therefore \quad v_C(t) = E_0\left(1 - e^{-t/\tau_0}\right) \tag{2.37}$$

Equations (2.35) and (2.37) describe how voltage and current across a capacitor follow abrupt changes in DC voltage level across the capacitor terminals. Term used to denote this type of changes is *transient* and it is, obviously, very nonlinear process. Points to note are:

1. a capacitor is very good in passing fast, abrupt voltage changes, while at the same time it presents open circuit for DC currents;

2. theoretically, a capacitor never reaches the E_0 voltage level, it only keeps tending to it forever. Because of that, practical decision that a capacitor is "fully charged" is usually made at about $t = 5\tau_0$, because at that moment the capacitor voltage v_C is at over 99% of the maximum level set by E_0, which is easily proved by (2.37).

2.3.3.6 Energy Stored in Capacitor

Important question regarding the charging and discharging process is "where does the power go?" At the beginning, during the capacitor charging period, when the voltage across the capacitor abruptly jumps from the low to the high voltage, almost the whole power is being dissipated in the resistor. As the transition current lowers due to the increase of the capacitor voltage, portion of the power being stored in the capacitor in form of electrostatic field increases. When the polarity at the capacitor terminals is reversed (at the falling edge of the pulse), the capacitor serves as the source of energy, which now flows into the resistor where it is dissipated. At the end of the charge–discharge cycle, the whole energy initially provided by the voltage source has been dissipated in the resistor.

By definition, we find

$$P(t) \stackrel{\text{def}}{=} i(t)\,v(t) = v\,C\frac{dv}{dt} \quad \therefore \quad E = \int_0^t P\,dt = \int_0^V C\,v\,dv = \frac{1}{2}CV^2 \tag{2.38}$$

which is a commonly used expression for the amount of energy in a capacitor.

2.3.4 Inductor

As a physical component, most often an inductor takes shape of a cylindrical coil that is built by winding low-resistance wire around a cylindrical body, while modern integrated circuits implementations use planar form, Fig. 2.20. Approximate formula commonly used for short cylindrical air-core inductor is

$$L = \frac{\pi r^2 \mu_0 N^2}{l} \quad (r \ll l) \tag{2.39}$$

where L is desired inductance, r is the coil radius, l is the coil length, N is the number of turns, and μ_0 is permeability in vacuum. Inductors are not that often used in low-frequency electronic circuits.

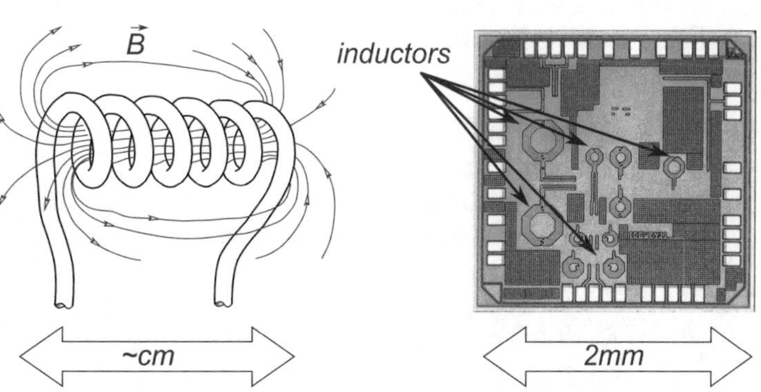

Fig. 2.20 Inductor: cylindric coil wire (left), and planar IC coil (right) versions

However, in wireless RF designs they are absolutely essential components. What is more, frequency behaviour of RF inductors limits the final specifications of RF circuits (arguably) more than any other component.

2.3.4.1 Inductive Reactance

Similarly to capacitors, inductors are two-terminal devices that are capable of storing energy. This time the energy is stored in the form of its internal magnetic field. Voltage–current relationship at inductor's terminals is described as

$$v(t) = L\,\frac{d}{dt}i(t) \tag{2.40}$$

where voltage $v(t)$ and the change of current di/dt are connected by the proportionality constant L, which is defined as *inductance*. Hence, voltage generated at terminals of an inductor is proportional to *rate of change* of the current flowing through.

1. **DC case**: if there is no current change, i.e. $i_L(t) = \texttt{const}$, then the term $di/dt = 0$ in (2.40) becomes zero, with direct consequence that the inductive voltage $v_L = 0$ must be zero,

$$Z_L(DC) = \frac{v_L}{i_L} = \frac{0}{i_L} = 0 \tag{2.41}$$

2. **AC case**: by using the same methodology as in Sect. 2.3.3.2, periodic current $i_L(t) = I_m \cos\omega t$, and after expanding (2.40) we show that in complex algebra notation inductive reactance Z_L is defined as

$$Z_L(j\omega) \stackrel{\text{def}}{=} \frac{V_L}{I_L} = j\omega L \tag{2.42}$$

showing that inductor's impedance $Z_L(j\omega)$ is *directly* proportional to frequency, starting with zero resistance at DC and tending to infinity as the frequency increases. Thus, once the transition process is over (i.e. $\omega \to 0$), an inductor presents zero resistance to DC current.

As a side note, there is important version of an inductor known as *RF choke* (RFC) whose main characteristic is that it is made intentionally large. Consequently, at higher frequencies it serves as AC blocking device while allowing DC current to flow. It is used extensively in RF circuits for providing DC biasing to active devices while minimally (ideally, not at all) interfering with the AC signal. To conclude, we note that an RFC has the same relationship to AC signals as a capacitor to DC signals, and vice versa.

Example 19: Inductor Impedance $Z_L(j\omega)$
To illustrate how inductor impedance Z_L changes relative to frequency, calculate impedance of a $L = 159\,\text{nH}$ inductor at the following frequencies: 100 Hz, 10 kHz, 1 MHz, 100 MHz, and 10 GHz.

Solution 19: Direct implementation of (2.42) results in $|Z_L| = 2\pi f L$, therefore we write frequency to impedance ($f \to Z_L$) pairs of this inductor as: (100 Hz → 100 μΩ), (10 kHz → 10 mΩ), (1 MHz → 1 Ω), (100 MHz → 100 Ω), and (10 GHz → 1 kΩ).

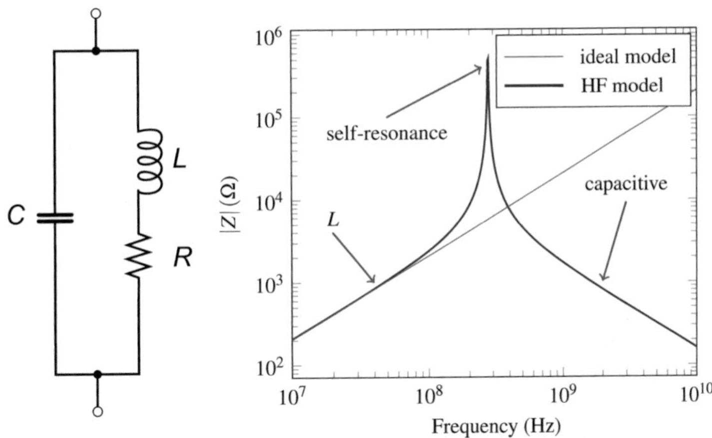

Fig. 2.21 Equivalent electrical circuit model of inductor (left), impedance vs. frequency of a typical wire-wound inductor (right)

Example 20: Inductor Impedance $Z_L(N, l, r)$
Estimate inductance L of a coil formed by $N = 50$ turns of a copper wire with radius $a = 80\,\mu\text{m}$, radius of air core is $r = 2\,\text{mm}$, length of the coil is $l = 10\,\text{mm}$. Note that distance between two adjacent turns is $d = l/N = 100\,\mu\text{m}$.

Solution 20: Direct implementation of (2.39) gives a close estimate of the coil inductance as

$$L = \frac{\pi r^2 \mu_0 N^2}{l} \approx 3.948\,\mu\text{H}$$

2.3.4.2 HF Inductor Model

It should be noted that the internal resistance of the wire is always present together with the parasitic capacitances between: (a) the neighbouring turns and (b) the inductor and the surrounding environment, which means that in reality it is only possible to achieve a close approximations of the ideal inductor behaviour, but never the "real ideal inductor". Therefore, a real inductor resembles a relatively complex RLC network behaviour, where dominant inductive behaviour is only within a limited range of operation, while quickly losing its inductive property outside of the optimal range. One of the limiting factors is phenomena known as *self-resonance* (see Chap. 10).

One of possible equivalent circuits of a high-frequency inductor, Fig. 2.21, includes the desired inductance L, the wire serial resistance R_s, and C_s is the parasitic shunt capacitance between the inductor's terminals.

A typical realistic frequency behaviour of an inductor clearly shows three distinct regions. For purposes of comparison, an impedance amplitude of an ideal inductor is shown in the same plot. Although the numbers shown in Fig. 2.21 are specific only for this particular numerical example, the curve shape is similar for other examples as well.

1. *Below* 100 MHz: inside this frequency region the intended function, i.e. the inductance, closely follows the one for ideal inductor, which is to say that parasitic components are negligible.
2. *Around* 250 MHz: A sharp, resonant behaviour of real inductor is clearly visible with self-resonant frequency.

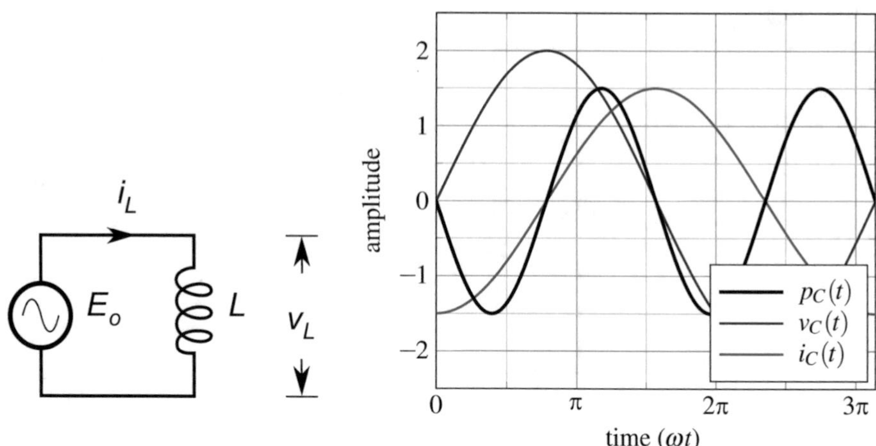

Fig. 2.22 Schematic diagram of a circuit with purely inductive load and AC voltage generator, (left), and the corresponding voltage–current–power time domain plot (right)

3. *Above* 1 GHz: in this frequency region capacitive parasitics are dominant turning this inductor into a capacitor, the desired function is almost completely suppressed.

This example illustrates complexity of behaviour associated with real components, in this case an inductor, in frequency domain due to their internal parasitics, which directly limits their useful frequency range of operation.

2.3.4.3 AC Steady State of a Circuit with Inductor

Parallel connection of a single-tone generator and purely inductive load is shown in Fig. 2.22 (left). Let us take a closer look at the voltage–current relationship in Fig. 2.22. At the beginning of the voltage waveform cycle, i.e. at $t = 0$, voltage is at its maximum rate of change, which means that the rate of current change is *the highest* (because $Z_L(t_0) = 0$), however it is *negative*. The current is opposing the large rate of voltage change. On the other hand, when the current is at its maximum, with its first derivative equal zero, the inductor voltage must be at zero value as well. This point in time is followed by reduction in current amplitude, which causes negative voltage, and the cycle keeps repeating. Once this analysis is done for all points in time, we reach the conclusion that the current waveform also takes sinusoidal shape, however it is out of phase, i.e. 90° *behind* (i.e. it "lags") the voltage waveform. This voltage–current relationship is conveniently described by using the phrase that "current lags the voltage by 90°".

For the numerical example in Fig. 2.22 important points to observe are:

(a) because the voltage and current through an inductor are out of phase, the power changes from its most negative value, through the zero, to the most positive value, and back. Its waveform also follows a sinusoidal shape. That means, for half of its cycle the power flows *out of the generator* into the inductor where it is stored in the form of magnetic field. For the other half the power flows *out of the inductor* back into the voltage generator;
(b) the power cycle is half the signal cycle;
(c) the average power is zero, i.e. it keeps bouncing back and forth between the source and the inductor. In short, there is no thermal power dissipation in an ideal inductor.

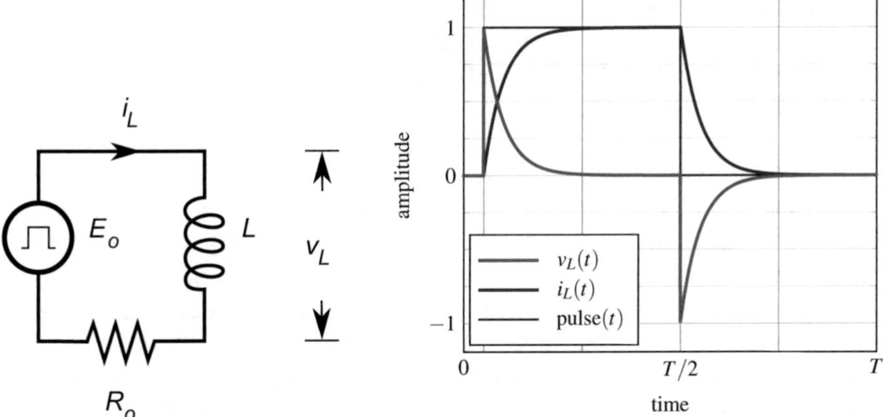

Fig. 2.23 Schematic diagram of a circuit with an inductor, limiting resistor R_0, and `pulse` voltage generator E_0, (left), and the corresponding voltage–current time domain plot (right)

2.3.4.4 Transient Inductive Current

Under given initial conditions, i.e. voltage across the inductor L is zero in the beginning, and the abrupt voltage change is introduced by the pulse function, Fig. 2.23, the time domain, non-steady-state behaviour of the capacitor is analysed as follows.

At any given moment in time, the source voltage E_0 is split between voltages across the resistor v_R and inductor v_L. At the moment $t = t_0$ when the voltage E_0 level becomes abruptly high, the source voltage is distributed only over the inductor (because the initial voltage $v_R(t_0) = 0$) and the current starts to approach the maximal $i(t) = E_0/R$ level (when Z_L becomes zero). However, as soon as the current starts to flow, the resistor's voltage v_R starts to rise at very high rate. As a consequence, less voltage is left across the inductor, which by itself further increases the current, while the rate of the current change is being constantly reduced.

Hence, after writing the KVL equations we conclude that

$$\left.\begin{aligned} v_R &= i\,R \\ v_L &= L\frac{di}{dt} \end{aligned}\right\} \quad \therefore \quad Ri + L\frac{di}{dL} = E_0 \quad \therefore \quad i(t) = \frac{E_0}{R}\left(1 - e^{-\,(R/L)\;t}\right) \tag{2.43}$$

where

$$\tau_0 = \frac{L}{R} \tag{2.44}$$

is the time constant of LR circuit and the current follows the natural growth law.

Equation (2.43) describes how current and voltage across an inductor follow transient changes in DC voltage level across the inductor terminals. Points to note are:

1. an inductor is very good in passing fast, abrupt current changes, while at the same time it presents short circuit for DC currents;

2. theoretically, an inductor never reaches the E_0/R current level, it only keeps tending to it forever. Because of that, practical decision that voltage across inductor drops to zero at about $t = 5\tau_0$, because at that moment the inductor current i_L is at over 99% of the maximum level set by E_0/R, which is easily proved by (2.43).

2.3.4.5 Energy Stored in Inductor

Same as for a capacitor, we pose question "where does the power go?" At the beginning, during the inductor charging period, when the voltage across the inductor abruptly jumps from the low to the high voltage, almost the whole power is being stored in the inductor ($V_R = 0$ thus the resistor current is zero). As the transition voltage $v_L(t)$ lowers due to the increase of the resistor voltage, portion of the power being stored in the inductor in form of electromagnetic field decreases. When the polarity at the inductor terminals is reversed (at the falling edge of the pulse), the inductor is fully charged and it serves as the source of energy, which now flows into the resistor where it is dissipated. At the end of the charge–discharge cycle, the whole energy initially provided by the voltage source has been dissipated in the resistor. These two phases of the full cycle must contain exactly the same amounts of energy that add up to the total available energy, thus we can say that the total charge

By definition, we find

$$P(t) \stackrel{\text{def}}{=} i(t)\,v(t) = i\,L\frac{di}{dt} \quad \therefore \quad E = \int_0^t P\,dt = \int_0^I L\,i\,di = \frac{1}{2}LI^2 \tag{2.45}$$

which is a commonly used expression for the amount of energy in an inductor.

Example 21: Impedance—Series RL

Derive expression for the equivalent impedance Z_{RL} if a resistor $R = 1\,\text{k}\Omega$ and an inductor $L = 1\,\text{mH}$ are connected in series.

1. Calculate absolute value of the impedance $|Z_{RL}(\omega)|$ and the associated phase $\varphi_{RL}(\omega)$ at the following two frequencies: (a) $\omega_1 = 2\pi\,f_1 = 2\pi \times 1\,\text{MHz}$; (b) $\omega_2 = 2\pi\,f_2 = 2\pi \times 1\,\text{kHz}$
2. What are limits of $|Z_{RL}(\omega)|$ and $\varphi_{RL}(\omega)$ when: (a) $\omega \to 0$ (i.e. at zero frequency, a.k.a. *DC*) and (b) $\omega \to \infty$, i.e. at very high frequencies.

Solution 21:

1. According to definitions of impedances for these two elements (2.13) and (2.42), it follows

$$Z_{RL}(\omega) = Z_R + Z_L = R + j\omega L = 1\,\text{k}\Omega + j\omega \times 1\,\text{mH}$$

Thus, by using the Pythagorean theorem (i.e. complex algebra), it follows

$$|Z_{RL}(\omega)|^2 = R^2 + (\omega L)^2 \Rightarrow |Z_{RL}(\omega)| = \sqrt{R^2 + (\omega L)^2} \tag{2.46}$$

$$\therefore$$

$$|Z_{RL}(\omega_1)| = \sqrt{(1\,\text{k}\Omega)^2 + (2\pi \times 1\,\text{MHz} \times 1\,\text{mH})^2} = 6.36\,\text{k}\Omega$$

$$|Z_{RL}(\omega_2)| = \sqrt{(1\,\text{k}\Omega)^2 + (2\pi \times 1\,\text{kHz} \times 1\,\text{mH})^2} = 1\,\text{k}\Omega$$

(continued)

We note that at 1 kHz impedance of this inductor is negligible relative to 1 kΩ resistance. Phase φ of a complex number is calculated by definition with the help of $\tan(\varphi)$ function. Both real and imaginary part of Z_{RL} are positive, therefore the associated phase φ is in the first quadrant, i.e.

$$\tan\varphi = \frac{\Im(Z_{RL})}{\Re(Z_{RL})} \tag{2.47}$$

$$\therefore$$

$$\varphi_1(\omega_1) = \arctan\left[\frac{\omega_1 L}{R}\right] = \arctan\left[\frac{2\pi \times 1\,\text{MHz} \times 1\,\text{mH}}{1\,\text{k}\Omega}\right] = 81^\circ$$

$$\varphi_2(\omega_2) = \arctan\left[\frac{\omega_2 L}{R}\right] = \arctan\left[\frac{2\pi \times 1\,\text{kHz} \times 1\,\text{mH}}{1\,\text{k}\Omega}\right] = 0.3^\circ$$

which is to say that phase of this RL network at 1 MHz is dominated by the inductor, while at 1 kHz it is close to zero (thus dominated by resistor).

2. When $\omega \to 0$, from (2.46) we find $|Z_{RL}(0)| = 1\,\text{k}\Omega$ (i.e. the network is resistive), and when $\omega \to \infty$ we find $|Z_{RL}(\infty)| = \infty$ (i.e. the network is inductive).
   When $\omega \to 0$ from (2.47) we find $\varphi(0) = 0$ (i.e. the network is resistive), and when $\omega \to \infty$ we find $\varphi(\infty)| = +(\pi/2)$ (i.e. the network is inductive).

In summary, serial RL network is dominated by inductor at high frequencies and by resistor at low frequencies (when reactance of the inductor is very low).

Example 22: Impedance—Parallel RL
Repeat calculations from Example 21, but this time in the case of parallel RL network.

Solution 22:

1. According to definitions of impedances for these two elements (2.13) and (2.42) their parallel impedance $Z_{R||L}$ is found as[3]

$$\frac{1}{Z_{R||L}(\omega)} = \frac{1}{Z_R} + \frac{1}{Z_L} \Rightarrow Z_{R||L}(\omega) = \frac{1}{\frac{1}{Z_R} + \frac{1}{Z_L}} \Rightarrow |Z_{R||L}(\omega)|$$

$$= \frac{1}{\sqrt{\frac{1}{Z_R^2} + \frac{1}{Z_L^2}}} = \frac{1}{\sqrt{\frac{1}{R^2} + \frac{1}{(\omega L)^2}}}$$

$$\therefore$$

$$Z_{R||L}(1\,\text{MHz}) = \frac{1}{\sqrt{\frac{1}{(1\,\text{k}\Omega)^2} + \frac{1}{(2\pi \times 1\,\text{MHz} \times 1\,\text{mH})^2}}} = 987.57\,\Omega$$

(continued)

[3] See complex algebra rules.

$$Z_{R||L}(1\,\text{kHz}) = \frac{1}{\sqrt{\frac{1}{(1\,\text{k}\Omega)^2} + \frac{1}{(2\pi \times 1\,\text{kHz} \times 1\,\text{mH})^2}}} = 6.28\,\Omega$$

which shows that at low frequencies this impedance is dominated by inductor's low resistance.
This time, phase φ is inverted relative to the frequency, i.e.

$$\varphi(1\,\text{MHz}) = \arctan\left[\frac{R}{\omega_1 L}\right] = \arctan\left[\frac{1\,\text{k}\Omega}{2\pi \times 1\,\text{MHz} \times 1\,\text{mH}}\right] = 0.3^\circ$$

$$\varphi(1\,\text{kHz}) = \arctan\left[\frac{R}{\omega_2 L}\right] = \arctan\left[\frac{1\,\text{k}\Omega}{2\pi \times 1\,\text{kHz} \times 1\,\text{mH}}\right] = 81^\circ$$

which is to say that phase of this RL network at 1 kHz is dominated by the inductor, while at 1 MHz it is close to zero (thus dominated by resistor).

2. When $\omega \to 0$, $|Z_{R||L}(0)| = 0$, and when $\omega \to \infty$, then $|Z_{R||L}(\infty)| = R$.
When $\omega \to 0$, $\varphi(0) = +\pi/2$, and when $\omega \to \infty$, then $\varphi(\infty)| = 0)$.
In summary, parallel RL network is dominated by inductor at low frequencies and by resistor at high frequencies (when reactance of the inductor is very high).

2.4 Summary

In this chapter we reviewed transfer functions of basic building blocks used in circuit design. We find that two groups of devices, namely voltage/current sources and passive RLC components, can be precisely described by their idealized behavioural models. These analytical models set "back of the envelope" limits on functionality that these devices could theoretically achieve, either alone or in combination with the other devices. Often, these models assume that the associated devices are capable of handling infinite amounts of power, which in reality we know very well is not possible. Nevertheless, behavioural models enable us to do very fast "back of the envelope" hand analysis and evaluations and to find theoretical limits of various parameters.

On the other hand, behaviour of real physical devices resembles functionality of their ideal counterparts only within a limited range of specifications. Outside of the given range, real devices almost always mimic behaviour of one or more other ideal components. For this reason, we use multiple equations to describe the functionally of real devices, each equation being valid only within a certain range. We found that, for example, a simple resistor shows behaviour of a real resistor only at relatively low frequencies, then, as the frequency increases, it morphs into a capacitive behaviour, only to suddenly switch its personality and to become almost perfect inductor. Accepting and understanding physical reality and limitations of devices is one of the keys to a successful circuit design.

Problems

2.1 For the current waveform in Fig. 2.24, sketch the graph of its corresponding voltage $v(t)$ across an inductor.

Data: $L = 3\,\mu\text{H}$.

2.2 A capacitor C and a resistor R are connected in parallel. Initially, at time $t_0 = 0\,\text{s}$ the capacitor was charged up to voltage V_0. Sketch a graph of the voltage $v_C(t)$ across the capacitor over the time interval t. (HINT: timing constant is $\tau = RC$)

Data: $C = 1\,\mu\text{F}$, $R = 1\,\text{k}\Omega$, $V_0 = 10\,\text{V}$, $t = 5\,\text{ms}$.

2.3 Calculate equivalent resistance R_{AB} for the four networks Fig. 2.25. First, find the equivalent resistances at the function of frequency, then estimate the resistances at DC and $f = \infty$.

2.4 For each of the two networks given in Fig. 2.26, derive the output voltage gain $A_V = V_{AB}/V_{out,B}$, then estimate the gain at DC and $f = \infty$.

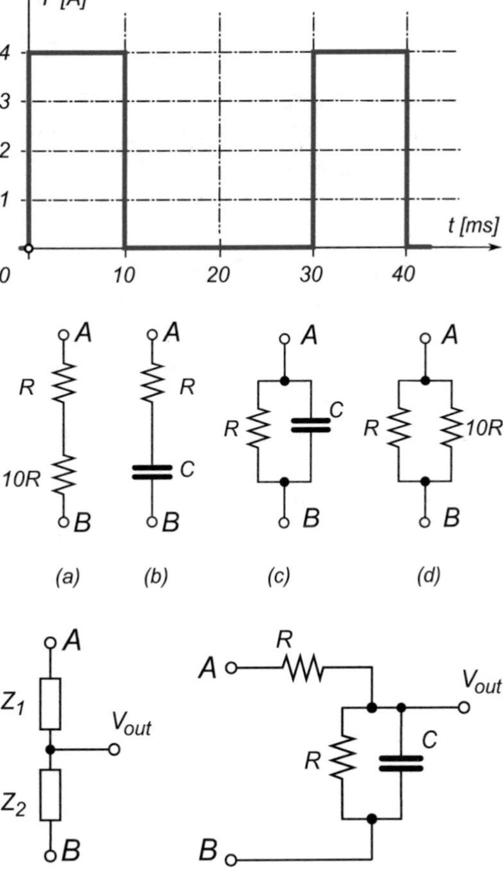

Fig. 2.24 Illustration for Problem 2.1

Fig. 2.25 Resistive networks for problem 2.3

Fig. 2.26 Networks for problem 2.4

Data: (1) $Z_1/Z_2 = 10:1$, and (2) $Z_1/Z_2 = 1:10$.

2.5 A typical bonding wire has a diameter of $d = 75\,\mu\text{m}$. Calculate how long is the wire whose resistance is $R = 1\,\Omega$ if the material used is: (a) copper, (b) aluminium, or (c) iron.

Data: (a) $\rho_{Cu} = 1.68 \times 10^{-8}\,\Omega\text{m}$, (b) $\rho_{Al} = 2.65 \times 10^{-8}\,\Omega\text{m}$, (c) $\rho_{Fe} = 9.70 \times 10^{-8}\,\Omega\text{m}$.

2.6 A rotary plate capacitor in Fig. 13.15 consists of five pairs of semicircular metallic plates; five static plates and five rotating plates. All static plates are connected to the same potential V_1, equally, all rotating plates are connected to the same potential V_2.

Initially, the capacitor plates are fully overlapping, calculate the total capacitance when the plates are: (a) at the initial position, (b) when the rotating plates are rotated by $\alpha = \pi/4$, and (c) when the rotating plates are rotated by $\alpha = \pi/2$.

Data: the plate radius is $r = 10\,\text{mm}$, distance between each two neighbouring plates is $d = 1\,\text{mm}$, they are separated by air, vacuum permittivity $\epsilon_0 = 8.854\,187\,812 \times 10^{-12}\,\text{F/m}$.

2.7 In reference only to the graph in Fig. 2.13 (right), for this realistic resistor, estimate value of the dominant equivalent capacitance C and inductance L. Then, write three expressions for the resistor's complex impedance in the form of $Z_R = \Re(Z_R) + j\Im(Z_R)$: (a) when the frequency is $f < 10\,\text{MHz}$; (b) when the frequency is in the range $20\,\text{MHz} < f < 1\,\text{GHz}$; and (c) when the frequency is $f > 5\,\text{GHz}$. Show these impedance vectors in the complex plane.

2.8 In reference only to the graph in Fig. 2.17 (right), for this realistic capacitor, estimate values of the capacitance C and its parasitic inductance L. Then, write two expressions for the capacitor's complex impedance in the form of $Z_C = \Re(Z_C) + j\Im(Z_C)$: (a) when the frequency is $f < 20\,\text{MHz}$; (b) when the frequency is $f > 200\,\text{MHz}$. Show these impedance vectors in the complex plane.

2.9 In reference only to the graph in Fig. 2.21 (right), for this realistic inductor, estimate values of its inductance L and its parasitic capacitance C. Then, write two expressions for the inductor's complex impedance in the form of $Z_L = \Re(Z_L) + j\Im(Z_L)$: (a) when the frequency is $f < 100\,\text{MHz}$; (b) when the frequency is $f > 1\,\text{GHz}$. Show these impedance vectors in the complex plane.

2.10 In reference to Fig. 2.19, recommend highest frequency of the square waveform that should be used in the RC circuit, given: (a) $R = 1\,\text{k}\Omega, C = 1\,\text{pF}$; (b) $R = 1\,\text{k}\Omega, C = 1\,\text{nF}$; and (c) $R = 1\,\text{k}\Omega, C = 1\,\mu\text{F}$.

Multistage Interface

3

Over the time we have developed what is known as "top-to-bottom then bottom-to-top design flow", that is to say a complicated system is designed hierarchically; during its development stage systems are often split into more than ten levels of hierarchy. Once the hierarchy chain is established and each of the stages is replaced by its equivalent Thévenin or Norton model, each of the blocks is considered to be a "black box" described by its input and output impedances and its transfer function. In this section we review basic interface models.

3.1 System Partitioning Concept

We appreciate the elegance and efficiency of the system level approach once we realize that each output signal generated by "driving stage" (or simply "the driver") is received as the input signal by the "loading stage" (or simply "the load"). It is important to understand that except for the first and the last element in the chain, by itself, each stage is both driver (relative to its neighbour down the signal path) and load (relative to its neighbour up the signal path). For the purpose of the signal transmission, the internal structure of each stage is not relevant; indeed, it is only important to know the following:

1. V_{out}: amplitude of voltage (or current) signal generated by the driver,
2. Z_{out}: output impedance of the driver,
3. Z_{in}: input impedance of the load stage, and
4. A: gain of each individual stage.

Conceptually, these driver–load relationships are used to model the signal transfer at each of the interface points. By doing so, at the conceptual level, analysis of a complicated system is reduced to repeated calculations of a simple voltage/current divider at each of the interfaces.

The system partitioning enables us to calculate, for example, the total gain of a system, Fig. 3.1, as a product of the individual gains of each stage, as

$$A = A_1 \times A_2 \times A_3 \times \cdots \times A_n \tag{3.1}$$

where gain of each stage is found by definition as the output to input signal ratio

R. Sobot, *Wireless Communication Electronics*,
https://doi.org/10.1007/978-3-030-48630-3_3

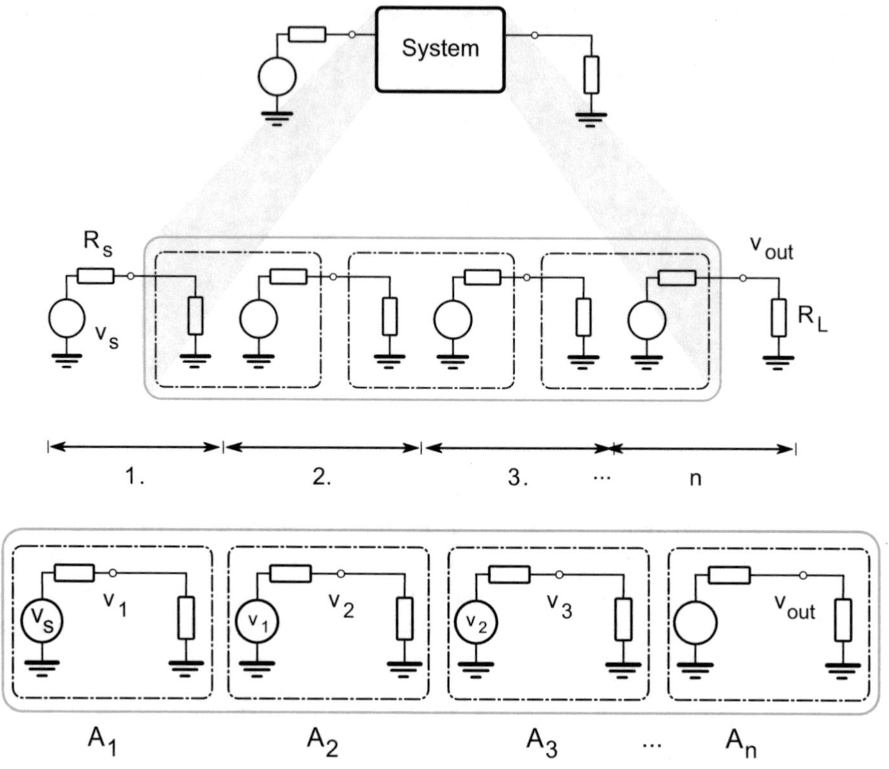

Fig. 3.1 System partitioning based on voltage divider model at interface planes between two subsequent stages

$$A_1 = \frac{v_1}{v_S}; \qquad A_2 = \frac{v_2}{v_1}; \qquad A_3 = \frac{v_3}{v_2}; \qquad \cdots \qquad A_n = \frac{v_{OUT}}{v_{n-1}} \tag{3.2}$$

which is to say that

$$A = \frac{\cancel{v_1}}{v_S} \times \frac{\cancel{v_2}}{\cancel{v_1}} \times \frac{\cancel{v_3}}{\cancel{v_2}} \times \cdots \times \frac{v_{OUT}}{\cancel{v_{n-1}}} = \frac{v_{OUT}}{v_S} \tag{3.3}$$

In addition, if the gain of each stage is expressed in decibels, then the total gain is simply the sum of the gains, as

$$A_{\text{dB}} = A_{1_{\text{dB}}} + A_{1_{\text{dB}}} + \cdots \times A_{n_{\text{dB}}} \tag{3.4}$$

This general system partitioning idea enables us to perform rapid analysis "by inspection", which gives us not only fast answers but also the insight into fundamental limits of the circuit topology under consideration. Routinely, the back of the envelope "rough" estimates are within less than 10% difference relative to the "exact" results derived by simulation or experiment.

3.2 Voltage Transfer Interface

Two main applications of the voltage divider model can be loosely described as the literal and the conceptual. A voltage divider in its basic form of two impedances connected in series is probably the simplest circuit model. Since only two of the three basic devices (i.e. R,L, and C) are used to build a voltage divider then there are nine possible serial configurations (e.g. R-R, R-C, R-L, etc.), each having slightly different behaviour.

When applied literally, a voltage divider is implemented as two physical components connected in series and used simply to scale down the voltage amplitude (either DC or AC) applied at its two end terminals. When one of the two impedances is complex, then due to its frequency dependance such voltage divider is capable of modifying "frequency profile" of the voltage signal applied to its terminals, i.e. it serves as a first-order RC or RL *filter*.

When applied conceptually, a voltage divider is used to model the signal transfer between any two system-level blocks. In Fig. 3.1, we introduce the concept of system partitioning based on voltage dividers. It cannot be emphasized enough that a clear understanding of simple voltage divider behaviour and its applications are of vital importance to all electrical engineers.

Question that needs to be answered is: in reference to Fig. 3.1, starting with a voltage source v_S whose resistance is R_S, how large are the internal voltages v_1, v_2, etc., until v_{out} at R_L? In other words, what is the gain of this system? In order to facilitate answer to this question, in the following sections, we review three of most important voltage divider structures.

3.2.1 Resistive Voltage Divider

One of the most important simple networks consists of one ideal voltage source and two resistors connected in series, see Fig. 3.2. It is assumed that all the elements are ideal and, aside from the conversion of electrical energy into thermal energy, there is no energy loss.

Analysis of this circuit structure is simple and the objective is to find the relationship between the source voltage v_{in} and the output voltage v_{out} at node① (the connecting point between the two resistors) in Fig. 3.2. In the ideal case, except for the current i_R flowing through the two resistors, there is no other current flow in or out of node①, and the two resistors present $(R_1 + R_2)$ load to the ideal voltage source. A direct application of Ohm's law leads to expression for the voltage gain A_V as

$$\left.\begin{aligned} i_R &= \frac{v_{in}}{R_1 + R_2} \\ v_{out} &= i_R\, R_2 \end{aligned}\right\} \quad \therefore \quad A_V \stackrel{\text{def}}{=} \frac{v_{out}}{v_{in}} = \frac{R_2}{R_1+R_2} = \frac{1}{1+\dfrac{R_1}{R_2}} \tag{3.5}$$

In other words, the ratio of the source voltage v_{in} and output voltage v_{out} is the same as one plus the ratio of their respective resistances. The output voltage v_{out} is measured across R_2, while the source voltage v_{in} is distributed across two resistors in series $(R_1 + R_2)$ (see Fig. 3.2). Equation (3.5) clearly shows that theoretical limit of the resistive voltage divider's gain A_V equals to one, i.e. a passive resistive circuit is not capable of amplifying signal. At best, voltage v_{in} generated by the voltage

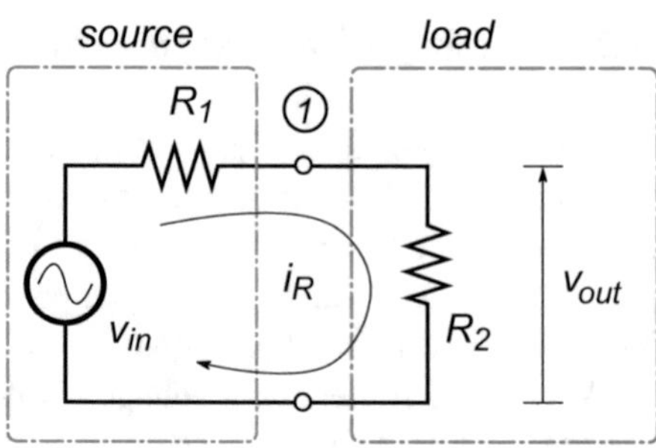

Fig. 3.2 Simple resistive voltage divider

source is delivered unattenuated to the receiving load R_2, i.e. $v_{out} = v_{in}$. This is possible only in two cases: either $R_2 \to \infty$ and/or $R_1 = 0$, then $v_{out} = v_{in}$, as shown here

$$\lim_{R_1 \to 0} A_V = \frac{1}{1 + \cancelto{0}{\frac{0}{R_2}}} = 1 \Rightarrow v_{out} = v_{in}$$

$$\lim_{R_2 \to \infty} A_V = \frac{1}{1 + \cancelto{0}{\frac{R_1}{\infty}}} = 1 \Rightarrow v_{out} = v_{in}$$

However, in practice we approximate the above two conditions as

$$R_2 \gg R_1 \Rightarrow v_{out} \approx v_{in} \tag{3.6}$$

so that (3.6) applies to realistic circuits. Therefore, this equation serves as the "back of the envelope" guideline for design of resistive voltage divider interfaces. In addition, resistive voltage divider is extensively used as a "voltage reference" circuit to derive biasing DC voltage from the power supply voltage. As a concluding remark, we note that being frequency independent circuit, resistive voltage divider indiscriminately scales both DC and AC voltage signals.

Example 23: Resistive Voltage Divider—Maximum Power
Derive an expression for the maximum possible power $P_{\max}$ that can be delivered by a realistic voltage generator, i.e. with non-zero internal resistance, to a resistive load. This case is modelled with the circuit network in Fig. 3.2 where the ideal generator v_{in} and resistance R_1 represent the realistic voltage source, while resistance R_2 represents the load.

Solution 23: Using Eqs. (3.5) and the definition of power, with reference to Fig. 3.2 we write

$$P \stackrel{\text{def}}{=} i_R\, v_{out} = I_R^2\, R_2 = \left[\frac{v_{in}}{R_1 + R_2}\right]^2 R_2 = \frac{v_{in}^2\, R_2}{(R_1 + R_2)^2}$$

(continued)

$$= \frac{v_{in}^2}{R_1} \frac{\frac{R_2}{R_1}}{\left(1 + \frac{R_2}{R_1}\right)^2} = \frac{v_{in}^2}{R_1} \frac{x}{(1+x)^2} \tag{3.7}$$

after substitution of $R_2/R_1 = x$. We find that function $f(x)$

$$f(x) = \frac{x}{(1+x)^2} \quad \therefore \quad f'(x) = \frac{1-x}{(1+x)^3} \tag{3.8}$$

where $x \geq 0$, has a maximum for $x = 1$ (found when derivative $f'(x) = 0$), leading into $\max[(f(x))] = 1/4$, hence

$$P_{\max} = \frac{v_{in}^2}{4\,R_1} \tag{3.9}$$

The conclusion is that the maximum power (3.9) that can be generated by a voltage generator v_{in} whose internal resistance is R_1 is achieved for $x = 1$, i.e. $R_1 = R_2$. Generalized complex form of this conclusion represents one of the key design guidelines in RF circuit design.

3.2.2 RC Voltage Divider

In this section we analyze RC serial network where capacitor's impedance Z_C represents the load, see Fig. 3.3 (left). As oppose to resistive voltage divider in Sect. 3.2.1, which is frequency independent, RC voltage divider includes a capacitive element C whose impedance (2.31). Therefore, RC voltage divider alters the frequency spectrum of the output signal.

Steady-state analysis of an RC voltage divider is done by complex algebra. By doing so, all three variables (the amplitude, the phase, and the frequency of the output signal) are calculated using the same equation. Moreover, after resistance R_2 is replaced with impedance $Z_C = 1/j\omega C$ and as long as complex algebra is used then (3.5) still holds. By inspection of schematic diagram in Fig. 3.3 (left), we derive an expression for impedance Z_{RC} seen by voltage source V_{in} as

$$Z_{RC} = Z_R + Z_C = R + \frac{1}{j\omega C} = R - \frac{j}{\omega C} \tag{3.10}$$

Therefore, similarly to (3.5), the transfer function of RC voltage divider becomes

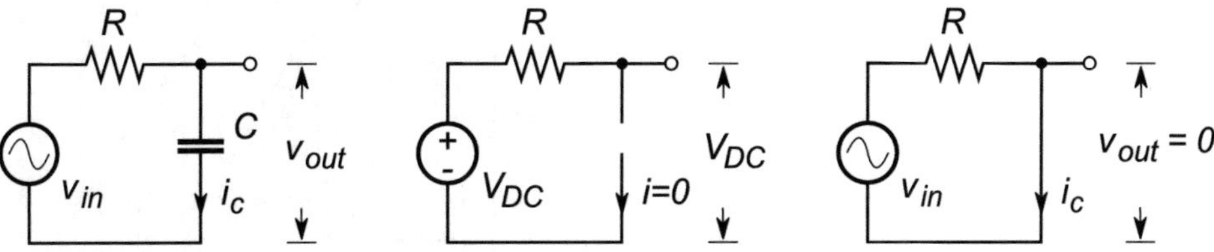

Fig. 3.3 RC voltage divider at AC (left), at DC (middle); and at $\omega = \infty$

$$A_V = \frac{V_{out}}{V_{in}} = \frac{Z_C}{Z_R + Z_C} = \frac{\frac{1}{j\omega C}}{R + \frac{1}{j\omega C}} = \frac{1}{1 + j\omega RC} = \frac{1}{1 + j\frac{\omega}{1/RC}} = \frac{1}{1 + j\frac{\omega}{\omega_0}} \tag{3.11}$$

where $\omega_0 = 1/RC$ and we note that (3.11) is in the form of (1.84). Therefore, amplitude of (3.11) is

$$A_V = \left|\frac{1}{1 + j\omega RC}\right| = \frac{1}{\sqrt{1 + (\omega RC)^2}} \tag{3.12}$$

and we repeat the analysis presented in Sect. 1.6.4, which helps us write the phase expression as

$$\phi = -\arctan\frac{\omega}{\omega_0} = -\arctan(\omega RC) \tag{3.13}$$

For the sake of discussion, we analyze transfer function (1.84) again, but this time in the context of RC voltage divider. Using the linear approximation method, we evaluate the extremes of (3.12) and find that, for DC (i.e. $\omega = 0$), the capacitor has an infinite impedance, i.e. it becomes an open connection, see Fig. 3.3 (middle), therefore $A_V = 1$ or, in other words, $v_{out} = v_{in}$. At the opposite end of the spectrum, for $\omega = \infty$, the capacitor has zero resistance, i.e. it is a short connection, see Fig. 3.3 (right), therefore $A_V = 0$ or $v_{out} = 0$. The output amplitude in the frequency domain between these two extremes is, therefore, described in accordance with (3.12) (see Fig. 3.4).

Following analysis in Sect. 1.6.4 and by inspection of (3.11) we write expression for "−3 dB" frequency as

$$\omega_0 = \frac{1}{RC} \tag{3.14}$$

This type of RC voltage divider is commonly referred to as a "low-pass filter" (LPF) because it attenuates high-frequency components of multi-tone signal, while at the same time components close

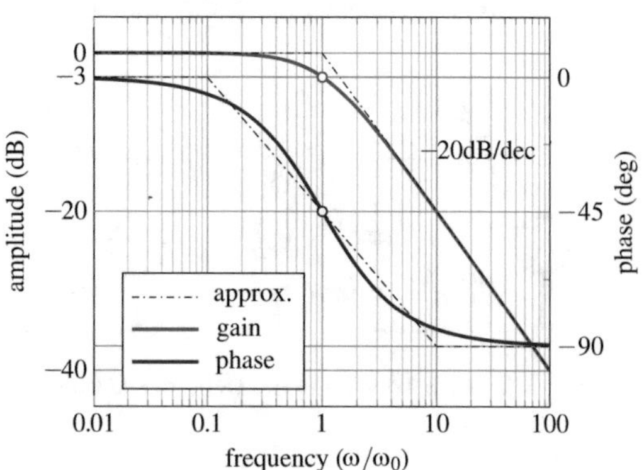

Fig. 3.4 Frequency domain plots of an LP RC filter: amplitude and phase response

to DC pass unaffected. Frequency ω_0 that corresponds to the -3 dB amplitude point is the frequency parameter that determines its pass-band frequency and $-45°$ phase shift, as in Fig. 3.4. Again, for RC filters, we note that phase of the output voltage lags phase of the input voltage.

3.2.3 RL Voltage Divider

Voltage divider that consists of serial RL network and voltage source is similar to the RC network of Sect. 3.2.2, with the capacitor being replaced by an inductor, see Fig. 3.5 (left). Due to the inductor's frequency dependence, this network also alters the frequency spectrum profile of the output signal. We find its transfer function as follows. The serial RL network's impedance is

$$Z_{RL} = Z_R + Z_L = R + j\omega L = R\left(1 + j\frac{\omega}{R/L}\right) = R\left(1 + j\frac{\omega}{\omega_0}\right) \tag{3.15}$$

where $\omega_0 = R/L$. By inspection of Fig. 3.5 (left), after substituting (3.15) we write RL network's gain transfer function as

$$A_V = \frac{v_{out}}{v_{in}} = \frac{Z_L}{Z_R + Z_L} = \frac{j\omega L}{R + j\omega L} = \frac{j\dfrac{\omega}{\omega_0}}{1 + j\dfrac{\omega}{\omega_0}} \tag{3.16}$$

Therefore, logarithmic form of (3.16) becomes

$$20\log A_V = 20\log\left(\frac{j\dfrac{\omega}{\omega_0}}{1 + j\dfrac{\omega}{\omega_0}}\right) = \underbrace{\underbrace{+20\log\left(j\frac{\omega}{\omega_0}\right)}_{\text{Sect. 1.6.2}}}_{①}\ \underbrace{\underbrace{-20\log\left(1 + j\frac{\omega}{\omega_0}\right)}_{\text{Sect. 1.6.4}}}_{②} \tag{3.17}$$

Thus, we find that transfer function of this RL network (3.17) is equivalent to the sum of functions in Sects. 1.6.2 and 1.6.4, thus its phase is similarly written as

$$\angle H(\omega) = \underbrace{\underbrace{\frac{\pi}{2}}_{\text{Sect. 1.6.2}}}_{①}\ \underbrace{\underbrace{-\arctan\frac{\omega}{\omega_0}}_{\text{Sect. 1.6.4}}}_{②} \tag{3.18}$$

With this conclusion, the gain and phase functions are derived by linear approximations as well as by numerical calculations, see Fig. 3.6. Frequency profile, as this one for serial RL network, is commonly referred to as high-pass filter (HPF), whose gain function is

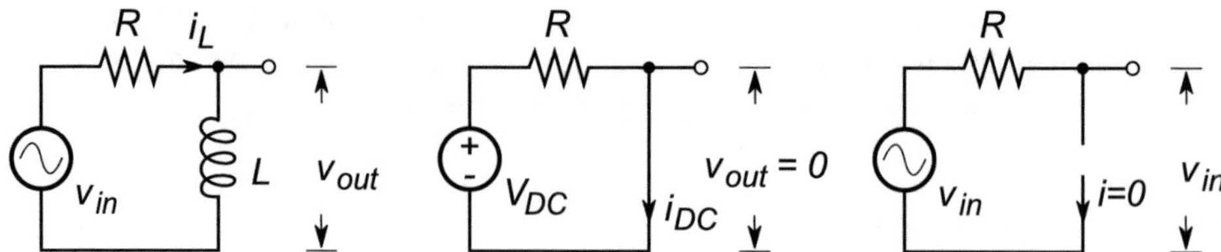

Fig. 3.5 Serial RL voltage divider at AC (left), at DC (middle), and at $\omega = \infty$ (right)

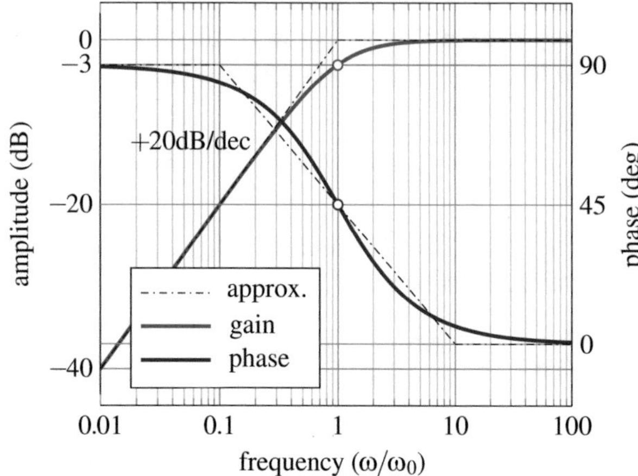

Fig. 3.6 Frequency domain plots of an HP RL filter: amplitude and phase response

$$A_V = \left|\frac{v_{out}}{v_{in}}\right| = \left|\frac{j\omega L}{R + j\omega L}\right| = \left|\frac{1}{1 - j\dfrac{R}{\omega L}}\right| = \frac{1}{\sqrt{1 + \left(\dfrac{R}{\omega L}\right)^2}} \tag{3.19}$$

For the sake of discussion, we analyze transfer function (3.16) again, but this time in the context of RL voltage divider, see Fig. 3.5. At zero frequency the inductor's reactance becomes zero (i.e. a short connection) causing the output voltage to drop to zero as well, Fig. 3.5 (middle). At the other end of the spectrum, at infinite frequency, the inductor becomes an open connection because its reactance also becomes infinite, which effectively stops AC through its branch. Stated differently, as per (3.6), the output voltage becomes equal to the input voltage, Fig. 3.5 (right). We find the −3 dB frequency by inspection of (3.15) as

$$\omega_0 = \frac{R}{L} \tag{3.20}$$

and, again, we note that the output voltage phase is "leading" the input phase by 90° at low frequencies. The phase lead reduces to 45° at the −3 dB point and, naturally, at high frequencies the input and output signals align their phase simply because they become equal.

3.3 Current Transfer Interface

A current divider in its basic form of two impedances connected in parallel is the second simple circuit model. If only two of the three basic devices (i.e. R,L, and C) are used to build a current divider, then there are six possible parallel configurations that are different. As oppose to the voltage divider where order of "first" and "second" impedance in the serial configuration is important, in the parallel configuration there is no "first" or "second" impedance on the path. That is to say, for example, that $Z_R||Z_L$ is equivalent to $Z_L||Z_R$.

When applied literally, a current divider is used simply to divide current i_{in} (either DC or AC) generated by the current source into two parallel current paths i_1 and i_2, see Fig. 3.7. Current path i_1 is considered the "internal path" when associated with non-infinite resistance of the current source (see Sect. 2.2.4). Therefore, portion of the source current that is actually delivered to the "load" through the interfacing node① is denoted as i_2 and it depends on relationship between the two resistance.

When applied conceptually, a current divider is used to model the signal transfer process between any two system-level blocks, the source and the load, Fig. 3.7.

Question that needs to be answered is: starting with input current source i_{in} whose output resistance is R_1, how much current is delivered to the node①, i.e. to the load R_2? In other words, how large is the current gain between the source and the load?

In the following sections, we review three of most important current divider structures.

3.3.1 Resistive Current Divider

It is assumed that all elements are ideal and, aside from the conversion of electrical energy into thermal energy, there is no energy loss. Analysis of this network structure is simple and the goal is to find the relationship between the source current i_{in} and the output current i_2 crossing the node①, see Fig. 3.7. The two resistors are seen by the ideal current source as parallel resistance $R_1||R_2$. Direct application of Kirchhoff's laws leads to expression for the current gain A_i as

$$\left.\begin{aligned} i_{in} &= i_1 + i_2 \\ v_{R_1} &= v_{R_2} = v_1 = i_{in}\,(R_1||R_2) \end{aligned}\right\} \quad \therefore \quad i_{R_2} = \frac{v_{R_2}}{R_2} = \frac{i_{in}\,(R_1||R_2)}{R_2} = i_{in}\,\frac{R_1\,\cancel{R_2}}{R_1+R_2}\,\frac{1}{\cancel{R_2}}$$

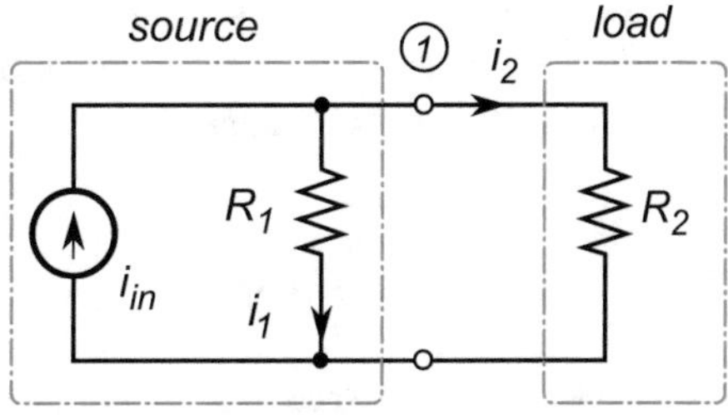

Fig. 3.7 Simple resistive current divider

$$A_i = \frac{i_2}{i_{in}} = \frac{1}{1 + \frac{R_2}{R_1}} \tag{3.21}$$

In other words, the ratio of the source current i_{in} and delivered current i_2 is the same as one plus the ratio of the two resistances. Equation (3.21) clearly shows that theoretical maximum current gain A_i of resistive divider equals to one. At best, the source current i_{in} is delivered unattenuated to the receiving load R_2, i.e. $i_2 = i_{in}$. This is possible only in two cases: either if $R_1 \to \infty$ and/or $R_2 = 0$, then $i_2 = i_{in}$, as we find by the following limits:

$$\lim_{R_1 \to \infty} A_i = \frac{1}{1 + \frac{R_2}{\infty}\nearrow^{0}} = 1 \Rightarrow i_2 = i_{in}$$

$$\lim_{R_2 \to 0} A_i = \frac{1}{1 + \frac{0}{R_1}\nearrow^{0}} = 1 \Rightarrow i_2 = i_{in}$$

In practice, the above two limits are summarized by the following approximation:

$$R_1 \gg R_2 \Rightarrow i_2 \approx i_{in} \tag{3.22}$$

where (3.22) is used as the “back of the envelope” guideline for design of current interfaces.

It is important to notice that in this case of current divider the relationship (3.22) between source and loading resistors is exactly opposite to the result (3.6) in the case of voltage divider. This conclusion is essential to understanding fundamental difference between voltage and current interface models. Consequently, input and output resistances of an amplifier, and therefore the type of the amplifier, are dependent on the signal interface type at the input and output sides. As a concluding remark, we note that being frequency independent circuit, resistive current divider indiscriminately scales both DC and AC voltage signals.

3.3.2 RC Current Divider

When the load impedance in a current divider takes the form of a capacitor, i.e. $R_2 = Z_C = 1/j\omega C$ then, relative to the frequency of the input signal, there are three distinct possibilities, Fig. 3.8.

1. *DC case,* ($\omega = 0$): when DC current source is used, the capacitive load impedance becomes infinite (see Sect. 2.3.3.2) and the equivalent schema of the current divider is shown in Fig. 3.8 (left). Consequently, no current is delivered to the load, i.e. $i_2 = 0$, and 100% of the source current keeps circulating through R.

2. *AC case,* $(0 < \omega \ll \infty)$: as per (3.21), for the intermediate values of frequencies, AC source current takes both available parallel paths in proportion to their respective resistances, see Fig. 3.8 (centre). The equivalent impedance seen by the current source is $Z_{eq} = R||Z_C$ is

$$Z_{R||C} = Z_R||Z_C \Rightarrow \frac{1}{Z_{R||C}} = \frac{1}{R} + j\omega C = \frac{1 + j\omega RC}{R}$$

$$\therefore$$

$$Z_{R||C}(\omega) = R\,\frac{1}{1 + j\omega RC} \Rightarrow \left|Z_{R||C}(\omega)\right| = \frac{R}{\sqrt{1 + (\omega RC)^2}} = \frac{R}{\sqrt{1 + \left(\frac{\omega}{\omega_0}\right)^2}} \quad (3.23)$$

where $\omega_0 = 1/RC$ is the -3 dB frequency.

3. *AC case,* $(\omega \to \infty)$: for high frequencies, the capacitive load impedance tends $Z_C \to 0$, which is to say that now 100% of the source current must take path through the capacitor, Fig. 3.8 (right).

Equation (3.23) also illustrates the limiting values of the equivalent RC impedance as shown in Fig. 3.8: when $\omega \to 0$ then (3.23) results in $|Z_{R||C}(0)| = R$; when $\omega \to \infty$ then (3.23) results in $|Z_{R||C}(\infty)| = 0$. With this simple back of the envelope analysis, we can already conclude that RC current dividing interface behaves as a HP filter: it suppresses tones close to DC while, at the same time, permits HF components to reach the load without attenuation.

Starting from the general expression for transfer function of current divider, see (3.21), substitution $R_2 = Z_C = 1/j\omega C$ gives

$$A_i = \frac{i_2}{i_{in}} = \frac{1}{1 + \dfrac{Z_C}{R}} = \frac{1}{1 + \dfrac{1}{j\omega RC}} = \frac{j\omega RC}{1 + j\omega RC} \quad (3.24)$$

Thus, limits of the current gain are found as

$$\lim_{\omega\to 0} |A_i| = \lim_{\omega\to 0} \frac{\omega RC}{\sqrt{1 + (\omega RC)^2}} = 0$$

$$\lim_{\omega\to\infty} |A_i| = \lim_{\omega\to\infty} \frac{\omega RC}{\sqrt{1 + (\omega RC)^2}} = \lim_{\omega\to\infty} \sqrt{\frac{(\omega RC)^2}{1 + (\omega RC)^2}}$$

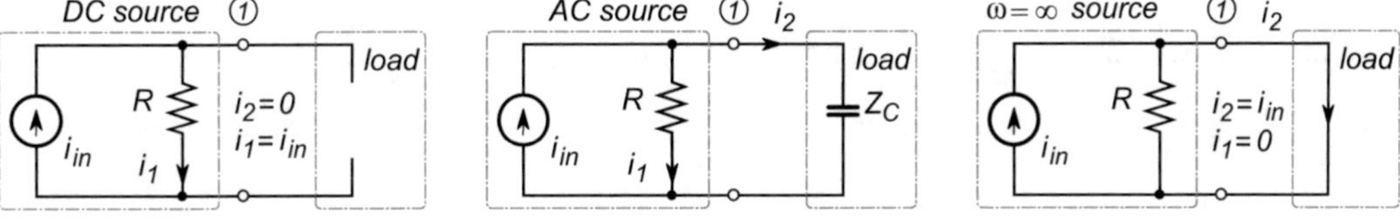

Fig. 3.8 RC current divider at DC (left), at AC (middle); and at $\omega = \infty$ (right)

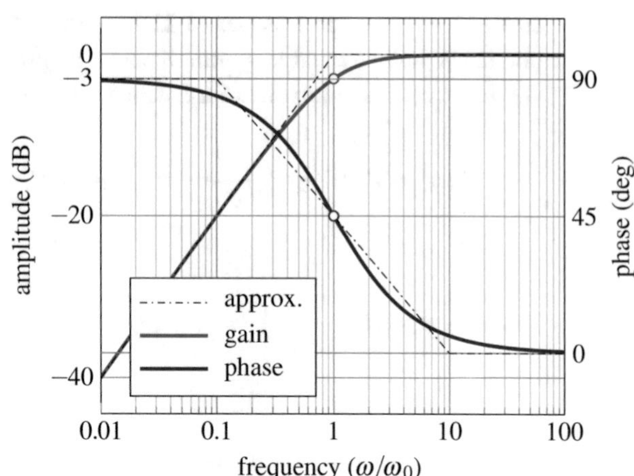

Fig. 3.9 Frequency domain plots of a current divider HP RC filter ($\omega_0 = 1/RC$): amplitude and phase response

$$= \lim_{\omega\to\infty} \sqrt{\frac{1}{\frac{1}{(\omega RC)^2} \stackrel{\nearrow 0}{} + 1}} = 1 \tag{3.25}$$

where general form of (3.24) is introduced in Sect. 3.3.1, therefore by inspection we conclude that

$$\omega_0 = \frac{1}{RC} \tag{3.26}$$

AC simulation of RC current divider once again confirms already achieved conclusions, see Fig. 3.9.

3.3.3 RL Current Divider

When the load impedance in a current divider takes the form of an inductor, i.e. $R_2 = Z_L = j\omega L$ then, relative to the frequency of the input signal, there are three distinct possibilities, Fig. 3.10.

1. *DC case,* ($\omega = 0$): when DC current source is used, then the inductive load impedance becomes zero (see Sect. 2.3.4) and the equivalent schema of the current divider is as in Fig. 3.10 (left). Consequently, 100% of the source current must take path through the inductor, $i_2 = i_{in}$.

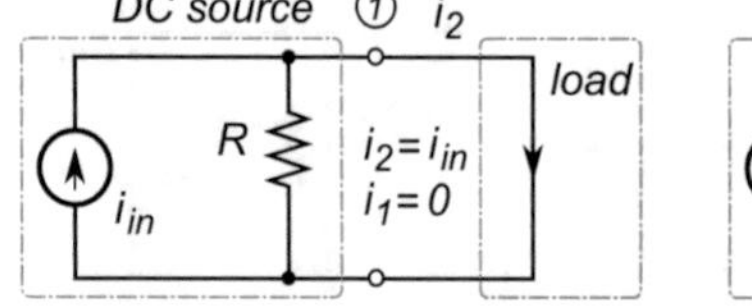

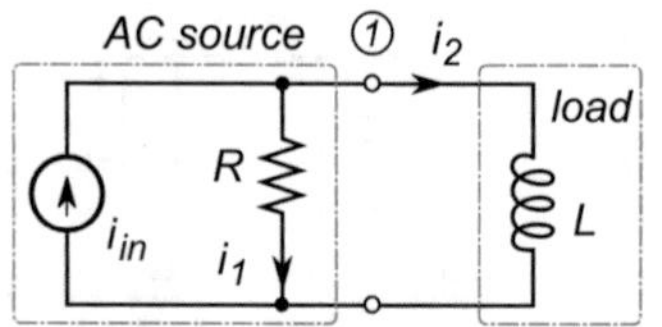

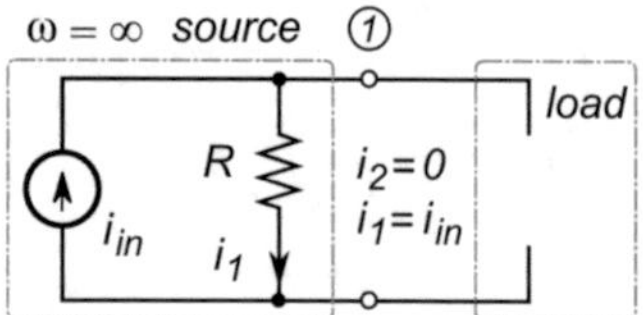

Fig. 3.10 RL current divider at DC (left), at AC (middle); and at $\omega = \infty$ (right)

2. *AC case,* ($0 < \omega \ll \infty$): at intermediate frequencies, AC current takes two available parallel paths, as per (3.21), see Fig. 3.10 (centre). Equivalent impedance seen by the current source is $Z_{\mathrm{eq}} = R||Z_L$. That is to say,

$$Z_{R||L} = Z_R||Z_L \Rightarrow \frac{1}{Z_{R||L}} = \frac{1}{R} + \frac{1}{j\omega L} = \frac{j\omega L + R}{j\omega RL}$$

$$\therefore$$

$$Z_{R||L}(\omega) = \frac{j\omega RL}{j\omega L + R} \Rightarrow \left|Z_{R||L}(\omega)\right| = \frac{|j\omega L|}{\left|1 + j\omega \dfrac{L}{R}\right|} = \frac{\omega L}{\sqrt{1 + \left(\dfrac{\omega L}{R}\right)^2}} \tag{3.27}$$

3. *AC case,* ($\omega \to \infty$): at high frequencies the inductor's impedance $Z_C \to \infty$, which is to say that $i_2 = 0$, Fig. 3.10 (right).

Equation (3.27) also illustrates the limiting values of the equivalent RL impedance as shown in Fig. 3.10: when $\omega \to 0$ then (3.27) results in $|Z_{R||L}(0)| = 0$; when $\omega \to \infty$ then (3.27) results in $|Z_{R||L}(\infty)| = R$, as

$$\lim_{\omega\to\infty} |Z_{R||L}| = \lim_{\omega\to\infty} \frac{\omega L}{\sqrt{1 + \left(\dfrac{\omega L}{R}\right)^2}} \approx \lim_{\omega\to\infty} \frac{\omega L}{\sqrt{\left(\dfrac{\omega L}{R}\right)^2}} = \lim_{\omega\to\infty} \frac{R\cancel{\omega L}}{\cancel{\omega L}} = R$$

With this simple back of the envelope analysis, we can already conclude that RL current dividing interface behaves as a LP filter: it suppresses HF components while, at the same time, permits LF components to pass without attenuation.

After substitution of $R_2 = Z_L = j\omega L$ into (3.21) we find out that RL current divider has the general transfer function form that was already introduced in Sect. 2.3.4, as

$$A_i = \frac{i_2}{i_{in}} = \frac{1}{1 + \dfrac{Z_L}{R}} = \frac{1}{1 + j\dfrac{\omega L}{R}} = \frac{1}{1 + j\dfrac{\omega}{\omega_0}} \tag{3.28}$$

where

$$\omega_0 = \frac{R}{L} \tag{3.29}$$

Furthermore, the limits of the current gain are found as

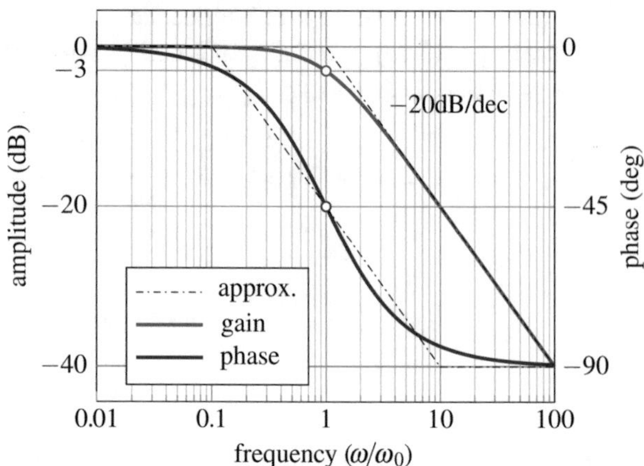

Fig. 3.11 Frequency domain plots of a current divider LP RL filter ($\omega_0 = R/L$): amplitude and phase response

$$\lim_{\omega\to 0} A_i = \lim_{\omega\to 0} \frac{1}{\sqrt{1+\left(\frac{\omega L}{R_1}\right)^2}} = 1$$

$$\lim_{\omega\to \infty} A_i = \lim_{\omega\to \infty} \frac{1}{\sqrt{1+\left(\frac{\omega L}{R_1}\right)^2}} = 0 \tag{3.30}$$

Numerical AC simulation of RL current divider again confirms our conclusions, see Fig. 3.11.

3.4 Maximum Power Transfer

Objective of circuit analysis at low frequencies is to rapidly deliver "good enough" voltage and current calculation. This approach is acceptable because the parasitic elements, which inherently have small RLC values, usually do not have a significant impact on circuit performance at low frequencies. Therefore, for low-frequency designs, first-order (i.e. back of the envelope) approximations of parasitic reactances with short/open connections provide a handy methodology.

At RF frequencies, however, voltage and current levels internal to the active circuit elements are, in general, not equal to the ones at the circuit terminals. Consequently, there is a non-negligible amount of wasted energy that is caused by the parasitic components associated with the circuit elements. Because of that unavoidable waste of energy, instead of evaluating the internal voltages and currents separately, it is more important to evaluate how the "instantaneous signal power" ($p = vi$) is transferred from one stage to another, with the implication that all internal impedances need to be accounted for. That is, important questions to answer are: how the power transfer between any two stages is influenced by the divider's impedances? How do we minimize the power transfer losses?

We have already concluded that there are two extreme cases to be considered, i.e. when the load impedance is either $Z_{\text{load}} = 0$ or $Z_{\text{load}} = \infty$. In the case of $Z_{\text{load}} = 0$, the voltage across the load terminals is forced to zero, which means that the power delivered to the load must be $p_{\text{load}} = v\,i = 0 \times i = 0$. In the case of $Z_{\text{load}} = \infty$, however, no current is delivered to the load, hence the delivered power must be zero again. Considering that electronic circuits do transfer signal power for the load impedances in between these two extremes, we conclude that there must be at least one non-zero maximum power transfer condition.

Let us assume complex source impedance $Z_S = R_S + jX_S$ and complex load impedance $Z_L = R_L + jX_L$ driven by ideal voltage source V_S. As already shown in (1.44), the average power $\langle P_L \rangle$ is dissipated in the resistive part of the load, while the current is complex, i.e.

$$\langle P_L \rangle = I_{\text{rms}}^2 R_L = \frac{1}{2}|I|^2 R_L = \frac{1}{2}\left[\frac{|V_S|}{|Z_S + Z_L|}\right]^2 R_L = \frac{1}{2}\frac{|V_S|^2}{(R_S + R_L)^2 + (X_S + X_L)^2} R_L \quad (3.31)$$

By inspection of (3.31), we note that the power P_L increases when the reactive term of the denominator is at a minimum. The minimum of a square function (which always has a non-negative value) is, of course, zero. That is, the minimum value of $(X_S + X_L)^2$ is achieved when the source and load reactances are equal and with opposite sign, i.e. $X_0 = -X_L$.

That leaves (3.31) with the resistive terms only, hence

$$P_L = \frac{1}{2}\frac{|V_S|^2}{(R_S + R_L)^2} R_L = \frac{1}{2}\frac{|V_S|^2}{\dfrac{R_S^2}{R_L} + 2R_S + R_L} \quad (3.32)$$

Therefore, the problem of finding the maximum value of P_L is reduced to the problem of finding the minimum value of the denominator in (3.32) in respect to the load resistance R_L, i.e.

$$\frac{d}{dR_L}\left(\frac{R_S^2}{R_L} + 2R_S + R_L\right) = -\frac{R_S^2}{R_L^2} + 1 = 0 \quad \therefore \quad R_S = R_L \quad (3.33)$$

because resistive values are always positive. The two derived conditions for reactances, $X_S = -X_L$, and resistances, $R_S = R_L$, are combined and written as

$$Z_S = Z_L^* \quad (3.34)$$

which is called “conjugate matching”. This condition guarantees the maximum power transfer; however, only at the one frequency for which source and load reactances are conjugate, i.e. $jX_L = -jX_S$.

In addition, by substituting $R_L = x\,R_S$ in (3.32) we can show how the ratio x of load and source resistances influences power matching and efficiency,

$$P_L = \frac{1}{2}\frac{x\,R_S}{(R_S + x\,R_S)^2}|V_S|^2 = \frac{1}{2}\frac{x}{(1+x)^2}\frac{|V_S|^2}{R_S} = \frac{1}{2}\frac{x}{(1+x)^2} \quad (3.35)$$

after normalizing to $V_S = 1\,\text{V}$ and $R_S = 1\,\Omega$; for $x = 1$ we write

$$P_L(\text{max}) = \frac{1}{2}\cdot\frac{1}{4} \quad (3.36)$$

where $P_L(\text{max})$ is the maximum power dissipated in the load, i.e. under the condition $R_L = R_0$ (i.e. $x = 1$). The normalized plot in Fig. 3.12 shows the delivered power at its maximum at the load ratio

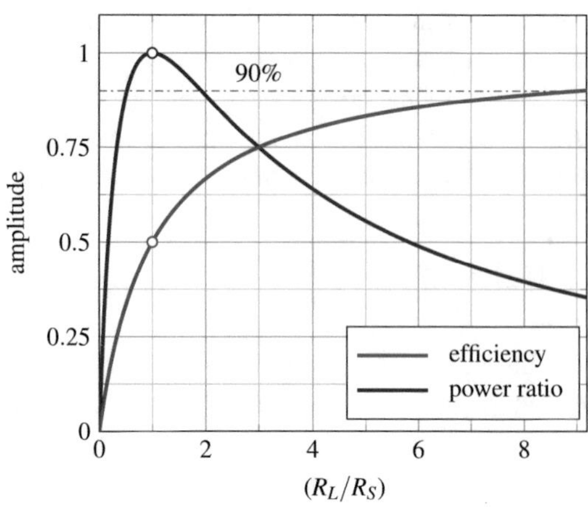

Fig. 3.12 Diagram showing maximum power transfer (ratio of Eqs. (3.31) and (3.36)), and maximum power efficiency (3.37), normalized to $V_S = 1\,\text{V}$; $R_S = 1\,\Omega$

$(R_L/R_L(\text{max}))$ when $R_L = R_S$. In addition, if we define power transfer efficiency as

$$\eta = \frac{R_L}{R_L + R_S} = \frac{1}{1 + \dfrac{R_S}{R_L}} = \frac{1}{1 + \dfrac{1}{x}} \tag{3.37}$$

which shows, Fig. 3.12, that when the maximum power is transferred to the load, the efficiency is only 50%, which is intuitively correct for the case of matched impedances. It is interesting to note that to reach 90% efficiency it is necessary to have $R_L/R_S = 9$.

Alternatively, a non-conjugate matching or broadband matching condition,

$$Z_S = Z_L \tag{3.38}$$

called *reflectionless match*, is also used. It is not as efficient as conjugate matching but it does offers broader maxima. In practice, the choice of matching condition depends upon the application.

3.4.1 Power Loss Due to Mismatch

When the maximum power transfer is not achieved, we have to quantify amount of the power loss. In case of two arbitrary source/load impedances Z_1 and Z_2, we define the "reflection coefficient" Γ as

$$\Gamma = \frac{Z_2 - Z_1}{Z_2 + Z_1} \tag{3.39}$$

where $0 \leq |\Gamma| \leq 1$. In theory, the power transfer is represented as a sum of two power waves: the incident power originating from the source and reflected power that was not delivered to the load. For good power transfer, the reflection should be as small as possible. In the case of perfect matching, the impedances Z_1 and Z_2 are equal, i.e. $\Gamma = 0$. In communication systems, the associated standards specify the maximum allowed value of the reflection coefficient. More often, the reflection coefficient, which is a unitless number, is converted to dBs and referred to as the "return loss"

$$RL = 10\log(|\Gamma|^2) = 20\log|\Gamma| \tag{3.40}$$

where $0\,\text{dB} \le RL \le \infty$. Loosely stated, the return loss quantifies the difference in power delivered to the two interfacing impedances. To find out how much power is wasted at the interface, the "mismatch loss" (ML) is defined a

$$ML = \frac{1}{1-|\Gamma|^2} \tag{3.41}$$

or, after conversion to dBs

$$ML_{\text{dB}} = -10\log(1-|\Gamma|^2) \tag{3.42}$$

where it is assumed that the signal source itself is ideal. In the case of arbitrary impedances at the network ports, calculation of the mismatch loss is a bit more complicated and we leave it for another occasion. To summarize this section, the return loss represents the difference between the reflected and the incident powers and a good match is indicated by a low return loss value. At the same time, mismatch loss represents the maximum possible power gain improvement relative to the case of a perfect match. Therefore, close to unity value of mismatch loss ML indicates a good match.

In order to visually illustrate the concept of complex power matching, it helps to introduce an analogy to the power flow in the form of a water flow through two pipes with diameters d_1 and d_2 (see Fig. 3.13). Pipe diameters and water flow have a relationship similar to that between resistance and current flow. We intuitively know that the most efficient water flow (i.e. no spills) happens when the two pipes have the same diameters, $d_1 = d_2$. To make the analogy even closer, let us note that if the two connecting pipes are cut at a right angle $\varphi = 90°$, then it does not matter how the pipes are rotated along their axes; they always make a good connection, i.e. they "match". However, if the pipes are cut at some other angle $\varphi \neq 90°$, the most efficient water flow is when the two angles are complementary, i.e. $\varphi_1 + \varphi_2 = \pi$. Non-perpendicular angles are equivalent to complex impedances Z_S and Z_L in the voltage divider, where the positive slope is equivalent to "inductive" and the negative slope to "capacitive" impedance, while the right angle is the special (and simpler) resistive case.

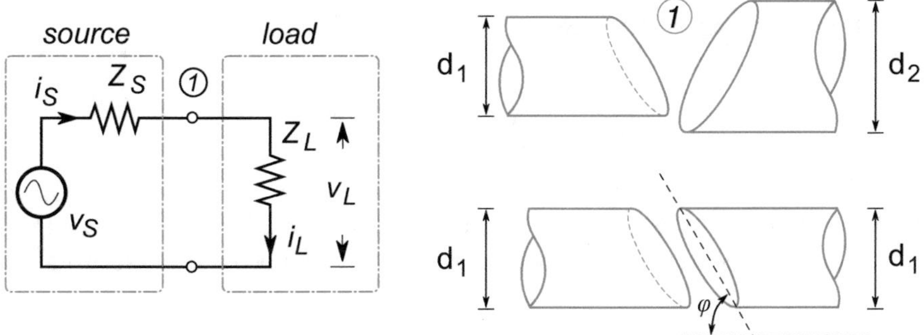

Fig. 3.13 Complex impedance matching network (left) and the pipe analogy showing the non-matched and matched cases (right)

3.5 Case Study: The Need for Signal Buffering

To illustrate the loss of signal energy in the case when there is large difference between source and load impedances, let us consider the following practical example. A voltage reference based on a resistive voltage divider is used to drive a $R_L = 100\,\Omega$ load, Fig. 3.14. As a numerical example, let us assume that $v_S = 10\,\text{V}$, $R_1 = R_2 = R = 10\,\text{k}\Omega$, and loading resistance is $R_L = 100\,\Omega$. Before connecting the load, the available voltage at node① is calculated as

$$v_1 = i_S\, R_2 = \frac{v_S}{R_1 + R_2}\, R_2 = \frac{v_S}{2\not{R}}\, \not{R} = \frac{v_S}{2} = 5\,\text{V}$$

However, when node① is directly connected to node② without the buffer in between, then we calculate

$$R_2||R_L = 10\,\text{k}\Omega||100\,\Omega = 99\,\Omega \approx 100\,\Omega$$

which is very close to the load resistance. Thus, the available voltage v_1 becomes

$$v_1 = i_S\,(R_2||R_L) = \frac{v_S}{R_1 + (R_2||R_L)}\,(R_2||R_L) = \frac{v_S}{\dfrac{R_1}{R_2||R_L} + 1} = \frac{10\,\text{V}}{\dfrac{10\,\text{k}\Omega}{99\,\Omega} + 1} = 98\,\text{mV} \approx 100\,\text{mV}$$

which illustrates consequence of connecting small load to this kind of voltage reference. Under these conditions, the total power delivered to R_L is

$$P_{R_L} = v\,i = \frac{v_1^2}{R_L} = \frac{(98\,\text{mV})^2}{100\,\Omega} = 96\,\mu\text{W} \approx 100\,\mu\text{W}$$

This result is to be compared with ideal case when the full $v_1 = 5\,\text{V}$ is available at node②,

$$P_{R_L} = v\,i = \frac{v_1^2}{R_L} = \frac{(5\,\text{V})^2}{100\,\Omega} = 250\,\text{mW}$$

that is to say, the ratio of delivered powers is

$$P = \frac{250\,\text{mW}}{99\,\mu\text{W}} \approx 2500 = 68\,\text{dB}$$

This large difference demonstrates why it is important to "isolate" the reference voltage level and maximize the power delivered to the load. Practical engineering solution to this problem is shown in Fig. 3.14, where an intermediate stage is added to serve as a buffer between the source and the load.

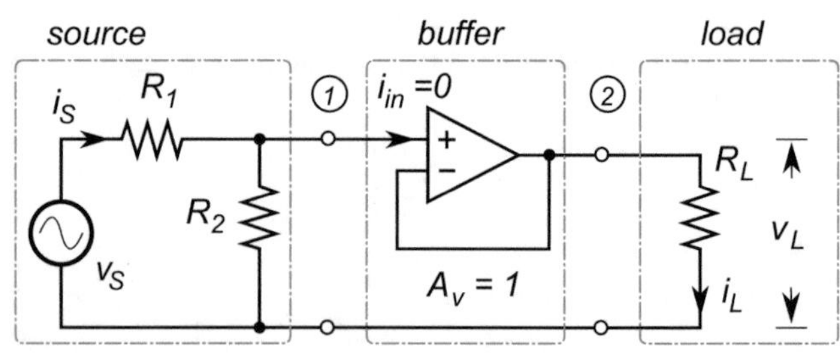

Fig. 3.14 Illustration of buffering stage to maximize power delivered to the load

In this case, the input impedance of the buffer must be very high (ideally infinite) so that no current is drawn from the source $i_{in} = 0$. At the output, however, this amplifier must serve as ideal voltage source with gain $A_v = 1$. Under these conditions, $v_2 = v_1$ and the maximum power is delivered to the load. In a real circuit, however, there is always a small attenuation of v_1, therefore $v_2 \approx v_1$ is the physical result that can be achieved in real circuits.

3.6 Summary

In this chapter we reviewed basic types of interfaces between subsequent stages in a multistage system. Two main models, namely voltage divider and current divider, are used to evaluate power transfer between the stages. Depending whether voltage or current is used to transmit the signal, the interface specifications are very different. What is more, a simple resistive interface behaves very different than, for example, RC or RL interfaces that create LPF or HPF. Based on the ratio of the source and load impedances, the designer aims to maximize the power transfer by applying the impedance matching and buffering techniques.

Problems

3.1 A three stage communication system is driven by an ideal Thévenin source and it consists of:

1. *RF amplifier*: with input resistance $R_{in} = 5\,\mathrm{k}\Omega$, voltage gain $A_V = 10$, and $R_{out} = 25\,\Omega$,
2. *Mixer*: with input resistance $R_{in} = 1\,\mathrm{k}\Omega$, voltage gain $A_V = 1$, and $R_{out} = 100\,\Omega$,
3. *IF amplifier*: with input resistance $R_{in} = 10\,\mathrm{k}\Omega$, voltage gain $A_V = 6\,\mathrm{dB}$, and $R_{out} = 50\,\Omega$,

At the output of IF amplifier there is a load resistor $R_L = 1\,\mathrm{k}\Omega$

Calculate the total voltage gain A_V of this system and express it in both V/V and dB units.

3.2 In reference to Fig. 3.2, assuming $V_{in} = 5\,\mathrm{V}$ is the power supply, design a $V_1 = V_{out} = 2\,\mathrm{V}$ voltage reference using a resistive divider. Second specification is that the equivalent resistance looking into node① node is $R_{out} = R_{①} = 6\,\mathrm{k}\Omega$.

3.3 In reference to Fig. 3.2, assuming $V_{in} = 10\,\mathrm{V}$ is the power supply, design a voltage reference at node① using a resistive divider so that (a) $V_{out} = 0.1\,V_{in}$, (b) $V_{out} = 0.5\,V_{in}$, (c) $V_{out} = 0.9\,V_{in}$.

For each of the cases in this problem, how would you categorize the (V_{in}, R_1) source: as the voltage or current source? Which (if any) of the cases could be declared as “the best voltage source”, and which one as “the best current source”?

3.4 In reference to Fig. 3.3, assuming $v_{in}(t) = 10\,\mathrm{V}\,\sin\omega t$ is the input signal, calculate the capacitor value C so that maximum voltage amplitudes V_{out} are

1. Given $R = 1\,\mathrm{k}\Omega$ and $f_1 = 10\,\mathrm{kHz}$ and: (a) $V_{out} = 0.1\,V_{in}$, (b) $V_{out} = 0.5\,V_{in}$, (c) $V_{out} = 0.9\,V_{in}$,
2. Given $R = 1\,\mathrm{k}\Omega$ and $f_2 = 1\,\mathrm{MHz}$ and: (a) $V_{out} = 0.1\,V_{in}$, (b) $V_{out} = 0.5\,V_{in}$, (c) $V_{out} = 0.9\,V_{in}$,
3. Given $R = 1\,\mathrm{k}\Omega$ and $f_3 = 1\,\mathrm{GHz}$ and: (a) $V_{out} = 0.1\,V_{in}$, (b) $V_{out} = 0.5\,V_{in}$, (c) $V_{out} = 0.9\,V_{in}$.

For each of the cases in this problem, calculate the $-3\,\mathrm{dB}$ frequency.

3.5 In reference to Fig. 3.5, assuming $v_{in}(t) = 10\,\text{V} \sin\omega t$ is the input signal, calculate the inductor value so that maximum voltage amplitudes V_{out} are

1. Given $R_1 = 1\,\text{k}\Omega$ and $f_1 = 10\,\text{kHz}$ and: (a) $V_{out} = 0.1\,V_{in}$, (b) $V_{out} = 0.5\,V_{in}$, (c) $V_{out} = 0.9\,V_{in}$,
2. Given $R_1 = 1\,\text{k}\Omega$ and $f_2 = 1\,\text{MHz}$ and: (a) $V_{out} = 0.1\,V_{in}$, (b) $V_{out} = 0.5\,V_{in}$, (c) $V_{out} = 0.9\,V_{in}$,
3. Given $R_1 = 1\,\text{k}\Omega$ and $f_3 = 1\,\text{GHz}$ and: (a) $V_{out} = 0.1\,V_{in}$, (b) $V_{out} = 0.5\,V_{in}$, (c) $V_{out} = 0.9\,V_{in}$.

For each of the cases in this problem, calculate the $-3\,\text{dB}$ frequency.

3.6 In reference to Fig. 3.7, assuming $i_{in}(t) = 10\,\text{mA} \sin\omega t$ is the input signal, calculate the resistor R_2 value so that maximum current amplitudes I_2 are

1. Given $R_1 = 100\,\text{k}\Omega$ and $f_1 = 10\,\text{kHz}$ and: (a) $I_2 = 0.1\,I_{in}$, (b) $I_2 = 0.5\,I_{in}$, (c) $I_2 = 0.9\,I_{in}$,
2. Given $R_1 = 100\,\text{k}\Omega$ and $f_2 = 1\,\text{MHz}$ and: (a) $I_2 = 0.1\,I_{in}$, (b) $I_2 = 0.5\,I_{in}$, (c) $I_2 = 0.9\,I_{in}$,
3. Given $R_1 = 100\,\text{k}\Omega$ and $f_3 = 1\,\text{GHz}$ and: (a) $I_2 = 0.1\,I_{in}$, (b) $I_2 = 0.5\,I_{in}$, (c) $I_2 = 0.9\,I_{in}$.

Comment on the calculated results.

3.7 In reference to Fig. 3.8, assuming $i_{in}(t) = 10\,\text{mA} \sin\omega t$ is the input signal, calculate the capacitor C value so that maximum current amplitudes I_2 are

1. Given $R = 100\,\text{k}\Omega$ and $f_1 = 10\,\text{kHz}$ and: (a) $I_2 = 0.1\,I_{in}$, (b) $I_2 = 0.5\,I_{in}$, (c) $I_2 = 0.9\,I_{in}$,
2. Given $R = 100\,\text{k}\Omega$ and $f_2 = 1\,\text{MHz}$ and: (a) $I_2 = 0.1\,I_{in}$, (b) $I_2 = 0.5\,I_{in}$, (c) $I_2 = 0.9\,I_{in}$,
3. Given $R = 100\,\text{k}\Omega$ and $f_3 = 1\,\text{GHz}$ and: (a) $I_2 = 0.1\,I_{in}$, (b) $I_2 = 0.5\,I_{in}$, (c) $I_2 = 0.9\,I_{in}$.

For each of the cases in this problem, calculate the $-3\,\text{dB}$ frequency and comment on the calculated results.

3.8 In reference to Fig. 3.10, assuming $i_{in}(t) = 10\,\text{mA} \sin\omega t$ is the input signal, calculate the inductor L value so that maximum current amplitudes I_2 are:

1. Given $R = 100\,\text{k}\Omega$ and $f_1 = 10\,\text{kHz}$ and: (a) $I_2 = 0.1\,I_{in}$, (b) $I_2 = 0.5\,I_{in}$, (c) $I_2 = 0.9\,I_{in}$,
2. Given $R = 100\,\text{k}\Omega$ and $f_2 = 1\,\text{MHz}$ and: (a) $I_2 = 0.1\,I_{in}$, (b) $I_2 = 0.5\,I_{in}$, (c) $I_2 = 0.9\,I_{in}$,
3. Given $R = 100\,\text{k}\Omega$ and $f_3 = 1\,\text{GHz}$ and: (a) $I_2 = 0.1\,I_{in}$, (b) $I_2 = 0.5\,I_{in}$, (c) $I_2 = 0.9\,I_{in}$.

For each of the cases in this problem, calculate the $-3\,\text{dB}$ frequency and comment on the calculated results.

3.9 An antenna, modelled as $Z = 50\,\Omega$, is connected to the input node of RF amplifier whose input resistance is (a) $R_{in} = 50\,\Omega$, (b) $R_{in} = 100\,\Omega$, and (c) $R_{in} = 1\,\text{k}\Omega$.

In each case, calculate the reflection coefficient, the return loss, and the mismatch loss.

3.10 Design LP and HP passive filters whose $-3\,\text{dB}$ frequency is $f_0 = 1\,\text{kHz}$. Comment on your solutions.

Basic Semiconductor Devices

4

A general electrical network theory is based on four fundamental functions of ideal passive devices, namely resistance (R), capacitance (C), inductance (L), and memristance (M). In order to amplify signal, however, passive elements alone are not sufficient. On the other hand, three-terminal active devices are capable of controlling large current flow at its output terminal if a small input signal is applied to its input terminal. That is to say, large waveform at the output terminal is a faithful replica of the small input side waveform, thus the amplification. Key point, however, is that the device merely *controls* the output current flow. The signal energy used both at the input and output side is provided by the *external* energy sources, battery for example. In a mechanical analogy, function of three-terminal devices is similar to a water tap that controls water flow (it does not create it).

In this chapter, we review the properties of the fundamental active electronic devices, namely diodes and three types of transistors: BJT, FET, and JFET.

4.1 Active Devices

Basic passive devices, see Chap. 2, alone are not capable of amplifying an electrical signal. Being *passive devices* they inherently can only form voltage/current dividers along the signal path, thus progressively reduce amplitude of the input signal. In order to have the gain larger than one, a network must include *active components*, namely diodes and transistors. The increased signal power, however, is provided by the external energy source, i.e. battery, and was not magically created inside a transistor (the energy conservation law still holds). That is, a transistor merely serves as a valve that controls the flow of a large current through the transistor's output terminals, where the current is drawn from the energy source, by means of low power signal at the input terminals. Simply put, we use flow of small amount of energy (the control signal) to control flow of large amount of energy (the external energy source, e.g. battery), hence, the "amplification" effect. As a comparison, for example, a transformer by itself can also increase amplitude of either voltage or current at its output terminals, but not both at the same time, which is definition of power. Hence, transformer is not a power amplifying device.

In the rest of this section, basic properties of PN junctions, diodes, and transistors are reviewed for the purposes of their application for amplification of weak RF signals.

R. Sobot, *Wireless Communication Electronics*,
https://doi.org/10.1007/978-3-030-48630-3_4

4.2 Diode

Using the black box method, Fig. 4.1, we find that transfer function of a diode shows behaviour similar to an ideal ON/OFF switch that is controlled by voltage at its terminals. Transition between ON/OFF diode states happens at voltage called "threshold voltage", usually denoted V_t. Relative to the threshold voltage, we find two distinct modes of operation:

1. When diode voltage V_D is in the range $(V_D \leq V_t)$, the diode behaves as an open connection, in other words as a switch in OFF state. Within this voltage interval the current passing through the black box is (ideally) zero, thus the conclusion would be that the box contains either an open switch or, equivalently, an infinite resistance.
2. When diode voltage is $V_D \geq V_T$, the relation between diode voltage V_D and current I_D follows exponential function law over remarkably large range of currents. In this interval of the terminal voltages, the conclusion would be that black box contains a nonlinear resistance. Therefore, it is necessary to find values of both DC and AC resistances.

We note that maximum negative diode voltage is not infinite, it is specific to the diode type (for example, in Fig. 4.1 the negative voltage is shown only up to $V_D = -1$). On the other hand, for most commercial diodes the threshold voltage is typically set to $V_t \approx 0.6 - 0.7\,\text{V}$, which is strictly a technological parameter.

4.2.1 Mathematical Model

It was found both theoretically and experimentally that voltage–current characteristics of this single p-n junction device obey "exponential law", which subsequently was curve-fitted by using very sophisticated mathematical algorithms into the following mathematical model

$$I_D(V_D) = I_S\left[\exp\left(\frac{V_D}{n\,V_T}\right) - 1\right] = I_S\left[\exp\left(\frac{q\,V_D}{n\,k\,T}\right) - 1\right] \tag{4.1}$$

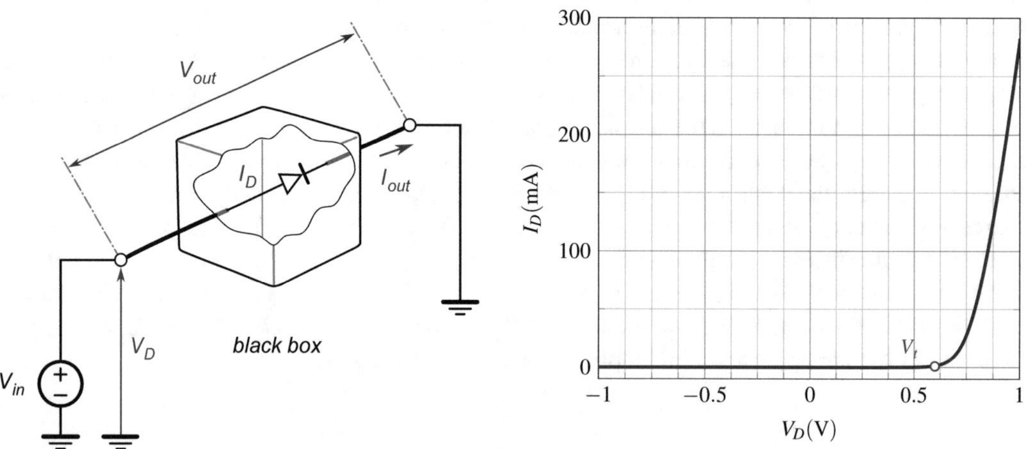

Fig. 4.1 The black box setup (left), diode's voltage to current nonlinear characteristics (right)

where

I_D is the current flowing through the diode
I_S is the diode leakage current
V_D is the voltage across the diode, i.e. biasing voltage
V_T is the thermal voltage ($V_T = kT/q \approx 25\,\text{mV}$) at room temperature
k is Boltzmann constant ($k = 1.380\,650 \times 10^{-23}$ [J/K])
T is the temperature in degrees Kelvin [K], ($K = °\text{C} + 273.15$)
q is the elementary charge ($q = 1.602\,176\,487 \times 10^{-19}$ [C])
n is the emission coefficient, unless specified it is assumed $n = 1$

Equation (4.1) shows that amplitude of diode current I_D is exponential function of diode voltage V_D. However, instead of focusing on the absolute value of the diode voltage V_D, it is actually the ratio V_D/V_T of the biasing voltage and the thermal voltage that matters. Direct consequence is that a diode current has a strong temperature dependence. As an illustration, a diode characteristic in Fig. 4.2 shows that, when temperature changes from −55 to 125 °C, the diode current I_D varies between zero and 125 mA for the same diode voltage $V_D = 750\,\text{mV}$. Therefore, calculation of the diode current I_D is valid only at one specific temperature and biasing point (V_D, I_D).

A very important limitation of (4.1) is that it implies when the temperature *increases* the diode current I_D *reduces*. Nevertheless, simulation results in Fig. 4.2 clearly show that *increase* in temperature results in the *increase* of diode current. This trend is due to the fact that (4.1) does not include very strong temperature dependence of I_S current, which dominates the overall temperature response.

Exponential function is considered a "strong" function from a mathematical perspective, thus it is a common practice to use its approximated V–I expressions for two distinct regions of diode operation:

1. ($V_D \gg V_T$): when the diode biasing voltage is much larger (even three to four times or more, for example) than the thermal voltage V_T, the exponential term in (4.1) becomes much larger than the "−1" term. In that case, (4.1) degenerates into a plain exponential function that is much easier to handle from the mathematical perspective (think of its derivatives and integrals), i.e.

$$I_D(V_D) = \left[\exp\left(\frac{V_D}{V_T}\right) - 1\right] \approx I_S \exp\left(\frac{V_D}{V_T}\right) \qquad (V_D \gg V_T) \tag{4.2}$$

In this case, the diode is said to be fully "forward biased", i.e. it fully conducts current and its behaviour is similar to a wire or a closed switch in series with an ideal voltage source with V_t

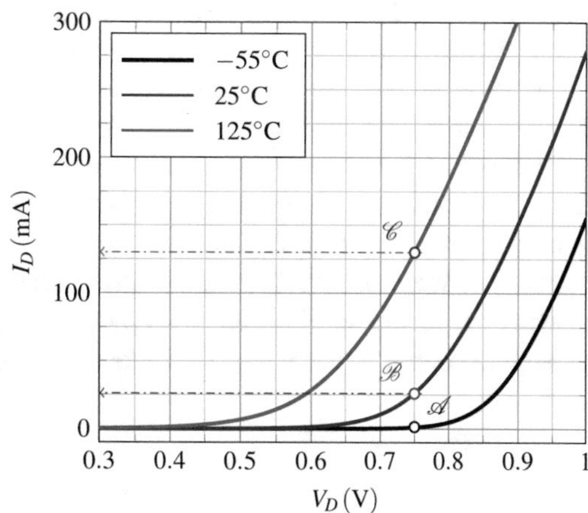

Fig. 4.2 Diode voltage-current characteristics (4.1) for three distinct temperatures, shown for the full military temperature range of operation (−55 °C ≤ T ≤ 125 °C)

volts at its terminals, i.e., its internal resistance is very low (see Sect. 2.2.3 for the ideal voltage source model). Once a diode is forward biased, for all practical purposes it could be approximated with an ideal voltage source V_t (thus its internal resistance is small or, ideally, zero).

We note that (4.2) is bidirectional: voltage V_D controls not only diode current I_D, but also current I_D that is forced through the diode by the external current source sets V_D across the diode. In the latter case, a diode is used as a voltage reference device.

2. ($V_D \ll V_T$): when the biasing voltage V_D is much smaller (ten times or more, for example) than the thermal voltage V_T, the exponential term in (4.1) becomes very close to one. In that case, (4.1) degenerates into the following expression:

$$I_D(V_D) = I_S \left[\exp\left(\frac{V_D}{V_T}\right) - 1\right] \approx 0 \qquad (V_D \ll V_T) \tag{4.3}$$

The diode is said to be "reverse biased", i.e. it is fully turned off and its behaviour is similar to that of an open switch—only a small portion of the leakage current I_S flows through the p-n junction boundary. We note that if the anode and cathode terminals are shorted (or, set to the same potential), in other words $V_D = 0$, then from (4.1) it follows that $I_D = 0$. This relationship is often used in circuits when there is a need to guarantee that a diode is turned off; it is sufficient to enforce $V_D = 0$ by the external circuit.

In practice, there are a number of ways to design diodes optimized for a particular behaviour. For example, a Schottky diode is designed to have very fast switching times; a Zener diode (i.e. an avalanche diode) is designed for a specific reverse-bias breakdown voltage that is useful as a reference in voltage-stabilizing circuits; a varactor diode is designed specifically for its voltage-controlled capacitance; a PiN diode is designed with a region of intrinsic silicon between p-type and n-type regions (hence the PiN name) to enable its linear voltage-controlled resistance behaviour, which is especially useful in microwave systems; and a light-emitting diode (LED) is designed so that the charge recombination process results in the release of photons of light—by controlling the free carrier's energy levels, thus frequency of the released photons, i.e. the emitted light colour is controlled.

4.2.2 Biasing Point

Normally, the first step in the circuit design process is to decide on "reference" biasing currents for all active devices. It is practical to choose value of I_D to be a "nice number". Later, in the circuit design process, I_D reappears frequently in the equations, thus a nice number reduces the analytical effort. However, as soon as a specific biasing current $I_D = I_0$ is set, consequently, the biasing voltage $V_D = V_0$ is also set. By doing so, unavoidably, the voltage-to-current gain "g_m" of this diode is also set. We note this "domino effect" in design process of any analog circuit.

To illustrate the voltage–current relationship, a typical diode transfer characteristic that is obtained either by simulation or experiment is shown in Fig. 4.3 (right). In this setup, an ideal voltage source is used to precisely control the diode voltage V_D which in return controls its corresponding diode current I_D. We note that if instead of the voltage source an ideal current source were used, then the diode current is the one that controls V_D voltage. That is to say, (4.1) is bidirectionally valid regardless of which variable, either V_D or I_D, is chosen to be independent.

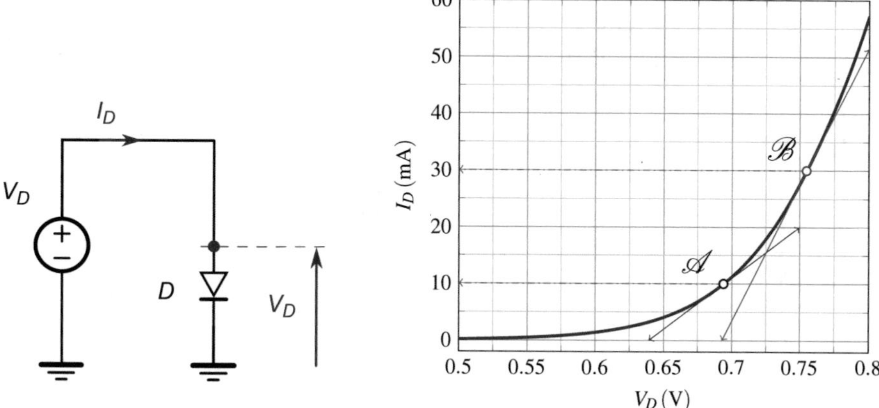

Fig. 4.3 A typical simulation setup (left) to determine voltage-current transfer characteristics (right) of a diode. Two different biasing points $\mathscr{A}$ and $\mathscr{B}$ illustrate its nonlinear nature

Example 24: Diode Biasing Point

Referring to Fig. 4.3, estimate and compare DC biasing (I_0, V_0) at two points $\mathscr{A}$ and $\mathscr{B}$. In addition, determine diode's DC resistance at these two points.

Solution 24: Zoom-in of Fig. 4.3 around these two bias points, see Fig. 4.4, can be used to evaluate setup of this diode.

Biasing points $\mathscr{A}$ and $\mathscr{B}$: are estimated by reading their respective coordinates, i.e.

$\mathscr{A} : (I_0, V_0) = (10\,\text{mA}, 695\,\text{mV})$, and $\mathscr{B} : (I_0, V_0) = (30\,\text{mA}, 755\,\text{mV})$.

Therefore, the choice of biasing current I_0 (which is motivated by application and technology constraints) leads into its associated biasing voltage V_0.

DC resistance: By Ohm's law, DC resistance is the ratio of the associated voltage and current, thus

$$R_{DC}(\mathscr{A}) \stackrel{\text{def}}{=} \left.\frac{V_0}{I_0}\right|_{\mathscr{A}} = \frac{695\,\text{mV}}{10\,\text{mA}} = 69.5\,\Omega\,; \qquad R_{DC}(\mathscr{B}) \stackrel{\text{def}}{=} \left.\frac{V_0}{I_0}\right|_{\mathscr{B}} = \frac{755\,\text{mV}}{30\,\text{mA}} = 25.1\,\Omega$$

which illustrates that, as opposed to linear resistors whose resistance is always constant, resistance of a nonlinear device is not constant, i.e. it depends on DC current flowing through the device.

Let us consider the following practical problem. Given the circuit setup in Fig. 4.5 (left) and the diode's voltage-current transfer characteristics in Fig. 4.5 (right), what resistor value R is needed to set the diode's biasing point $\mathscr{A} : (I_0, V_0)$?

Resistor and diode are in serial connection, thus they share same current and we write

$$I_D = I_R \tag{4.4}$$

$$V_{CC} = V_R + V_D \tag{4.5}$$

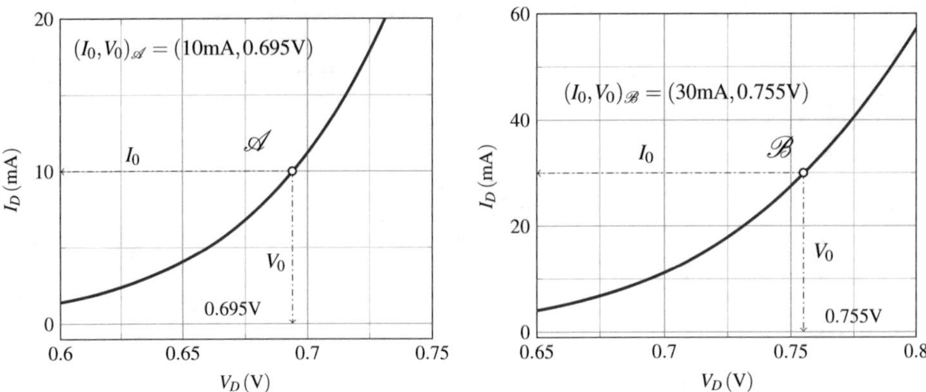

Fig. 4.4 Zoom-ins of Fig. 4.3 showing details around the biasing points

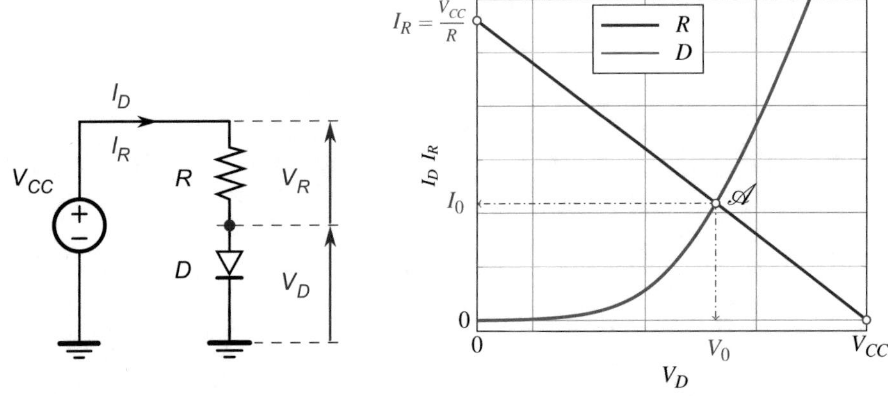

Fig. 4.5 Practical application of a diode, the accompanying resistor R serves as the diode's "local' current source that limits the maximum current in the R-D branch

$$I_R = \frac{V_{CC} - V_D}{R} \tag{4.6}$$

$$I_D = I_S \left[\exp\left(\frac{V_D}{V_T}\right) - 1 \right] \tag{4.7}$$

Mathematically, this is a classic example of a system that consists of linear and nonlinear equations, therefore either graphical or iterative numerical methods are used to solve it for I_D and V_D.

Figure 4.5 (right) illustrates the graphical method, which is a direct implementation of (4.4)–(4.7), where $\mathscr{A}$ is the only point that belongs both to (4.6) and (4.7). Therefore, point (V_0, I_0) is the solution of this system, i.e. the only point where $I_D = I_R$. By Ohm's law, voltage–current relationship of a resistor is a linear function.[1] From (4.5) voltage across resistor is

$$V_R = V_{CC} - V_D \tag{4.8}$$

[1]See Sect. 2.3.1.

where we draw I_R vs. V_D graph as follows. When $V_D = V_{CC}$ both terminals of R are at the same potential, then from (4.8) it follows that $V_R = 0$, which substituted into (4.6) gives that $I_R = 0$. Similarly, when $V_D = 0$ then from (4.8) we find $V_R = V_{CC}$, which after substituting into (4.6) gives $I_R = V_{CC}/R$. Two points are sufficient to determine a linear function. We conclude that all possible current-voltage values of this resistor are therefore found on the solid line, which is often referred to as the "load line", and is determined by two points: $(V_R, I_R) : (V_{CC}, 0)$ and $(V_R, I_R) : (0, V_{CC}/R)$.

Plotting the load line together with the diode's transfer function enables us to graphically determine biasing point $\mathscr{A} : (V_0, I_0)$, as in Fig. 4.5 (right). We note that changing the resistor value would change the slope of its linear function, which would cross the diode's transfer function at different biasing point. Therefore, we conclude that this resistor R indeed controls the diode's biasing point.

Example 25: Diode Biasing Point
Referring to Fig. 4.5, given that $\mathscr{A} : (10\,\text{mA}, 0.695\,\text{V})$ calculate the required R.

Data $V_{CC} = 5\,\text{V}$

Solution 25: With respect to (4.4), direct implementation of (4.6) results in

$$I_D = I_R = \frac{V_{CC} - V_D}{R} \Rightarrow R = \frac{V_{CC} - V_D}{I_D} = \frac{5 - 0.695\text{V}}{10\,\text{mA}} = 430.5\,\Omega$$

Example 26: Diode Biasing Point
Referring to Fig. 4.5, calculate biasing point $\mathscr{A} : (I_0, V_0)$

Data $V_{CC} = 5\,\text{V}$, $R = 430.5\,\Omega$, $V_T = 25\,\text{mV}$, $I_S = 8.445 \times 10^{-15}\text{A}$ (assumed constant)

Solution 26: From (4.4), (4.6), and (4.7) we write

$$\frac{V_{CC} - V_D}{R} = I_S \left[\exp\left(\frac{V_D}{V_T}\right) - 1\right]$$

which we rearrange to set up the iterative equation as

$$V_{D_{(n+1)}} = V_T \ln\left[\frac{V_{CC} - V_{D_{(n)}}}{R\, I_S} + 1\right] \tag{4.9}$$

where $(n + 1)$th index refers to the subsequent value that is calculated after (n)th value is used. Choosing an arbitrary number $V_{D_{(n)}} = 1\,\text{V}$ as the initial value and substituting into (4.9) gives

$$V_{D_{(n+1)}} = 25\,\text{mV} \ln\left[\frac{5 - 1\,\text{V}}{430.5\,\Omega \times 8.445 \times 10^{-15}\text{A}} + 1\right] = 0.693\,\text{mV} \tag{4.10}$$

(continued)

Therefore, result of the first iteration is $V_D = 0.693\,\text{mV}$. We now use this value in the second iteration by substituting $V_{D_{(n)}} = 0.693\,\text{mV}$ into (4.9), which produces

$$V_{D_{(n+1)}} = 25\,\text{mV}\ \ln\left[\frac{5\,\text{V} - 0.693\,\text{mV}}{430.5\,\Omega \times 8.445 \times 10^{-15}\text{A}} + 1\right] = 0.695\,\text{mV} \qquad (4.11)$$

We repeat again this process by setting $V_{D_{(n)}} = 0.695\,\text{mV}$ and substituting into (4.9) as

$$V_{D_{(n+1)}} = 25\,\text{mV}\ \ln\left[\frac{5\,\text{V} - 0.695\,\text{mV}}{430.5\,\Omega \times 8.445 \times 10^{-15}\text{A}} + 1\right] = 0.695\,\text{mV} \qquad (4.12)$$

Since there is no change in the last two iterations, we accept this 1 mV resolution in the result and stop the iterative process. Repeat this exercise with some other initial values, and confirm that the iterative process is indeed rapid. For example, $V_{D_n} = 4.999\,\text{V}$[2] converges to the same result.

4.2.3 Small Signal g_m Gain

By definition, a diode voltage-to-current gain g_m equals to a *small change* of current I_D caused by *small change* of voltage V_D, i.e. it is calculated as the first derivative of (4.1) at diode's biasing point, which in ideal case gives

$$g_m \overset{\text{def}}{=} \left.\frac{dI_D}{dV_D}\right|_{(I_0,V_0)} = \frac{d}{dV_D}\left[I_S \exp\left(\frac{V_D}{V_T}\right) - I_S\right]_{(I_0,V_0)} = \frac{1}{V_T}\underbrace{\left[I_S \exp\left(\frac{V_0}{V_T}\right)\right]}_{I_0}$$

$\therefore$

$$g_m = \frac{I_0}{V_T}\ [\text{S}] \qquad (4.13)$$

It should be noted however that (4.13) is ideal as it calculates the upper g_m limit. Specifically, we assumed $n = 1$ in (4.1), as well as strong temperature dependences of I_S and V_T. Furthermore, the diode small signal resistance r_d to small changes of V_D is by definition the inverse of g_m, i.e.

[2]Logarithmic function is defined only for the positive arguments.

$$r_d \overset{\text{def}}{=} \frac{1}{g_m} = \frac{V_T}{I_0} \; [\Omega] \tag{4.14}$$

Rapid estimate of "small-signal" resistance shows that if $I_D = 1\,\text{mA}$, assuming room temperature $T = 290\,\text{K}$ and $V_T \approx 25\,\text{mV}$, then $r_d = 25\,\Omega$, but if $I_D = 2\,\text{mA}$ then $r_d = 12.5\,\Omega$, etc. Even though being only a rough approximate, estimates produced by (4.13) and (4.14) are very useful.

By knowing the exact shape of I_D vs. V_D transfer characteristics, for example as in Fig. 4.3, an estimate of g_m may be obtained by using graphical interpretation of the first derivative, i.e. by constructing tangent at the biasing point. For example, slope of tangent at point $\mathscr{A}$ is smaller than the one at point $\mathscr{B}$, that is to say $g_m(\mathscr{B}) > g_m(\mathscr{A})$. Furthermore, by using the right-angled triangles at the biasing points, a fairly accurate estimate of g_m is also achieved.

Once the biasing point is chosen, therefore the associated g_m gain, we calculate small signal gain of the input signal. Following definition (4.13) we conclude that small variations of diode voltage produce small variations of diode current. In order to clarify what exactly "small" means, we accept rather liberal engineering definition that "small signal" assumes a signal whose amplitude is much smaller than the DC value of the associated biasing point. In engineering practice, ratios of $1/10$ and larger (i.e. 10% or less) are assumed to be "much smaller". For example, if the biasing voltage is $V_0 = 1\,\text{V}$ then amplitudes of $100\,\text{mV}$ or less are accepted as "small". Therefore, if diode current/voltage stays close to DC biasing point (I_0, V_0), we can assume that $g_m = \text{const.}$ and write

$$V_D(t) = V_0 + v_S(t) \quad \text{and} \quad I_D(t) = I_0 + i_D(t) \tag{4.15}$$

Since the assumption is that (I_0, V_0) stays constant, (therefore its derivative is zero) we focus only on small AC diode current and conclude that

$$i_D = g_m\, v_S \tag{4.16}$$

where i_D, v_S are both small AC signals. Relation (4.16) is fundamental to operation of active devices based on pn junctions. Simulation setup, Fig. 4.6 (left), illustrates small signal amplitude of the output current i_D caused by small changes of the input voltage v_S (which is in effect identical to AC diode voltage v_D), Fig. 4.6 (right).

In order to illustrate small-signal gain of a nonlinear device by simulaiton, we reuse results already found for a diode whose transfer characteristics are shown in Fig. 4.3 with $\mathscr{A} : (I_0, V_0) = (10\,\text{mA}, 695\,\text{mV})$ and $\mathscr{B} : (I_0, V_0) = (30\,\text{mA}, 755\,\text{mV})$. In the simulation setup, DC voltage source is used to set biasing point V_0, while an AC voltage source is used as the "small signal" $v_S(t)$ source.

Simulation confirms, Fig. 4.6 (right), that for a small sine signal whose amplitude is, for example, $\pm 10\,\text{mV}$ (thus "small" relative to either 695 or 755 mV), the diode current amplitudes are $i_d(\mathscr{B}) = \pm 5\,\text{mA}$ and $i_d(\mathscr{A}) = \pm 2\,\text{mA}$. Time domain plots of the output sinusoidal current levels i_D show that the two small-signal output currents are added to their respective biasing levels, i.e. $I_D(t) = I_0 + i_D(t)$.

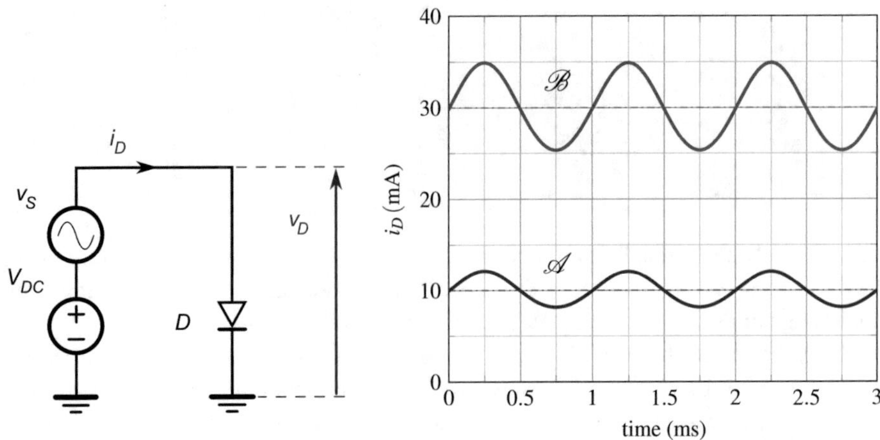

Fig. 4.6 AC simulation setup (left) and diode AC currents at two biasing points (right)

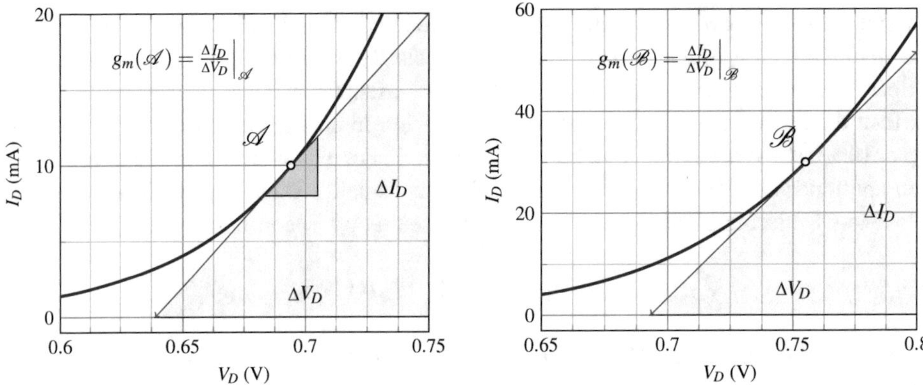

Fig. 4.7 Geometrical interpretation of the first derivative and the similarity of triangles

Example 27: Diode's g_m Gain and AC Resistances
Referring to Fig. 4.3, estimate: $g_m(\mathscr{A})$, $g_m(\mathscr{B})$, $r_d(\mathscr{A})$, and $r_d(\mathscr{B})$.

Solution 27: The zoom-in of Fig. 4.3 around the two bias points, see Fig. 4.7, enables us to estimate AC parameters of a nonlinear device, such as a diode in this and Example 24. Practical and quick estimation method shown here is based on geometrical interpretation of the first derivative, while precise results are obtained by numerical simulations.

g_m gain: practical estimate of g_m at the biasing point is done graphically as in Fig. 4.7, where at biasing point $\mathscr{A}$ we find: $\Delta I_D = 10\,\text{mA} - 0 = 10\,\text{mA}$ and $\Delta V_D = 695 - 640\,\text{mV} = 55\,\text{mV}$. Similarly, at biasing point $\mathscr{B}$: $\Delta I_D = 30\,\text{mA}{-}0 = 30\,\text{mA}$ and $\Delta V_D = 755{-}695\,\text{mV} = 60\,\text{mV}$. Thus,

(continued)

$$g_m(\mathscr{A}) \approx \left.\frac{\Delta I_D}{\Delta V_D}\right|_{\mathscr{A}} = \frac{10\,\text{mA}}{55\,\text{mV}} = 180\,\text{mS}; \quad g_m(\mathscr{B}) \approx \left.\frac{\Delta I_D}{\Delta V_D}\right|_{\mathscr{B}} = \frac{30\,\text{mA}}{60\,\text{mV}} = 500\,\text{mS}$$

AC resistance: small signal resistances r_d (a.k.a. AC resistance), i.e. resistance to *change* of voltage V_D (i.e. ΔV_D) around the chosen biasing point V_0. Thus, by definition we write

$$r_d(\mathscr{A}) \stackrel{\text{def}}{=} \left.\frac{1}{g_m}\right|_{\mathscr{A}} \approx \left.\frac{\Delta V_D}{\Delta I_D}\right|_{\mathscr{A}} = \frac{55\,\text{mV}}{10\,\text{mA}} = 5.5\,\Omega\,; \quad r_d(\mathscr{B}) \stackrel{\text{def}}{=} \left.\frac{1}{g_m}\right|_{\mathscr{B}} \approx \left.\frac{\Delta V_D}{\Delta I_D}\right|_{\mathscr{B}} = \frac{60\,\text{mV}}{30\,\text{mA}} = 2\,\Omega$$

It is useful to note that, in any case, resistance of a conducting diode is relatively low, thus often approximated to zero.

Example 28: A Small Signal g_m Gain
Referring to Fig. 4.3 and solutions in Examples 24 and 27, assuming a small signal variation of the diode voltage $v_D(t) = 10\,\text{mV}\ \sin(2\pi \times 1\,\text{kHz} \times t)$, estimate the total diode current $i_D(t)$ at the biasing points $\mathscr{A}$ and $\mathscr{B}$.

Solution 28: At biasing point $\mathscr{A}$: (10 mA, 695 mV), we calculate AC current as:

$$i_D(t) = \frac{v_D(t)}{r_d(\mathscr{A})} \stackrel{\text{def}}{=} g_m(\mathscr{A}) \times v_D(t) = 180\,\text{mS} \times 10\,\text{mV}\ \sin(2\pi \times 1\,\text{kHz} \times t)$$

$$= 1.8\,\text{mA}\ \sin(2\pi \times 1\,\text{kHz} \times t)$$

$$\therefore$$

$$I_D(t) = I_0 + i_D(t) = 10\,\text{mA} + 1.8\,\text{mA}\ \sin(2\pi \times 1\,\text{kHz} \times t)$$

At biasing point $\mathscr{B}$: (30 mA, 755 mV), we calculate AC current as:

$$i_D(t) = \frac{v_D(t)}{r_d(\mathscr{B})} \stackrel{\text{def}}{=} g_m(\mathscr{A}) \times v_D(t) = 500\,\text{mS} \times 10\,\text{mV}\ \sin(2\pi \times 1\,\text{kHz} \times t)$$

$$= 5\,\text{mA}\ \sin(2\pi \times 1\,\text{kHz} \times t)$$

$$\therefore$$

$$I_D(t) = I_0 + i_D(t) = 30\,\text{mA} + 5\,\text{mA}\ \sin(2\pi \times 1\,\text{kHz} \times t)$$

Simulation results, see Fig. 4.6, confirm these hand calculated results. As a comparison, instead of calculating g_m as in this example, use (4.13) and recalculate diode currents.

4.2.4 Varicap Diode

Inherently, by its physical structure a p-n junction is also a capacitor, not much different than the geometrical structure shown in Fig. 2.16. Once the depletion region is established between p-type and n-type junction layers, the region serves as any other non-conducting layer between two charged plates. Thus, in accordance with (2.26) it is the overlapping plate surface and distance between the two plates that determine the capacitor's value.

As opposed to the linear capacitors made of metallic plates, where distance d between the two plates is fixed, the width of diode's depletion region is dependent on the external voltage $d = f(V_D)$ being applied to the diode terminals. Thus, internal capacitance of a diode C_D is also function of the applied voltage, $C_D = f(V_D)$. In standard diodes the internal capacitance is minimized by design, i.e. by making the diodes physically small. However, we can exploit this voltage to capacitance relationship to create a diode with large C_D that is electronically *tuneable*, thus it can be used to design a tuneable LC resonator. This type of a diode is called "varicap diode" (a.k.a. "varactor") to emphasize its internal capacitance C_D as the primary parameter.

As with any device, it is necessary to know its C-V characteristics in advance. This characteristic is found either by experiment or by simulation. In case of a varicap diode de-embedding its internal capacitance is somewhat challenging. Here, we show two simulation methods that can be used: (1) TRAN simulation permits measurements of 5τ constant for a given R by monitoring the RC discharge time, and (2) AC simulation permits measurements of the resonant frequency ω_0 of LC resonator where a varicap diode is used along with known inductor L. Both methods enable us to de-embed the diode's internal capacitance C_D, Fig. 4.8. In reference to Fig. 4.9, the equivalent LC circuit is derived by the circuit reduction technique. After the serial branch of two capacitors $2C_p$ is replaced by its equivalent capacitance C_p, the total equivalent capacitance is

$$\frac{1}{C} = \frac{1}{C_p} + \frac{1}{C_D} \quad \therefore \quad C = \frac{C_p C_D}{C_p + C_D} \tag{4.17}$$

To illustrate analysis of a typical varicap diode, a typical CV characteristic of a reverse biased diode is shown in Fig. 4.10 for ($-7\,\text{V} \geq V_D \geq -0.5\,\text{V}$). In practice, capacitance associated with zero-volt biasing is extrapolated, in this example to $C_{D0} \approx 140\,\text{pF}$, Fig. 4.10 (left). We note that since the diode is *reverse* biased its current is $I_D \approx 0$, therefore voltages before and after the source resistor R_S are practically equal, i.e. $V_{ctrl} = V_D$ (there is no DC path for V_{ctrl}. It is of practical value to

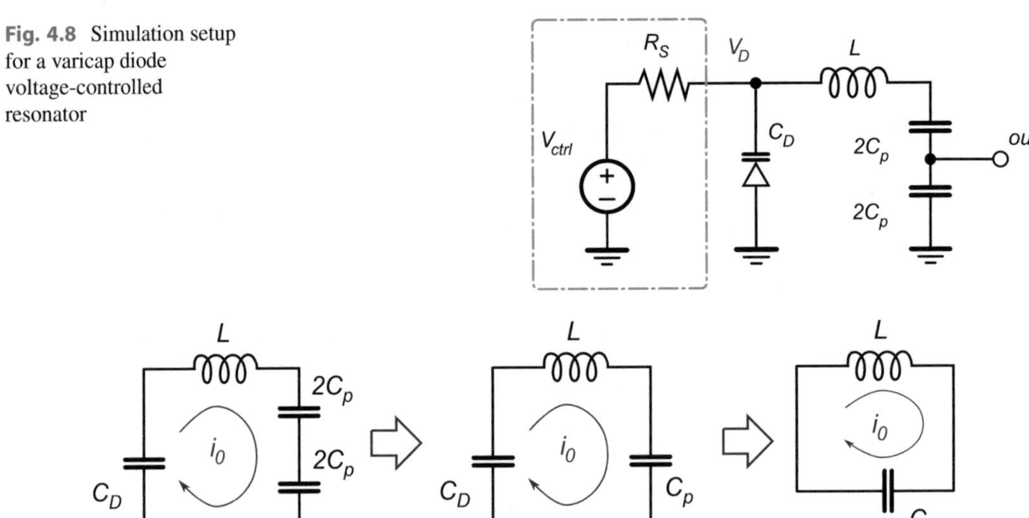

Fig. 4.8 Simulation setup for a varicap diode voltage-controlled resonator

Fig. 4.9 AC model of a varicap diode based LC voltage-controlled resonator in Fig. 4.8

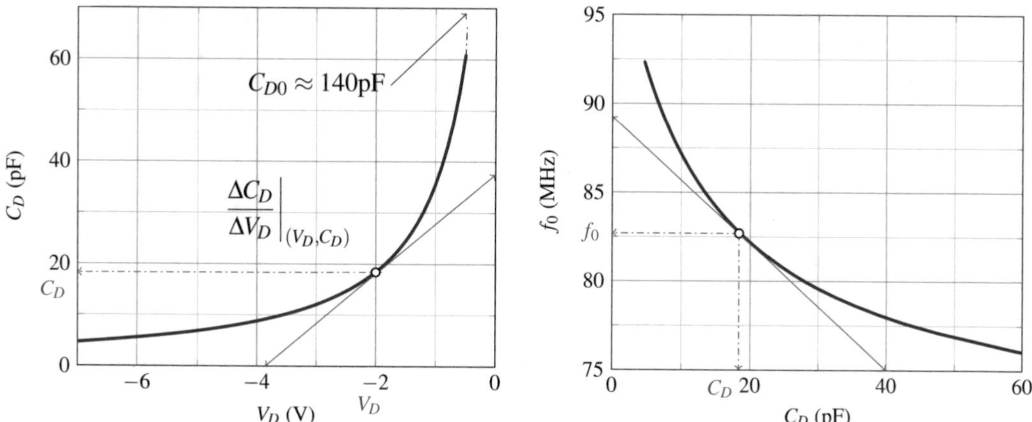

Fig. 4.10 Simulated transfer characteristics of tuneable LC resonator using a varicap diode showing C_D vs. V_D and f_0 vs. C_D curves

estimate sensitivity of diode's CV characteristics at the biasing point, for example at $V_D = -2\,\text{V}$. By using the graphical method for a tangent at a point, Fig. 4.10, for the given example we write

$$\left.\frac{\Delta C_D}{\Delta V_D}\right|_{(V_D=-2\,\text{V})} = \frac{(0-38)\,\text{pF}}{(-3.8-0)\text{V}} \approx 10\,\text{pF/V} \tag{4.18}$$

which illustrates significant and useful variation of the internal diode's capacitance.

In addition to the numerical (or experimental) transfer function in Fig. 4.10 (left), it is necessary to have some form of analytical function as well. There are various nonlinear models of a varicap diode found in the literature, here we use curve fitting method to derive the following model

$$C_D = \frac{C_{D0}}{(1-2V_D)^{\frac{5}{4}}} \tag{4.19}$$

which is very good approximation of this diode's transfer function. For example, substituting $V_D = -2\,\text{V}$ in (4.19) we calculate $C_D(-2\,\text{V}) = 18.7\,\text{pF}$, which is in agreement with the graph in Fig. 4.10 (left).

With this model in (4.19), we can also derive analytically the sensitivity found in (4.18) as follows:

$$\frac{\partial C_D}{\partial V_D} = \frac{5}{2}\frac{C_{D0}}{(1-2V_D)^{\frac{9}{4}}} = \frac{5}{2}\underbrace{\frac{C_{D0}}{(1-2V_D)^{\frac{5}{4}}}}_{C_D}\frac{1}{(1-2V_D)} = \frac{5}{2}\frac{C_D}{(1-2V_D)} \tag{4.20}$$

$$\therefore$$

$$\left.\frac{\partial C_D}{\partial V_D}\right|_{(V_D=-2\,\text{V})} = \frac{5}{2}\frac{18.7\,\text{pF}}{[1-2(-2\,\text{V})]} = 9.35\,\text{pF/V} \tag{4.21}$$

which is very close to result in (4.18) that is obtained graphically by the tangent method.

Transfer characteristics in Fig. 4.10 (right), which is obtained by AC simulation of circuit in Fig. 4.8, relation between its resonant frequency f_0 and C_D shows that, indeed, this circuit is a voltage-controlled resonator that is suitable for the FM radio frequency range. Voltage-controlled resonator circuit in Fig. 4.8 consists of an inductance L, and the equivalent capacitance C_{eq} that is made of C_p and C_D connected in series. Thus by definition, the resonant frequency ω_0 is found as

$$\omega_0 = \frac{1}{\sqrt{LC_{eq}}} = \frac{1}{\sqrt{L\frac{C_pC_D}{C_p+C_D}}} = \sqrt{\frac{C_p + C_D}{L\,C_pC_D}} \quad \therefore \quad L = \frac{C_p + C_D}{\omega_0^2\,C_pC_D} \tag{4.22}$$

With the help of (4.22) we find the ω_0 vs. C_D sensitivity is as follows:

$$\begin{aligned}\frac{\partial\omega_0}{\partial C_D} &= \frac{\partial}{\partial C_D}\left(\sqrt{\frac{C_p + C_D}{L\,C_p\,C_D}}\right) = \frac{1}{2}\frac{1}{\sqrt{\frac{C_p+C_D}{L\,C_pC_D}}}\,\frac{L\,C_pC_D - (C_p + C_D)\,L\,C_p}{(L\,C_p\,C_D)^2}\\ &= \frac{1}{2}\sqrt{\frac{L\,C_pC_D}{C_p + C_D}}\frac{(-1)}{L\,C_D^2} = -\frac{1}{2}\sqrt{\frac{C_pC_D}{L(C_p + C_D)}}\frac{1}{C_D^2}\end{aligned} \tag{4.23}$$

It is convenient to express (4.23) relative to its ω_0 by substituting L as found in (4.22), thus we write

$$\frac{\partial\omega_0}{\partial C_D} = -\frac{1}{2}\sqrt{\frac{C_pC_D}{\frac{C_p + C_D}{\omega_0^2\,C_pC_D}(C_p + C_D)}}\frac{1}{C_D^2} = -\frac{1}{2}\sqrt{\frac{\omega_0^2\,C_pC_D\,C_pC_D}{(C_p + C_D)\,(C_p + C_D)}}\frac{1}{C_D^2}$$

$$\therefore$$

$$\frac{\partial\omega_0}{\partial C_D} = -\frac{1}{2}\frac{\omega_0\,C_p}{(C_p + C_D)}\frac{1}{C_D} = -\frac{1}{2}\frac{\omega_0}{C_D}\frac{1}{1 + \frac{C_D}{C_p}} = -\frac{1}{2}\frac{\omega_0}{C_D}\frac{1}{1 + n}$$

where it is convenient to use the ratio substitution $n = C_D/C_p$. Therefore, after further substituting $\omega_0 = 2\pi f_0$ on both sides of the last equation we find

$$\frac{\partial f_0}{\partial C_D} = -\frac{1}{2}\frac{f_0}{C_D}\frac{1}{1 + n} \tag{4.24}$$

For example, SPICE simulation of an LC resonator in Fig. 4.8 shows f_0 vs. C_D function of a typical varicap diode, Fig. 4.10 (right). By Pythagorean triangle rule for a tangent at a point, for the given example we write

$$\left.\frac{\Delta f_0}{\Delta C_D}\right|_{(C_D=18.7\,\text{pF})} = \frac{(102.5 - 75)\,\text{MHz}}{(0 - 35)\,\text{pF}} \approx -0.786\,\text{MHz/pF} \tag{4.25}$$

For circuit shown in Fig. 4.8, evaluation of (4.24) gives

$$\left.\frac{\partial f_0}{\partial C_D}\right|_{(C_D=18.7\,\text{pF})} = -\frac{1}{2}\frac{f_0}{C_D}\frac{1}{1 + n} = -\frac{1}{2}\frac{88.4\,\text{MHz}}{18.7\,\text{pF}}\frac{1}{1 + 1.87} = -0.823\,\text{MHz/pF}$$

which illustrates that, given a good estimate of C_D, the graphical method is reasonably good relative to the analytical result.

In practice, a VCO is viewed as a voltage-to-frequency converter. That being the case, sensitivity of the output frequency relative to the variations of varicap's biasing voltage V_D must be derived. Using (4.20) and (4.24) we find the frequency deviation constant k as derivative of the frequency with respect to the varicap biasing voltage V_D at the point V_0 (that is to say, its corresponding varicap diode capacitance C_0) at the resonant frequency f_0, i.e.

$$k = \left.\frac{\partial\omega_0}{\partial V_D}\right|_{V_0,C_0} = \frac{\partial\omega_0}{\partial C_D}\frac{\partial C_D}{\partial V_D} = \left[\frac{5}{2}\frac{\cancel{C_D}}{(1-2V_D)}\right]\left[-\frac{1}{2}\frac{f_0}{\cancel{C_D}}\frac{1}{1+n}\right]$$
$$= -\frac{5}{4}\frac{f_0}{(1-2V_D)(1+n)} \tag{4.26}$$

which for our numerical example results in $k = -7.7\,\text{MHz/V}$.

Varicap diode is among key devices in all modern electronics because, even though its capacitance is nonlinear function of voltage, it enables *electronic* tuning of circuit parameters.

Example 29: Voltage-Controlled LC Resonator

For a given voltage-controlled LC oscillator, Fig. 4.8 calculates the range of control voltages V_{ctrl} that would enable this oscillator to tune its resonant frequency for the commercial FM radio range, i.e. f_0 between 87.5 and 108 MHz. Transfer functions of varicap diode are as in Fig. 4.10.

Data: $C_p = 10\,\text{pF}$, $L = 500\,\text{nH}$, $C_{D0} = 140\,\text{pF}$, $C_p = 10\,\text{pF}$, $V_{ctrl} = (0.5 - 0.7\,\text{V})$, $R_S = 20\,\text{k}\Omega$

Solution 29: Equation (4.22) gives

$$C_{eq} = \frac{1}{L\omega_0^2} \;\therefore\; \frac{C_pC_D}{C_p+C_D} = \frac{1}{L\omega_0^2} \;\therefore\; C_D = \frac{C_p}{C_pL\omega_0^2 - 1} = \frac{C_p}{C_pL(2\pi f_0)^2 - 1}$$

For the given data, numerical solutions are: at $f_0 = 87.5\,\text{MHz}$ capacitance $C_D = 19.559\,\text{pF}$, and for $f_0 = 108\,\text{MHz}$ capacitance $C_D = 7.678\,\text{pF}$.

To set these two capacitance values, in accordance with (4.19) we derive that

$$V_D = \frac{1}{2}\left[1 - \left(\frac{C_{D0}}{C_D}\right)^{\frac{4}{5}}\right]$$

Thus, to set $C_D = 7.678\,\text{pF}$ the diode must be biased with $V_D = -4.601\,\text{V}$, and to set $C_D = 19.559\,\text{pF}$ the diode must be biased with $V_D = -1.914\,\text{V}$. AC simulation of circuit in Fig. 4.8 confirms validity of these results, see Fig. 4.11.

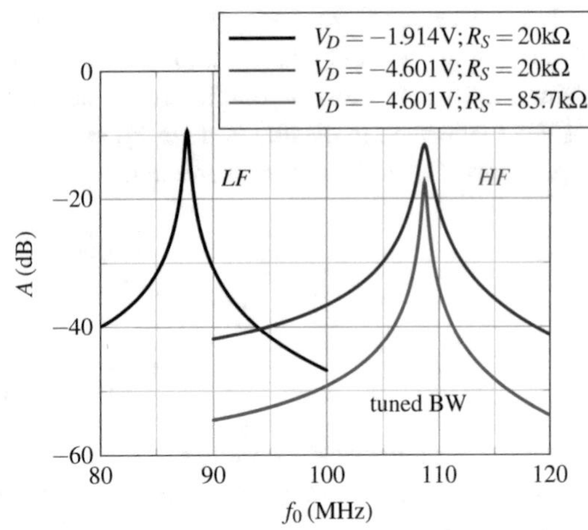

Fig. 4.11 AC simulation of tuneable LC resonator for the range of 87.5–108 MHz that is based on varicap diode.

4.3 Bipolar Junction Transistor (BJT)

In comparison with two-terminal devices, as in Sect. 4.2, analysis of a three-terminal device requires significantly greater experimental effort. Nevertheless, by using "divide and conquer" approach, with systematic methodology we are able to create working models of three-terminal devices.

General scientific and engineering strategy is to reduce the unknown complicated problem to several simpler ones that are already well understood. In this case it means that by fixing one of the three I/O terminals the analysis of a three-terminal device is reduced to analysis of a number of cases of two-terminal devices (while keeping track of the imposed limitation on the third terminal). In addition, one of the three terminals is declared to be the reference point, i.e. it is "shared" between the input and output side of the black box. In the two-terminal cases, it was the "ground" that served as the common I/O reference point. Without analysing all possible I/O combinations, here we review only two important cases while the analysis of the rest follows the same idea.

In the first typical experiment, Fig. 4.12 (left), the black box is connected as follows:

1. on its right side, call it "output", a voltage source is connected between terminals labeled "C" and "E". This source, labeled V_{CE}, generates a fixed DC voltage, while its current I_C is measured as the output (i.e. controlled) variable. Voltage V_{CE} itself is considered a technological parameter. Here, we assume that V_{CE} is a non-zero positive voltage that is safe to use with this particular device (e.g. 10 V).
2. on its left side, call it "input", a voltage source is connected between terminals labeled "B" and "E". This source, labeled V_{BE}, generates a variable DC voltage. This voltage is declared the input (i.e. controlling) variable and it is swept within a certain range of voltage values.

We note that terminal labeled "E" is fixed to the ground level and therefore serves as the reference point as well as "shared" I/O terminal. Consequently, from the perspective of terminals B and C this setup is equivalent to the two-terminal setup for a diode, see Fig. 4.1, because the third terminal E is temporarily fixed to the reference point and it does not affect the experiment.

After sweeping the input variable V_{BE}, for example over -1 to 1 V range,[3] the output I_C as a function of the input voltage V_{BE} is found to be as in Fig. 4.12 (right). Without knowing anything else about the content of the black box, this result is indistinguishable from the one we already found to

[3]This voltage range is not known in advance, we gradually zoom in.

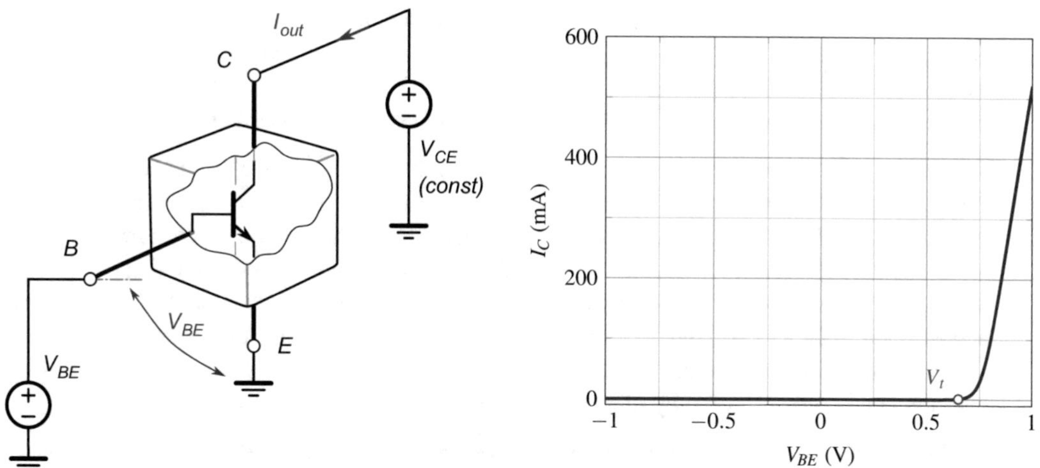

Fig. 4.12 Three-terminal element in a black box (left), its input side voltage–current characteristics (right)

be "signature" of a diode (see Fig. 4.1). The conclusion of this experiment is that the output current I_C is an exponential function of the input voltage V_{BE}, i.e. it appears as if there is a diode connected between the output terminals.

Therefore, the same procedure already used in Sects. 4.2.2 and 4.2.3 (in the case of a diode in the black box) is repeated to set up biasing points and g_m gains of a BJT transistor, when terminal "B" is used to accept the input voltage V_{BE} that controls the output current I_C. It is noted that, as same as for any other diode, there is a minimum voltage $V_t \approx 0.7\,\text{V}$.

If voltage source V_{BE} in Fig. 4.12 is replaced with a current source I_B, then we can establish current I/O relationship. After sweeping $I_B \geq 0$, by experiment we find these two relationships

$$I_C = \beta\, I_B \;\; \text{and} \;\; I_E = I_C + I_B \tag{4.27}$$

where β is the current multiplication factor between terminals B and C. Additional measurements show that two currents entering the black box through B and C terminals are found at E terminal, see (4.27). In other words, three terminals of this black box obey Kirchhoff's current law: the exiting current from this node is the sum of the entering currents.

In modern technologies the current multiplication factor β is approximately 50 and 300. Thus, in accordance with (4.27), the following approximations are very often used

$$I_C \gg I_B \;\; \text{and} \;\; I_E \approx I_C \tag{4.28}$$

This experiment on its own still does not give a complete picture what is in the black box. Before making the final conclusion and the appropriate mathematical (i.e. behavioural) model we need to find the other I/O relationships as well.

In the second experiment, Fig. 4.13 (left), the black box is connected as follows:

1. on its right side, a voltage source is connected between terminals labeled "C" and "E". This source, labeled V_{CE}, generates a DC voltage that is declared the input (i.e. controlling) variable, while its current I_C is measured as the output (i.e. controlled) variable. During each measurement, this controlling variable V_{CE} is swept within a certain range of voltage values, e.g. 0–10 V.
2. on its left side, a voltage source is connected between terminals labeled "B" and "E". This source, labeled V_{BE}, generates a fixed DC voltage; during each experiment, this source is kept at a constant value (i.e. "parametric constant"). This parametric constant is changed between each measurement to produce a family of parametric I/O characteristics.

A typical family of I/O characteristics that is produced, for example, after four measurements is given in Fig. 4.13 (right). Here, starting with $V_{BE} = 0$ V, in each subsequent measurement this parametric variable is increased[4] to 0.600, 0.650, and 0.700 V. Produced experimental characteristics in Fig. 4.13 (right) are interpreted as follows:

1. ($V_{BE} = 0$): regardless of V_{CE} the output current $I_C = 0$, which is switch behaviour in OFF state; there is no current flow between C and E terminals.
2. ($V_{BE} > 0$, V_{CE} is small), e.g. $0 \leq V_{CE} \leq 100$ mV: for each value of V_{BE} the transfer function I_C vs. V_{CE} is similar to the one associated with a linear resistor. This resistance value is small (see the slope of this function in this region, note that the vertical axis is a current). However, this slope (thus the resistance) changes with each value of V_{BE}, in other words this small resistance is *programmable*.
3. ($V_{BE} > 0$, V_{CE} is large), e.g. $100\text{ mV} \leq V_{CE} \leq 10$ V: at around $V_{CE} = 100$ mV the transfer function rapidly changes and becomes similar to the one associated with a current source between C and E terminals—the output current is almost constant over large range of V_{CE} voltages. What

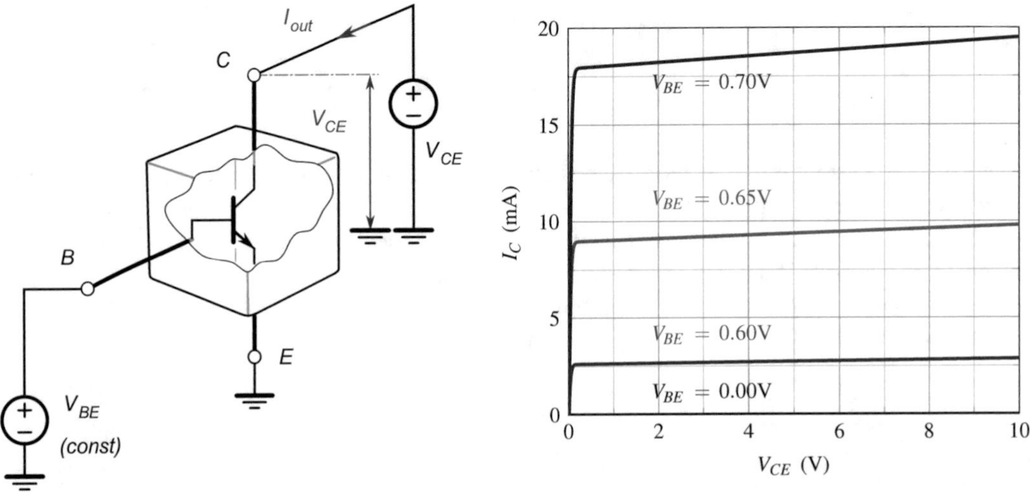

Fig. 4.13 Three-terminal element in a black box (left), its output side voltage–current characteristics (right)

[4]This is only a typical subset of all possible V_{BE} values shown in Fig. 4.12 (right).

is more, current level of this current source is also programmable by V_{BE}, and its resistance is high but not infinite (i.e. it is a realistic current source).

In conclusion, BJT transistor is a rather sophisticated and versatile device that is primarily used as a programmable current source, where the output current I_C is controlled either by the input voltage V_{BE} or by the input current I_B. This constant current mode is possible under condition that the output voltage V_{CE} is greater than a certain minimum voltage that is technology dependent (call it $V_{CE}(min)$, in modern technologies typically $V_{CE}(min) \approx$ 100–200 mV).

4.3.1 Mathematical Model

A mathematical description of the relationship between collector current I_C and base–emitter voltage V_{BE} in a BJT is similar to the voltage–current relationship of a diode, see (4.1).[5] In addition, as per (4.29), a transistor may be looked at as a three-terminal current node, where the three currents must obey KCL. In addition, the alternative interpretation of base to collector current gain amplification is by the factor β, see (4.30). Therefore, for a given V_{BE} voltage we write

$$I_E = I_C + I_B \tag{4.29}$$

$$I_C = \beta I_B \tag{4.30}$$

$$I_C = I_S \left[\exp\left(\frac{V_{BE}}{n\, V_T} \right) - 1 \right] \tag{4.31}$$

where

I_C is the collector current
I_B is the base current
I_E is the emitter current
I_S is the BJT leakage current, a.k.a. "dark current" which is in order of pA
V_{BE} is the base–emitter voltage
β is the current gain factor
n is the emission coefficient, unless specified otherwise, it is assumed $n = 1$

Equation (4.30) illustrates the basic current-amplifying property of a BJT—the collector current is β times larger than the base current. The current amplification factor β is usually of the order of 100–300 and it is controlled by the manufacturing process. However, a given β is not constant either, indeed it is a strong function of the temperature, the transistor type, collector current, and the collector-emitter voltage V_{CE}. All in all, circuit designers try hard to design circuits with gains that are independent of β.

The relationship between the emitter and collector currents is derived by substituting (4.30) into (4.29) as follows:

$$I_E = I_C + \frac{I_C}{\beta} \quad \therefore \quad I_C = \frac{\beta}{\beta + 1} I_E = \alpha\, I_E \approx I_E \tag{4.32}$$

[5]Under conditions of forward-biased base–emitter diode, and $V_{CE} \geq V_{CE}(min)$.

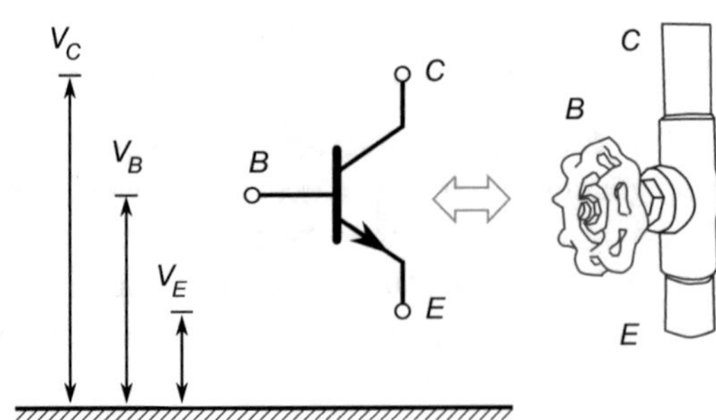

Fig. 4.14 Electrical symbol of a BJT (left) and its functional valve analogy (right) showing the relative potential levels of the three terminals in the case of an "open" valve

where α is the ratio of the collector and emitter currents and the last approximation is valid when $\beta \gg 1 \Rightarrow \beta + 1 \approx \beta \Rightarrow \alpha \approx 1$ (which is valid for virtually all modern transistors). It is also useful to keep these relationships

$$\alpha = \frac{\beta}{\beta + 1} = \frac{1}{1 + \dfrac{1}{\beta}} \quad \therefore \quad \frac{1}{\alpha} = 1 + \frac{1}{\beta} \quad \therefore \quad \beta = \frac{\alpha}{1 - \alpha} \tag{4.33}$$

A physical analogy that may be helpful to understand functionality of a transistor is that, in its basic function, a transistor can be described as a simple valve that controls the current flow in the CE branch. A BJT transistor is like a unidirectional pipe with a third terminal (base) that controls the amount of current flow through the CE "pipe", from fully closed to fully open (see Fig. 4.14).

Example 30: NPN BJT Biasing Point
Estimate ΔV_{BE} if the collector current is increased ten times. Assume the room temperature.

Solution 30: Direct implementation of (4.34) for the two currents gives

$$I_C = I_S \exp\left(\frac{V_{BE1}}{V_T}\right) \quad \text{and} \quad 10 \times I_C = I_S \exp\left(\frac{V_{BE2}}{V_T}\right)$$

$$\therefore$$

$$\frac{10 \times \cancel{I_C}}{\cancel{I_C}} = \frac{\cancel{I_S} \exp\left(\frac{V_{BE2}}{V_T}\right)}{\cancel{I_S} \exp\left(\frac{V_{BE1}}{V_T}\right)} = \exp\left(\frac{V_{BE2} - V_{BE1}}{V_T}\right) = \exp\left(\frac{\Delta V_{BE}}{V_T}\right)$$

$$\therefore$$

$$\Delta V_{BE} = V_T \ln 10 = 25\,\text{mV} \times 2.3026 = 57.567\,\text{mV} \approx 60\,\text{mV}$$

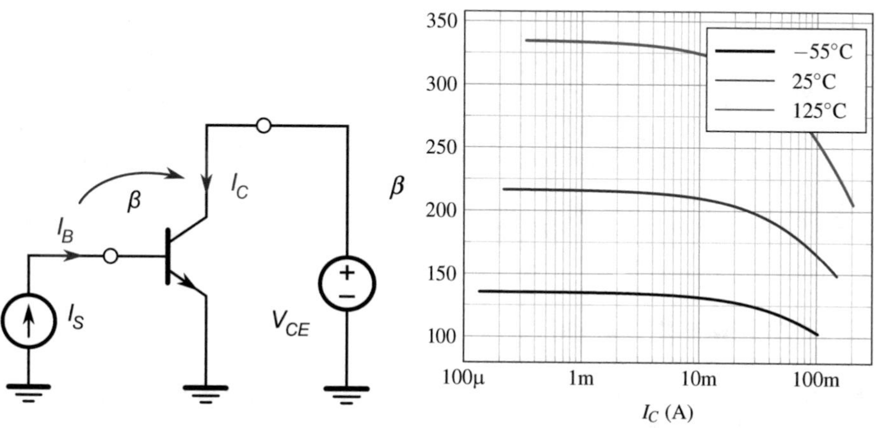

Fig. 4.15 Simulation setup for measuring β gain of a typical NPN BJT transistor (left), and simulated characteristics (right)

4.3.2 Current Gain β

BJT transistors serve both as current-to-current amplifier whose current gain is $\beta = I_C/I_B$ and voltage-to-current amplifier whose gain is $g_m = i_C/v_{BE}$, where β gain is assumed constant. In reality, however, β is strongly dependent upon both temperature and the biasing current I_C. Simulation technique used to evaluate β of a typical NPN transistor is shown in Fig. 4.15 (left). Base current is forced by an ideal current source, and the collector current is measured. All SPICE simulators based simulators are capable of performing mathematical operation on the simulated waveforms, thus graph in Fig. 4.15 (right) is created by calculating $\beta = I_C/I_B$, see its definition (4.30).

Example 31: NPN BJT—β Dependence on Temperature
In reference to Fig. 4.15, if the collector current is held constant $I_C = 1$ mA estimate how many times n the base current changes over the full temperature range from −55 to 125 °C?

Solution 31: By inspection of Fig. 4.15 (right), we find that $\beta(-55\,°\text{C}, 1\,\text{mA}) = 135$ and $\beta(125\,°\text{C}, 1\,\text{mA}) = 335$. Direct implementation of (4.30) leads into conclusion that

$$I_B(-55\,°\text{C}, 1\,\text{mA}) = \frac{1\,\text{mA}}{\beta(-55\,°\text{C}, 1\,\text{mA})} = \frac{1\,\text{mA}}{135} = 7.41\,\mu\text{A}$$

and

$$I_B(125\,°\text{C}, 1\,\text{mA}) = \frac{1\,\text{mA}}{\beta(-55\,°\text{C}, 1\,\text{mA})} = \frac{1\,\text{mA}}{335} = 2.98\,\mu\text{A}$$

that is, if the collector current is to be held constant, the base current must change

$$n = \frac{I_B(-55\,°\text{C}, 1\,\text{mA})}{I_B(125\,°\text{C}, 1\,\text{mA})} = \frac{7.41\,\mu\text{A}}{2.98\,\mu\text{A}} = 2.48 \text{ times } = 248\%$$

(continued)

which is significant change that illustrates the need for using the temperature compensation techniques in high precision applications (e.g. A/D converters), or in circuits intended for harsh environments (e.g. space, military). However, if the environment temperature is stable (e.g. biomedical implants), then β factor is inherently stable over relatively wide range of biasing currents.

4.3.3 BJT Small Signal Models

In the subsequent analysis we consider only the effects related to the small variable signals (i.e. AC), while assuming the biasing point (i.e. DC) is already set. To indicate whether DC or AC values of voltages and currents are used, in accordance with the most textbooks, the following notification is used whenever possible:

1. *AC voltages and currents:* indexes written in small letters indicate AC values, e.g. i_{in}, v_{out}
2. *DC voltages and currents:* capital letters indicate hard DC values, e.g. I_C, V_{BE}, I_0
3. *Total voltages and currents:* to indicate that given variable is the sum of its DC and AC values we use capitalized indexes, e.g. i_{IN}, v_{OUT}, i_C, v_{BE}

Linear circuit analysis employs the traditional small-signal BJT model, Fig. 4.16 (top), based on controlled current sources (either VCCS or CCCS, see Fig. 2.6), which emphasizes the two main BJT functions: either voltage to current $i_C = g_m v_{BE}$ or current to current amplifier $i_C = \beta i_B$. A

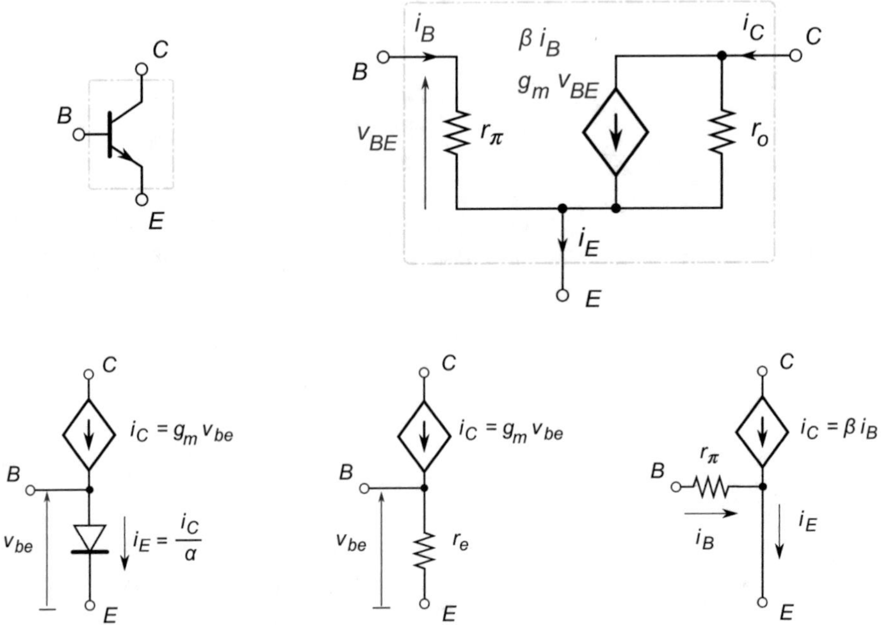

Fig. 4.16 NPN BJT transistor model based on controlled current sources (top). Three somewhat simplified and often used BJT model versions are T-models (below, assuming $r_o = \infty$, i.e. r_o is omitted). All controlled current sources are either VCCS for $i_C = g_m i_B$ or CCCS for $i_C = \beta i_B$ calculations

forward-biased base–emitter diode controls the overall i_C current in accordance with its exponential function (4.31); the β factor controls the ratio between the collector and base currents as in (4.30).

We note that the complete models of active devices (i.e. BJT, MOS, diodes) are very sophisticated and complicated, thus suitable only for numerical simulations. For hand analysis, therefore, there is no one single model that is used in every situation. Instead, depending on what aspect of active devices is calculated at the given moment, we use multiple simplified versions of BJT transistor models, sometimes called "T-models", Fig. 4.16 (bottom). For example, the first two BJT models in Fig. 4.16 (bottom) show the voltage to current conversion property (thus, $i_C = g_m\, i_B$) while distinguishing whether the base-emitter branch is considered a diode or the equivalent emitter resistance r_e. Similarly, T-model on the right emphasizes current amplification property $i_C = \beta\, i_B$ and the input resistance r_π at the base terminal. We note that if $r_o = \infty$, which is the case of an ideal current source connected between C and E terminals, this T-model is identical to the model in Fig. 4.16 (top).

4.3.4 Small Signal g_m Gain

Equation (4.29) illustrates that a BJT obeys KCL, while (4.31) emphasizes the fact that a BJT device is fundamentally a g_m (i.e. "transconductance") amplifier. The total collector current i_C (the output variable) is controlled by the base–emitter voltage v_{BE} (the input variable). Similar to diode approximations (4.2) and (4.3), the BJT's base–emitter diode expression is approximated as

$$I_C = I_S \left[\exp\left(\frac{V_{BE}}{V_T}\right) - 1\right] \tag{4.34}$$

Then, by definition, we write

$$V_T \overset{\text{def}}{=} \frac{kT}{q} \approx 25\,\text{mV} \quad \text{at room temperature } T = 290.22\,\text{K} \tag{4.35}$$

which means that, at room temperature, as soon as V_{BE} voltage is greater than about $4V_T$ or so, the "-1" term in (4.34) can be ignored. We keep in mind that in this version of (4.34) it is base–emitter voltage V_{BE} that controls the collector current. However, this relationship is bidirectional, i.e. if the collector current I_C is forced by the external source, then the associated V_{BE} is generated—in this case, we create a "diode voltage reference". At room temperature ($T = 290.22$ K), (4.34) becomes

$$I_C \approx I_S\, e^{40\, V_{BE}} \tag{4.36}$$

At a given biasing point (I_{C0}, V_{BE0}), the transconductance gain g_m of a BJT is found by definition[6] as

$$g_m \overset{\text{def}}{=} \left.\frac{\partial I_C}{\partial V_{BE}}\right|_{(I_{C0}, V_{BE0})} = \frac{\partial}{\partial V_{BE}}\left[I_S \exp\left(\frac{V_{BE}}{V_T}\right)\right] = \frac{I_S \exp\left(\frac{V_{BE}}{V_T}\right)}{V_T} = \left.\frac{I_C}{V_T}\right|_{(I_{C0}, V_{BE0})} = \frac{I_{C0}}{V_T}$$

$\therefore$

[6]See Sect. 4.2.3.

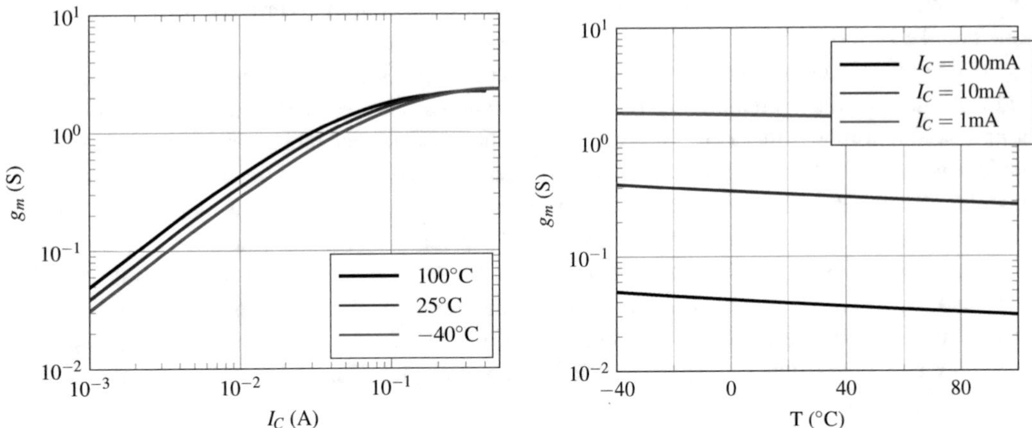

Fig. 4.17 g_m dependence on I_C (temperature is a parameter), and g_m dependence on T (I_C is a parameter)

$$g_m = \frac{I_{C0}}{V_T} \tag{4.37}$$

Again, we note that (4.37) assumes ideal conditions ($n = 1$) thus it produces upper limit value. By inspection of (4.37), we conclude that at the given temperature the transconductance gain of a BJT is set strictly by its biasing current. Zoom-in of transfer characteristics in Fig. 4.12 is identical to the one in Fig. 4.7, thus the same graphical method is used to estimate g_m at a given biasing point.

In practice, once the external network is designed it sets the biasing point to BJT device. In the subsequent small signal analysis the details of this biasing network are simply assumed through g_m value. Or, equivalently, the first step in the design process is to decide the collector current I_{C0} which subsequently determines the associated V_{BE0}, g_m, and r_e (see (4.37)).

Example 32: NPN BJT—g_m Temperature Dependence
By simulation, of a typical NPN BJT create plots of: (a) its g_m dependence relative to the collector current, at three typical temperatures 100, 25, −40 °C; and (b) its g_m dependence relative to the temperature, at three typical collector currents 100, 10, 1 mA.

Solution 32: Equation (4.37) implies that g_m gain is directly proportional to I_C and inversely proportional to V_T. By themselves, both IC and VT are temperature dependent.

The simulation shows that g_m dependence is not exactly as it would be implied by (4.37), see Fig. 4.17. This is because there is also a strong $I_S(T)$ function involved. Nevertheless, with a careful design, it is possible to achieve very good temperature compensation of $g_m(T)$ function.

4.3.5 Emitter Resistance

The "small-signal resistance" r_e of BJT is perceived by "looking into" the emitter. Assuming that the collector and emitter currents are approximately the same, i.e. $I_C \approx I_E$, which is close enough if $\beta \gg 1$, or equivalently the base current is negligible relative to the collector current, this resistance is calculated by definition as

$$r_e \stackrel{\text{def}}{=} \frac{\partial V_{BE}}{\partial I_E} \approx \frac{\partial V_{BE}}{\partial I_C} \quad \therefore \quad \frac{1}{r_e} = \frac{\partial I_C}{\partial V_{BE}} \stackrel{\text{def}}{=} g_m \tag{4.38}$$

as well as by the less accurate but practical relationship

$$\begin{aligned} r_e &= \frac{1}{gm} \approx \frac{25\,\text{mV}}{I_C} \quad \text{at the room temperature } T = 290.22\,\text{K} \\ &= \frac{25}{I_C\ [\text{mA}]}\ [\Omega] \end{aligned} \tag{4.39}$$

that helps us to quickly estimate the emitter's output impedance of a BJT in the active mode of operation. For example, a typical biasing current $I_C = 1\,\text{mA}$ results in emitter output resistance of $r_e = 25\,\Omega$; $I_C = 0.5\,\text{mA}$ leads to $r_e = 50\,\Omega$, etc. We note that this resistance is an intrinsic property of a BJT transistor, i.e. of its base–emitter diode.

Functional diagrams of an NPN BJT, which are based purely on BJT's behaviour at the three terminals, Fig. 4.16, are very useful for hands-on rapid analysis. Relation between the base and emitter resistances is derived as follows:

$$v_{be} = r_e\, i_e \tag{4.40}$$

therefore,

$$v_{be} = r_\pi\, i_B = r_\pi\, \frac{i_C}{\beta} = r_\pi\, \frac{\alpha\, i_E}{\beta} = \frac{r_\pi}{\beta + 1}$$

$$\therefore$$

$$\boxed{r_e = \frac{r_\pi}{\beta+1} \approx \frac{r_\pi}{\beta}} \tag{4.41}$$

where (4.41) shows again the bilateral relationship between the base and emitter resistances. It is very handy to interpret (4.41) as the "magnifying effect" between the base and emitter resistances: the resistance associated with the emitter node is perceived at the base node as being *multiplied* $(\beta + 1)$ times and the resistance associated with the base is perceived at the emitter node as being *divided* $(\beta + 1)$ times.

This relatively small emitter resistance r_e is important because it acts as a minimum resistance in the emitter branch, thus it prevents the emitter branch resistance from becoming zero. Therefore, it determines the maximum voltage gain of an amplifier, see Sect. 6.2.

On the other hand, a detailed analysis shows that, contrary to what (4.31) would suggest, the base–emitter voltage has a positive temperature coefficient (TC). It is useful to remember that the base–emitter voltage V_{BE} increases by approximately $2\,\text{mV}/°\text{C}$. This consequence is due to very strong temperature dependence of "dark current" I_S (which is not explicitly shown in (4.31)). In practice, this relationship is important for the design of temperature-independent voltage references, known as "bandgap references".

4.3.6 Base Resistance

It is already established in (4.41) that base resistance is perceived as "reflection" of the emitter resistance. Here, in reference to the two T-models in Fig. 4.18, by definition, we write

$$r_\pi \stackrel{\text{def}}{=} \frac{v_{BE}}{i_B} = \frac{v_{BE}}{i_C/\beta} = \frac{\cancel{v_{BE}}}{g_m\,\cancel{v_{BE}}/\beta} \quad \therefore \quad r_\pi = \frac{\beta}{g_m} \tag{4.42}$$

In addition, (4.42) can be expressed in terms of DC base current (i.e. at the biasing point) after (4.37),

$$r_\pi = \frac{\beta}{g_m} = \frac{\beta}{I_C/V_T} = \frac{\cancel{\beta}}{\cancel{\beta} I_B/V_T} = \frac{V_T}{I_B} \tag{4.43}$$

Furthermore, we write

$$i_C = g_m\,v_{BE} \quad \text{and} \quad i_B = \frac{v_{BE}}{r_\pi} \quad \text{and} \quad i_E = \frac{v_{BE}}{r_e} \tag{4.44}$$

$$\therefore$$

$$i_E \stackrel{\text{def}}{=} i_C + i_B = g_m\,v_{BE} + \frac{v_{BE}}{r_\pi} = \frac{v_{BE}}{r_\pi}\,(g_m\,r_\pi + 1)$$

$$= \frac{v_{BE}}{r_\pi}\left(\cancel{g_m}\frac{\beta}{\cancel{g_m}} + 1\right) = \frac{v_{BE}}{\dfrac{r_\pi}{\beta+1}} \stackrel{(4.44)}{} \stackrel{\text{def}}{=} \frac{v_{BE}}{r_e}$$

$$\therefore$$

$$r_e = \frac{r_\pi}{\beta + 1} \quad \text{or,} \quad r_\pi = (\beta + 1)\,r_e \tag{4.45}$$

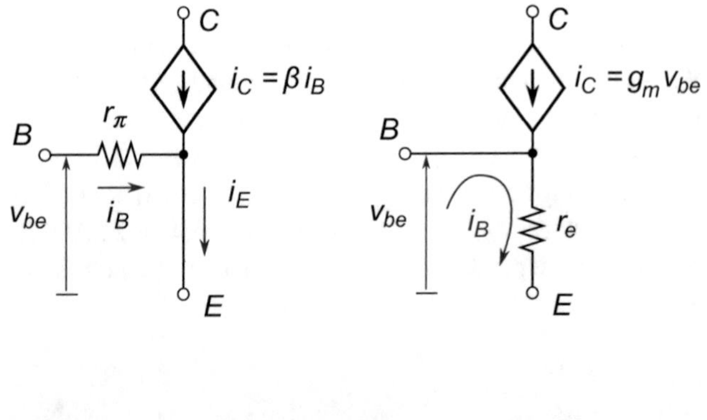

Fig. 4.18 T-models used to calculate r_e and r_π

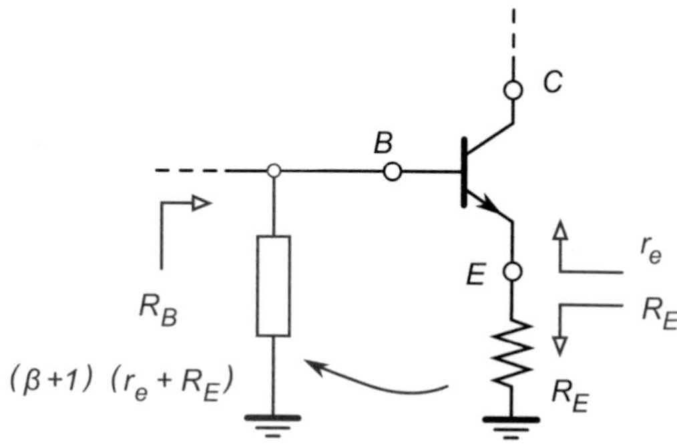

Fig. 4.19 Base resistance that includes "projected" emitter resistance R_E, as perceived by "looking into" technique

which again shows that the base side resistance may be perceived as a "virtual" resistance whose value is the emitter resistance "magnified" by large multiplication factor $(\beta + 1)$. It is to say that even though r_e is relatively small it is multiplied by large β and projected to the base terminal giving it relatively large base resistance. Equally, the relatively large base side resistance r_π is perceived at the emitter node as a small resistance r_e because r_π is divided by large β, i.e. is "magnified" by very small multiplication factor $1/(\beta + 1)$.

This "magnification" interpretation of emitter and base resistances is very useful for rapid estimates. What is more, even if there are discrete resistors connected to each of the two sides, this magnification effect is still valid for evaluating the total resistances. For example, in very important case when there is the external discrete resistor R_E connected to the emitter (a.k.a. "degenerated emitter"), Fig. 4.19, first we deduce that there is the total emitter resistance consisting of $R_E + r_e$, then it is projected to the base side as

$$R_B = (\beta + 1)(r_e + R_e) \tag{4.46}$$

which explains relatively large resistance perceived at the base terminal, see Sect. 4.3.8.

4.3.7 Collector Resistance

Voltage/current characteristics in Fig. 4.13 show that the collector current is very stable over wide range of V_{CE} voltages, from $V_{CE}(min)$ to V_{CC}, nevertheless it is not perfectly constant, Fig. 4.20. By Ohm's law, therefore, the collector resistance parameter $r_o \neq \infty$ implies that I_C current is dependent on V_{CE} voltage. To account for this relatively small but important modulation of collector current by the collector-emitter voltage, we explicitly add term in (4.31) so that the linear dependence of I_C relative to V_{CE} is accounted for as

$$I_C = \underbrace{I_S \left[\exp\left(\frac{V_{BE}}{V_T}\right) - 1\right]}_{I_{C0}=\text{const.}} \underbrace{\left(1 + \frac{V_{CE}}{V_A}\right)}_{I_{C0}=f(v_{CE})}$$

$$\therefore$$

$$\frac{1}{r_o} \stackrel{\text{def}}{=} \left.\frac{\partial I_C}{V_{CE}}\right|_{V_{BE}=\text{const.}} = I_S \left[\exp\left(\frac{V_{BE}}{V_T}\right) - 1\right] \frac{1}{V_A} = \frac{I_{C0}}{V_A} \quad \therefore \quad r_o = \frac{V_A}{I_{C0}} \tag{4.47}$$

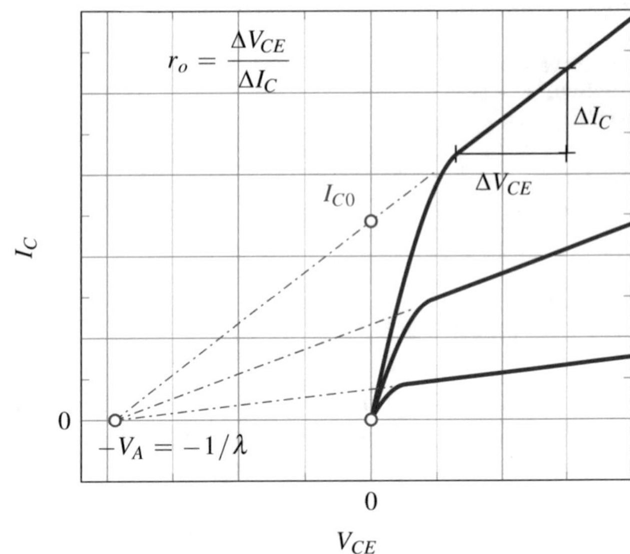

Fig. 4.20 Collector current with exaggerated slopes to illustrate r_o and V_A parameters for various currents I_{C0}

therefore, E:IcVA shows collector resistance r_o as perceived by "looking into" collector, where I_{C0} is the constant collector current level that is not dependent upon V_{CE} voltage. The technological parameter V_A is also known as Early voltage. This parameter is found by extrapolating slopes of several I_C vs. V_{CE} characteristics, where these extrapolated lines converge at one single point, Fig. 4.20 (left), at Early voltage. In conclusion, the collector terminal provides practical implementation of a current source, see Sect. 2.2.4, which implies that higher output resistance r_o translates into better approximation of the ideal current source.

4.3.8 Collector Resistance: "Degenerated Emitter" Case

Realizing that collector terminal of a BJT shows behaviour that is very close to the current source model, natural question is: can we improve this particular aspect of BJT and make even better approximation of an ideal current source? To do so, it is necessary to increase collector's finite resistance r_o so that I_C is less dependent on the voltage V_{CE} at its terminals (in the ideal case $r_o \to \infty$, see Sect. 2.2.4).

Special transistor configuration known as "degenerated emitter" is created by adding the external resistor R_E between BJT emitter and the ground nodes, Fig. 4.21 (left and centre). Addition of this resistor causes significant increase in the perceived "looking into collector" resistance R_C, which is exactly what is required to approach the ideal current source model.

In order to illustrate the achieved increase in collector's resistance, first, we run simulations and show qualitatively, then we develop analytical formulas to quantify this effect. In the case of non-degenerated emitter case, i.e. $R_E = 0$, resistance seen at the collector terminal is estimated by looking at the slope of graphs in Fig. 4.21 (right) as $r_o \approx 100\,\text{k}\Omega$ for this specific transistor. However, as R_E value is increased, so is the collector resistance which in the limiting case $R_E \to \infty$ approaches $R_C \to (\beta + 1)\, r_o$ value.

For the purpose of developing the intuition and better understanding of transistor's internal functionality, here we develop analytical expression for the collector resistance R_{OUT} in the case of degenerated BJT emitter. We already found that resistance at one port of BJT is affected by resistances

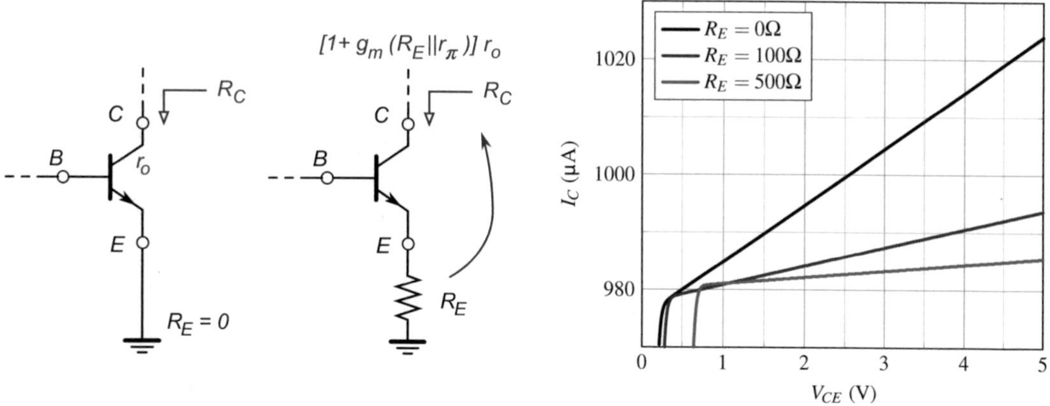

Fig. 4.21 BJT transistor with and without R_E resistor (left), and the collector resistance as the function of R_E, (right)

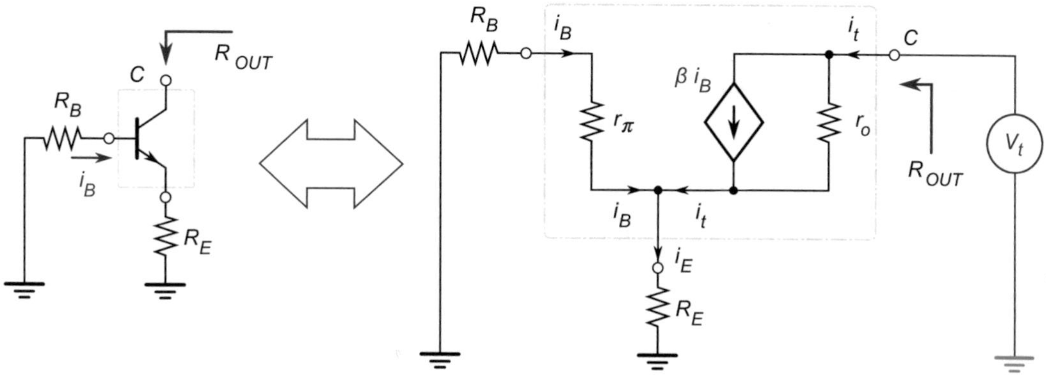

Fig. 4.22 Setup for delivering the collector resistance in the case of degenerated emitter

connected to the other ports (for example, see the relation between r_π and r_e showing the "reflected resistance" as in (4.45)). Thus, we consider transistor setup in Fig. 4.22 (left), where R_B represents the total resistance connected to the base terminal, for example, the equivalent resistance of biasing network and/or the signal source resistance.

For the moment and without loss of generality let us assume $R_E \gg r_e$, thus in this analysis we omit the explicit use of emitter's r_e resistance and we assume R_E to be the total emitter resistance. After replacing BJT with its small signal model, Fig. 4.22 (right) and applying a test signal v_t connected to the collector terminal, we calculate the perceived collector resistance R_{OUT} by definition, i.e.

$$R_{OUT} \stackrel{\text{def}}{=} \frac{v_t}{i_t} \tag{4.48}$$

where schematic in Fig. 4.22 (right) is further reorganized as in Fig. 4.23 to explicitly show serial/parallel relationships among devices, as well as the internal current/voltage relationships.

We note that relative orientation between i_B and v_E is negative, that the serial connection $(R_B + r_\pi)$ appears in parallel with the external resistor R_E, and that the test current i_t is initially split between CCCS and r_o branches, however it recombines again before entering the equivalent R_{eq} serial/parallel restive branch (R_E, r_π, R_B). By inspection of Fig. 4.23 we write the following system of equations

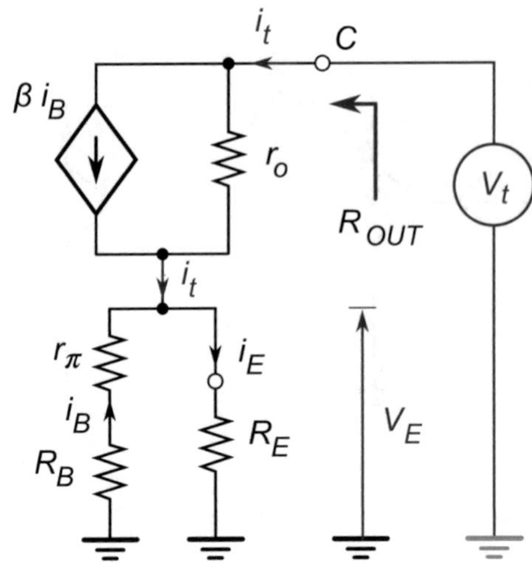

Fig. 4.23 The equivalent schematic for the case of degenerated emitter in Fig. 4.22

$$i_B = -\frac{v_E}{R_B + r_\pi} \quad \text{where} \quad v_E = i_t\, R_{eq} = i_t\, R_E||(R_B + r_\pi)$$

$\therefore$

$$i_B = -\frac{i_t\, R_E||(R_B + r_\pi)}{R_B + r_\pi} = -i_t\,\frac{R_E\,(R_B + r_\pi)}{R_E + R_B + r_\pi}\,\frac{1}{R_B + r_\pi} = -i_t\,\frac{R_E}{R_E + R_B + r_\pi} \tag{4.49}$$

as well as

$$i_t = \beta i_B + \frac{v_t - v_E}{r_o} \tag{4.50}$$

Substitution of (4.49) into (4.50) leads into

$$i_t = \beta\left(-i_t\,\frac{R_E}{R_E + R_B + r_\pi}\right) + \frac{v_t - i_t\, R_E||(R_B + r_\pi)}{r_o}$$

$$\therefore \frac{v_t}{r_o} = i_t\left[1 + \frac{\beta R_E}{R_E + R_B + r_\pi} + \frac{R_E||(R_B + r_\pi)}{r_o}\right]$$

$$\therefore R_{OUT} = \frac{v_t}{i_t} = r_o\left[1 + \frac{\beta R_E}{R_E + R_B + r_\pi} + \underbrace{\frac{R_E||(R_B + r_\pi)}{r_o}}_{\approx 0}\right] \tag{4.51}$$

where the third term in (4.51) is very small and often ignored because $r_o \gg R_E||(R_B + r_\pi)$.[7] The conclusion is that collector resistance depends on resistances connected to the other two ports, namely the signal source resistance, the equivalent biasing network, and emitter resistances.

We show several useful simplifications and approximations derived from (4.51):

[7]The equivalent resistance of multiple parallel resistors is smaller than the resistance of the smallest resistor in the group, while r_o is in general large resistance.

1. *Non-degenerated case,* $R_E = 0$: in this case (4.51) reduces to the already known solution[8]

$$R_{OUT} = r_o \tag{4.52}$$

2. *Ideal voltage source driving the base,* $R_B = 0$, $R_E \neq 0$: this substitution into (4.51) gives

$$\begin{aligned} R_{OUT} &= r_o \left[1 + \frac{\beta R_E}{R_E + r_\pi}\frac{r_\pi}{r_\pi} + \frac{R_E||r_\pi}{r_o}\right] = r_o \left[1 + \frac{\beta}{r_\pi}(R_E||r_\pi) + \frac{R_E||r_\pi}{r_o}\right] \\ &= r_o \left[1 + g_m\,(R_E||r_\pi) + \underbrace{\frac{R_E||r_\pi}{r_o}}_{\approx 0}\right] \qquad (4.53) \\ &\approx r_o\left[1 + g_m\,(R_E||r_\pi)\right] \qquad (4.54) \end{aligned}$$

where we substitute (4.42), and because ($r_o \gg R_E||r_\pi$) we can ignore their ratio in the third term in (4.53). We note that in this case, for the given $R_E||r_\pi$ resistance, R_{OUT} is the linear function of g_m, in other words it is the function of the biasing current. Equation (4.54) implies that by increasing g_m we should be able to indefinitely increase the collector's resistance.

3. *Ideal voltage source driving the base,* $R_B = 0$, $R_E \gg r_\pi$: in this case (4.51) (or equally, (4.54)) simplifies into

$$R_{OUT} \approx r_o \left[1 + g_m\, r_\pi + \cancelto{0}{\frac{r_\pi}{r_o}}\right] = r_o\,(1 + \beta) \tag{4.55}$$

where we keep in mind that the equivalent parallel resistance of small resistor connected in parallel to much larger resistor is approximately same as the small resistance, therefore $R_E||r_\pi \approx r_\pi$ when ($R_E \gg r_\pi$). Subsequently, because $r_o \gg r_\pi$ we ignore their ratio too, which after substituting (4.42) produces the final form (4.55). Equation (4.55) is the limiting case for an "open emitter" circuit, that is to say for unloaded common-collector amplifier ($R_E \to \infty$, $R_L = \infty$). Therefore, (4.55) shows the theoretical limit of BJT's collector resistance.

4. *Ideal voltage source driving the base,* $R_B = 0$, $R_E \ll r_\pi$: in this case (4.51) simplifies into

$$R_{OUT} \approx r_o \left[1 + g_m\, R_E + \cancelto{0}{\frac{R_E}{r_o}}\right] = r_o\,(1 + g_m\, R_E) \tag{4.56}$$

after assuming $r_o \gg R_E$. We note that in this case the collector's resistance is similar to the result found in (4.54), thus same comments in regard to its relation to g_m.

[8]Resistor r_o is in parallel with the ideal current source whose impedance is infinite, see Sect. 2.2.4, thus only r_o value counts as the equivalent resistance.

4.3.9 Brief Summary

For the sake of emphasizing versatility of a BJT transistor, we summarize again its three-terminal resistances. Estimating the three-terminal resistance by "looking into" the transistor ports (i.e. "by inspection") is a very efficient practical technique for simplifying circuits that contain transistors. Subsequently, hand-analysis is much simpler and conclusions can be reached quickly because the transistor itself can be replaced by its equivalent port resistance. Visual interpretation of these three port equivalent terminal resistances is illustrated in Fig. 4.24.

1. *Emitter resistance* r_e: looking into the emitter terminal of BJT, in (4.37) we derived approximate upper limit expression for ideal g_m, which is by definition inverse to emitter's equivalent resistance r_e, as shown in (4.39). Thus,

$$r_e \stackrel{\text{def}}{=} \left. \frac{dV_{BE}}{dI_C} \right|_{(I_{C0}, V_{BE0})} = \frac{1}{g_m} = \frac{V_T}{I_C} \tag{4.57}$$

   where $V_T = k\,T/q$ is approximately 25–26 mV at room temperature.
2. *Base resistance* r_π: looking into the base terminal of BJT, in (4.42) and (4.43) we derived multiple versions of its equivalent resistance r_π:

$$r_\pi = \frac{\beta}{g_m} = (\beta + 1)\, r_e = \frac{V_T}{I_B} \tag{4.58}$$

   A practical way of looking at (4.58) is to imagine that the emitter resistance (in this case only r_e) is "projected" to the base terminal after being *magnified* by β factor to create $r_\pi = (\beta + 1)\, r_e$.

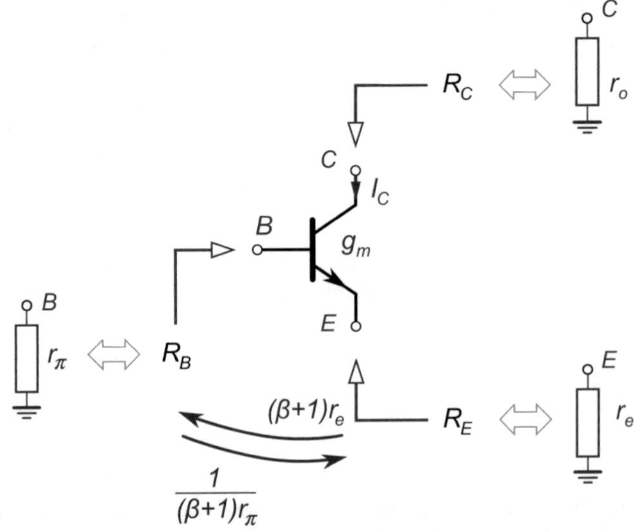

Fig. 4.24 Electrical symbol of a BJT with its main parameters g_m, I_C, and its equivalent port resistances, as interpreted by "looking into" technique

Thus, we note that the base resistance is function of the emitter's *total* resistance, that is to say even if the external resistor R_E is connected to emitter node.

3. *Collector resistance* r_o: looking into the collector terminal of BJT, in (4.51)–(4.55) we derived expressions for r_o resistance. Practical simulation technique in Fig. 4.13, however, is used to determine r_o while the transistor is in its constant current mode as in (4.47). That is, calculation of the equivalent resistance is reduced to calculation of a simple ratio between *change* of collector-emitter voltage relative to the *change* of collector current[9]

$$r_o \approx \left.\frac{\Delta V_{CE}}{\Delta I_C}\right|_{V_{BE}=\text{const.}} = \frac{V_A}{I_{C0}} \tag{4.59}$$

4. *Collector resistance with the degenerated emitter*: We learned that collector terminal shows properties that, if looked at as in the black box, make it appear very similar to ideal current source. One of the characteristic properties of ideal current source is its infinite resistance.[10] In practice, we found that by using the degenerated source technique collector resistance is increased relative to the transistors' r_o resistance as

$$R_{OUT} \approx r_o \left[1 + g_m \left(R_E || r_\pi\right)\right] \tag{4.60}$$

which increases linearly with R_E, however this increase has its limit

$$R_{OUT} = r_o \left(1 + \beta\right) \tag{4.61}$$

when $R_E \gg r_\pi$ and the source generator is ideal.

In this short review we conclude that, after establishing its biasing point, i.e. (I_{C0}, V_{BE0}) in other words its g_m, transistor is then replaced by one of its terminal resistances. In addition, the collector node with its large resistance is a very good implementation of current source (Norton generator), while at the same time the emitter node with its low resistance is a very good model of voltage source (Thévenin generator).

[9]Derivative of a linear function is simply $\Delta y/\Delta x$ ratio.

[10]See Sect. 2.2.4.

Example 33: R_E Magnifying Effect
If a BJT transistor with current gain factor $\beta = 100$ has $R_E = 1\,\text{k}\Omega$ connected between its emitter node and the ground, estimate the input impedance perceived at the input node. For the sake of simplicity, $R_E \gg r_e$ resistance.

Solution 33: By using the magnification effect reasoning, resistance associated with the emitter node, $R_E = 1\,\text{k}\Omega$ is seen from the base node as

$$R_{in} = (\beta + 1)\,(R_E + r_e) \approx \beta\, R_E = 100\,\text{k}\Omega$$

which illustrates how the emitter resistance is projected to the base node.

4.4 MOSFET Transistor

From the functional perspective, a field-effect transistor (FET) is equivalent to a BJT. In its discrete version, it is a three-terminal device (see Fig. 4.25), the three terminals being drain (D), gate (G), and source (S), whose roles are equivalent to the collector, base, and emitter of a BJT.[11] Similarly to a BJT, its main role is to control the flow of current through its drain-source branch.

4.4.1 Mathematical Model

The black box method uncovers transfer functions of a MOS device that are similar to the ones of BJT transistor, Fig. 4.26. In its saturation mode, i.e. when $I_D = \text{const.}$, MOS transistor is also used as a programmable current source whose mathematical model obeys a quadratic function law

$$I_D = \frac{1}{2}\,\mu_n C_{ox}\,\frac{W}{L}\,(V_{GS} - V_t)^2\;(1 + \lambda V_{DS}) = \frac{1}{2}\,\mu_n C_{ox}\,\frac{W}{L} V_{OV}^2\;(1 + \lambda V_{DS}) \tag{4.62}$$

$$I_G = 0 \tag{4.63}$$

$$I_D = I_S \tag{4.64}$$

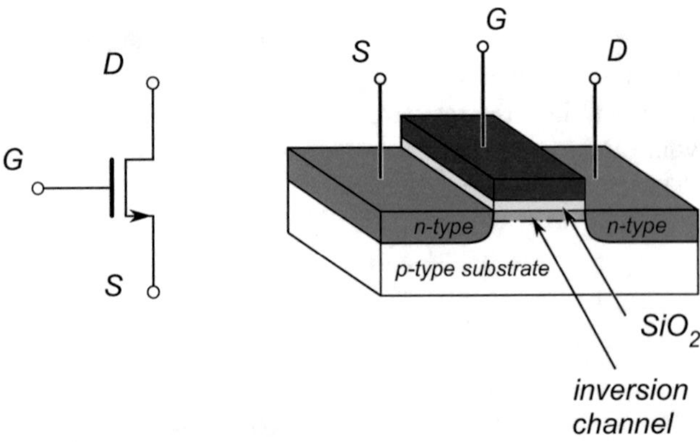

Fig. 4.25 Electrical symbol of a NMOS and its physical geometry

[11] In its IC version, it is a four-terminal device, the fourth terminal being the body, i.e. the substrate.

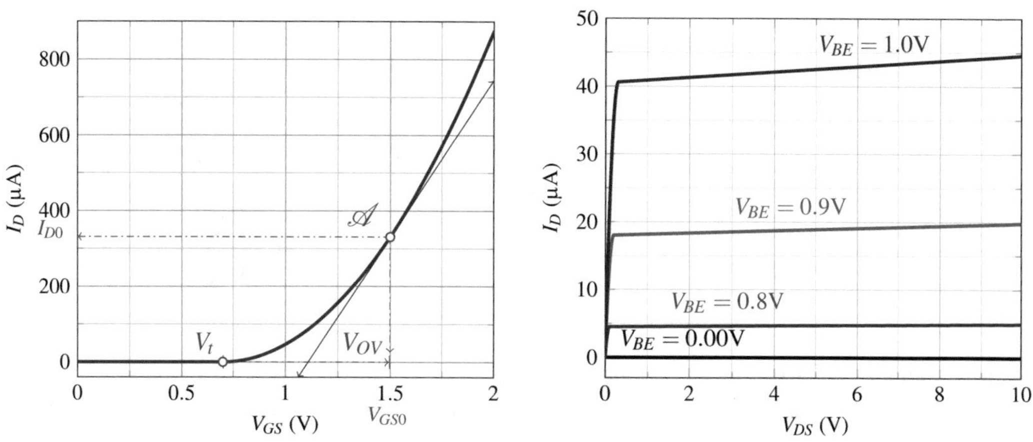

Fig. 4.26 Voltage–current characteristics of a NMOS transistor: I_D vs. V_{GS} (left), and I_D vs. V_{DS} (right)

where

I_D is the drain current
I_G is the gate current
I_S is the source current
μ_n is the mobility of the charges in the induced n channel
C_{ox} is the capacitance of parallel plate capacitor per unit gate area
W is the width of the gate
L is the length of the gate
V_{GS} is the gate-source voltage
V_t is the threshold voltage of NMOS transistor (it is negative for PMOS)
λ is the channel-length modulation parameter, $\lambda = 1/V_A$
V_{OV} is the effective voltage or the overdrive voltage, defined as $V_{OV} \stackrel{\text{def}}{=} V_{GS} - V_t$

Figure 4.26 (left) shows explicitly I_D vs. V_{GS} transfer curve with $V_t \approx 0.7$ V and, for example at the biasing point $\mathscr{A}$: $(I_D, V_{GS}) = (330\,\mu\text{A}, 1.5\,\text{V})$, we find that static $V_{OV} = 1.5 - 0.7\,\text{V} = 0.8\,\text{V}$. Similar to BJT devices, drain current I_D vs. V_{DS} characteristics in Fig. 4.26 (right) shows that there is a minimum $V_{DS}(min)$ voltage required for MOSFET to be in the constant current mode region. Depending on the specific transistor technology, $V_{DS}(min) \approx 100$ mV is similar to its bipolar counterpart.

4.4.2 MOS Small Signal Model

From the black box perspective, functionality of an NMOS transistor is very similar to the one already found by NPN BJT, as the transfer characteristics in Fig. 4.26 clearly illustrate. However, there are several important details typical for MOS transistors that we point out in the following sections.

Major difference between BJT and MOS transistors is found in (4.63) and (4.64). Under normal circumstances, there is no DC current flowing through the gate terminal. By taking a look at the physical structure of a MOS transistor, Fig. 4.25, we find that there is SiO_2 isolation layer between the conductive gate plate above and the conductive inversion layer underneath. In other words,

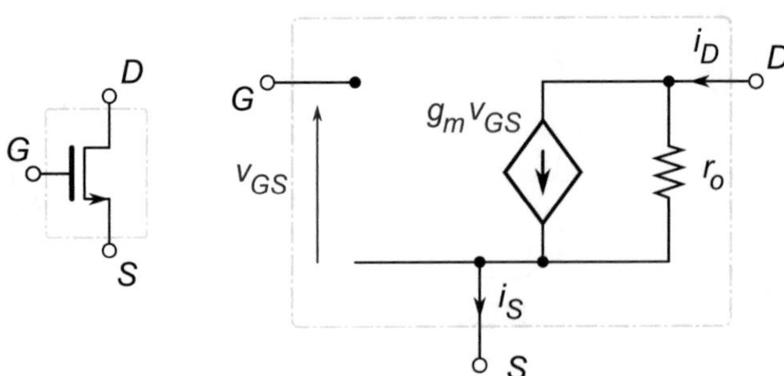

Fig. 4.27 NMOS transistor small signal model based on controlled current source

the gate terminal is structured as a *plate capacitor*, consequently there is no DC current flowing through. In order to emphasize this property of MOS devices, the gate terminal of electrical symbol in Fig. 4.27 (left) is drawn similarly to a capacitor. Zero current is interpreted as an open switch or infinite resistance, thus the small signal model in Fig. 4.27 (right) shows the gate terminal as a "floating" node. There is no equivalent r_π resistance between the gate and source terminal. Therefore, by definition of input/output gain relation,[12] the drain current is found as

$$g_m \stackrel{\text{def}}{=} \left.\frac{dI_D}{dV_{GS}}\right|_{(I_{D0},V_{GS0})}$$

$$\therefore$$

$$i_D = f(g_m\, v_{GS}) \tag{4.65}$$

which is to say that the small signal current i_D is controlled strictly by v_{GS} voltage, there is no current controlled mode as the one related to β factor of BJT devices. In other words, MOS transistors represent almost perfect implementation of a g_m amplifying device. In all other respects, functionality of the two small signal models (BJT and MOS devices) is similar.

4.4.3 Small Signal g_m Gain

Assuming $\lambda = 0$, then g_m gain is obtained by derivative of (4.62) at a given DC point, i.e.

$$\begin{aligned} g_m \stackrel{\text{def}}{=} \left.\frac{dI_D}{dV_{GS}}\right|_{(I_{D0},V_{GS0})} &= \frac{d}{dV_{GS}}\left[\frac{1}{2}\,\mu_n C_{ox}\,\frac{W}{L}\,(V_{GS}-V_t)^2\right] \\ &= \mu_n C_{ox}\,\frac{W}{L}\,(V_{GS}-V_t) = \mu_n C_{ox}\,\frac{W}{L}\,V_{OV} \end{aligned} \tag{4.66}$$

[12]See Sect. 6.1.

where we use $V_{OV} = (V_{GS} - V_t)$ to annotate only the gate voltage above the minimum V_t voltage (see Fig. 4.26), which indicates that, in the given technology, to increase g_m it is necessary to use shorter and wider transistors (i.e. to increase the W/L ratio), in addition to increasing the V_{OV}. With (4.66), assuming $\lambda = 0$, we can rewrite (4.62) as

$$I_D = \frac{1}{2}\,\mu_n C_{ox}\,\frac{W}{L} V_{OV}\,V_{OV} = \frac{1}{2}\,g_m\,V_{OV} \tag{4.67}$$

Another practical expression for g_m is obtained from substituting (4.67) as

$$g_m = \left.\frac{2\,I_D}{V_{OV}}\right|_{I_D=I_{D0}} = \frac{2\,I_D}{V_{GS} - V_t} \tag{4.68}$$

which shows that by setting MOS biasing current I_{D0}, and therefore the overdrive voltage V_{OV}, we in effect set g_m gain of the MOS transistor.

4.4.4 Source Resistance

While in Sect. 4.3.5 we used approximation $i_E \approx i_C$ due to assumption of ($\beta \gg 1$), in the case of MOS transistor there is no need to make similar assumption—drain and source currents are equal, see (4.63) and (4.64), and the gate current is zero.

Even though the current entering gate is zero, current leaving the source is not because it is equal to the drain current (4.64) thus, after (4.65) we write

$$i_S = g_m\,v_{GS} \Rightarrow i_D = g_m\,v_{GS}$$

$$\therefore$$

$$\frac{1}{r_S} \overset{\text{def}}{=} \frac{di_D}{dv_{GS}} = g_m$$

$$\therefore$$

$$r_S = \frac{1}{g_m} \tag{4.69}$$

We note that r_S is not a real resistance between gate and source, instead it is a small signal model of resistance "as seen by looking" into source terminal that enables us to calculate the source current which happens to be $i_s = g_m\, v_{GS}$ as a consequence of property (4.64) of MOS transistors. We found the same result as the approximative value of r_e in the case of BJT, which is to say the functionality of the two types of transistors is practically same.

4.4.5 Drain Resistance

In order to analytically determine output resistance r_o at the drain terminal, it is necessary to include dependence of I_D relative to V_{DS}, as shown in (4.62). Thus, for that purpose it is necessary to use the complete equation (4.62) for I_D that includes $\lambda \neq 0$ parameter.

Dependence of the drain current upon the output voltage V_{DS}, that is to say a finite r_o, is illustrated in Fig. 4.28. Slope of I_D is found as the first derivative of (4.62),

$$\frac{1}{r_o} \stackrel{\text{def}}{=} \left.\frac{dI_D}{dV_{DS}}\right|_{V_{GS}=\text{const.}} = \frac{d}{dV_{DS}}\left[\underbrace{\frac{1}{2}\mu_n C_{ox}\frac{W}{L}V_{OV}^2}_{\text{const}}\,(1+\lambda v_{DS})\right]$$

$$= \left[\frac{1}{2}\mu_n C_{ox}\frac{W}{L}V_{OV}^2\right]\frac{d}{dV_{DS}}(1+\lambda v_{DS})$$

$$= \lambda\,\frac{1}{2}\mu_n C_{ox}\frac{W}{L}V_{OV}^2 = \lambda\, I_{D0}$$

$$\therefore$$

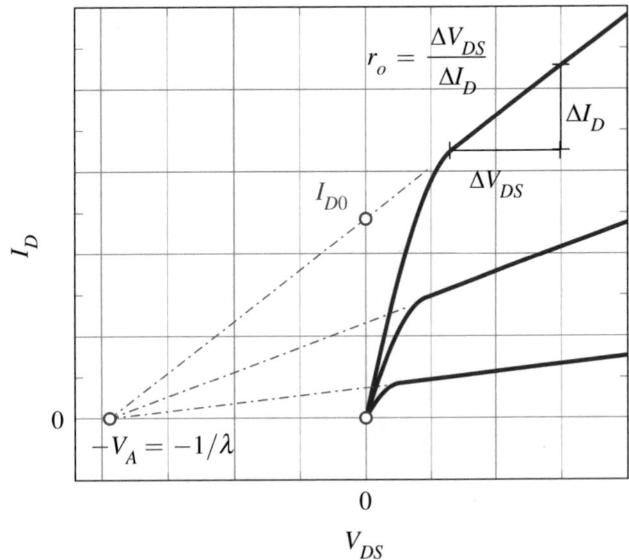

Fig. 4.28 Practical method to estimate r_o is by calculating slope of the drain current in the saturation region. (Slopes in this graph are exaggerated to illustrate the λ parameter)

$$r_o = \frac{1}{\lambda\, I_{D0}} = \frac{V_A}{I_{D0}} \tag{4.70}$$

where I_{D0} is the ideal drain current, i.e. when $\lambda = 0$. Equation (4.70) is easily verified geometrically by the right-angled triangle rules and Fig. 4.28, it is practically same as in BJT case (see (4.47)).

Example 34: NFET—g_m and r_o
Estimate g_m of NMOS transistor whose biasing point is set as $I_D = 1\,\text{mA}$ and $V_{OV} = 1\,\text{V}$. Then, estimate its r_o if $V_A = -100\,\text{V}$.

Solution 34: Direct implementation of (4.68) gives

$$g_m = \left.\frac{2\, I_D}{V_{OV}}\right|_{I_D = I_{D0}} = \frac{2 \times 1\,\text{mA}}{1\,\text{V}} = 2\,\text{mS}$$

Direct implementation of (4.70) gives

$$r_o = \frac{V_A}{I_{D0}} = \frac{|-100\,\text{V}|}{1\,\text{mA}} = 100\,\text{k}\Omega$$

4.4.6 Drain Resistance: "Degenerated Source" Case

Thus, we consider transistor setup in Fig. 4.29 (left), where R_G represents the total resistance connected to the base terminal, for example, the equivalent resistance of biasing network and/or the signal source resistance. Assuming $R_E \gg r_S$, in this analysis we can omit the explicit use

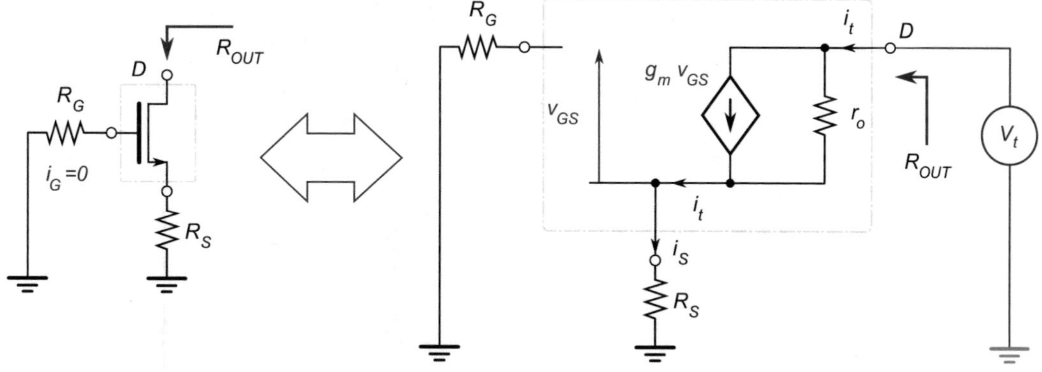

Fig. 4.29 Setup for calculating the drain resistance in the case of degenerated source

of r_S resistance. The first step in this procedure is to replace NMOS with its small signal model, Fig. 4.29 (right). Next, a test signal v_t is connected to the drain terminal, thus the perceived drain resistance R_{OUT} is found by definition (using black-box approach) as the ratio between the test voltage and the test current entering the drain terminal, i.e.

$$R_{OUT} = \frac{v_t}{i_t} \tag{4.71}$$

To make the following analysis more obvious and easier to follow, schematic in Fig. 4.29 (right) is reorganized as in Fig. 4.30 to show explicitly serial/parallel relationships among the resistors, and also relative orientations of the internal currents and voltages.

We note that the test current i_t is initially split between $g_m v_{GS}$ VCCS and r_o branches, however it recombines again before entering R_S branch. Therefore, the initial system of equations is as follows:

$$v_S = i_t\, R_S = -v_{GS} \quad \text{(because gate current is zero)}$$

$$\therefore$$

$$i_t = g_m\, v_{GS} + \frac{v_t - v_S}{r_o} = -g_m\, i_t\, R_S + \frac{v_t}{r_o} - \frac{i_t R_E}{r_o}$$

$$\therefore$$

$$\frac{v_t}{r_o} = i_t \left(1 + g_m R_S + \frac{R_E}{r_o}\right)$$

$$\therefore$$

$$R_{OUT} \stackrel{\text{def}}{=} \frac{v_t}{i_t} = r_o \left(1 + g_m R_S + \frac{R_S}{r_o}\right) \tag{4.72}$$

We show useful simplifications and approximations derived from (4.72) that are found in textbooks:

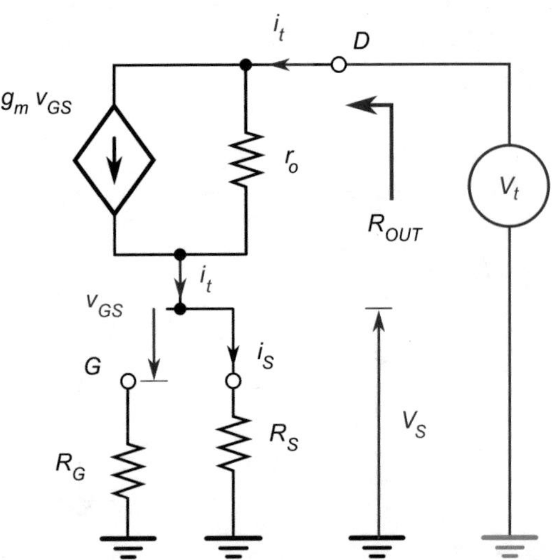

Fig. 4.30 Equivalent schematic for the case of degenerated source in Fig. 4.29

1. *Non-degenerated case,* $R_S = 0$: in this case the output resistance in (4.72) reduces to the drain resistance r_o,

$$R_{OUT} = r_o \tag{4.73}$$

2. *Ideal current source model, i.e.* $r_o \to \infty$ *(or equally* $r_o \gg R_S$*)*: in this case (4.72) simplifies into

$$R_{OUT} = r_o \left(1 + g_m R_S + \cancelto{0}{\frac{R_S}{r_o}} \right) = r_o \, (1 + g_m R_S) \tag{4.74}$$

   where (4.74) implies that, given g_m, drain resistance is linearly increased for larger R_S, which gives us another mean to approach the ideal current source model at the drain terminal. Often, it is assumed that $(g_m R_S \gg 1)$, which allows for $R_{OUT} \approx g_m R_S$ approximation.

4.4.7 Brief Summary

We summarize MOS terminal resistances so that further hand-analysis is simplified and conclusions are reached faster, see Fig. 4.31.

1. *Source resistance* r_S: looking into the source terminal of NMOS, in (4.69) we derived expression for g_m, which is by definition inverse to source equivalent resistance r_S. Thus, in summary

$$r_S = \frac{1}{g_m} \tag{4.75}$$

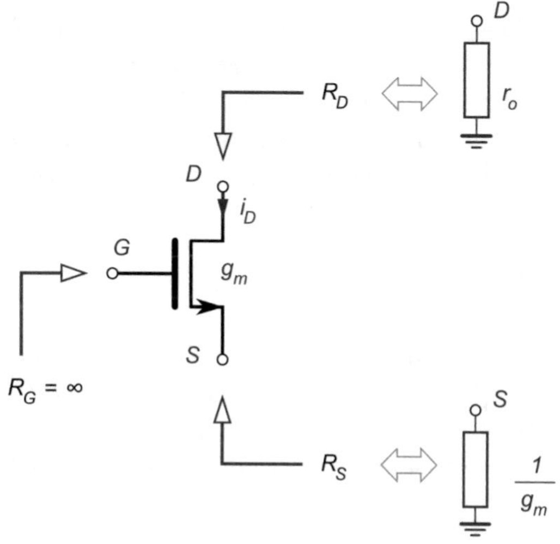

Fig. 4.31 Electrical symbol of a NMOS with its main parameters g_m, i_D, and its equivalent port resistances, as interpreted by "looking into" technique

2. *Gate resistance*: looking into the gate terminal of MOS we found out that, under normal circumstances, DC current entering this port is equal to zero. In other words, we say that the gate resistance of MOS transistor is infinite because its physical structure is equivalent to a capacitor.
3. *Drain resistance* r_o: looking into the drain terminal of MOS, in (4.70) we derived expressions for r_o resistance. Family of functions in Fig. 4.28 is used to determine r_o while the transistor is in its constant current mode. That is, calculation of the equivalent resistance is found by a simple ratio between *change* of drain-source voltage relative to the *change* of drain current.

$$r_o \stackrel{\text{def}}{=} \left.\frac{\Delta V_{DS}}{\Delta I_D}\right|_{V_{GS}=\text{const.}} = \frac{1}{\lambda\, I_{D0}} = \frac{V_A}{I_{D0}} \tag{4.76}$$

4. *Drain resistance with the degenerated source*: In the case of ideal controlled current source its resistance r_o is infinite.[13] Thus, by increasing resistance looking into the drain source we create better approximation of an ideal current source. Equation (4.72) shows that the degenerated source technique increases the drain resistance as

$$R_{OUT} = r_o\ (1 + g_m R_S) \tag{4.77}$$

which in comparison with BJT limitation (4.61) is theoretically not limited but instead linearly increases with the increase of R_S resistance.

4.5 Junction Field-Effect Transistor

Although the junction field-effect transistor (JFET) structure was conceived before the structures presented in the previous sections,[14] technology limitations delayed its realization until years after the other structures. Unlike the other two types of device, there is no p-n junction on the drain-source path. Instead, a JFET is built from a slab of, for example, n-type doped material with metallic contacts (see Fig. 4.32), thus JFETs are exclusively voltage-controlled devices because the gate current is zero.

4.5.1 Small Signal Model

The main difference between JFET and other FET devices is that, when the gate voltage $V_{GS} = 0$, it is said that JFET operates in the "ohmic region" and the depletion layer is very narrow, which means that the JFET behaves as a voltage-controlled resistor. In order to increase the depletion region, the gate potential must be *negative* relative to the source potential.

[13] See Sect. 2.2.4.

[14] US 1745175 Julius Edgar Lilienfeld: "Method and apparatus for controlling electric current" first filed in Canada on 22.10.1925 (original document CA272437 (A)), describing a device similar to a MESFET.

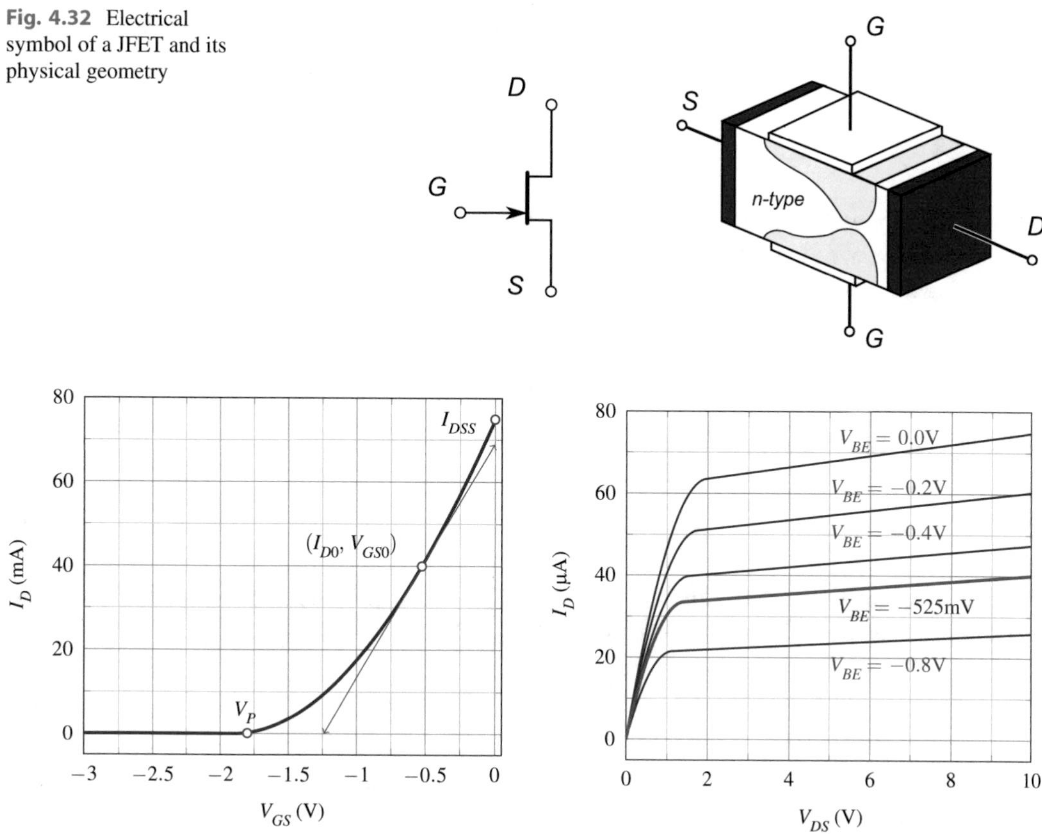

Fig. 4.32 Electrical symbol of a JFET and its physical geometry

Fig. 4.33 JFET $I_D - V_{GS}$ function showing a typical biasing point set at approximately 50% of the I_{DSS} current (left); and $I_D - V_{DS}$ family of curves, where VCCS current set by biasing point in the left graph is shown in red (right)

When in constant current mode, it is said that JFET operates in the "pinch-off region" and it behaves as a VCCS. In pinch-off mode, the relationship of the gate voltage V_{GS} to the drain current I_D is given by

$$I_D = I_{DSS}\left[1 - \frac{V_{GS}}{V_P}\right]^2 \tag{4.78}$$

where V_P is the pinch-off voltage, i.e. the V_{GS} voltage when drain current drops to $I_D = 0$, and I_{DSS} is the drain current when $V_{GS} = 0$, This parabolic relationship is shown in Fig. 4.33 (left). We note that the characteristic curve does not extend in positive direction much beyond $V_{GS} = 0$. This is because a JFET must be operated with a reverse-biased gate, otherwise the gate diode is turned on and (4.78) is not valid any more. For a specific JFET device, its values of I_{DSS} and V_P are set by technology. Then, the parabolic function (4.78) provides a means to calculate the biasing point (I_{D0}, V_{GS0}). In the example given in Fig. 4.33 (left) $I_{D0} = 40\,\text{mA}$ when $V_{DS0} = 525\,\text{mV}$.

The transconductance g_m of a JFET operating in pinch-off region is found by calculating the derivative of (4.78) as

$$g_m \stackrel{\text{def}}{=} \left.\frac{dI_D}{dV_{GS}}\right|_{V_{DS}=\text{const.}} = \frac{2\,I_{DSS}}{|V_P|}\left(1 - \frac{V_{GS}}{V_P}\right) \tag{4.79}$$

which is *linear function* relative to the gate voltage V_{GS}. Note that due to quadratic function, in (4.79) only positive (i.e. absolute value) g_m is used. This JFET linearity is useful in, for example, JFET-based multiplying circuits. In the example given in Fig. 4.33 (left) we estimate g_m as

$$g_m \approx \frac{\Delta I_D}{\Delta V_{GS}} = \frac{68\,\text{mA}}{1250\,\text{mV}} = 54.4\,\text{mS} \tag{4.80}$$

Finite output resistance r_o is found from Fig. 4.33 (right) as

$$r_o = \frac{\Delta V_{DS}}{\Delta I_D} = \frac{V_A}{I_{D0}} \tag{4.81}$$

In the linear region, shown in Fig. 4.33 (right) for $V_{DS} < 1\,\text{V}$, JFET behaves as a voltage-controlled resistor, which is controlled by gate voltage V_{GS} as

$$I_D = \frac{2\,I_{DSS}}{|V_P|}\left(1 - \frac{V_{GS}}{V_P}\right) V_{DS} \tag{4.82}$$

In general, JFET devices are capable of dissipating power, thus they are used in RF circuits where the power transfer is more important than voltage/current gains.

4.5.2 BJT to MOSFET Transistor Comparison

Although they are made to serve essentially the same function, there are several fundamental differences between MOS Field-Effect Transistor (MOSFET) and BJT devices in terms of how the current flow control function is implemented:

1. Current flow in a BJT device is caused by the movement of electrons and holes at the same time (hence, it is "bipolar"). In a FET device, only one type of carrier makes the current, i.e. either electrons or holes, hence a FET device is "unipolar".
2. The current conduction mechanism in a BJT device is based on the principle of injection of minority carriers, while the current conduction mechanism in a FET device is based on the "inversion channel" principle. The minimum gate-source voltage that is needed to create the inversion layer is called the "threshold voltage" V_t. Subsequently, even a small horizontal electric field caused by a voltage difference between the drain and source potentials causes current flow.
3. A BJT device is asymmetrical by design, hence the current always flows from the collector to the emitter. FET devices are symmetrical and the roles of drain and source are determined only by their potentials within the specific circuit for each device separately.
4. A BJT is a "vertical device"—it is manufactured so that the NPN (or PNP) sandwich is vertical relative to the substrate surface. A FET is considered to be a "lateral device"—its conducting inversion layer is parallel to the substrate surface.

5. For all practical purposes a FET gate represents one plate of a *capacitor* (the substrate itself being the second), see Fig. 4.25. In contrast to the base current of a BJT, under normal operational conditions there is no DC flow into the gate.[15] Therefore, while a BJT is considered as a current-amplifying device (i.e. base current in becomes collector current out), a FET device is a true *transconductance device* (i.e. voltage input controls current output).
6. A BJT device has one forward-biased diode (the base–emitter diode). Inside a FET device, however, under normal working conditions all internal diodes are reverse biased.
7. The collector current in a BJT is controlled by the exponential function (4.31). A FET device is controlled by the square function (4.62) between its drain current I_D and its gate voltage V_{GS}.

Once the above differences are accounted for, analysis of circuits using FET devices is, in the first approximation, almost identical to that with BJT circuits.

Example 35: JFET Biasing Point
Given $I_D = 26\,\text{mA}$, give approximate estimates of V_{GS}, V_P, I_{DSS}, g_m, and r_o for a JFET whose characteristics are shown in Fig. 4.33. By using the estimated values, evaluate I_D as per (4.78) and compare estimated g_m with the analytical definition (4.79).

Solution 35: By inspection of graph in Fig. 4.33 (left) we estimate: $I_D = 26\,\text{mA} \rightarrow V_{GS} = 800\,\text{mV}$, as well as $V_P \approx -1.8\,\text{V}$, and $I_{DSS} \approx 75\,\text{mA}$. In order to estimate g_m we draw a tangent at (I_D, V_{GS}) : (26 mA, 800 mV) biasing point, see Fig. 4.34 (left), which enables us to estimate

$$g_m \approx \frac{\Delta I_D}{\Delta V_{GS}} \approx \frac{62\,\text{mA}}{1.35\,\text{V}} \approx 46\,\text{mS}$$

By magnifying graph in Fig. 4.33 (right) as shown in Fig. 4.34 (right) we estimate the output resistance as

$$r_o \approx \frac{\Delta V_{DS}}{\Delta I_D} \approx \frac{8\,\text{V}}{4\,\text{mA}} = 2\,\text{k}\Omega$$

Calculation of I_D as per (4.78) gives

$$I_D = I_{DSS}\left[1 - \frac{V_{GS}}{V_P}\right]^2 = 75\,\text{mA}\left[1 - \frac{-0.8\,\text{V}}{-1.8\,\text{V}}\right]^2 \approx 23\,\text{mA}$$

which is found on the output characteristics in Fig. 4.34 (right), where the maximum value is 26 mA due to the finite output resistance r_0. Similarly, when g_m gain is calculated per (4.79) we find

$$g_m = \frac{2\,I_{DSS}}{|V_P|}\left(1 - \frac{V_{GS}}{V_P}\right) = \frac{2 \times 75\,\text{mA}}{1.8\,\text{V}}\left(1 - \frac{0.8\,\text{V}}{1.8\,\text{V}}\right) = 46.3\,\text{mS}$$

which illustrates applicability of the graph method.

[15] Modern FET devices have a very thin gate layer and, therefore, there is visible "current leakage" which is ignored in the first approximation.

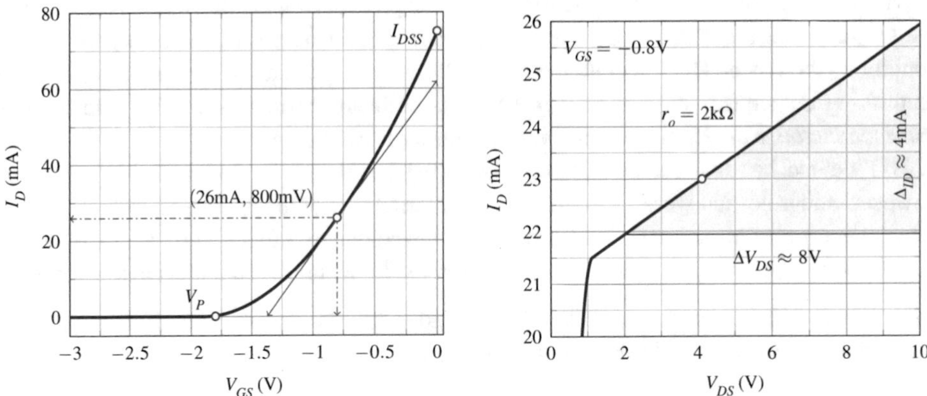

Fig. 4.34 I_D dependence on V_{GS} showing biasing point (left), and zoom-in of I_D dependence on V_{DS} (at the same biasing point) given in Example 35

4.6 Summary

In this chapter, we have reviewed the basic active devices that are used in RF circuit design. This review is by no means complete and thorough; it merely serves the purpose of being a reminder to the reader about the very basic and approximate facts describing the functionality of devices. Detailed treatment of each of the devices mentioned would cover a book similar to this one. The reader is advised to follow the literature and expand on the new concepts learned in this chapter; without knowledge of fundamental device behaviour, any attempt to design an RF circuit is futile.

Problems

4.1 Calculate the current I_D of a typical diode if: (a) the forward biasing voltage is $V_D = 200\,\text{mV}$. Estimate the calculation error in percents if (4.2) is used instead of (4.1); and (b) the forward biasing voltage is $V_D = 70\,\text{mV}$. Estimate the calculation error in percents if (4.2) is used instead of (4.1).

Data: $I_S = 72.2\,\text{nA}$, $T = 28\,°\text{C}$.

4.2 Assuming an ideal diode, i.e. $r_D = 0\,\Omega$ and $V_t = 0\,\text{V}$, in reference to Fig. 4.35, describe the diode's state in terms of "ON" or "OFF" under the following conditions:

(a) $R_3 = R_4$, $R_1 = R_2$;
(b) $R_3 = R_4$, $R_1 = 9R_2$;
(c) $R_3 = R_4$, $R_2 = 9R_1$;
(d) $R_1 = R_2$, $R_3 = 9R_4$;
(e) $R_1 = R_2$, $R_4 = 9R_3$;

What is your conclusion?

4.3 In reference to Fig. 4.35 (right), assuming both diodes D_1 and D_2 have $V_t = 0.6\,\text{V}$, determine the relationship between V_1 and V_{CC} so that:

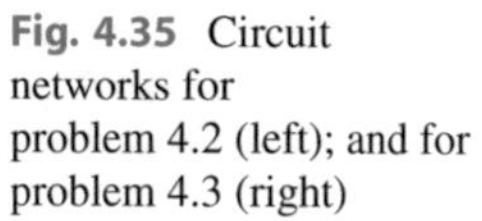

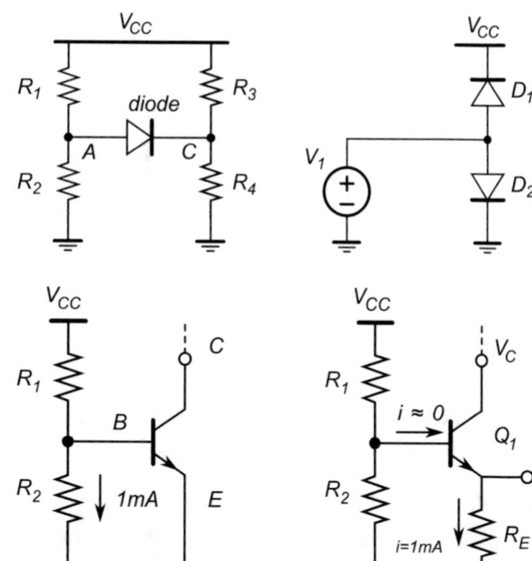

Fig. 4.35 Circuit networks for problem 4.2 (left); and for problem 4.3 (right)

Fig. 4.36 Schematic of a network for Problems 4.7 (left) and 4.8 (right)

(a) D_1 is ON and D_2 is OFF; (b) D_1 is ON and D_2 is ON; (c) D_1 is OFF and D_2 is OFF;

Comment on the proposed solutions.

4.4 A typical diode is connected in series with an ideal current source $I = 1\,\text{mA} \pm 10\%$. Calculate voltage V_D and its variation for the given current source.

Data: $I_S = 72.2\,\text{nA}$, $T = 28\,°\text{C}$.

4.5 Plot the I_D vs. V_{GS} curve for a typical JFET transistor. Estimate its g_m gain if $I_D = I_{DSS}/2$.

Data: $V_P = 4.5\,\text{V}$, $I_{DSS} = 1\text{A}$.

4.6 For a BJT transistor whose $I_S = 5 \times 10^{-15}\text{A}$ at room temperature, the biasing current is $I_C = 1\,\text{mA}$. Calculate its biasing point (I_C, V_{BE}).

(a) Recalculate the collector currents I_C for the range of base–emitter voltages from $V_{BE} = 0.500\,\text{V}$ to $V_{BE} = 0.800\,\text{V}$ in steps of 50 mV.
(b) Calculate g_m gain at each of the V_{BE} voltages.

4.7 For a network shown by schematic diagram in Fig. 4.36 (left),

(a) assuming the base–emitter diode threshold voltage is $V_{th}(BE) = 0\,\text{V}$, i.e. ideal BE diode, find value(s) of R_2 so that the transistor Q_1 is turned on. What potential V_C is required at the collector node C to maintain the saturation mode of operation?
(b) assuming the base–emitter diode threshold voltage is $V_{th}(BE) = 0.7\,\text{V}$, i.e. realistic BE diode, find value(s) of R_2 so that the transistor Q_1 is turned on. What potential is required at the collector node V_C to maintain the saturation mode of operation?

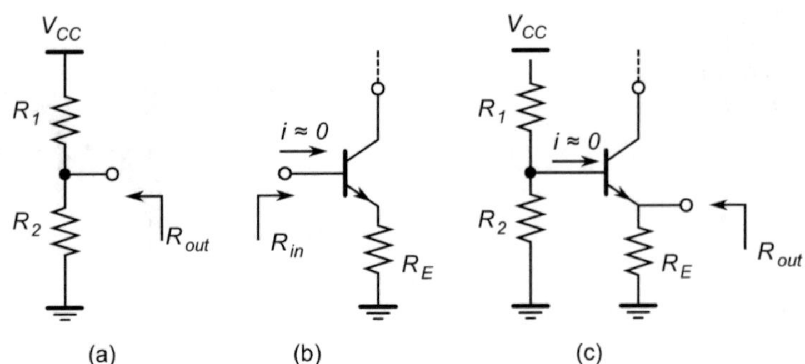

Fig. 4.37 Schematic of a network for Problem 4.9

4.8 What is the required resistor ratio R_1/R_2 for the network in Fig. 4.36 (right), so that the transistor Q_1 operates in saturation, assuming:

(a) the base–emitter diode threshold voltage is $V_{th}(BE) = 0\,\text{V}$, i.e. ideal BE diode, find value(s) of R_2 so that the transistor Q_1 is turned on. What potential is required at the collector node V_C to maintain the saturation mode of operation?
(b) the base–emitter diode threshold voltage is $V_{th}(BE) = 0.7\,\text{V}$, i.e. realistic BE diode, find value(s) of R_2 so that the transistor Q_1 is turned on. What potential is required at the collector node V_C to maintain the saturation mode of operation?

Data: $V_{CC} = 10\,\text{V}$, $R_E = 1\,\text{k}\Omega$.

4.9 Estimate resistances looking into networks (a), (b), and (c), Fig. 4.37.

4.10 Calculate thermal voltage V_T under the following conditions: (a) $T = -55\,°\text{C}$, (b) $T = 25\,°\text{C}$, and (c) $T = 125\,°\text{C}$. Then calculate the diode's current at the room temperature only. (As a side note, these three temperatures are commonly used to characterize military grade electronic equipment.)

Data: $V_D = 200\,\text{mV}$, $I_S = 72.2\,\text{nA}$.

4.11 Assuming the following conditions for the diode voltage V_D: (a) $V_D = 0.1V_T$, (b) $V_D = V_T$, and (c) $V_D = 10V_T$, calculate the diode current I_D at room temperatures.

Data: $I_S = 72.2\,\text{nA}$.

4.12 A simple voltage reference is built using a resistor and a diode as in schematic Fig. 4.38 (left). Calculate voltage across the diode at room temperature.

Data: $V_{CC} = 5\,\text{V}$, $R = 1\,\text{k}\Omega$, $I_S = 1\,\text{fA}$.

4.13 Calculate biasing voltage V_{BE} at room temperature, if BJT collector current is set to $I_C = 1\,\text{mA}$ and $I_S = 100\,\text{fA}$, Fig. 4.38 (right). Repeat the calculations for $I_S = 200\,\text{fA}$.

4.14 Assuming room temperature, estimate unknown collector current I_C, Fig. 4.38 (right), that is required to force biasing voltage $V_{BE} = 768.78\,\text{mV}$ if: (a) $I_S = 100\,\text{fA}$ and (b) $I_S = 200\,\text{fA}$.

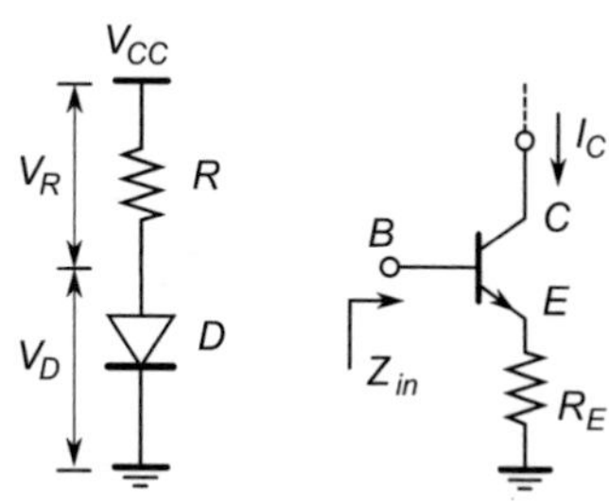

Fig. 4.38 Voltage reference network for problem 4.12 (left); and BJT biasing for problems 4.13–4.17 (right)

4.15 For BJT transistor in Fig. 4.38 (right) estimate biasing voltage V_B required at the base node so that the collector biasing current is set to $I_C = 1\,\text{mA} \approx I_E$.

Data: $I_S = 100\,\text{fA}$, $R_E = 100\,\Omega$, $T = 25\,°\text{C}$.

4.16 Estimate the input impedance Z_{in} looking into the base node, Fig. 4.38 (right), if the forward gain β_F is assumed as: (a) $\beta_F = 99$, and (b) $\beta_F \to \infty$.

Data: $R_E = 100\,\Omega$.

4.17 Design a resistive voltage divider to set the base biasing voltage V_{BE} for circuit in Fig. 4.38 (right), assume: (a) $\beta_F = 99$, and (b) $\beta_F \to \infty$. Calculate the difference in percents between the two V_{BE} solutions (a) and (b).

Data: $R_E = 100\,\Omega$, $V_{CC} = 9\,\text{V}$.

4.18 Calculate transconductance I_C, g_m and the intrinsic emitter resistance r_e of a BJT transistor. What happens to r_e if $I_C = 2, 3\,\text{mA}, \ldots$? Comment on the observed results.

Data: $I_S = 100\,\text{fA}$, $V_{BE} = 768.78\,\text{mA}$, $T = 25\,°\text{C}$.

Transistor Biasing

5

Active devices pose a design challenge due to their non-linear voltage–current characteristics. Inherently, there are two very different resistances that are found at each point of the V–I transfer characteristics, one static and one dynamic. When fixed voltage/current stimulus is applied at the terminals of an active device then by Ohm's law we find static (i.e. DC) resistance at the given point simply by calculating V/I ratio. However, when the input signal varies relative to its average level, resistance to that change is then calculated by using the first derivative mathematical operation. Depending on the specific shape of a non-linear function, this dynamic (i.e. AC) resistance is very much dependent on the function's curvature at the given static point, that is to say on its derivative. In this section we review basic technique to choose and setup DC and AC resistances of active devices, in a process known as the biasing setup.

5.1 The Biasing Problem

Fundamentally, a BJT device behaves as a *current controlled* current source, where the output current at collector is controlled by the input current at the base terminal through $i_C = \beta\, i_B$ relationship, where β is the current gain of the device. In addition, a BJT is used as g_m amplifier, that is to say, for the given input voltage v_{BE} it delivers the output current i_C. Unfortunately, our manufacturing technology is not ideal and, therefore, there are at least three main problems relevant to circuit design:

(a) first, a BJT current gain factor β is not constant, it depends on both device geometry and on the manufacturing parameters. Unavoidably, the two have certain processing variations, which leads to the large variations of current factor β around its nominal value. What is more, β has strong temperature dependance and to a certain extent dependance on the collector current. The final consequence is that the overall realistic gain, which is expected to be *fixed*, if left uncorrected would have large variations, thus would render BJT devices practically useless.

(b) second consequence is that g_m also depends on collector current as well as on temperature. Detailed device analysis shows that base–emitter voltage V_{BE} changes at the rate of $2.5\,\text{mV}/^\circ\text{C}$. On the other hand, high-reliability electronic circuits must satisfy, for example, military and space standard which is specified within the environmental temperature range of $T_1 = -55\,^\circ\text{C}$ to $T_2 =$

R. Sobot, *Wireless Communication Electronics*,
https://doi.org/10.1007/978-3-030-48630-3_5

+125 °C. Unavoidably, this wide temperature range of $\Delta T = T_2 - T_1 = 180\,°\text{C}$ causes the biasing V_{BE} voltage to change by $\Delta V_{BE} = 180\,°\text{C} \times 2.5\,\text{mV}/°\text{C} = 450\,\text{mV}$. At the same time, thermal voltage V_T changes as

$$V_T(T_2) = \frac{kT_2}{q} = 34.31\,\text{mV} \quad \text{and} \quad V_T(T_1) = \frac{kT_2}{q} = 18.8\,\text{mV}$$

As a consequence, taking a typical BJT transistor with (for example) $V_{BE}(T_2) = 0.925\,\text{mV}$ and $V_{BE}(T_1) = 0.475\,\text{mV}$, while assuming constant saturation current I_S, leads into

$$I_{C2}(T_2) \approx I_S \exp\frac{V_{BE2}}{V_{T2}} \quad \text{and} \quad I_{C1}(T_1) \approx I_S \exp\frac{V_{BE1}}{V_{T1}}$$

$$\therefore$$

$$\frac{I_{C2}(T_2)}{I_{C1}(T_1)} = \exp\left(\frac{V_{BE2}}{V_{T2}} - \frac{V_{BE1}}{V_{T1}}\right) \approx 5.4$$

which directly translates into the overall gain variations. Conclusion is that BJT is not much useful as an amplifier without some form of the external mechanism to stabilize the current.

(c) third, the combination of component aging, leakage currents in active devices, and other secondary effects of the integrated circuit technology amounts to a non-consistent and unpredictable variation of the current gain, which must also be compensated by the external mechanism.

Obviously, setting up a stable biasing current is first and foremost task in the circuit design process that must be done. In the following sections we review basic techniques for setting biasing point of transistors, then we briefly compare relative stability of these techniques.

5.1.1 Biasing Point Setup

To setup biasing point of a non-linear device means that a fixed voltage/current pair (V_0, I_0) is chosen based on the targeted specifications of the circuit about to be designed. To do so, first, transistor must be characterized for its input and output side I_{DC} vs. V_{DC} characteristics, see Fig. 5.1.

Modern low-power IC works well with biasing currents in order of microamperes, while discrete transistor works well with biasing currents in order of milliamperes. Practical approach is to choose the collector current as a "nice" round number, for example "1,2,3, ..." or "10, 20, 30, ..." of basic units, so that the subsequent calculations are simplified (biasing current shows up directly or indirectly in almost all design equations). The choice of biasing point is driven by the need for a specific

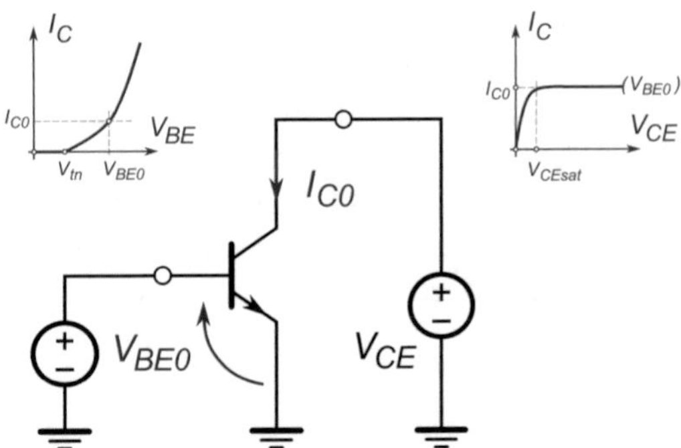

Fig. 5.1 NPN BJT biasing point setup by using two voltage sources. The input side graph shows the position of the biasing point (V_{BE0}, I_{C0}). The output side graph shows that, for the given V_{BE0}, $I_{C0}(V_{CE})$ is a constant function under condition of $V_{CE} \geq V_{CE}(min)$

g_m gain. Once the biasing collector current I_{C0} is chosen, then the transfer function in Fig. 4.12 determines the required biasing V_{BE0} voltage and, at the same time, g_m gain at this particular biasing point (I_{C0}, V_{BE0}). Finally, knowing g_m the emitter's AC resistance is also univocally determined as $r_e = 1/g_m$.

We note that, from the perspective of a non-linear device, it is the external circuitry that controls its biasing point—not the device itself. For example, in the literal implementation, the biasing circuit consists of two external voltage sources that are used to directly control the input and output terminals of transistor, Fig. 5.1. In most applications, transistors are biased so that the collector current is constant, i.e. "saturation" for MOS and "active" mode for BJT. The constant current mode of operation is achieved by simultaneously setting up the following two conditions:

1. Emitter diode (source in the case of MOS) is *forward* biased, i.e. set to conduct current, thus "low-resistance" mode that is modelled by r_e looking into emitter (the diode is ON).
2. Collector diode (drain in the case of MOS) is *reverse* biased, i.e. set to block current, thus "high-resistance" mode that is modelled as r_o when looking into collector (the diode is OFF).

A "back-to-back diodes" model is a practical method to visualize these two conditions for setting up the constant current mode of a transistor, Fig. 5.2. We keep in mind though that although this model is very useful to summarize the two conditions for biasing a transistor in the constant current mode, it is not completely accurate. Details of the actual physical mechanisms for transistor operation are found in books on semiconductors physics.

The two diodes are controlled separately by two external voltages. For example, in the case of NPN transistor, to turn on the emitter diode it is necessary to set $V_B \geq V_E + V_t$ (this is first condition of BJT's active mode), and at the same time in order to keep the collector diode OFF it is necessary to set $V_C \geq V_B - V_t$ (this is second condition). That is, to keep the collector diode OFF, potential at the collector's diode cathode may be lowered as low as V_t below its anode that is at V_B potential (which is the condition for any diode to start conducting). In the case of PNP transistor, the two conditions are symmetrical, Fig. 5.2 (right). In the first approximation, the same back-to-back diode models are also valid in the cases of NMOS and PMOS transistors.

5.1.1.1 Collector–Emitter Saturation Voltage $V_{CE}(min)$

Another way to look at the two conditions for setting up the biasing point of N-type transistor is to note that the output side characteristics, see, for example, Fig. 5.1 (right), shows that the constant collector current I_{C0} that is "requested" by the input side $V_{BE} \geq V_t$ voltage (the first condition) is delivered only if the collector–emitter voltage is above certain minimum level

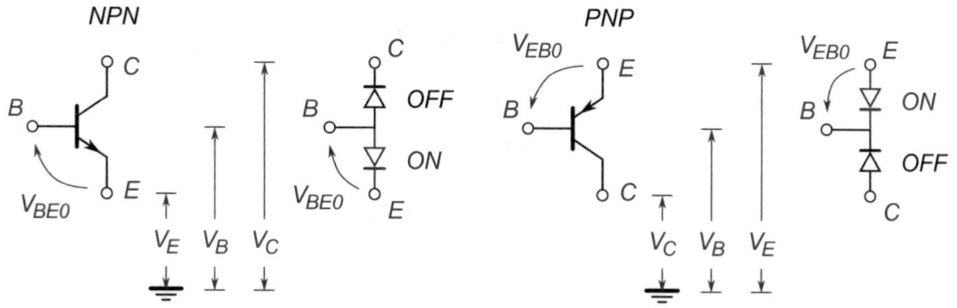

Fig. 5.2 Back-to-back diode models for NPN (or NMOS) (left), and for PNP (or PMOS) transistors (right), which illustrate the two necessary conditions to set NPN transistor in the constant current mode of operation

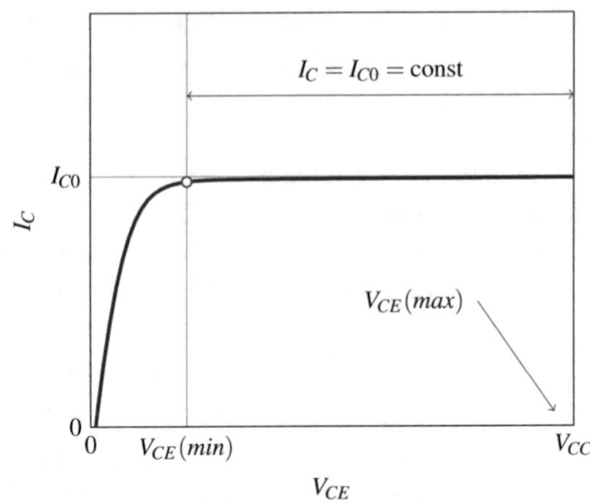

Fig. 5.3 NPN transistor collector current characteristics

$$V_{CE} \geq V_{CE}(min) \tag{5.1}$$

which is the second condition of setting up the constant current mode, Fig. 5.3. This technological parameter is referred to as the "saturation voltage" $V_{CE}(min)$, and depending on the specific transistor it is approximately in $0.1\,\text{V} \leq V_{CE}(min) \leq 0.2\,\text{V}$ range for modern transistors. Important consequence of this limit is that we must always verify if condition (5.1) is satisfied or not, by calculating the collector–emitter voltage V_{CE} that is imposed by the external circuit. The upper limit $V_{CE}(max)$ for linear circuit (e.g. linear amplifiers) is set by the power supply voltage, which is not case for non-linear circuits (e.g. oscillators and other circuits containing LC components).

Therefore, question is how to implement practical circuit to replace the ideal model in Fig. 5.1 so that the desired biasing point is set. In the following sections, practical evolution of a typical biasing network for NPN and PNP transistors is shown. While the current mode biasing setup is applicable only to BJT transistors (due to its non-zero base current), both MOS and BJT transistors employ same techniques for design of biasing circuits, thus they will not be repeated.

5.1.2 Voltage Divider Biasing Technique

Literal implementation of (4.31), where the collector current I_C is controlled by V_{BE}, requires two ideal voltage sources, see Fig. 5.1. Although it does provide the ideal conditions, main practical drawback of this literal implementation is that it requires two large and expensive voltage sources (e.g. batteries): one source to provide V_{BE0} that controls the collector current, and one source to provide V_{CE} and to serve as the energy source for the large collector current. Indeed, in practice the voltage source that controls V_{BE0} is replaced with a simple resistive voltage divider (see Sect. 3.2.1). We calculate the voltage divider ratio as

$$V_{BE0} = I_{R_2} R_2 = \frac{V_{CE}}{R_1 + R_2} R_2 = V_{CE} \frac{1}{1 + \dfrac{R_1}{R_2}} \tag{5.2}$$

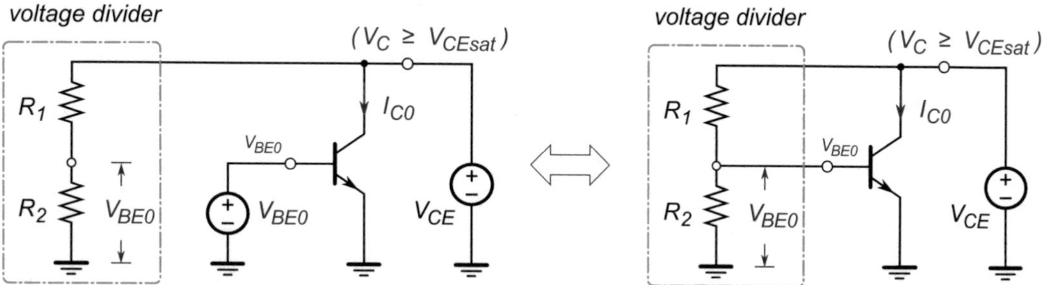

Fig. 5.4 Evolution of the biasing network required to setup a NPN device

where it is the *ratio* R_1/R_2 that controls V_{BE0} voltage—not their respective absolute values. Once the voltage divider (5.2) is designed, see Fig. 5.4 (left), it can safely replace V_{BE0} source, Fig. 5.4 (right). This technique is applicable to both BJT and MOS transistors.

Example 36: Voltage Divider Biasing Technique
For BJT in Fig. 5.4 (right), calculate the biasing voltage (V_{BE0}).

Data: $V_{CE} = 10\,\text{V}$, $R_1 = 143.85\,\text{k}\Omega$, $R_2 = 10\,\text{k}\Omega$.

Solution 36: Direct implementation of (5.2), after neglecting the base current, results in

$$V_{BE0} = V_{CE}\,\frac{1}{1+\dfrac{R_1}{R_2}} = 10\,\text{V}\,\frac{1}{1+\dfrac{143.85\,\text{k}\Omega}{10\,\text{k}\Omega}} = 10\,\text{V} \times 65 \times 10^{-3} = 650\,\text{mV}$$

Also, we always verify that $V_{CE} > V_{CE}(min)$, which in this case is true because $10\,\text{V} \gg 100\,\text{mV}$, thus we conclude that BJT is in the constant current mode.

Example 37: Voltage Divider Biasing Technique—Analysis
For BJT transistor in Fig. 5.4 (right), and assuming $V_{CE} = 10\,\text{V}$, $V_{BE0} = 0.650\,\text{V}$, design the required voltage-divider biasing circuit. Then, calculate variation of V_{BE0} if discrete resistors are used that have standard industrial values and $\pm 10\%$ tolerance.

Solution 37: Direct application of (5.2) results in

$$1 + \frac{R_1}{R_2} = \frac{V_{CE}}{V_{BE0}} \Rightarrow \frac{R_1}{R_2} = \frac{V_{CE}}{V_{BE0}} - 1 \quad \therefore \quad \frac{R_1}{R_2} = \frac{10\,\text{V}}{0.650\,\text{V}} - 1 = 14.385$$

Without additional constrains, this ratio is set by choosing, for example, $R_2 = 1\,\text{k}\Omega$ which results in $R_1 = 14.385\,\text{k}\Omega$. Later in this book we introduce additional constraints for the two voltage divider resistors. The standard discrete resistor values with $\pm 10\%$ tolerance are based on these numbers (10, 12, 15, 18, 22, 27, 33, 39, 47, 56, 68, 82)[1] followed by the exponent.

(continued)

[1]Search the external resources for this standard.

Therefore, $1 \times 10^3\,\Omega \pm 10\%$ is a standard resistor value, while the number that is closest to $14.385\,\text{k}\Omega$ is $R = 15 \times 10^3\,\Omega \pm 10\%$. Recalculation of (5.2) results in biasing voltage of

$$V_{BE0} = 10\,\text{V}\,\frac{1}{1+\dfrac{15\,\text{k}\Omega}{1\,\text{k}\Omega}} = 625\,\text{mV} \tag{5.3}$$

Thus, if made of standard resistors, this biasing circuit reduces V_{BE0} by 25 mV relative to the initial value of 650 mV. By following up Example 30, we find that the change $\Delta V_{BE0} = 25\,\text{mV}$ at room temperature would results in

$$\Delta V_{BE} = V_T \ln x \Rightarrow \ln x = \frac{\Delta V_{BE}}{V_T} \quad \therefore \quad x = \exp\left(\frac{\Delta V_{BE}}{V_T}\right) = \exp\left(\frac{25\,\text{mV}}{25\,\text{mV}}\right) = e = 2.718\ldots$$

that is, after substituting the standard resistor values the actually realized biasing current is increased 2.7 times relative to the ideal case. In addition, after accounting for the $\pm 10\%$ resistor variations, we find the extreme variations by calculating minimum and maximum[2] of (5.3) as

$$V_{BE0}(min) = 10\,\text{V}\,\frac{1}{1+\dfrac{15\,\text{k}\Omega + 10\%}{1\,\text{k}\Omega - 10\%}} = 517\,\text{mV};\; V_{BE0}(max) = 10\,\text{V}\,\frac{1}{1+\dfrac{15\,\text{k}\Omega - 10\%}{1\,\text{k}\Omega + 10\%}} = 753\,\text{mV}$$

That is, after accounting for worst case scenario with standard $\pm 10\%$ resistor tolerances, the actual biasing voltage variance is $\Delta V_{BE0} = 753\,\text{mV} - 517\,\text{mV} = 236\,\text{mV}$. Thus, the actually achieved biasing voltage is $V_{BE0} = 625 \pm 118\,\text{mV}$. Furthermore, accounting for this variation, the collector biasing current may change by the factor of $x \approx 12 \times 10^3$. In addition, we must account for the power supply, temperature and β factor variations and the component aging.

In conclusion, this example illustrates that the design of biasing networks must be undertaken with appreciation for all approximations and assumptions made in the initial "ideal" design. Nevertheless, in practice, the voltage divider reference is used very often for electronic circuits that are not intended for high precision applications. Otherwise, more expensive low-tolerances components must be used, including the custom "trimming" technique used in IC designs.

5.1.3 Base-Current Biasing Technique

Literal implementation of (4.30), where the collector current I_C is controlled by I_B through β parameter, requires one current (I_B) and one voltage (V_{CE}) source. However, the current source may be implemented by a single resistor. Thus, it is possible to further simplify biasing circuit in Fig. 5.4 (right) by eliminating R_2 resistor and using R_1 (renamed to R_{BB}) to control the base current, Fig. 5.5. With careful setup, V_{CE} and R_{BB} produce desired V_{BE0} biasing voltage. Consequently, from the transistor's perspective nothing changed relative to the biasing technique introduced in Sect. 5.1.2.

We note that voltage across R_{BB} is generated by the base current I_B, which is the only current in this branch. Thus, DC voltage equation is written by inspection

[2] A rational function increases when its denominator decreases, and vice versa.

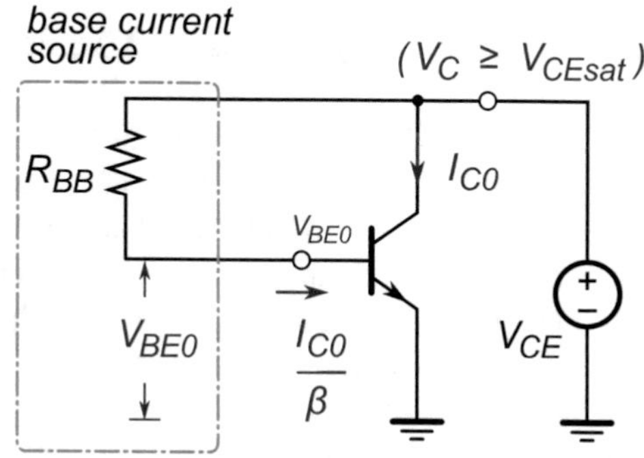

Fig. 5.5 Biasing technique of BJT transistor that controls the base current

$$V_{CE} = V_{R_{BB}} + V_{BE0} = R_{BB}\, I_{B0} + V_{BE0}$$

$$\therefore$$

$$R_{BB} = \frac{V_{CE} - V_{BE0}}{I_{B0}} \tag{5.4}$$

where

$$I_{C0} = \beta\, I_{B0} \qquad \therefore \qquad I_{B0} = \frac{I_{C0}}{\beta} \tag{5.5}$$

In this biasing technique, it is assumed that β of BJT is a priori known, which determines the relationships between all three terminal currents. In this interpretation, a BJT transistor is physical realization of Kirchhoff's current node: two currents enter the node (I_C and I_B) and one current exists (I_E). In other words, $i_E = i_C + i_B = \beta i_B + i_B = (\beta + 1) i_B$, where the input current i_b is multiplied by current gain factor β to generate the output collector current.

Example 38: Biasing Circuit—Base-Current Technique
In reference to Fig. 5.5, given $\beta = 100$, $V_{CE} = 10\,\text{V}$, calculate R_{BB} that is required to setup biasing point (650 mV, 1 mA). Then replace the initial ideal resistance value with the classiest 10% standard value and evaluate the actually implemented biasing point.

Solution 38: Direct implementation of (5.5) and (5.4) results in

$$I_{B0} = \frac{I_{C0}}{\beta} = \frac{1\,\text{mA}}{100} = 10\,\mu\text{A} \Rightarrow R_{BB} = \frac{V_{CE} - V_{BE0}}{I_{B0}} = \frac{10\text{V} - 0.65\,\text{V}}{10\,\mu\text{A}} = 935\,\text{k}\Omega$$

Although there are various possibilities to implement 935 kΩ resistance, for the sake of argument let us use two 470 kΩ resistors in series to create $R_{BB} = 940\,\text{k}\Omega \pm 10\%$ resistor. Thus, in order to achieve the required $I_{C0} = 1\,\text{mA}$, according to (5.4) we calculate the equivalent V_{BE0} voltage as

$$V_{BE0} = V_{CE} - I_{B0}\, R_{BB} = 10\,\text{V} - 10\,\mu\text{A} \times 940\,\text{k}\Omega = 600\,\text{mV}$$

Similar to the voltage divider biasing technique in Example 37, biasing technique based on the base current setup also shows similar sensitivity to the resistor values.

5.2 Sensitivity of Biasing Circuits

In the review of biasing techniques so far we assumed that voltage sources are constant, as well as β parameter of BJT devices. In reality, that is not the case, as introduced in Sect. 5.1, and it is important to evaluate *sensitivity* S_x^y of these biasing techniques to the variations of circuit and environment parameters. Detailed studies of sensitivity relative to various environmental and device parameters are found in the literature, here we briefly illustrate sensitivity of basic biasing techniques relative to voltage supply variations and β parameter.

Calculation of sensitivity S_x^y gives answer to the following question: if the input variable changes x percents,[3] by how many percents the output variable y changes? Sensitivity calculation is given in the form of ratio, thus the answer $S_x^y = 1$ is interpreted as, for example, 10% increase in x causes 10% increase of y, that is, both input and output variable change by the same percentage. Result $S_x^y = 0$ indicates that y is not dependent on x variable. Formal definition of this ratio is written as

$$S_x^y = \frac{\frac{dy}{y}}{\frac{dx}{x}} = \frac{x}{y}\frac{dy}{dx} \tag{5.6}$$

which is widely used in science and engineering. A sensitivity analysis enables us to identify variables with significant impact to the overall circuit functionality. In modern circuit design flow we use numerical simulations to estimate the overall distribution of various specifications by assigning extreme variance to the individual components. This approach is known as "corner analysis" and it is main tool that enables us to do preliminary cost analysis.

Example 39: Biasing Circuit—Voltage Divider Sensitivity
Calculate $S_{V_{BE0}}^{V_{CE}}$ for a voltage divider reference.

Solution 39: Using result (5.2), for a given ratio R_1/R_2 by definition (5.6) we write

$$\frac{dV_{BE0}}{dV_{CE}} = \frac{d}{dV_{CE}}\left(V_{CE}\,\frac{R_2}{R_1+R_2}\right) = \frac{R_2}{R_1+R_2} \stackrel{(5.2)}{=} \frac{V_{BE0}}{V_{CE}}$$

$$\therefore$$

$$S_{V_{CE}}^{V_{BE0}} = \frac{\frac{dV_{BE0}}{V_{BE0}}}{\frac{dV_{CE}}{V_{CE}}} = \frac{V_{CE}}{V_{BE0}}\frac{dV_{BE0}}{dV_{CE}} = \frac{\cancel{V_{CE}}}{\cancel{V_{BE0}}}\frac{\cancel{V_{BE0}}}{\cancel{V_{CE}}} = 1$$

This result is important because it shows that voltage divider generates voltage reference V_{BE0} whose variation directly follows the variation of voltage source that is used. In accordance to (4.34) this ΔV_{BE0} variation then translates to collector current ΔI_{C0} variation. We keep in mind that standard commercial power supply voltages vary $VCC \pm 10\%$, which directly results in $V_{BE0} \pm 10\%$, which may not be acceptable for more demanding applications.

[3]Recall definition of percentage calculation, $\Delta x\,[\%] = (\Delta x)/x \times 100$.

Example 40: Biasing Circuit—Sensitivity to β variations
Calculate $S_{I_{C0}}^{\beta}$ for biasing technique in Fig. 5.5.

Solution 40: After substituting (5.4) into (5.5) we write

$$I_{C0} = \beta \, \frac{V_{CE} - V_{BE0}}{R_{BB}}$$

which leads into

$$\frac{dI_{C0}}{d\beta} = \frac{d}{d\beta}\left(\beta \, \frac{V_{CE} - V_{BE0}}{R_{BB}}\right) = \frac{V_{CE} - V_{BE0}}{R_{BB}} \overset{(5.4)}{=} \frac{I_{C0}}{\beta}$$

$$\therefore$$

$$S_{\beta}^{I_{C0}} = \frac{\dfrac{dI_{C0}}{I_{C0}}}{\dfrac{d\beta}{\beta}} = \frac{\beta}{I_{C0}} \, \frac{dI_{C0}}{d\beta} = \frac{\beta}{\cancel{I_{C0}}} \, \frac{\cancel{I_{C0}}}{\beta} = 1$$

This result is important because it shows that in this technique variation I_{C0} directly follows the variation of transistor's β factor. We keep in mind that commercial BJT transistors have β that is very strong function of process (i.e. large spread among components) and the temperature.

Example 41: Biasing Circuit—Voltage Divider Sensitivity Due to β Variations
Calculate $S_{I_{C0}}^{\beta}$ for voltage divider biasing technique in Fig. 5.4 (right).

Solution 41: To find expression for collector current, first, circuit Fig. 5.4 (right) is transformed into its equivalent version using Thévenin generator (Fig. 5.6)
where $R_{Th} = R_1 || R_2$ and $V_{Th} = V_{CE} \, R_2/(R_1 + R_2)$. After this transformation, by inspection we write

$$V_{Th} = V_{R_{Th}} + V_{BE0} = I_{B0} \, R_{Th} + V_{BE0} = \frac{I_{C0}}{\beta} \, R_{Th} + V_{BE0}$$

$$\therefore$$

(continued)

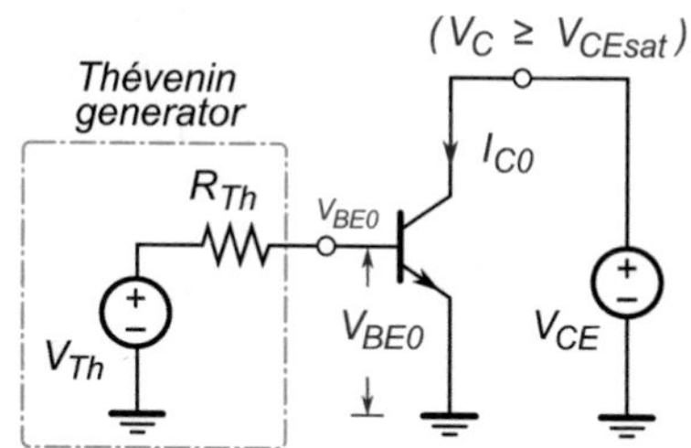

Fig. 5.6 BJT with Thévenin equivalent of voltage divider reference (Example 41)

$$I_{C0} = \beta \frac{V_{Th} - V_{BE0}}{R_{Th}}$$

which leads into

$$\frac{dI_{C0}}{d\beta} = \frac{d}{d\beta}\left(\beta \frac{V_{Th} - V_{BE0}}{R_{Th}}\right) = \frac{V_{Th} - V_{BE0}}{R_{Th}} = \frac{I_{C0}}{\beta}$$

$$\therefore$$

$$S_{\beta}^{I_{C0}} = \frac{\dfrac{dI_{C0}}{I_{C0}}}{\dfrac{d\beta}{\beta}} = \frac{\beta}{I_{C0}}\frac{dI_{C0}}{d\beta} = \frac{\beta}{\cancel{I_{C0}}}\frac{\cancel{I_{C0}}}{\beta} = 1$$

This result shows that sensitivity $S_{\beta}^{I_{C0}}$ of voltage divider biasing technique is as same as the sensitivity for base-current technique, as shown in Example 40.

In conclusion, a voltage reference derived from voltage source through means of passive resistors directly follows variations of the voltage source itself. In addition, variations of the components themselves, i.e. resistance, and β factor, as well as process and temperature variations present significant challenge to the designer who must apply some compensation techniques to stabilize the reference voltage and currents.

5.3 Stabilization of Biasing Current by "Degenerated Emitter" Technique

In order to rectify some of the biasing point sensitivity issues pointed out in the previous sections, let us take a look at "degenerated emitter" version of BJT circuit, Fig. 5.7. Distinct feature of this topology is that as oppose to emitter being connected directly to ground, there is the external R_E added between the emitter and ground terminals. As a consequence, the potential of emitter terminal is elevated to $V_E = I_E R_E$.

Using the circuit version in Fig. 5.7 (right) we write by inspection

$$V_{Th} = V_{R_{Th}} + V_{BE0} + V_E = I_{B0} R_{Th} + V_{BE0} + I_{E0} R_E = \frac{I_{C0}}{\beta} R_{Th} + V_{BE0} + \frac{\beta + 1}{\beta} I_{C0} R_E$$

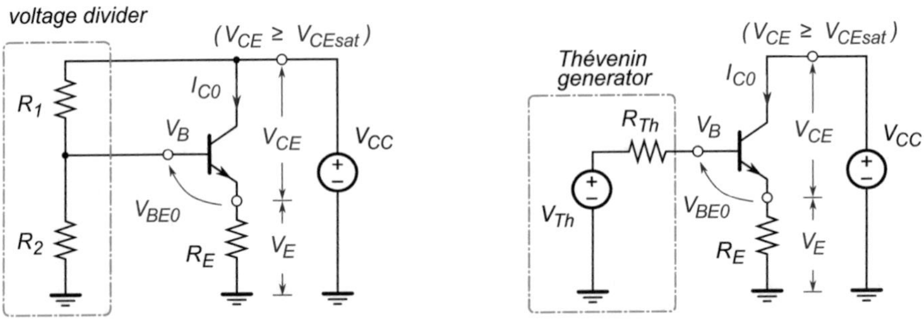

Fig. 5.7 "Degenerated emitter" with voltage divider (left) and its equivalent Thévenin circuit (right)

$\therefore$

$$I_{C0} = \frac{V_{Th} - V_{BE0}}{\dfrac{R_{Th}}{\beta} + \dfrac{\beta + 1}{\beta} R_E} = (V_{Th} - V_{BE0}) \frac{\beta}{R_{Th} + (\beta + 1) R_E}$$

which leads into

$$\frac{dI_{C0}}{d\beta} = (V_{Th} - V_{BE0}) \frac{d}{d\beta} \left(\frac{\beta}{R_{Th} + (\beta + 1) R_E} \right) = (V_{Th} - V_{BE0}) \frac{R_{Th} + R_E}{(R_{Th} + (\beta + 1) R_E)^2}$$
$$= \frac{I_{C0}}{\beta} \frac{R_{Th} + R_E}{R_{Th} + (\beta + 1) R_E}$$

$\therefore$

$$S_{\beta}^{I_{C0}} = \frac{\dfrac{dI_{C0}}{I_{C0}}}{\dfrac{d\beta}{\beta}} = \frac{\beta}{I_{C0}} \frac{dI_{C0}}{d\beta} = \frac{\cancel{\beta}}{\cancel{I_{C0}}} \frac{\cancel{I_{C0}}}{\cancel{\beta}} \frac{R_{Th} + R_E}{R_{Th} + (\beta + 1) R_E} < 1 \tag{5.7}$$

because (R_{Th}, R_E, β) are positive numbers and $(\beta \gg 1)$, thus the denominator in (5.7) is greater than its numerator. The conclusion is that relative to the previous biasing techniques where $S_{\beta}^{I_{C0}} = 1$, degenerated emitter technique is always less sensitive, thus preferred option.

What is more, from (5.7) we can say that

$$S_{\beta}^{I_{C0}} = \frac{R_{Th} + R_E}{R_{Th} + (\beta + 1) R_E} \frac{\beta}{\beta} = \frac{\dfrac{R_{Th}}{\beta} + \dfrac{R_E}{\beta}}{\dfrac{R_{Th}}{\beta} + \dfrac{(\beta + 1)}{\beta} R_E} \tag{5.8}$$

Assuming $(\beta \gg 1)$ it follows that

$$\frac{(\beta + 1)}{\beta} \approx 1 \Rightarrow \frac{R_{Th}}{\beta} + \frac{(\beta + 1)}{\beta} R_E \approx \frac{R_{Th}}{\beta} + R_E$$

then, if

$$\frac{R_{Th}}{\beta} \ll R_E \Rightarrow \frac{R_{Th}}{\beta} + R_E \approx R_E$$

which, for $(\beta \gg 1)$, leads into

$$S_{\beta}^{I_{C0}} \approx \frac{\cancelto{0}{\dfrac{R_{Th}}{\beta}} + \cancelto{0}{\dfrac{R_E}{\beta}}}{R_E} \approx 0 \tag{5.9}$$

which gives us important guideline for design of biasing network that is independent of β variations. Under condition

$$\frac{R_{Th}}{\beta} \ll R_E \tag{5.10}$$

we conclude that, when degenerated emitter technique and modern transistors are used, the sensitivity $S_{\beta}^{I_{C0}}$ can be practically eliminated. A reminder, in practice "much smaller" means ten times or more.

In summary of this section, in order to achieve maximum voltage gain we are motivated to rely on emitter resistance r_e that is the minimal resistance we can have at the emitter node. Subsequently, voltage gain is set by ratio of collector resistance and r_e. However, large temperature dependance of a diode resistance r_e translates into biasing current variations. Practical solution is to use external $R_E \gg r_e$ in series so that influence of r_e on the biasing current is reduced. Subsequently, the external C_E is used to bypass R_E for AC signals, and therefore restore the maximum voltage gain.

Example 42: Biasing Circuit—Degenerated Emitter Technique
For BJT in Fig. 5.7, design biasing network to set biasing point to (650 mV, 1 mA).

Data: $V_{CC} = 10\,\text{V}, \beta = 100, V_{CEsat} = 0.1\,\text{V}, R_E = 1\,\text{k}\Omega$.

Solution 42: By inspection of Fig. 5.7, we find DC voltage at the base as

$$V_B = V_{R_E} + V_{BE} = R_E\,I_E + V_{BE} = R_E\,\frac{\beta+1}{\beta} I_C + V_{BE}$$
$$= 1\,\text{k}\Omega\,\frac{100+1}{100} \times 1\,\text{mA} + 650\,\text{mV} = 1.660\,\text{V}$$

The voltage divider reference needed to generated V_B is found from (5.2) as

$$\frac{R_1}{R_2} = \frac{10\,\text{V}}{1.660\,\text{V}} - 1 = 5.024 \approx 5$$

To minimize the influence of β variations, by following (5.10) and $R_{th} = R_1||R_2$, we write

$$\frac{R_{Th}}{\beta} \ll R_E \quad \therefore \quad \frac{R_1||R_2}{\beta} \ll R_E \quad \therefore \quad \frac{R_1\,R_2}{R_1+R_2} \ll \beta\,R_E$$

which gives us the following system of equations:

$$R_1 = 5\,R_2 \quad \text{and} \quad \frac{R_1\,R_2}{R_1+R_2} \ll 100\,\text{k}\Omega \quad \therefore \quad R_2 \ll \frac{6}{5}\,100\,\text{k}\Omega$$

Following the rule of thumb that much larger is ten times or more, it is reasonable to choose a standard value $R_2 = 10\,\text{k}\Omega$, which impels $R_1 = 50\,\text{k}\Omega$. These two values give $R_{th} = 8.33\,\text{k}\Omega$, which after substituting in (5.8) results in the sensitivity of

$$S_{\beta}^{I_{C0}} = \frac{\dfrac{R_{Th}}{\beta} + \dfrac{R_E}{\beta}}{\dfrac{R_{Th}}{\beta} + \dfrac{(\beta+1)}{\beta} R_E} = \frac{\dfrac{8.33\,\text{k}\Omega}{100} + \dfrac{1\,\text{k}\Omega}{100}}{\dfrac{8.33\,\text{k}\Omega}{100} + \dfrac{(100+1)}{100}\,1\,\text{k}\Omega} = 85.3 \times 10^{-3} \approx 9\%$$

(continued)

that is to say that the variation of collector current is almost ten times lower than the β factor variation, which illustrates validity of (5.8) and (5.9).

The introduction of R_E, depending on the emitter current, introduces a possibility that V_E voltage is raised too high and violated the condition for minimum $V_{CEmin} = 0.1\,\text{V}$. We verify this possibility

$$V_{CC} = V_E + V_{CE} \Rightarrow V_{CE} \approx V_{CC} - R_E\, I_C = 10\,\text{V} - 1\,\text{k}\Omega \times 1\,\text{mA} = 9\,\text{V} > 0.1\,\text{V}$$

and confirm that the transistor is still operated in the constant current mode.

Example 43: Biasing Circuit—I_{C0} Temperature Dependance
Using a typical BJT transistor biased as in Fig. 5.4, then its "degenerated emitter" version as in Fig. 5.7 sets both biasing currents to $I_{C0} = 1\,\text{mA}$. Compare the temperature dependance of the two collector's current by SPICE simulation. Use the standard industrial temperature range interval, i.e. ($-40\,°\text{C}$, $+80\,°\text{C}$). Then, compare with the case of $I_{C0} = 5\,\text{mA}$.

Solution 43: SPICE simulations demonstrate the effect of two fundamental relations (4.31) and (4.35), with and without emitter resistor, for two values of collector current.

Adding R_E in series with the emitter resistance r_e shows that instead of the exponential dependance the collector current moderately and linearly changes as $I_C \approx I_{C0} \pm 200\,\mu\text{A}$ for the two biasing currents at room temperature, Fig. 5.8 (left).

In addition, a larger R_E is beneficial because the approximation $R_E + r_e \approx R_E \neq f(T)$ is better Fig. 5.8 (right). Here, "better" means higher output resistance, i.e. better current source.

This example illustrates multiple compromises and trade-offs that are possible during the design process. If even tighter voltage dependance is required by the final application, then additional more sophisticated temperature compensation techniques must be used.

(continued)

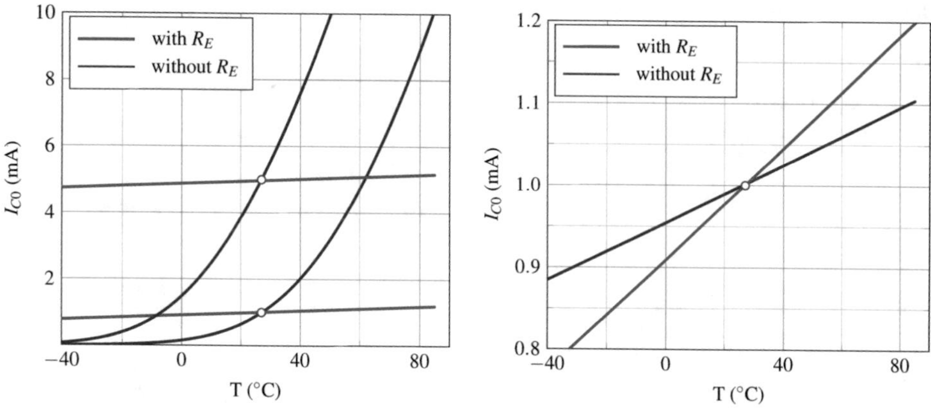

Fig. 5.8 Temperature dependance of I_D, with and without emitter resistor (Example 43)

Again, we keep in mind that the collector current increases with the increase of the temperature, which is opposite to what (4.31) implies. This trend is due to very strong temperature dependance of I_s current.

5.4 The Collector Resistance R_C

So far, in our discussion of biasing circuits we omitted collector resistance (a.k.a. "load") R_C because the biasing current is set by the input side resistive network and there is a wide range of V_{CE} voltages available. Therefore, by setting the load resistance to $R_C = 0$ we calculate biasing point (I_{C0}, V_{BE0}) under the condition $V_{CE} > V_{CE}(min)$. Then, ideally the collector current is constant over wider range of V_{CE} including $V_{CE} = V_{CC}$, Fig. 5.9.

We now discuss the second condition for setting BJT into its constant current mode (the first condition was to set up $V_{BE} \geq V_t$, see Fig. 4.12), namely the condition that

$$V_{CE} \geq V_{CE}(min) \tag{5.11}$$

where $V_{CE}(min)$ is a technological parameter of a transistor. In reference to the back-to-back diode model, see Fig. 5.2, if (5.11) is not satisfied then collector–base diode starts to conduct and the transistor crosses the edge of constant current mode, therefore it enters the linear region mode of operation, i.e. $V_{CE} \leq V_{CE}(min)$, see Fig. 5.9.

Therefore, the voltage–current characteristics in Fig. 5.9 show that the load resistance can have values within the interval $R_C(max) \geq R_C \geq 0$. We now find $R_C(max)$ that defines the edge of the linear mode region and still permits the constant current mode of operation. By inspection of Fig. 5.9 (left) we write

$$V_{CC} = V_{R_C} + V_{CE} = I_{R_C}\, R_C + V_{CE} = I_{C0}\, R_C + V_{CE} \tag{5.12}$$

which is valid as long as (5.11) is satisfied, in other words I_{C0} = const and $R_C < R_C(max)$. In the boundary case, when $V_{CE} = V_{CE}(min)$, we write

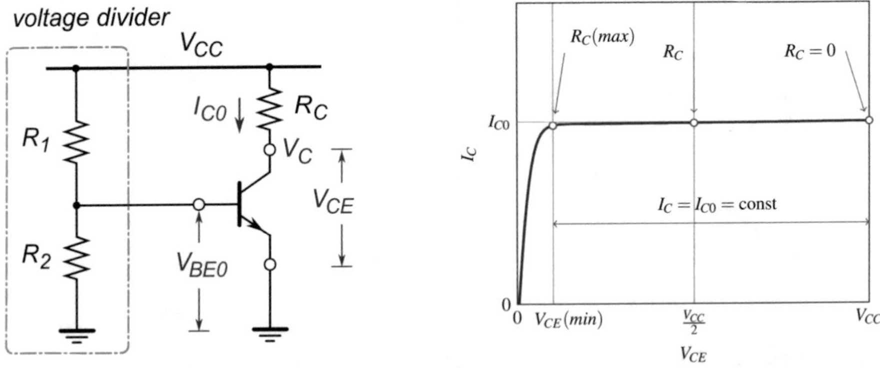

Fig. 5.9 BJT transistor with its biasing setup and R_C (left), and its constant collector current characteristics (right)

$$V_{CC} = I_{C0}\, R_C(max) + V_{CE}(min)$$

$$\therefore$$

$$R_C(max) = \frac{V_{CC} - V_{CE}(min)}{I_{C0}} \tag{5.13}$$

which is referred to as “maximal load” that permits the constant current mode. In practice, the initial value of R_C is chosen so that the collector voltage is $V_C \approx V_{CC}/2$, as shown in Fig. 5.9 (right). This condition allows for maximal voltage swing at the collector node (between $V_{CE}(min)$ and V_{CC}) while I_{C0} is kept constant.

Example 44: Biasing Circuit—$R_C(max)$ Calculation
Assuming that biasing current of transistor in Fig. 5.9 is set to $I_{C0} = 1\,\text{mA}$, given that $V_{CC} = 10\,\text{V}$, $V_{CE}(min) = 0.1\,\text{V}$, find the range of values of resistor R_C.

Solution 44: Direct implementation of (5.13) gives maximum R_C as

$$R_C(max) = \frac{V_{CC} - V_{CE_{sat}}}{I_{C0}} = \frac{10\,\text{V} - 0.1\,\text{V}}{1\,\text{mA}} = 9.9\,\text{k}\Omega$$

therefore, the load resistance must be $0 \leq R_C \leq 9.9\,\text{k}\Omega$. As discussed earlier, “optimal” load is calculated at $V_{CE} = V_{CC}/2$, that is

$$R_C = \frac{V_{CC} - V_{CC}/2}{I_{C0}} = \frac{10\,\text{V} - 5\,\text{V}}{1\,\text{mA}} = 5\,\text{k}\Omega$$

This value of R_C sets DC voltage at collector node as $V_C = V_{CC}/2$ that allows a sinusoidal output voltage to have maximum possible amplitude to swing up and down, while the transistor is still in the constant current mode.

5.5 BJT Biasing

Regardless what type of transistor is used, polarization of biasing voltages must follow the same idea: in the case of BJT transistors, base–emitter diode must be conducting (forward biased) and at the same time collector–base diode must be turned off (reverse biased). In addition, collector–emitter voltage must be kept above the minimum $V_{CE}(min)$ voltage, see Fig. 5.9. In the case of FET transistors, the equivalent relationships are required for gate–source, drain–gate, and drain–source voltages. By providing these biasing conditions we setup a transistor in its constant current mode of operation. Last point to note is that the polarity of biasing voltages for NPN and NFET devices is opposite relative to the one required by their PNP and PMOS counterparts. One practical way to see the biasing voltage orientation is to use the ground as reference for N type devices, and positive power supply rail V_{CC} as the reference level for P type devices.

5.5.1 Biasing Setup for a Single N-Type Transistor Amplifier

Details of more complete and advanced biasing techniques are found in the literature, nevertheless the understanding of techniques presented so far is considered mandatory. Again, only circuits based on transistors in the constant current mode are used in this book. Modern and more advanced analog design techniques for low-power IC are subject of advanced courses.

With the introduction of load resistor R_C we have completed the basic practical setup of a single transistor, Fig. 5.10. Note that, although the voltage divider biasing technique is shown, when R_2 is removed this biasing setup morphs into the base-current biasing technique. Circuit in Fig. 5.10 (right) is *static*, i.e. there is no "signal" injected into any of the three transistor terminals. As is, until we decide which terminal is the "input" and which terminal is the "output" there is no "amplifier".

Example 45: Biasing Circuit—$R_C(max)$ Calculation
Assuming that biasing setup of transistor in Fig. 5.10 is as same as in Example 44, with the addition of $R_E = 1\,\text{k}\Omega$, calculate the maximum value of resistor R_C.

Solution 45: Addition of R_C does not influence the biasing point calculations that are already done in Example 44, but this time we write the voltage equation in the collector–emitter branch as

$$V_{CC} = V_{R_C} + V_{CE} + V_{R_E} \approx R_C\, I_C + V_{CE} + R_E\, I_C \Rightarrow R_C \approx \frac{V_{CC} - V_{CE} - R_E\, I_C}{I_C}$$

$$\therefore$$

$$R_C(max) \approx \frac{V_{CC} - V_{CE}(min) - R_E\, I_C}{I_C} = \frac{10\,\text{V} - 0.1\,\text{V} - 1\,\text{k}\Omega \times 1\,\text{mA}}{1\,\text{mA}} = 8.9\,\text{k}\Omega$$

which forces the collector voltage to

$$V_C = V_{CC} - V_{R_C} = V_{CC} - R_C\, I_C = 10\,\text{V} - 8.9\,\text{k}\Omega \times 1\,\text{mA} = 1.1\,\text{V}$$

(continued)

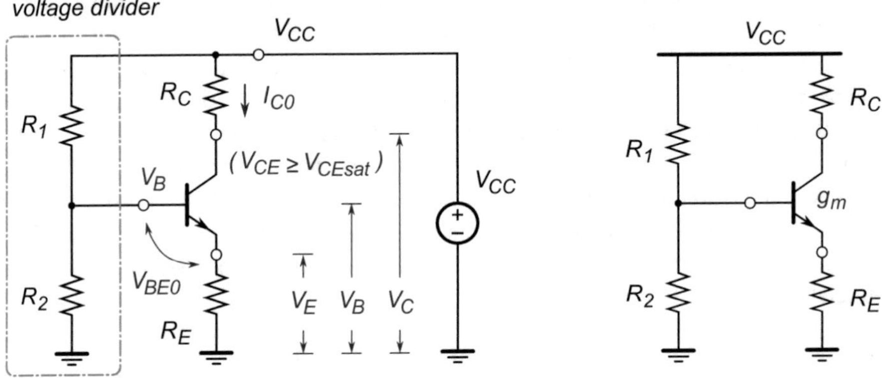

Fig. 5.10 Complete static schematic of NPN BJT biasing setup based on a voltage divider

In this case, the average collector voltage that allows for the maximum signal swing is between 10 V and 1.1 V, that is when $V_{C0} = (10\,\text{V} - 1.1\,\text{V})/2 = 5.55\,\text{V}$. This setup requires that the "optimal" load resistance is set to $R_C = (10\,\text{V} - 5.55\,\text{V})/1\,\text{mA} = 4.45\,\text{k}\Omega$.

5.5.2 Biasing of PNP Type Transistors

For the sake of completeness, we briefly review the biasing schemas used for P-type (PNP BJT or PMOS) transistors, that are symmetrical to the ones used for N-type devices. Figure 5.11 shows the two-voltage-source biasing setup required by P-type devices. We note that, although the graph forms are identical, directions of current and voltages are negative relative to N-type devices. This fact is also reflected by the used notations, e.g. V_{EB} instead of V_{BE} and $V_{EB} = -V_{BE}$. Convenient way to visualize schematics with P-type devices is to use power supply voltage level as the reference, instead of usually used ground level. By doing so, schematics for P-type devices appear as if they were mirrored versions for N-type devices schematics. Accordingly, the network equations are also symmetric relative to the respective plus/minus signs.

Following this line of thinking, the implementation of the voltage divider and base-current biasing techniques result in circuits shown in Fig. 5.12. We note that base current is now directed towards the ground node through R_{BB} resistor.

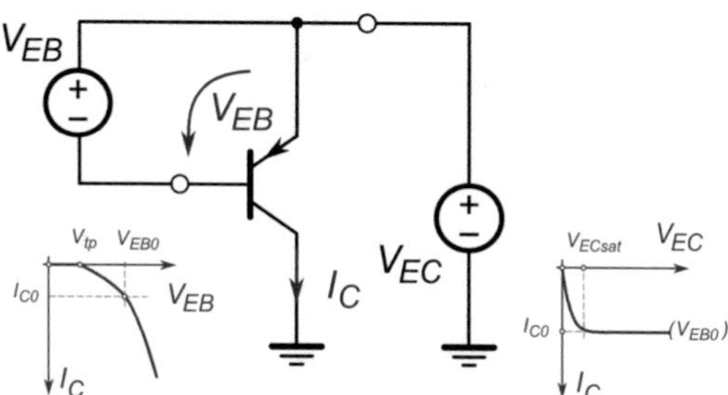

Fig. 5.11 PNP BJT biasing point setup by using two voltage sources. The input side graph shows the position of the biasing point (V_{EB0}, I_{C0}). The output side graph shows that, for the given V_{EB0}, $I_{C0}(V_{EC})$ is a constant function under condition of $V_{EC} \geq V_{ECsat}$

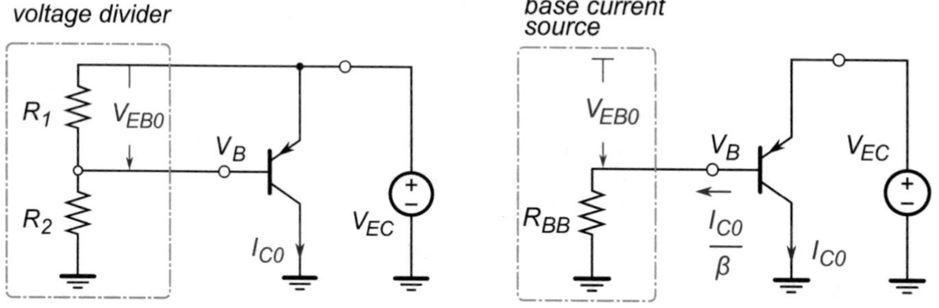

Fig. 5.12 PNP BJT transistor biased using the voltage divider (left) and base-current (right) technique

Example 46: Biasing Circuit—PNP BJT Biasing
In reference to Fig. 5.12 (left), design voltage divider reference to set $V_{EB0} = 650\,\text{mV}$ with $V_{EC} = 10\,\text{V}$. Then, in reference to Fig. 5.12 (right), design base-current reference to setup biasing point (650 mV, 1 mA) given that PNP transistor has $\beta = 100$.

Solution 46: By inspection, after neglecting the base current, we write

$$V_{EC} - V_{EB0} = V_{R_2} = \frac{V_{EC}}{R_1 + R_2} R_2 = V_{EC} \frac{1}{\frac{R_1}{R_2} + 1}$$

$$\therefore$$

$$\frac{R_1}{R_2} = \frac{V_{EC}}{V_{EC} - V_{EB0}} - 1 = \frac{\cancel{V_{EC}} - \cancel{V_{EC}} + V_{EB0}}{V_{EC} - V_{EB0}}$$

$$\therefore$$

$$\frac{R_2}{R_1} = \frac{V_{EC} - V_{EB0}}{V_{EB0}} = \frac{10\,\text{V} - 0.650\,\text{V}}{0.650\,\text{V}} = 14.385$$

Since there are no additional constrains, we choose $R_1 = 1\,\text{k}\Omega$, which forces $R_2 = 14.385\,\text{k}\Omega$. Thus, the voltage divider generates $V_{EB0} = 650\,\text{mV}$ and $V_B = 9.35\,\text{V}$.

By inspection of Fig. 5.12 (right) we write

$$I_{R_{BB}} = I_B = \frac{I_{C0}}{\beta} = \frac{V_B}{R_{BB}} = \frac{V_{EC} - V_{EB0}}{R_{BB}}$$

$$\therefore$$

$$R_{BB} = \beta \frac{V_{EC} - V_{EB0}}{I_{C0}} = 100 \frac{10\,\text{V} - 0.650\,\text{V}}{1\,\text{mA}} = 935\,\text{k}\Omega$$

which sets $V_B = 9.35\,\text{V}$, i.e. $V_{EB0} = 650\,\text{mV}$. We note symmetry of this circuit relative to N-type in Example 37.

5.5.3 Biasing Setup for a Single P-Type Transistor Amplifier

Again, with the introduction of load resistor we have developed the basic practical setup of a single transistor, Fig. 5.13. Relative voltage levels among the three transistor terminals are shown explicitly in Fig. 5.13 (left), the equivalent practical schematic diagram is shown in Fig. 5.13 (right). Although the voltage divider technique is shown, when R_2 is removed it morphs into the base-current biasing technique.

Finally, we note that circuit in Fig. 5.13 (right) is *static*, i.e. there is no "signal" injected into any of the three transistor terminals. For this reason, until we define which terminal in the circuit serves as the "input" and which terminal serves as the "output", we cannot say anything more about this circuit. At this moment, the only certain information that circuit conveys is that the transistor is biased at the certain g_m point. Everything else is the subject of discussion in the following chapter.

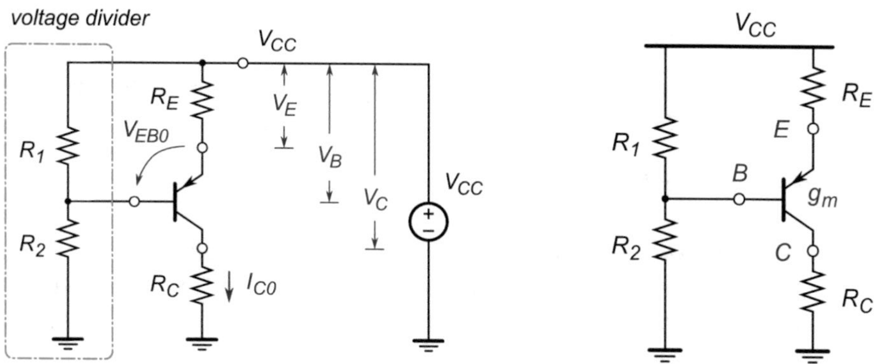

Fig. 5.13 Completed static schematic of PNP BJT biasing setup based on voltage divider and including loading resistor R_C, with and without voltage annotation

5.6 Summary

Setting up DC operating point of active devices (diodes, BJTs, or FETs) is first step in circuit design, which in itself sets their respective gains. The subsequent AC signal analysis is then simplified by omitting details of DC biasing circuit while preserving g_m values of active devices instead. Additionally, in the first approximation, a BJT device is seen either as a current amplifier, where the current gain is β, or g_m amplifier where v_{BE} is the input and i_C is the output variable. Accordingly, a BJT is biased either in voltage-to-current configuration to implement g_m amplification, or in current-to-current mode to implement current amplification. On the other hand, MOS devices can be set only in voltage-to-current configuration to implement g_m amplification.

As a side note, the base–emitter voltage V_{BE} still depends on the temperature, i.e. $V_{BE} = f(T)$. Therefore, in high precision applications, some of the temperature compensation techniques are employed to further stabilize the biasing point, while in the first approximation all circuits are assumed to operate at "room temperature".

Problems

5.1 For a single NPN BJT transistor, draw the schematic symbol and indicate potentials at the three terminals, i.e. the V_C, V_B, and V_E, and their relationship assuming the transistor is turned on, i.e. it is operating in the forward active region. Repeat the exercise using a PNP BJT transistor.

5.2 If V_{BE} voltage of a BJT transistor changes by 18 mV what is change of I_C in expressed in dB? What if V_{BE} changes by 60 mV ? (note: $V_T = kT/q = 25\,\text{mV}$)

5.3 Resistance seen by looking into a BJT emitter is $R_{\text{out}} = 100\,\Omega$. Resistance looking into the base is $R_{\text{in}} = 100\,\text{k}\Omega$. For $\beta = 99$. Calculate the total resistances R_B and R_E at the base–emitter nodes.

5.4 Given a simple CE amplifier with non-generated emitter and resistive load R_C, estimate its voltage gain when (a) $V_{\text{out}} = 7.5\,\text{V}$, (b) $V_{\text{out}} = 5\,\text{V}$, (c) $V_{\text{out}} = 0.2\,\text{V}$. Comment on the calculated results.

Data: $V_{CC} = 10\,\text{V}$, $R_C = 5.1\,\text{k}\Omega$, $V_T = 25\,\text{mV}$.

5.5 Assuming that BJT in set in its constant current mode of operation, estimate voltage gain A_v for circuit in Fig. 5.14 (1). Express the gain in [dB].

Data: $R_C = 10\,\text{k}\Omega$, $C_E = 0\,\text{F}$, $R_E = 1\,\text{k}\Omega$.

5.6 Calculate the voltage gain of circuit in Fig. 5.14 (1). Estimate the voltage gain A_v for: (a) $C_E = 1.6\,\text{pF}$, (b) $C_E = 160\text{nF}$, and (c) $C_E \to \infty$. How large are the gain differences in comparison with the gain calculated in Problem 5.5? Comment on the calculated results.

Data: $f = 10\,\text{MHz}$, $T = 25\,°\text{C}$, $R_C = 10\,\text{k}\Omega$, $R_E = 1\,\text{k}\Omega$, $I_S = 100\,\text{fA}$, and $V_{BE} = 650.6\,\text{mV}$.

5.7 An ideal signal generator is connected to the input terminal of the CE amplifier through a capacitor C, see Fig. 5.14 (2). Assuming large voltage gain and ignoring the base current, estimate the frequency range where this CE amplifier should be used.

Data: $C = 1\,\mu\text{F}$, $R_1 = 2\,\text{k}\Omega$, $R_2 = 2\,\text{k}\Omega$.

5.8 Further to the calculations in Problem 5.6, estimate the Miller capacitance C_M.

Data: $C_{CB} = 1\,\text{pF}$.

5.9 In the amplifier, Fig. 5.14 (2), resistors R_1 and R_2 make Q_1 base voltage divider which should be set such that their current is $I_{R_{1,2}} = 1/10\, I_E$ (for this calculation only ignore the base current) and $V_B = 1/3\, V_{CC}$. At the room temperature, estimate base voltage V_B, R_1, R_2, R_E, I_C, g_m, r_e, R_C and the amplifier's input resistance at the base node if the signal source resistance is R_{sig}.

Data: $A_V = -8$, $V_{CC} = 9\,\text{V}$, $I_E = 2\,\text{mA}$, $\beta = 100$, $V_{BE} = 0.7\,\text{V}$, $R_{\text{sig}} = 10\,\text{k}\Omega$, $C = C_E = \infty$.

5.10 Given data, state the necessary assumptions and estimate the frequency range where CE amplifier in Fig. 5.14 (3) should be used. What would be the consequence if current gain β is not infinite?

Data: $\beta \to \infty$, $L = 2.533\,\mu\text{H}$, $R_C = 9.9\,\text{k}\Omega$, $R_E = 100\,\Omega$, $C_{CB} = 1\,\text{pF}$.

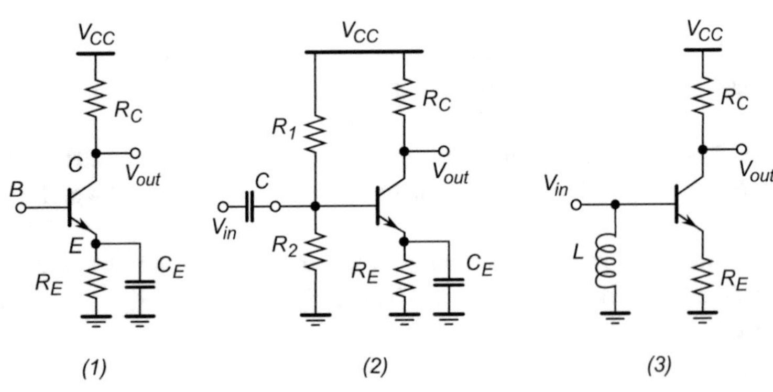

Fig. 5.14 Schematics for Problems 5.5–5.9

5.11 For amplifier, Fig. 5.15 (left), estimate value of the inductor L so that the input stage accepts a signal frequency f_0. Assume base current to be zero.

Data: $A = -99$, $C = 1\,\text{pF}$, $f_0 = 15.915\,\text{MHz}$.

5.12 The circuit shown in Fig. 5.15 (right) is a common-base amplifier. Assuming $\beta \gg 1$, estimate: (a) DC voltage at the collector, (b) $g_m(Q_1)$, (c) AC voltage gain, $A = v_C/v_i$.

Data: $\beta = \infty$, $V_T = 25\,\text{mV}$, $R_C = 7.5\,\text{k}\Omega$, $I = 0.5\,\text{mA}$, $C = \infty$, $V_{CC} = 5\,\text{V}$.

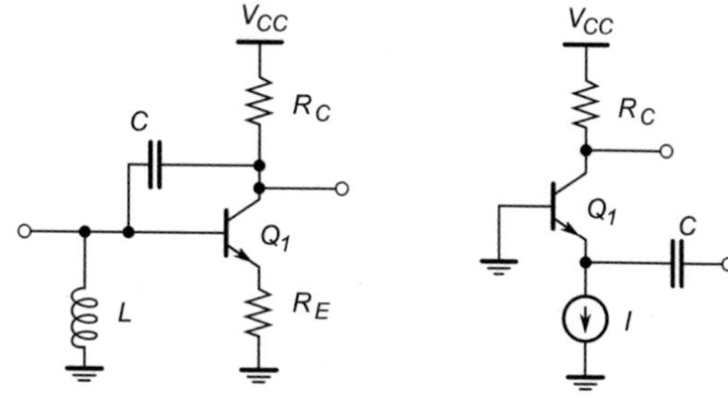

Fig. 5.15 Schematic for Problems 5.11 (left) and 5.12 (right)

Review of Basic Amplifiers

6

After a weak radio frequency (RF) signal has arrived at the antenna, it is channelled to the input terminals of the RF amplifier through a passive matching network that enables maximum power transfer of the receiving signal by equalizing the antenna impedance with the RF amplifier input impedance. Then, it is job of the RF amplifier to increase the power of the received signal and prepare it for further processing. In the first part of this chapter, we review the basic principles of linear baseband amplifiers and common circuit topologies. In the second part of the chapter, we introduce RF and IF amplifiers. Aside from their operating frequency, for all practical purposes, there is not much difference between the schematic diagrams of RF and IF amplifiers. In this book, unless we need to specifically separate the two functions, we refer to all tuned amplifiers as RF amplifiers.

6.1 General Amplifiers

The topic of linear baseband amplifiers is usually covered in introductory undergraduate courses in electronics, thus there is a large number of excellent textbooks available with a thorough treatments of the subject, some of them listed in the reference section. Here, however, we introduce the "back of the envelope" circuit analysis approach with the intent to develop intuition for the circuits and to adopt rapid techniques for circuit analysis. Even though the "back of the envelope" approach is more than often based on *very* rough approximations, and it leads into conclusions that are sometimes order of magnitude off the "correct" numerical solution, its usefulness is in enabling circuit designer to focus on the underlying principles of the circuit operation instead on the fine and obscuring numerical details. As a result, the amount of time spend to reach correct conclusions is often measured in seconds. By practicing rapid circuit analysis the designers, eventually, develop intuition for the underlying principles and ability to spot possible problems, the circuit limitations, or improvements. Indeed, the machines and simulators still *cannot solve problems* and improve existing solutions, they merely produce numbers that may or may not have anything to do with the problem at hand. Until we reach the age of intelligent machines, our brain is still the only tool we have that is capable of creative reasoning.

In addition, we note that the very notion of "correct" answer is more than often fuzzy. The point is that circuit's internal states keep changing in both time and frequency domains. That is, even before the signal-processing operation is finished, the circuit's internal voltage and current levels have changed many times. Therefore, it is valid question to ask "which of the states *is* the "correct"

R. Sobot, *Wireless Communication Electronics*,
https://doi.org/10.1007/978-3-030-48630-3_6

one?" The answer is "all of them, but not all of the time", and that is why the numerical simulators are useful. They enable designers to observe the ever-changing internal states of the circuit, which is otherwise not possible to do by hand analysis.

6.1.1 General Amplifier Classification

A simple, general classification of amplifiers is done in respect to the nature of their input and output signals. In the world of electronic circuits the signals are in form of either voltage or current, which are not two independent variables that we could separate at will. Instead, they are two representations of the same phenomena, i.e. position of the charge carriers in time and space. At the highest level of abstraction, voltage and current relationship is described by Maxwell's equations that also include media where the charges are located. Kirchhoff's and Ohm's laws are simply the low-frequency approximations of Maxwell's equations derived under assumption that the wavelength λ of the signal being observed is much longer than the distance d it needs to travel, i.e. $\lambda \gg d$.

Naturally, often asked question is "so, how do we decide whether to use voltage or current signal?". In fact, deciding whether to process a signal in form of voltage or current is a challenge on its own because, except in purely abstract mathematical world, there is no such thing as pure "voltage amplifier" or pure "current amplifier", or any other "pure" signal-processing circuit for that matter. Instead, we *approximate* a circuit function as "voltage amplifier" or "current amplifier" based on two main circuit characteristics, namely, input and output impedances.

From the purely mathematical perspective, function of an ideal linear amplifier is written as

$$y(x(t)) = K\, x(t) \tag{6.1}$$

where $x(t)$ is the time-dependent signal variable that is presented at the amplifier's input terminals, $y(x(t))$ is the time-dependent variable at the amplifier's output terminals, K is the multiplication factor between the $x(t)$ and $y(t)$ variables, which is called *gain*. Strictly speaking, although the word "gain" implies a number larger than one, gain K can take any value, i.e. $-\infty \leq K \leq \infty$. Negative gain indicates that variables x and y have opposite phase while their amplitude relation is still controlled by the absolute value of K. Although gain less than one, i.e. $y < x$, is sometimes referred to as *loss*, the term gain assumes both "gain" and "loss". Additionally, in (6.1) it is assumed that $K = \texttt{const}$, which translates into $y(x)$ being, in the ideal case, linear function.

In the real material world, the two abstract variables (x, y) are given physical meaning. Electronic amplifiers have two sets of terminals, i.e. the input and the output, that are capable of accepting two forms of signals, i.e. voltage and current, hence there are only four possible amplifier variants:

1. *Voltage amplifier*: a circuit is classified as a "voltage amplifier" if voltage signal v_{in} at its input causes proportional voltage signal v_{out} at its output, thus $v_{out} = A_v\, v_{in}$, where the multiplication constant A_v is referred to as *voltage gain* in units of V/V, (or in [dB]).
2. *Current amplifier*: a circuit is classified as a "current amplifier" if current signal i_{in} at its input causes proportional current signal i_{out} at its output, thus $i_{out} = A_i\, i_{in}$, where the multiplication constant A_i is referred to as *current gain* in units of [$^{A}/_{A}$], (or in [dB]).
3. *Transconductance amplifier*: a circuit is classified as a "transconductance (G_m) amplifier" if voltage signal v_{in} at its input causes proportional current signal i_{out} at its output, thus $i_{out} = G_m\, v_{in}$, where the multiplication constant G_m is referred to as *voltage to current gain* in units of [S] ("siemens", $S = {}^{A}/_{V} = {}^{1}/_{\Omega}$). Accepted convention is to use notification with the capital

letter G_m to indicate transconductance of a *circuit*, e.g. amplifier, and lower case g_m to indicate transconductance of a single *device*, e.g. BJT transistor.

4. *Transresistance amplifier*: a circuit is classified as a "transresistance amplifier" if current signal i_{in} at its input causes proportional voltage signal v_{out} at its output, thus $v_{out} = A_R\, i_{in}$, where the multiplication constant A_R is referred to as *current to voltage gain* in units of [Ω] ("ohm", $\Omega = V/A$). Trivial example of transresistance amplifier is a linear resistor, $v_R = R\, i_R$.

By combining these four possible ideal amplifying functions we are able to both synthesize and analyse any complicated multistage amplifying circuit that may be optimized to process either voltage or current form of signals, or even to keep switching the signal form along the way. As a first step in moving from the ideal mathematical concept of amplifiers into the real world, for each of the four ideal amplifiers, we take a closer look at their characteristics and consequences of interfacing them with the signal source and the subsequent loading stages.

6.1.2 Voltage Amplifier

Behavioural model of an ideal voltage amplifier, Fig. 6.1 (left), shows the literal implementation of (6.1), which is based on ideal voltage controlled voltage source (VCVS)[1] whose voltage gain is A_v.

Important characteristics of an ideal voltage amplifier to observe are as follows. Beginning from the left side of the ideal voltage amplifier model, any voltage v_{in} presented at the input terminals *outside* of the amplifier is immediately transferred to the *inside* of the amplifier without any either loss or change. This is because the input current is zero, due to the *input impedance* Z_i of the ideal voltage amplifier being equivalent to *open connection*. That is, the input impedance of the ideal voltage amplifier is $Z_i = \infty$, which is another way of saying that the input current is $i_{in} = v_{in}/Z_i = v_{in}/\infty = 0$.

At the right side of the ideal voltage amplifier symbol, the output voltage v_{out} is generated by the internal VCVS that simply takes the input voltage value v_{in} as seen at the *internal* nodes of the amplifier, and multiplies it with the multiplication constant A_v before delivering it to the output nodes. Because the internal impedance of an ideal voltage source is zero, see Sect. 2.2.3, the output impedance of the ideal voltage amplifier in Fig. 6.1 (left) must also be zero, i.e. $Z_o = 0$ (looking into the output terminals of an ideal voltage amplifier VCVS is the only element connected).

Consequently, an element or circuit aspiring to be classified as "voltage amplifier" must be as close as possible to the ideal model, Fig. 6.1 (left). The criteria for quantifying the success of this aspiration

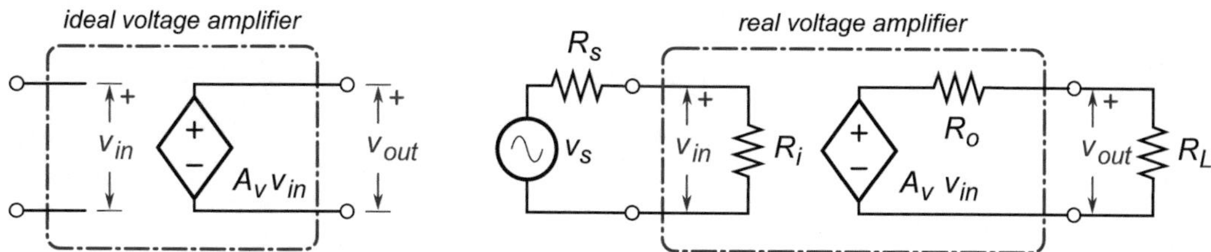

Fig. 6.1 Ideal voltage amplifier (left), and realistic voltage amplifier connected with the input signal source and the output load (right)

[1] Instead of circular symbols, controlled sources are usually indicated by diamond shaped symbols.

are: (1) the input impedance Z_i has to be very high, ideally infinite; (2) the output impedance Z_o has to be very low, ideally zero; and (3) the voltage gain A_v has to be uniquely defined and constant.

In reality, the amplifier's input resistance is explicitly modelled by resistor R_i, and at the same time, the realistic VCVS at the output side exhibits its internal resistance R_o. Following the same idea, the input signal source (i.e. driver) is modelled as an ideal voltage source v_S in series with a non-zero resistance $R_S > 0$. The loading circuit (i.e. the load) that receives the amplifier's output signal v_{out} is modelled only by its input impedance, here annotated as a simple load resistor R_L.

At the input side of the amplifier, there is a voltage divider created by the voltage source v_S, source resistance R_S, and the amplifier's input resistance R_i. Therefore, portion of the source signal level v_S that is transferred to the amplifier's internal nodes v_{in} (and subsequently multiplied by gain A_v) is calculated as

$$v_{in} = i_{in}\, R_i = \frac{v_S}{R_S + R_i}\, R_i \quad \therefore \quad A'_v = \frac{v_{in}}{v_S} = \frac{R_i}{R_S + R_i} = \frac{1}{\dfrac{R_S}{R_i} + 1} \tag{6.2}$$

where A'_v is the voltage gain of the input side voltage divider by itself (obviously less than one). In order to efficiently transfer the source *voltage* signal into the amplifier with no attenuation, this attenuation effect of this voltage divider must be minimized.

With passive components, theoretical maximum that we can hope for is to transfer the full source signal level v_S to the inside of the amplifier, i.e. to achieve $v_{in} = v_S$ (or equivalently, $A'_v = 1$). By inspection of (6.2) we find two possibilities that would lead into $v_{in} = v_S$, the first one being

$$\lim_{R_S \to 0} A'_v = \frac{1}{\cancelto{0}{\dfrac{0}{R_i}} + 1} = 1 \tag{6.3}$$

that is, in the case of an ideal source signal generator (i.e. the one whose source impedance is $R_S = 0$).

The second limiting case is

$$\lim_{R_i \to \infty} A'_v = \frac{1}{\cancelto{0}{\dfrac{R_S}{\infty}} + 1} = 1 \tag{6.4}$$

that is, in the case of an ideal voltage amplifier (i.e. one whose input impedance is $R_i = \infty$).

In reality, the two conditions for theoretically perfect voltage transfer at the source side (6.3) and (6.4) can be combined together by stating that a *voltage* amplifier must have *large input impedance* R_i *relative to the signal source impedance* R_S, i.e.

$$R_i \gg R_S \tag{6.5}$$

An important case is that of matched impedances $R_i = R_S$ leading into the voltage gain of $A'_v = 1/2$, which means that only a half of the input voltage is transferred through matched networks. In other words, only *quarter* of the signal power is transferred due to the $P = f(V^2)$ relationship.

Similarly, the output voltage divider consists of the amplifier's output impedance R_o and the load impedance R_L. Thus, we write

$$v_{out} = i_{out}\,R_L = \frac{A_v\,v_{in}}{R_o + R_L}\,R_L \quad \therefore \quad A''_v = \frac{v_{out}}{v_{in}} = A_v\,\frac{R_L}{R_o + R_L} = A_v\,\frac{1}{\dfrac{R_o}{R_L} + 1} \tag{6.6}$$

where A''_v is voltage gain of the output side voltage divider. Again, non-zero impedances at the load side create resistive voltage divider which causes proportional reduction of the output signal v_o on its way to the load terminals.

By inspection of (6.6) we find two conditions that would lead into $v_{out} = A_v\,v_{in}$, the first one being

$$\lim_{R_o \to 0} A''_v = A_v\,\frac{1}{\cancelto{0}{\dfrac{\cancelto{0}{R_o}}{R_L}} + 1} = A_v \tag{6.7}$$

that is, in the case of zero output impedance R_o.

The second limiting case is

$$\lim_{R_L \to \infty} A''_v = A_v\,\frac{1}{\cancelto{0}{\dfrac{R_o}{\cancel{\infty}}} + 1} = A_v \tag{6.8}$$

that is, in the case of infinitely large loading impedance connected to a real amplifier. Physical interpretation of this condition is that the amplifier is *disconnected* from the load and it is appropriately referred to as "non-loaded gain" that is possible to achieve only in theory.

The two conditions for perfect voltage transfer at the load side (6.7) and (6.8) can be combined together by stating that a *voltage* amplifier must have *much smaller output impedance R_o relative to the load impedance R_L*, i.e.

$$R_L \gg R_o \tag{6.9}$$

We recognize that, under conditions of $R_i \to \infty$ and $R_o \to 0$ the real voltage amplifier model, Fig. 6.1 (right), degenerates into the ideal voltage amplifier model, Fig. 6.1 (left). Another way to look at the realistic voltage amplifier is put together conditions (6.2) and (6.6). Then, the total gain of real source, real amplifier, and real load chain is found as if the two gains A'_v and A''_v follow each other. Thus, the total gain A_v from the source to the load terminals is

$$A_v = \frac{v_{out}}{v_s} = A'_v\,A''_v = \underbrace{\frac{R_i}{R_S + R_i}}_{<1} \times \underbrace{\qquad A_v \qquad}_{\text{voltage gain}} \times \underbrace{\frac{R_L}{R_o + R_L}}_{<1} \tag{6.10}$$

which clearly shows the three terms contributing to the total gain of a real amplifiers. The first and third terms show the attenuations due to the input and the output side voltage divider, and each is less than one. The second term is the only term possibly larger than one, and it represents the maximum possible voltage gain that could be achieved under the ideal condition of zero loss at the input and output interfaces. This reasoning is the one used in rapid back of the envelope evaluations.

6.1.3 Current Amplifier

Behavioural model of an ideal current amplifier, Fig. 6.2 (left), shows the literal implementation of (6.1), which is based on ideal current controlled current source (CCCS) whose gain is A_i.

Important characteristics of an ideal current amplifier to observe are as follows. Beginning at the left side of the ideal current amplifier symbol, any input current i_{in} presented at the input terminals *outside* of the amplifier is immediately transferred to the *inside* of the amplifier without any either loss or change. This is because the *input impedance* Z_i of an ideal current amplifier is equivalent to *short connection*. That is, the input impedance of ideal current amplifier is $Z_i = 0$, which is another way of saying that the input voltage is $v_{in} = i_{in}\, Z_i = i_{in} \times 0 = 0$.

At the right side of the ideal current amplifier symbol, the output current i_{out} is generated by the CCCS which simply takes value of the input current i_{in} as seen at the *internal* branch of the amplifier, and multiplies it with the multiplication constant A_i before delivering it to the output nodes. Because the internal resistance of an ideal current source is infinite (see Sect. 2.2.4) the output resistance of the ideal current amplifier is also infinite, i.e. $R_o = \infty$ (looking into the output terminals CCCS is the only element connected, and its impedance is infinite).

Overall, an element or circuit aspiring to be called "current amplifier" must be as close as possible to the ideal current amplifier model, Fig. 6.2 (left). The criteria for quantifying the success of this aspiration are: (1) the input impedance R_i has to be zero; (2) the output impedance R_o has to be infinite; and (3) the current gain A_i has to be uniquely defined and constant.

Following the same analytical steps as in Sect. 6.1.2, we derive conditions required for the efficient transfer of the source *current* signal i_S to the output current i_{out} entering the load. Realistic current source is modelled using its equivalent Norton model, hence we include resistances R_S and R_o, Fig. 6.2 (right). At the same time, the following stage is modelled as resistance R_L.

As oppose to the voltage amplifier circuit network that was analysed by using a voltage divider model, current divider is created at the input terminals of a realistic current amplifier that is driven by realistic current source, Fig. 6.2 (right), where a non-zero input voltage v_{in} develops across $R_S||R_i$. Therefore, at the input side interface, by inspection, we write

$$i_s = \frac{v_{in}}{R_i||R_S} = \frac{R_S + R_i}{R_S}\frac{v_{in}}{R_i} = \frac{R_S + R_i}{R_S}\, i_{in} \quad \therefore \quad A_i' = \frac{i_{in}}{i_S} = \frac{R_S}{R_S + R_i} = \frac{1}{1 + \dfrac{R_i}{R_S}} \tag{6.11}$$

where A_i' is the current gain of the input side current divider.

Non-zero impedances at the source side create resistive current divider which causes proportional reduction of the current signal i_S on its way to the amplifiers' internal nodes. The minimum current

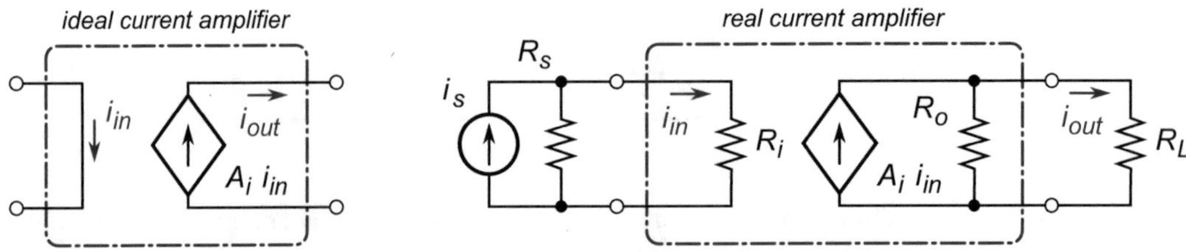

Fig. 6.2 Ideal current amplifier (left), and realistic current amplifier connected with the input signal source and the output load (right)

signal loss condition translates into $i_{in} = i_S$ (or equivalently, $A'_i = 1$). By inspection of (6.11) we find two conditions that would lead into the lossless current transmission through the input side amplifier terminals, the first one being

$$\lim_{R_i \to 0} A'_v = \frac{1}{1 + \cancelto{0}{\frac{0}{R_S}}} = 1 \tag{6.12}$$

that is, in the case of zero input resistance R_i.

The second limiting case is

$$\lim_{R_S \to \infty} A'_i = \frac{1}{1 + \cancelto{0}{\frac{R_i}{\infty}}} = 1 \tag{6.13}$$

that is, in the case of the ideal current source (i.e. one whose output impedance is $R_S = \infty$).

The two conditions for perfect current transfer at the source side (6.12) and (6.13) can be combined together by stating that a *current* amplifier must have *small input impedance relative to the signal source impedance*, i.e.

$$R_i \ll R_S \tag{6.14}$$

Similarly, by inspection of the current amplifier's output terminals, Fig. 6.2 (right), we write

$$i_{out} = \frac{R_o}{R_o + R_L} A_i \, i_{in} \quad \therefore \quad A''_i = \frac{i_{out}}{i_{in}} = A_i \frac{R_o}{R_o + R_L} = A_i \frac{1}{1 + \frac{R_L}{R_o}} \tag{6.15}$$

where A''_i is the current gain of the output side current divider. Non-zero impedances at the load side create resistive current divider which causes proportional reduction of the output current i_{out} on its way to the load terminals.

By inspection of (6.15) we find two conditions that would lead into $i_{in} = A_i \, i_{out}$, the first being

$$\lim_{R_L \to 0} A''_i = A_i \frac{1}{1 + \cancelto{0}{\frac{0}{R_o}}} = A_i \tag{6.16}$$

that is, in the case of zero load resistance R_L.

The second limiting case is

$$\lim_{R_o \to \infty} A''_i = A_i \frac{1}{1 + \cancelto{0}{\frac{R_L}{\infty}}} = A_i \tag{6.17}$$

that is, in the case of infinite output impedance R_o. Under conditions of $R_i \to 0$ and $R_o \to \infty$ the real current amplifier model, Fig. 6.2 (right), degenerates into the ideal current amplifier model, Fig. 6.2 (left).

The two conditions for perfect current transfer at the load side (6.16) and (6.17) can be combined together by stating that a *current* amplifier must drive *small load impedance relative to its own output impedance*, i.e.

$$R_L \ll R_o \tag{6.18}$$

The total gain A_i from the source to the load terminals, as per (6.11) and (6.15), is

$$A_i = \frac{i_{out}}{i_s} = A_i' \; A_i'' = \underbrace{\frac{R_S}{R_S + R_i}}_{<1} \times \underbrace{\qquad\qquad A_i}_{\text{current gain}} \times \underbrace{\frac{R_o}{R_o + R_L}}_{<1} \tag{6.19}$$

which clearly shows how the three terms contribute to the total gain of real amplifiers. The first and third terms show the attenuations due to the input and the output side current dividers, and each is less than one. The second term is the only term possibly larger than one, and it represents the maximum possible current gain that could be achieved under the ideal condition of zero loss at the input and output terminals. This reasoning is the one used in rapid back of the envelope evaluations.

6.1.4 Transconductance Amplifier

By definition, a transconductance (G_m) amplifier converts the *input voltage* signal v_{in} into the *output current* signal i_{out}, and it looks as if we took the input stage of a voltage amplifier and merged it with the output stage of a current amplifier. Behavioural model of an ideal G_m amplifier, Fig. 6.3 (left), shows the literal implementation of (6.1), which is based on ideal voltage controlled current source (VCCS) element.

Therefore, all comments and conclusions that we have made about the input side of a voltage amplifier and about the output side of a current amplifier in the previous sections of this chapter still apply, which simplifies our analysis of this kind of amplifier.

By combining results in (6.2) to (6.19), enables us to directly write expression for a real G_m amplifier gain as

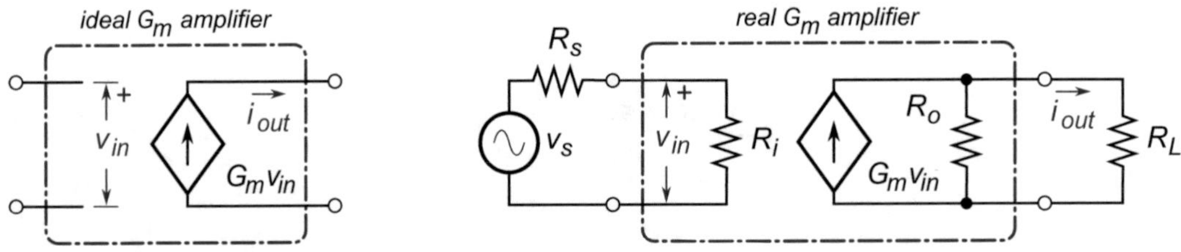

Fig. 6.3 Ideal transconductance (G_m) amplifier (left), and realistic transconductance amplifier connected with the input signal source and the output load (right)

$$G_m = \frac{i_{out}}{v_s} = \underbrace{\frac{R_i}{R_i + R_S}}_{<1} \times \underbrace{\qquad G_m \qquad}_{g_m \text{ gain}} \times \underbrace{\frac{R_o}{R_o + R_L}}_{<1} \tag{6.20}$$

and state that, in order to make an amplifier that would efficiently control the output current by means of the input voltage, its input impedance must be much larger than the source impedance and, at the same time, its output impedance must be much larger than the load impedance, i.e.

$$R_i \gg R_S \quad \text{and} \quad R_o \gg R_L \tag{6.21}$$

That is, an element or circuit aspiring to be called "G_m amplifier" must be as close as possible to the ideal model, Fig. 6.3 (left). The criteria for quantifying the success of this aspiration are: (1) the input impedance R_i has to be infinite; (2) the output impedance R_o has to be infinite; and (3) the transconductance gain G_m has to be uniquely defined and constant.

Let us note that a single MOSFET transistor device is as close to the ideal g_m stage as the technology allows. The reason is that FET device has extremely high input impedance (in order of $M\Omega$s), and in active mode its drain current i_D is controlled by the overdrive voltage at the input side $v_{OV} = (v_{GS} - V_{Tn})$, where V_{GS} is the gate–source voltage and V_{Tn} is the threshold voltage of N-type device. In other words, a FET device is the literal implementation of transconductance gain definition, see (4.67). By convention, we use a small letter in g_m to indicate transconductance of a single device, as opposed to the capital letter version G_m used to indicate gain of an amplifying circuit.

6.1.5 Transresistance Amplifier

The fourth kind of amplifier is a transresistance (A_R) amplifier that, by definition, converts the *input current* signal i_{in} into the *output voltage* signal v_{out}, and it looks as if we took the input stage of a current amplifier and merged it with the output stage of a voltage amplifier. Behavioural model of an ideal A_R amplifier, Fig. 6.4 (left), shows the literal implementation of (6.1), which is based on ideal current controlled voltage source (CCVS) element whose current gain is A_R.

From (6.2) to (6.19) we conclude that

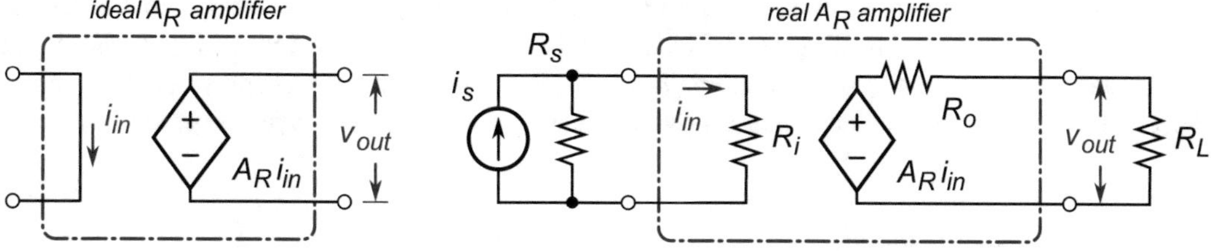

Fig. 6.4 Ideal transresistance (A_R) amplifier (left), and realistic transresistance amplifier connected with the input signal source and the output load (right)

$$A_R = \frac{v_{out}}{i_s} = \underbrace{\frac{R_S}{R_i + R_S}}_{<1} \times \underbrace{\quad A_R \quad}_{\text{R gain}} \times \underbrace{\frac{R_L}{R_o + R_L}}_{<1} \tag{6.22}$$

and state that, in order to make an amplifier that would efficiently control the output voltage signal by means of the input current signal, its input impedance must be much lower than the source impedance and, at the same time, its output impedance must be much smaller than the load impedance, i.e.

$$R_i \ll R_S \quad \text{and} \quad R_o \ll R_L \tag{6.23}$$

That is, an element or circuit aspiring to be called "transresistance amplifier" must be as close as possible to the ideal model, Fig. 6.4 (left). The criteria for quantifying the success of this aspiration are: (1) the input impedance R_i has to be zero; (2) the output impedance R_o has to be zero; and (3) the transresistance gain A_R has to be uniquely defined and constant.

6.2 Single-Stage BJT/MOS Amplifiers

As already introduced in Sect. 3.1, main idea of rapid circuit analysis is to decompose a complicated circuit into a series of simple interfacing circuits, either voltage or current dividers, see Fig. 3.1. Then, we use the "chain rule" to calculate the total gain, where the equivalent source/load resistance are calculated by serial/parallel network transformations.

In Sect. 6.1, we considered idealized amplifying functions from the most abstract perspective. Our goal was to determine how the external parameters influence the overall amplifier behaviour and to derive general rules of circuit interaction with the external world. It turns out that knowing the input, output, source and load impedances, and the internal gain factors is sufficient to specify conditions for four possible ways of amplifying the input signal. The exact details of the circuit's internal structure and the ways it may be implemented did not play any role in the analysis.

In this section, we analyse the three main single-transistor amplifier topologies, i.e., common base (or common gate), common emitter (or common source), and common collector (or common drain), shown in Fig. 6.5, using either BJT or FET devices. In addition, we introduce the cascoded amplifier that is one among many multistage amplifiers. In all cases, the assumptions are that all transistors are in the constant current mode and there are no feedback loops except the degenerated emitter/source case. Further, we use the "by inspection" technique to derive expressions for the input and output impedances, and the overall gain of the three basic amplifier architectures. General strategy for either analysis or synthesis of circuits is to use a series of circuit reductions to derive one of the basic four

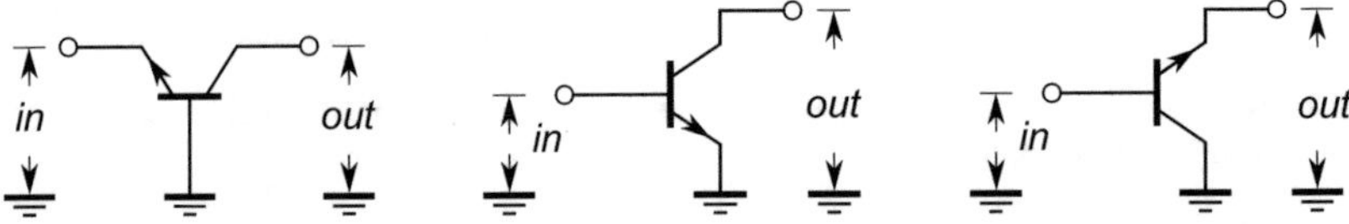

Fig. 6.5 Basic single-stage amplifiers, common base (left), common emitter (centre), and common collector (right). Details of biasing NPN BJT are not shown, i.e. ground symbols used represent the *small signal grounds*

fundamental models, see Sect. 6.1.1. Having said that, in the following sections we consider only the circuits without feedback path.

6.3 Common-Base/Gate Amplifier

Static circuit in Fig. 5.10 whose biasing point is set is used as the starting point in developing all three types of amplifiers. As per Fig. 6.5, common-base amplifier is identified by fixed non-perturbed voltage of the base terminal that is shared between the input and output ports, Fig. 6.6 (centre). In the following discussion we use Thévenin generator model, which can be always replaced with its equivalent Norton generator. From the perspective of the signal source, the whole amplifier circuit is perceived as a simple resistance, which is to say the amplifier's input resistance R_i, see Fig. 6.6 (left). With this insight, the input side interface is then reduced to a simple voltage divider, see Sect. 3.3.1. Equally, from the loading resistor's perspective, the amplifier is perceived as a current source with its associated R_o resistance (this is in effect the amplifier's output resistance), thus the output side interface is also reduced either to a simple voltage divider, or to its equivalent Norton generator to emphasize that the output variable is current, see Fig. 6.6 (right).

Practical implementation of the CB amplifier, Fig. 6.7, shows a Thévenin source connected to the emitter terminal through the coupling capacitor C_0, which protects the already established biasing point but permits the small AC (i.e. sinusoidal) signal to pass. Another coupling capacitor C_0 is used at the output side, which is at the BJT collector, to enable AC signal to reach the loading resistor R_L without disturbing the collector's DC operating point. For the moment, we ignore C_B capacitance and we introduce it later in this section.

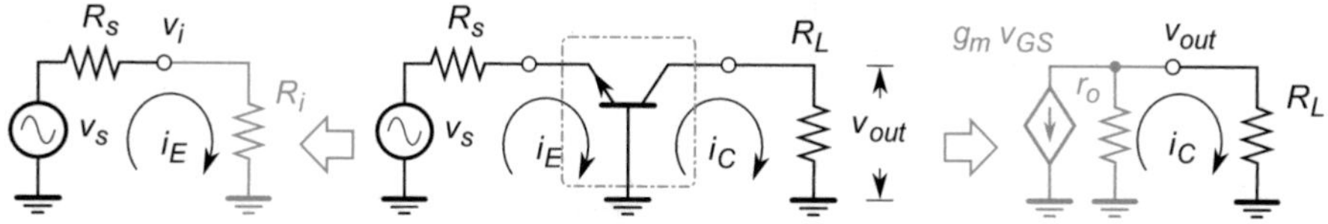

Fig. 6.6 Basic single-stage CB amplifier (centre), it is driven by a voltage source v_S, whose internal resistance is R_S, and it drives resistive load R_L. Equivalent voltage divider schematics for CB amplifier, looking into the input resistance (left) and the output resistance (right)

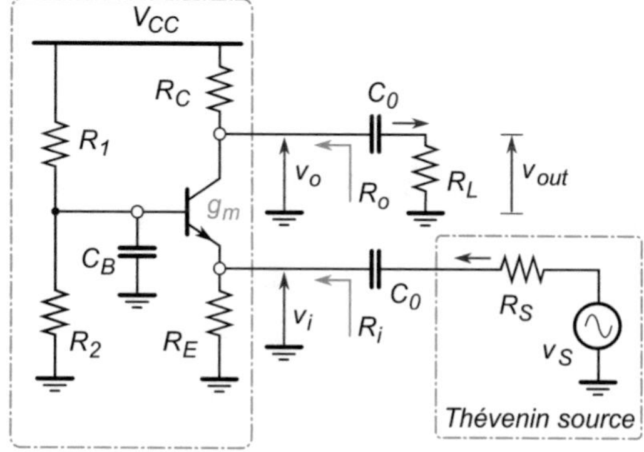

Fig. 6.7 Common-base amplifier stage derived from BJT biasing setup in Fig. 5.10 (right) by adding Thévenin source at the emitter terminal and load resistance at the collector terminal

6.3.1 AC Equivalent Circuit

Circuit in Fig. 6.7 must be first transformed into its equivalent AC model: i.e. power supply source is "folded" to the ground,[2] and C_0 capacitances are shorted. At the same time, two voltage divider biasing resistors R_1 and R_2 are replaced with their equivalent AC resistance $R_B = R_1||R_2$. This transformation results in the equivalent AC circuit, Fig. 6.8.

6.3.2 Input Resistance

From the perspective of the signal source, there is R_i resistance in series with R_S, Fig. 6.9. Resistance R_i consists of R_E in parallel with R_{IN} looking into the BJT emitter, therefore, we write

$$R_{IN} = r_e + \frac{R_B}{\beta + 1}$$

$$\therefore$$

$$R_i = R_E||R_{IN} = R_E||\left(r_e + \frac{R_B}{\beta + 1}\right) \approx R_E||\left(r_e + \frac{R_B}{\beta}\right) \quad (\beta \gg 1) \tag{6.24}$$

which in series with R_S creates the input side voltage divider interface, Fig. 6.9 (right). Obviously, this is voltage divider interface, thus

$$\left.\begin{aligned} i_E &= \frac{v_S}{R_S + R_i} \\ v_i &= v_E = R_i\, i_E \end{aligned}\right\} \Rightarrow v_i = R_i\, \frac{v_S}{R_S + R_i}$$

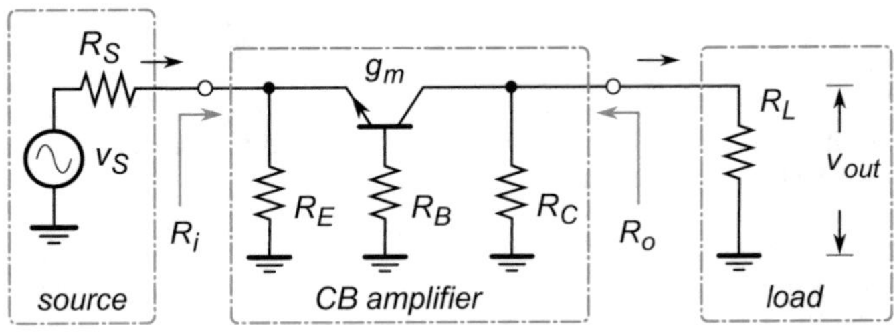

Fig. 6.8 Equivalent AC circuit of a single-stage CB amplifier in Fig. 6.7

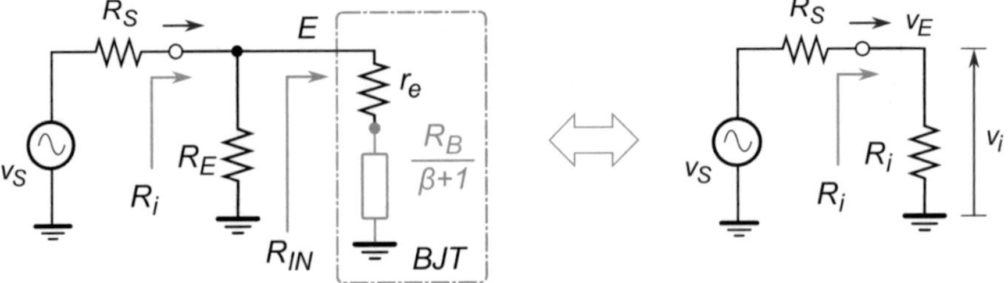

Fig. 6.9 The input side interface network of the CB amplifier

[2]Resistance of ideal voltage sources is zero, thus R_C and R_1 are connected to AC ground node.

$$\therefore$$

$$A_1 \stackrel{\text{def}}{=} \frac{v_i}{v_S} = \frac{R_i}{R_S + R_i} = \frac{1}{1 + \dfrac{R_S}{R_i}} \tag{6.25}$$

which as expected shows that it is necessary to make the source resistance much smaller than the input resistance. However, resistance r_e is relatively small and it is comparable to R_B/β leading into conclusion that it is difficult to use voltage source because its resistance must be much smaller than the already small R_i. Therefore, to achieve the same result, it is easier to force input current by using Norton source that should have its resistance as high as possible relative to R_i.

6.3.3 Output Resistance

Transformation steps in Fig. 6.10 show reduction of the CB amplifier in Fig. 6.8 so that its output side interface is reduced to the equivalent current divider. We start by the equivalent parallel resistance $R_S||R_E$, which is then perceived as being in parallel with the projected base resistance R_B.

Since our objective is to find the equivalent resistance looking into collector, we focus on BJT itself (see Fig. 6.10 zoom-in in the middle). We conclude that from the collector's perspective, this is BJT with degenerated emitter, where $R_B = R_1||R_2$ and the equivalent emitter resistor is $R_S||R_E$. We already analysed this case in (4.51), thus we write

$$R_{OUT} \approx r_o \left[1 + \frac{\beta R_S||R_E}{R_S||R_E + R_B + r_\pi} \right] \tag{6.26}$$

Knowing (6.26) resistance we transform AC schematic in Fig. 6.8 to show only the output side of the circuit where BJT is replaced with its equivalent voltage controlled current source $g_m\, v_{GS}$ and its

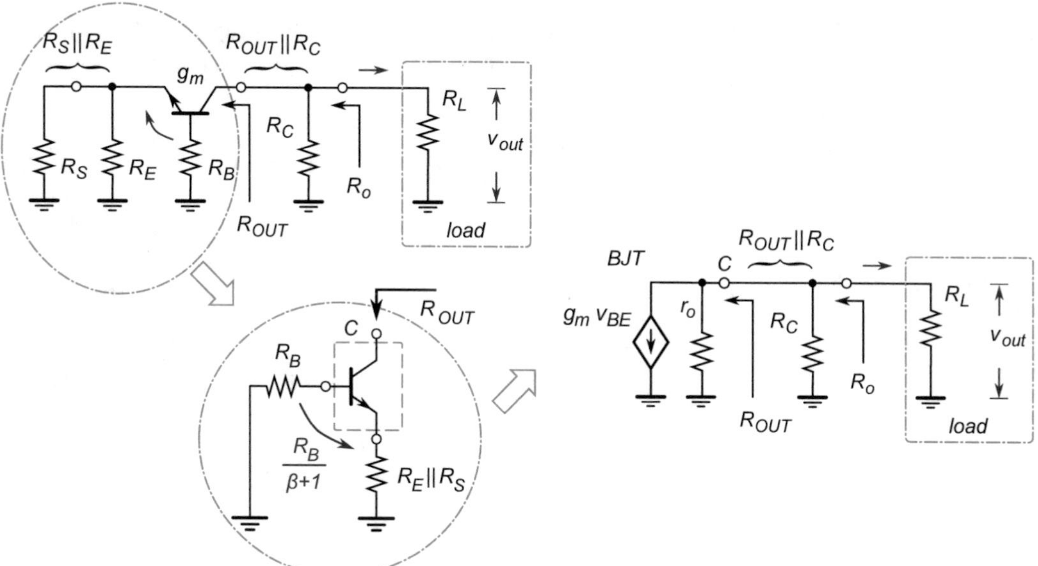

Fig. 6.10 Transformation of common-base amplifier to derive the output side interface

perceived R_{OUT}, see Fig. 6.10 (right). By inspection, we write

$$R_o = R_C||R_{OUT} = R_C||r_o \left[1 + \frac{\beta R_S||R_E}{R_S||R_E + R_B + r_\pi}\right] \tag{6.27}$$

where R_o is the resistance looking into the unloaded CB amplifier output that consists of R_C and collector's resistance in the case of degenerated emitter.

6.3.4 Voltage Gain

The output voltage v_{OUT} is generated by $g_m\, v_{GS}$ current source driving the total equivalent resistance connected to its terminals, which consists of three resistances in parallel, see Figs. 6.10 and 6.11, so we write

$$\begin{aligned} v_{OUT} &= g_m\, v_{GS} \times R_L||R_o = g_m\, v_{GS} \times R_L||R_C||R_{OUT} \\ &= g_m\, v_{GS} \times R_L||R_C||r_o \left[1 + \frac{\beta R_S||R_E}{R_S||R_E + R_B + r_\pi}\right] \end{aligned} \tag{6.28}$$

therefore, we estimate voltage gain A_2 of the output interface stage itself as

$$A_2 \overset{\text{def}}{=} \frac{v_{OUT}}{v_{GS}} = g_m\,(R_L||R_o) \tag{6.29}$$

The total gain is the product of chained gains, therefore for the complete CB amplifier we write

$$A_v = A_1 \times A_2 = \frac{R_i}{R_S + R_i} \times g_m\,(R_L||R_o) = \frac{1}{\dfrac{R_S}{R_i} + 1} \times g_m\,(R_L||R_o) \tag{6.30}$$

In the case when $R_B = 0$ for AC signals, which is possible by connecting capacitor C_B in parallel with R_B (see Fig. 6.7) therefore shorting the base node to ground for AC signals, then (6.24) becomes

$$R_i = R_E|| \left(r_e + \frac{\cancelto{0}{R_B}}{\beta} \right) = R_E||r_e \approx r_e \quad (R_E \gg r_e) \tag{6.31}$$

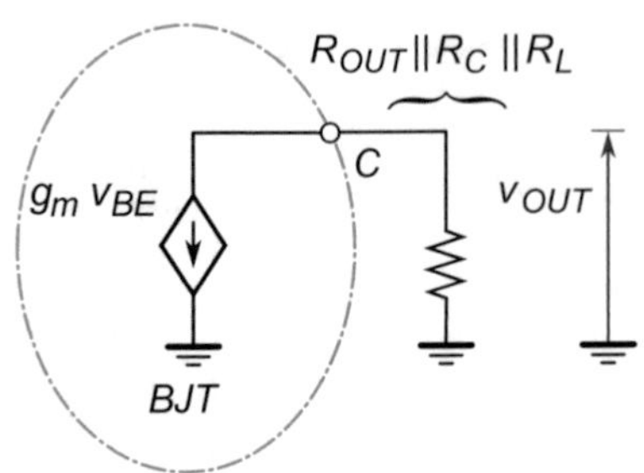

Fig. 6.11 The equivalent output side network of the loaded common-base amplifier

which shows that the input AC resistance of the CB amplifier is lower than r_e, in other words "low input impedance" that is suitable to interface with the current generator whose impedance is very large.

At the same time, the output impedance is also influenced by R_B, thus (6.27) becomes

$$R_o = R_C||r_o\left[1 + \frac{\beta R_S||R_E}{R_S||R_E + \cancelto{0}{R_B} + r_\pi}\right] = R_C||\underbrace{r_o\left[1 + \frac{\beta R_S||R_E}{R_S||R_E + r_\pi}\right]}_{>r_o}$$

$$\approx R_C \quad (r_o \gg R_C) \tag{6.32}$$

which is to say that for large output resistance r_o of the collector (i.e. when close to the ideal current source model), the overall output impedance of the CB amplifier is determined by the external R_C resistor.

Furthermore, the expression for voltage gain (6.30) is approximated as

$$A_v = \frac{R_i}{R_S + R_i} \times g_m\left(R_L||R_o\right) \approx \frac{r_e}{R_S + r_e} \times g_m\left(R_L||R_C\right)$$

which, after $r_e = 1/g_m$ substitution, can be written in the following form:

$$A_v = \underbrace{\frac{r_e}{R_S + r_e}}_{A_1} \times \underbrace{\frac{R_L||R_C}{r_e}}_{A_2} \tag{6.33}$$

A very useful interpretation of (6.33) is to say that the voltage gain of the CB amplifier in Fig. 6.7 is very close to the ratio of the total resistance at the collector (i.e. $R_L||R_C$) divided by the total resistance at the emitter (r_e) multiplied by the gain voltage divider interface at the input side. With a little practice, voltage gain evaluation is done by inspection of these resistances, where R_C and R_L are controlled by the designer, while r_e is consequence of the chosen biasing point.

6.3.5 A Brief CB Amplifier Summary

We briefly summarize most important properties of this basic amplifier architecture.

1. *Current gain*: because emitter terminal serves as the input port and collector serves as the output, due to relationship (4.32) and because modern transistors have $\alpha \approx 1$, we say that

$$i_C = \alpha\, i_E \quad \therefore \quad i_C \approx i_E \quad \therefore \quad A_i(CB) = \frac{i_{out}}{i_{in}} = \frac{i_C}{i_E} \approx 1 \tag{6.34}$$

   in other words, CB amplifier is not really capable of current amplification. Instead, it is often referred to as a "current buffer".

2. *Input resistance*: we found that the input resistance of the CB amplifier is approximately r_e

$$R_i \approx r_e = \frac{1}{g_m} = \frac{V_T}{I_C} \quad (R_B \to 0) \tag{6.35}$$

where $V_T = kT/q \approx 25\,\text{mV}$ at room temperature. In other words, it is about $25\,\Omega$ when $I_C = 1\,\text{mA}$, about $12.5\,\Omega$ when $I_C = 2\,\text{mA}$, etc. It is shown in Sect. 6.1.3 that low input resistance is property either of current or transresistance amplifier. Conclusion is that CB amplifier should be used with a current signal source (whose output resistance is much higher than r_e, which is not difficult requirement), even though CB current gain is approximately one.

3. *Output resistance*: analysis of the CB output resistance revealed that if high resistance signal sources is used, then the total output node resistance R_{tot} becomes

$$R_{tot} = R_o || R_C || R_L \approx R_C || R_L \quad (r_o \gg R_C) \tag{6.36}$$

which is in general high. It is shown in Sect. 6.1.3 that high output resistance is characteristics of a current amplifier. Conclusion is that CB amplifier should be used to drive relatively low resistance load ($R_L \ll R_C$), in other words to transmit a current signal.

4. *Voltage gain*: CB transistor with a certain g_m followed by a resistive load R_L gives rise to voltage gain as shown in (6.29). It is important to note that in order to amplify voltage, a g_m device (whose input is voltage and the output is current) must be followed by transresistance device (whose input is current and the output is voltage). Thus, referring to Sect. 6.1.1 where the four basic amplifying topologies are introduced, a practical CB amplifier could be imagined as being made of two general amplifiers, where R_L resistor is used as a trivial transresistance amplifier that accepts current output of a g_m amplifier.

6.4 Common Emitter Amplifier

At the fundamental level we identified four types of basic amplifiers, see Sect. 6.1.1. Thus, the first inclination may be that the practical voltage amplifier (i.e. "voltage to voltage") implementation is as simple as using a transformer, which is after all the literal realization of model in Fig. 6.1. However, that is not the case because transformer is a passive device, that is to say it cannot amplify the power—the increase of its output voltage, by definition, goes with the decrease of its output current.

In order to achieve the "power amplification" we must use an active device in addition to the external source of energy. If the intent is to design voltage amplifier, we face first practical problem: transistors are not capable of amplifying voltage to voltage, instead they are "g_m devices", which is to say voltage to current amplifiers. Then, the question is: how do we make practical implementation of model in Fig. 6.1?

6.4.1 Fundamental Principle of CE Amplifier

In order to resolve this problem, we are forced to use combination of two basic amplifiers: first g_m type to convert voltage to current, (see Fig. 6.1 in Sect. 6.1.2), which is immediately followed by second amplifier of transresistance type to convert current to voltage (see Fig. 6.4 in Sect. 6.1.5). We can, therefore, see a practical voltage amplifier as a *two-stage* amplifier, Fig. 6.12 and conclude that the total gain is by definition the product of two gains, i.e.

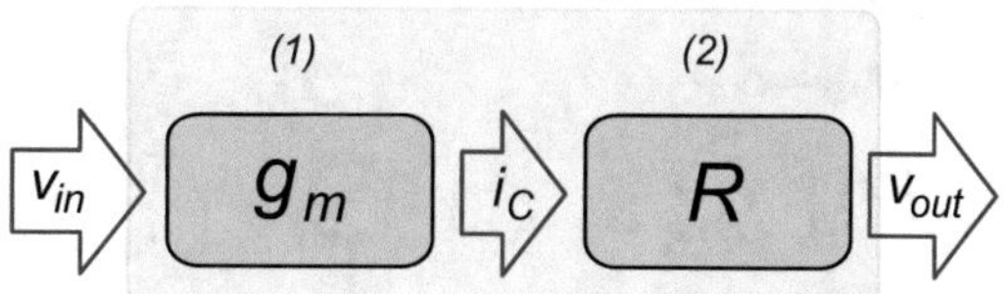

Fig. 6.12 Fundamental principle of a BJT CE amplifier

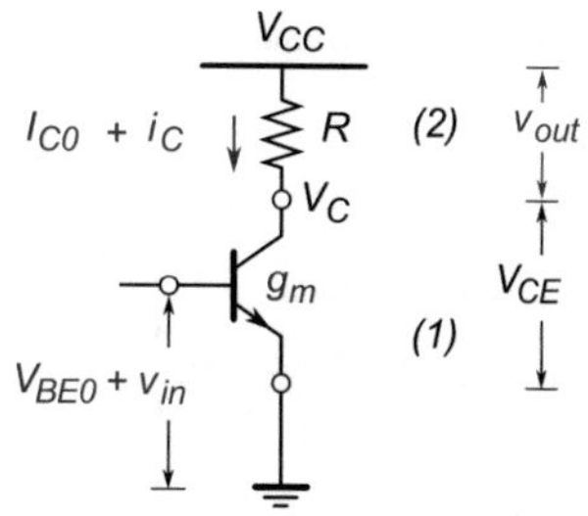

Fig. 6.13 "Bare bones" literal implementation of BJT CE amplifier

$$A_V = g_m\, R \tag{6.37}$$

In this interpretation we realize that (6.37) provides back of the envelope estimate of the voltage gain. That estimate gives us theoretical gain limit of this two-stage architecture, which sets our design goal that we approach as close as possible. What is more, we now realize that we can make literal implementation of Fig. 6.12 model that is based on NPN BJT transistor and a resistor, as in Fig. 6.13.

Example 47: BJT Voltage Gain Mechanism
In reference to Fig. 6.13, in order to illustrate the voltage gain A_V mechanism of BJT, let us assume that a small AC signal v_{in} is applied across its input terminals and the output resistance is R. Calculate the voltage gain A_V.

Data: $\beta \gg 1$, $r_e = 25\,\Omega$, and $R = 10\,\text{k}\Omega$.

Solution 47: Small AC signal v_{in} is applied across the emitter diode whose resistance is r_e, thus

$$i_E = \frac{v_{in}}{r_e}$$

Due to $\beta \gg 1$ the assumption $I_C \approx I_E$ is valid, thus

$$v_{out} = i_C \times R \approx v_{in}\frac{R}{r_e}$$

$$\therefore$$

$$A_V = \frac{v_{out}}{v_{in}} = \frac{R}{r_e} \stackrel{\text{def}}{=} g_m \times R = \frac{10\,\text{k}\Omega}{25\,\Omega} = 400 = 52\,\text{dB}$$

(continued)

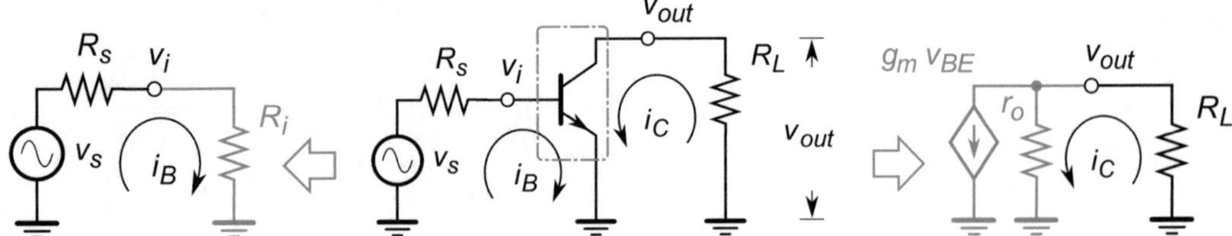

Fig. 6.14 Common emitter amplifier (centre), its input side interface model (left), and its output side model (right)

Note that the output collector current i_C is converted into voltage by means of the loading resistance R that serves as the second stage amplifier.

Another way to show the same result is by definition as

$$A_V \stackrel{\text{def}}{=} \frac{v_{out}}{v_{in}} = \frac{v_{out}}{v_{BE}} = \frac{i_C\, R}{i_E\, r_e} \approx \frac{\cancel{i_C}\, R}{\cancel{i_C}\, r_e} = \frac{R}{r_e}$$

This last form or voltage gain is very useful when interpreted as follows: CE amplifier voltage gain A_V is evaluated as a simple ratio between the *total* resistance found at the collector terminal and the *total* resistance found at the emitter node. As this statement is general, thus we use it to evaluate voltage gain "by inspection".

Basic single-transistor common emitter (CE) amplifier is created when the base terminal serves as the input port and collector terminal serves as the output port. The third transistor's terminal—emitter—is therefore shared between the input/output ports. A realistic voltage source v_S with the internal resistance R_S provides small signal perturbations, Fig. 6.14, and the loading resistance R_L is connected to the collector node.

At this level of simplification, it becomes clearly visible that from the signal source perspective, the input side of the amplifier is perceived as its equivalent R_i resistance, Fig. 6.14 (left). Similarly, the collector current (as modelled by $g_m v_{GS}$ source and R_o resistance) delivers the AC signal to R_L resistance, Fig. 6.14 (right). Therefore, a series of circuit reductions is used to simplify circuit until the expressions for R_i and R_o are found by inspection.

Once the biasing point of BJT is set, a realistic CE amplifier is achieved by connecting the signal source and load, as in Fig. 6.15. In order to protect DC voltage levels, coupling capacitors C_0 are used at both interfaces. We also note that there is C_E capacitor added to bypass R_E, however for the moment we ignore it. Later in this section we evaluate its role in more details. First, we derive expressions for the input resistance, the output resistance, and the voltage gain of the CE amplifier.

6.4.2 AC Equivalent Circuit

First circuit reduction is to create AC equivalent circuit, which is circuit as perceived by the time-varying (AC) input signal, as in Fig. 6.16. Power supply source is shorted (its internal resistance is zero) and its terminals "folded" to the ground reference node (which now becomes the small signal

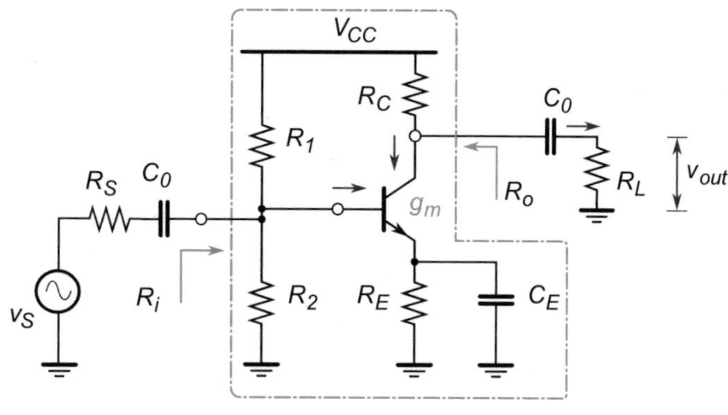

Fig. 6.15 Common emitter amplifier stage

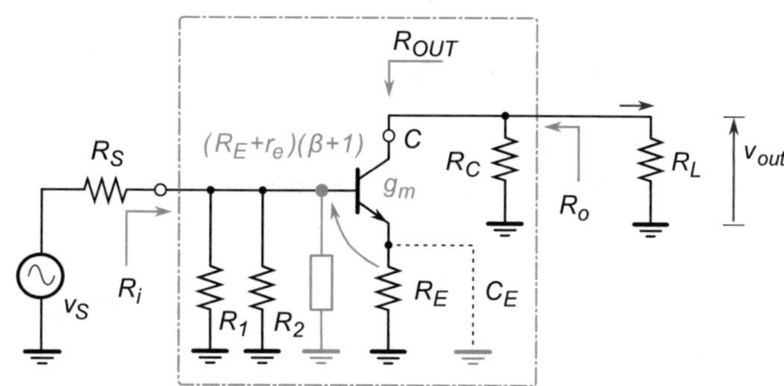

Fig. 6.16 AC model of a common emitter amplifier

ground). In addition, capacitors are shorted (they also present zero resistance to AC signals). As a consequence, at the input side there is the parallel connection $R_B = (R_1)||R_2$ in parallel with the virtual base resistance (i.e. projected emitter resistance). Equally, the total resistance at the output side becomes parallel connection $(R_{OUT}||R_C||R_L)$.

6.4.3 Input Resistance

In reference to Fig. 6.14 (left), by inspection we write

$$i_B = \frac{v_S}{R_S + R_i} \tag{6.38}$$

Then, focusing on the input side interface only, we reduce the input side of circuit in Fig. 6.15 to circuit shown in Fig. 6.17, which enables us to write by inspection

$$R_i = R_B||(R_E + r_e)(\beta + 1) \approx R_B||\beta(R_E + r_e) \tag{6.39}$$

which in series with R_S creates the input side voltage divider interface, Fig. 6.14 (left). In practical versions of CE amplifiers, often it is used $R_E \gg r_e$, so that (6.39) becomes

$$R_i \approx R_B||\beta R_E \tag{6.40}$$

which further is set to $\beta R_E \gg R_B$ leading into

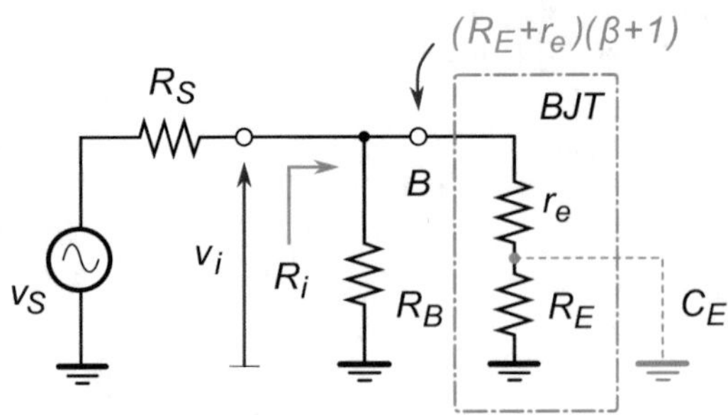

Fig. 6.17 The input side equivalent circuit for a common emitter amplifier

$$R_i \approx R_B||\beta R_E \approx R_B \tag{6.41}$$

and, therefore, reduces dependence of BJT biasing point (as set by ratio of R_1, R_2, thus R_B) as already discussed in Sect. 5.2.

Following the idea of the chained interface gains, we take this input side interface as first gain A_1 in the chain and we calculate the actual voltage that reached the base terminal as

$$\left.\begin{aligned} i_B &= \frac{v_S}{R_S + R_i} \\ v_i &= R_i\, i_B \end{aligned}\right\} \Rightarrow v_i = R_i\,\frac{v_S}{R_S + R_i}$$

$$\therefore$$

$$A_1 \stackrel{\text{def}}{=} \frac{v_i}{v_S} = \frac{R_i}{R_S + R_i} = \frac{1}{1 + \dfrac{R_S}{R_i}} \tag{6.42}$$

In conclusion, when using the Thevenin source its resistance should be as small as possible relative to the amplifier's input resistance.

6.4.3.1 Output Resistance

Focusing to the output side interface, we replace the output side of circuit in Fig. 6.15 with its equivalent circuit in Fig. 6.18. The total resistance attached to base is, therefore, $R_b = R_S||R_B$, which after division by $(\beta + 1)$ factor is projected to the emitter side as a very small resistance in series with r_e. Subsequently, both resistances appear in series with the external R_E. In total, from the collector node perspective, BJT is in the degenerated emitter configuration (4.51), so we write

$$R_{OUT} \approx r_o \left[1 + \frac{\beta R_E}{R_E + R_B + r_\pi}\right]$$

$$\therefore$$

$$R_o = R_C||r_o \left[1 + \frac{\beta R_E}{R_E + R_B + r_\pi}\right] \tag{6.43}$$

where R_o is resistance looking into the CB amplifier output that consists of R_C and collector's resistance in the case of degenerated emitter. However, when C_E is connected in parallel with R_E (as annotated by dash-dot line in Fig. 6.18), it effectively enables AC signal to bypass R_E on its way

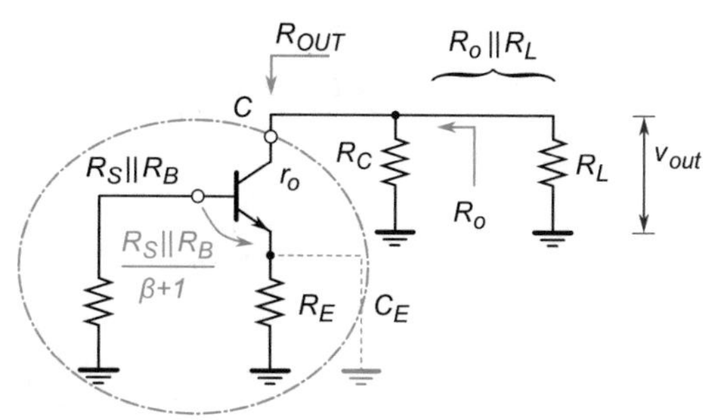

Fig. 6.18 The output side equivalent circuit for a common emitter amplifier

to ground, therefore from the AC signal perspective (while DC is not affected with this connection) we write

$$R_o = R_C||r_o \left[1 + \underbrace{\left(\frac{\beta \cancelto{0}{R_E}}{\cancelto{0}{R_E} + R_B + r_\pi} \right)}_{\to 0} \right] \approx R_C||r_o \tag{6.44}$$

which illustrates one of the reasons to add the bypass capacitor R_E and its importance in CE amplifier.

6.4.4 Voltage Gain

Further reduction of circuit in Fig. 6.18 leads into same output side interface as already found in Fig. 6.11, so by inspection we write

$$\begin{aligned} v_{OUT} &= -g_m\, v_{BE} \times R_L||R_o \\ &= -g_m\, v_{BE} \times R_L||R_C||r_o \left[1 + \frac{\beta R_E}{R_E + R_B + r_\pi} \right] \end{aligned} \tag{6.45}$$

where the negative sign is due to the signal inversion.[3] Therefore, we estimate voltage gain A_2 of the output side network itself as

$$A_2 \stackrel{\text{def}}{=} -\frac{v_{OUT}}{v_{BE}} = -g_m\,(R_L||R_o) \tag{6.46}$$

The total gain of the two chained stages is

$$A_v = A_1 \times A_2 = -\frac{R_i}{R_S + R_i} \times g_m\,(R_L||R_o) = -\frac{R_i}{R_S + R_i} \times \frac{(R_L||R_o)}{r_e} \tag{6.47}$$

General expression (6.47) shows that the voltage gain is determined by the total resistance at the collector divided by the total resistance at the emitter, then multiplied by gain of the input side voltage divider.

[3] CE is a typical inverting amplifier.

1. For $R_E = 0$, which is possible when the bypass capacitor C_E is connected to the emitter node, as shown in (6.44). Consequently, after $r_e = 1/g_m$ substitution the total CE amplifier gain (6.47) becomes

$$A_v = -\frac{R_i}{R_S + R_i} \times g_m \left(R_L || R_o\right) \approx \frac{R_B}{R_S + R_B} \times g_m \left(R_L || R_C || r_o\right)$$

$$\therefore$$

$$A_v = -\underbrace{\frac{1}{\frac{R_S}{R_B} + 1}}_{A_1} \times \underbrace{\frac{R_L || R_C || r_o}{r_e}}_{A_2} \tag{6.48}$$

Often, $R_L || R_C \ll r_o$ so that (6.48) is further approximated

$$A_v = -\frac{1}{\frac{R_S}{R_B} + 1} \times \frac{R_L || R_C}{r_e} \tag{6.49}$$

and is interpreted as the voltage gain being equal to ratio of the total resistance at the collector node (i.e. $R_L || R_C$) versus the total resistance at the emitter node (i.e. r_e when C_E is used), after accounting for the input side voltage divider interface gain (R_S, R_B voltage divider).

6.4.5 A Brief CE Amplifier Summary

We briefly summarize most important properties of CE amplifier architecture.

1. *Input resistance*: we found that input resistance of the CE amplifier is approximately R_B when βR_E is much larger. As a result, input resistance of the CE amplifier is medium to large, which makes it suitable to accept voltage type input signals.
2. *Output resistance*: in the case of CE amplifier, the total output resistance is made of parallel $R_C || R_L$, both of which may affect BJT biasing point if $v_{CE} < V_{CE}(sat.)$. Given R_C, the range of loading resistances is very important specification. "Unloaded" amplifier case (i.e. $R_L \to \infty$, in other words open connection) delivers maximum theoretical voltage gain (but zero power because the output current is zero). Shorted output (i.e. $R_L = 0$, in other words, output node shorted to ground) pulls the collector voltage below $V_{CE}(sat.)$ and, effectively, forces collector current to zero. In conclusion, it is important to specify the minimum load resistance.
3. *Voltage gain*: CE transistor with a certain g_m followed by a resistive load $R_L || R_C$ gives rise to voltage gain as shown in (6.49). It is important to note that voltage amplification is achieved in two steps, g_m of BJT device (whose input is voltage and the output is current) is followed by transresistance device (whose input is current and the output is voltage).

6.5 Common Collector Amplifier

Common collector (CC) amplifier is realized when collector terminal serves as the small signal ground (i.e. shared terminal), Fig. 6.19, and the emitter terminal is used to recover amplified signal. As with the other two basic amplifiers, circuit analysis is based on reduction techniques that produce the input and output interface voltage dividers, Fig. 6.19. We note, the output interface is modelled as Thévenin generator because the emitter resistance is small, see Sect. 4.3.5, thus closer to the ideal voltage source model. Practical CC amplifier circuit is shown in Fig. 6.20 where it should be noted that R_C resistor is removed because in CC amplifier it serves no purpose.[4]

6.5.1 AC Circuit Model

Reduction of practical CC amplifier circuit results in schematic shown in Fig. 6.21. We note that in this circuit, R_C is not critical and in practice is replaced with a short connection, i.e. $R_C = 0$. In addition, voltages at base and emitter terminals are related as

$$v_E = v_B - v_{BE0} \tag{6.50}$$

in other words, the output signal simply "follows" the input at v_{BE0} distance (thus, "emitter follower" amplifier).

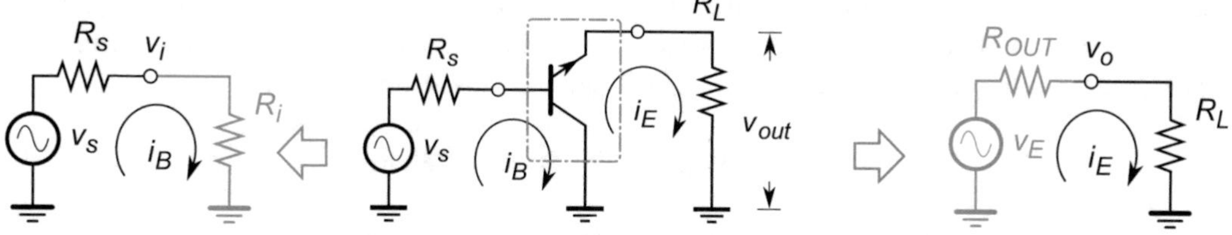

Fig. 6.19 Common collector amplifier (centre), its input side interface model (left), and its output side model (right)

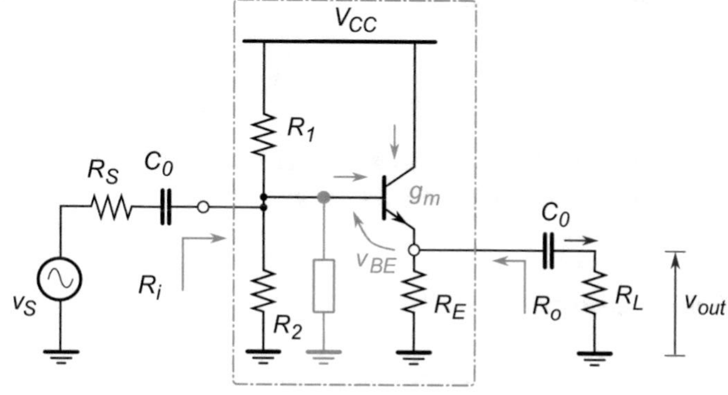

Fig. 6.20 Common collector amplifier

[4]We recall the range of R_C resistances that permit constant current mode, see Sect. 5.4.

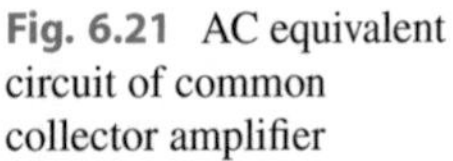

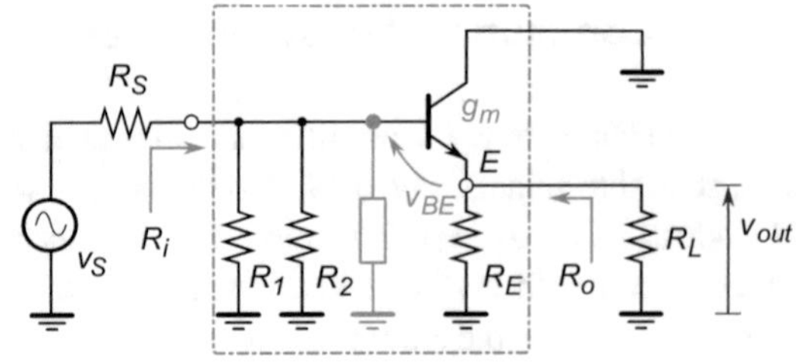

Fig. 6.21 AC equivalent circuit of common collector amplifier

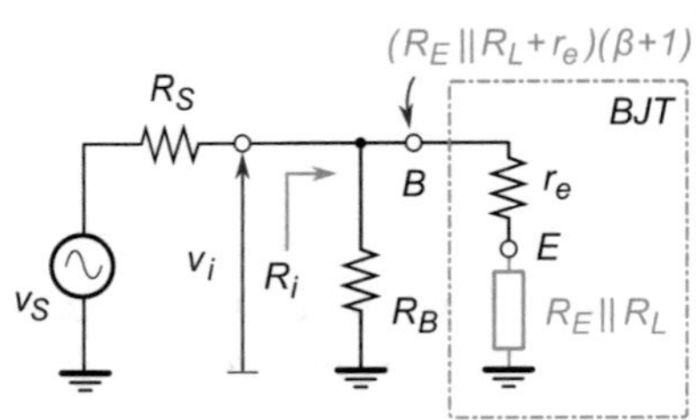

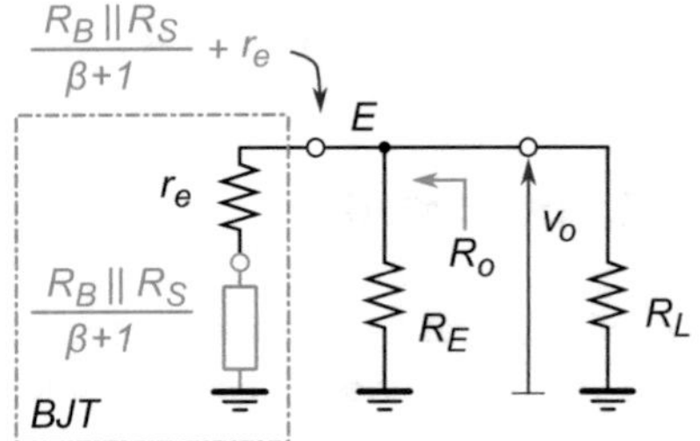

Fig. 6.22 Common emitter amplifier stage, its input side (left) and output side (right) equivalent circuits

6.5.2 Input Resistance

In order to evaluate the input side resistance of the CC amplifier, first transformation is it to replace the two biasing voltage divider resistor with their equivalent parallel resistance $R_B = R_1||R_2$, see Fig. 6.22 (left). Then, two parallel resistors at the emitter node are also replaced with their equivalent $R_E||R_L$ to form serial resistance with r_e. After that, this branch is projected to the base terminal as virtual resistance in parallel with R_B, see Fig. 6.22 (left). By inspection we write

$$R_i = R_B||(\beta + 1)(R_E||R_L + r_e) \approx R_B||\beta(R_E||R_L + r_e) \tag{6.51}$$

where other approximations depend on each specific case, because in general R_E is not too large, while the actual value of R_L also influences the input side resistance.

The input side voltage divider consists of R_S and R_i, accordingly gain at the input side interface is

$$A_1 = \frac{v_B}{v_S} = \frac{R_i}{R_S + R_i} \tag{6.52}$$

which approaches unity if source resistance is much smaller than R_i, that is to say if the source generator is close to the ideal voltage generator.

6.5.3 Output Resistance

Circuit transformation at the emitter node, see Fig. 6.22 (right), clearly illustrates virtual BJT base resistance that is perceived by the emitter node, and it appears in series with r_e. However, by looking into the emitter node, this branch appears connected in parallel with R_E to form R_o. By inspection we write

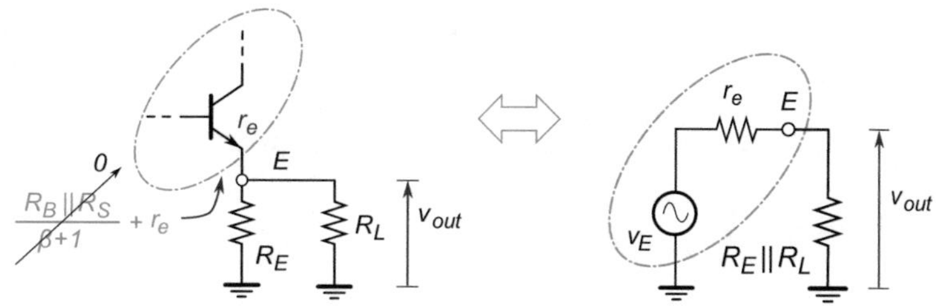

Fig. 6.23 Common collector amplifier: the output side equivalent Thévenin generator

$$R_o = R_E|| \left(r_e + \frac{R_B||R_S}{\beta+1} \right) \approx R_E|| \left(r_e + \frac{R_B||R_S}{\beta} \right) \tag{6.53}$$

which is further simplified either for the case of ideal voltage source $R_S = 0$ or, equally, for $\beta \gg 1$. In both cases the second term in the brackets is very small, thus we write

$$R_o \approx R_E|| \left(r_e + \frac{R_B||R_S \to 0}{\beta} \right) \approx R_E||r_e \approx r_e \tag{6.54}$$

where the last approximation is in the case of $R_E \gg r_e$, which is often the case. For example, if $I_C = 1\,\text{mA}$, then $r_e = V_T/I_C \approx 25\,\Omega$ at the room temperature. Consequently, if $R_E = 100\,\Omega$ or more, then approximation $R_o \approx 20\,\Omega \approx r_e$ is already very close.

6.5.4 Voltage Gain

Voltage at the emitter closely "follows" voltage at the base terminal at v_{BE0} distance, see (6.50), in other words there is a gain of approximately one between the base and emitter nodes. We already derived gain A_1 of the input side voltage divider (6.52), now we derive gain A_2 of the output side voltage divider. Following result (6.54) and Fig. 6.22 (right), we find that the output voltage v_{out} is across $R_L||R_E$. Looking into the emitter node, explicit circuit transformation shown in Fig. 6.23 illustrates functionality of the BJT emitter as a voltage source generator driving $R_E||R_L$ loading resistance. Due to its small output resistance r_e, the emitter node is very good practical implementation of Thévenin type source generator.

Therefore the output voltage divider consists of r_e and $R_L||R_E$, see Fig. 6.23, by inspection we write

$$A_2 = \frac{v_{out}}{v_E} = \frac{R_L||R_E}{R_L||R_E + r_e} \tag{6.55}$$

There are no other gains to consider, accordingly we write expression for the total voltage gain of the CC amplifier as

$$A_v = A_1 \times A_2 = \frac{R_i}{R_S + R_i} \frac{R_L||R_E}{R_L||R_E + r_e} \tag{6.56}$$

Expression (6.56) clearly shows that the maximum possible voltage gain of the CC amplifier is when $R_i \gg R_S$ and $R_L||R_E \gg r_e$, which reduces (6.56) to

$$A_v \approx 1 \tag{6.57}$$

In conclusion, under normal operation, CC amplifier has voltage gain of one, which makes it suitable to serve as "voltage buffer", or impedance converter (high input impedance, low output impedance).

6.5.5 A Brief CC Amplifier Summary

We briefly summarize most important properties of the CC amplifier architecture.

1. *Voltage gain*: because emitter terminal voltage "follows" the base at fixed distance of v_{BE0}, this type of amplifier is in effect limited to $A_v = 1$ by the input output voltage divider interfaces. Instead, it is often referred to as a " voltage buffer" or "emitter follower".
2. *Input resistance*: we found that the input resistance of the CC amplifier is medium to large, which makes it suitable to accept voltage type input signals.
3. *Output resistance*: in the case of CC amplifier, looking into the emitter terminal we find r_e that forces this node resistance to be low, even after adding R_E and R_L resistances because, in the end, from the perspective of AC signal all three resistances are connected in parallel. Inevitably, the output node resistance is low.

6.6 Cascode Amplifier

Important amplifier configuration, so much that it has its own name, *cascode* amplifier, is in effect a combination of two single-stage amplifiers, CE followed by CB amplifier, Fig. 6.24, (or their respective equivalent MOS versions). The two amplifiers are connected serially, that is to say the output CE current passes through CB current buffer before reaching the load resistance. Since the current does not change while passing through CB buffer, which is to say that from the perspective of R_L the overall setup is very similar to the one already analysed in the case of CE amplifier.

Practical schematic of cascode amplifier is derived by merging Figs. 6.7 and 6.15, as shown in Fig. 6.25. We note that from the perspective of Q_1, instead of passive R_C component as its load, it perceives the input resistance $r_{in2} = r_e$ of Q_2. At the same time, from the perspective of Q_2 emitter, it perceives the input current i_C that is in reality generated by Q_1 and its collector resistance r_{o1}. Consequently, Q_2 is in degenerated emitter configuration with r_{o1} being perceived by its collector terminal.

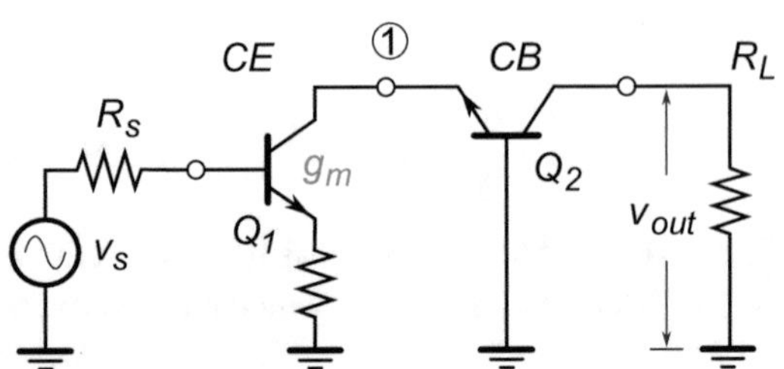

Fig. 6.24 Architecture of cascode amplifier illustrating serial connection of CE and CB amplifiers

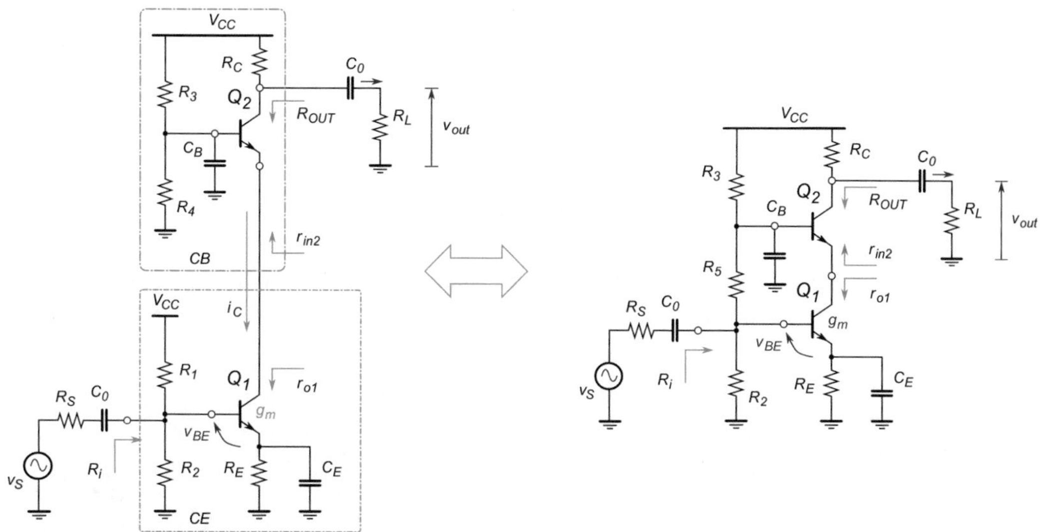

Fig. 6.25 Cascode amplifier, practical circuit schematic

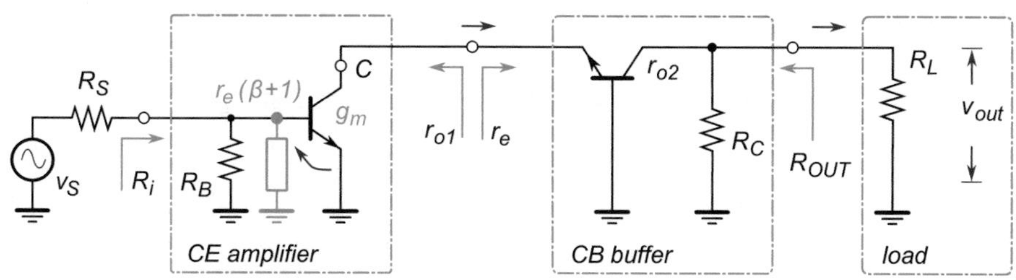

Fig. 6.26 AC model of cascode amplifier stage

Furthermore, instead of using two separate resistors R_1 and R_4 to derive biasing voltages, they are combined in one R_5 resistor as in Fig. 6.25 (right). In addition, in order to maximize gain of the CE amplifier, R_E connected to Q_1 is bypassed by C_E (see (6.39)), which creates virtual AC ground.

6.6.1 AC Circuit Model

Small signal AC model of a cascode amplifier is derived from Fig. 6.25 resulting in circuit in Fig. 6.26. Following analysis from the previous sections we write by inspection:

1. *Input resistance* of the CE amplifier, in the case of $R_E = 0$ due to the bypass capacitor C_E, is found as parallel connection of R_B and $r_\pi = r_e(\beta + 1)$ (see (6.39)), i.e.

$$R_i = R_B || r_\pi \tag{6.58}$$

2. *Output resistance* at the collector of the CB amplifier consists of R_C in parallel with collector resistance r_{o2} of Q_2 with degenerated emitter, see (4.60), where the virtual emitter resistance is in effect the collector resistance r_{o1} of Q_1, hence

$$R_{OUT} = R_C||r_{o2}\left[1 + \frac{\beta\, r_{o1}}{r_{o1} + r_\pi}\right] \tag{6.59}$$

which can be further approximated in the case of $r_{o1} \gg r_\pi$ and subsequently $\beta \gg 1$, as

$$R_{OUT} \approx R_C||r_{o2}\left[1 + \frac{\beta\, \cancel{r_{o1}}}{\cancel{r_{o1}}}\right] \approx R_C||\beta\, r_{o2} \tag{6.60}$$

which illustrates large increase in the output resistance at BJT's collector that is consequence of cascoding transistors. The total resistance R_o at the output node of a cascoded amplifier is therefore

$$R_o = \beta\, r_{o2}||R_C||R_L \tag{6.61}$$

3. *Voltage gain:* by inspection, we write

$$v_{out} = -i_C \times R_C||R_L = -g_m\, v_{BE} \times R_C||R_L$$

$$\therefore$$

$$A_v = \frac{v_{out}}{v_{BE}} = -g_m(R_C||R_L) \tag{6.62}$$

which is as same as the voltage gain of a CE amplifier by itself. However, relative to CE amplifier the output resistance is increased orders or magnitude, which is interpreted as a current source that is much closer to the ideal case (whose output resistance equals infinity). Therefore, almost by default, realistic current sources are made out of cascode stages.

6.6.2 Output Resistance of FET Cascode Transistor

Motivation of using cascode transistors is mostly due to orders of magnitude higher output resistance in comparison with a single transistor, as in (6.61). In this section we quickly evaluate output resistance of a JFET cascode transistor, Fig. 6.27.

By applying a small signal test current i_x, we measure v_x after all other voltage sources are shorted including v_{sig}. by inspection of the small signal model schematic, Fig. 6.27 (right), we write

$$v_{\text{sig}} = 0 \Rightarrow v_{gs1} = 0 \Rightarrow g_{m1}v_{gs1} = 0 \Rightarrow v_{gs2} = -v_1 = -i_x r_{o1}$$

$$\therefore$$

$$i_x = \frac{v_x - v_1}{r_{o2}} + g_{m2}v_{gs2} = \frac{v_x - i_x r_{o1}}{r_{o2}} - g_{m2}\, i_x r_{o1}$$

Therefore, assuming two identical transistors (i.e. $r_{o1} = r_{02} = r_o$, and also $g_{m1} \approx g_{m2} = g_m$) we write the expression for a cascoded FET transistor as

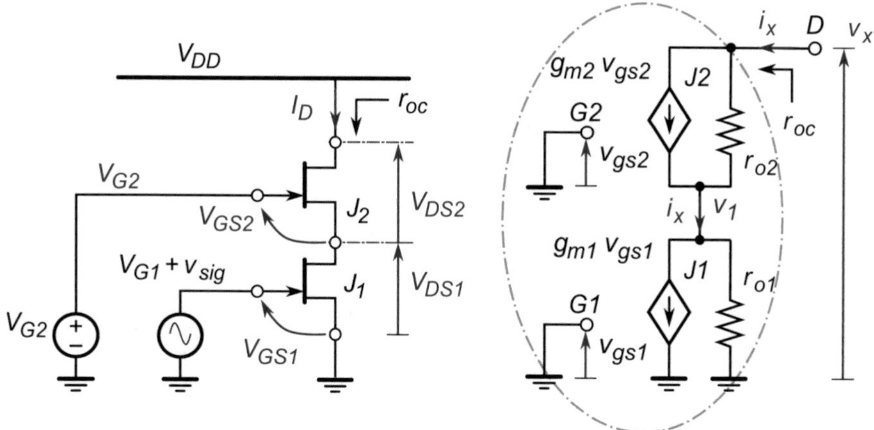

Fig. 6.27 JFET MOS cascode transistor

$$r_{oc} \stackrel{\text{def}}{=} \frac{v_x}{i_x} = (1 + g_{m2}r_{o1})r_{o2} + r_{o1} = r_{o1} + r_{o2} + g_{m2}r_{o1}r_{o2} \approx 2r_o + g_m r_o^2 \approx g_m r_o^2 \tag{6.63}$$

Assuming, for example, $g_m = 6\,\text{mS}$ and $r_o = 75\,\text{k}\Omega$ for a single JFET transistor, Eq. (6.63) shows that the output of cascoded transistor is $r_{oc} \approx 34\,\text{M}\Omega$, which is to say that, in the current technology, the cascoded JFET output is as close as it can get to the ideal current source. We note that from the perspective of J_2 the resistance at its drain is similar to the case of degenerated emitter (or source).

6.7 Case Study: Parameters of BJT CE Amplifier

In this section, based on the discussion in the previous sections, we show two examples of design procedure that is used to determine parameters of an N-type CE amplifier.

Development of an amplifier (or any other circuit for that matter) starts with establishing the collector current I_C vs. base–emitter voltage V_{BE} transfer function, Fig. 6.28. This function is found by simulation, or it may be provided in the transistor's datasheet. In this example, biasing current was arbitrary chosen to be $I_{C0} = 1\,\text{mA}$. We note that this is the first decision to be made in the design process. Modern IC electronics is usually biased with $1\,\mu\text{A}$, $10\,\mu\text{A}$, or some other similar "nice number". Similarly, general discrete components circuits may be set to 1 mA, 10 mA, or some other nice round number. This kind of choices for the biasing currents greatly simplifies all further calculations.

Schematic diagram in Fig. 6.28 shows DC simulation setup used to generate transfer functions in Fig. 6.29. As a result of DC simulation of a BJT device we obtain first set of design parameters:

(a) *The biasing point*: for example, if $I_{C0} = 1\,\text{mA}$, for this particular transistor, then the biasing voltage must be $V_{BE0} = 653\,\text{mV}$. Due to the exponential relationship (4.31) resolution of 1 mV is appropriate for the biasing voltage. Therefore, as per transfer function in Fig. 6.29 (left), the biasing point is set to

$$(I_{C0}, V_{BE0}) = (1\,\text{mA}, 653\,\text{mV}) \tag{6.64}$$

(b) *BJT g_m gain*: the biasing point is also interpreted by its associated g_m gain. Either by using SPICE numerical functions, or manually we estimate the tangent at the biasing point, Fig. 6.29 (left). Once the biasing point is chosen g_m gain is already fixed. Manual technique is based on applying

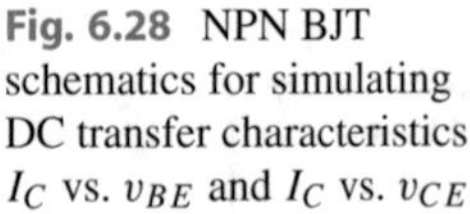

Fig. 6.28 NPN BJT schematics for simulating DC transfer characteristics I_C vs. v_{BE} and I_C vs. v_{CE}

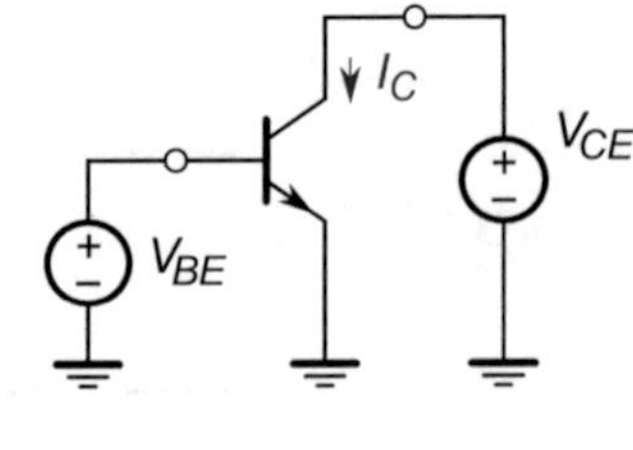

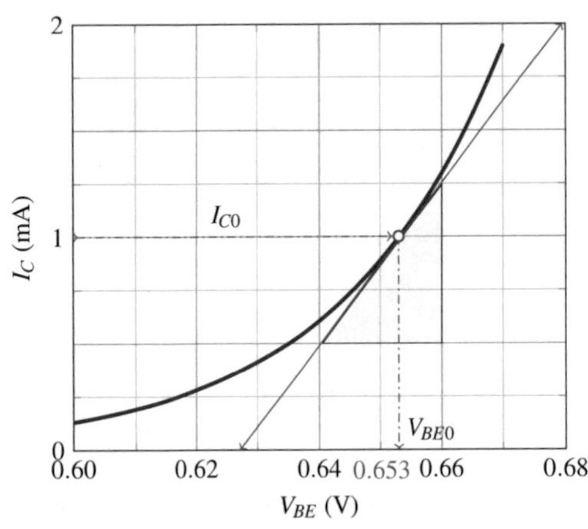

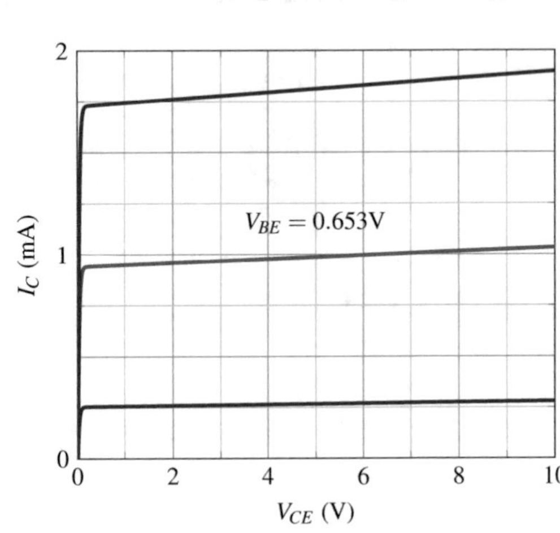

Fig. 6.29 NPN BJT: DC transfer characteristics

Pythagorean theorem on a convenient right angle triangle, as for example yellow triangle in Fig. 6.29 (left) (or any other similar triangle), that results in

$$g_m = \left.\frac{dI_C}{dV_{BE}}\right|_{(I_{C0},V_{BE0})} = \frac{\Delta I_C}{\Delta V_{BE}} = \frac{(1-0)\,\text{mA}}{(0.653-0.628)\,\text{V}} = 40\,\text{mS} \tag{6.65}$$

or, equally from (4.35) and (4.37), at room temperature, we calculate

$$g_m = \frac{I_C}{V_T} = \frac{1\,\text{mA}}{25\,\text{mV}} = 40\,\text{mS} \tag{6.66}$$

(c) *Emitter r_e resistance*: direct consequence of I_{C0} choice is that by (4.39) we find that

$$r_e = \frac{1}{g_m} = \frac{25\,\text{mV}}{1\,\text{mA}} = 25\,\Omega \tag{6.67}$$

Note that the above result is easily evaluated for $I_C = 1\,\text{mA} \rightarrow r_e = 25\,\Omega$, $I_C = 2\,\text{mA} \rightarrow r_e = 12.5\,\Omega$, etc., which is easily calculated without a calculator.

(d) *Collector resistance r_o*: set of the output characteristics, Fig. 6.29 (right), is used to determine collector r_o resistances for various collector currents. This characteristics is obtained by experiment or simulation, and for this particular transistor with $I_{C0} = 1\,\text{mA}$, we estimate the collector resistance by using the linear region of the graph (when the current is in ideal case constant), Fig. 6.30 (left). Then, as per (4.59) and Pythagorean theorem we write

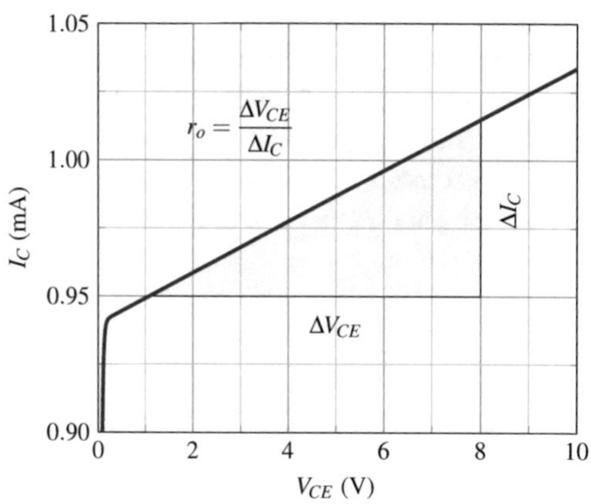

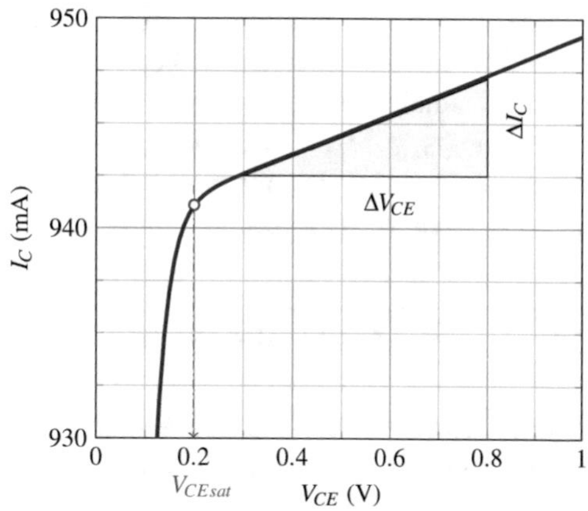

Fig. 6.30 NPN BJT schematics for simulating DC transfer characteristics I_C vs. v_{BE} and I_C vs. v_{CE}

$$r_o \overset{\text{def}}{=} \left.\frac{\Delta V_{CE}}{\Delta I_C}\right|_{V_{BE}=\text{const.}} = \frac{(10-1.1)\,\text{V}}{(1010-950)\,\mu\text{A}} = 115\,\text{k}\Omega \approx 100\,\text{k}\Omega \tag{6.68}$$

It is important to note that lower values of I_{C0} have smaller current variations, i.e. higher r_o. This relationship provides another argument for choosing one biasing current over the other and permits us to make trade-offs if necessary.

(e) *Early voltage* V_A: this technological parameter is calculated by extrapolation of curves in Fig. 6.29. Focusing on a single characteristics that corresponds to $I_{C0} = 1\,\text{mA}$, Fig. 6.30 (left), and properties of similar triangles, we write equation for proportions of the yellow triangle Fig. 6.30 (left) and triangle formed by I_{C0} and V_A (see, for example, Fig. 4.20) as

$$\frac{\Delta V_{CE}}{\Delta I_C} = \frac{V_A}{I_{C0}} \quad \therefore \quad V_A = I_{C0} \times \frac{\Delta V_{CE}}{\Delta I_{C0}} = 1\,\text{mA} \times 100\,\text{k}\Omega = 100\,\text{V} \tag{6.69}$$

(f) *Saturation voltage* $V_{CE}(min)$: zoom to the region close to the end of the (almost) constant current mode, Fig. 6.30 (right), shows that the collector resistance stays constant until the point when V_{CE} drops below $V_{CE}(min) \approx 0.2\,\text{V}$. Further reduction in V_{CE} voltage forces the transistor to start behaving as a linear resistor, and to enter its "linear resistor" mode of operation where the output resistance r_o is direct function of V_{CE}. It is important to keep this relationship in mind

$$V_{CE} \geq V_{CE}(min) \tag{6.70}$$

and verify its validity once the circuit design is finished. Unless intentionally setting up a transistor in this mode of operation, if the last check-up of any transistor in the circuit shows violation of (6.70) the circuit must be redesigned again. The technological parameter $V_{CE}(min)$ is typically in the range of 100–200 mV for modern transistors.

The initial choice of biasing current I_{C0} is constrained by the specifications of targeted application, current technology, device size, power consumption, price, etc., and the preferred trade-offs. Thus, with DC simulation we characterize BJT transistor and subsequently by "domino effect" find the ensuing V_{BE0}, g_m, r_e, r_o, V_A, and $V_{CE}(min)$ parameters. Once these small signal BJT parameters are determined, we can start the circuit development. However, the circuit design is an iterative process and very often we are forced to return to the step one and choose another value of the biasing current.

6.8 Amplifier Design Flow

In this section we briefly review a simple design flow (among many possible) for a simple BJT CE amplifier circuit in Fig. 6.13. The objective is to illustrate some of the practical knottiness in the circuit synthesis process. Main source of these difficulties is the iterative nature of the underlying set of equations. In order to see how the design knot is created, let us start with a goal to create CE amplifier with voltage gain A_V.

Naturally, we could start with (6.37) that gives the upper limit gain as $A_V = g_m R$. Therefore, there are two design variables, g_m and R, that must be known a priori. Since there is an infinity of possibilities to choose from so that their product equals to A_V, without other constrains we are forced to choose one of the two. So, to start with, first we may choose a specific transistor to give us a certain range g_m values.

Then, unavoidably, for the given transistor's g_m the following parameters are also determined: r_e, I_{C0}, V_{BE0}. At the same time, in order to satisfy (6.37) the value of total collector's resistance is enforced too. We note that the total collector resistance consists of $R = R_L||r_o||R_C$, where r_o is the BJT collector resistance, R_C is the external collector resistor, and R_L is the external loading resistance.

However, the total collector's resistance R in combination with biasing current I_{C0} sets DC voltage at the collector as

$$V_C = V_{CE} = V_{CC} - V_{out} = V_{CC} - R\, I_{C0}$$

which goes back to the gain A_V, see Fig. 6.31. The problem is that the resulting V_{CE} may be either too close to the ground, thus less than $V_{CE}(min)$, or too close to V_{CC}. Both cases are not good outcome: the transistor is either shut down (because $V_{CE} < V_{CE}(min)$) or there is not enough headroom for the signal. Ideally, V_C should be around $V_{CC}/2$ level so that there is maximal symmetric headroom for the amplified signal.

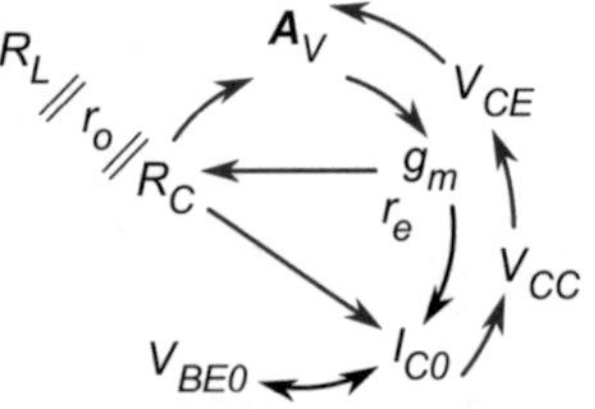

Fig. 6.31 Relation among design variables

6.9 Summary

In this section we reviewed fundamental concepts of LF amplifiers and developed intuitive views of the internal amplifier operation. In our review, we concluded that important parameters of any amplifier are its input and output resistance, and its gain. We also realized that two basic electrical variables, voltage and current, determine the total of four possible amplifier transfer functions: voltage gain A_v, current gain A_i, voltage to current gain G_m, and current to voltage gain A_R.

As the first step in amplifier design, gain of the active devices (either BJT or FET) is set by their DC operating point, and the subsequent signal analysis is simplified by omitting details of the biasing circuit, i.e. it is simply assumed that the g_m gain is already set. We reviewed basic circuit configurations for setting up a stable DC operating points for the active devices. In the first approximation, a BJT device is seen as a current amplifier, where the current gain β serves as the multiplication factor in the relationship between the base and collector currents. After the collector current is passed through a resistive load R_{ct}, which is effectively seen as a current to voltage amplifier, i.e. input base current is amplified into voltage across the loading resistor. The alternative view of BJT and FET devices is the one of Gm amplifier, i.e. the input voltage is converted into the ouput current.

Transformation of LF baseband amplifier into an RF amplifier is done by replacing the collector resistance with RLC resonator.

Problems

6.1 Design the model of a voltage amplifier whose gain is, for example, $A_V = 40\,\text{dB}$. The ideal amplifier model is to be made as a two-stage amplifier, where the first stage must be a G_m amplifier. The input voltage signal v_S is generated by an antenna and is to be delivered the load resistance R_L. Propose two "back of the envelope" model solutions by assuming that both gain stages are:

(a) ideal, i.e. their respective input/output resistances are assumed to be ideal as needed,
(b) more realistic, i.e. their respective input/output resistances are non-ideal.

Comment on your two solutions and the necessary constrains resulting from the above assumptions.

6.2 Design a simple NPN BJT amplifier in Fig. 6.13, given the only objective is to have $A_V = 40\,\text{dB}$. Use transistor whose characteristics are given in Fig. 6.32.

6.3 As a follow up to Problem 6.2, calculate maximum resistance at the collector node.

6.4 Given the results and experience encountered in Problem 6.2, propose a modification of the initial specifications and a possible design flow to achieve the new goal.

6.5 Design biasing circuit to support CE BJT amplifier designed in Problem 6.4. How many possible solutions can you propose?

6.6 By SPICE simulations verify functionality of the amplifier designed in Problem 6.5 over the temperature range specified by aerospace and military standards, i.e. from $-55°$ to $+125°$. What are the result showing? What do you think is the explanation for the results? and can you propose a possible remedy ?

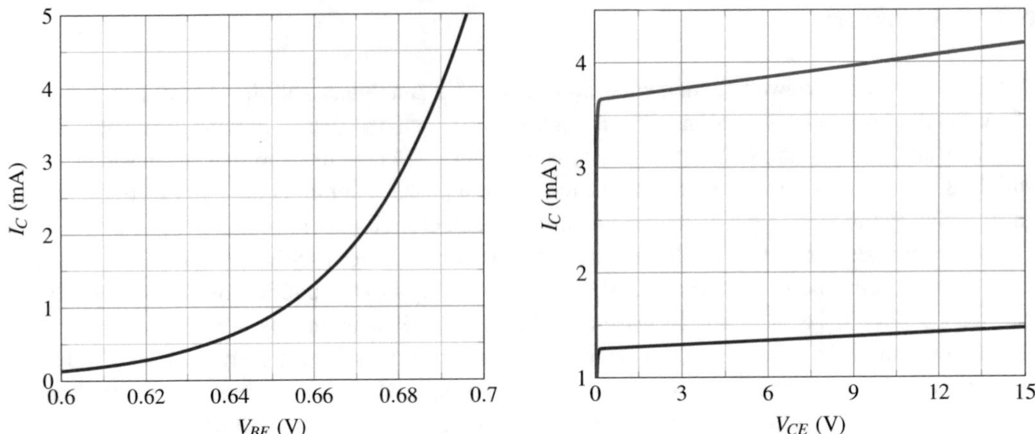

Fig. 6.32 NPN BJT characteristics for Problem 6.2

6.7 Modify design solution from Problem 6.4 by adding the external emitter resistor R_E while preserving the achieved gain A_V. Comment on your new circuit's functionality.

6.8 Compare performances of two amplifiers designed in Problems 6.5 and 6.7 over the same temperature range $-55°$ to $+125°$. How do you explain the observed simulation results.

6.9 Design similar amplifiers based on NFET and JFET transistors of your choice, and compare with the your best solution based on NPN BJT. Comment on the three amplifier versions.

Introduction to Frequency Analysis of Amplifiers 7

Frequency independent analysis that is used so far in this book provides fast and relatively simple way to analyse and design amplifiers and other electronic circuits. This method is based on a simple assumption that the circuit is capable to accept and process signals whose frequency spectrum includes all frequencies, from minus infinity to plus infinity. Nonetheless, we already know that elements capable to store energy need a finite amount of time to change their internal states. For slow changes this time delay is negligible, thus "low-frequency" approximation produces acceptable results. However, as the signal frequency increases, the impedances of frequency dependent components drastically change. In this chapter we review basic techniques of "frequency analysis" that is appropriate for low to medium frequencies.

7.1 Amplifier Bandwidth

A simple first-order RC (or RL) network shows frequency profile that corresponds to either HPF or LPF, see Chap. 3 and they are simplest frequency dependent circuit models (LC network comprises two energy storage elements, thus it is a second-order network). In principle, to the first approximation any circuit could be reduced and approximated by its equivalent RC or RL network. In other words, practical technique to determine frequency domain response of an amplifier involves a sequence of circuit reductions until a simple RC or RL network is achieved and analysed.

Complex components (i.e. C or L) are in effect *frequency controlled resistances* that create voltage/current dividers whose gains are also frequency dependent. At the same time, realistic communication systems are based on processing of multi-tone signals (see Sect. 9.1.2) such as square and other modulated waveforms, not only a single sinusoidal forms. For this reason, it is important to determine "frequency bandwidth", that is, the range of acceptable frequencies (or, frequency spectrum of the I/O signal), of amplifiers intended for RF applications.

A typical electronic circuit shows the frequency characteristics as in Fig. 7.1, where three distinct regions are clearly visible:

1. *LF band:* this region is characteristic for DC frequency component being completely blocked by infinite resistance of HPF path (see Sects. 3.2 and 3.3). On decibel scale, DC gain limits to negative infinity (i.e. DC means $\omega = 0$, thus $\log(0) = -\infty$). Very low-frequency tones close to DC experience HPF attenuation of $+20^{\text{dB}}/_{\text{dec}}$ from very low frequency until ω_L. This boundary

R. Sobot, *Wireless Communication Electronics*,
https://doi.org/10.1007/978-3-030-48630-3_7

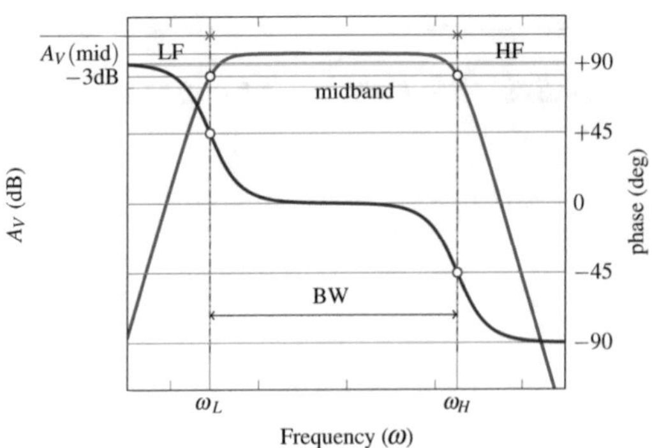

Fig. 7.1 A typical frequency bandwidth profile and its definitions

frequency is set at the point of -3dB attenuation relative to the maximal amplitude as measured in decibels. This region is consequence of large capacitive/inductive components in the circuit, and we note that the phase is $\pi/2$ for most of this region. However, at ω_L the phase is $\pi/4$.

2. *Midband:* this region is characteristic for its flat gain over the wide range of frequencies from ω_L to ω_H. Here, all external and internal capacitances are neglected: the external capacitances are shorted, and internal parasitic capacitances are omitted. This range of frequencies is important, it is often referred to as "bandwidth" (BW), and in the case of linear circuits is used in the definition of "gain-bandwidth" (GBW) product as

$$GBW \stackrel{\text{def}}{=} A_{V_{\text{mid}}} \times BW \tag{7.1}$$

which is (in the case of linear circuits) a constant quantity. Relationship in (7.1) indicates that we can reduce gain to increase BW and vice versa. This trade-off is routinely used in amplifier design to achieve, for example, either desired high narrow-band gain or low wide-band gain. The -3dB frequency at the end of midband is set by ω_H. We note that the phase is zero (i.e. the circuit transfer function is real) for most of the midband region.

3. *HF band:* this HF region is characteristic for its LPF attenuation that starts at ω_H frequency, which is -3dB point of LPF created by the output resistance and small parasitic capacitances in BJT/MOS devices. When the circuit is reduced to the first order LPF, the attenuation rate is $-20^{\text{dB}}/_{\text{dec}}$ and the phase follows accordingly as in Fig. 7.1. We note that the phase is $-\pi/2$ for most of this region, however, at ω_H the phase is $-\pi/4$.

The objective of doing frequency analysis is to determine bandwidth of an amplifier, i.e. possible frequency range of signals being processed by the amplifier. In practice, this analysis is done in four steps: (a) the midband gain is calculated with all capacitances appropriately neglected; (b) LF analysis takes into account only the external impedances; (c) HF analysis includes only the internal parasitic impedances; and (d) Body plot summary is created.

7.2 Frequency Analysis: Definitions

Discussion on the subject of frequency analysis of single-transistor amplifiers is based on the following "vocabulary":

1. *Time constant* τ: it is remarkable to note that the product of two basic electrical variables, resistance (whose unit is Ω) and capacitance (whose unit is F), results in the physical variable of *time* (whose unit is s)

$$\tau = RC \text{ [s]} \tag{7.2}$$

Similarly, in the case of RL network, the time constant is

$$\tau = \frac{L}{R} \text{ [s]} \tag{7.3}$$

2. *The* 3dB *frequency*: main significance of the time constant τ is that it equals to the inverse of the 3dB frequency ω_0, either in the case of LPF or HPF, and is calculated as

$$\tau = \frac{1}{\omega_0} = \frac{1}{2\pi f_0} \tag{7.4}$$

which enables us to set and calculate boundaries of the associated frequency range by choosing values of R and C components. In the case of multiple poles the total time constant is calculated as $\tau_H \cong \sum \tau_k$, where each τ_k is calculated independently.

3. *Zeros*: as a consequence of the reactive component properties, the resulting transfer function $A(s)$ is complex and rational function of the complex variable $s = j\omega$, whose factorized form is written as

$$A(s) = A_0 \frac{(s - z_1)(s - z_2)(s - z_3)\cdots}{(s - p_1)(s - p_2)(s - p_3)\cdots} \tag{7.5}$$

where $z_1, z_2, z_3, \ldots,$ are referred to as the "zeros" of function $A(s)$. This term comes from algebra, where we learned that if any of the factors in the numerator equals zero, then amplitude of $A(z_i)$ is forced to zero. That is, if

$$s - z_1 = 0 \quad \therefore \quad s = z_1 \quad \Rightarrow \quad A(s) = 0$$

$$s - z_2 = 0 \quad \therefore \quad s = z_2 \quad \Rightarrow \quad A(s) = 0$$

etc.

By setting $s = j\omega = 2\pi f$ we give physical interpretation to the above algebra reasoning as follows. In the frequency domain, gain of transfer function $A(j\omega)$ becomes zero at frequencies $f_1, f_2, \ldots,$ that are calculated as

$$s = z_1 \quad \therefore \quad j\omega_1 = j\,2\pi f_1 = z_1 \quad \Rightarrow \quad f_1 = \frac{z_1}{j\,2\pi}$$

$$s = z_2 \quad \therefore \quad j\omega_2 = j\,2\pi f_2 = z_2 \quad \Rightarrow \quad f_2 = \frac{z_2}{j\,2\pi}$$

etc.

In short, the factorized form (7.5) of $A(j\omega)$ conveys numerical values of z_i (which are mathematical numbers) that are easily converted into ω_i thus f_i (which are physical values).
In the piecewise linear approximation of a frequency response function (see Sect. 1.6) we find that at frequencies f_1, f_2, etc. the gain function $A(s)$ starts to *increase* at a rate of +20dB for each subsequent ten-fold increase in frequency, thus "+20dB/dec" expression. Precise calculation shows the gain of $1/sqrt2 = -3$dB, thus "+3dB pole frequency" expression (see, for example, Fig. 1.26).

4. *Poles*: The other extreme of $A(s)$ function (7.5) is when its denominator equals zero, i.e. $s_i = p_i$, which forces amplitude of $A(s) \to \infty$, thus "pole" of the transfer function at each respective frequency. In the piecewise linear approximation of a frequency response function (see Sect. 1.6) at frequencies f_1, f_2, etc. corresponding to the respective poles of the frequency response (i.e. the gain) function $A(s)$ starts to *decrease* at a −20dB for each subsequent ten-fold increase in frequency, thus "−20dB/dec" expression. Precise calculation shows the gain of $1/sqrt2 = -3$dB, thus "−3dB zero frequency" expression (see, for example, Fig. 1.28).
5. *Dominant pole* f_0: In the case of circuits with multiple poles and zeros, as in (7.5), there are two possible situations. If the pole and zero frequencies are sufficiently separated, then we assume that there is no significant interaction among them. In that case, the highest-frequency pole is declared *dominant* and the others are neglected. We note that in the single-pole networks there is already only one pole, thus it is dominant. In the case of multiple poles and zeros, however, it can be shown that the dominant pole frequency f_0 is found to be very close to

$$f_0 \cong \frac{1}{\sqrt{\dfrac{1}{f_{p1}^2} + \dfrac{1}{f_{p2}^2} + \cdots - \dfrac{2}{f_{z1}^2} - \dfrac{2}{f_{z2}^2} - \cdots}} \tag{7.6}$$

which, again, helps us to approximate a multi-poles circuit by using a single-pole model.

Example 48: Dominant Pole Estimate
Estimate dominant pole if transfer function of an amplifier contains two pole and one zero frequency: $f_{p1} = 10\,\text{kHz}$, $f_{p2} = 50\,\text{kHz}$, and $f_{z1} = 100\,\text{kHz}$.

Solution 48: Direct application of (7.6) gives

$$f_0 = \frac{1}{\sqrt{\dfrac{1}{f_{p1}^2} + \dfrac{1}{f_{p2}^2} - \dfrac{2}{f_{z1}^2}}} = \frac{1}{\sqrt{\dfrac{1}{10\,\text{kHz}^2} + \dfrac{1}{50\,\text{kHz}^2} - \dfrac{2}{100\,\text{kHz}^2}}} = 9.76\,\text{k}\Omega$$

which is very close to $f_{p1} = 10\,\text{kHz}$ because the second pole and the zero are "sufficiently" far away, thus we could make the approximation without the above calculation.

7.3 Frequency Analysis of Single-Stage Amplifiers

In this section we review circuit reduction techniques that enable us to estimate time constants, that is to say poles and zeros, found in the three basic single-transistor amplifiers. With that information we can further estimate frequency range of the input signal frequencies that can be amplified, or frequency range of the input signal frequencies that are attenuated. The technique by inspection is based on approximation of time constants associated with each RC branch separately.

7.3.1 Time Constants in CB Amplifier

In the frequency independent analysis we assume that all internal discrete capacitors are approximated with a short connection (i.e. either a large capacitor or high frequency). In other words, we assume an ideal frequency independent amplifier that equally amplifies all input signals from DC to infinite frequency. In this section we estimate frequency limitations of realistic amplifiers, that is to say, we determine range of frequencies where the signal is amplified and where it is attenuated. In the first approximation, we consider only the external discrete capacitors that are large relative to the internal parasitic capacitances found in the active devices. Therefore, this approximation is referred to as a "low-frequency analysis".

In this technique, the objective is to reduce given circuit to a simple equivalent R that is "seen" by each C in the network. The equivalent resistances connected to three discrete capacitors C_1, C_2, C_3 found in CB amplifier, Fig. 7.2, are determined by circuit reduction techniques. Thus, we find three associated time constants as follows.

1. *The input side node:* we focus only on C_1 that connects the voltage source v_S to CB input terminal, and search the equivalent resistance R_{eq} that is perceived by this capacitor. Resistance of the voltage source generator v_S is zero therefore it is shorted, which connects the bottom node of R_S to ground while its other node stays connected to C_1. The second terminal of C_1 perceives resistance equivalent to R_i of CB amplifier, Fig. 7.3 (left). We already found (see (6.24)) that the input side resistance of CB amplifier is equivalent to

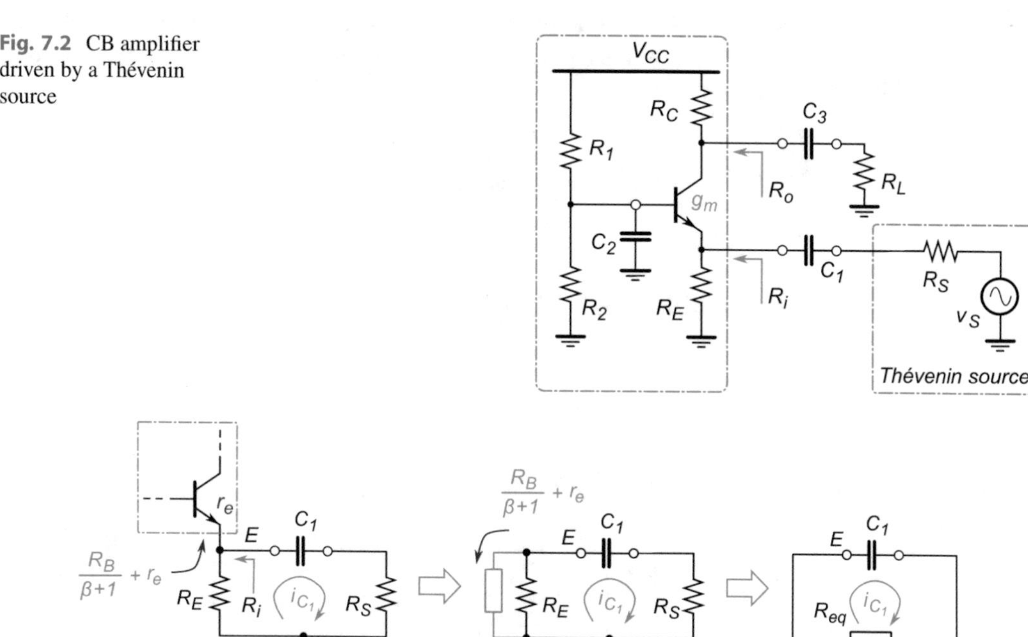

Fig. 7.2 CB amplifier driven by a Thévenin source

Fig. 7.3 The equivalent RC network associated with the CB input terminal

$$R_i = R_E || \left(r_e + \frac{R_B}{\beta + 1} \right) \tag{7.7}$$

Following the capacitor's current i_C flow, by inspection of Fig. 7.3 we write

$$R_{eq} = R_S + R_i = R_S + R_E || \left(r_e + \frac{R_B}{\beta + 1} \right)$$

which after substitution in (7.2) leads into the expression for the input side time constant τ_1 as

$$\tau_1 = C_1 \left[R_S + R_E || \left(r_e + \frac{R_B}{\beta + 1} \right) \right] \tag{7.8}$$

The fact that one of the internal nodes in $R_{eq}\, C_1$ loop is connected to DC ground does not change the conclusion that R_S and R_i are connected in series. It is the path of AC current i_C that helps us determine whether resistances are connected in series or in parallel.

2. *The base node:* capacitor C_2 creates AC ground at the base terminal. We find that, see Fig. 7.4, on the one side C_2 is connected to ground, and on the other side to the equivalent resistance R_{eq} that consists of R_B and virtual projected resistance in parallel, as

$$R_{eq} = R_B || (\beta + 1)(R_E + r_e)$$

which after substitution in (7.3) leads into the second time constant τ_2 as

$$\tau_2 = C_2 \left[R_B || (\beta + 1)(R_E + r_e) \right] \tag{7.9}$$

3. *The output node:* Following the signal to the output side, capacitor C_3 is connected on the one side to R_L and on the other side to the equivalent R_{OUT}resistance attached to the collector node, Fig. 7.5. Thus, by inspection we write

$$R_{OUT} = R_C || r_o \left(1 + \frac{\beta\, R_E}{R_E + R_B + r_\pi} \right)$$

$$\therefore$$

$$R_{eq} = R_L + R_{OUT} = R_L + R_C || r_o \left(1 + \frac{\beta\, R_E}{R_E + R_B + r_\pi} \right)$$

$$\therefore$$

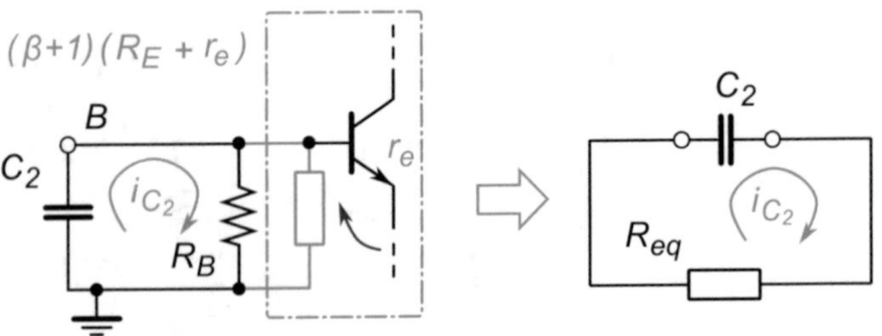

Fig. 7.4 The equivalent RC network associated with the CB base terminal

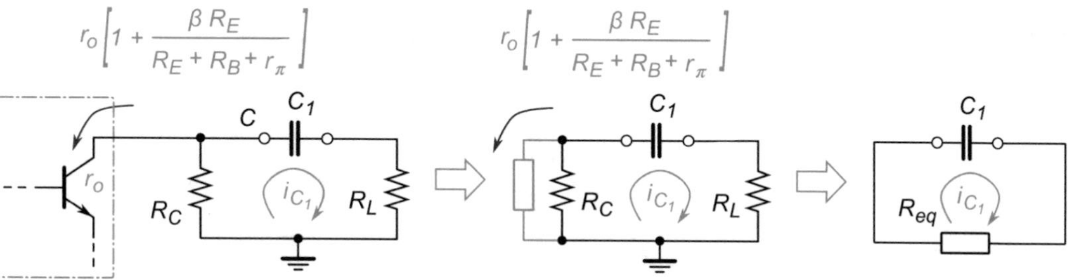

Fig. 7.5 The equivalent RC network associated with the CB output terminal

$$\tau_3 = C_3\left[R_L + R_C||r_o\left(1+\frac{\beta\, R_E}{R_E+R_B+r_\pi}\right)\right] \tag{7.10}$$

The nature of the time constants, i.e. whether it belongs to pole or zero, is determined by the equivalent RC circuit network. If the capacitor is on the signal's path it creates a zero (i.e. blocks DC component of the signal, because if $f \to 0$, then $Z_C = 1/(2\pi f C) \to \infty$), thus it is HPF. If the capacitor provides path to the ground (because if $f \to \infty$, then $Z_C = 1/(2\pi f C) \to 0$), i.e. it shorts HF component of the signal, thus creates LPF interface.

Example 49: CB Time Constants
Assuming $\beta \to \infty$ and $R_E \gg r_e$ estimate τ_1, τ_2, and τ_3 time constants of CB amplifier.

Solution 49: Given data, we write the following approximations

$$\tau_1 = C_1\left[R_S + R_E||\left(r_e+\frac{R_B}{\beta+1}\right)\right] \approx C_1\,(R_S + R_E||r_e) \approx C_1 R_S$$

$$\tau_2 = C_2\,[R_B||(\beta+1)(R_E+r_e)] \approx C_2 R_B$$

$$\tau_3 = C_3\left[R_L + R_C||r_o\left(1+\frac{\beta\, R_E}{R_E+R_B+r_\pi}\right)\right] \approx C_3\;(R_L+R_C)$$

7.3.2 Time Constants in CE Amplifier

For each of the three main capacitors in CE amplifier (see Fig. 7.6), which are constantly charged and discharged during the amplifier operation, there is an equivalent resistance connected to their respective terminals. These resistances provide the charge/discharge current paths that, in effect, determine the total amplifier's frequency response.

With the approximation that there is no interference between these three RC networks, by inspection we find their associated time constants. Following the usual procedure for creating a small signal AC model, we can isolate the equivalent RC circuits at the input and output sides, as well as the network connected to the emitter terminal.

1. *Time constant due to C_1:* the sequence of circuit transformations in Fig. 7.7 enables us to use "by inspection" technique to determine time constant due to $R_{eq}C_1$ network and to write

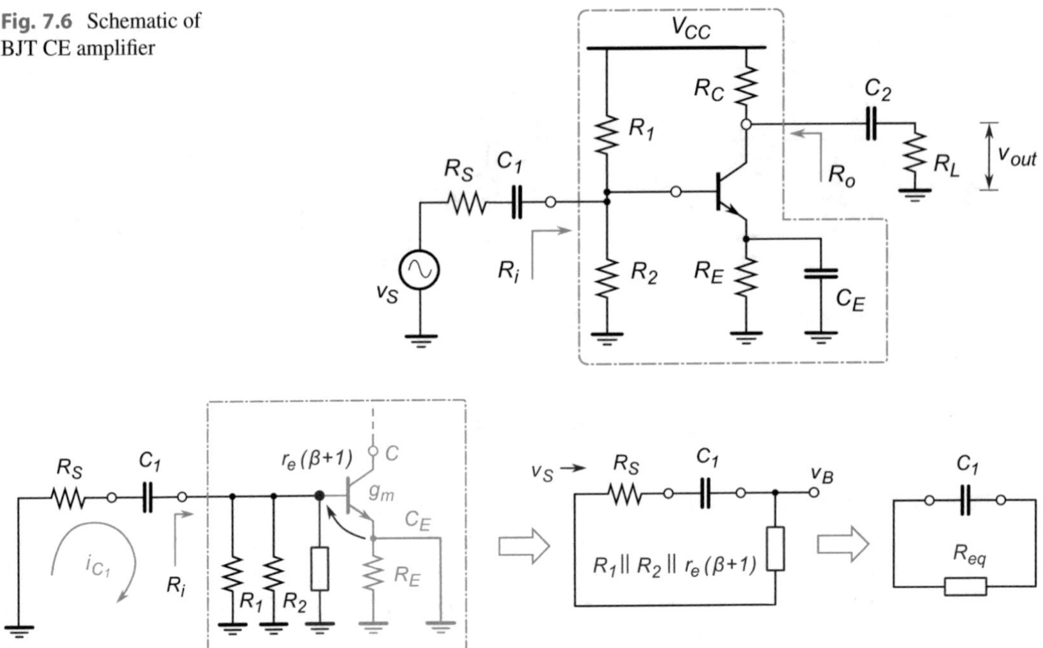

Fig. 7.6 Schematic of BJT CE amplifier

Fig. 7.7 The equivalent RC network at the input terminal of CE amplifier

$$R_{eq_1} = R_S + R_1||R_2||r_e(\beta + 1) \approx R_S + R_1||R_2||\beta r_e$$

$$\therefore$$

$$\tau_1 = R_{eq_1} C_1 = [R_S + R_B||(\beta + 1)r_e]\ C_1, \qquad (R_B = R_1||R_2) \tag{7.11}$$

therefore,

$$\omega_{\tau_1} = \frac{1}{[R_S + R_B||(\beta + 1)r_e]\ C_1} \tag{7.12}$$

Time constant found by the circuit reduction technique gives frequency of poles and/or zeros of the overall gain transfer function due to C_1. It is therefore necessary to deduce the nature of this constant in this specific case.

Capacitor C_1 presents a serial impedance inserted into the input signal's path, therefore it blocks DC component of the input signal current i_{C_1}—in other words, the signal gain is zero at DC frequency. Gain of RC voltage divider formed by (R_S, C_1) on the one side and the total equivalent base resistance $R'_B = R_B||r_e(\beta + 1)$ on the other side (see Fig. 7.7 (middle)) is obviously frequency dependent. Thus, we observe that DC gain of this voltage divider is zero because in that case $Z_{C_1} = \infty$, which blocks DC signal current. On the other side of the spectrum, at high frequencies the capacitor impedance becomes $Z_{C_1} = 0$, which transforms this RC network into a simple resistive voltage divider. By inspection of Fig. 7.7 (middle) we write

$$v_B = \frac{v_S}{R_S + Z_{C_1} + R'_B} R'_B$$

$$\therefore$$

$$A_1 \stackrel{\text{def}}{=} \frac{v_B}{v_S} = \frac{R'_B}{R_S + \frac{1}{j\omega C_1} + R'_B} = \frac{j\omega\, C_1\, R'_B}{1 + j\omega\, C_1 (R_S + R'_B)} = \underbrace{\frac{R'_B}{R_S + R'_B}}_{a_{01}} \frac{j\frac{\omega}{\omega_{\tau_1}}}{1 + j\frac{\omega}{\omega_{\tau_1}}} \tag{7.13}$$

where DC gain is determined by the equivalent resistive voltage divider at the base node

$$a_{01} = \frac{R'_B}{R_S + R'_B} \tag{7.14}$$

and it is always true that $a_{01} < 1$ (i.e. it is a negative number in log scale). The total gain transfer function (7.13) is shown in log scale as

$$20\log(A_1) = \underbrace{+\,20\log(a_{01})}_{\text{(1) Sect. 1.6.1}} + \underbrace{20\log\left(j\frac{\omega}{\omega_{\tau_1}}\right)}_{\text{(2) Sect. 1.6.2}} - \underbrace{20\log\left(1 + j\frac{\omega}{\omega_{\tau_1}}\right)}_{\text{(3) Sect. 1.6.3}} \tag{7.15}$$

This is a first order network (there is only one energy storing element used), where the three terms in (7.15) clearly show: (1) DC gain a_0 term (i.e. $\omega = 0$) due to the resistive divider; (2) zero forming term; and (3) pole forming term. Therefore, in log/log scale, the total frequency response (4) is derived as a simple sum[1] of the three terms in (7.15). We note that pole (contributing -20dB/dec) and zero (contributing $+20$dB/dec) are found at the same frequency ω_{τ_1}. Consequently, they cancel each other to produce 0dB/dec for frequencies above ω_{τ_1} and the overall HPF frequency response, see Fig. 7.8. As a side note, zero-pole cancelation technique is often used in circuit design.

2. *Time constant due to C_2:* Decoupling capacitor C_2 at the output side of CE amplifier provides AC signal path to the load R_L resistance while at the same time it blocks DC component of the output current i_{C_2}. Therefore this capacitor also sets a zero at the output side of CE amplifier.

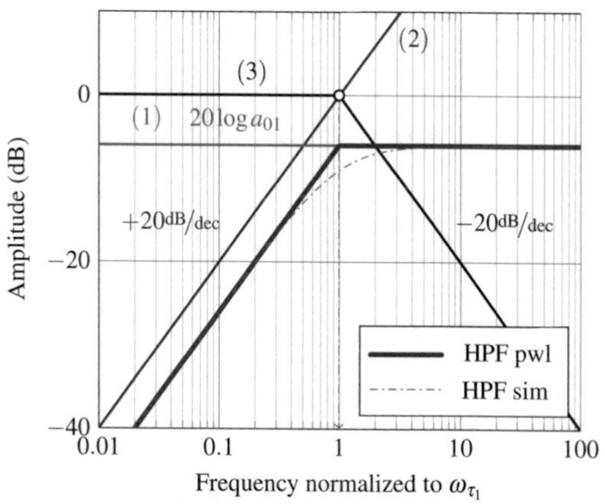

Fig. 7.8 HPF gain function of CE amplifier's equivalent input side network: piecewise approximation (4) found as the sum of three terms in (7.15), and its simulated AC curve (red)

[1]See Sect. 1.6.

By inspection of Fig. 7.9 (middle) we find RC voltage divider formed by $r_o||R_C$ and Z_{C_2} on the one side, and R_L on the other. At high frequencies this RC voltage divider also transforms into a simple resistive voltage divider formed by $r_o||R_C$ and R_L. Therefore, we conclude that frequency response of this RC network has the same form as in (7.15) and Fig. 7.8. This time, however, by inspection of Fig. 7.9 (middle) we write

$$R_{eq} = r_o||R_C + R_L \approx R_C + R_L \text{ when } (r_o \gg R_C) \tag{7.16}$$

$$\therefore$$

$$\tau_2 = (r_o||R_C + R_L)\, C_2 \approx (R_C + R_L)\, C_2 \tag{7.17}$$

which gives us pole/zero frequency at

$$\omega_{\tau_2} = \frac{1}{\tau_2} = \frac{1}{(r_o||R_C + R_L)\, C_2} \tag{7.18}$$

Bypass capacitor C_E is connected between the emitter terminal and ground, therefore there is no emitter degeneration, in other words the collector resistance is simply r_o. In addition, gain a_0 of the resistive voltage divider (when $Z_{C_2} = 0$) is found by inspection of Fig. 7.9 (middle) as

$$a_{02} = \frac{R_L}{r_o||R_C + R_L} \tag{7.19}$$

3. *Time constants due to C_E:* as opposed to C_1 and C_2, detailed analysis shows that capacitor C_E creates zero and pole pair that does not cancel each other, they are found at two different frequencies (Fig. 7.10).

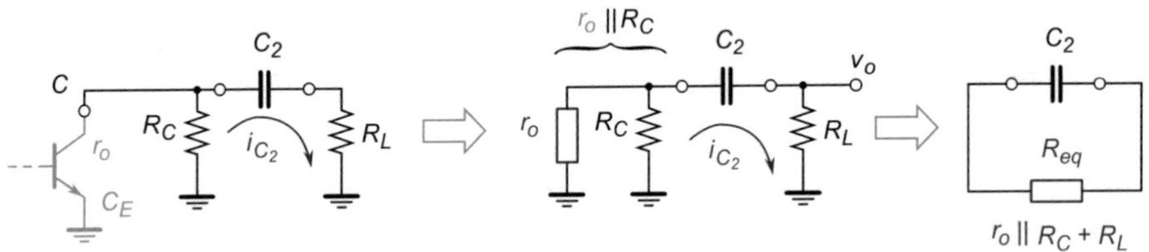

Fig. 7.9 The equivalent RC network at the output terminal of CE amplifier

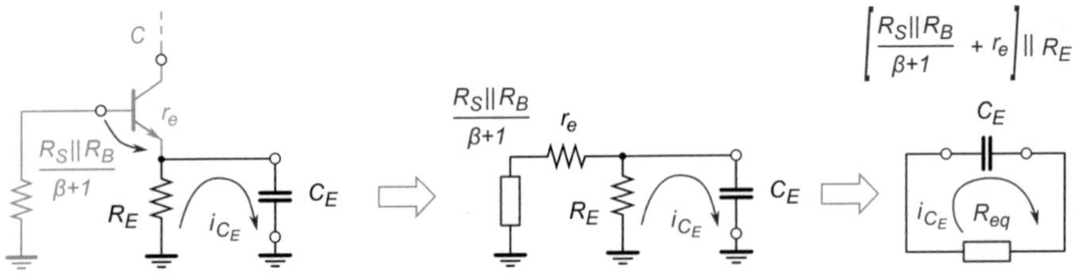

Fig. 7.10 The equivalent RC network at the emitter node of CE amplifier that sets pole of the transfer function

(a) *Zero due to* C_E*:* at DC, the input signal current leaving the emitter node can reach the ground node only through R_E (because $Z_{C_E}(DC) = \infty$). Consequently, in this case DC gain is not zero. Instead, as already found in the previous chapters, CE amplifier voltage gain[2] A_V is at its minimum theoretical value

$$a_{03} \approx \frac{R_C}{R_E} \qquad (r_0,\, R_L \to \infty,\, R_E \gg r_e) \tag{7.20}$$

where, in general, $a_{03} \geq 1$. Therefore in log/log scale DC gain is a positive number shown as a constant function (1) in Fig. 7.11.

Nevertheless, as the signal frequency moves from DC and increases, impedance of C_E decreases, which causes the total emitter impedance $R'_E = R_E || Z_{C_E}$ to also decrease. As a consequence, the voltage gain $|A_V| \approx R_C/R'_E$ gradually increases at $+20^{\text{dB}}/_{\text{dec}}$ rate. Therefore, there is a zero in the transfer function whose time constant τ_z is simply

$$\tau_z = R_E\, C_E \tag{7.21}$$

$$\therefore$$

$$\omega_z = \frac{1}{R_E\, C_E} \tag{7.22}$$

In log/log scale, this zero forming term is a piecewise function (2) whose slope is $+20^{\text{dB}}/_{\text{dec}}$, see Fig. 7.11. It starts as zero until ω_z frequency, then it introduces $+20^{\text{dB}}/_{\text{dec}}$ slope.[3]

(b) *Pole due to* C_E*:* in reference to Fig. 7.10 we note that one terminal of C_E is connected to ground, while it perceives a signal being delivered by the total emitter resistance on its other side. Therefore, this network is equivalent to RC voltage divider[4] whose gain is zero at hight frequency when $Z_{C_E} = 0$. In other words, this RC network is equivalent to LPF whose pole frequency $\omega_p(C_E)$ is found by inspection as

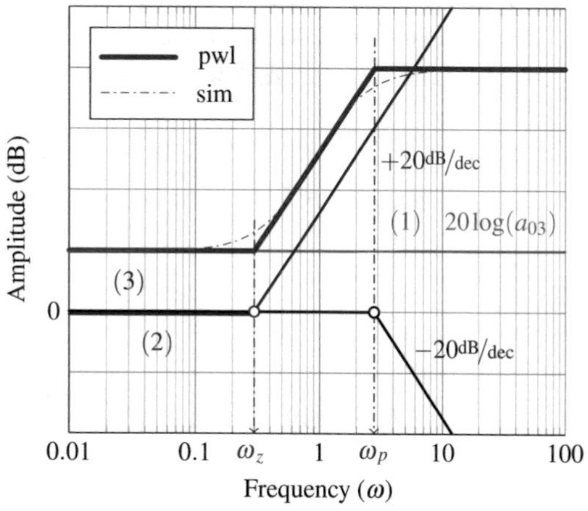

Fig. 7.11 Frequency response at the emitter node of CE amplifier showing: (1) DC gain term; (2) zero term; (3) pole term; and (4) piecewise linear sum (black solid) as well as the simulated AC curve (solid thin red)

[2] Voltage gain of CE amplifier is equivalent to the total collector and total emitter resistances ratio.

[3] See Sect. 1.6.3.

[4] See Sect. 3.2.2.

$$\tau_p = R_{eq}\, C_E = \left[\left(\frac{R_S||R_B}{\beta+1} + r_e \right) ||R_E \right] C_E \tag{7.23}$$

$\therefore$

$$\omega_p = \frac{1}{\left[\left(\frac{R_S||R_B}{\beta+1} + r_e \right) ||R_E \right] C_E} \tag{7.24}$$

Even though it may seem complicated, (7.24) is derived simply by inspection while looking into emitter node of CE amplifier circuit and mentally performing the circuit transformations in Fig. 7.10.

7.3.3 Case Study: CE Amplifier Poles-Zeros

Problem: Estimate piecewise linear Bode plot for the CE amplifier in Fig. 7.6. Compare the hand derived result with AC simulation, estimate the simulated GBW, then illustrate how a wider bandwidth could be achieved. **Data:** $R_1 = 2.5\,\text{k}\Omega$, $R_2 = 1.0\,\text{k}\Omega$, $R_S = 50\,\Omega$, $R_C = 1250\,\Omega$, $R_E = 500\,\Omega$, $R_L = 100\,\text{k}\Omega$, $r_o = 5\,\text{k}\Omega$, $C_1 = 10\,\mu\text{F}$, $C_2 = 80\,\text{nF}$, $C_E = 200\,\text{nF}$, $g_m = 160\,\text{mS}$, $\beta = 200$.

Solution: Knowing g_m gives us the emitter resistance as $r_e = 1/g_m = 6.250\,\Omega$, see (4.39). Although not explicitly asked in this example, it is useful to quickly estimate upper limit of the biasing current at the room temperature, see (4.14) and (4.37), as

$$I_{C0} = g_m\, V_T = g_m \frac{kT}{q} \approx 160\,\text{mS} \times 25\,\text{mV} = 4\,\text{mA}$$

which is a reasonable current level for discrete transistors that is found at biasing point where g_m is measured. (We keep in mind the $n = 1$ assumption in BJT model.)

1. *Midband gain calculations:* Amplifier gain A_{V0} in the midband range is calculated assuming C_E capacitor is not ignored, i.e. it bypasses R_E and brings AC ground to the emitter node.
   In reference to (6.46) we estimate CE amplifier midband gain. In order to illustrate possible approximations, the following progressively more accurate midband gains are calculated as follows:

$$(R_L, \beta, r_o \to \infty) \quad |A_{V0}| = |-g_m\, R_C| = \frac{R_C}{r_e} = \frac{1250\,\Omega}{6.250\,\Omega} = 200\,\text{V/v} = 46\,\text{dB}$$

$$(R_L, \beta \to \infty) \quad |A_{V0}| = |-g_m\, R_C||r_o| = \frac{R_C||r_o}{r_e} = \frac{1000\,\Omega}{6.250\,\Omega} = 158.4\,\text{V/v} = 44\,\text{dB}$$

$$(R_L \to \infty) \quad |A_{V0}| = \left| -\frac{R_C||r_o}{r_e + \frac{R_B}{\beta}} \right| = \frac{1\,\text{k}\Omega}{9.821\,\Omega} = 101.8\,\text{V/v} = 40.2\,\text{dB}$$

$$|A_{V0}| = \left| -\frac{R_C||r_o||R_L}{r_e + \frac{R_B}{\beta}} \right| = \frac{990.1\,\Omega}{9.821\,\Omega} = 100.8\,\text{V/v} = 40.1\,\text{dB}$$

The above results illustrate how the limiting midband gain of 46 dB (for unloaded amplifier with the ideal transistor) is reduced to $A_{V0} = 40\,\text{dB}$ when realistic parameters and the loading resistance are taken into account. This is the gain level found after ω_p frequency.

2. *LF gain calculations:* at DC impedance $Z(C_E)$ equals infinity, which means that R_E is not bypassed. Consequently, the amplifier gain is reduced because the total emitter resistance is increased relative to its minimal value of r_e, see Sect. 6.4.4.

   In order to illustrate possible approximations, the following progressively more accurate DC gains are calculated as follows:

$$(R_L, \beta, r_o \to \infty) \qquad |A_{DC}| = \frac{R_C}{R_E} = \frac{1250\,\Omega}{500\,\Omega} = 2.5\,\text{V/v} = 8\,\text{dB}$$

$$(R_L, \beta \to \infty) \qquad |A_{DC}| = \frac{R_C||r_o}{R_E} = \frac{1000\,\Omega}{500\,\Omega} = 2\,\text{V/v} = 6.02\,\text{dB}$$

$$(\beta \to \infty) \qquad |A_{DC}| = \frac{R_C||r_o||R_L}{R_E} = \frac{990.1\,\Omega}{500\,\Omega} = 1.98\,\text{V/v} = 5.93\,\text{dB}$$

$$|A_{DC}| = \frac{R_C||r_o||R_L}{R_E + \dfrac{R_B}{\beta}} = \frac{990.1\,\Omega}{503.6\,\Omega} = 1.97\,\text{V/v} = 5.87\,\text{dB}$$

   The above results illustrate that, for the given R_C, LF gain is limited to 8dB due to R_E itself, while the inclusion of other realistic parameters further reduces LF gain to approximately 5.9dB. We also note that this gain is a result of the sum of LF gains (7.14), (7.19), and (7.20).

3. *Pole-zero calculations:* The three main discrete capacitors determine time constants of poles and zeros in the transfer function, as per (7.12) (7.18) (7.22), and (7.24).

   (a) *Time constant f_{τ_1} and DC gain due to C_1:* evidently, the input DC signals are blocked due to C_1 being connected in series on their path, thus C_1 creates HPF function (see Sect. 7.3.2 and Fig. 7.8) whose transition frequency is found as

$$(\beta \to \infty) \qquad f_{\tau_1} = \frac{1}{2\pi\, C_1(R_S + R_B)} = \frac{1}{2\pi \times 10\,\mu\text{F} \times (50\,\Omega + 714.3\,\Omega)} = 20.8\,\text{Hz}$$

$$(\beta \neq \infty) \qquad f_{\tau_1} = \frac{1}{2\pi\, C_1(R_S + R_B||\beta r_e)} = \frac{1}{2\pi \times 10\,\mu\text{F} \times (50\,\Omega + 526.2\,\Omega)} = 30.2\,\text{Hz}$$

   As per (7.14), first we calculate $R'_B = R_B||r_e(\beta+1) = 455.4\Omega$ so that DC gain is calculated as

$$a_{01} = \frac{R'_B}{R_S + R'_B} = \frac{455.4\,\Omega}{50\,\Omega + 455.4\,\Omega} = 0.9\,\text{V/v} = -0.9\,\text{dB}$$

   This HPF function is labeled as (1) in Fig. 1.33 (left). It shows +20 dB per decade slope until the transition frequency f_{τ_1} where it reaches $a_{01} = -0.9\,\text{dB}$ gain level.

   (b) *Time constant f_{τ_2} and DC gain due to C_2:* time constant due to the collector side resistance and C_2 also causes HPF profile

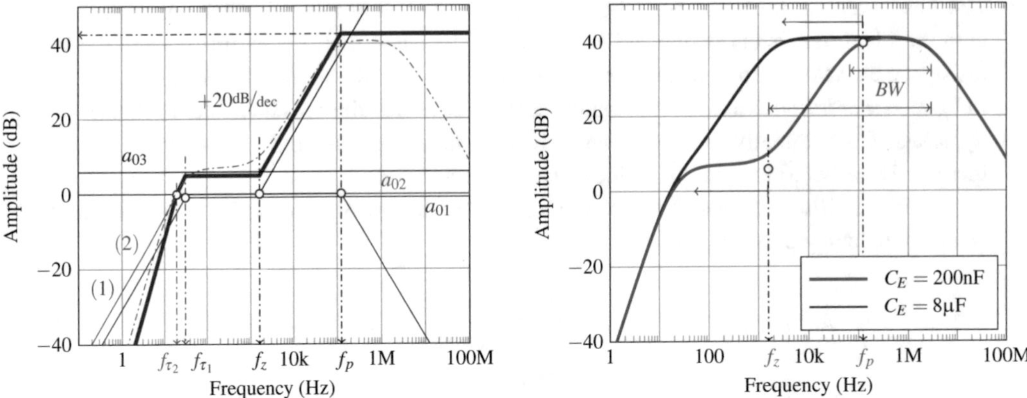

Fig. 7.12 Frequency characteristics of CE amplifier used in the case study

$$(r_o \to \infty) \quad f_{\tau_2} = \frac{1}{2\pi C_2(R_C + R_L)} = \frac{1}{2\pi \times 80\,\text{nF} \times (1250\,\Omega + 100\,\text{k}\Omega)} = 19.6\,\text{Hz}$$

$$(r_o \neq \infty) \quad f_{\tau_2} = \frac{1}{2\pi C_2(R_C||r_o + R_L)} = \frac{1}{(2\pi \times 80\,\text{nF} \times 101\,\text{k}\Omega)} = 19.7\,\text{Hz}$$

As per (7.19), we calculate DC gain as

$$a_{02} = \frac{R_L}{r_o||R_C + R_L} = \frac{100\,\text{k}\Omega}{5\,\text{k}\Omega||1250\,\Omega + 100\,\text{k}\Omega} = 0.99\,\text{V/V} = -0.1\,\text{dB}$$

This HPF function is labeled as (2) in Fig. 7.12 (left). It also shows +20 dB per decade slope until the transition frequency f_{τ_2} where it reaches $a_{02} = -0.1$ dB gain level.

(c) *Pole and zero pair due to* R_E*:* the two time constants at the emitter node are calculated per (7.22) and (7.24), thus the zero frequency is calculated as

$$f_z = \frac{1}{2\pi R_E C_E} = \frac{1}{2\pi \times 500\,\Omega \times 200\,\text{nF}} = 1.6\,\text{kHz}$$

and the pole frequency is

$$(\beta \to \infty) \quad f_p = \frac{1}{2\pi C_E(R_E||r_e)} = \frac{1}{2\pi \times 200\,\text{nF} \times (500\,\Omega||7.14\,\Omega)} = 66.7\,\text{kHz}$$

$$(\beta \neq \infty) \quad f_p = \frac{1}{2\pi C_E\left[\left(\frac{R_S||R_B}{\beta+1} + r_e\right)||R_E\right]} = \frac{1}{2\pi \times 200\,\text{nF} \times 6.4\,\Omega} = 124.3\,\text{kHz}$$

Zero at f_z frequency introduces +20 dB per decade gain slope, which increases the amplifier gain until the next pole at $f_p = 124.3$ kHz frequency when it reaches approximately 42 dB. When calculated as ratio of resistances, depending on the level of approximation, the midband gain is in the range of 40–46 dB while the simulation shows maximum at 40.1 dB.

Hand calculated piecewise approximation of gain transfer function is close to the simulated gain curve, Fig. 7.12 (left). In this approximation, there is no additional HF pole to limit the upper frequency of this amplifier. Simulation, however, clearly shows that there must be an additional pole at around 2 MHz, see Fig. 7.12 (right). This additional HF pole is the consequence of parasitic BJT capacitance, to be introduced in Sect. 7.4. Emitter capacitor controls position of the pole-zero pair, thus if the emitter capacitor value is increased to, for example, $C_E = 8\,\mu$F, then the emitter's zero-pole frequencies are shifted lower and bandwidth is increased approximately from $BW = 2.93$ MHz to $BW = 3$ MHz, Fig. 7.12 (right).

7.4 High-Frequency Analysis of Single-Stage Amplifiers

In our discussion so far, it was assumed that active devices are ideal and only the external components determine the circuit's functionality. In other words, we used a *low to medium frequency approximation*. In this section, we include internal parasitic capacitances into our analysis. Inherently, these capacitances are small, thus they contribute HF zeros and poles.

7.4.1 HF Transistor Models

In Sects. 4.3 and 4.4 we introduced BJT and MOS transistors and their respective models, Figs. 4.16 and 4.27 without taking into account the inherent p-n junction parasitic capacitances which, for example, were exploited in Sect. 4.2.4 to create voltage-controlled capacitance. Here, we include explicitly the internal BJT and MOS capacitances as in Figs. 7.13 and 7.14. While we ignore some of the others, we focus on two p-n junction capacitances in a BJT transistor (namely C_{BE}, C_{BC}), and two principal parasitic capacitances in a MOS device, that is (C_{GD}, C_{GB}).

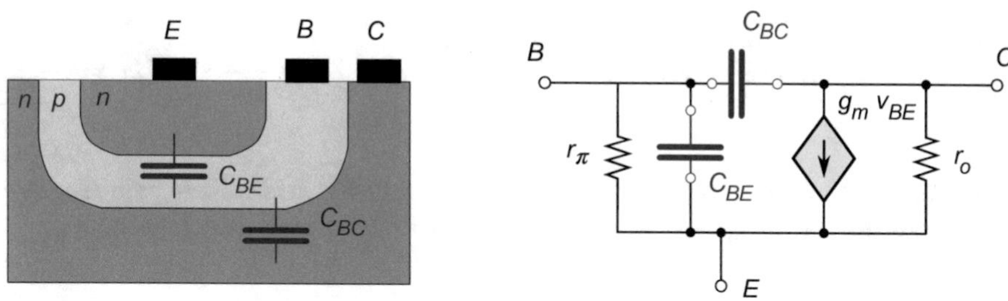

Fig. 7.13 HF BJT cross-section (vertical axis is exaggerated) and AC model

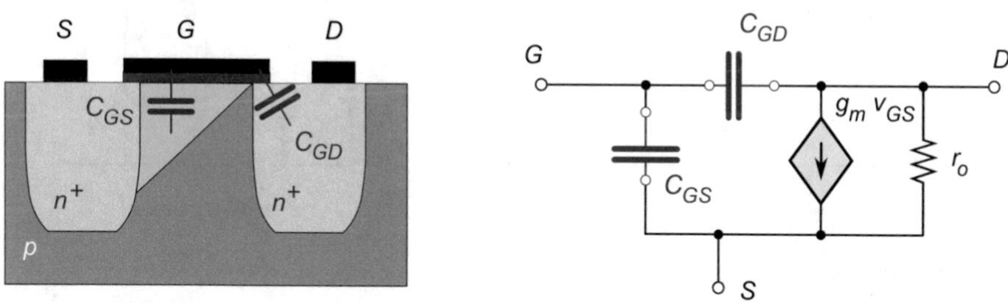

Fig. 7.14 HF MOS cross-section (vertical axis is exaggerated) and AC model

Being in the order of a few fF to a pF, these parasitic capacitances are relatively small in the absolute terms; however, at sufficiently high frequencies even their respective impedances become small and significantly affect the overall functionality of the circuit.

In particular, even if one of these small capacitances creates a feedback path for a signal, the overall circuit could become unstable. On the other hand, if used appropriately the feedback paths enable design of important circuits, for example, oscillators and phase-frequency locked loops (PLLs). In order to understand this particular phenomenon, in the following section we review one of very important theorems in engineering, the Miller theorem.

7.4.2 Miller Theorem

A theorem that is very often used in circuit frequency analysis is applicable in the case of "floating" (i.e. not connected to ground node) impedance Z_{AB} that is branching two nodes A and B as in Fig. 7.15. Miller theorem derives transformation of equality between networks in Fig. 7.15 (left) and (right). Main advantage of this transformation is that the floating impedance Z_{AB}, which is difficult to analyse, is replaced with impedances Z_A and Z_B that are connected to ground. Therefore they are easily included in analysis as will be demonstrated shortly.

Looking into node A both on the left and right side circuits, if the two circuits are to be identical, then the same current is being drawn in both cases, that is to say

$$i_A = \frac{v_A - v_B}{Z_{AB}} \quad \text{and} \quad i_A = \frac{v_A}{Z_A} \quad \therefore \quad \frac{v_A - v_B}{Z_{AB}} = \frac{v_A}{Z_A} \quad \Rightarrow \quad Z_A = \frac{v_A}{v_A - v_B} Z_{AB} = \frac{1}{1 - \dfrac{v_B}{v_A}} Z_{AB}$$

Looking into node B we apply the same reasoning, so we write

$$i_B = \frac{v_B - v_A}{Z_{AB}} \quad \text{and} \quad i_B = \frac{v_B}{Z_B} \quad \therefore \quad \frac{v_B - v_A}{Z_{AB}} = \frac{v_B}{Z_B} \quad \Rightarrow \quad Z_B = \frac{v_B}{v_B - v_A} Z_{AB} = \frac{1}{1 - \dfrac{v_A}{v_B}} Z_{AB}$$

By definition, the ratio of voltages at the "end" and the "start" nodes is voltage gain A_V, in this case $A_V = v_B/v_A$, thus the above expressions derived by Miller theorem are usually written as

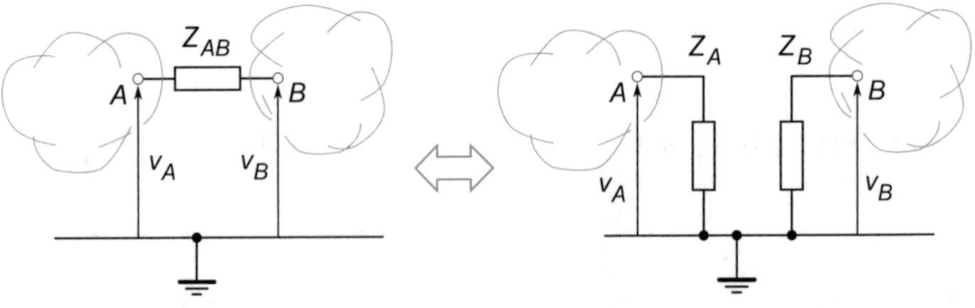

Fig. 7.15 A general case of Miller theorem

$$Z_A = \frac{1}{1 - A_V} Z_{AB} \quad \text{and} \quad Z_B = \frac{1}{1 - \dfrac{1}{A_V}} Z_{AB} \tag{7.25}$$

Physical interpretation of (7.25) is that known bridging impedance Z_{AB} is perceived (or, "projected") at the input and output sides of the circuit after being multiplied by their respective factors, which are functions of the voltage gain between the two branching nodes.

7.4.3 Miller Capacitance and Inverting Amplifier

An amplifying network configuration, which satisfies the following three conditions, is quite common in electronics and in the nature:

1. the amplifier is *inverting* voltage amplifier
2. the amplifier's voltage gain is larger than one, i.e. $|A_v| \gg 1$
3. a capacitor C is connected *between its input and output* terminals

General network that illustrates the three conditions listed above is shown in Fig. 7.16 (we note C_f connecting the output and input nodes of the amplifier), and it is analysed as follows. Assuming an inverting voltage amplifier with infinite input impedance, the output and input voltages are related as $v_{out} = -A_v\, v_{in}$. Under the condition of no current flowing into the amplifier's input terminal, the input impedance Z_{in} of the network is calculated as

$$i_{in} = \frac{v_{in} - v_{out}}{Z_{C_f}} = \frac{v_{in} + A_v\, v_{in}}{Z_{C_f}} = \frac{v_{in}(1 + A_v)}{Z_{C_f}} \tag{7.26}$$

$$\therefore$$

$$Z_{in} \stackrel{\text{def}}{=} \frac{v_{in}}{i_{in}} = \frac{Z_{C_f}}{1 + A_v} \tag{7.27}$$

which, in the case of capacitive bridging impedance $Z_C = 1/s\,C$, takes the form of

$$Z_{in} = \frac{1}{j\omega \underbrace{C\,(A_v + 1)}_{\text{capacitance}}} = \frac{1}{j\omega\, C_M} \tag{7.28}$$

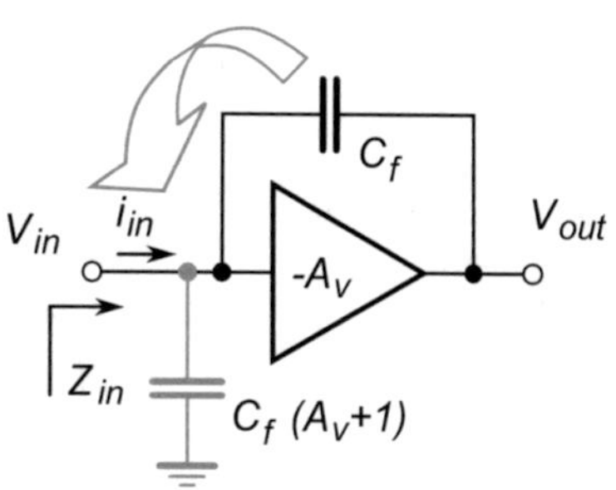

Fig. 7.16 A general inverting voltage amplifier with Miller capacitance

where the effective Miller capacitance is defined as

$$C_M = C\,(A_v + 1) \tag{7.29}$$

We note that C_M is not a physical capacitor, instead it is "folded over and magnified" feedback capacitor C_f (which is a real capacitor). Effectively, in combination with the source resistance, Miller capacitance C_M creates a low-pass filter. Therefore, an inverting amplifier, e.g. CE, is not a good choice for RF applications because its frequency bandwidth is limited by Miller effect.

Example 50: Miller Theorem

Assuming an amplifier with the infinite input impedance, Fig. 7.17 (left), calculate the input Z_{in} and output Z_{out} side perceived impedances, Fig. 7.17 (right), in the following cases: (a) $A_V = 100$, $R_f = 1\,\text{k}\Omega$; (b) $A_V = -100$, $R_f = 1\,\text{k}\Omega$; (c) $A_V = -100$, $C_f = 1\,\text{pF}$, $\omega = 10\,\text{Mrad}$.

Solution 50: By direct implementation of (7.25), we write:

(a) At the input and output nodes of a non-inverting $A_V = 100$ amplifier, the feedback resistor $R_f = 1\,\text{k}\Omega$ is perceived as real resistors whose values are

$$Z_A = \frac{R_f}{1 - A_V} = \frac{1\,\text{k}\Omega}{1 - 100} = -\frac{1\,\text{k}\Omega}{99} \approx -10\,\Omega$$

$$Z_B = \frac{R_f}{1 - \frac{1}{A_V}} = \frac{1\,\text{k}\Omega}{1 - \frac{1}{100}} = \frac{1\,\text{k}\Omega}{0.99} \approx 1\,\text{k}\Omega$$

We note that at the input side a non-inverting amplifier converts a real feedback resistor into a *negative* small resistor, which by itself does have useful applications. The impedance at the output node of a non-inverting amplifier, however, is very close to R_f.

(b) At the input and output nodes of an inverting $A_V = -100$ amplifier, the feedback resistor $R_f = 1\,\text{k}\Omega$ is perceived as real resistors whose values are

$$Z_A = \frac{R_f}{1 - A_V} = \frac{1\,\text{k}\Omega}{1 - (-100)} = \frac{1\,\text{k}\Omega}{99} \approx 10\,\Omega$$

$$Z_B = \frac{R_f}{1 - \frac{1}{A_V}} = \frac{1\,\text{k}\Omega}{1 - \frac{1}{-100}} = -\frac{1\,\text{k}\Omega}{1.01} \approx 1\,\text{k}\Omega$$

(continued)

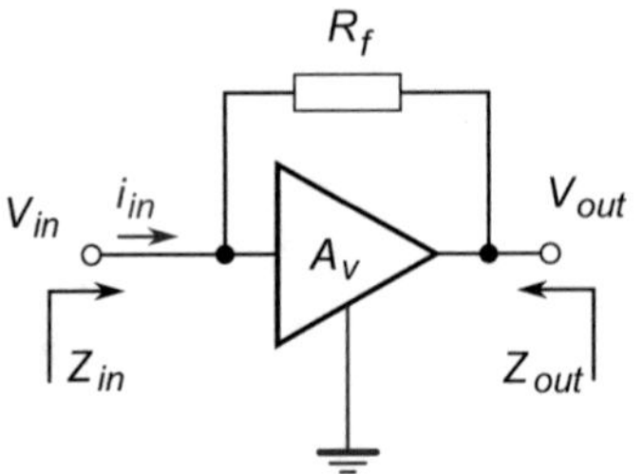

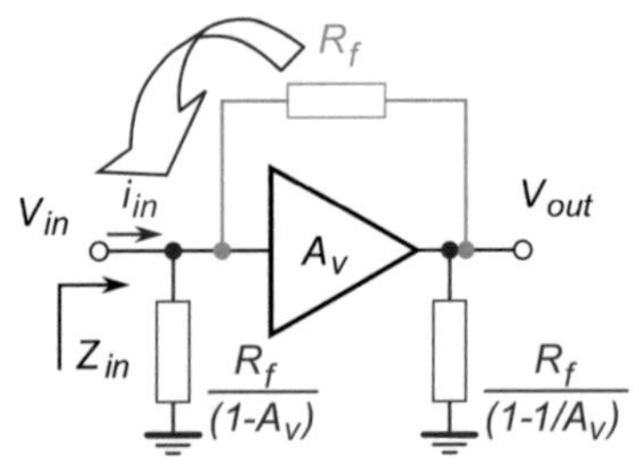

Fig. 7.17 Voltage amplifier with feedback impedance (left), and the equivalent circuit after Miller transformation (right)

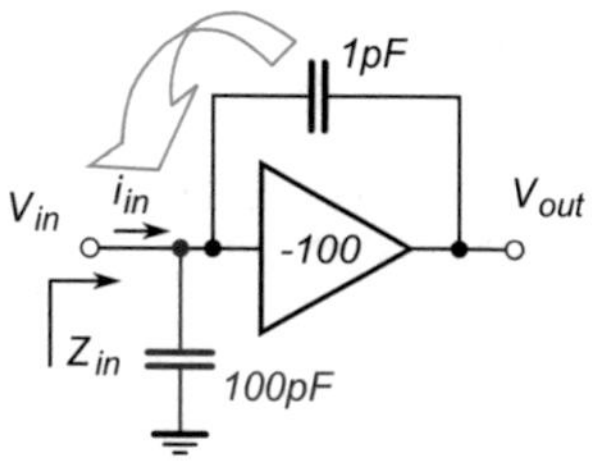

Fig. 7.18 Voltage amplifier with Miller capacitance at the input node

This time, at the input side an inverting amplifier converts a relatively large real feedback resistor into a small resistor perceived as Z_{in}. At the output node, again, high-gain non-inverting amplifier presents resistance of approximately R_f.

(c) If, instead of a resistor, a 1 pF feedback capacitor is connected, then at the input and output nodes of an inverting $A_V = -100$ amplifier it is perceived as a capacitor. First, it is necessary to calculate the impedance at the given frequency as

$$Z_f = \left|\frac{1}{j\omega C_f}\right| = \frac{1}{10\,\text{Mrad} \times 1\,\text{pF}} = 100\,\text{k}\Omega \tag{7.30}$$

Now, we calculate the input and output impedances as

$$Z_A = \frac{R_f}{1 - A_V} = \frac{100\,\text{k}\Omega}{1-(-100)} = \frac{100\,\text{k}\Omega}{101} \approx 1\,\text{k}\Omega \Rightarrow C_A \cong \frac{1}{10\,\text{Mrad} \times 1\,\text{k}\Omega} = 100\,\text{pF}$$

$$Z_B = \frac{R_f}{1 - \frac{1}{A_V}} = \frac{100\,\text{k}\Omega}{1 - \frac{1}{-100}} = \frac{100\,\text{k}\Omega}{1.01} \approx 100\,\text{k}\Omega \Rightarrow C_B \cong 1\,\text{pF} = C_f$$

Looking at the amplifier's output node the perceived impedance did not change, which is to say that capacitance seen at the output node stays equal to C_f. At the input node, however, an inverting high-gain amplifier reduces the feedback impedance by the factor equal to the voltage gain A_V, which results in *increased* capacitance by the same factor.

This case is of particular interest because, everything else being equal, an inverting amplifier creates large Miller capacitance at its input node, which consequently reduces the amplifier's bandwidth, thus making it not a good choice for RF applications (Fig. 7.18).

Fundamental reason for a CE amplifier's sensitivity to Miller effect is that base-collector capacitance C_{CB} is real and unavoidable. That parasitic capacitance exists due to the reverse biased base-collector p-n junction that behaves as a voltage-controlled capacitor. By inspection of the equivalent BJT transistor model, Fig. 7.13, in the case of CE amplifier, we find that capacitance C_{BE} is connected across the input terminals and is safely grounded on one side, therefore introduces only minor frequency limitation to the overall amplifier behaviour. However, the floating collector–base C_{CB} capacitance provides feedback path to the signal, which gives rise to the Miller effect. To complete the set of three conditions needed for Miller effect, CE stage inherently inverts signal and has large voltage gain.

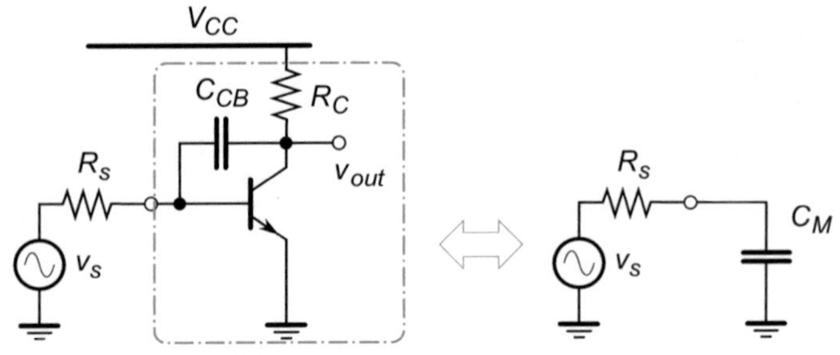

Fig. 7.19 CE voltage amplifier with feedback capacitance C_{CB} (left), and its equivalent input side LP filter network (right)

Because of this CE amplifier weakness, a CB amplifier is often used in RF designs in cases where its low input resistance is compatible with the driving stage. Everything else being equal, in the case of CB amplifier there is no significant capacitance that connects the CB stage input–output terminals.

Example 51: Miller Effect: CE Amplifier
Assume an ideal, single-stage CE amplifier (i.e. whose input resistance $R_{in} = \infty$) with voltage gain of $A_v = -99$, Fig. 7.19 (left). The amplifier is driven by a voltage source whose output resistance is $R_S = 50\,\Omega$. In addition, there is a $C_{CB} = 1\,\text{pF}$ capacitor connected between the transistor's collector and base. It is also assumed that the voltage gain is constant, i.e. $A_v \neq f(\omega)$, and that collector–base capacitance C_{CB} is independent of collector–base voltage, i.e. $C_{CB} \neq f(V_{CB})$. Biasing details of the circuit are not shown.

Estimate CE amplifier's frequency bandwidth.

Solution 51: This CE amplifier satisfies all three conditions required for Miller effect. It is an inverting amplifier, its voltage gain is greater than one, and it has a capacitive component that creates feedback path between the output and input terminals. Therefore, the equivalent input side schematic diagram, Fig. 7.19 (right), is analysed as a voltage divider that consists of the source resistance $R_S = 50\,\Omega$ and the Miller capacitance $C_M = C_{CB}(|A_V| + 1) = 100\,\text{pF}$.

The frequency bandwidth defined by this (R_S, C_M) LP filter is in effect the "useful range" of signal frequencies, i.e.

$$f_{3\,\text{dB}} = \frac{1}{2\pi\, R_S\, C_M} = \frac{1}{2\pi \times 50\,\Omega \times 100\,\text{pF}} = 31.831\,\text{MHz} \tag{7.31}$$

Considering that a simple $(50\,\Omega, 1\,\text{pF})$ LP filter would allow the bandwidth of 3.183 GHz, this example illustrates significant reduction in CE amplifier's bandwidth due to Miller effect.

7.4.4 HF CB Amplifier Model

High-frequency CB amplifier model in Fig. 7.20 is derived from Fig. 7.2 after replacing BJT with its HF BJT model and shorting (C_1, C_2, C_3) capacitances. Then, we rearrange schematic Fig. 7.20 into a more structured form shown in Fig. 7.21. Now, it becomes more evident that it is the current source $g_m v_{BE}$ that creates a feedback path between the output node (collector) and the input node (emitter), see Fig. 7.21. In addition, we find the feedback path through serial connection of C_{BE} and C_{BC}. In this case, however, the equivalent serial capacitance is smaller than any of the two already

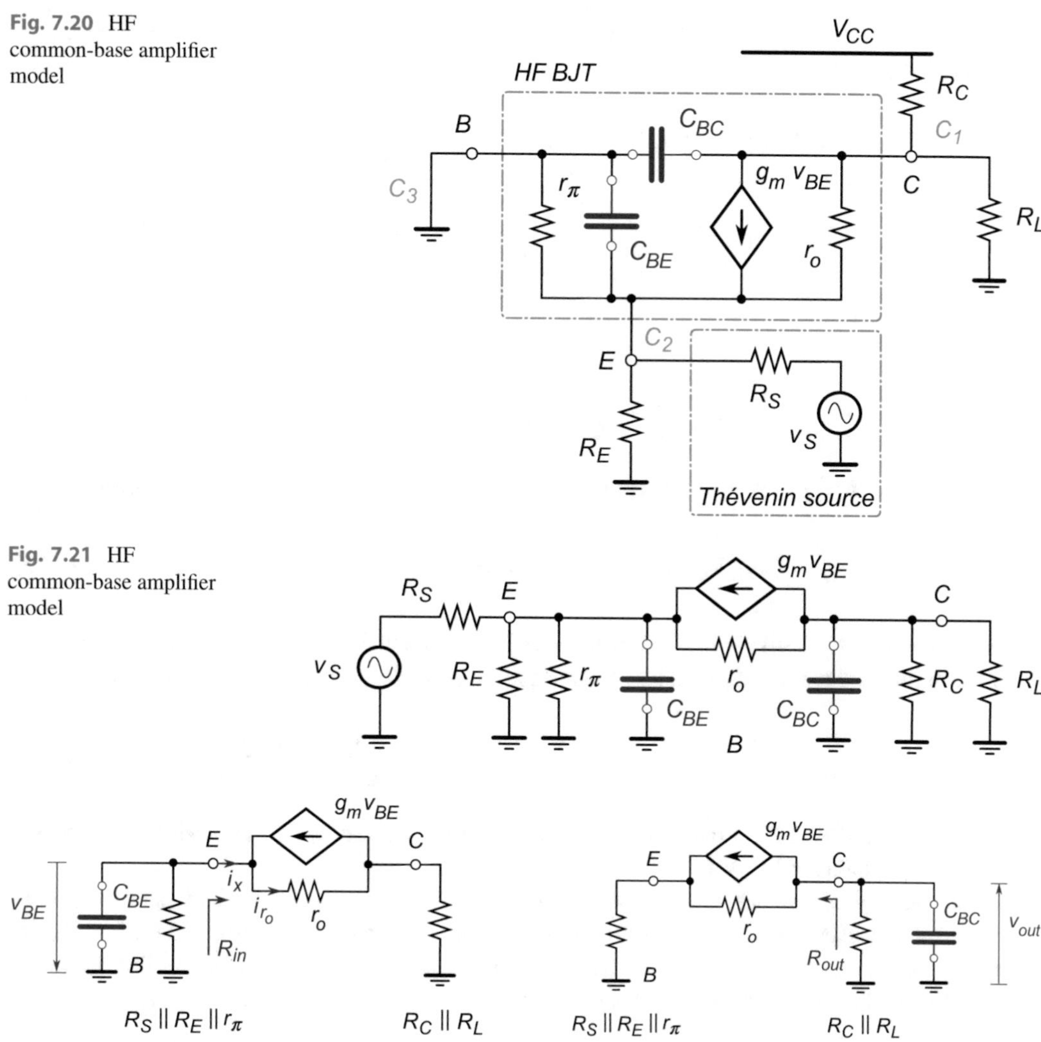

Fig. 7.20 HF common-base amplifier model

Fig. 7.21 HF common-base amplifier model

Fig. 7.22 HF common-base amplifier model

small capacitances C_{BE} and C_{BC}. Thus, the influence of this capacitive path becomes non-negligible only in the VHF/UHF range and in that case it is commonly known as a "feed-through" path.

Lastly, CM base amplifier is not the inverting type because the signal does not change its phase between the input and output nodes. Therefore, the conclusion is that there is no Miller effect in CB amplifier case. Consequently, in that respect CB amplifier is considered a wide-band amplifier that is suitable for RF applications.

Time constants of CB amplifier are found by circuit reduction technique that enables estimates of the equivalent resistance seen by each of the capacitors. In the following analysis we consider two capacitors, thus we calculate two time constants—one associated with the input and one associated with the output side of CB amplifier.

1. *Time constant at the input side:* by inspection we find that the total equivalent resistance associated with C_{BE} consists of $(R_S||R_E||r_\pi)$ in parallel with the R_{in}, as shown in Fig. 7.22 (left). In Sect. 4.3.5 we found that at emitter node $R_{in} \approx 1/g_m$, for the sake of clarity we show it again.

In order to simplify writing we denote $R'_C = R_C||R_L$. First, we note that $v_x = v_E = -v_{BE}$, so by applying KCL we write:

$$i_x + g_m v_{BE} = i_{r_o} \Rightarrow i_x - g_m v_x = \frac{v_x}{R'_C + r_o}$$

$$\therefore$$

$$i_x = v_x \left(g_m + \frac{1}{R'_C + r_o} \right) = v_x \left(\frac{g_m (R'_C + r_o) + 1}{R'_C + r_o} \right)$$

$$\therefore$$

$$R_{in} \stackrel{\text{def}}{=} \frac{v_x}{i_x} = \frac{R'_C + r_o}{g_m (R'_C + r_o) + 1} \approx \frac{\cancel{R'_C + r_o}}{g_m \cancel{(R'_C + r_o)}} = \frac{1}{g_m} \tag{7.32}$$

At HF capacitor C_{BE} creates a short to ground, therefore it defines a HF pole at frequency

$$\tau_{C_{BE}} = (R_S||R_E||r_\pi||1/g_m)C_{CE} \Rightarrow \omega_{C_{BE}} = \frac{1}{(R_S||R_E||r_\pi||1/g_m)C_{BE}} \tag{7.33}$$

By inspection of (7.33) we find that resistance $1/g_m$ is much lower than $(R_S||R_E||r_\pi)$, therefore we approximate the overall parallel resistance as $(R_S||R_E||r_\pi||1/g_m) \approx 1/g_m$ and write

$$\omega_{C_{BE}} = \frac{1}{(R_S||R_E||r_\pi||1/g_m)C_{BE}} \approx \frac{1}{(1/g_m)C_{BE}} = \frac{g_m}{C_{BE}} \tag{7.34}$$

which is indeed very high frequency due to a numerical value of g_m that is divided by orders of magnitude smaller value of C_{BE}.

2. *Time constant at the output side:* at high frequencies C_{BC} provides a short path to ground, therefore it also creates a HF pole whose time constant is found by inspection of Fig. 7.22 (right). Since there is resistor connected at the emitter node, when looking into BJT collector we see "degenerated emitter". In Sect. 4.3.8 we derived equation (4.54), which is applied as

$$R_{out} = r_o \left[1 + g_m (R_S||R_E||r_\pi)\right] \quad \therefore \quad \tau_{C_{BC}} = \{r_o \left[1 + g_m (R_S||R_E||r_\pi)\right]\}||R_C||R_L \times C_{BC} \tag{7.35}$$

In the case of degenerated emitter R_{out} is very high, which makes the collector node a very good current source to drive a low impedance R_L load (see Sect. 6.3). Thus, we conclude that R_L resistance is the lowest among parallel resistances in (7.35). Accordingly we write

$$\begin{aligned} \tau_{C_{BC}} &= \{r_o \left[1 + g_m (R_S||R_E||r_\pi)\right]\}||R_C||R_L \times C_{BC} \\ &\approx R_L C_{BC} \end{aligned}$$

$$\therefore$$

$$\omega_{C_{BC}} \cong \frac{1}{R_L C_{BC}} \tag{7.36}$$

Again, relatively small R_L is multiplied by a very small number C_{BC} resulting in very high frequency $\omega_{C_{BC}}$, thus it confirms the wide-band property of CB amplifier.

Due to contribution of two VHF poles, we conclude that after the first one there is $-20\,^{\text{dB}}/_{\text{dec}}$ gain slope that increases to $-40\,^{\text{dB}}/_{\text{dec}}$ after the second pole.

7.4.5 HF CE Amplifier Model

High-frequency CE amplifier model in Fig. 7.23 is derived from circuit in Fig. 7.6 where BJT is replaced with its HF BJT model, capacitances (C_0, C_E) are shorted and final circuit is rearranged as in Fig. 7.24. Clearly, it is the capacitor C_{BC} that creates a feedback path between the output node (collector) and the input node (base). By intent, the voltage gain of CE amplifier is made large, and in addition CE amplifier inverts the signal's phase between the input/output nodes. Therefore, all three conditions for Miller theorem are satisfied, see Fig. 7.25 (right).

By inspection of Fig. 7.25 we derive the total equivalent capacitance C_{eq} and resistance R_{eq} attached to the input side of CE amplifier as

$$\begin{aligned} C_{eq} &= C_{BE} + (A_V + 1)C_{CB} = C_{BE} + C_M \\ R_{eq} &= R_S||R_1||R_2||r_\pi \end{aligned} \tag{7.37}$$

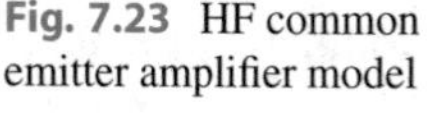

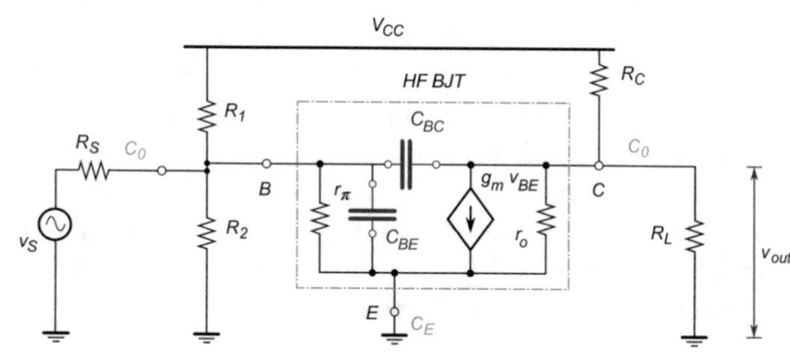

Fig. 7.23 HF common emitter amplifier model

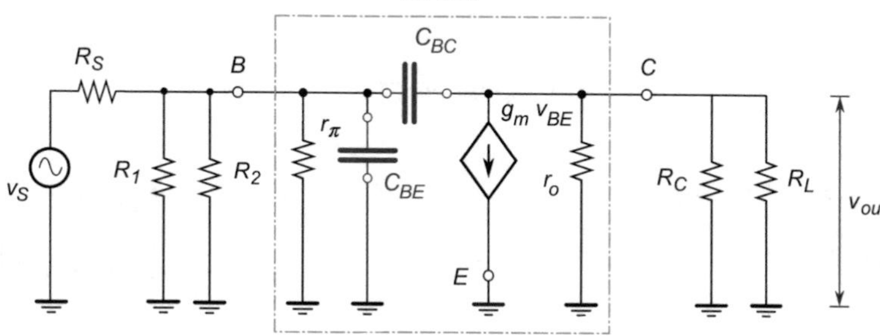

Fig. 7.24 AC model of HF CE amplifier in Fig. 7.23

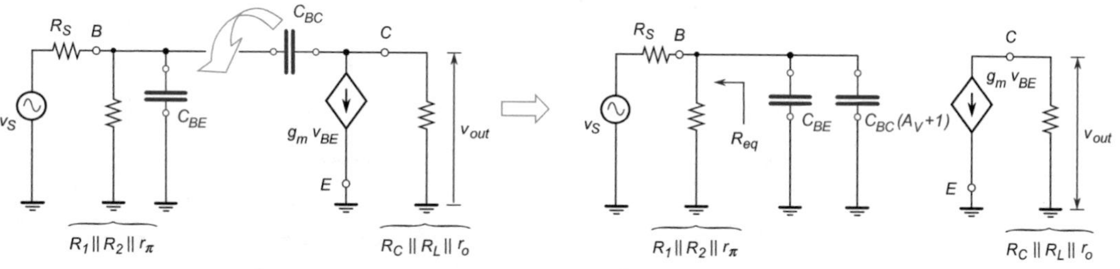

Fig. 7.25 Simplified AC model of a HF common emitter showing Miller effect folding of C_{BC} capacitor

What is more, as the frequency increases the C_{eq} impedance reduces, thus it shorts the input signal. In other words, due to Miller effect this capacitance creates a HF pole in the transfer function of CE amplifier, i.e. it creates LPF. By inspection of Fig. 7.25 (right) time constant and frequency of this pole are written as

$$\tau_M = R_{eq}C_{eq} \qquad \therefore \qquad f_M = \frac{1}{2\pi(R_S||R_1||R_2||r_\pi)\,(C_{BE} + C_M)} \tag{7.38}$$

In conclusion, at high frequencies the internal capacitances of BJT transistor form a pole with the input side resistance of CE amplifier, i.e. the inherently formed LPF determines the amplifier's bandwidth. The real problem is that due to Miller effect this bandwidth is limited to the frequency range that is typically well below radio frequencies, which is to say that due to this reduced bandwidth a CE amplifier is not best option for RF amplifiers.

7.4.6 HF CC Amplifier Model

After explicitly setting $R_C = 0$ and substituting HF BJT model, schematic of emitter-follower amplifier in Fig. 6.21 is rearranged as in Fig. 7.26. We observe that the collector terminal is grounded, thus C_{BC} is not subject to Miller effect. It is, however, connected between the base and ground terminals, thus it contributes to the total input side capacitance.

On the other hand, C_{BE} does provide a path between the input (base) and output (emitter) terminals. In reference to Fig. 7.26 we develop HF model of the input side network as in Fig. 7.27 where $R_B = R_1||R_2$ and $R'_L = R_E||R_L$. We have already found that the voltage gain of a CC amplifier is $A_V \approx 1$, which leads into conclusion that the influence of C_{BE} may be minimal. For the sake of clarity we verify our intuitive conclusion.

1. *Time constant at the input side:* by looking into the base, we find that voltages at the emitter and base terminals are

$$v_E = g_m v_{BE}(R'_L||r_o)$$

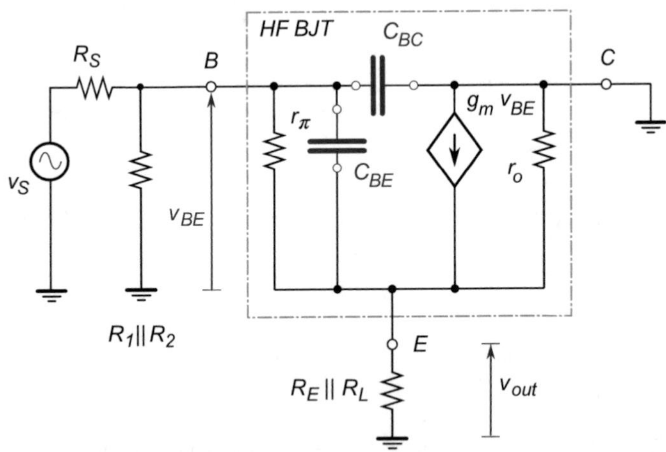

Fig. 7.26 HF model of CC amplifier shown in Fig. 6.21

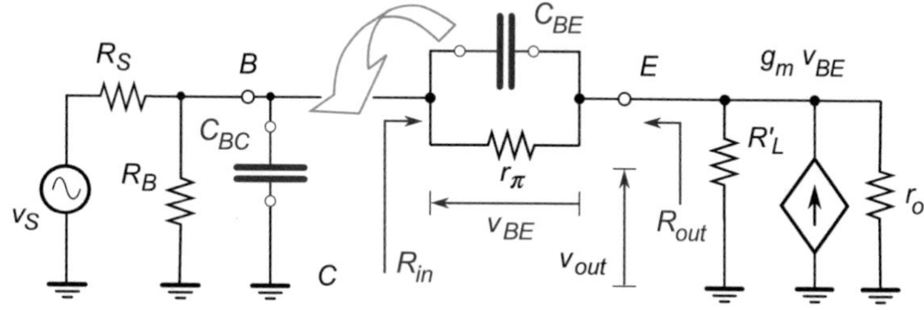

Fig. 7.27 HF model of CC amplifier for calculating the input side time constant, derived from Fig. 7.26

$$v_B = v_{BE} + v_E = v_{BE} + g_m v_{BE}(R'_L||r_o) = v_{BE}[1 + g_m(R'_L||r_o)]$$

$$\therefore$$

$$A_V(BE) \stackrel{\text{def}}{=} \frac{v_E}{v_B} = \frac{\cancel{v_{BE}}\, g_m(R'_L||r_o)}{\cancel{v_{BE}}[1 + g_m(R'_L||r_o)]} \approx \frac{g_m R'_L}{1 + g_m R'_L} \quad (R'_L \ll r_o) \tag{7.39}$$

With the help of Miller theorem, see Sect. 7.4.2, we write the expression for the total (real and projected) capacitance seen at the base terminal as[5]

$$\begin{aligned} C_{eqB} &= C_{BC} + C_{BE}(1 - A_V(BE)) = C_{BC} + C_{BE}\left(1 - \frac{g_m R'_L}{1 + g_m R'_L}\right) \\ &= C_{BC} + C_{BE}\frac{1 + \cancel{g_m R'_L} - \cancel{g_m R'_L}}{1 + g_m R'_L} = C_{BC} + \frac{C_{BE}}{1 + g_m R'_L} \end{aligned} \tag{7.40}$$

where the second term clearly shows that only a fraction of the original C_{BE} (the denominator is greater than one) is added to the total sum, hence it is often approximated to zero. Impedance of the equivalent capacitance C_{eqB} reduces as the frequency increases, consequently the input signal is shorted to ground. In conclusion, this capacitance creates HF pole in CC amplifier.

Further, the equivalent resistance R_{eqB} associated with the base node is found by inspection of Fig. 7.27 (the total emitter resistance is multiplied by $(\beta + 1)$ factor and is projected to the base node), which results in[6]

$$R_{eqB} = R_S||R_B||[r_\pi + (\beta + 1)R'_L] \tag{7.41}$$

It is now straightforward to write the expression for the input side time constant

$$\tau_B = R_{eqB}C_{eqB} = R_S||R_B||[r_\pi + (\beta + 1)R'_L]\left(C_{BC} + \frac{C_{BE}}{1 + g_m R'_L}\right)$$

$$\therefore$$

$$\omega_B = \frac{1}{R_S||R_B||[r_\pi + (\beta + 1)R'_L]\left(C_{BC} + \frac{C_{BE}}{1 + g_m R'_L}\right)} \tag{7.42}$$

---

[5]Pay attention to the voltage signs while calculating Miller capacitance.

[6]we keep in mind that $(\beta + 1)(r_e + R'_L) = (\beta + 1)r_e + (\beta + 1)R'_L = r_\pi + (\beta + 1)R'_L$.

Indeed, because C_{BE} is a small capacitance, and the total parallel resistance is smaller than the relatively low resistance R_S, the input side time constant is very small which makes the pole frequency very high.

2. *Time constant at the output side:* by inspection of the emitter node we find the equivalent capacitance and resistance[7] as

$$C_{eqE} = C_{BE}$$

$$R_{eqE} = R_{out}||R'_L = \left(\frac{R_S||R_B}{\beta+1} + \frac{1}{g_m}\right)||\,R'_L$$

$$\therefore$$

$$\omega_E = \frac{1}{R_{eqE}\,C_{eqE}} = \frac{1}{\left[\left(\dfrac{R_S||R_B}{\beta+1} + \dfrac{1}{g_m}\right)||\,R'_L\right] C_{BE}} \approx \frac{1}{r_e\,C_{BE}} \tag{7.43}$$

which again illustrates that the output side pole frequency is indeed very high.

Therefore, because it does not suffer from Miller effect, CC amplifier also has two VHF/UHF poles that are function of the load impedance as well as β factor. In addition, the feed-through path provided by C_{BE} creates a zero at the emitter node whose time constant is approximately

$$\tau_z \cong r_e\,C_{BE} = \frac{1}{g_m}C_{BE} \Rightarrow \omega_z \cong \frac{g_m}{C_{BE}} = \frac{1}{r_e\,C_{BE}} \tag{7.44}$$

In conclusion, CC amplifier is similar to CB in terms of wide-band operation, for that reason it is also suitable for HF application where it serves as “impedance converter”, a.k.a voltage buffer stage.

7.4.7 Cascoded Amplifier HF Model

Low-frequency AC model of a cascode amplifier in Fig. 6.26 is modified to explicitly show the internal parasitic capacitors of BJT transistor, Fig. 7.28, while the external capacitances are shorted. Although the analysis of this circuit by substituting full HF models of active devices is possible, a

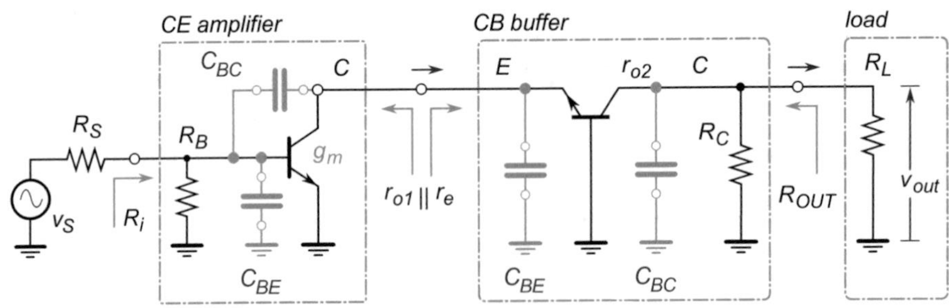

Fig. 7.28 HF model of cascode amplifier stage

[7]The total base resistance $R_S||R_B$ is projected to the emitter side after division by $(\beta+1)$ factor, then it appears in series with $r_e = 1/g_m$. In addition, r_o is not seen at the emitter node, and C_{BE} is not multiplied.

simpler approach is to keep the transistor symbols and explicitly show only the internal capacitances of interest. Because cascode amplifier consists of two stages it can be analysed by applying the already derived results for CE and CB amplifiers and of the interface circuits.[8] This rapid approach gives us good intuition into the circuit behaviour and enables us to do hand analysis of complicated circuits, while the subsequent numerical simulations produce more refined results.

By looking into the collector-emitter interface in Fig. 7.28 we deduce that the equivalent resistance seen at that node is low. Indeed, looking left into the output of CE stage we see the collector output resistance r_{o1} (which is relatively high, in order of kΩ), while looking right into the input of CB stage we see r_e resistance (which is relatively low, in order of Ω). Thus, the node equivalent resistance is $r_{o1}||r_e \approx r_e$. With this low resistance load, by inspection we find that CE amplifier's gain is[9]

$$A_V = \frac{R_{Ceq}}{R_{Eeq}} = \frac{r_{o1}||r_e}{r_e} \cong \frac{r_e}{r_e} \approx 1 \qquad (7.45)$$

This unity gain is the consequence of reduced collector loading resistance, which is not the case in a single-stage CE amplifiers. However, this low voltage gain in effect eliminates Miller effect by removing one of the three conditions, namely the large multiplying factor for C_{BC} feedback capacitor.

We proceed with HF hand analysis of cascade amplifier by separately analysing the input side, the interface between CE and CB amplifiers, and the output side nodes.

1. *The input side node:* the equivalent HF model of the input node, Fig. 7.29, shows that collector–base feedback capacitor C_{BC} is projected both to the base node (C_{Mi}) and the collector node (C_{Mo}), thus we apply Miller theorem to calculate the two effective capacitances. In addition, the input resistance and time constant are found by inspection as

$$\begin{aligned}
C_i &= C_{BE} + C_{Mi} = C_{BE} + C_{BC}(1 + \overset{\nearrow 1}{\cancel{A_V}}) \cong C_{BE} + 2C_{BC}\\
R_i &= R_S||R_B||r_\pi\\
&\therefore\\
\tau_i &= R_i C_i = (C_{BE} + 2C_{BC})(R_S||R_B||r_\pi)\\
&\therefore\\
\omega_i &= \frac{1}{(C_{BE} + 2C_{BC})(R_S||R_B||r_\pi)}
\end{aligned} \qquad (7.46)$$

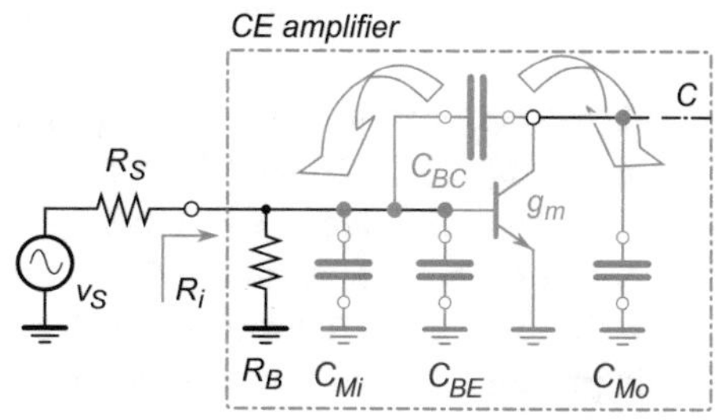

Fig. 7.29 HF model of cascode amplifier stage: the input side node

[8] See Chap. 3.

[9] CE voltage gain is calculated as the ratio of total resistance at collector versus total resistance at emitter, see Sect. 6.4.4.

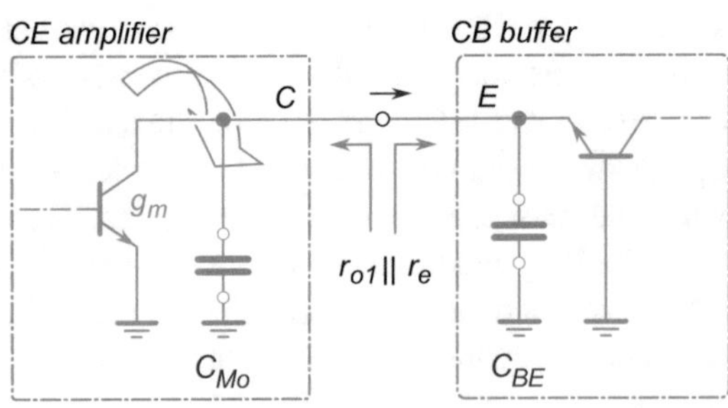

Fig. 7.30 HF model of cascode amplifier stage: CE to CB interface node

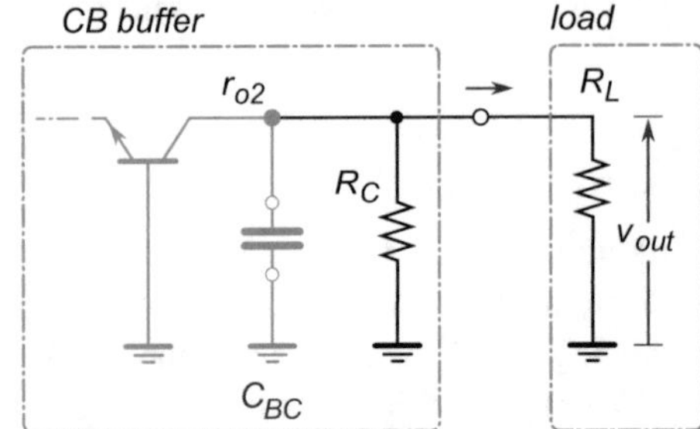

Fig. 7.31 HF model of cascode amplifier stage: the output side node

Due to relatively low total resistance R_i (lower than R_S) and very small capacitors (C_{BE}, C_{BC}) this time constant is very small which forces the pole frequency into UHF/VHF range.

2. *CE to CB interface node:* following the signal path we already deduced resistance at this node to be approximately r_e. HF model in Fig. 7.30 shows that Miller projection C_{Mo} appears in parallel with C_{BE} capacitance of CB amplifier. It is therefore straightforward to write

$$C_1 = C_{BE} + C_{Mo} = C_{BE} + C_{BC}\left(1 + \frac{1}{\cancel{A_V}^{\,1}}\right) \cong C_{BE} + 2C_{BC}$$

$$R_1 = r_{o1}||r_e \approx r_e$$

$$\therefore$$

$$\tau_1 = R_1 C_1 = (C_{BE} + 2C_{BC})r_e$$

$$\therefore$$

$$\omega_1 = \frac{1}{(C_{BE} + 2C_{BC})r_e} \tag{7.47}$$

Again, considering that all three quantities in (7.47) are very small, as a result this pole frequency ω_1 is also in the UHF/VHF range.

3. *The output side node:* CB amplifier does not suffer from Miller effect, thus HF model of the output node is shown in Fig. 7.31 only with C_{BC} capacitance. On the other hand, looking into collector of CB amplifier we see degenerated emitter resistance r_{o2} that is orders of magnitude higher than r_o, hence by inspection we write

$$C_o = C_{BC}$$

$$R_o = r_{o2}||R_C||R_L \approx R_C||R_L$$

$$\therefore$$

$$\tau_o = R_o C_o = (R_C||R_L)C_{BC}$$

$$\therefore$$

$$\omega_0 = \frac{1}{(R_C||R_L)C_{BC}} \tag{7.48}$$

Finally, although a little bit larger than the other two, the output side time constant is nevertheless very small, which creates another pole in the VHF range.

This back of the envelope HF analysis of cascode amplifier illustrates its wideband property. We found out that the fundamental reason for this extension in the frequency range is due to reduction of CE amplifier voltage gain that removed Miller effect. However, because CE amplifier output current is passed to CB amplifier (whose current gain is very close to one), the same current reaches load R_L, which is to say that the overall result is unchanged CE voltage gain while the bandwidth is increased.

Example 52: CE Amplifier: HF Band-Limiting Poles
Assuming that $C_{BC} = 15\,\text{pF}$ and $C_{BE} = 5\,\text{pF}$, continue the case study in Sect. 7.3.3 and estimate frequency of HF pole that is not predicted by LF model. However, this bandwidth limiting pole is clearly visible in the simulation, Fig. 1.33 (right).

Solution 52: For the given data in the case study we calculate $R_{eq} = R_S||R_B||r_\pi = 1.256\,\text{k}\Omega$, and $C_M = (A_V + 1)C_{BE} = 101 \times 15\,\text{pF} = 1.5\,\text{nF}$
then by direct implementation of (7.38) we calculate and compare with the simulations

$$f_M = \frac{1}{2\pi(R_S||R_1||R_2||r_\pi)\,(C_{BE} + C_M)} = \frac{1}{2\pi \times 1.256\,\text{k}\Omega \times (5\,\text{pF} + 1.5\,\text{nF})}$$
$$\approx 2.3\,\text{MHz} \tag{7.49}$$

7.5 Summary

In this section we reviewed fundamental concepts of LF and HF hand analysis and developed intuitive views of the amplifier's internal operation in the frequency domain. In our review, we concluded that important parameters of any amplifier are its input and output resistances, the internal and external capacitances, and its gain. For the sake of simplicity, we used HF version of BJT/MOS devices with only two major internal capacitors (see Sect. 7.4.1) while ignoring all the other relatively less important parasitic capacitances.

Additionally, in this chapter we used strictly resistive model as the loading impedance R_L. Having said that, realistic HF loads should be modelled as a parallel RC network, and further at VHF and above as a parallel RLC network. Consequently, for example, in (7.48) the loading capacitance would appear in parallel to C_{BC} and would increase the total capacitive load to $(C_{BC} + C_L)$. Consequently, the corresponding HF pole frequency would be much lower. Thus, we apply various levels of simplification when performing frequency analysis by hands.

Problems

7.1 Given a transfer function of an amplifier

$$(a)\quad H(s) = \frac{1}{1 + \dfrac{s}{2\pi \times 10^6}} \qquad\qquad (b)\quad H(s) = \frac{1 - \dfrac{s}{10^6}}{\left(1 + \dfrac{s}{10^5}\right)\left(1 + \dfrac{1}{5 \times 10^5}\right)}$$

sketch the Bode plot and estimate its bandwidth.

7.2 Given RC and RL networks in Fig. 7.32, derive their respective $Z_{AB}(j\omega)$ functions relative to the terminals AB.

7.3 Given voltage networks in Fig. 7.33, derive their respective $H(j\omega)$ functions relative to the terminals V_1,V_2 then sketch their Bode plots and estimate the bandwidths.

7.4 We wish to design CE amplifier whose -3 dB dominant frequency is set to f_L by C_E capacitor, see Fig. 7.6. Calculate the required C_E, C_1, and C_2 capacitors.

Data: $f_L = 100\,\text{kHz}$, $R_B = R_1||R_2 = 20\,\text{k}\Omega$, $R_C = R_L = 10\,\text{k}\Omega$, $R_E = 1\,\text{k}\Omega$, $R_{\text{sig}} = 10\,\text{k}\Omega$, $g_m = 400\,\text{mS}$, $r_o \to \infty$, $\beta = 100$.

7.5 Given data, calculate relevant pole/zero frequencies in a CE amplifier.

Data: $C_1 = C_2 = C_E = 100\,\text{pF}$, $R_B B = 100\,\text{k}\Omega$, $R_C = R_L = 10\,\text{k}\Omega$, $r_o \to \infty$, $R_{\text{sig}} = 1\,\text{k}\Omega$, $g_m = 40\,\text{mS}$.

7.6 Given data for a CE amplifier, calculate its midband gain $A_{V_{\text{mid}}}$ and the upper -3 dB frequency f_H. Then, if the upper frequency f_H needs to be doubled and everything else being equal, what would be the maximum allowed value of C_{BC}?

Data: $R_S = 10\,\text{k}\Omega$, $R_i = 100\,\text{k}\Omega$, $R_C = R_L = 10\,\text{k}\Omega$, $r_o = 20\,\text{k}\Omega$, $C_{BE} = 1\,\text{pF}$, $C_{BC} = 0.5\,\text{pF}$.

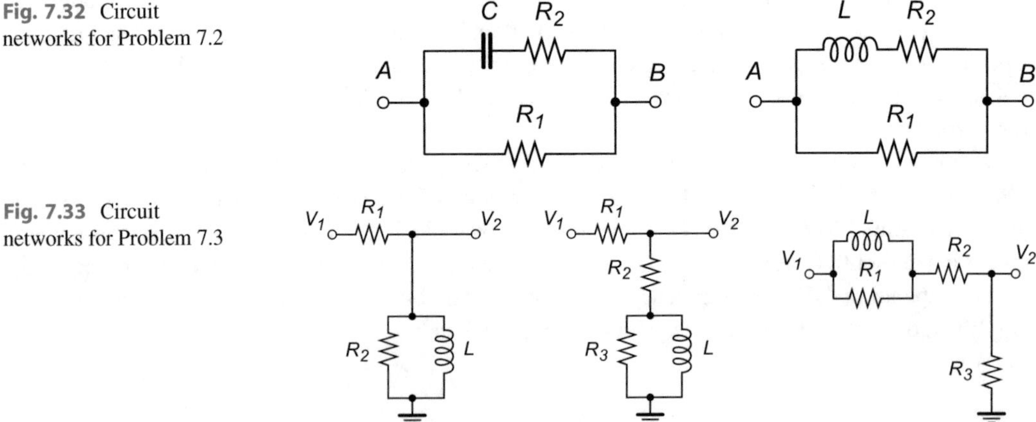

Fig. 7.32 Circuit networks for Problem 7.2

Fig. 7.33 Circuit networks for Problem 7.3

7.7 Given amplifier in Problem 7.6, everything else being equal, what is the maximum value of the load resistance R_L so that the midband gain is doubled? What happened to the GBW?

7.8 For a given cascode BJT CE amplifier, assuming room temperature calculate its input resistance, midband gain, the output resistance.

Data: $I_C = 1\,\text{mA}$, $\beta = 100$, $r_o = 50\,\text{k}\Omega$

Electrical Noise

8

Any electrical signal that makes recovery of the information signal more difficult is considered noise. For example, "white snow" on a TV picture and "hum" in an audio signal are typical electrical noise manifestations. Noise mainly affects receiving systems, where it sets the minimum signal level that it is possible to recover before it becomes swamped by the noise. We note that amplifying a signal already mixed with noise does not help the signal recovery process at all. Once it enters the amplifier, noise is also amplified, which is to say that the ratio of signal-to-noise (S/N) power does not improve and that is what matters. When the power of the noise signal becomes too large relative to the power of the information signal, information content may be irreversibly lost. In this chapter, we study the basic classification of noise sources and methods for evaluation of noise effects.

8.1 Thermal Noise

At the fundamental level, the numerical value of electrical current is just an average number of electrons coming out of the conductor per unit of time. This movement is caused by the external field generated by energy source, for example a battery. However, even without any external electric field, an electron cloud moves inside a material and interacts with the vibrating ions, each electron moving in Brownian motion (i.e. similar to a pinball). The random motion of each individual electron makes a micro current that, together with all the other micro currents in the given volume, adds up to a macro current with zero average value. Due to its random nature, this current does not contain information, therefore we consider it "noise", Fig. 8.1 (left). This motion is responsible for the conductor's temperature, hence it is known as "thermal" noise; in real conductors, it is what constitutes the conductor's resistance. Given that the movement of electrons produces current, and current through a resistor creates voltage across its terminals, we also consider a resistor as a random noise generator. Both experiments and theory have found that the power spectrum of thermal noise is flat, which (loosely) means that each frequency component in the noise spectrum has the same power level, as shown in Fig. 8.1 (right). This conclusion is valid over a very wide range of frequencies (up to approximately 10^{13}Hz). Similarly to white light, which contains all colours (i.e. light frequencies), a noise signal that contains single tones at all possible frequencies is called, appropriately, *white noise*. Of course, it is only a very good approximation, because the implication is that, theoretically, if measured over all possible frequencies, the total noise energy would add up to be infinite.

R. Sobot, *Wireless Communication Electronics*,
https://doi.org/10.1007/978-3-030-48630-3_8

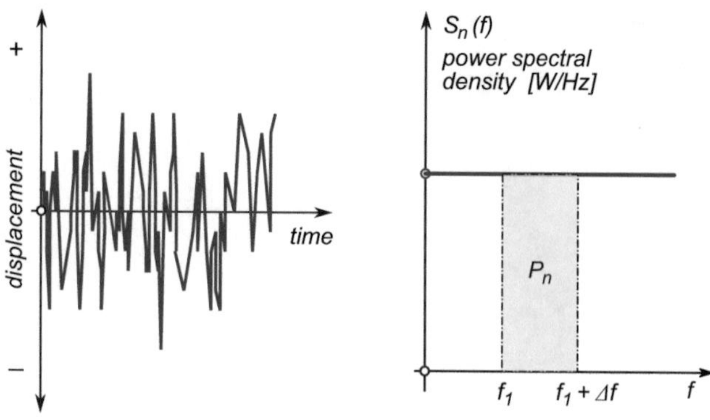

Fig. 8.1 Thermal noise: in the time domain (left); and the noise power spectrum density (right)

Variables that have zero average, which is the case with thermal noise, are much better evaluated by measuring their RMS value as in Sect. 1.5.2. Using methods from statistical thermodynamics and quantum mechanics, it has been shown that the noise spectrum density S_n (sometimes referred to as the available noise power) within a 1Hz bandwidth is

$$S_n(f) = k\,T \;[\mathrm{W/Hz}] \tag{8.1}$$

which is not a function of frequency (because it is already normalized to 1 Hz), i.e. it is constant, Fig. 1.1 (right). Therefore, the noise power generated within the frequency bandwidth Δf is, by definition

$$P_n = \int_{f_1}^{f_1+\Delta f} S_n(f)\,df = S_n(f) \int_{f_1}^{f_1+\Delta f} df = k\,T\,\Delta f \quad [\mathrm{W}] \tag{8.2}$$

where

k is the Boltzmann's constant (1.380×10^{-23} [J/K])
T is the absolute temperature of the conductor [K]
Δf is the frequency bandwidth in which the noise measured [Hz]

It is interesting to note that, even though it is modelled with a resistor, the noise power does not depend on resistance of the conductor. Equation (8.2) is also known as Johnson's law, and it implies that since the noise power is proportional to the system bandwidth, it is desirable to reduce the bandwidth of a receiver to the minimum.

Because the real conductor with resistance R generates the electrical noise power, it is modelled as the equivalent voltage (or the equivalent current) generator circuit consisting of an ideal voltage source[1] E_n and an ideal resistor R, Fig. 8.2. The average power delivered by a voltage generator of internal RMS voltage V_S and internal resistance R_S to a load R_L is at maximum, assuming matched impedances (i.e. $R_S = R_L$),[2] as shown in Fig. 8.2 (right). After substituting (8.2), we have

$$P_{Lmax} = \frac{(V_n/2)^2}{R_S} \quad \therefore \quad k\,T\,\Delta f = \frac{V_n^2}{4\,R} \quad \therefore \quad V_n = \sqrt{4\,R\,k\,T\,\Delta f} \tag{8.3}$$

[1]We use the letter E instead of V to indicate noise voltage generator.

[2]See Chap. 11.

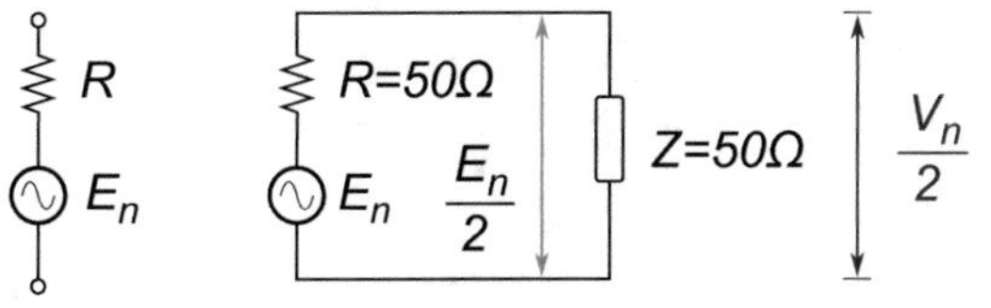

Fig. 8.2 Equivalent noise generator V_n with internal resistance R (left), and noise power delivered to a system whose input impedance is $R_L = Z$ (right)

Equation (8.3) is one of the most often used representations of electrical noise and is therefore widely used in calculating the system noise performance. Sometimes, because of the square root, it is more convenient to work with V_n^2 instead of V_n. The equivalent noise voltage of combinations of resistors in series and in parallel is calculated after finding the equivalent resistance R or conductance G as

$$V_n^2 = 4(R_1 + R_2 + \cdots)k\,T\Delta f = V_{n1}^2 + V_{n2}^2 + \cdots \tag{8.4}$$

$$I_n^2 = \frac{V_n^2}{R} = 4(G_1 + G_2 + \cdots)k\,T\Delta f = I_{n1}^2 + I_{n2}^2 + \cdots \tag{8.5}$$

where R is the equivalent noise resistance, $G = 1/R$ is the equivalent noise conductance, V_n is the equivalent noise voltage, I_n is the equivalent noise current.

Example 53: Thermal Noise Definitions

Find: (a) spectrum density for thermal noise at room temperature ($T = 300\,\text{K}$); (b) Available noise power within bandwidth of 1 MHz; (c) available signal power for 1 μV signal from 50 Ω source delivered to the matched load; (d) SNR for noise in (b) and signal in (c).

Solution 53:

$$\text{(a) } S_n = kT = 1.38 \times 10^{-23} \times 300 = 4.14 \times 10^{-21}\ \text{W/Hz}$$

$$\text{(b) } P_n = kT\Delta f = 4.14 \times 10^{-21} \times 10^6 = 4.14 \times 10^{-15}\ \text{W}$$

$$\text{(c) } P_s = \frac{(v_s/2)^2}{R_s} = \frac{\left(1 \times 10^{-6}/2\right)^2}{50} = 5 \times 10^{-15}\ \text{W}$$

$$\text{(d) } SNR = \frac{P_s}{P_n} = \frac{5 \times 10^{-15}}{4.14 \times 10^{-15}} = 0.82\,\text{dB}$$

Example 54: Thermal Noise Definitions

Resistors $R_1 = 20\,\text{k}\Omega$ and $R_2 = 50\,\text{k}\Omega$ are at room temperature $T = 290\,\text{K}$. For a given bandwidth of $BW = 100\,\text{kHz}$ find: (a) the thermal noise voltage for each resistor; (b) for their serial combination; (c) for their parallel combination.

Solution 54:

(continued)

(a) For the two resistors separately, from (8.3) follows that

$$V_n^2(R_1) = 4 \times 20\,\text{k}\Omega \times 1.38 \times 10^{-23} \times 290\,\text{K} \times 100\,\text{kHz} = 32 \times 10^{-12}\,\text{V}^2$$

$$V_n^2(R_2) = 4 \times 50\,\text{k}\Omega \times 1.38 \times 10^{-23} \times 290\,\text{K} \times 100\,\text{kHz} = 80 \times 10^{-12}\,\text{V}^2$$

$$\therefore$$

$$V_n(R_1) = 5.658\,\mu\text{V}$$

$$V_n(R_2) = 8.946\,\mu\text{V}$$

(b) Serial resistance is $R_s = R_1 + R_2 = 70\,\text{k}\Omega \quad \therefore \quad V_n(R_s) = 10.59\,\mu\text{V}$.
(c) Parallel resistance is $R_p = R_1 || R_2 = 14.286\,\text{k}\Omega \quad \therefore \quad V_n(R_p) = 4.78\,\mu\text{V}$.

8.2 Equivalent Noise Bandwidth

Although reactive components do not generate thermal noise because they do not dissipate thermal power, it is important to estimate the noise power of networks that contain inductive and capacitive reactances. This is because both capacitive and inductive components do influence the frequency bandwidth, hence the effect of reactances on the kT noise spectrum must be taken into account. We consider two important network cases for thermal noise calculations: resistor–capacitor (RC) networks and resistor–inductor–capacitor (RLC) networks.

8.2.1 Noise Bandwidth in an RC Network

It can be shown that, when noise passes through a passive filter which has a complex transfer function $H(\omega)$, the noise output spectrum density S_{no} for the given input spectrum density (8.1) is (in general) calculated as

$$S_{no} = |H(\omega)|^2\, k\, T \tag{8.6}$$

In the case of capacitive load (see Fig. 8.3 (left)), the LP filter with noise generator V_s^2 and output voltage V_n^2 taken across the capacitor has the transfer function[3] $H(\omega)$ as

$$|H(\omega)| = \frac{1}{\sqrt{1 + (\omega RC)^2}} \quad \therefore \quad S_{no} = \frac{k\,T}{1 + (\omega RC)^2} \tag{8.7}$$

$$P_{no} = \int_0^\infty S_{no}\, df = \int_0^\infty \frac{k\,T}{1 + (2\pi RCf)^2}\, df = \frac{k\,T}{2\pi RC} \int_0^\infty \frac{1}{1 + x^2}\, dx \tag{8.8}$$

[3]See Sects. 1.6.4 and 3.2.2.

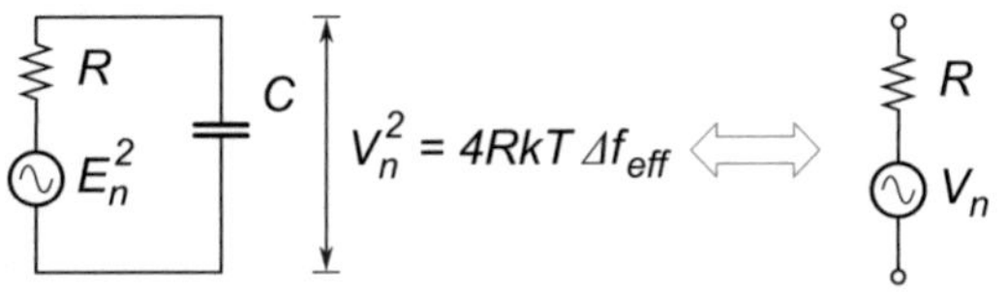

Fig. 8.3 Equivalent noise voltage in RC circuit

The output spectrum, therefore, decreases as the frequency increases due to the bandwidth limiting of the low-pass filter. The total noise power available at the output is obtained by integrating[4] (8.8) from zero to infinity. The integral is tabular and the result is

$$P_{no} = \frac{k\,T}{2\pi\,RC}\ \arctan x\,\Big|_0^{\infty} = \frac{\cancel{k}\cancel{T}}{2\cancel{\pi}\,RC}\frac{\cancel{\pi}}{2} \stackrel{\text{def}}{=} \cancel{k}\cancel{T}\,\Delta f_{\text{eff}} \quad \therefore \quad \Delta f_{\text{eff}} = \frac{1}{4RC} \tag{8.9}$$

where we introduced Δf_{eff} as "effective noise bandwidth". This definition allows introduction of equivalent circuit in Fig. 8.3 (right). Thus, the noise spectrum only within the effective bandwidth Δf_{eff} is considered to be equal to $k\,T$, and zero everywhere else.

Furthermore, the equivalent noise voltage V_n (see (8.3)) can be written as

$$V_n^2 = \cancel{4}\cancel{R}k\,T\frac{1}{\cancel{4}\cancel{R}C} = \frac{k\,T}{C} \tag{8.10}$$

which shows that even though the noise was generated by the resistor R, the output noise voltage is not a function of the resistor R. Instead, it is determined by the capacitor C, which does not generate the thermal noise by itself.

Example 55: Thermal Noise Definitions

Calculate equivalent noise voltage V_n generated by a resistor R in series with $C = 100\,\text{pF}$ capacitor at room temperature $T = 300\,\text{K}$

Solution 55: Straightforward implementation of (8.10) yields

$$V_n^2 = \frac{k\,T}{C} = \frac{1.38\times10^{-23}\times300\,\text{K}}{100\,\text{pF}} = 4.14\times10^{-11}\ V^2 \qquad \therefore \qquad V_n = 6.434\,\mu\text{V}$$

8.2.2 Noise Bandwidth in RLC Network

The RLC tuned circuit in Fig. 8.4 consists of the ideal lossless capacitor C and a real inductor L whose wire resistance r generates noise. We consider the noise voltage E_n as the input to the network and V_n as the output from the network, thus the modulus of the transfer $H(\omega)$ function is found (using the voltage divider rule with $(r + Z_L)$ impedance on one side and Z_C on the other) as

$$|H(\omega)| = \frac{|X_C|}{|Z_S|} \tag{8.11}$$

[4]Use of the substitution $(2\pi\,RCf = x)$ leads into the tabular integral $\int \frac{1}{1+x^2} = \arctan x$.

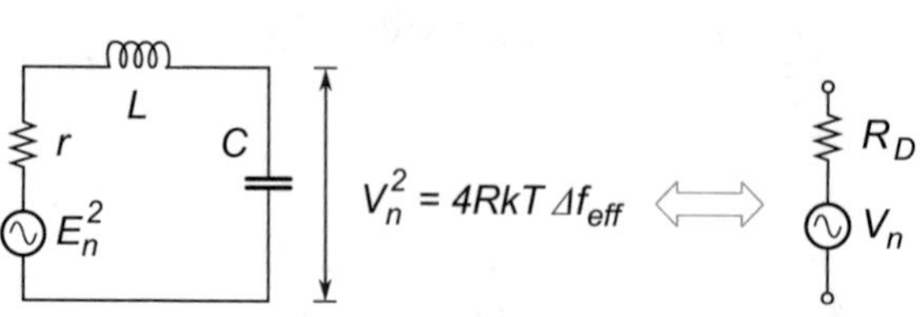

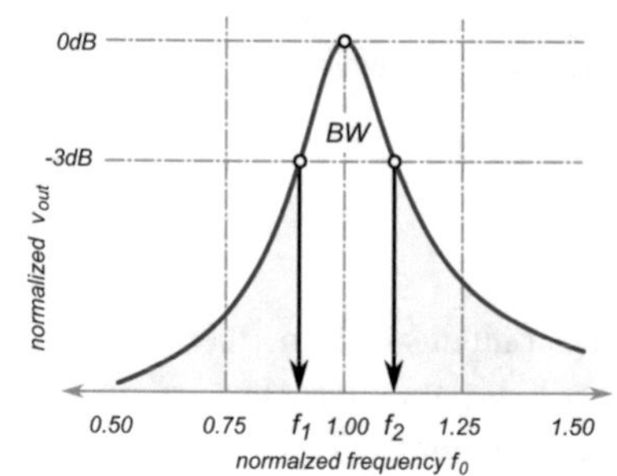

Fig. 8.4 Equivalent noise voltage in RLC tuned circuit (left), and its amplitude function (right)

where Z_S is the series impedance of the resonant RLC circuit (10.37)[5] and X_C is the reactance of capacitor C. We note the existence of two energy storing elements in the network, namely the inductor and capacitor, which means that the amplitude function (see Sect. 1.6.5) has a form of bandpass filter (BPF) centred around its resonant frequency.

If the noise calculation is limited to a narrow bandwidth $\Delta f \ll f_0$ around the resonant frequency f_0, then the transfer function $H(\omega_0)$ (8.11) is approximated as $H(\omega_0) \approx Q$, i.e. the Q factor of the RLC network, where one of Q definitions has the form of

$$Q = \frac{\omega_0}{BW}, \qquad (\omega_0 = 2\pi f_0) \tag{8.12}$$

where BW is defined strictly by the -3dB points, see Fig. 8.4 (right). Solving integral (8.6) in the case of $H(\omega_0) \approx Q$, thus $S_{no} = Q^2\,kT$, the noise power generated within the frequency bandwidth $BW = \Delta f$ is found to be

$$P_{no} = \int_0^\infty S_{no}\,df = \int_{f_0-\Delta f/2}^{f_0+\Delta f/2} Q^2\,kT\,df = Q^2\,\cancel{kT}\,\Delta f \stackrel{\text{def}}{=} \cancel{kT}\,\Delta f_{\text{eff}} \quad \therefore \quad \Delta f_{\text{eff}} = Q^2\,\Delta f \tag{8.13}$$

which gives the equivalent noise voltage V_n as

$$V_n^2 = 4\,r\,k\,T\,Q^2\,\Delta f = 4\,R_D\,k\,T\,\Delta f \tag{8.14}$$

where $R_D = Q^2\,r$ is the "dynamic impedance" of the RLC circuit at resonance. This result is very important for practical calculations because the noise bandwidth in RLC tuned networks is indeed limited to a narrow bandwidth around the resonant frequency.

By combining (8.2), (8.10), and (8.14), we can express the total noise power in reference to the capacitance C as

$$P_{no} = \frac{k\,T}{4\,R_D\,C} = k\,T\,\Delta f_{\text{eff}} \tag{8.15}$$

where $\Delta f_{\text{eff}} = 1/4R_D C$ is the effective noise bandwidth of the RLC network at resonance. In practice, however, the effective noise bandwidth Δ_{eff} of an RLC network is approximated by

[5]For more details see Sect. 10.2 on serial RLC resonance.

$$\Delta f_{\text{eff}} = \frac{\pi}{2} BW_{3\text{dB}} \tag{8.16}$$

to account for the “fringe” power found in the region outside of the $BW_{3\text{dB}}$ range, as indicated by coloured surface in Fig. 8.4 (right). The idea of an equivalent noise bandwidth can be extended to amplifiers and receivers as well.

Example 56: Thermal Noise Definitions
A tuned parallel LC tank has the following data: $f_0 = 120\,\text{MHz}$, $C = 25\,\text{pF}$, $Q = 30$, bandwidth $\Delta f = 10\,\text{kHz}$. Knowing that one of the forms for dynamic resistance is $R_D = Q/\omega_0 C$, calculate the effective noise voltage of the LC tank at room temperature within the given bandwidth.

Solution 56: From (10.79), dynamic resistance of LC resonator at resonance is calculated as

$$R_D = \frac{Q}{\omega_0 C} = \frac{30}{2\pi \times 120\,\text{MHz} \times 25\,\text{pF}} = 1.59\,\text{k}\Omega$$

then from (8.14)

$$V_n^2 = 4Q^2\, R_L\, k\, T \Delta f = 4R_D\, k\, T \Delta f = 0.254 \times 10^{-12}\,\text{V}^2$$

$$\therefore$$

$$V_n = 0.50\,\mu\text{V}$$

8.3 Signal-to-Noise Ratio

One of the most (arguably, the most) important quantitative measures of a signal’s “noisiness” is the signal-to-noise ratio (SNR), which is defined as the ratio of the signal and noise powers,

$$SNR = \frac{P_s}{P_n} \tag{8.17}$$

where P_s is the signal power and P_n the noise power. As defined, it shows how many times more powerful is the signal than the noise; it is a *relative* measure of the two powers. Note that SNR is a unitless number that merely shows the value of the signal-to-noise power ratio.

It is practical to express the power ratios in units of dB, defined as follows:

$$SNR = 10 \log \frac{P_s}{P_n} \quad [\text{dB}] \tag{8.18}$$

$$= 10 \log \frac{V_s^2/\cancel{R}}{V_n^2/\cancel{R}} = 10 \log \left(\frac{V_s}{V_n}\right)^2 = 20 \log \frac{V_s}{V_n} \quad [\text{dB}] \tag{8.19}$$

where V_s the signal voltage and V_n is the noise voltage, measured across the resistive load R terminals. Note the multiplication constants in the expressions for power (8.18) and voltage (8.19). It is trivial to derive an expression similar to (8.19) for currents instead of voltages. Although it may appear a bit counterintuitive to introduce the cumbersome logarithmic function to replace the clean ratio, it turns out that calculations in units of dB are much simpler because the ratios become differences,[6] which is a much simpler arithmetic operation.

The relative measure of power (8.18) tells us only that, for example, P_1 is double P_2 in the case of $SNR = 3\,\text{dB}$. It does not tell us whether we compared 6 to 3 mW, or 6 to 3 kW, i.e. it does not tell us anything about the *absolute* power levels. In order to convey that information as well, we need to define the absolute unit of power P_{dBm} as

$$P_{\text{dBm}} = 10 \log \frac{P_1}{1\,\text{mW}} \quad [\text{dBm}] \tag{8.20}$$

where P_1 power is normalized to 1 mW. In the following sections, we will show examples of how to use dBm units. Its unity step is identical to the dB unity step, which means that adding dB and dBm units is a perfectly valid mathematical operation.

Example 57: SNR Definitions

Convert signal power levels of; (a) $P_1 = 1\,\text{mW}$, (b) $P_2 = 1\,\text{W}$, and (c) $P_3 = 10\,\text{W}$ into dBm units. Then, find SNR of the same three signals if the noise power is $P_n = 1\,\text{mW}$

Solution 57: As per (8.20) leads into: (a) $P_1 = 0\,\text{dBm}$, (b) $P_2 = 30\,\text{dBm}$, and (c) $P_3 = 40\,\text{dBm}$. As per (8.18) leads into: (a) $SNR = 0\,\text{dB}$, (b) $SNR = 30\,\text{dB}$, and (c) $SNR = 40\,\text{dB}$. Note the difference between the *absolute* values in dBm and the *relative* values in *deci*B.

8.4 Noise Figure

Knowing SNR_{in} of the signal presented to the input terminals of a circuit network is only one step in the circuit design process. For the purposes of measuring the "noisiness" of the circuit itself, i.e. of finding out how much noise was generated by the circuit's internal components, SNR is measured both at the input and output terminals (see Fig. 8.5),

$$SNR_{in} = \frac{P_{si}}{P_{ni}} \quad \text{and} \quad SNR_{out} = \frac{P_{so}}{P_{no}} \tag{8.21}$$

$$\therefore$$

$$F = \frac{SNR_{in}}{SNR_{out}} = \frac{P_{si}\,P_{no}}{P_{ni}\,P_{so}} = \frac{P_{no}}{A_P\,P_{ni}} \tag{8.22}$$

where noise factor F is the ratio of the output and input SNRs, $A_P = P_{so}/P_{si}$ is the signal power gain. In practice, any of the three forms in (8.22) is used to calculate F. If, for example, $SNR_{out} = SNR_{in}$, then $F = 1$, which is to say that there was no additional noise contribution between the input and output terminals, hence the circuit itself is noiseless. Note that, as defined, noise factor F is a unitless number. Again, it is practical to introduce *noise figure* (NF) as

[6] Some of the basic logarithmic identities are: $\log(x/y) = \log(x) - \log(y)$; $\log(xy) = \log(x) + \log(y)$; $\log(x^n) = n\,\log(x)$.

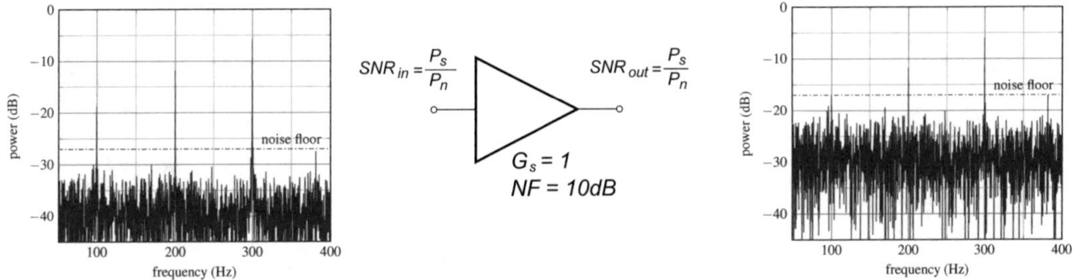

Fig. 8.5 An amplifier whose signal gain is $G_s = 1$ and $NF = 10\,\text{dB}$, with its input and output SNRs

$$NF = 10 \log F \text{ [dB]} \tag{8.23}$$

where the noiseless circuit (i.e. $F = 1$) has noise figure $NF = 10 \log 1 = 0\,\text{dB}$, which is the ideal case, of course it is not achievable in real systems.

Example 58: Thermal Noise Definitions

An amplifier has signal-to-noise ratio of $SNR(out) = 5$ at its input and $SNR(in) = 10$ at its output. Calculate its F and NF.

Solution 58: By direct application of (8.22) we simply write

$$F = \frac{SNR(in)}{SNR(out)} = \frac{10}{5} = 2 \quad \therefore \quad NF = 10 \log 2 = 3\,\text{dB}$$

that is, the amplifier itself contributes 3 dB of noise inside the overall system.

8.5 Noise Temperature

Thermal noise power in (8.2) can be rearranged to define the noise temperature T_n as

$$T_n = \frac{P_n}{k\,\Delta f} \tag{8.24}$$

where index n is added to the temperature T to indicate that the noise temperature T_n is referring to the noise power P_n.

For a given amplifier, however, its thermal noise is generated by the internal components and it can be measured at the output terminal. It is convenient in noise analysis to refer the noise back to the input terminal of the circuit and imagine that it is generated by the equivalent external noise source, while the circuit itself is assumed noiseless (see Fig. 8.6). If the circuit's power gain is A_P and if the equivalent noise power at the input is P_{ni}, then the output noise power P_{no} is calculated simply as

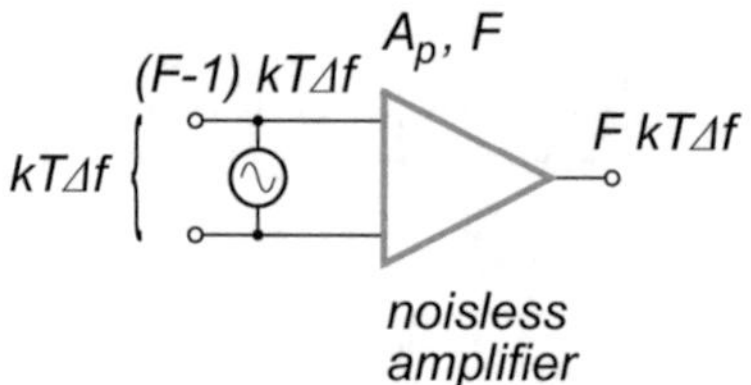

Fig. 8.6 Equivalent input referred noise power for a noiseless amplifier

$$P_{no} = A_P\, P_{ni} \quad \therefore \quad P_{ni} = \frac{P_{no}}{A_P} \tag{8.25}$$

On the other hand, if the input signal power is P_{si} and the input noise power is $P_{ni} = k\,T\,\Delta f$, then from (8.21) the input side signal-to-noise ratio SNR_{in} is

$$SNR_{in} = \frac{P_{si}}{k\,T\,\Delta f} \tag{8.26}$$

then, while keeping in mind that both signal and noise are amplified with the same gain A_P, (8.22) can be formatted as

$$F = \frac{P_{no}}{A_P\,P_{ni}} = \frac{P_{no}}{A_P\,k\,T\,\Delta f} \tag{8.27}$$

Substituting (8.25) into (8.27), it follows that the total available input-referenced noise is

$$F = \frac{P_{no}}{A_P\,k\,T\,\Delta f} = \frac{P_{ni}}{k\,T\,\Delta f} \quad \therefore \quad P_{ni} = F\,k\,T\,\Delta f \tag{8.28}$$

Therefore, the amplifier's noise contribution P_{na} is simply the difference between the output and input noise powers

$$P_{na} = F\,k\,T\,\Delta f - k\,T\,\Delta f = (F - 1)\;k\,T\,\Delta f \tag{8.29}$$

Substituting (8.29) into (8.24), (in the case of an amplifier for which $P_n = P_{na}$), we write

$$T_n = (F - 1)\;T \quad \text{or} \quad F = 1 + \frac{T_n}{T} \tag{8.30}$$

where T_n is the noise temperature, T is the ambient temperature. The significance of (8.30) is that it shows the equivalence between noise factor F and equivalent noise temperature T_n (which is not as same as the temperature of the noise source); if one is known, the other is known too. In addition, in cases of low noise power levels, noise temperature turns out to be more sensitive than noise factor, which makes the measurements easier. Because of that, noise temperature is used mostly at higher frequencies and in radio astronomy.

Example 59: Noise Figure and Temperature Definitions
Equivalent noise temperature of an amplifier is $T_n = 50\,\text{K}$. Calculate the amplifier's noise factor F at room temperature $T = 300\,\text{K}$. As a comparison, calculate F if $T_n = 300\,\text{K}$ and if $T_n = 0\,\text{K}$.

Solution 59: Direct implementation of (8.30) leads into

$$T_n = (F - 1)\,T \quad \therefore \quad 50\,\text{K} = (F - 1) \times 300\,\text{K} \quad \therefore \quad F = \frac{50\,\text{K}}{300\,\text{K}} + 1 = 1.167$$

$$\therefore$$

$$NF = 10 \log 1.167 = 0.669\,\text{dB}$$

If the equivalent noise temperature of the amplifier is as same as the environment $T_n = 300\,\text{K}$, $T_n = 0\,\text{K}$ and we find

$$NF = 10 \log 2 = 3\,\text{dB} \quad \text{and} \quad NF = 10 \log 1 = 0\,\text{dB}$$

which is to say that noiseless amplifier is equivalent to $T_n = 0\,\text{K}$ (it does not add noise to the input signal), and when T_n is equal to the environment noise produced and added by this amplifier is equal to the noise of the input signal.

8.6 Noise Figure of Cascaded Networks

Analysis in Sects. 8.4 and 8.5 demonstrated that any noise signal P_{ni} presented at the input terminals of an amplifier (or any general circuit, for that matter) is multiplied with its gain A_P and produces the output noise signal P_{no}, as shown by (8.25). In addition, the amplifier itself generates internal noise P_{na}, which is quantified by its noise factor F, as shown by (8.29). Therefore, a single-stage amplifier generates total output noise power P_1 as the sum

$$P_1 = P_{no} + P_{na} = A_P\, P_{ni} + (F - 1)\, k\, T\, \Delta f \tag{8.31}$$

or, in general, rearranging (8.27) we can also write for the total output noise power $P_{(no)(tot)}$

$$P_{(no)(tot)} = F_{(tot)}\, A_{P(tot)}\, k\, T\, \Delta f \tag{8.32}$$

where "(tot)" indicates that the internal structure of the amplifier may consist of multiple stages.

Let us now evaluate the noise factor of a cascade of networks, each stage with its own noise factor F_i, $(i = 1, \ldots, n)$. Considering that system-level analysis is based on a cascade of driver–load pairs, it is important to find an expression for the total noise factor of the cascaded system (see Fig. 8.7).

In its simplest, very important case, the system consists of only two stages ($i = 1, 2$), so that the noise factor F_{12} of the combination is calculated as follows. The input to the first stage is connected to a resistor R_{eq}, which is used to model the thermal noise injected into the two-stage system. For the sake of simplicity, let us assume that the two noise bandwidths Δf of the stages are identical and therefore equal to the noise bandwidth Δf of the cascaded combination.

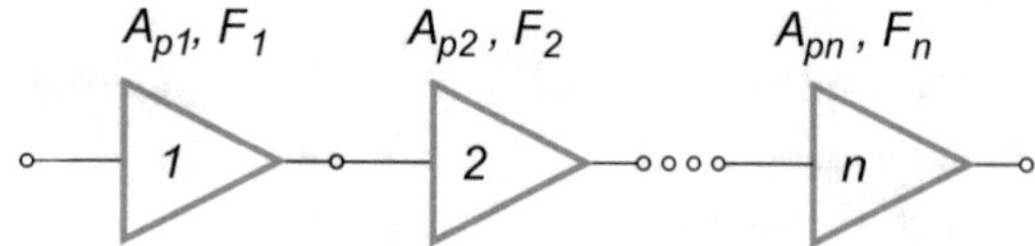

Fig. 8.7 Cascaded system of n stages, each with its own noise factor F_i, $(i = 1, \dots, n)$

The total gain A_{P12} of the two subsequent stages, obviously, must be their product

$$A_{P12} = A_{P1}\, A_{P2} \tag{8.33}$$

and, according to (8.27), the noise output $P_{(no)(1)}$ after the first stage is

$$P_{(no)(1)} = F_1\, A_{P1}\, k\, T\, \Delta f \tag{8.34}$$

which, after being multiplied by the second stage gain A_{P2} is

$$P_{(no)(2)} = A_{P2}\, P_{(no)(1)} = F_1\, A_{P1}\, A_{P2}\, k\, T\, \Delta f \tag{8.35}$$

if the second stage were noiseless. However, it amplifies its own input referred thermal noise (8.29)

$$P_{i2} = A_{P2}\, (F_2 - 1)\, k\, T\, \Delta f \tag{8.36}$$

Therefore, the total noise output from the second stage is the sum of (8.35) and (8.36)

$$\begin{aligned} P_2 &= A_{P2}\, (F_2 - 1)\, k\, T\, \Delta f + F_1\, A_{P1}\, A_{P2}\, k\, T\, \Delta f \\ &= \left(F_1 + \frac{F_2 - 1}{A_{P1}}\right) A_{P1}\, A_{P2}\, k\, T\, \Delta f \\ &= \left(F_1 + \frac{F_2 - 1}{A_{P1}}\right) A_{P12}\, k\, T\, \Delta f \end{aligned} \tag{8.37}$$

By comparison of (8.32) and (8.37) we have

$$F_{(tot)} = F_1 + \frac{F_2 - 1}{A_{P1}} \tag{8.38}$$

which is the noise factor expression for a two-stage cascaded network. By reiterating same reasoning, we generalize (8.38) to a cascaded network of n stages, resulting in

$$F_{(tot)} = F_1 + \frac{F_2 - 1}{A_{P1}} + \frac{F_3 - 1}{A_{P1} A_{P2}} + \cdots + \frac{F_n - 1}{A_{P1} A_{P2} \cdots A_{P(n-1)}} \tag{8.39}$$

Equation (8.39) is known as Friis's formula and is widely used for evaluating the NF of cascaded networks. Friis's formula suggests that in a cascaded network, the noise factor of the very first stage, i.e. F_1, is the most critical because noise factors of the subsequent stages are divided by the combined gain of all previous stages.

Example 60: NF of Cascaded Networks
A three-stage amplifier has the following specifications: gain of the first stage is $A_{P1} = 14\,\text{dB}$, and its noise figure is $NF_1 = 3\,\text{dB}$, the second stage has $A_{P2} = 20\,\text{dB}$, and its noise figure is $NF_2 = 8\,\text{dB}$, and the third-stage amplifier is identical to the second stage. Calculate the overall noise figure NF of the system.

Solution 60: Using Friis's formula we write:

$$A_{P1} = 14\,\text{dB} = 25.1, \quad A_{P2} = A_{P3} = 20\,\text{dB} = 100$$

$$NF_1 = 3\,\text{dB}, \quad F_1 = 2, \quad NF_2 = NF_3 = 8\,\text{dB}, \quad F_2 = F_3 = 6.31$$

therefore,

$$F_{(tot)} = 2 + \frac{6.31 - 1}{25.1} + \frac{6.31 - 1}{25.1 \times 100} = 2.212 \quad \therefore \quad NF = 10 \log 2.212 = 3.448\,\text{dB}$$

This example illustrates that NF of the first stage is the one that is most important, even if the subsequent stages have terrible NF, the overall NF of the system is close to the NF of the first stage. For that reason, most of the effort in design of RF front-end circuits is given to design the first-stage amplifier, also known as "low-noise amplifier" (LNA).

Example 61: The Input Referred Noise
For the amplifier in Fig. 8.8 (left), calculate the signal voltage V_s and the equivalent noise voltage V_n appearing at the input terminals.

Data: bandwidth $\Delta f = 10\,\text{kHz}$, room temperature $T = 290\,\text{K}$, equivalent internal noise resistance $R_n = 400\,\Omega$, amplifier input resistance $R_i = 600\,\Omega$, source resistance $R_s = 50\,\Omega$, and source voltage $V_s = 1\,\mu\text{V}$.

Solution 61: Application of Thévenin theorem on the V_s, R_s, and R_i network results in the following:

$$R_t = \frac{R_s\,R_i}{R_s + R_i} = 46.15\,\Omega$$

$$V_t = V_s\,\frac{R_i}{R_s + R_i} = 0.923\,\mu\text{V}$$

The equivalent noise voltage at the amplifier input is calculated for the case of serial resistors as

$$R = R_t + R_n = 400\,\Omega + 46.15\,\Omega = 446.15\,\Omega$$

which after applying (8.3), results in

(continued)

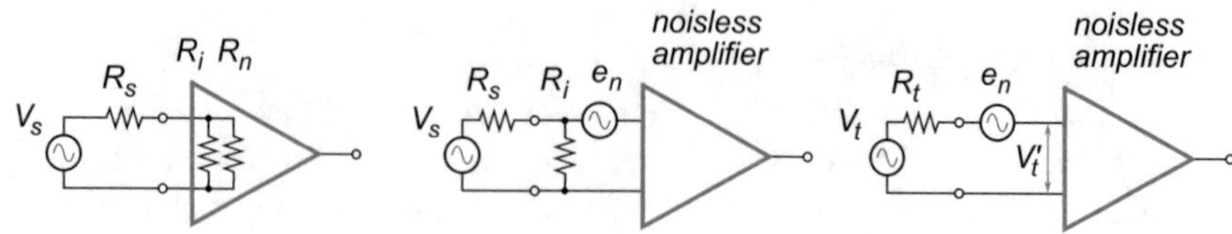

Fig. 8.8 An amplifier with internal noise and input resistances (left); the noise equivalent voltage generator (middle), and equivalent Thévenin representation (right)

$$V_n = \sqrt{4\,R\,k\,T\,\Delta f} = \sqrt{4 \times 446.15\,\Omega \times 1.38 \times 10^{-23} \times 290 \times 10\,\text{kHz}} = 0.267\,\mu\text{V}$$

8.7 Summary

The topic of noise analysis is much broader than presented in this short chapter. A large number of research publications and textbooks are available for further study. In this chapter, we reviewed the most important basic definitions and applications that are considered essential for further discussion. The reader is encouraged to become fluent with the terminology and principles related to noise analysis.

Problems

8.1 Calculate:

(a) spectrum density for thermal noise at room temperature ($T = 300$ K);
(b) Available noise power within the bandwidth of 1 MHz;
(c) available signal power for 1 μV signal from 50 Ω source delivered to the matched load;
(d) SNR for noise in (b) and signal in (c).

8.2 Determine the noise voltage generated by 50 Ω, 1 kΩ, and 1 MΩ resistors at room temperature 290 K and within 20 kHz bandwidth.

8.3 Resistors $R_1 = 20\,\text{k}\Omega$ and $R_2 = 50\,\text{k}\Omega$ are at room temperature $T = 290$ K. For a given bandwidth of $BW = 100$ kHz find: (a) the thermal noise voltage for each resistor; (b) for their serial combination; (c) for their parallel combination.

8.4 A tuned parallel LC tank has the following data: $f_0 = 120$ MHz, $C = 25$ pF, $Q = 30$, bandwidth $\Delta f = 10$ kHz. Find the effective noise voltage of the LC tank at room temperature within the given bandwidth.

8.5 An oscilloscope probe is specified as $R = 1\,\text{M}\Omega$ and $C = 20$ pF, with the bandwidth of $BW =$ 200 MHz. Determine noise voltage generated due to the probe at the room temperature.

8.6 A television set consists of the following chain of sub-blocks: two RF amplifiers with 20 dB gain and 3 dB noise figure each, a mixer with a gain of −6 dB and noise figure of 8 dB, two additional amplifiers with 20 dB gain and noise figure of 10 dB each. Assuming room temperature, calculate: (a) the system noise figure; and (b) the system noise temperature.

8.7 An amplifier with the input signal power of 5×10^{-6} W, noise input power 1×10^{-6} W, has output signal power of 50×10^{-3} W and the output noise power 40×10^{-3} W. Determine the noise factor F and the nose figure of this amplifier.

8.8 Calculate noise current and equivalent noise voltage for a diode biased with $I_{DC} = 1\,\text{mA}$ at the room temperature 300 K and within the bandwidth of 1 MHz.

8.9 A voltage source V_S with the internal resistance of R_S provides a signal to amplifier whose input resistance is R_{in}. Under those conditions the equivalent DC shot noise current is I_{DCn} at the room temperature. Calculate the input $SNR(in)$ assuming the bandwidth BW.

Data: $v_S = 100\,\mu V_{\texttt{rms}}$, $R_S = 50\,\Omega$, $R_{in} = 1\,\text{k}\Omega$, $I_{DCn} = 1\,\mu\text{A}$, $BW = 10\,\text{MHz}$, $T = 300\,\text{K}$.

8.10 A front-end RF amplifier whose gain is 50 dB and noise temperature 90 K provides signal to a receiver that has a noise figure of 12 dB. Calculate the noise temperature of the receiver by itself, and the overall noise temperature of the amplifier plus the receiver system at the room temperature $T = 300\,\text{K}$.

Part II

Radio Receiver Circuit

Radio Receiver Architecture

9

Wireless communication systems are result of multidisciplinary research that exploits various mathematical, scientific, and engineering principles in a very creative manner. The inner structure of signal waveforms (such as a voice, for example) is revealed by Fourier transformations, which enables us to design appropriate filters and amplifiers. By using Fourier theory we are able to both synthesize and decompose waveforms that are continuous (i.e. analog) or sampled (i.e. digital). By using Maxwell's theory we explain the creation and propagation of EM waves. By using mathematical theorems and techniques, such as basic trigonometry identities, for example, we are able to manipulate signals at the system level. By using circuit theory and techniques, we are able to practically implement the underlying theoretical equations and therefore create "wireless communication system". In this chapter we briefly introduce the basic RF receiver architecture, a.k.a. "heterodyne receiver" that is principal architecture exploited in all modern RF systems.

9.1 Electromagnetic Waves

In a very similar way to acoustic waves creating periodic pressure disturbance in the air, electric waves create periodic disturbance of the EM field. In order for EM induction to generate a meaningful potential, the Rx/Tx antennas must be similar length to the wavelength λ of the EM disturbance.

Consequently, to radiate radio waves at audio frequencies, an antenna would have to be of the order of kilometres, so one might just as well (and more conveniently) use it at ground level as a telephone line. What is more, it turns out that RF waves do not need air to propagate. If anything, they travel much easier through a vacuum. In short, the fundamental difference between sound waves and RF waves is that sound waves propagate by *mechanical* vibrations, while RF waves *radiate* from antenna in the form of electromagnetic waves.

Relative to the complete frequency spectrum known to exist in nature, our natural wave receptors (ears and eyes) cover only two minor frequency ranges (i.e. audio and optical). In addition, there is a relatively big gap between the two frequency bands. It is no surprise that most of our engineering efforts go into building artificial wave receptors that operate in our "blind spots" and enable us to "see" the full sound and EM spectrums.

R. Sobot, *Wireless Communication Electronics*,
https://doi.org/10.1007/978-3-030-48630-3_9

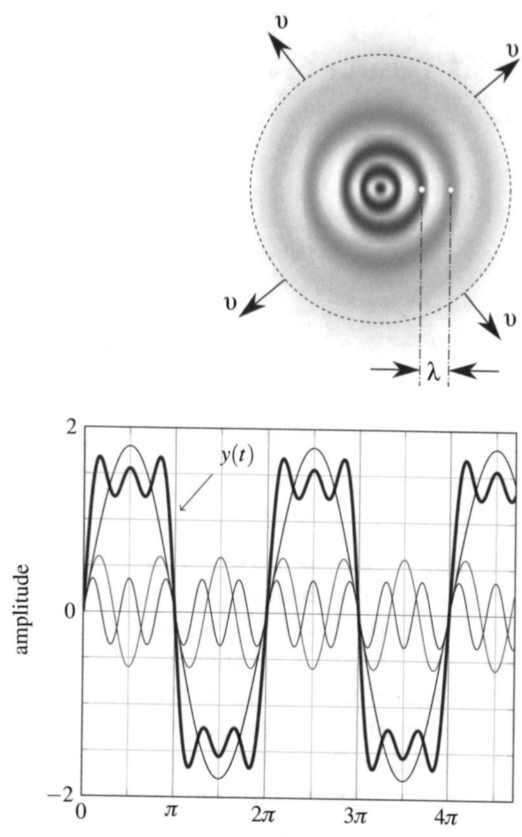

Fig. 9.1 A spherical wave spreading in all directions. The imaginary expanding sphere (represented by the blue circle) that separates space still not affected from the wave inside is the wavefront. Its velocity determines the wave propagation speed ν. The wavelength λ is the spatial distance between two subsequent crests

Fig. 9.2 A multi-tone waveform (solid dark line), created by linear addition of its first, third, and fifth harmonics. Increasing the number of harmonics to 15 or more creates an almost-perfect, square-pulse waveform

9.1.1 Multi-tone Waveform

A brilliant intuition led Fourier to speculate that an arbitrary waveform, which is a typical shape found in nature, such as spherical wave in Fig. 9.1, can be decomposed into the sum of single-tone waveforms. Eventually, he proved the idea and earned his space in history by developing the "Fourier transform", which is known to virtually every engineer and scientist in the world.

A very liberal interpretation of the Fourier transform is that any arbitrary waveform can be synthesized from an infinite number of harmonics added together in a certain proportion, as prescribed by a formula that was delivered for that particular waveform. We start with a single tone whose frequency is ω, referred to as the first harmonic (or, the fundamental), the second harmonic is a single-tone sinusoidal waveform whose frequency is $2\,\omega$, the third harmonic is a single-tone sinusoidal waveform whose frequency is $3\,\omega$, and so on. All single-tone terms in a Fourier transform (referred to as "harmonics") are then appropriately scaled in amplitude and added together as in, for example, (9.1).

Using Fourier transform, the squarish looking waveform $y(t)$ in Fig. 9.2 is synthesized by adding only the first three odd harmonics as

$$y(t) = \frac{4}{\pi}\left[\sin\omega\, t + \frac{1}{3}\sin 3\,\omega\, t + \frac{1}{5}\sin 5\,\omega\, t + \cdots\right] \tag{9.1}$$

where sin terms, together with their respective frequencies $n \times \omega$ ($n = 1, 3, 5, \ldots$) and amplitude scaling coefficients ($1, 1/3, 1/5, \ldots$) multiplied by $4/\pi$ represent harmonics of the waveform $y(t)$. One way to interpret (9.1) is to say that the waveform $y(t)$ is constructed using the three single-tone signals as its basic building blocks. In a way, the Fourier transform serves a similar role for a waveform as an X-ray machine does for a body: it shows what a complicated waveform is made of.

9.1.2 Frequency Spectrum

The shape of a sinusoid is always the same and all that we need to describe it are the three numbers representing its amplitude, frequency, and phase. If the sine plot axes are labelled, the three numbers are found by inspection.

However, a more complicated waveform, such as $y(t)$ in Fig. 9.2, cannot be analysed by visual inspection only, because it was created by adding multiple harmonics and it is defined by its total amplitude, frequency, and phase parameters.

Instead, a wave frequency spectrum clearly shows that the wave may contain many tones in an infinite number of combinations, as implied by (9.1). It is very useful to create a plot that shows the relationship among all the harmonics in the frequency–power domain, as for example in Fig. 9.3 that shows how harmonics are used to create two common waveforms: square and sawtooth. To illustrate the point, in order to transform a waveform into its equivalent frequency spectrum function, a fast Fourier transform (FFT) numerical algorithm is applied to the time domain waveform data, Fig. 9.4. In frequency domain plots, it is common practice to convert units of amplitude (for instance, volts or amperes) into decibels (dB) (unit of a relative gain).

A signal's time domain graph in Fig. 9.4 (left) is converted into its equivalent frequency domain plot, Fig. 9.4 (right), which is interpreted as follows. Starting from the zero frequency point (i.e. DC) on the left side and moving along the horizontal axis (scaled in units of Hz), each point of the graph symbolizes its respective $(x, y) =$ (frequency, power) pair of numbers. In other words, each pixel of the curve shows (in the ideal case) individual power levels for each of the infinite possible single tones within this frequency band. The three distinct vertical lines represent the three waveform harmonics, each with its dB power level quantified by the highest vertical point. In addition, there is a "sea of noise" (i.e. a "noise floor"), caused by various random sources that exist all around us, which is relatively constant.

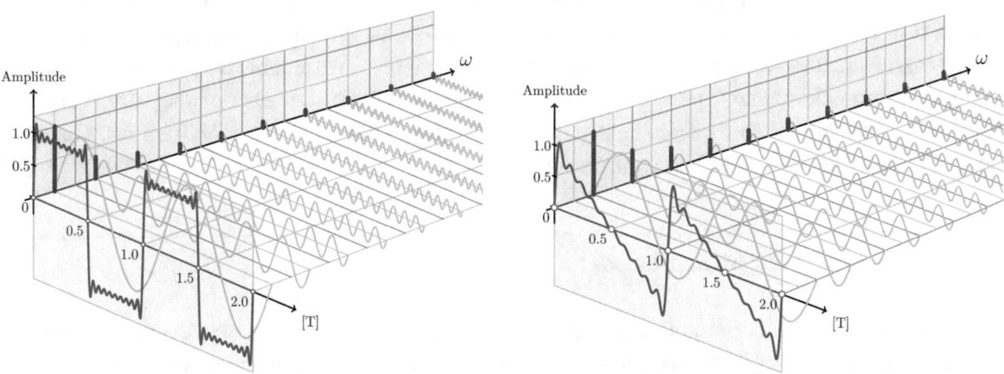

Fig. 9.3 Fourier synthesis of a square waveform (left) and sawtooth waveform (right) using first nine harmonics

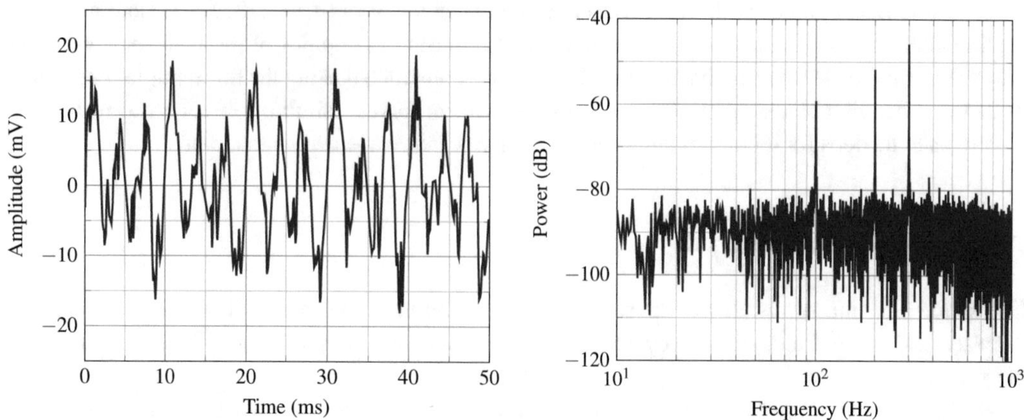

Fig. 9.4 Time domain waveform (left), and its power spectrum plot (right) relative to the noise floor at around −80 dB

Table 9.1 Classification of radio frequency bands

Frequency band	Abbreviation	Frequency range	Typical application
Extremely low	ELF[a]	3–30 Hz	Military underwater communications
Super low	SLF[a]	30–300 Hz	Military underwater communications
Ultra low	ULF[a]	0.3–3 kHz	Military underground communications
Very low	VLF	3–30 kHz	Submarine navigation
Low	LF	30–300 kHz	LORAN, time signals
Medium	MF	0.3–3 MHz	AM broadcasting, radio beacons
High	HF	3–30 MHz	Amateur radio
Very high	VHF	30–300 MHz	Short-distance terrestrial communication
Ultra high	UHF	0.3–3 GHz	TV broadcasting, cell phones
Super high	SHF	3–30 GHz	Wireless LAN, satellite links
Extremely high	EHF	30–300 GHz	Radio astronomy, research, military

[a]The whole Earth may serve as antenna

It should be emphasized that detailed examination of a complicated waveform includes both time and frequency domain analysis. To help the process, an oscilloscope is a test instrument that serves as a time domain waveform plotter and spectrum analyzer is a test instrument that performs real-time Fourier transformation of the given waveform and displays its "power spectrum plot". It is assumed that all engineers and scientists are familiar with these two test instruments that enable us to see two distinct, yet complementary, perspectives of the same waveform. Because of the very wide range of useful frequencies used by radio communication systems, the most common frequency bands are categorized into sub-bands as shown in Table 9.1.

9.2 The Need for Modulation

The main purpose of a communication system is to transmit a message from one point in space to another. It would be very inefficient, to say the least, if we had a single communication system that is capable of transmitting only one message at a time. Just imagine if the whole world's phone system used a single metallic wire and users had to line up at the two ends for a chance to communicate with the other side. An evolutionary improvement is the development of a network with multiple

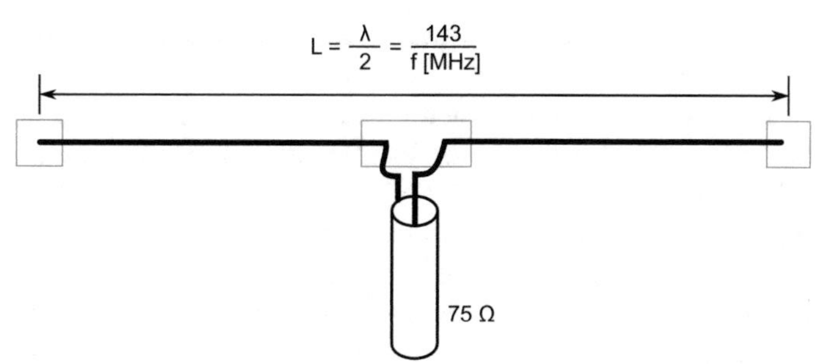

Fig. 9.5 Dipole antenna diagram

communication points. Indeed, the phone system is based on the existence of a temporary physical wire connection between the sending and the receiving points, which, of course, necessitates the existence of switching circuits. Today, direct transmission of relatively LF signals, such as our voice, is routinely done through the phone network.

However, we note that the wire-based communication network is rather expensive to build and maintain. That is mostly because the wire itself needs a supporting medium, in this case the Earth's surface, which presented huge technical challenges, for instance, when the intercontinental cables were laid at the bottom of the Atlantic and Pacific oceans.

A wireless transmission system does not have this issue, however, its own problem becomes visible in the case of direct transmission of audio signals. One commonly used antenna is known as a "half-wave dipole antenna",[1] a name derived from the requirement that the antenna wire length L is approximately equal to half the wavelength λ (Fig. 9.5), which is calculated as

$$L = \frac{1}{2}\lambda = \frac{1}{2}c\,T = \frac{1}{2}\frac{c}{f} \approx \frac{300 \times 10^6 \mathrm{m/s}}{2}\;\frac{1}{f\ [^1/\mathrm{s}]} = \frac{150}{f\ [\mathrm{MHz}]}\ [\mathrm{m}] \tag{9.2}$$

where λ is the wavelength of the incoming EM wave, T is its period, and c is the speed of light. Antenna designers usually fudge (9.2) by approximately 5%, so that the commonly cited rule of thumb for the length L of the wire intended to receive a signal at the frequency f is written as

$$L = \frac{143}{f\ [\mathrm{MHz}]}\ [\mathrm{m}] \tag{9.3}$$

which, for a simple example of audio frequency $f = 1\,\mathrm{kHz} = 0.001\,\mathrm{MHz}$ leads to a required antenna $L = 143\,\mathrm{km}$ long. For all practical purposes, we already have that antenna in the form of the phone cables laid in trenches all around the world. Which is to say that direct radio transmission of audio signals is not practical. A straightforward solution to this problem is to apply the frequency shifting principle and move the audio signals higher in the frequency domain. For instance, if the signal frequency is $f = 10\,\mathrm{MHz}$, then we calculate, from (9.3), that the required antenna must be approximately $L = 143/10 = 14.3\,\mathrm{m}$, while for a signal $f = 1\,\mathrm{GHz}$, the required dipole antenna is $L = 14.3\,\mathrm{cm}$ long. One of the first generations of cellphones used 850 MHz frequency, thus its retractable antenna was a wire of approximately that length. Obviously, the use of higher frequency signals results in more practical antenna sizes. A further study of EM wave propagation properties showed that transmission losses through various materials are dependent on the frequency. Hence, for the given transmission medium (air in this case), not all wavelengths travel the same distance for the

[1]For a detailed theory of quarter-wave and dipole antennas, see, for example, Antennas and Propagation for Wireless Communication Systems by S. Saunders and A. Aragón-Zavala.

same initial signal power, which means that the choice of operating frequency is very important in regard to how much energy is used for the transmission.

An efficient communication system needs to be capable of transmitting multiple transmissions at the same time. Considering that the audio bandwidth requires approximately 20 kHz, if the RF equipment is capable of working from, say, 1 GHz–2 GHz, then the $(2 - 1)\,\text{GHz} = 1\,\text{GHz}$ frequency bandwidth can be viewed as a wide cable that consists of $1\,\text{GHz}/20\,\text{kHz} = 50 \times 10^3$ parallel "wires", i.e. 50 000 separate "channels", where each channel can carry one full HiFi audio signal. If each of the 50 000 audio sources is precisely frequency shifted and aligned next to each other within the 1 GHz bandwidth, then by means of frequency shifting and filtering, the wireless system is enabled to transmit multiple signals at the same time. In practical terms, a wirelessly transmitted signal consists of two signals: a high-frequency signal that serves as the carrier and a low-frequency information signal that is somehow embedded into the carrier by the transmitter circuitry and de-embedded by the receiver.

We can now summarize the reasons why modulation is needed:

1. To enable practical wireless transmission of audio signals.
2. To enable power-efficient transmission that depends on the carrier frequency.
3. To serve as the mechanism of embedding low-frequency information into the high-frequency carrier.

For a given periodic signal, a natural question is what exactly we can modulate. A general, time domain, periodic signal $c(t)$ is described as

$$c(t) = C \sin(\omega t + \varphi) \tag{9.4}$$

where C is its maximum amplitude, ω is its radial frequency, and φ is its initial phase. By inspection of (9.4), we conclude that there are three possible ways to embed information into the carrier:

1. Vary the amplitude C in time so that $C(t)$ becomes equal to the time variation of the information signal (this method is called "amplitude modulation" (AM)).
2. Vary the frequency ω in time so that $\omega(t)$ becomes equal to the time variation of the information signal (this method is called "frequency modulation" (FM)).
3. Vary the phase φ in time so that $\varphi(t)$ becomes equal to the time variation of the information signal (this method is called "phase modulation" (PM)).

Although, theoretically any combination of the three parameters C, ω, and φ could be used to modulate the carrier $c(t)$, including modulating all three at the same time, in practice a communication system is designed to work with only one type of modulation. Hence, we only talk about either amplitude (AM), frequency (FM), or phase modulation (PM) systems.

9.3 RF Communication System

The goal is to transmit an audio signal using RF waves and faithfully reproduce it at the receiving end. Based on the principles introduced so far, a rough block diagram of one possible system architecture is shown in Fig. 9.6.

At the beginning of the transmission chain, the mechanical sound wave produced by vocal cords (1) must be converted into its equivalent electrical signal (2) by a microphone. This electrical signal contains the complete information that needs to be transmitted and it is now ready to take a ride on its assigned carrier. It is the job of the modulator (3) to accept the signal and mix it with the carrier,

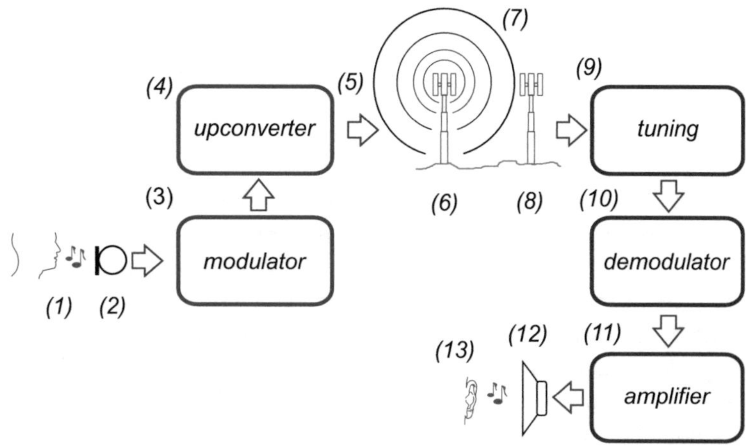

Fig. 9.6 A basic block diagram of a wireless communication system and its required components

which is enabled by the upconverter (4), so that the signal is imprinted as the carrier's envelope. The modulated carrier (5), with the information "riding" as its envelope, is now pushed into the transmitting antenna (6) and radiated into open space in the form of an EM wave (7). At this point, the information is available to anyone who is within the receiving range and whose "electrical length" matches that of the carrier. It should be noted that, at the same time, space is very busy and filled with many other carriers trying to reach their respective destinations. For the time being, the most important condition is that within the given space there must be no more than one carrier of any given frequency, i.e. each carrier must use its assigned and separate channel. Otherwise, two information packages travelling on separate carriers with indistinguishable carrier frequencies would unintentionally mix with each other and would be lost forever. In Sect. 14.7, we expand this condition to include one more frequency that, for now, we refer to as a "ghost frequency". There is no restriction on the number of receivers within the receiving range; in fact, radio and TV broadcasting companies spend vast amounts of money and resources to keep increasing the number of receivers for their broadcasting signals within the receiving distance range. This receiving distance range is limited by the power of the transmitted signal (its "loudness") and the sensitivity of the receiver (the quality of its "hearing system").

A receiver expecting a message must first adjust its "electric length" to match the frequency of the carrier. Under that condition, the receiving antenna (8) and the tuning section (9) start to oscillate in synchronicity with the incoming carrier, while (ideally) ignoring all other carriers. Using a simple analogy, we can visualize the receiver and the tuning section as a wall with a number of doors in various colours. At any given time, only one door is open (i.e. tuned) and only carriers of the matching colour get through. All other carriers face the wall with their matching colour doors closed.

Now there is no need for the carrier signal itself (it is like a car that needs to be parked at the end of the trip, while the transmitted information is similar to a passenger getting out of the parked car)—it is the job of the demodulator (10) to extract the envelope and discard the carrier. After travelling a long way, the incoming wave is very weak (it is not economical to place receivers closer to the transmitter than needed), hence there is significant amplification (11) present in the receiving path. Stated differently, it is beneficial to design receivers with high sensitivity and maximize the distance between the transmitter and the receiver so that the total number of users is maximized.

At the end of the receiving chain, the signal carrying the information is ready to be converted from electrical to mechanical form by a speaker (12) and finish the last leg of the journey the way it started—as a sound wave understandable by humans (13). The magic is done and the virtual distance between two humans becomes independent of the physical distance.

9.4 Heterodyne AM Radio Receiver Architecture

Depending on specific application and its frequency range, there are many practical implementations of communication systems that follow the principle explained in Sect. 9.3. For the purpose of introduction to design of practical RF systems, in this book we follow a design of a heterodyne AM receiver, Fig. 9.7. This receiver is based on the circuit architecture referred to as "heterodyne", which implies that there is only one mixer on the signal path, either for upconversion or downconversion of the carrier's frequency. Logically, the term "super-heterodyne" implies that more than one mixer is used. To explain this terminology, first we note that a problem arises when the carrier frequency is much higher than the one used in the rest of the receiver's circuits. As an analogy, when a street is not too wide, pedestrians can cross it easily in a single "jump". However, in the middle of a wide avenue we find "pedestrian refuge island" that serves as a temporarily stop point before the pedestrians can continue their trip to the other side, thus making "two-jump" crossing. Similarly, for example, when the carrier frequency is 10 MHz and the rest of receiver's circuit work at 445 kHz it is not difficult to design a mixer that is capable to make this frequency shift (i.e. heterodyne architecture). Just the same, if the carrier frequency is 2.4 GHz it is not possible to design a practical mixer that is capable of shifting this carrier's frequency in one jump, instead two or more mixers are used in series until the end frequency is reached (i.e. super-heterodyne architecture). For the purpose of introduction to practical RF systems, obviously heterodyne architecture is a good choice.

Reason to choose relatively low 10 MHz frequency (by today's standards) for design of a demo heterodyne AM receiver presented in this book is that we can focus on the fundamental principles of RF circuit design while, for the moment, we can ignore more advanced and more rigorous constrains associated with HF circuits. These HF issues are appropriately left for more advanced RF courses.

Being the oldest and simplest among the three modulation techniques, study and design of an AM receiver is a logical first step. As a practical demonstration of the knowledge acquired in this book, we

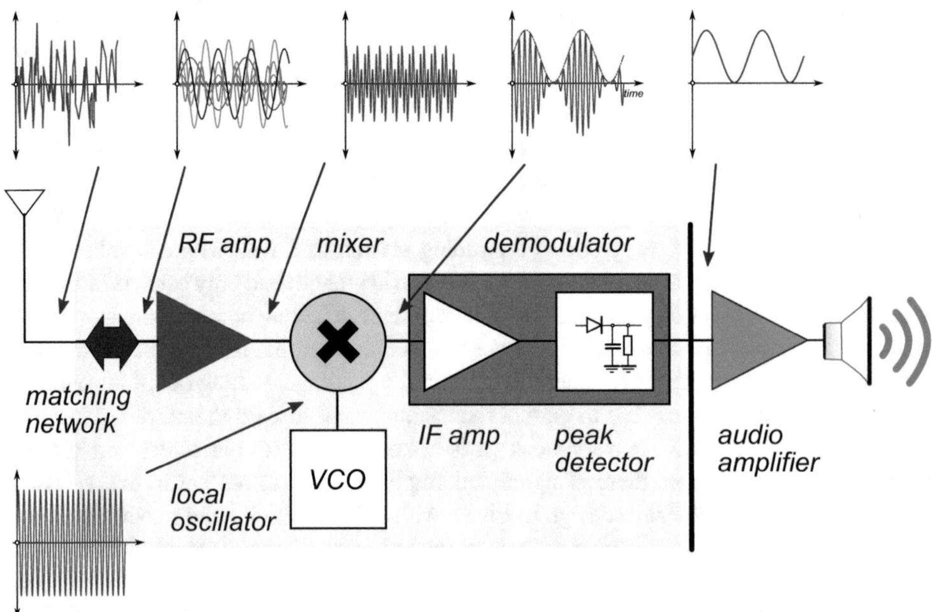

Fig. 9.7 A basic block diagram of an AM receiver and its subcircuits

can design a working demo RF receiver that, for example, receives time reference RF signals emitted by the state owned transmitters. Despite advances of GPS technology, these terrestrial transmitters are still used in many countries in the world because their time reference is derived from atomic clocks. For example, 14.67 MHz time reference signal is transmitted by NRC in Canada, 10 MHz signal is transmitted by NIST in the USA, and the TDF time signal at 162 kHz is transmitted by the Allouis long-wave transmitter in France.

Due to very large number of the existing RF communication systems that are continuously transmitting, e.g. radio and TV stations, cellphones, satellite and GPS systems, etc., for all practical purposes the overall frequency spectrum around us looks as a white noise.

At the front-end side of the AM receiver, Fig. 9.7, among all existing signals the antenna preselects the desired carrier waveform that is at ω_{RF} frequency that is same as the resonant frequency ω_0 (see Chap. 10) of the front-end circuits. The incoming RF signal is accepted by the matching network (see Chap. 11) on its way to RF amplifier (see Chap. 12). The combined BPF effect of the antenna, the matching network, and RF amplifier permits only waveform whose frequency is within the BPF bandwidth[2] to enter the receiver.

Amplitude-modulated signal at the RF amplifier's output is centred at ω_{RF} and it must be downconverted to the standard intermediate frequency (IF) of 455 kHz used in AM receivers. In the mathematical sense, downconversion process is simply a multiplication of the received RF signal and the local reference signal. This precise local reference is generated in the form of sinusoidal signal by the local oscillator (LO) circuit (see Chap. 13). In the following chapters we find that it is simply the difference between frequencies of the incoming RF and the local oscillator that must be set to $IF = 455$ kHz. In order to be able to select and receive more than one single radio station (as we take for granted for our radio receivers), the LO must be able to generate a range of reference frequencies, in other words it must be designed as a tuneable voltage-controlled oscillator (VCO). In the following step of the signal processing, the frequency shifting, it is the role of mixer to perform the multiplication and to produce IF signal at its output node. This upconversion/downconversion frequency operation performed by a mixer is referred to as "frequency shifting" (see Chap. 14).

The process of de-embedding the transmitted message is known as "demodulation" (see Chap. 16). Signal entering the demodulator circuit is found at ω_{IF} frequency with clearly visible envelope (for example, a single sinusoid in Fig. 9.7). In order to compensate for the inevitable amplitude loss that occurs between RF amplifier's output and demodulator's input nodes, there is an IF amplifier inserted on the signal path. The internal structure of an IF amplifier is practically the same as the one found in the RF amplifier, with the main difference in the setup of LC loads: the resonant frequency in the RF amplifier is set to ω_{RF} (which is a high, RF frequency), while the resonant frequency in the IF amplifier is set to ω_{IF} (which is much lower, downconverted frequency).

In the last stage of the AM receiver, just before the input node of the audio amplifier, the audio message embedded as the envelope of AM IF waveform is recovered by "peak detector" circuit, see Sect. 16.2. This little circuit that consists of a diode, resistor, and capacitor is certainly one of the most versatile circuits in electronics. Except for the inevitable distortions, waveform generated at its output is almost identical to the one generated by the microphone. The quality and limitations of the received signal is measured by the set of standard specifications, some of the most important ones are introduced in Chap. 17. With this closing step of the transmission/reception the AM receiver design flow is completed. And all of this effort, at the first glance seemingly unnecessary, is needed simply because it is not practical to directly transmit an audio signal without involving the upconversion/downconversion procedure.

[2]The Q factor of this real BPF is not infinite, i.e. $Q = \Delta\omega/\omega_0$ is set by design.

9.5 Summary

In this short chapter we introduced the architecture of the AM receiver that is at the core of all modern RF communication systems. Of course, over the time we have developed many different circuits and systems, both analog and digital, both at high and at very high frequencies for various wireless communication applications. Nevertheless, the fundamental principles and ideas that are introduced so far and applied in the following chapters are general, thus should be mastered before continuing studies of HF communication systems.

Problems

9.1 Given two sine waveforms, $s_1(t) = \sin(2\pi \times 2\,\text{kHz})$ and $s_2(t) = \sin(2\pi \times 3\,\text{kHz})$ using a simulation software show frequency spectrum of: (a) $s_1(t)$ and $s_2(t)$ as stand-alone waveforms; (b) $s(t) = s_1(t) + s_2(t)$; and (c) $p(t) = s_1(t) \times s_2(t)$. For this product, derive the analytical form of $p(t)$ waveform.

Compare the three plots and comment on the results.

9.2 Given three sine waveforms, $s_1(t) = \sin(2\pi \times 1\,\text{kHz})$, $s_2(t) = \sin(2\pi \times 10\,\text{MHz})$, and $s_3(t) = \sin(2\pi \times 9.45\,\text{MHz})$, first derive analytical forms of $s_1(t)$, $s_2(t)$, and $s_3(t)$, then using a simulation software show their respective frequency spectrums.

Compare the three plots and comment on the results.

9.3 Given $\omega = 2\pi \times 10^3$ Hz and a sine waveform

$$p(t) = \frac{4}{\pi}\left[\sin(\omega t) + \frac{1}{3}\sin(3\omega t) + \frac{1}{5}\sin(5\omega t) + \cdots + \frac{1}{33}\sin(33\omega t)\right] \tag{9.5}$$

using a simulation software show their time and frequency domain plots. Then,

(a) randomly choose and remove one or more of the harmonics in (9.5) and observe again the time and frequency domain plots;
(b) repeat the exercise but this time randomly change some of the harmonic amplitudes;
(c) repeat the exercise but this time randomly add a phase $0 \le \phi \le 2\pi$ to some of the harmonics;
(d) plot $f(t) = (1 + p(t))/2$; $f(t) = 2 + p(t)$; $f(t) = -3 + p(t)$;
(e) add a simple RC filter and observe the $f(t)$ waveforms at its output if: (1) $\tau = 1.59\,\mu\text{s}$, (2) $\tau = 15.9\,\mu\text{s}$, (3) $\tau = 159\,\mu\text{s}$, (4) $\tau = 1.59\,\text{ms}$.

Comment on the observed waveforms.

9.4 Given $\omega = 2\pi \times 10^3$ Hz and a sine waveform

$$w(t) = \frac{2}{\pi}\left[\sin(\omega t) - \frac{1}{2}\sin(2\omega t) + \frac{1}{3}\sin(3\omega t) - \cdots + \frac{1}{17}\sin(17\omega t)\right] \tag{9.6}$$

using a simulation software show their time and frequency domain plots. Then,

(a) randomly choose and remove one or more of the harmonics in (9.6) and observe again the time and frequency domain plots;
(b) repeat the exercise but this time randomly change some of the harmonic amplitudes;

(c) repeat the exercise but this time randomly add a phase $0 \leq \phi \leq 2\pi$ to some of the harmonics;
(d) plot $f(t) = (1 + w(t))/2$; $f(t) = 2 + w(t)$; $f(t) = -3 + w(t)$.
(e) add a simple RC filter and observe the $f(t)$ waveforms at its output if: (1) $\tau = 1.59\,\mu s$, (2) $\tau = 15.9\,\mu s$, (3) $\tau = 159\,\mu s$, (4) $\tau = 1.59\,ms$.

Comment on the observed waveforms.

9.5 Given waveforms $p(t)$ and $w(t)$ in problems 9.3 and 9.4 and $w = 2\pi \times 10^3$, plot the following waveforms:

(a) $f(t) = p(t) - {}^4/_{\pi} \sin(\omega t)$, $f(t) = p(t) - {}^4/_{\pi} \sin(\omega t) - {}^4/_{3\pi} \sin(3\omega t)$, etc.
(b) $f(t) = p(t) - {}^4/_{33\pi} \sin(33\omega t)$, $f(t) = p(t) - {}^4/_{33\pi} \sin(33\omega t) - {}^4/_{31\pi} \sin(31\omega t)$, etc.
(c) $f(t) = w(t) - {}^2/_{\pi} \sin(\omega t)$, $f(t) = p(t) - {}^2/_{\pi} \sin(\omega t) + {}^2/_{2\pi} \sin(2\omega t)$, $f(t) = p(t) - {}^2/_{\pi} \sin(\omega t) + {}^2/_{2\pi} \sin(2\omega t) - {}^2/_{3\pi} \sin(3\omega t)$, etc.
(d) $f(t) = w(t) - {}^2/_{17\pi} \sin(17\omega t)$, $f(t) = p(t) - {}^2/_{17\pi} \sin(17\omega t) + {}^2/_{16\pi} \sin(16\omega t)$, $f(t) = p(t) - {}^2/_{17\pi} \sin(17\omega t) + {}^2/_{16\pi} \sin(16\omega t) - {}^2/_{15\pi} \sin(15\omega t)$, etc.

Comment on the observed waveforms.

9.6 In reference to Table 9.1, estimate the length of either real or superficial dipole antenna that would have to be used by:

(a) military underwater communication systems;
(b) amateur radios;
(c) GSM-850 cellphones (carrier frequency $f_c = 850\,\text{MHz}$), UMTS-FDD cellphones (carrier frequency $f_c = 2.1\,\text{GHz}$);
(d) radio astronomy systems.

10 Electrical Resonance

In the most familiar form of mechanical oscillations, the pendulum, the total system energy constantly bounces back and forth between the kinetic and potential forms. In the absence of friction (i.e. energy dissipation), a pendulum would oscillate forever. Similarly, after two ideal electrical elements capable of storing energy (a capacitor and an inductor) are connected in parallel then the total initial energy of the system bounces back and forth between the electric and magnetic energy forms. This process is observed as electrical oscillations and the parallel LC circuit is said to be "in resonance". The phenomenon of electrical resonance is essential to wireless radio communications technology because without it, simply put, there would be no modern communications. In this chapter, we study behaviour and derive the main parameters of electrical resonant circuits.

10.1 The LC Circuit

The simplest electrical circuit that exhibits oscillatory behaviour consists of an inductor L and a capacitor C connected in parallel (see Fig. 10.1). Let us assume the initial condition where the capacitor contains q amount of charge, hence the initial voltage V across the LC parallel network is related to the charges as $q = C\,V_C = C\,v(\max)$.

10.1.1 Qualitative Description of LC Resonance

At time $t = 0$, the voltage across the charged capacitor is at its maximum $v(\max)$, its associated electric field and stored energy are also at maximum, and the network current is still at zero value. That is, at time $t = 0$, the inductor is still "seen" by the capacitor charge as an ideal wire. Naturally, due to the electric field, the capacitive charge is forced to move through the only available path, the inductive wire. However, as soon as the first electron leaves the capacitor plate, this movement qualifies as a change of current in time and, according to (2.40), this "ideal wire" starts to show strong inductive properties accompanied by an appropriate magnetic field. Hence, while this rising current flows through the inductor, it must obey Lenz's law and create the magnetic field that opposes the change that produced it. Eventually, the current reaches its maximum value $i(\max)$ (at $t = T/4$) when the capacitor is fully discharged; the complete energy of the LC system is now stored in the inductor's magnetic field.

R. Sobot, *Wireless Communication Electronics*,
https://doi.org/10.1007/978-3-030-48630-3_10

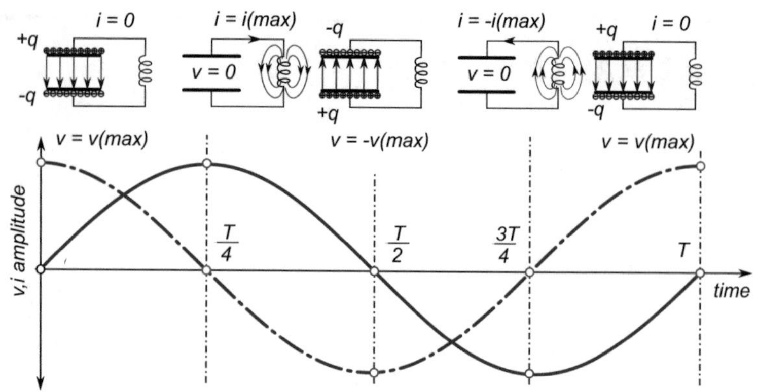

Fig. 10.1 Ideal LC resonance, the first cycle

It is now up to the inductor to serve as the energy source in the circuit and to push charges inside the wire while gradually passing the magnetic energy into the capacitive electrostatic energy. The uninterrupted flow of the current continues to cause the charges to keep accumulating at the other capacitor plate and along the way to create an electric field in the opposite direction relative to the initial state. This process continues until the capacitor is fully charged again (at $t = T/2$), (see Fig. 10.1), this time the voltage across the capacitor is at its minimum $v(\text{min}) = -v(\text{max})$. Because the system is assumed to be ideal, there is no thermal dissipation in wires, capacitor, and inductor. Consequently, the energy conservation law must be maintained, which is the condition for a sustained repetitive exchange of energy between the inductor and capacitor.

10.1.2 Formal Derivation of LC Resonance

In time domain, ideal LC circuit in Fig. 10.1 indeed creates electrical current that follows a sinusoidal waveform. We write the KVL equation around the loop as

$$v_C - v_L = 0 \quad \therefore \quad \frac{q}{C} + L\frac{di}{dt} = 0 \tag{10.1}$$

$$\text{(by definition)} \quad i = \frac{dq}{dt} \tag{10.2}$$

therefore, after differentiating (10.1) we have

$$\frac{i}{C} + L\frac{d^2 i}{dt^2} = 0 \quad \therefore \quad \frac{d^2 i}{dt^2} + \frac{1}{LC} i = 0 \tag{10.3}$$

hence its solution takes the form,[1]

$$i = I_0 \cos(\omega_0 t + \phi) \quad \text{or} \quad i = I_0 \sin(\omega_0 t + \theta) \tag{10.4}$$

where (10.4) is the standard solution of the second-order differential equation (10.3), and

[1]This is the second-order differential equation with a standard form of the solution.

$$\omega_0 = \frac{1}{\sqrt{LC}} \quad \Rightarrow \quad f_0 = \frac{1}{2\pi\sqrt{LC}} \tag{10.5}$$

where ω_0 is the frequency of oscillation of an ideal LC resonating circuit (i.e. with no thermal losses). We duly note that (10.4) is indeed a sinusoidal form that applies both to the resonating current and to the voltage (Fig. 10.1). Angular frequency ω_0, defined in (10.5), is one of the key variables in RF design, so much so that it was given its own name, the *resonant frequency*. The resonant frequency is calculated either as ω_0 in rad/s or as f_0 in Hz, where $\omega_0 = 2\pi f_0$. The physical definition of resonance is the tendency of a system to oscillate at maximum amplitude at a certain frequency. This frequency is known as the system's resonant frequency. It is very important to distinguish the resonant frequency from other modes of oscillations. While a system can oscillate at many frequencies, only the frequency associated with the maximal amplitude of oscillation (which in theory tends to infinity) is named the resonant, or natural, frequency.

We recall that the expression for the total energy $W = W_C + W_L$ contained in the LC resonator network is the sum of energies stored in the capacitor W_C and the inductor W_L, i.e.

$$W_C = \frac{1}{2}\frac{q^2}{C} = \frac{1}{2} v_C\, q = \frac{1}{2} C\, v_C^2 \tag{10.6}$$

$$W_L = \frac{1}{2} L\, i^2 \tag{10.7}$$

where, assumption is that at time $t = 0$ there is no initial energy stored in the inductor $W_L(t = 0) = 0$; that is, the complete initial energy of the LC network is stored in the capacitor.

10.1.3 Damping and Maintaining Oscillations

In the ideal resonating system, once started, the sinusoidal oscillations would maintain forever the amplitude of its waveform. Of course, that would have been a kind of perpetual motion machine, because we ignored the internal energy losses due to the system's internal resistance. The analogical mechanical system would be, for example, a swing that once pushed keeps swinging forever. In reality, this situation does not happen because of energy losses caused by internal friction in the swing's joints and air resistance to the body movement. As a result, the amplitude of oscillations becomes smaller with every passing cycle until, eventually, the movement completely stops.

For that reason we should determine under what conditions oscillations of a realistic oscillator are maintained and at what rate energy is lost from the oscillator due to the internal and external imperfections in the system. A harmonic oscillator of which the oscillations lose amplitude over time is referred to as a "damped harmonic oscillator", which is a very general and common mode of behaviour in nature, exhibited by many seemingly unrelated systems: an imperfect LC resonator, a pendulum, a guitar string, or a bridge, to name a few.

The general mathematical treatment of a damped harmonic oscillator is found in many textbooks on mathematics and physics; for completeness of our topic, we repeat the basic definitions. The second-order linear differential equation is

$$a_2 \frac{d^2x}{dt^2} + a_1 \frac{dx}{dt} + a_0\, x = 0 \tag{10.8}$$

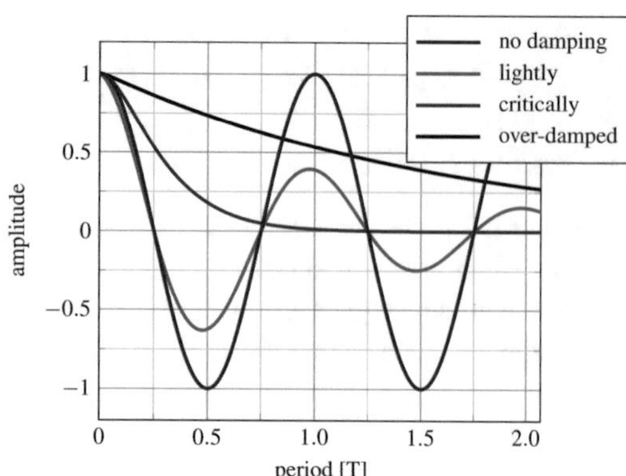

Fig. 10.2 Oscillation waveforms for various types of damping: ideal oscillations with no energy loss (no damping); oscillations with light energy loss (lightly); a system on the verge of starting oscillations (critically); and an over-damped system that cannot start oscillations (over-damped)

where a_2, a_1, a_0 are constants and x is the variable. It is more convenient to rewrite (10.8) in a form where the constant a_2 associated with the second derivative is normalized to one, hence we have

$$\frac{d^2x}{dt^2} + \frac{a_1}{a_2}\frac{dx}{dt} + \frac{a_0}{a_2}\,x = 0 \quad \therefore \quad \frac{d^2x}{dt^2} + \gamma\,\frac{dx}{dt} + \omega_0^2\,x = 0 \tag{10.9}$$

where constant γ is the ratio a_1/a_2 and ω_0 is the natural frequency of the damped harmonic oscillator, defined as the ratio a_0/a_2. In general, (10.9) has three possible solutions depending on how much damping is applied to the oscillator (see Fig. 10.2):

1. Lightly damped oscillators have minimal energy loss (i.e. they are very close to the case of a non-damped oscillator). This class of oscillators, if left on its own, is able to sustain oscillations for an appreciable amount of time.
2. Critically damped oscillators are on the verge of being able to start and sustain oscillations.
3. Over-damped oscillators cannot start the oscillation process because of large energy losses.

Let us focus on a lightly damped oscillator because it is the only one of the three cases that can start oscillations. Intuitively, we conclude that the solution of (10.9) must include a term that reduces the initial amplitude over time; hence, we multiply the solution from (10.4) with an exponentially decaying function and adopt the solution of (10.9) in the following form:

$$x = \exp\left(-\frac{t}{\tau}\right)\,A_0\,\cos\omega\,t \tag{10.10}$$

where t is the time variable, ω is the oscillation frequency, and τ is the timing constant that controls the rate of amplitude decay. For example, if $\tau = \infty$, then there is no reduction in the amplitude A_0 because the exponential term becomes equal to one at all times. At the other extreme, if $\tau = 0$, then the exponential term becomes zero, that is, the cosine function is completely suppressed. For any other value of τ, there will be natural decay in the initial amplitude A_0.

The first and second derivatives of (10.10) are

$$\frac{dx}{dt} = -A_0\exp\left(-\frac{t}{\tau}\right)\left(\omega\sin\omega t + \frac{1}{\tau}\cos\omega t\right) \tag{10.11}$$

$$\frac{d^2x}{dt^2} = A_0\exp\left(-\frac{t}{\tau}\right)\left[\frac{2\omega}{\tau}\sin\omega t + \left(\frac{1}{\tau^2} - \omega^2\right)\cos\omega t\right] \tag{10.12}$$

After substituting (10.10)–(10.12) into (10.9), we have

$$A_0 \exp\left(-\frac{t}{\tau}\right)\left[\left(\frac{2\omega}{\tau} - \gamma\omega\right)\sin\omega t + \left(\frac{1}{\tau^2} - \omega^2 - \frac{\gamma}{\tau} + \omega_0^2\right)\cos\omega t\right] = 0 \tag{10.13}$$

Equation (10.13) is possible at all times if the two multiplying constants of both sine and cosine terms are zero, i.e.

$$\left(\frac{2\omega}{\tau} - \gamma\omega\right) = 0 \quad \therefore \quad \tau = \frac{2}{\gamma} \tag{10.14}$$

$$\therefore$$

$$\frac{1}{\tau^2} - \omega^2 - \frac{\gamma}{\tau} + \omega_0^2 = 0 \quad \Rightarrow \quad \omega = \sqrt{\omega_0^2 - \left(\frac{\gamma}{2}\right)^2} \tag{10.15}$$

after eliminating τ from the cosine coefficient. Now, we can rewrite solution (10.10) for the case of a lightly damped oscillator with zero initial phase as

$$x = A_0 \exp\left(-\frac{\gamma}{2}\,t\right)\cos\omega t \tag{10.16}$$

The solution (10.16) is valid for ω as found in (10.15) and represents oscillatory motion if ω is real, i.e. if

$$\omega_0^2 > \frac{\gamma^2}{4} \tag{10.17}$$

which is the condition for lightly damped harmonic oscillations. In addition, the frequency of a lightly damped oscillator is close to its natural resonant frequency if

$$\omega_0^2 \gg \frac{\gamma^2}{4} \quad \therefore \quad \omega = \sqrt{\omega_0^2 - \left(\frac{\gamma}{2}\right)^2} \approx \omega_0 \tag{10.18}$$

It is important to note that parameters γ and ω_0 are solely determined by the physical parameters of the circuit. An example of lightly damped harmonic oscillation is shown in Fig. 10.3. Let us now determine how the amplitude of a decaying cosine function changes along the envelope function by finding the ratio of the two maxima of the cosine function.

We write (10.16) at time $t = t_0$ and at $t = t_0 + nT$, where T is the cosine period, and n represents the index of the n-th maxima away from the one at t_0. Hence, we have expressions for the two amplitudes as

$$A_k = x(t_0) = A_0 \exp\left(-\frac{\gamma}{2}\,t_0\right)\cos\omega t_0 \tag{10.19}$$

$$A_{k+n} = x(t_0 + nT) = A_0 \exp\left[-\frac{\gamma}{2}\,(t_0 + nT)\right]\cos\omega(t_0 + nT) \tag{10.20}$$

$$= A_0 \exp\left(-\frac{\gamma}{2}\,t_0\right)\exp\left(-\frac{\gamma}{2}\,nT\right)\cos\omega t_0 \tag{10.21}$$

$$\therefore$$

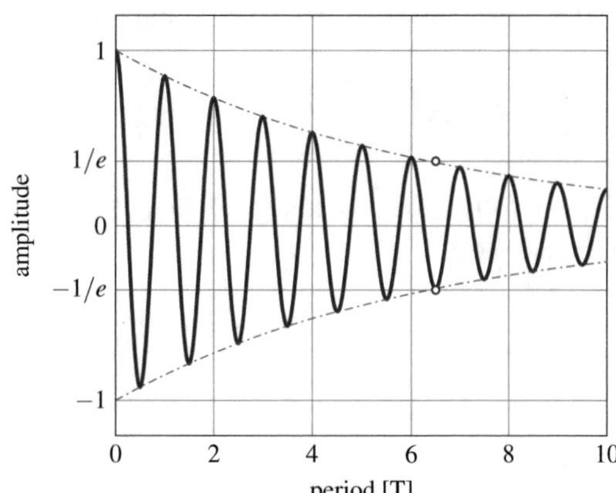

Fig. 10.3 Normalized amplitude and period of a decaying oscillator whose $Q \approx 20$. It can be found from the plot by counting the number of oscillation periods n until the amplitude drops to $^1/_e$, then $Q = n\,\pi$. In this case, $n = 6.5$, then $6.5 \times \pi \approx 20$. The red lines are the exponential envelope function $\exp(-\gamma\, ^t/_2)$

$$\ln\left(\frac{A_k}{A_{k+n}}\right) = \frac{\gamma\, nT}{2} = \frac{\gamma\, n\, 2\pi}{2\omega} = \frac{\gamma\, n\, \pi}{\omega} \approx \frac{\gamma\, n\, \pi}{\omega_0} = \frac{n\,\pi}{Q} \tag{10.22}$$

because $\cos\omega(t_0 + T) = \cos\omega t_0$. In (10.22), we introduced the ratio of the natural resonant frequency and γ as the figure of merit for the quality of oscillations, the Q factor,

$$Q = \frac{\omega_0}{\gamma} \tag{10.23}$$

Obviously, for finite frequencies, $Q \to \infty$ implies that $\gamma \to 0$, which enforces terms $A_0 \exp(-\gamma\, t/2) \to A_0$, in other words, there is no damping.

Example 62: LC Resonance—Q Factor
It was determined by measurement that the amplitude of a decaying cosine function at approximately 6.5 periods from $t = 0$ is e times smaller than the initial amplitude value A_0. Estimate the Q factor of this resonator.

Solution 62: From (10.22) we write

$$\ln(e) = \frac{6.5\,\pi}{Q} \quad \therefore \quad Q = 6.5\,\pi \approx 20.4$$

which is shown in Fig. 10.3.

Now that we have determined the behaviour of a lightly damped harmonic oscillator, we conclude that condition (10.17) sets boundaries for the other two damping conditions as

$$\omega_0^2 > \frac{\gamma^2}{4} \quad \text{lightly dumped} \tag{10.24}$$

$$\omega_0^2 = \frac{\gamma^2}{4} \quad \text{critically dumped} \tag{10.25}$$

$$\omega_0^2 < \frac{\gamma^2}{4} \quad \text{over-dumped} \tag{10.26}$$

The critically damped oscillator has the fastest time to equilibrium and still does not start oscillations. The over-damped system is dominated by the exponential decay function and slowly follows the envelope path (see Fig. 10.2).

Example 63: LC Resonance—Q Factor
The amplitude of a decaying oscillation is $E(t) = E_0 \exp(-t/\tau)$, where E_0 is the initial amplitude and τ is the decay time. For a guitar string that produces a tone at $f_0 = 334\,\text{Hz}$, the sound decayed by factor 2 after 4 s. Estimate the decaying time τ and the quality factor Q.

Solution 63: We write

$$E(t) = E_0 \exp\left(-\frac{t}{\tau}\right) \quad \therefore \quad \tau = \frac{t}{\ln\left(\dfrac{E_0}{E(t)}\right)}$$

$$\tau = \frac{4s}{\ln(2)} = 5.77s$$

and, from (10.14) and (10.23) we write

$$Q = \frac{\omega_0}{\gamma} = \frac{\omega_0\,\tau}{2} = \frac{2\pi \times 334\,\text{Hz} \times 5.77\,\text{s}}{2} \approx 6 \times 10^3$$

This relatively high Q indicates that the guitar string produces very narrowband[2] tone, i.e. very clean sinusoid, which is what is expected from a musical instrument.

A realistic electrical resonator consists of a serial RLC loop, similar to that shown in Fig. 10.6 (left) except that nodes a and b are connected. Under those conditions, we write KVL around the loop, as

$$L\frac{di}{dt} + iR + \frac{q}{C} = 0 \tag{10.27}$$

$$\therefore$$

$$\frac{d^2q}{dt^2} + \frac{R}{L}\frac{dq}{dt} + \frac{1}{LC}\,q = 0 \tag{10.28}$$

where (10.28) has an identical form to (10.9), hence (after including (10.18)) we write by inspection that

$$\omega_0^2 = \frac{1}{LC}; \quad \gamma = \frac{R}{L} \tag{10.29}$$

[2]Recall that also $Q = f_0/\Delta f$.

$$\therefore$$

$$q(t) = q_0 \exp\left(-\frac{R}{2L}t\right) \cos\left(\sqrt{\frac{1}{LC} - \frac{R^2}{4L^2}}\, t\right) \tag{10.30}$$

therefore, the Q factor is found from (10.23) and (10.29) as

$$Q = \frac{1}{R}\sqrt{\frac{L}{C}} \tag{10.31}$$

We observe that the presence of resistive element R is the fundamental cause of damping factor γ and a finite value of Q factor. In the case of $R = 0$, we have again the ideal resonator, i.e. $\gamma = 0$ and $Q \to \infty$, with no damping and $\omega = \omega_0$.

10.1.4 Forced Oscillations

Understanding (10.16) for a free running resonator is important in order to be able to control conditions that are favourable for maintaining oscillations in a real system. Going back to the swing analogy, in order to maintain the oscillations, at the end of every cycle the swing needs to receive just the right amount of push in the right direction. This action causes just the right amount of energy to be regularly injected into the system so that the energy loss due to friction is compensated. The key points are that the compensating energy must be injected at the right moment in time and with the right phase, i.e. synchronized with the oscillations in the right direction.

As opposed to a simple mechanical system, such as a swing, it is not as practical to manually compensate for thermal losses in electronic systems. The good news, however, is that it is not difficult to synchronize electronic systems so that the losses are correctly compensated for and to maintain the oscillations. For example, a hypothetical realistic LC resonator with internal losses, Fig. 10.4 (left), is connected to a transistor Q_1 that serves as a current source (the biasing details are omitted for simplicity). If at the end of each cycle we manually press the switch for a short period of time, given right conditions we could inject just about right amount of energy to compensate for the losses in each cycle and force the resonator to maintain constant amplitude of oscillations. Still hypothetical,

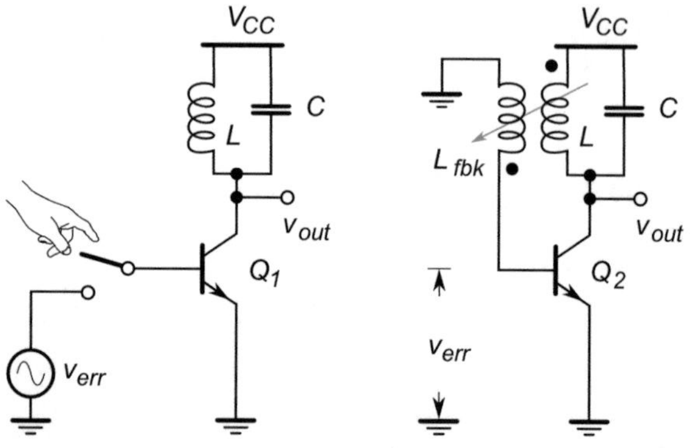

Fig. 10.4 LC resonating circuits with (left) a manual compensation mechanism for the internal thermal losses and (right) an automatic compensation mechanism

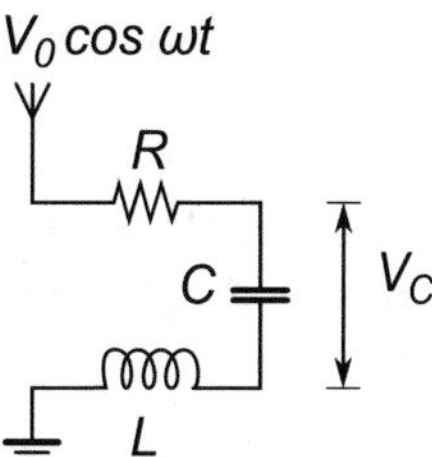

Fig. 10.5 RLC resonating circuit driven by RF voltage source $V_0 \cos \omega t$

but more practical method would be to replace the manual switch with some form of feedback system that would perform the same operation automatically, Fig. 10.4 (right).

The case of a forced RLC resonator is essential to RF communication systems and we are going to take a closer look by rewriting (10.28) as

$$\frac{d^2q}{dt^2} + \frac{R}{L}\frac{dq}{dt} + \frac{1}{LC}\, q = V_0 \cos \omega t \tag{10.32}$$

where $V_0 \cos \omega t$ represents a source of forced oscillations. The mathematical procedure for solving a non-homogeneous, linear differential equation (10.32) is a bit more involved, however it is easily found in calculus textbooks, hence we only write the solution for voltage across the capacitor $v_C(t)$ in the RLC forced resonator as

$$v_C(t) = V_C(\omega)\, \cos(\omega t - \delta) \tag{10.33}$$

$$V_C(\omega) = \frac{\dfrac{V_0}{LC}}{\sqrt{(\omega_0^2 - \omega^2)^2 + \left(\dfrac{\omega R}{L}\right)^2}} \tag{10.34}$$

where δ is the phase difference between the voltage source and the oscillator's frequency. At resonance, when $\omega = \omega_0$, then (10.34) becomes

$$V_C(\omega_0) = \frac{V_0}{\omega_0\, RC} = Q\, V_0 \tag{10.35}$$

We observe the very important fact, that when the local RLC resonator is designed for the resonant frequency ω_0 and the frequency of the external driving voltage source $V(\omega)$ coincides, i.e. $\omega = \omega_0$, there is significant amplification of the incoming tone (Q is usually very large). The simplified RLC circuit is driven by the incoming radio signal provided by the antenna shown in Fig. 10.5.

10.2 RLC Circuit

In real systems, there is always a small resistance R associated with the connection wires and the inductor, as well as a small leakage current in the capacitor (due to the less than infinite resistance of the capacitor's dielectric material) which, all combined, cause a small amount of energy to be lost each cycle in the form of heat. As a result, if generated by a real RLC circuit with no external compensation, the waveform amplitude exponentially decays (Fig. 10.3). That amplitude decay is the main reason for having an external energy source that compensates for the internal thermal losses in real oscillating RLC circuits (Fig. 10.4). This statement is confirmed experimentally by LC resonators made of superconductive materials; once the internal current is induced, the superconductive resonator oscillates for a very long time (measured in days and months) without any external energy source (strictly speaking this is not correct—a large amount of external energy is spent on keeping the resonator cool, however that is not the point).

10.2.1 Serial RLC Network

In Sects. 2.3.3.4 and 2.3.4.3, we already learned that, in a capacitive network, the capacitive voltage V_C lags the current by 90° and that, in an inductive network, the inductive voltage leads the current by 90°. Intuitively, we conclude that if the two elements are put in the same network the two voltages must, therefore, have the phase difference of 180°, which leads to an interesting question: what happens if the two voltages are equal in amplitude? Obviously, one voltage must be subtracted (remember differential signals?) from the other, which leads to interesting conclusions. To illustrate the point, let us take a look at the following example.

Example 64: LC Resonance—Q Factor
An AC voltage source V is connected across a serial LC connection. Find the capacitive X_C and inductive X_L reactances and voltages V_C and V_L across their respective terminals.

Data: $V = 5\,\text{V}$, $f = 10\,\text{MHz}$, $C = 1\,\text{nF}$, and $L = 1\,\mu\text{H}$.

Solution 64: The two reactances are calculated as

$$X_L = 2\pi f L = +62.832\Omega \quad \text{and} \quad X_C = 1/2\pi f C = -15.915\Omega$$

$$\therefore$$

$$X_{LC} = X_L + X_C = +46.916\Omega \quad \Rightarrow \quad L_{eq} = X_{LC}/2\pi\, f = 746.697\,\text{nH}$$

That is,[3] the total reactance at this frequency is equivalent to the reactance of an inductor, which further implies that, from the perspective of the voltage generator, at this particular frequency the serial LC connection could be replaced with a single 746.697 nH inductor without disturbing the rest of the circuit. The total branch current is, therefore, $I = V/X_{LC} = 106.573\,\text{mA}$ with $-90°$ phase relative to the voltage.

While keeping in mind the phase relationships, the voltage across the inductor is calculated as

$$V_L = I \times X_L = 106.573\,\text{mA} \times 62.832\Omega = 6.696\,\text{V}$$

and,

$$V_C = I \times X_C = 106.573\,\text{mA} \times 15.915\Omega = 1.696\,\text{V}$$

(continued)

[3]This is a relative comparison, thus it is agreed convention that the negative reactance is associated with a capacitance because $X_C = \frac{1}{j\omega C} = -j\frac{1}{\omega C}$.

Note that the inductor voltage is much higher than the one provided by the voltage source. However, the difference between the voltages is $V_L - V_C = 6.696\,\text{V} - 1.696\,\text{V} = 5\,\text{V}$, as it should be in order to agree with the applied voltage. As a consequence, we must be careful about the operational range of components used to build high-Q RLC resonators.

The addition of resistance R into a serial LC network, turning it into an RLC network (see Fig. 10.6), changes the overall circuit behaviour. The resistive component becomes responsible for thermal dissipation of the energy that was originally stored in the electrostatic and magnetic fields. As a consequence, the sinusoidal resonating current calculated in (10.4) cannot sustain its maximum value indefinitely. With each cycle, some of the electrical energy dissipates into heat, while the output voltage decays (Fig. 10.2). The absolute value of the total impedance is calculated as

$$Z = R + j\omega L + \frac{1}{j\omega C} = R + j\underbrace{\left(\omega L - \frac{1}{\omega C}\right)}_{X_{RLC}} \tag{10.36}$$

$$|Z| = \sqrt{R^2 + (X_L - X_C)^2} = \sqrt{R^2 + \left(\omega L - \frac{1}{\omega C}\right)^2} \tag{10.37}$$

where the two reactances, X_L and X_C, determine the equivalent total reactance X_{RLC} of the RLC network, which becomes *zero* if the two reactances are equal in their absolute values. The equality

$$X_L = X_C \Rightarrow \omega L = \frac{1}{\omega C} \quad \therefore \quad X_{RLC} = 0 \tag{10.38}$$

is considered the required condition for resonance.

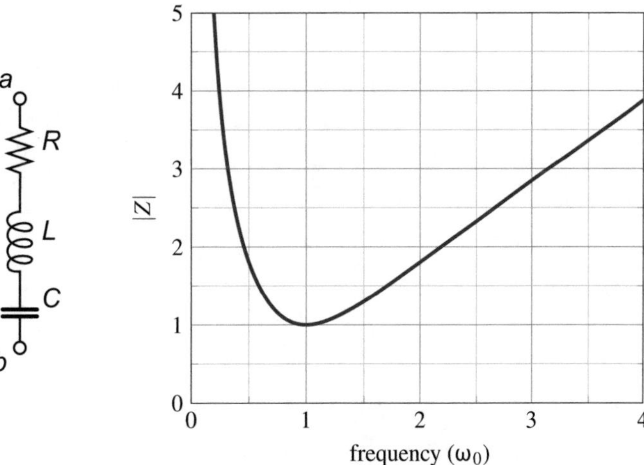

Fig. 10.6 Serial RLC circuit network with normalized resonant frequency $\omega_0 = 1$ (left) and its total impedance plot versus frequency (right)

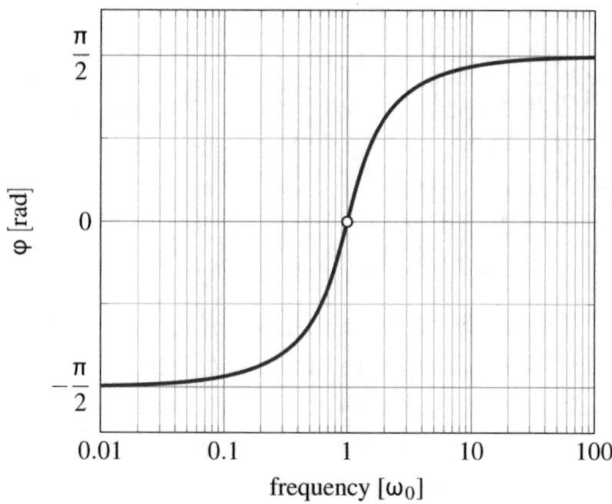

Fig. 10.7 Phase plot of a serial RLC network, normalized to the resonant frequency ω_0

Under the condition of resonance, the absolute value of serial impedance is at its minimum, i.e. it is equal to R (Fig. 10.6), and it is, therefore, real, which means that the total phase angle equals zero (Fig. 10.7). The resonance condition (10.38) is satisfied at the one and only frequency ω_0 found as

$$\omega_0 = \frac{1}{\sqrt{LC}} \tag{10.39}$$

where ω_0 indicates the resonant frequency of the LC network under the $X_L = X_C$ condition, which is the same conclusion as the result (10.5) from the differential equation. At other frequencies, the total reactance is greater than zero, (10.37), which implies that R is the minimum value of the absolute impedance Z. The phase plot is easily obtained, after using definitions (3.13), (3.18), and (10.36) as

$$\phi = \arctan\frac{\Im(Z)}{\Re(Z)} = \arctan\frac{\left(\omega L - \dfrac{1}{\omega C}\right)}{R} \tag{10.40}$$

which is given in Fig. 10.7 after normalizing to the ω_0 resonant frequency.

Example 65: LC Resonance—Serial RLC
The RLC circuit in Fig. 10.8 is connected to a realistic voltage source V_{in}, whose internal resistance is R_S. Find: (a) output voltage $V_{out} = V_R$ as measured across the resistor R, at the resonant frequency $f = f_0$; (b) output voltage $V_{out} = V_C$ as measured across the capacitor C at $f = 1\,\text{kHz}$.

Data: $C = 1\,\text{nF}$, $L = 1\,\mu\text{H}$, $R = 1\,\text{m}\Omega$, $R_S = 50\,\Omega$, $V_{in} = 1\,\text{mV}$

Solution 65:

(a) At the resonant frequency f_0, the condition (10.38) gives $Z_L = Z_C \Rightarrow Z_{RLC} = R$. Therefore, after applying the voltage divider rule, it follows:

(continued)

$$\frac{V_{out}}{V_{in}} = \frac{V_R}{V_{in}} = \frac{R}{R + R_S} = \frac{1\,\text{m}\Omega}{1\,\text{m}\Omega + 50\Omega} \approx 20 \times 10^{-6}\text{V/v}$$

$$\therefore$$

$$V_{out} = 20 \times 10^{-6}\text{V/v} \times 1\,\text{mV} = 20\,\text{nV}$$

(b) At the frequency $f = 1\,\text{kHz}$, inductor and capacitor reactances are

$$X_L = |j\omega L| = 2\pi \times 1\,\text{kHz} \times 1\,\mu\text{H} = 6.283\,\text{m}\Omega$$

$$X_C = \left|\frac{1}{j\omega C}\right| = \frac{1}{2\pi \times 1\,\text{kHz} \times 1\,\text{nF}} = 159.2\,\text{k}\Omega$$

which helps us to calculate the total impedance seen by the ideal voltage source as

$$Z_{tot} = \sqrt{(R_S + R)^2 + (X_L - X_C)^2} = \sqrt{(50\,\Omega + 1\,\text{m}\Omega)^2 + (6.283\,\text{m}\Omega - 159.2\,\text{k}\Omega)^2}$$
$$\approx 159.2\,\text{k}\Omega$$

Voltage across the capacitor V_C is found again by voltage division rule as

$$\frac{V_{out}}{V_{in}} = \frac{V_C}{V_{in}} = \frac{X_C}{Z_{tot}} \approx \frac{159.2\,\text{k}\Omega}{159.2\,\text{k}\Omega} \approx 1\text{V/v} \quad \therefore \quad V_{out} = 1\text{V/v} \times 1\,\text{mV} = 1\,\text{mV}$$

which illustrates how impedances, thus voltages across their respective components, drastically change as a function of frequency.

Example 66: LC Resonance—Serial RLC
For the circuit in Fig. 10.8, find the resonant frequency f_0 and calculate the total impedance Z_{tot} at: (a) 1 kHz, (b) 7.335 MHz, (c) 1 GHz.

Data: $R_S = 0\,\Omega$, $R = 1\,\text{m}\Omega$, $L = 4.708\,\text{nH}$, and $C = 100\,\text{nF}$.

(continued)

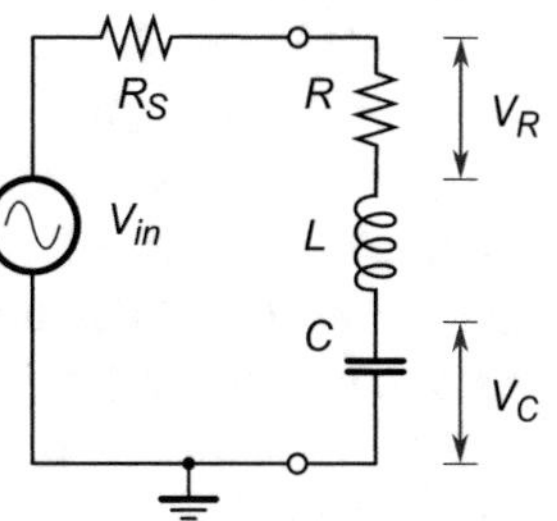

Fig. 10.8 An RLC network driven by a realistic non-ideal voltage source

Solution 66: The resonant frequency is

$$f_0 = \frac{1}{2\pi\sqrt{LC}} = \frac{1}{2\pi\sqrt{4.708\,\text{nH}\,100\,\text{nF}}} \approx 7.335\,\text{MHz}$$

(a) at $f = 1\,\text{kHz}$:

$$X_L = 2\pi f L = 2\pi \times 1\,\text{kHz} \times 4.708\,\text{nH} = 29.581\,\mu\Omega$$

$$X_C = \frac{1}{2\pi f C} = \frac{1}{2\pi \times 1\,\text{kHz} \times 100\,\text{nF}} = 1.592\,\text{k}\Omega$$

therefore,

$$Z_{tot} = \sqrt{R^2 + (X_L - X_C)^2} = \sqrt{1\,\text{m}\Omega^2 + (29.581\,\mu\Omega - 1.592\,\text{k}\Omega)^2} \approx 1.592\,\text{k}\Omega$$

that is to say, at $f = 1$ kHz this RLC network is dominated by the capacitor's reactance.

(b) at $f = 7.335$ MHz: this is the resonant frequency, hence $Z_{tot} = R = 1\,\text{m}\Omega$, i.e. only the real resistance R is "visible" to the source.

(c) at $f = 1$ GHz: at this frequency capacitor reactance is $X_C \approx 1.6\,\text{m}\Omega$, while $X_L = 29.581\,\Omega$, thus $Z_{tot} \approx X_L = 29.581\,\Omega$, i.e. dominated by the inductor's reactance.

10.2.2 Parallel RLC Network

A parallel RLC network has a few subtle differences from the serial version which was discussed in the previous paragraphs. It also represents a frequency controlled impedance which has the same expression for the resonant frequency. However, the impedance behaves slightly differently.

In a parallel RLC circuit (see Fig. 10.9), the voltage is equal across all three components, where each component defines the branch whose current is described by the Ohm's law and each component keeps its own voltage, current, and phase relationship. For a parallel RLC network, the total current I_{tot} is written as

$$I_{tot} = \sqrt{I_R^2 + (I_L - I_C)^2} \tag{10.41}$$

By inspection of Fig. 10.9 (left) and examining the two extreme cases, at DC and very high frequency, we conclude the following. At DC, the inductor has zero impedance (i.e. it becomes a short connection), the resistor holds the R value, and the capacitor has infinite impedance (i.e. it becomes an open connection). The three components are in parallel, hence the equivalent impedance is zero due to short connection by the inductor. At very high frequencies (i.e. $\omega \to \infty$), the inductor has infinite impedance (i.e. it becomes an open connection), the resistor still holds R, and the capacitor has zero impedance (i.e. it becomes a short connection). Again, the equivalent impedance is zero, this time due to short connection by the capacitor. This behaviour implies the existence of at least one maxima between the two extreme points. For the time being, let us stay with this conclusion and only roughly plot the frequency dependence of the impedance, as shown in Fig. 10.9 (right).

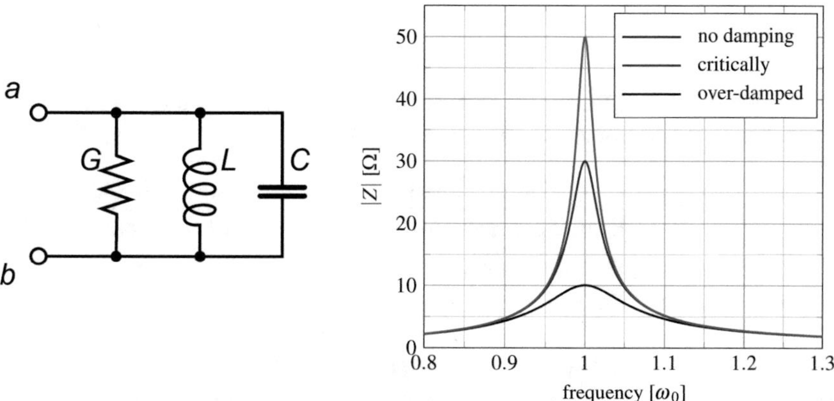

Fig. 10.9 Parallel RLC circuit network, $G = 1/R$, (left) and the plot of impedance $|Z_{ab}|$ against frequency, normalized at $\omega_0 = 1$ Hz, (right). The maximum impedance value is $Z_{\omega_0} = Q^2 R$

The following example illustrates how the three components in parallel share the current branches.

Example 67: LC Resonance—Parallel RLC
For the circuit in Fig. 10.9 (left) estimate the total current I_{tot} supplied by the source and the circuit impedance Z_{tot}.

Data: $R_S = 0\Omega$, $V_{in} = 12\,\text{V}$, $R = 400\Omega$, $X_L = 500\Omega$, and $X_C = 200\Omega$,

Solution 67: The three branch currents are

$$I_R = \frac{V_{in}}{R} = \frac{12\,\text{V}}{400\Omega} = 30\,\text{mA}; \quad I_L = \frac{V_{in}}{X_L} = \frac{12\,\text{V}}{500\Omega} = 24\,\text{mA}, \quad \text{and}$$

$$I_C = \frac{V_{in}}{X_C} = \frac{12\,\text{V}}{200\Omega} = 60\,\text{mA}$$

Then, from (10.41) it follows that

$$I_{tot} = \sqrt{I_R^2 + (I_L - I_C)^2} = \sqrt{(30\,\text{mA})^2 + (24\,\text{mA} - 60\,\text{mA})^2} = 46.862\,\text{mA}$$

$$\therefore$$

$$Z_{tot} = \frac{V_{in}}{I_{tot}} = \frac{12\,\text{V}}{46.862\,\text{mA}} = 256.074\Omega$$

We note that the current through the capacitive branch is larger than the total current I_{tot} provided by the signal generator. We recall that we must use vector algebra for complex components, which means that module must be larger than either real or complex vector projection.

10.3 Q Factor

A more general definition than the one used in Sect. 10.1.3 would be that the Q factor is the ratio of the energy stored in the resonator and the energy given by a generator, i.e. it is evaluated in each cycle as

$$Q = 2\pi \times \frac{\text{Energy Stored}}{\text{Energy dissipated per cycle}} = \omega_0 \times \frac{\text{Energy Stored}}{\text{Power Loss}} \tag{10.42}$$

where, in electrical systems, the stored energy is the sum of energies initially stored in lossless inductors and capacitors and the lost energy is the sum of the energies dissipated in resistors per cycle.

In the ideal case, energy stored in the magnetic field of the inductor is eventually converted without loss into energy of the electrostatic field of the capacitor. At the resonant frequency, the maximum energy stored in the network keeps bouncing back and forth between the inductor and the capacitor without loss and, therefore, is calculated either at the moment when the capacitor is fully discharged (and therefore the inductor holds the full amount of energy W_L) or when the capacitor is fully charged (and therefore temporarily holds the full amount of the energy W_C), i.e.

$$W_L = \int_0^T i(t)v(t)\,dt = \int_0^T i(t)L\frac{di(t)}{\cancel{dt}}\cancel{dt} = L\int_0^{I_p} i\,di = \frac{1}{2}L\,I_p^2 = L\,I_{\mathrm{rms}}^2 \tag{10.43}$$

or, similarly

$$W_C = \int_0^T v(t)i(t)\,dt = \int_0^T v(t)\,C\frac{dv(t)}{\cancel{dt}}\cancel{dt} = C\int_0^{V_p} v\,dv = \frac{1}{2}CV_p^2 = CV_{\mathrm{rms}}^2 \tag{10.44}$$

where $I_p = \sqrt{2}I_{\max}$ is the peak current through the inductor, and $V_p = \sqrt{2}V_{\max}$ is the peak voltage across the capacitor.[4]

10.3.1 Q Factor of a Series RLC Network

During one full resonant cycle

$$T_0 = \frac{1}{f_0} = \frac{2\pi}{\omega_0} \tag{10.45}$$

the energy dissipated in the resistor W_R is simply, by definition, power multiplied by the time, i.e.

$$W_R = P_R \times T_0 = R\,I_{\mathrm{rms}}^2 \times T_0 = \frac{2\pi}{\omega_0}R\,I_{\mathrm{rms}}^2 \tag{10.46}$$

which means that (10.42) becomes (using either W_L or W_C) for serial RLC

$$Q_S = 2\pi\frac{W_L}{W_R} = \cancel{2\pi}\frac{L\,\cancel{I_{\mathrm{rms}}^2}}{\cancel{2\pi}/\omega_0\,R\,\cancel{I_{\mathrm{rms}}^2}} = \frac{\omega_0 L}{R} \tag{10.47}$$

[4]Make an example to show the integrals for Vp of sin and square functions.

At resonance, the resonant frequency ω_0, inductance L, and capacitance C are related as in (10.39); therefore, the three equivalent formulations of Q_S are

$$\omega_0 = \frac{1}{\sqrt{LC}} \quad \therefore \quad Q_S = \frac{\omega_0 L}{R} = \frac{1}{\omega_0 RC} = \frac{1}{R}\sqrt{\frac{L}{C}} \tag{10.48}$$

where (10.48) shows all three variants of the expression for the Q factor of an RLC network in series. Expressions in (10.48) for the quality factor Q are very important and are used to quantify a number of other specifications in radio design.

It is important to note that for the ideal inductor, i.e. $R = 0$, the Q factor becomes $Q = \infty$. It is desirable to keep control over the Q factor for many reasons that are mentioned throughout this book. In addition, it should be noted that in serial configurations, the Q factor is inversely proportional to the resistance R.

Example 68: LC Resonance—Q Factor
Estimate the resonant frequency and the Q factor for a typical serial RLC network.

Data: $L = 1\,\text{mH}$, $C = 25.33\,\text{pF}$, and $R = 15\Omega$ is the total resistance at the resonant frequency.

Solution 68: The resonant frequency is

$$f_0 = \frac{1}{2\pi\sqrt{(LC)}} \approx 1\,\text{MHz}$$

and the Q factor is

$$Q = \frac{1}{R}\sqrt{\frac{L}{C}} = \frac{1}{15\Omega}\sqrt{\frac{1\,\text{mH}}{25.33\,\text{pF}}} \approx 420$$

which are typical numbers in the current state of the art.[5]

10.3.2 Q Factor of a Parallel RLC Network

We now find the resonant frequency ω_{p0} of a realistic parallel LC network, Fig. 10.10, where the resistance r accounts for all thermal losses, i.e. the combined resistance of the inductor and wires and the effective series resistance (ESR) of the capacitor.

$$\begin{aligned} Y(\omega) &= \frac{1}{r + j\omega L} + j\omega C = \frac{r - j\omega L}{r^2 + (\omega L)^2} + j\omega C \\ &= \frac{r}{r^2 + (\omega L)^2} + j\left(\omega C - \frac{\omega L}{r^2 + (\omega L)^2}\right) \end{aligned} \tag{10.49}$$

[5]Make an example to show how to measure RD as in lab.

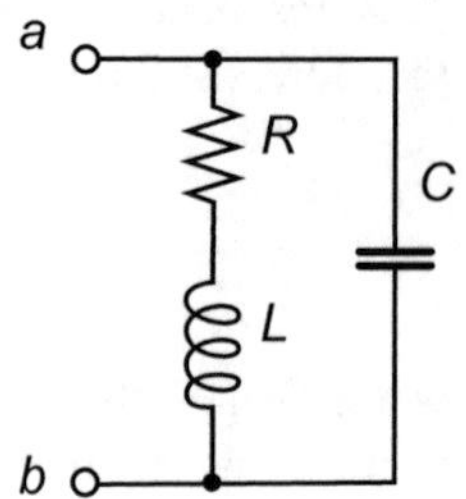

Fig. 10.10 Realistic parallel LC network

at resonance (i.e. $\omega = \omega_{p0}$), the two reactances are equal $|Z_L| = |Z_C|$, which is to say that the imaginary part is $\Im(Y) = 0$, hence we write

$$\omega_{p0} C = \frac{\omega_{p0} L}{r^2 + (\omega_{p0} L)^2} \quad \therefore \quad r^2 + (\omega_{p0} L)^2 = \frac{L}{C} \tag{10.50}$$

which leads to the conclusion

$$\omega_{p0} = \sqrt{\frac{1}{LC} - \frac{r^2}{L^2}} \tag{10.51}$$

We conclude that the resonant frequency ω_{p0} of a parallel LC network that includes realistic inductance has the additional term $(r/L)^2$ due to the finite wire resistance, which slightly reduces the resonant frequency relative to the case of the ideal LC resonator. When $r \to 0$, (10.51) becomes the same as (10.39) for the ideal LC resonator, i.e. $\omega_{p0} \to \omega_0$.

For parallel RLC configuration, as shown in Fig. 10.9 (left), however, it is desirable to have $R = 1/G$ as high as possible in order to reduce the power dissipation (i.e. to reduce the current through the R branch of the RLC network), which is to say that, using the principle of duality (resulting in mirroring (10.48)), the three equivalent quality factor Q_P formulations for a parallel RLC network are

$$Q_P = \frac{R}{\omega_0 L} = \omega_0 R C = R\sqrt{\frac{C}{L}} \tag{10.52}$$

To elaborate the point, it is useful to evaluate the size of the difference between resonant frequencies of series and parallel RLC networks. We already derived expression (10.51) for the resonant frequency ω_{p0} of a parallel RLC network, which can be reformulated with the help of (10.48) as

$$\omega_{p0} = \sqrt{\frac{1}{LC} - \frac{R^2}{L^2}} = \sqrt{\omega_{s0}^2 - \frac{R^2}{L^2}} = \omega_{s0}\sqrt{1 - \frac{R^2}{\omega_{s0}^2 L^2}} = \omega_{s0}\sqrt{1 - \frac{1}{Q_S^2}} \tag{10.53}$$

$$\therefore \quad \omega_{p0} \approx \omega_{s0} \text{ for } (Q_S > 10) \tag{10.54}$$

where we appropriately introduced series resonant frequency ω_{s0} through the serial Q_S factor. Equation (10.53) shows that for ideal or high Q networks (i.e. $Q > 10$) there is a very small error in calculating resonating frequencies ω_{s0} and ω_{p0} of the series and parallel circuits due to term $(1/Q_S^2 \rightarrow 0)$. Hence, ω_{s0} and ω_{p0} can be used interchangeably, as long as the Q factor is high.

To simplify the calculation, for high Q values, we assume that $(\omega L)^2 \gg R^2$ (this is justified because an inductor's wire resistance is relatively small) or, equivalently, the same condition is written as $R^2 + (\omega L)^2 \cong (\omega L)^2$. In addition, from (10.48) we find that

$$\frac{\omega_0 L}{R} \times L = \frac{1}{\omega_0 RC} \times L \quad \therefore \quad \frac{R}{(\omega_0 L)^2} = \frac{RC}{L} \tag{10.55}$$

Therefore, at resonance the admittance is resistive, which after applying condition (10.54) to the real part of (10.49) yields

$$Y_0 = \frac{R}{R^2 + (\omega_0 L)^2} \approx \frac{R}{(\omega_0 L)^2} = \frac{RC}{L} \quad \therefore \quad R_D = \frac{1}{Y_0} = \frac{L}{RC} \tag{10.56}$$

where R_D now represents the "dynamic resistance" of the LC tank at resonance. We note that this equivalent resistance is what an RLC circuit presents to the rest of the world under the condition of resonance. It is not the physical resistance of neither R, C nor L.

After having delivered expressions for both non-resonant admittance Y and admittance at resonance Y_0, it becomes straightforward to find out how LC tank admittance changes with frequency, relative to its resonant value. Then, from (10.49) and (10.56) for large Q factor we write

$$\frac{Y}{Y_0} \cong \frac{L}{RC}\left[\frac{R}{(\omega L)^2} + j\left(\omega C - \frac{1}{\omega L}\right)\right] \tag{10.57}$$

$$= \frac{1}{\omega^2 LC} + j\left(\frac{\omega L}{R} - \frac{1}{\omega RC}\right) \tag{10.58}$$

$$= \frac{\omega_0^2}{\omega^2} + j\,\delta\,Q \quad \text{where} \quad \delta = \frac{\omega}{\omega_0} - \frac{\omega_0}{\omega} \tag{10.59}$$

where after substituting (10.48) and rearranging (for high Q factor, the serial resonance ω_{s0} and parallel resonance ω_{p0} are equal). Therefore,[6]

$$|Y| = Y_0 \sqrt{\left(\frac{\omega_0}{\omega}\right)^4 + (\delta\,Q)^2} \tag{10.60}$$

Result (10.60) is an important relation that is used to estimate the amplitude of a signal located at ω that is not exactly at the resonant frequency ω_0. More applications of this formula are shown in Sect. 14.7.1.

[6]Becasue, $|Z| = \sqrt{\Re(Z)^2 + \Im(Z)^2}$.

Example 69: LC Resonance—Non-Ideal Resonator
For typical RLC components, $L = 1\,\text{mH}$, $C = 25.33\,\text{pF}$, and $R = 15\Omega$, find by how much the resonant frequency of the realistic resonator is off relative to the ideal resonator.

Solution 69: The ideal resonant frequency is simply

$$\omega_0 = \frac{1}{\sqrt{LC}} = \frac{1}{\sqrt{1\,\text{mH} \times 25.33\,\text{pF}}} = 1.00000584\,\text{MHz}$$

while the realistic resonant frequency is calculated as

$$\omega_0 = \sqrt{\frac{1}{LC} - \frac{R^2}{L^2}} = \sqrt{\frac{1}{1\,\text{mH} \times 25\,\text{pF}} - \frac{(15\Omega)^2}{(1\,\text{mH})^2}} = 1.00000299\,\text{MHz}$$

therefore, the difference of 2.85 Hz relative to 1×10^6 Hz is negligible for most practical purposes.

10.4 Self-resonance of an Inductor

As implied in the previous paragraphs, real inductors show characteristic properties of a complex RLC circuit. Based on the analysis of resonance, and knowledge that non-ideal inductors have parasitic capacitances related to the wire, we intuitively conclude that the circuit diagram shown in Fig. 10.10 could also be used to represent a real inductor by itself, where the inductor's wire resistance is $R_L = R$. It is only one of the possible ways to create a model of real inductor; it is also one of the most often used models. Following the same procedure as in the previous sections, and after applying the low wire resistance approximation, an expression for the admittance of a non-ideal inductor by itself is found as

$$\begin{aligned} Y_L &= \frac{1}{R_L + j\omega L} + j\omega C_L \approx \frac{R_L}{(\omega L)^2} + j\left(\omega C_L - \frac{1}{\omega L}\right) \\ &= \frac{R_L}{(\omega L)^2} - j\left(\frac{1 - \omega^2 L C_L}{\omega L}\right) = \Re(Y_L) + j\,\Im(Y_L) \end{aligned} \tag{10.61}$$

Because of the R_L, C_L, L component values are associated with physical realization of the inductor, obviously it must resonate on its own, i.e. the non-ideal inductor does have resonant frequency ω_{0L} of its own. It is very important for a designer to have at least some estimate of where this self-resonant frequency might be. Typically, the wire resistance is $R_L \leq 1\Omega$ and associated parasitic capacitance C_L is in the order of pF, which means that the self-resonant frequency is, typically, in the order of megahertz to a few hundreds of megahertz. That is, if a non-ideal inductor is to be used in an LC tank, the designer is forced to limit the intended signal frequency to no more than one decade (i.e. ten times) below the inductor's self-resonant frequency. This rule of thumb is most often used as a measure of how good an inductor is needed for the intended design.

A natural question one could ask would be why an external capacitor C is needed in parallel with a non-ideal inductor to set the resonant frequency. Why not just use the non-ideal inductor alone? This approach would simplify things, at least in terms of the component count. Indeed, that approach is used for the design of circuits working at very high frequencies, for example, in satellite communication systems. However, detailed analysis of such components and circuits is beyond the scope of this book. For purposes of designing circuits with discrete components working at frequencies in the order of up to a few 100 MHz, the limitation is that controlling the inductor's parasitic capacitance is not practical.

In this model, all wire resistances associated with the non-ideal LC are now merged with inductive resistance R_L. The effective parallel resistance R_P is represented by the real part of (10.61), i.e.

$$\frac{1}{R_P} = \frac{R_L}{(\omega L)^2} \qquad \therefore \qquad R_P = \frac{(\omega L)^2}{R_L} \tag{10.62}$$

while the effective parallel inductance of the coil Left is given by the imaginary part of (10.61), i.e.

$$\frac{1}{\omega L_{\text{eff}}} = \frac{1 - \omega^2 L C_L}{\omega L}$$

$$\therefore$$

$$L_{\text{eff}} = \frac{L}{1 - \omega^2 L C_L} = \frac{L}{1 - \left(\dfrac{\omega}{\omega_{0L}}\right)^2} \tag{10.63}$$

Next, we estimate the deviation from the ideal LC tank model at resonance ω_0 (which has to be at least one decade below the self-resonance ω_{0L} of the coil). Another way of stating this condition is that the resonator's external capacitor C_T has to be much larger than the parasitic capacitance C_L, i.e. $C_T \gg C_L$, (the ideal inductance L is always the same). At circuit resonance ω_0, the dynamic resistance R_D of the LC tank in Fig. 10.10, as same as in (10.48), is

$$R_D = Q\,\omega_0 L \tag{10.64}$$

At the same time, the dynamic resistance R_D of the equivalent circuit is described using effective values

$$R_D = Q_{\text{eff}}\,\omega_0 L_{\text{eff}} \tag{10.65}$$

Because of the equivalence of these two circuits and from (10.64) and (10.65) it follows that

$$Q\,\omega_0 L = Q_{\text{eff}}\,\omega_0 L_{\text{eff}} \tag{10.66}$$

therefore,

$$Q_{\text{eff}} = Q\left(1 - \omega^2 L C_L\right) = Q\left[1 - \left(\frac{\omega}{\omega_0}\right)^2\right]; \quad (\omega \ll \omega_0) \tag{10.67}$$

where $Q = \omega_0 L/R$, as defined in (10.48). This result shows that the effective Q factor Q_{eff} of a realistic LC tank decreases as the operating frequency ω approaches the self-resonating frequency ω_{0L} and eventually becomes zero when $\omega = \omega_0$.

How do we use the above results? It really depends. For the parallel circuit case in Fig. 10.10, the dynamic resistance can be calculated using either (10.64) or (10.65). The general definitions from (10.79) and (10.80) can be used as well, as long as either C is replaced with $(C + C_L)$ or Q is replaced with Q_{eff}. The bandwidth Δf from (10.68), however, must be calculated using Q and not Q_{eff} because, for a given resonant frequency, capacitance C is adjusted to absorb C_L, because they are in parallel connection.

Note that a serial tuned circuit is different, i.e. capacitor C is in serial connection with the inductive capacitance C_L. Effectively, C resonates with L_{eff} which means that instead of using Q, Q_{eff} is used, so that

$$Q_{\text{eff}} = \frac{f_0}{\Delta f} \tag{10.68}$$

Those little differences should be accounted for when analysing serial and parallel RLC resonant circuits.

10.5 Serial to Parallel Impedance Transformations

Often, it is useful to transform a serial RLC network into its equivalent parallel configuration or vice versa. This transformation must be done only at a single frequency, which does not affect the serial and parallel Q factors of the networks,

$$Q_S = \frac{X_S}{R_S} \tag{10.69}$$

$$Q_P = \frac{R_P}{X_P} \tag{10.70}$$

so that, assuming $Q_S = Q_P = Q$ with the help of (10.69) and (10.70), at the given frequency

$$Z_S = R_S + jX_S = R_S + jQ_S R_S = R_S(1 + jQ_S) \tag{10.71}$$

$$Y_p = \frac{1}{Z_S} = \frac{1}{R_S(1 + jQ)} = \frac{1}{R_S(1 + jQ)} \frac{1 - jQ}{1 - jQ} \tag{10.72}$$

$$= \frac{1}{R_S(1 + Q^2)} - j\frac{Q}{R_S(1 + Q^2)} \tag{10.73}$$

$$= \frac{1}{R_S(1 + Q^2)} - j\frac{Q}{\frac{X_S}{Q}(1 + Q^2)} = \frac{1}{R_P} - j\frac{1}{X_P} \tag{10.74}$$

therefore,

$$R_P = R_S(1 + Q^2) \tag{10.75}$$

$$X_P = X_S\left(1 + \frac{1}{Q^2}\right) \tag{10.76}$$

which, for large Q, i.e. $Q > 10$, becomes

$$X_P \approx X_S \tag{10.77}$$

and

$$R_P \approx Q^2 R_S \tag{10.78}$$

The last two expressions are often used approximations in resonant circuit network analysis.

10.6 Dynamic Resistance

The imaginary part $\Im(Y)$ controls the resonant frequency, while the real part $\Re(Y)$, from (10.49), determines the dynamic resistance R_D, i.e. the equivalent real resistance of the LC resonator at the resonant frequency ω_{p0}, as,

$$\Re(Y(\omega_{p0})) = \frac{R}{R^2 + (\omega_{p0}L)^2} = \frac{R}{R^2 + \left[\sqrt{\frac{1}{LC} - \frac{R^2}{L^2}}\right]^2 L^2} = \frac{R}{R^2 + \left(\frac{L}{C} - R^2\right)} = \frac{RC}{L}$$

$$\therefore$$

$$R_D = \frac{1}{\Re(Y(\omega_{p0}))} = \frac{L}{RC} \tag{10.79}$$

In the ideal case, i.e. $R = 0$, the dynamic resistance of the LC resonator (see Fig. 5.9) becomes $R_D = \infty$. It should be noted that, from the perspective of the resonant current, which circulates inside the RLC loop, the three elements are in series. Hence, reducing the resistance associated with the inductive branch is desirable in order to increase the dynamic impedance perceived by the network external to the RLC resonator.

Finally, the expression for dynamic resistance (10.79) can also be reformulated in terms of the Q factor ($Q > 10$), after using (10.48)), as:

$$R_D = \frac{L}{RC} = \omega_0 L Q = \frac{Q}{\omega_0 C} = Q^2 R \tag{10.80}$$

which is, again, the resistance of a realistic RLC tank in resonance, as perceived by the external network. An important distinction to make is that the resistance R is a physical entity: in a serial RLC network, it needs to be as small as possible; in a parallel configuration, it needs to be as large as possible. However, at resonance, this small resistance is perceived by the external network as if magnified by the Q^2 factor. In the ideal case, i.e. when serial the resistance $R = 0$, value of the Q factor becomes infinity. Hence, expression (10.80) is only a mathematical approximation, see Fig. 10.9.

The maximum impedance value happens at the resonant frequency ω_0, as shown in Fig. 10.9 (right), however this time it is calculated as

$$Z_{max} = Z(\omega_0) = Q^2 R \tag{10.81}$$

This is a very important property of parallel RLC networks, which indicates that, at the resonant frequency, an ideal LC parallel network (i.e. $R = \infty$) would have infinite Q factor and, therefore, infinite voltage output for any given current. One should note the obvious risk associated with the possible destruction of components used in high Q resonators.

10.7 General RLC Network

A truly realistic model of an LC resonator must include losses of both the inductor and capacitor, modelled by resistor r_1 and effective series resistance (ESR), (see Fig. 10.11). In this section, we derive expressions for the resonant frequency ω_0 and the associated dynamic resistance R_D of a general LC circuit as the function of both Q_1 (inductor) and Q_2 (capacitor) factors. Finally, we show how to transform the resonator in Fig. 10.11 into its equivalent parallel RLC network, assuming that the capacitor is lossless, i.e. $ESR = 0$.

By definition, Q factors of inductive and capacitive branches in the LC network (after substitution $ESR = r_2$) at resonant frequency ω_0 are[7]

$$Q_1 \stackrel{\text{def}}{=} \frac{X_L}{r_1} = \frac{\omega_0 L}{r_1} = \tan\theta_1 \quad \therefore \quad \theta_1 = \arctan Q_1 \tag{10.82}$$

$$Q_2 \stackrel{\text{def}}{=} \frac{X_C}{r_2} = \frac{1}{\omega_0 C\, r_2} = \tan\theta_2 \quad \therefore \quad \theta_2 = \arctan Q_2 \tag{10.83}$$

where θ_1 and θ_2 are the respective phase angles in the inductor and capacitor due to the thermal losses (resistances $r_{1,2}$ denote the internal resistances of the coil and the ESR of the capacitor, respectively). We also define, after including (10.48),

$$Z_1 = r_1 + j\omega_0 L = \frac{\omega_0 L}{Q_1} + j\omega_0 L$$

$$\therefore$$

$$|Z_1| = \sqrt{\left(\frac{\omega_0 L}{Q_1}\right)^2 + (\omega_0 L)^2} = \omega_0 L \sqrt{1 + \frac{1}{Q_1^2}} \tag{10.84}$$

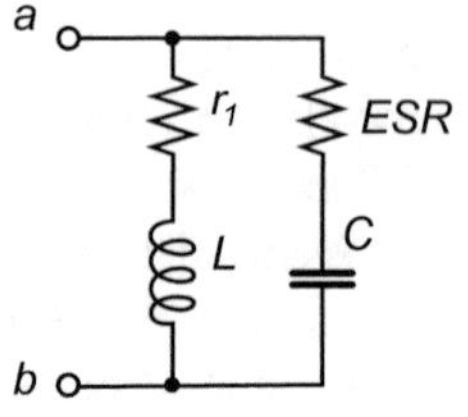

Fig. 10.11 Realistic, general parallel LC network

[7] Phase angle θ of a complex number z is found through the relation $\tan\theta \stackrel{\text{def}}{=} \Im(z)/\Re(z)$.

as well as,

$$Z_2 = r_2 + \frac{1}{j\omega_0 C} = \frac{1}{Q_2\omega_0 C} + \frac{1}{j\omega_0 C} \tag{10.85}$$

therefore,

$$|Z_2| = \sqrt{\frac{1}{(Q_2\omega_0 C)^2} + \frac{1}{(\omega_0 C)^2}} = \frac{1}{\omega_0 C}\sqrt{1 + \frac{1}{Q_2^2}} \tag{10.86}$$

From (10.82) and (10.83) with the help of trigonometric identities,[8] we write

$$\sin\theta_1 = \frac{Q_1}{\sqrt{1+Q_1^2}}\,;\ \cos\theta_1 = \frac{1}{\sqrt{1+Q_1^2}} \tag{10.87}$$

$$\sin\theta_2 = \frac{Q_2}{\sqrt{1+Q_2^2}}\,;\ \cos\theta_2 = \frac{1}{\sqrt{1+Q_2^2}} \tag{10.88}$$

10.7.1 Derivation for the Resonant Frequency ω_0

If an AC voltage source $V_{in} = V\cos\theta_1$ is connected to the resonator (see Fig. 10.12), the total current needed to compensate for the thermal losses $i = i_1 + i_2$ is split between the two branches. The inductive branch current i_1 has two components: one that is in phase with the source voltage V_{in}, i.e. $(V\cos\theta_1)/Z_1$, and one that is lagging by 90°, $(V\sin\theta_1)/Z_1$. At the same time, the capacitive branch current i_2 also has two components: one that is in phase with the source voltage V_{in}, i.e. $(V\cos\theta_2)/Z_2$, and one that is leading the source voltage V_{in} by 90°, i.e. $(V\sin\theta_2)/Z_2$.

At resonance, the two quadrature current components must be opposite and equal (so that the vector sum is zero), which leads to the following expressions (after using results (10.84)–(10.88))

$$(V\sin\theta_1)\frac{1}{Z_1} = (V\sin\theta_2)\frac{1}{Z_2} \tag{10.89}$$

therefore,

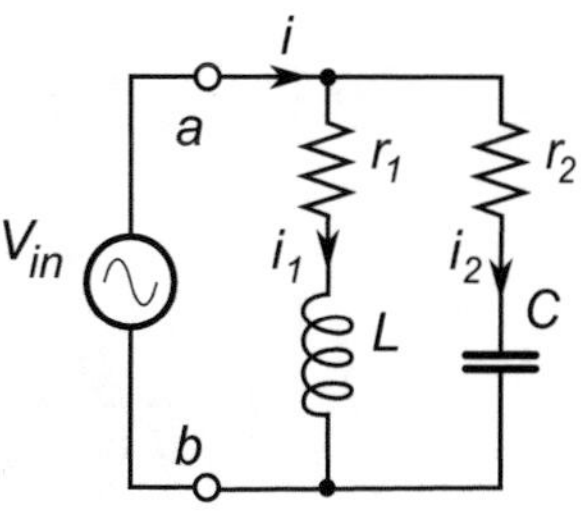

Fig. 10.12 Realistic LC resonator driven by the external signal source V_{in}

[8] $\cos[\arctan x] = 1/\sqrt{1+x^2}$ and $\sin[\arctan x] = x/\sqrt{1+x^2}$.

$$\frac{Q_1}{\sqrt{1+Q_1^2}}\frac{1}{\omega_0 L\sqrt{1+\frac{1}{Q_1^2}}} = \frac{Q_2}{\sqrt{1+Q_2^2}}\frac{\omega_0 C}{\sqrt{1+\frac{1}{Q_2^2}}}$$

$$\frac{Q_1}{\sqrt{\left(1+Q_1^2\right)\left(1+\frac{1}{Q_1^2}\right)}} = \frac{Q_2}{\sqrt{\left(1+Q_2^2\right)\left(1+\frac{1}{Q_2^2}\right)}}\,\omega_0^2 LC \tag{10.90}$$

where both the left and right side of (10.90) contain algebraic term that can be simplified as follows:

$$\frac{x}{\sqrt{(1+x^2)(1+\frac{1}{x^2})}} = \sqrt{\frac{x^2}{x^2+2+\frac{1}{x^2}}} = \sqrt{\frac{x^2}{\left(x+\frac{1}{x}\right)^2}} = \frac{x}{x+\frac{1}{x}} = \frac{1}{1+\frac{1}{x^2}} \tag{10.91}$$

Using (10.91) it is straightforward to rewrite (10.90) as

$$\frac{1}{1+\frac{1}{Q_1^2}} = \frac{1}{1+\frac{1}{Q_2^2}}\,\omega_0^2 LC$$

$$\therefore$$

$$\omega_0 = \frac{1}{\sqrt{LC}}\sqrt{\frac{1+\frac{1}{Q_2^2}}{1+\frac{1}{Q_1^2}}} \approx \frac{1}{\sqrt{LC}}; \qquad (Q_{1,2} \gg 1) \tag{10.92}$$

which is the solution for the resonant frequency of an LC resonator with a non-ideal inductor and a non-ideal capacitor. Naturally, for very good L and C components the thermal losses are negligible, in other words $Q_{1,2} \gg 1$, hence (10.92) can be approximated with the expression for the resonant frequency ω_0 that was defined earlier for the case of the ideal LC resonator. However, note that the assumption of high Q is not always valid, for example, in the case of the on-chip inductors manufactured in the standard CMOS process that are used in modern wireless devices.

10.7.2 Derivation for the Dynamic Resistance R_D

At resonance, the sum of complex quadrature components of the two branch currents is zero, which leaves only the two in-phase current components. Similarly to the previous derivation, we write

$$i = V\left(\frac{\cos\theta_1}{Z_1} + \frac{\cos\theta_2}{Z_2}\right) = V\left[\frac{1}{\sqrt{1+Q_1^2}}\frac{1}{\omega_0 L\sqrt{1+\frac{1}{Q_1^2}}} + \frac{1}{\sqrt{1+Q_2^2}}\frac{\omega_0 C}{\sqrt{1+\frac{1}{Q_2^2}}}\right]$$

$$\therefore$$

$$= V\left[\frac{Q_1}{\omega_0 L(1+Q_1^2)} + \frac{Q_2}{1+Q_2^2}\,\omega_0 C\right] \tag{10.93}$$

We introduce substitution for the $\omega_0\, C$ term, first by rewriting (10.92) as follows:

$$\omega_0^2\,LC = \frac{1+\frac{1}{Q_2^2}}{1+\frac{1}{Q_1^2}} \quad \therefore \quad \frac{Q_2^2\,\omega_0 C}{1+Q_2^2} = \frac{Q_1^2}{(1+Q_1^2)\omega_0 L} \quad \therefore \quad \omega_0 C = \frac{1+Q_2^2}{Q_2^2}\,\frac{Q_1^2}{(1+Q_1^2)\omega_0 L} \tag{10.94}$$

then, (10.93) becomes,

$$i = V\left[\frac{Q_1}{\omega_0 L(1+Q_1^2)} + \frac{Q_2}{1+Q_2^2}\,\frac{1+Q_2^2}{Q_2^2}\,\frac{Q_1^2}{(1+Q_1^2)\omega_0 L}\right] = V\frac{Q_1}{\omega_0 L(1+Q_1^2)}\left[1+\frac{Q_1}{Q_2}\right] \tag{10.95}$$

which now leads into the expression for dynamic resistance R_D as

$$R_D \stackrel{\text{def}}{=} \frac{V}{i} = \omega_0 L\,\frac{Q_1+\frac{1}{Q_1}}{1+\frac{Q_1}{Q_2}} \tag{10.96}$$

for the case of a non-ideal inductor and a non-ideal capacitor. As is, (10.96) shows the dependence of dynamic resistance versus the Q factors of L and C components. In case of very good (but still not perfect) inductors, i.e. $Q_1 \gg 1$ or in other words $(1/Q_1) \approx 0$, then (10.96) can be written as the very close approximation,

$$R_D = \omega_0 L\,\frac{Q_1}{1+\frac{Q_1}{Q_2}} = \omega_0 L\,\frac{Q_1\,Q_2}{Q_1+Q_2} \tag{10.97}$$

Modern capacitors are made using very good dielectrics, which is to say that Q_2 is not only large but could be approximated as $Q_2 \to \infty$, in other words $Q_2 \gg Q_1$, which leads into $Q_1/Q_2 \approx 0$. Therefore, in case of a lossless capacitor, (10.97) is further approximated as

$$R_D = \omega_0 L\,Q_1 \tag{10.98}$$

which is commonly used in practice because, in comparison with capacitors, inductors are much harder components to build.

Finally, further approximation that is good for fast "back-of-an-envelope" analysis, even the inductor is assumed to be perfectly lossless, i.e. $Q_1 \to \infty$, which means that (10.98) simplifies as

$$R_D \to \infty \tag{10.99}$$

which is what was concluded earlier, in (10.80). With the assumptions used in the above analysis, expressions (10.96)–(10.99), in addition to (10.107), for dynamic resistance R_D are also useful.

10.8 Selectivity

The ability of a resonating circuit to select and amplify a weak voltage signal at one specific frequency ω_0 is its core quality used in RF circuits, it is referred to as "selectivity". In the ideal case of $Q \to \infty$, the resonating circuit would pick one and only one frequency, ω_0, while all other tones would be

Fig. 10.13 Normalized output voltage across the inductor at normalized resonant frequency $\omega_0 = 1$ for various Q factors

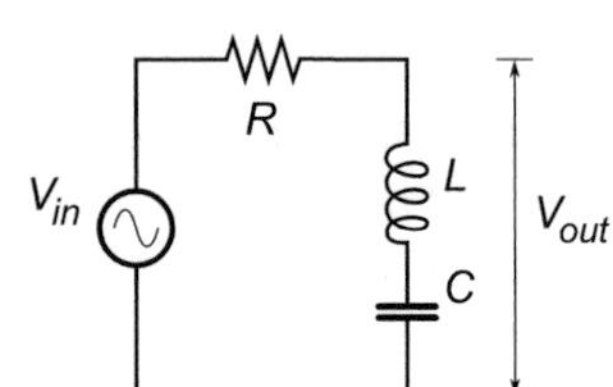

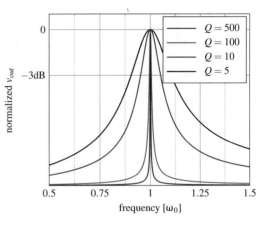

Fig. 10.14 LC network in series, driven by a sinusoidal voltage source

completely suppressed, i.e. their amplitudes multiplied by zero gain. However, in realistic circuits there is always some finite resistance causing the thermal loss, which is measured by the circuit's finite Q factor. A plot of selectivity curves as a function of Q factor is shown in Fig. 10.13. The plot indicates that, for good selectivity, we need high Q factor resonating circuits, in other words very narrow bandwidth around the resonant frequency.

An interesting question to answer is: in the case of resonators with finite Q, what is the range of frequencies that passes through the resonator without being significantly suppressed? In the following paragraphs, we examine the realistic case of RLC network behaviour regarding this important filtering property.

10.8.1 Bandpass Filters

Let us consider a serial RLC network from the perspective of voltage source with resistance R driving impedance $Z = j\,(\omega L - 1/\omega C)$ (see Fig. 10.14). Maximum power transfer, therefore, happens when the source is matched to the load, i.e. $R = |Z|$. Otherwise, at DC the capacitor becomes open while the inductor becomes a short connection; and at the other side of the frequency spectrum, at very high frequencies, the capacitor becomes short while the inductor becomes an open connection. In both extreme cases, there is no power transfer because the loop current must drop to zero due to the open connections at DC and infinite frequency.

Hence, the condition for maximum power transfer $R = |Z|$ leads to

$$V_{out} = \frac{V_{in}}{|R + Z|}\,|Z| = \frac{V_{in}}{|R \pm jR|}\,R = \frac{V_{in}}{|1 \pm j|} = \frac{V_{in}}{\sqrt{2}} \qquad (10.100)$$

which happens at two frequency points. Let us label them (for the time being) as ω_U and ω_L (for the "upper" and "lower" frequency, respectively), so that $R = |Z|$ is written as

$$R = \omega_U L - \frac{1}{\omega_U C} \tag{10.101}$$

$$-R = \omega_L L - \frac{1}{\omega_L C} \tag{10.102}$$

which, at resonance leads to

$$\begin{aligned} R &= \omega_U L - \frac{1}{\omega_U \frac{1}{\omega_0^2 L}} \;\therefore\; R = \omega_U L - \frac{\omega_0^2 L}{\omega_U} \;\therefore\; \frac{R}{\omega_0 L} = \frac{\omega_U}{\omega_0} - \frac{\omega_0}{\omega_U} \\ -R &= \omega_L L - \frac{1}{\omega_L \frac{1}{\omega_0^2 L}} \;\therefore\; -R = \omega_L L - \frac{\omega_0^2 L}{\omega_L} \;\therefore\; -\frac{R}{\omega_0 L} = \frac{\omega_L}{\omega_0} - \frac{\omega_0}{\omega_L} \end{aligned} \tag{10.103}$$

We substitute $Q = \omega_0 L/R$:

$$\frac{1}{Q} = \frac{\omega_U}{\omega_0} - \frac{\omega_0}{\omega_U} \tag{10.104}$$

$$-\frac{1}{Q} = \frac{\omega_L}{\omega_0} - \frac{\omega_0}{\omega_L} \tag{10.105}$$

After adding (10.104) and (10.105), it follows that

$$\frac{\omega_U}{\omega_0} + \frac{\omega_L}{\omega_0} = \frac{\omega_0}{\omega_U} + \frac{\omega_0}{\omega_L} \;\therefore\; \omega_0^2 = \omega_U \omega_L \tag{10.106}$$

and now, using (10.104) and (10.106), we write

$$\frac{1}{Q} = \frac{\omega_U}{\omega_0} - \frac{\omega_0}{\omega_U} = \frac{\omega_U^2 - \omega_0^2}{\omega_U \omega_0}$$

$$\therefore$$

$$Q = \frac{\omega_U \omega_0}{\omega_U^2 - \omega_0^2} = \frac{\omega_U \omega_0}{\omega_U^2 - \omega_U \omega_L} = \frac{\omega_0}{\omega_U - \omega_L} = \frac{\omega_0}{\Delta\omega} \tag{10.107}$$

The last part of expression (10.107)

$$Q = \frac{\omega_0}{\Delta\omega} \tag{10.108}$$

is very important, because the two frequencies ω_U and ω_L are used to define the resonator's bandwidth BW (see Fig. 10.15). The two frequencies are at -3 dB points relative to the maximum amplitude of the resonator (which is at ω_0). Also, (10.108) shows that a narrow band is achieved by using high Q components, see Fig. 10.13.

In serial RLC configuration, high Q also means very low resistance R and high inductance L, which implies that it is good for matching with a low impedance source, such as an antenna, for example, which usually has impedance in the order of 50Ω. Otherwise, if the source impedance is very high,

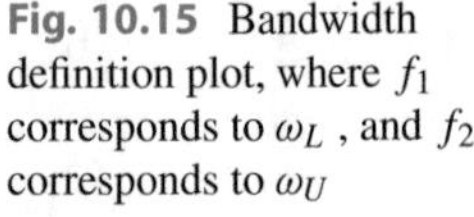

Fig. 10.15 Bandwidth definition plot, where f_1 corresponds to ω_L , and f_2 corresponds to ω_U

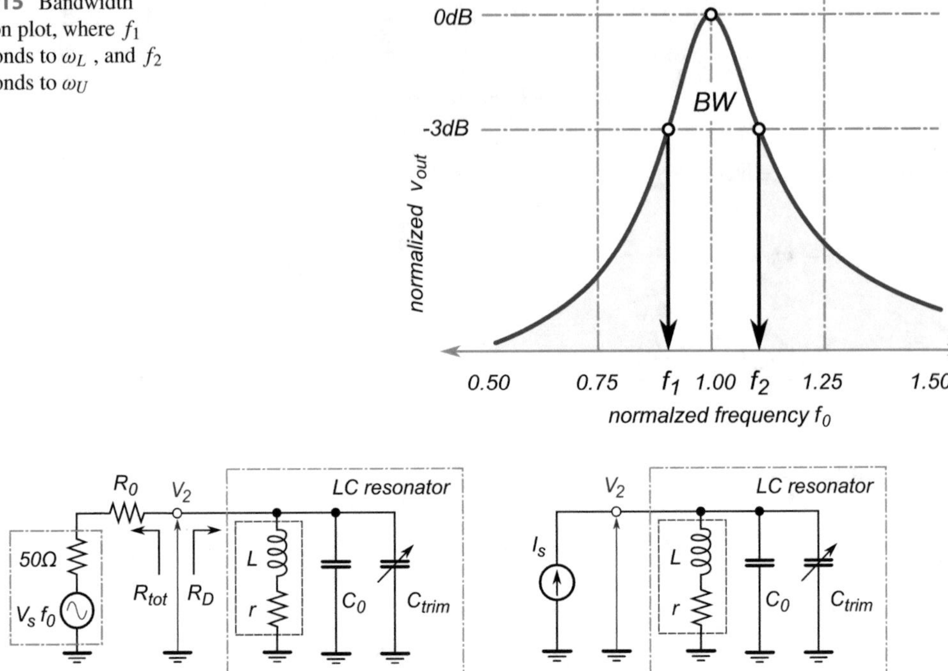

Fig. 10.16 LC resonator simulation setup with voltage source to measure R_D (left), and setup with current source to measure BW (right)

then a parallel RLC configuration must be used, where high Q means high resistance and very low inductance.

10.8.2 Measurement of LC Resonator's Dynamic Resistance

A practical method of measuring (or simulating) dynamic resistance R_D of a non-ideal LC resonator, which is the consequence of the inductor's non-zero serial DC resistance[9] r is based on the following idea. At its resonance, LC impedance is real because $X_L + X_C = 0$. That means, a Thevenin voltage generator with its internal real resistance R_S creates a voltage divider with $R_D = Q^2 r$. Since voltage source and its resistance are known, it is then sufficient to measure voltage across LC resonator and deduce R_D from the R-R voltage divider analysis (see Sect. 3.2.1). We note that due to sensitivity of LC resonator to the external (thus parasitic) capacitances associated with the measuring instruments an RF oscilloscope probe is used in real experiments. Simulation setup of this measuring is shown in Fig. 10.16.

In this example, a typical discrete $L = 3.3\,\mu\text{H}$ in parallel with $C_0 = 50\,\text{pF}$ and trim $C_{\text{trim}} = 0$-$50\,\text{pF}$ is tuned to set LC resonant frequency to $f_0 = 10\,\text{MHz}$. From (10.39) we know that it is the exact value of $C = C_0 + C_{\text{trim}} = 76.758\,\text{pF}$ that is needed to establish $f_0 = 10\,\text{MHz}$ resonant frequency. However, in the case of non-ideal inductor, see (10.51), there is slight offset in the resonant frequency. The R_D measurement then proceeds as follows.

[9]Most SPICE simulators default inductor's serial resistance to $r = 1\,\text{m}\Omega$.

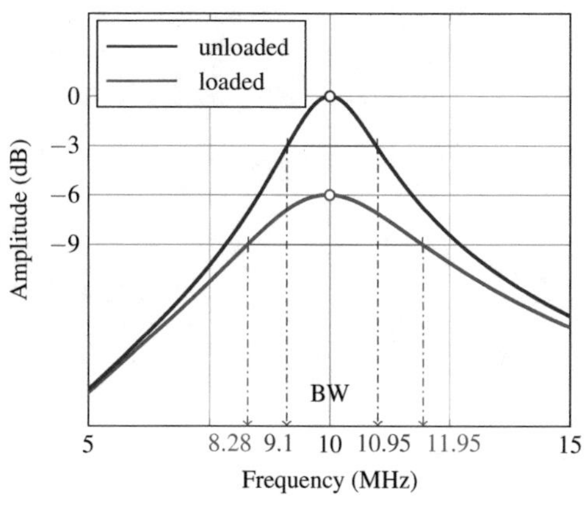

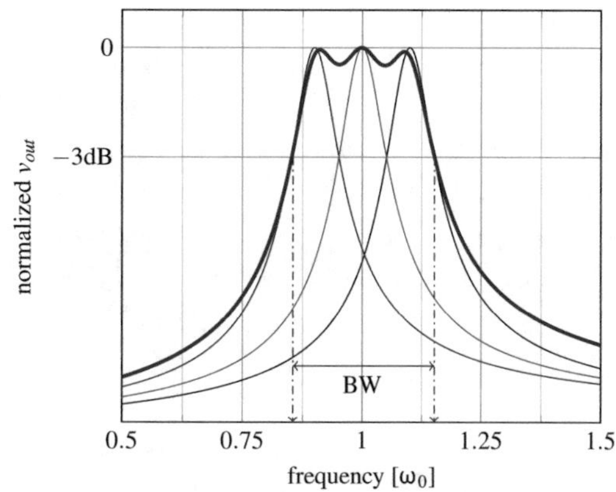

Fig. 10.17 LC resonator simulation: AC transfer characteristics for loaded and unloaded resonator (left), and coupled LC resonators (right)

1. *Capacitor tuning:* First, the signal generator is set to $f_0 = 10\,\text{MHz}$, then the trim capacitor is tuned until V_2 is at its maximum. Once that value is found, the LC resonator is in its state of resonance.
2. *Resistor tuning:* In general, R_D is much larger than the $R_{\text{sig}} = 50\Omega$ source resistance. Thus, it is practical to use $R_0 = R_{\text{trim}} = 0$–$10\,\text{k}\Omega$ trim resistor in series with the output of the signal generator. Then, the trim resistor is tuned until, for example, $V_2/V_S = 0.5$, which happens when $R_{\text{trim}} + R_{\text{sig}} = R_D$. Alternatively, a few kΩ fixed resistor R_0 is used and R_D is calculated from the voltage divider rule. In this example, $R_D = 1.1\,\text{k}\Omega$ is measured.

Frequency response in Fig. 10.17(left) shows the maximum voltage amplitude at $-6\,\text{dB}$ (i.e. when $V_2/V_S = 0.5$) and in accordance to (10.108) Q factor of this resonator is, therefore, found as

$$Q = \frac{f_0}{BW} = \frac{10\,\text{MHz}}{3.67\,\text{MHz}} = 2.72 \tag{10.109}$$

However, when $R_{\text{tot}} = R_{\text{trim}} + R_{\text{sig}} = R_D$, Q factor found in (10.109) is for the case of "loaded" LC resonator, that is to say with the signal source side resistance R_{tot} is included. We must realize that the effective resistance of "loaded" LC resonator consists of $R_{\text{tot}}||R_D = 550\Omega$ (see Sect. 11.3.1). Indeed, simulation shows, see Fig. 10.17(left), that unloaded $BW = 1.85\,\text{MHz}$, which is equivalent to $Q = 5.4$ for the LC resonator itself. In accordance to (10.81) we conclude that the internal resistance of this inductor is $r = R_D/Q^2 = 1100\Omega/5.4^2 = 37.6\Omega$. Note that ideal Norton generator (i.e. with infinite source resistance) is used to simulate LC resonator circuit.

Example 70: LC Resonance—Q Factor

A parallel LC tank consists of: $L = 2.533\,\text{nH}$ with internal wire resistance of $R_L = 1\,\text{m}\Omega$, and $C = 100\,\text{nF}$. Calculate: (a) the resonant frequency f_0 ; (b) the Q factor at resonance; (c) the impedance at resonance Z_{max}; and (d) the bandwidth BW.

Solution 70:

$$\text{(a)} \quad f_0 = \frac{1}{2\pi\sqrt{LC}} = \frac{1}{2\pi\sqrt{2.533\,\text{nH} \times 100\,\text{nF}}} = 10\,\text{MHz}$$

(continued)

(b) $Q = \frac{X_L}{R_L} = \frac{2\pi f_0 L}{R} = \frac{2\pi \times 10\,\text{MHz} \times 2.533\,\text{nH}}{1\,\text{m}\Omega} = 159.153$

(c) $Z_{\max} = Q^2 R_L = (159.153)^2 \times 1\,\text{m}\Omega = 29.330\Omega$

(d) $BW = \frac{f_0}{Q} = \frac{R_L}{2\pi L} = \frac{1\,\text{m}\Omega}{2\pi \times 2.533\,\text{nH}} = 62.833\,\text{kHz}$

10.9 Coupled Tuned Circuit

Although it may seem that increasing the Q factor of a resonating circuit is always desirable, that is not the case. In addition to improving the selectivity of a receiver, an increased Q factor helps with amplification of weak RF signals arriving to the antenna (i.e. with the sensitivity). However, higher Q also reduces the bandwidth, which may start cutting into the frequency content of the signal and, therefore, start introducing distortions.

For example, if a receiver is meant to receive the complete voice frequency spectrum, i.e. 20 Hz–20 kHz, using a 10 MHz carrier signal, then the minimum required bandwidth, calculated according to (10.108) as $Q = f_0/\Delta f = 100\,\text{MHz}/20\,\text{kHz} = 500$. Using a wider bandwidth would not benefit the quality of the received signal; instead, it would allow more noise into the system. If, for whatever reason, a resonator with higher Q is being used, in practical systems it is always possible to widen the overall bandwidth by staggering more than one resonator while maintaining the required sensitivity (see Fig. 10.17 (right)). Each of the resonators is tuned to a slightly different resonant frequency and the overall frequency response becomes equal to the sum of the individual responses

10.10 Summary

In this section we introduced serial and parallel resonant LC circuits. The LC resonant behaviour is very important for generating voltage and current variables that follow sinusoidal shape in time domain. We explored both ideal and realistic cases of LC resonators, and introduced Q factor as commonly used measure of the internal thermal losses. In the second important use of LC resonators, by controlling the Q factor, we are able to determine bandwidth of the bandpass LC resonating filter and, therefore, limit the frequency range of single-tone signals that pass through the LC resonator. These two functions are fundamental for RF circuit design and we use both of them.

Problems

10.1 For a given coil, $L = 2\,\mu\text{H}$, $Q = 200$, $f_0 = 10\,\text{MHz}$, calculate:

(a) its equivalent series resistance,
(b) its parallel resistance
(c) the value of the resonating capacitor,
(d) parallel resistance which, when added, provides bandwidth of 200 MHz.

10.2 If a single-tone signal $f_0 = 8\,\text{MHz}$ is passed through a low-pass RC filter followed by a high-pass RC filter:

1. Choose R and C values such that bandwidth around the f_0 is $BW = 10\,\text{kHz}$.
2. What would you choose for $BW = 5\,\text{kHz}$.
3. Design RLC filters with the same characteristics.

Note: pick component values at your will, do not have to be the standard values.

10.3 For a given inductor $L = 2.533\,\text{nH}$ and trimming capacitor whose range is $C = 80\,\text{nF}–120\,\text{nF}$ calculate the tuning range ($\Delta f = f_{\text{max}} - f_{\text{min}}$) of this LC resonator.

10.4 Design an LC resonator whose resonant frequency is $f_0 = 10\,\text{MHz}$ if only the following components are available:

(a) $L = 2.533\,\text{nH}$, $C_1 = 10\,\text{nF}$, $C_2 = 40\,\text{nF}$, and $C_3 = 50\,\text{nF}$
(b) $L = 2.533\,\text{nH}$, $C_1 = 20\,\text{nF}$, $C_2 = 30\,\text{nF}$, and $C_3 = 60\,\text{nF}$
(c) $L = 2.533\,\text{nH}$, $C_1 = 70\,\text{nF}$, $C_2 = 60\,\text{nF}$, and $C_3 = 60\,\text{nF}$

10.5 Calculate the Q factor of a serial RLC network if inductor $L = 2.533\,\text{nH}$ and the lumped wire resistance $r = (\pi)\text{m}\Omega$, at: (a) $f_1 = 10\,\text{MHz}$; and (b) $f_2 = 100\,\text{MHz}$.

10.6 For a series RLC network derive expression for bandwidth BW at the resonant frequency ω_0 as a function of the Q. What is the conclusion?

10.7 A $1\,\mu\text{H}$ inductive coil has wire resistance of $R = 5\,\Omega$ and self-capacitance of 5 pF. The inductor is used to create LC resonator with $f_0 = 25\,\text{MHz}$. Calculate the effective inductance and effective Q factor.

10.8 Calculate resonant frequency of a serial RLC network with $R = 30\,\Omega$, $L = 3\,\text{mH}$, and $C = 100\,\text{nF}$. Calculate its impedance at $f = 10\,\text{kHz}$.

10.9 A frequency response curve of an LC resonator looks as in Fig. 10.15. Assuming, $f_1 = 450\,\text{kHz}$, $f_2 = 460\,\text{kHz}$, $f_0 = 455\,\text{kHz}$. Determine the resonator bandwidth, Q factor, inductance L if capacitance is $C = 1\,\text{nF}$, the total internal circuit resistance R.

10.10 A parallel LC tank consists of $L = 1\,\text{mH}$ whose wire resistance is $R = 1\Omega$, and a capacitor $C = 100\,\text{nF}$. Determine, resonant frequency, the Q factor, dynamic resistance R_D, and bandwidth of this resonator.

10.11 A series RC branch consists of $R_S = 10\,\Omega$ and $C_S = 7.95\,\text{pF}$. Convert it into its equivalent parallel RC network form at $f = 1\,\text{GHz}$

11 Matching Networks

The main purpose of an electronic circuit is to process an electronic signal that has arrived at its input terminals. The circuit is then expected to modify (i.e. to process) the input signal in accordance with the intended mathematical function and to pass the result to the next stage. In addition, a real, well-designed system should perform the signal-processing operations efficiently with minimal waste of time and energy. This objective is achieved by using "power matching" techniques.

In this chapter, we study a simple basic methodology for interfacing two stages in the signal-processing chain that is commonly used in the design of RF electronic systems, with the main criterion being maximum power transfer between the stages. This approach is justified by the argument that wireless RF signals that have arrived at the system input terminals (e.g. at the antenna) are very weak, thus subsequent power loss would have broad consequences for the overall system performance.

11.1 Matching Networks

An optimal interface is designed with the assumption that driver/load impedances (see Sect. 3.4) are matched. In the case of the antenna and RF amplifier interface, it is the matching network that minimizes losses of the weak RF signal, Fig. 11.1. A general (and realistic) case is when the two matching impedances are not the same. Going back to the pipe analogy, when two pipes with unequal diameters need to be connected, we add a third pipe to serve as an adapter, Fig. 11.2 (left). Similarly, in order to enable efficient power transfer between two stages with non-matching impedances, an additional circuit network has to be designed and inserted at the interface to serve as an "impedance converter", Fig. 11.2 (right). Detailed coverage of the matching network design is beyond the scope of this book; nevertheless, some of the basic concepts of matching network design are introduced by means of examples.

For sake of clarifying the terminology, it is important that we distinguish between two similar, but often confused, circuit design activities: impedance transformation and impedance matching.

Impedance transformation is used to transform one impedance at the input node to a different value at the output. At the output node of the transformation network, only the new impedance is visible and it effectively masks the impedance connected to the input node of the transformation network. This interface is always unidirectional and is intended to interface only one impedance with the rest of the system.

R. Sobot, *Wireless Communication Electronics*,
https://doi.org/10.1007/978-3-030-48630-3_11

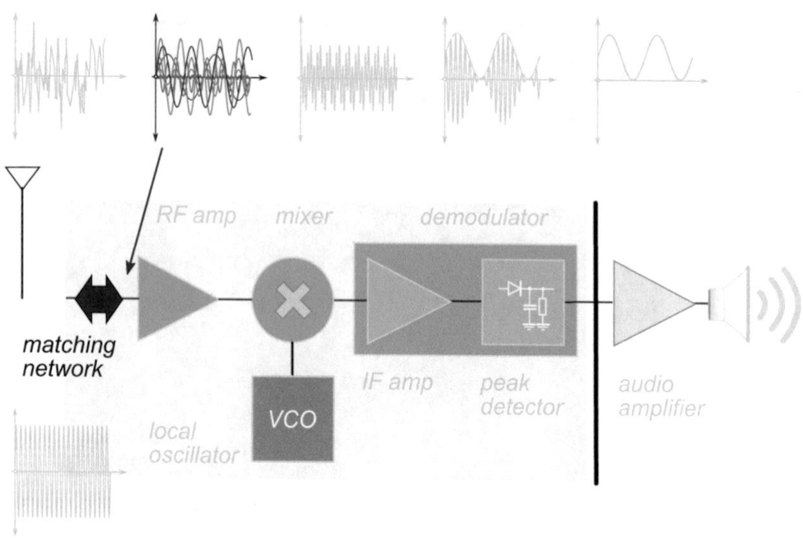

Fig. 11.1 Heterodyne AM radio receiver architecture: RF front end, matching network

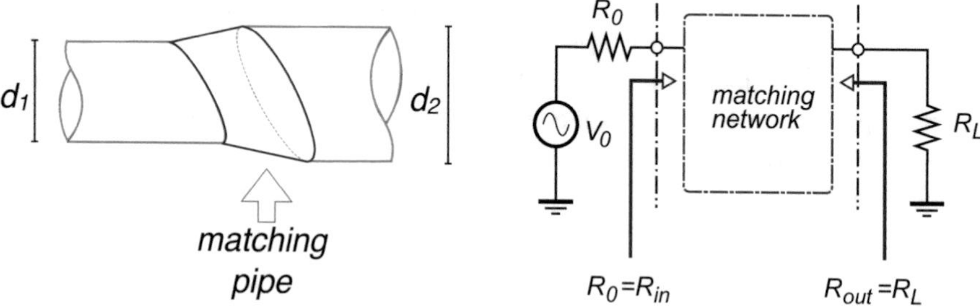

Fig. 11.2 An example of a typical case of non-matched source and load resistances, $R_0 \neq R_L$

Impedance matching is always performed between two non-equal impedances (real or complex). The interface is always bidirectional and intended to maximize power transfer between the two impedances. In this book, unless otherwise specified, we design the inserted matching network with the goal of maximizing signal power transfer.

11.2 The Q Matching Technique

This impedance matching technique is based on the idea that a single L-shaped (X_S, X_P) branch is sufficient to provide impedance transition between two real, unequal resistances R_0 and R_L, Fig. 11.3, with the limitation that perfect matching is achieved at the one and only frequency. When the two resistances are already equal, i.e. $R_0 = R_L$, there is no need for additional matching.

Because there are only two initial resistances to compare, R_0 and R_L, and two possible flavours of reactances that can be used $(j\omega L)$ and $(1/j\omega C)$, there are only four possible combinations that can be made. On the left of Fig. 11.3 either inductive or capacitive reactance X_S is used in series with $R_0 < R_L$. At the same time, either capacitive or inductive reactance X_P is used in parallel with load resistance $R_L > R_0$. The rule of Q matching technique is that the two reactive components X_S and

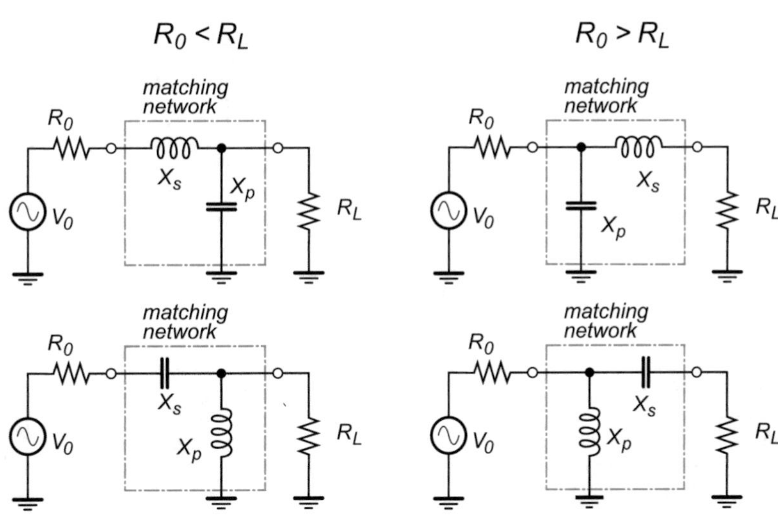

Fig. 11.3 Four possible cases of using a single X_S–X_P circuit branch as a matching network between the source resistance R_0 and the load resistance R_L. The left two cases are for $R_0 < R_L$, and the right two cases are for $R_0 > R_L$

X_P must not be of the same type, i.e. one must be inductive and the other capacitive. Thus, on the right of Fig. 11.3, either capacitive or inductive reactance is used in parallel with the source resistance $R_0 > R_L$ and either inductive or capacitive reactance is used in series with $R_L < R_0$.

At this point it is valid to ask why we use reactances when the same goal is achievable with a resistive network. Yes, it is possible to design a matching network using only resistors, however the power loss increases drastically and wideband networks are always much noisier, which reduces SNR. In addition, for each of the two relations between $R_0 \lessgtr R_L$, there are two possible matching networks; we may ask if there is any difference between the two. If there are no additional constraints, either solution is valid. For example, either of the two matching networks on the left of Fig. 11.3 is valid when $R_0 < R_L$. In practice, we usually have additional constraints, for example, if a DC connection needs to be maintained between the source and load resistance, then the serial reactance must be inductive, $X_S = j\omega L$ (the upper two cases in Fig. 11.3); if an AC connection is desired, the serial reactance must be capacitive, $X_S = 1/j\omega C$ (the lower two cases in Fig. 11.3). In the following sections, first we consider cases of matching real impedances, then we study the cases of matching complex impedances.

11.2.1 Matching Real Impedances

A typical matching problem involves only real source and load resistances that are not equal, $R_0 \neq R_L$. Let us find out how a general problem such as this one is solved using the Q matching technique. We distinguish two typical cases: $R_0 < R_L$ and $R_0 > R_L$ ($R_0 = R_L$ is already matched).

When the two matching resistances are not equal $R_0 \neq R_L$, we intuitively try to equalize the two resistances by adding serial reactance X_S to the lower resistance (thus increasing the equivalent resistance of the branch) and, at the same time, by adding parallel reactance X_P to the side with the higher initial resistance (thus lowering the equivalent resistance of that branch). Obviously, when the equivalent resistance of the serial branch is equal to the equivalent resistance of parallel branch, the matching is achieved. We distinguish two specific cases: $R_0 < R_L$ and $R_0 > R_L$. In the following section we use relationships (10.78) and (10.77) that are developed in Sect. 10.5.

1. Case ($R_0 < R_L$): to illustrate the matching technique we use a typical example in Fig. 11.4 where source resistance $R_0 = 5\,\Omega$ must drive a load of $R_L = 50\,\Omega$. After the matching network

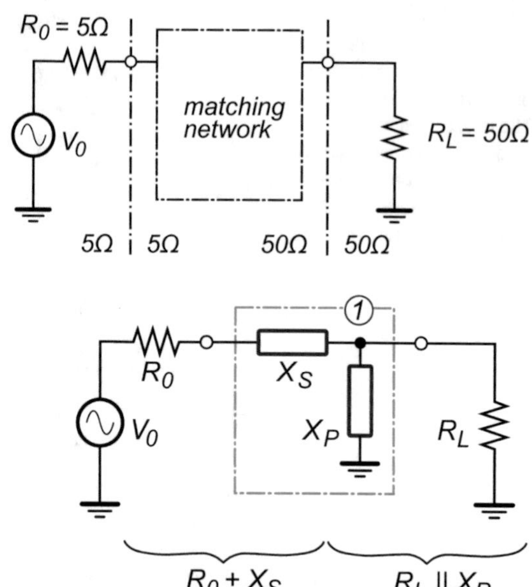

Fig. 11.4 An example of a typical case of non-matched source and load resistances, $R_0 < R_L$

Fig. 11.5 An L–C section placed between two resistive terminations creates a series and a parallel subnetwork. When the two subnetworks are conjugate matched to each other, their Qs are equal

is designed and inserted, the source should "see" a load value of, in this case, 5 Ω and, at the same time, the load resistor should "feel" as if it was driven by a source resistance equal to its own, in this example 50 Ω.

Design procedure consists of the following steps.

a. Add a series reactive element X_S next to R_0 and increase the impedance of the serial subnetwork branch. Add a parallel reactive element X_P next to R_L and reduce the impedance of the parallel subnetwork branch. We note that, if the serial element is an inductor, adding a parallel capacitor creates a LP topology (see Fig. 11.5); a serial capacitor in combination with a parallel inductor forms a high-pass section.
b. At the design frequency, the two newly created subnetworks, one in series and one in parallel (Fig. 11.5), must represent complex conjugate impedances to each other. Thus, the Q factors of these two subnetworks must be equal at the frequency where the match is computed. The serial Q factor Q_S and the parallel Q factor Q_P of the two subnetworks are (see Sect. 10.5)

$$Q_S = \frac{X_S}{R_0} \quad \text{and} \quad Q_P = \frac{R_L}{X_P} \tag{11.1}$$

c. Using (10.78) and (10.77) we calculate the serial and parallel Q factors of the two subnetworks as

$$Q_S = Q_P = Q = \sqrt{\frac{R_L}{R_0} - 1} \tag{11.2}$$

and we note that because $R_L > R_0$ the square root result is positive. For the example given in Fig. 11.4, $R_0 = 5\,\Omega$ and $R_L = 50\,\Omega$, we find

$$Q = \sqrt{\frac{R_L}{R_0} - 1} = \sqrt{\frac{50\,\Omega}{5\,\Omega} - 1} = 3 \tag{11.3}$$

d. Once the Q factors are calculated, the next step is to calculate the series and parallel reactances from (11.1) and to compute the inductor and capacitor values by using their respective impedance definitions for the given design frequency.

2. Case ($R_0 > R_L$): to illustrate the matching technique we use a typical example in Fig. 11.4 where source resistance $R_0 = 50\,\Omega$ must drive a load of $R_L = 5\,\Omega$. After the matching network is designed and inserted, the source should "see" a load value of, in this case, $50\,\Omega$ and, at the same time, the load resistor should "feel" as if it was driven by a source resistance equal to its own, in this example $5\,\Omega$.
Design procedure consists of the following steps.

   a. Add a series reactive element X_S next to R_L and increase the impedance of the serial subnetwork branch. Add a parallel reactive element X_P next to R_0 and reduce the impedance of the parallel subnetwork branch. We note that, if the serial element is an inductor, adding a parallel capacitor creates a LP topology (see Fig. 11.7); a serial capacitor in combination with a parallel inductor forms a high-pass section.
   b. At the design frequency, the two newly created subnetworks, one in series and one in parallel (Fig. 11.7), must represent complex conjugate impedances to each other. Thus, the Q factors of these two subnetworks must be equal at the frequency where the match is computed. The serial Q factor Q_S and the parallel Q factor Q_P of the two subnetworks are (see Sect. 10.5)

$$Q_S = \frac{X_S}{R_L} \quad \text{and} \quad Q_P = \frac{R_0}{X_P} \tag{11.4}$$

   c. Using (10.78) and (10.77) we calculate the serial and parallel Q factors of the two subnetworks as

$$Q_S = Q_P = Q = \sqrt{\frac{R_0}{R_L} - 1} \tag{11.5}$$

   and we note that because $R_0 > R_L$ the square root result is again positive. For the example given in Fig. 11.6, $R_0 = 50\,\Omega$ and $R_L = 5\,\Omega$, we find

$$Q = \sqrt{\frac{R_0}{R_L} - 1} = \sqrt{\frac{50\,\Omega}{5\,\Omega} - 1} = 3 \tag{11.6}$$

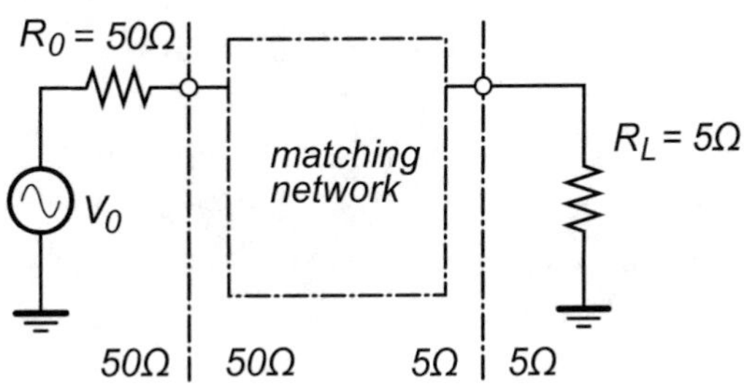

Fig. 11.6 An example of a typical case of non-matched source and load resistances, $R_0 > R_L$

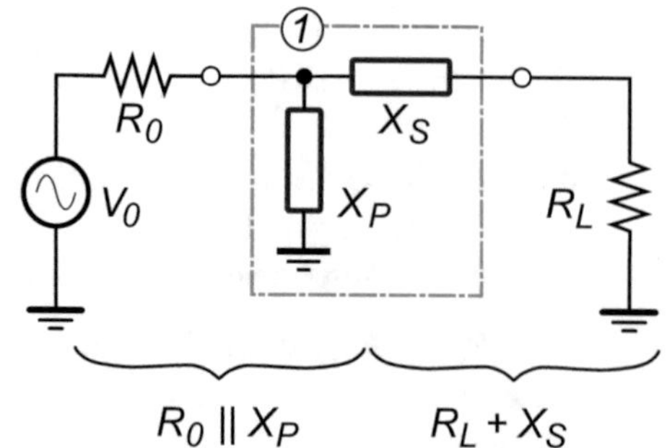

Fig. 11.7 An L–C section placed between two resistive terminations creates a series and a parallel subnetwork. When the two subnetworks are conjugate matched to each other, their Qs are equal

d. Once the Q factors are calculated, the next step is to calculate the series and parallel reactances from (11.4) and to compute the inductor and capacitor values by using their respective impedance definitions for the given design frequency (Fig. 11.7).

In summary, the Q matching methodology for the case of signal source V_0 with real source resistance R_0 that drives a load with real resistance R_L is a straightforward procedure because there are only four possible matching networks to consider. In order to make the solution unique, an additional constraint must be introduced to further determine the nature of serial and parallel impedances in the matching network. For example, if the matching network is to preserve a DC connection between the source and the load, then an inductor must be chosen as the serial element. Similarly, if an AC connection between the source and the load is to be preserved, a capacitor must be chosen as the serial element.

Example 71: Q Matching Technique, Real Resistances ($R_0 < R_L$)
Using the Q matching technique, design a single-section LC network to match a source resistance $R_0 = 5\,\Omega$ to a resistive load $R_L = 50\,\Omega$ at $f = 10\,\text{MHz}$ (see Fig. 11.4). Maintain a DC connection between the source and the load.

Solution 71: The source resistance is smaller than the load resistance, $R_0 < R_L$, (see Figs. 11.4 and 11.5). Therefore, serial reactance needs to be added to the source resistance R_0 and parallel reactance to the load resistance R_L. Adding a serial inductor to the 5Ω source side and a parallel capacitor to the 50 Ω load side keeps the DC connection and creates the LP matching configuration.

From (11.2) the required Q factors are calculated as

$$Q_S = Q_P = \sqrt{\frac{R_L}{R_S} - 1} = \sqrt{\frac{50}{5} - 1} = 3$$

From (11.1), it follows, first for the serial component,

$$X_S = Q_S\, R_0 = 3 \times 5\,\Omega = 15\,\Omega \quad \therefore \quad L = \frac{15\,\Omega}{2\pi \times 10\,\text{MHz}} = 238.732\,\text{nH}$$

and then for the parallel component,

(continued)

Fig. 11.8 Matching network for Example 71

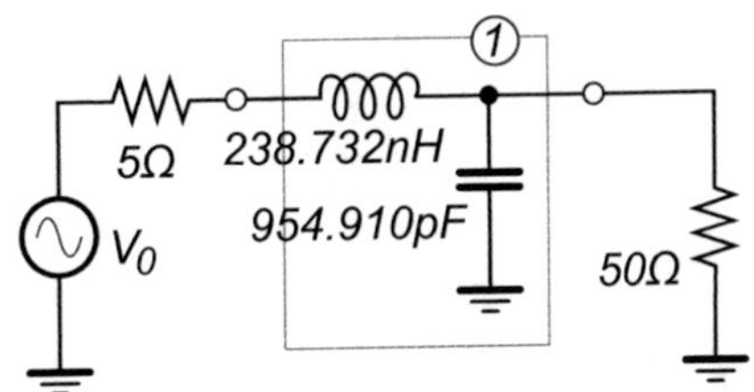

$$X_P = \frac{R_L}{Q_P} = \frac{50\,\Omega}{3} = 16.667\,\Omega \quad \therefore \quad C = \frac{1}{2\pi \times 10\,\text{MHz} \times 16.667\,\Omega} = 954.910\,\text{pF}$$

Let us verify the above result. After inserting the matching network and looking into the source side relative to node one, Figs. 11.4 and 11.8, there is a serial connection of the source resistance R_0 and the matching network's inductor X_S. Therefore, the total serial source side impedance is

$$|Z_0| = \sqrt{R_0^2 + X_S^2} = \sqrt{5^2 + 15^2}\,\Omega = 15.811\,\Omega$$

At the same time, looking into the load side, there is a parallel connection of R_L and X_P. Therefore, the parallel impedance at the load side is

$$|Z_L| = \frac{1}{\sqrt{\frac{1}{R_L^2} + \frac{1}{X_P^2}}} = \frac{1}{\sqrt{\frac{1}{50^2} + \frac{1}{16.667^2}}}\,\Omega = 15.811\,\Omega$$

Thus, the source side impedance increased and the load side impedance decreased, with the apparent matching of the two sides at 15.811 Ω at node one.

It is worth mentioning that design of matching networks results in non-standard component values, neither those of capacitors nor of inductors. As a consequence, the use of "trimer capacitors" is a standard practice, in addition to manufacturing of customized inductors.

Numerical simulation of matching circuit found in Example 71 shows that indeed maximum power level that the source can deliver is achieved, Fig. 11.9. By sweeping the source resistance R_0 over two decades and measuring RMS power dissipated in the two real resistors, graph in Fig. 11.9 (left) confirms that the source delivers maximal power when $R_0 = 5\,\Omega$ that is at the same time equal to the power received by the 50 Ω load. At the same time AC simulation shows that maximal power is achieved at the desired $f = 10\,\text{MHz}$, Fig. 11.9 (right). We note that this matching network has a non-symmetrical AC transfer function relative to the centre frequency (also, noted by slight offset of the phase function). Thus, the $-3\,\text{dB}$ bandwidth is approximated, for example in this case, to $BW \approx 7.5\,\text{MHz}$. This result is not surprising, because typically simple RLC networks result in a relatively low Q factor.[1]

[1] See Sect. 10.8.1.

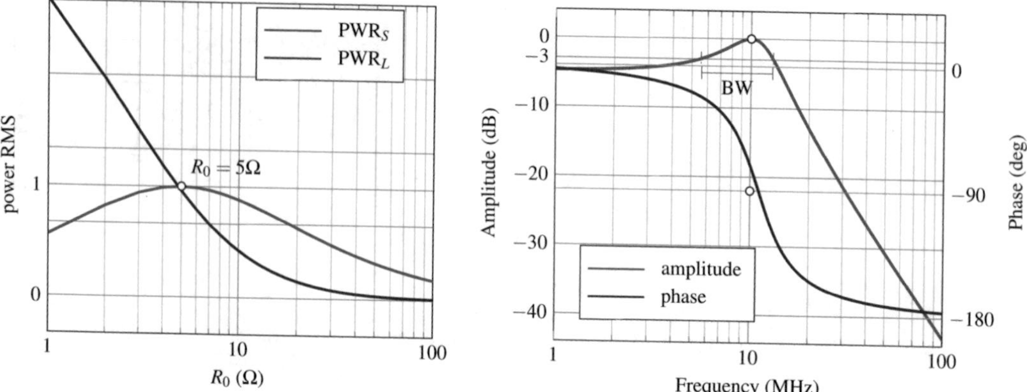

Fig. 11.9 SPICE simulation of matching network in Example 71 shows maximal power level (normalized) delivered by source is when $R_0 = 5\Omega$ and $R_L = 50\,\Omega$ (left). AC simulation confirms that power delivered at $f = 10\,\text{MHz}$ is indeed maximal (right)

Example 72: Q Matching Technique, Real Resistances ($R_0 > R_L$)
Using the Q matching technique, design a single-section LC network to match a source resistance $R_0 = 50\,\Omega$ to a resistive load $R_L = 5\,\Omega$ at $f = 10\,\text{MHz}$ (see Fig. 11.6). Maintain a DC connection between the source and the load.

Solution 72: The source resistance is greater than the load resistance, $R_0 > R_L$, see (Figs. 11.6 and 11.7). Therefore, parallel reactance needs to be added to the source resistance R_0 and serial reactance to the load resistance R_L. Adding a parallel capacitor to the $50\,\Omega$ source side and a serial inductor to the $5\,\Omega$ load side keeps the DC connection and creates the LP matching configuration.

From (11.2) the required Q factors are calculated as

$$Q_S = Q_P = Q = \sqrt{\frac{R_0}{R_L} - 1} = \sqrt{\frac{50}{5} - 1} = 3$$

From (11.4), it follows, first for the parallel component,

$$X_P = \frac{R_0}{Q_P} = \frac{50\,\Omega}{3} = 16.667\,\Omega \quad \therefore \quad C = \frac{1}{2\pi \times 10\,\text{MHz} \times 16.667\,\Omega} = 954.910\,\text{pF}$$

and then for the serial component,

(continued)

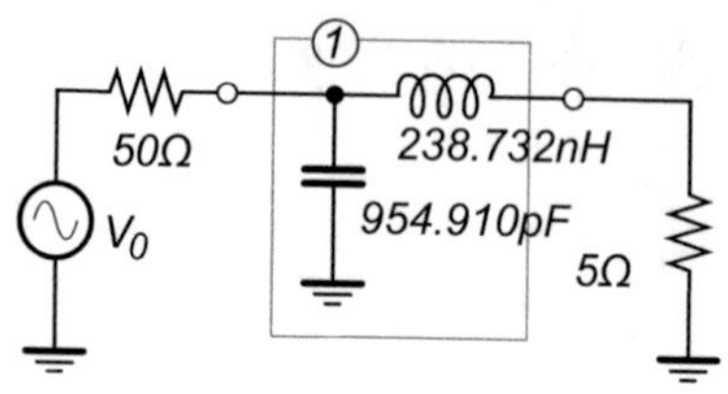

Fig. 11.10 Matching network for Example 72

$$X_S = Q_S\, R_L = 3 \times 5\,\Omega = 15\,\Omega \quad \therefore \quad L = \frac{15\,\Omega}{2\pi \times 10\,\text{MHz}} = 238.732\,\text{nH}$$

Let us verify the above result. After inserting the matching network and looking into the source side relative to node one, Figs. 11.7 and 11.10, there is a parallel connection of the source resistance R_0 and the matching network's capacitor X_P. Therefore, the total parallel source side impedance is

$$|Z_0| = \frac{1}{\sqrt{\frac{1}{R_0^2} + \frac{1}{X_P^2}}} = \frac{1}{\sqrt{\frac{1}{50^2} + \frac{1}{16.667^2}}}\,\Omega = 15.811\,\Omega$$

At the same time, looking into the load side, there is a serial connection of R_L and X_S. Therefore, the serial impedance at the load side is

$$|Z_L| = \sqrt{R_L^2 + X_S^2} = \sqrt{5^2 + 15^2}\Omega = 15.811\,\Omega$$

Thus, the source side impedance increased and the load side impedance decreased, with the apparent matching of the two sides at 15.811 Ω at node one.

11.2.2 Matching Complex Impedances

A general case of matching complex impedances follows the same design methodology presented in the previous sections, i.e. properly designed matching network must provide correct complex conjugate matching both at the input terminal plane and at the output terminal plane. When looking into the output terminals of the matching networks we need to see the complex conjugate value of the load impedance and when looking into the input terminals of the matching network we need to see the complex conjugate value of the source impedance (Fig. 11.11). Under those conditions, all of the signal power is delivered to the load without any reflection at the two ports.

The reactances associated with source and load impedances are referred to as "parasitics". If any of the two matching impedances Z_0 and Z_L already contains parasitics, the matching network design problem can be approached in two possible ways that may lead to the desired solution: we could try to "absorb the parasitics" into the matching network or to eliminate the parasitics by resonance, i.e. to "resonate them out". In both of these methods, the parasitic reactances may be eliminated either completely or partially. A design procedure for matching complex impedances starts by solving the

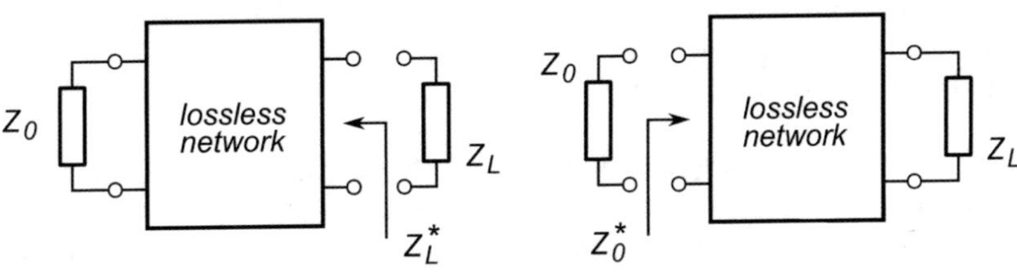

Fig. 11.11 Matching complex impedances by design of the lossless matching network simultaneously provides bidirectional complex conjugate matching

matching network for the real parts and then proceeds by absorbing the parasitics or resonating them out, either completely and partially.

11.2.2.1 Absorbing the Parasitics

Let us consider a case where source or load impedances include parasitic reactances. In addition, let us assume that values of the parasitic reactances are *lower* than the component values of the matching network that is required to match only the real parts of the two impedances. If that is the case, there is an opportunity to "absorb", i.e. to combine, these source or load parasitics with the matching network components. First, we solve the case of real resistance of the load and source, then we explore the possibility to absorb the complex impedances. Let us take a look at the following example.

Example 73: Q Matching Technique, Absorbing Parasitic Impedances, $(R_0 < R_L)$
Design a single-stage LC matching network at $f = 10\,\text{MHz}$ for the case of a source V_0 whose impedance consists of a resistor $R_0 = 5\,\Omega$ connected in series with an $L_S = 138.732\,\text{nH}$ inductor, which has to drive an $R_L = 50\,\Omega$ load resistance in parallel with $C_L = 454.910\,\text{pF}$, see Fig. 11.12. The matching network is expected to maintain a DC path between the source and the load.

Solution 73: In the case of a source or load with complex impedances, a general strategy is to first resolve the matching network only for the real parts of the two impedances. In Example 71, we already designed a matching network for the case of real $R_0 = 5\,\Omega$ source and $R_L = 50\,\Omega$ load at 10 MHz, which happen to be numerically equal to the real parts of the impedances in this example. Hence, we reuse the results and treat those calculations as the first phase of this example.

As we already found in Example 71, see Fig. 11.8, to match $R_0 = 5\,\Omega$ to $R_L = 50\,\Omega$ we need an $X'_S = 238.732\,\text{nH}$ inductor and an $X'_P = 954.910\,\text{pF}$ capacitor. However, the source impedance in this example already contains $L_S = 138.732\,\text{nH}$ inductance, which means that the additional $X_S = 238.732 - 138.732 = 100\,\text{nH}$ inductor in series is needed. We can imagine the matching network in Fig. 11.8 as its equivalent network in Fig. 11.13 (left).

Because the existing inductance is smaller than the required 238.732 nH it is possible to reuse the existing inductance and simply add a 100 nH serial inductance to make the total inductance, as shown in Fig. 11.13 (left). By doing this addition we "absorb" the existing 138.732 nH parasitic inductance into the total value of inductance required to match $5\,\Omega$ source to $50\,\Omega$ real load.

At the same time, in the case of 5–50 Ω matching, the loading impedance needs a total of $X'_P = 954.910\,\text{pF}$ capacitance, which means that the additional $X_P = 954.910 - 454.910 = 500\,\text{pF}$ capacitor is needed in parallel with the existing $C_L = 454.910\,\text{pF}$ parasitic capacitance. By adding the 500 pF difference in parallel, see Fig. 11.13 (left), we "absorb" the existing parasitic capacitance into the total value of capacitance required by the matching network.

Therefore, the required matching network consists of an $X_S = 100\,\text{nH}$ inductor and an $X_P = 500\,\text{pF}$ capacitor, as shown in Fig. 11.13 (right).

Numerical simulation of matching circuit found in Example 73 confirms the hand calculation that indeed maximum power level that source can deliver is achieved, Fig. 11.14. By sweeping serial X_S inductance over a range of values around nH and measuring RMS power dissipated in the load and the source real resistors, graph in Fig. 11.14 confirms that source delivers maximal power when $L_S = 100\,\text{nH}$ at the given frequency.

Fig. 11.12 Matching network in the case of complex source and load impedances

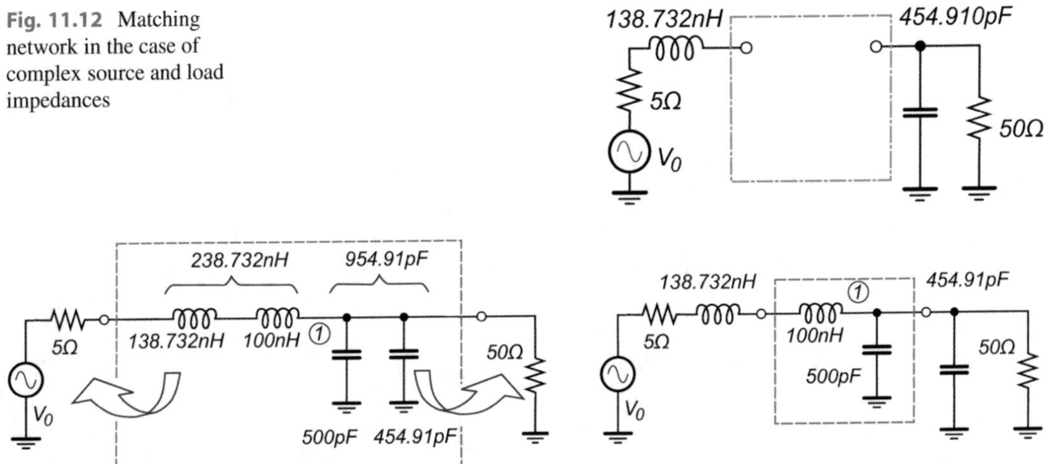

Fig. 11.13 Matching network in the case of complex source and load impedances

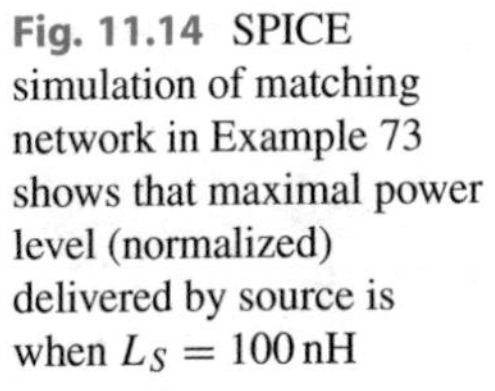
Fig. 11.14 SPICE simulation of matching network in Example 73 shows that maximal power level (normalized) delivered by source is when $L_S = 100\,\text{nH}$

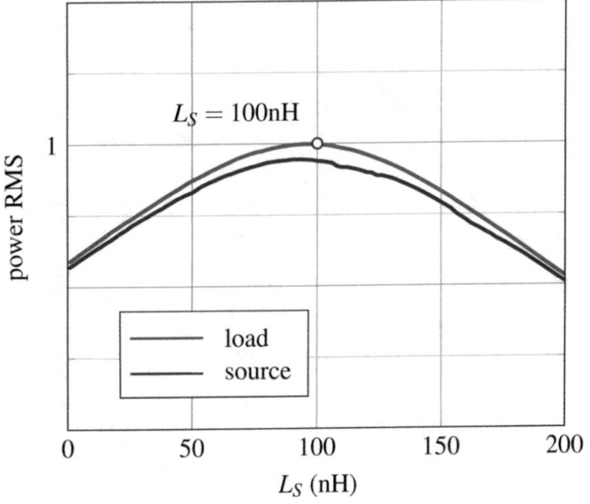

11.2.2.2 Resonating Out Excessive Parasitics

Let us consider a case where, for example, the load impedance includes parasitic reactance. In addition, let us assume that value of the parasitic reactance is *greater* than the value of the component of the matching network designed to match only the real parts of the two impedances (see Example 71). If that is the case, there is an opportunity to "resonate out", either fully or partially, the load's parasitic reactance with the matching network's components.

To illustrate the idea of "resonating out" an impedance, let us consider the following two examples: first, in Example 74 there is a parasitic capacitance in series with real load resistor; then, in Example 75 the parasitic capacitance is in parallel with the load resistor. In both cases we resonate out the total value of the load capacitance.

Example 74: Q Matching Technique, Resonating Out Series Capacitance ($R_0 = R_L$)
Match a 50 Ω resistive source at 10 MHz to a load that is a series connection of a 50 Ω resistor and a 100 pF capacitance (see figure below left).

Solution 74: This is a special case because the real parts of the source and load impedances are equal. That being the case, the idea is then to create a branch whose impedance at the given frequency is zero by using the technique of "resonating out" the existing 100 pF capacitance.

Because the otherwise already matched load has an extra capacitive (i.e. negative) reactance

$$X_{S_C} = \left|\frac{1}{j\omega C}\right| = -|j|\frac{1}{\omega C} = -\frac{1}{2\pi \times 10\,\text{MHz} \times 100\,\text{pF}} = -159.155\,\Omega$$

where negative signs in the result account for "$-j$" phase. In order to nullify the existing negative reactance we must add inductive (i.e. positive) serial reactance $X_{S_L} = +159.155\,\Omega$ towards the source side. In other words, at $f = 10\,\text{MHz}$, we need an inductor with

$$|j|\,L = \frac{X_{S_L}}{\omega} = \frac{+159.155\,\Omega}{2\pi \times 10\,\text{MHz}} = 2.533\,\mu\text{H}$$

where the module of j is accounted as positive inductor's reactance. At $f = 10\,\text{MHz}$ this serial L, C branch resonates and therefore presents the total of zero resistance between the source and load, see Fig. 11.15 (right).

Another way to look at this problem is to formally derive the same result by setting the serial impedance Z_{LC} of the inductor and capacitor to zero, which gives the required inductance. Thus we write

$$Z_{LC} = j\omega L + \frac{1}{j\omega C} = j\omega L - j\frac{1}{\omega C} = j\left(\omega L - \frac{1}{\omega C}\right)$$

Again, the above equation is remarkable because it shows that, at the given frequency and ideal L, C components, it is possible to make serial impedance Z_{LC} equal to *zero*. This is due to the complex nature of L and C components and destructive addition of signals. We recall that, in the complex plane, positive (i.e. "+j") and negative ("−j") directions of imaginary axis translate into phase difference of 180°, which is in effect the condition for destructive addition of two sinusoidal waveforms (see Sect. 1.4.3). That is to say,

$$Z_{LC} = 0 \Rightarrow \omega L - \frac{1}{\omega C} = 0$$

which is possible at a particular frequency $\omega = \omega_0$ that is calculated as

$$Z_{LC} = 0 \Rightarrow \omega L = \frac{1}{\omega C} \Rightarrow \omega_0 = \frac{1}{\sqrt{LC}} \tag{11.7}$$

As we already found in Chap. 10, frequency ω_0 in (11.7) is known as the "resonant frequency ω_0" and is not only one of the key properties of LC circuit but also oneof the fundamental

(continued)

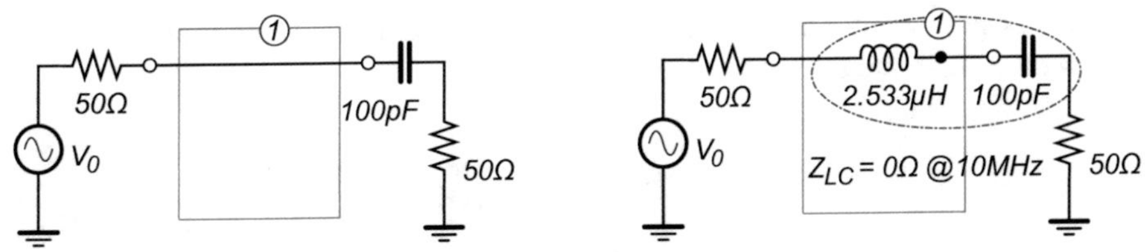

Fig. 11.15 Matching network in the case of serial C, R load impedance

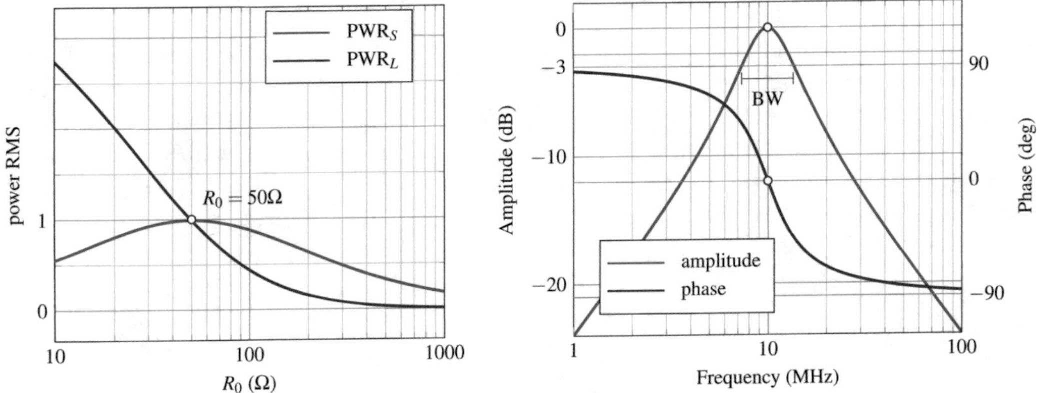

Fig. 11.16 SPICE simulation of matching network in Example 74 shows maximal power level (normalized) delivered by source is when $R_0 = 50\,\Omega$ and $R_L = 50\,\Omega$ (left). AC simulation confirms that power delivered at $f = 10\,\text{MHz}$ is indeed maximal (right)

relationships in physics. Therefore, for the given C, at the resonant frequency $\omega_0 = \omega = 10\,\text{MHz}$ we find

$$L = \frac{1}{\omega_0^2\, C} = \frac{1}{(2\pi\ 10\,\text{MHz})^2 \times 100\,\text{pF}} = 2.533\,\mu\text{H} \tag{11.8}$$

With (11.8) we formally confirm that the required matching network consists only of $X_S = 2.533\,\mu\text{H}$ inductor in series to 100 pF capacitor, see Fig. 11.15 (right), because at $f_0 = 10\,\text{MHz}$ it helps to create a zero impedance. This elegant one-additional-component solution illustrates the "resonating out" technique for matching network design in the case of serial RC load.

Finally, we note that in this case of serial RC complex load there is not much choice left; the original connection was already blocking DC signals, thus adding inductor in series does not change this situation.

Numerical simulation of matching circuit found in Example 74 shows that the use of resonating out technique enables short connection between source and load, Fig. 11.16. By sweeping the source resistance R_0 over two decades and measuring RMS power dissipated in the two real resistors, graph in Fig. 11.16 (left) confirms that the source delivers maximal power when $R_0 = 50\,\Omega$ that is at the same time equal to the power received by the 50 Ω load. At the same time AC simulation shows that maximal power is achieved at the desired $f = 10\,\text{MHz}$, Fig. 11.16 (right). We note that this time matching network has symmetrical AC transfer function relative to the centre frequency. Thus, the −3 dB bandwidth is measured as $BW = 7.5\,\text{MHz}$.

Example 75: Q Matching Technique, Resonating Out Parallel Capacitance ($R_0 = R_L$)
Match a 50 Ω resistive source at 10 MHz to a load that is a parallel connection of a 50 Ω resistor and a 100 pF capacitance (see Fig. 11.17 (left)).

Solution 75: Similar to Example 74 if somehow the reactance of 100 pF capacitor is removed, then there would be no need for matching because we would have the case of $R_0 = R_L$.

One way when that would be possible is if the capacitor's reactance were infinite, which would make the parallel (R_L, C) impedance equal to R_L. However, a finite capacitor's reactance equals to infinity only at DC, which is not the working frequency of this network.

We find another possible way to create an infinite impedance in parallel with R_L by recalling that an impedance of a parallel (L, C) network at resonance equals to infinity.[2]

Let us explore this idea of adding an inductor L in parallel to the loading capacitor C. The equivalent admittance $Y_{LC} = 1/Z_{LC}$ is calculated as the sum of parallel admittances, i.e.

$$Y_{LC} = \frac{1}{Z_{LC}} = \frac{1}{Z_L} + \frac{1}{Z_C} = \frac{1}{j\omega L} + j\omega C = \frac{1 + (j\omega C)(j\omega L)}{j\omega L} = \frac{1 - \omega^2 LC}{j\omega L}$$

$$\therefore$$

$$Z_{LC} = j\,\frac{\omega L}{1 - \omega^2 LC} \tag{11.9}$$

As we already found in Chap. 10, (11.9) is remarkable because it shows that it is possible to make impedance Z_{LC} of an ideal parallel LC branch equal to *infinity*. That is to say, if

$$1 - \omega^2 LC = 0 \Rightarrow Z_{LC} = \infty$$

which is possible at a particular frequency $\omega = \omega_0$ that is calculated as

$$1 - \omega^2 LC = 0 \Rightarrow \omega_0 = \frac{1}{\sqrt{LC}} \tag{11.10}$$

In comparison with result found in Example 74, as expected both serial and parallel connections of ideal L and C components, (11.7) and (11.10) resonate at same ω_0 frequency.

If (11.10) is satisfied, then the loading 50 Ω resistor perceives *infinite* resistance connected in parallel, that is to say, the total loading resistance becomes only $R_L = (50\,\Omega||\infty) = 50\,\Omega$. Given $f_0 = 10\,\text{MHz}$ and $C = 100\,\text{pF}$, from (11.10) we find

$$L = \frac{1}{(2\pi\, f_0)^2\, C} = \frac{1}{(2\pi \times 10\,\text{MHz})^2 \times 100\,\text{pF}} = 2.533\,\mu\text{H}$$

Therefore, the required matching network consists only of $X_P = 2.533\,\mu\text{H}$ inductor in parallel to 100 pF capacitor, as shown Fig. 11.17 (right), which at $f_0 = 10\,\text{MHz}$ effectively creates an infinite Z_{LC} impedance. This elegant one-additional-component solution illustrates the "resonating out" technique for matching network design in the case ofparallel RC load. We

(continued)

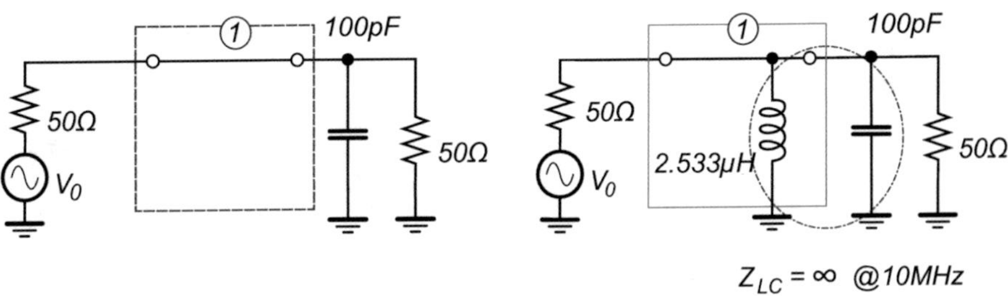

Fig. 11.17 Matching network in the case of parallel RC load impedance

note that with this solution, DC connection between source and load is inherently set by the short connection.

A slightly different case arises, for example, when the existing parasitic capacitance is greater than the one absorbed in the previous examples, and there is an opportunity to use either total or partial resonating out. In the following example we illustrate a case of 5 Ω source driving parallel RC load consisting of 50 Ω resistor and 1.055 nF parasitic capacitance (Fig. 11.18).

Example 76: Q Matching Technique, Resonating Out Parallel Capacitance ($R_0 < R_L$) Design a single-stage LC matching network at $f = 10\,\text{MHz}$ for the case of a source V_0 with a $R_0 = 5\,\Omega$ output resistance, which has to drive a $R_L = 50\,\Omega$ load resistance in parallel with $C_L = 1.055\,\text{nF}$, Fig. 11.19. The matching network is expected to maintain a DC path between the source and the load.

Solution 76: As we already found in Example 71, to match $R_0 = 5\,\Omega$ to $R_L = 50\,\Omega$ we need $X'_S = 238.732\,\text{nH}$ inductor and $X'_P = 954.910\,\text{pF}$ capacitor. However, the load impedance already includes parallel parasitic capacitance $C_L = 1.055\,\text{nF}$, which means that somehow we need to reduce it to the required $X'_P = 954.910\,\text{pF}$.

In general, there are two possible ways to approach this kind of problem. First possibility is to resonate out the total 1.055 nF value of the load capacitance, then to use the already calculated matching network from Example 71. Second possibility is to resonate out only the excess portion of 1.055 nF, i.e. 1.055 nF – 954.910 pF = 100 pF, and to leave only 954.910nF of the total capacitance. Let us explore the two possibilities.

1. *Total resonating out*: following same idea as in Example 75 and relationship (11.10), given $C_L = 1.055\,\text{nF}$ and $f_0 = 10\,\text{MHz}$, we calculate

$$L = \frac{1}{(2\pi\, f_0)^2\, C} = \frac{1}{(2\pi \times 10\,\text{MHz})^2 \times 1.055\,\text{nF}} = 240.098\,\text{nH} \approx 240.1\,\text{nH}$$

which creates condition that $Z_{LC} = \infty$ at $f_0 = 10\,\text{MHz}$. With this step we resonated out full value of load capacitance, thus creating situation equivalent to 5Ω real source driving

(continued)

[2]See Chap. 10.

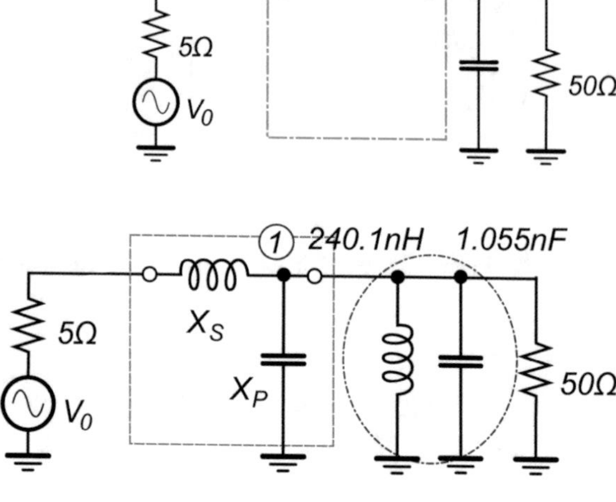

Fig. 11.18 Matching network in the case of parallel RC load when C is relatively large

Fig. 11.19 Matching network, the idea of fully resonating out the initial load capacitance

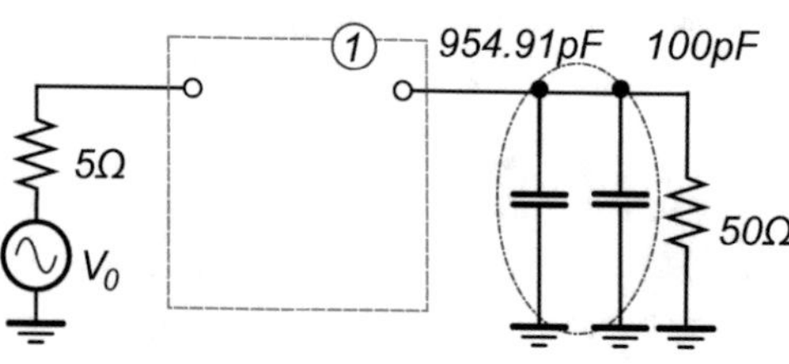

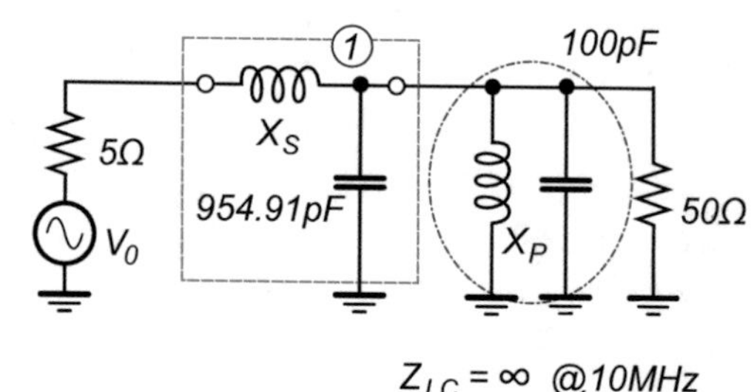

Fig. 11.20 Matching network, the idea of partial resonating out the initial load capacitance

50 Ω real load. Now we reuse results from Example 71, and add $X_S = 238.732\,\text{nH}$ and $X_P = 954.910\,\text{pF}$ capacitance as shown in Fig. 11.19. In total, three components are used: two inductors and one capacitor to create this matching network.

2. *Partial resonating out*: second idea that can be used is to imagine that the total 1.055 nF capacitance consists of two capacitors connected in parallel, that is, 954.910 pF + 100 pF, as shown in Fig. 11.20 (left).

   In order to create infinite impedance with the 100 pF portion of the loading capacitance, we calculate inductance

$$L = \frac{1}{(2\pi\, f_0)^2\, C} = \frac{1}{(2\pi \times 10\,\text{MHz})^2 \times 100\,\text{pF}} = 2.533\,\mu\text{H}$$

   that is shown as X_P in Fig. 11.20 (right). All that is left to do is to add serial inductance already calculated in Example 71, i.e. $X_S = 238.732\,\text{nH}$, as shown in Fig. 11.20 (right). Therefore this solution requires only two new components, inductors $X_S = 238.732\,\text{nH}$ and $X_P = 2.533\,\mu\text{H}$. Portion of the original loading capacitance is resonated out with X_P,

(continued)

while the leftover is used in combination with X_S inductance to create the same matching network as in Example 71.

Overall, both solutions are valid, while the differences are strictly practical. The decision between the three components for the first solution, or two components for the second solution is now a matter of practicality, size, price.

11.3 Bandwidth of LC Matching Network

So far in our discussion of single-stage LC matching networks, we have only focused on the main goal of matching the source side impedance to the load side impedance. We had no freedom to control the bandwidth of the overall network. We have learned by now that a general RLC network always behaves as a bandpass filter centred around the resonant frequency ω_0, which is determined by the LC components. We also have learned that, as a good approximation (assuming Q factor larger than ten or so), the serial and parallel RLC networks resonate at the same frequency (see Sect. 10.5). In any case, it is important to estimate the bandwidth of matching networks, because we may reach a wide-bandwidth solution (that allows too much noise into the system) or narrow bandwidth (that alters the frequency content of the passing signals, i.e. the matching network itself distorts the signal).

A more detailed network analysis, which is beyond the scope of this book, would have revealed that determining the network bandwidth using the standard definition based on the −3 dB points turns out to be problematic, to say the least. There are at least two good reasons for this statement. One has to do with the fact that some resonant networks may never reach the −3 dB attenuation points. For example, a low Q resonant curve is almost flat—it may not even have 3 dB difference between its maximum amplitude at the point of the resonant frequency and the side points. A second, and less obvious, reason for our difficulties in determining the 3 dB bandwidth of LC matching networks unambiguously is that, in general, resonant curves are not symmetrical around the resonant frequency. Nevertheless, in this book we assume high Q resonant networks (which is a reasonable assumption in the case of wireless radio) and we assume symmetrical resonant curves (which is a valid assumption in the narrow region around the resonant frequency point).

11.3.1 Hand-Calculation of Bandwidth

To illustrate the point, let us estimate bandwidth for matching network of Example 71, but instead of numerical simulation we use hand calculations, Fig. 11.21 (left). To do so, we use a technique of converting serial RL branches into parallel. The goal is to create the equivalent RLC network where all three components are in parallel. By doing so, it is then possible to use relation

$$Q = \frac{f_0}{\text{BW}} \tag{11.11}$$

Thus, we convert the serial source impedance $Z_0 = R_0 + j\omega L_S = (5 + j15)\,\Omega$ subnetwork into its equivalent parallel subnetwork (at $f = 10\,\text{MHz}$). Relationship between serial and parallel

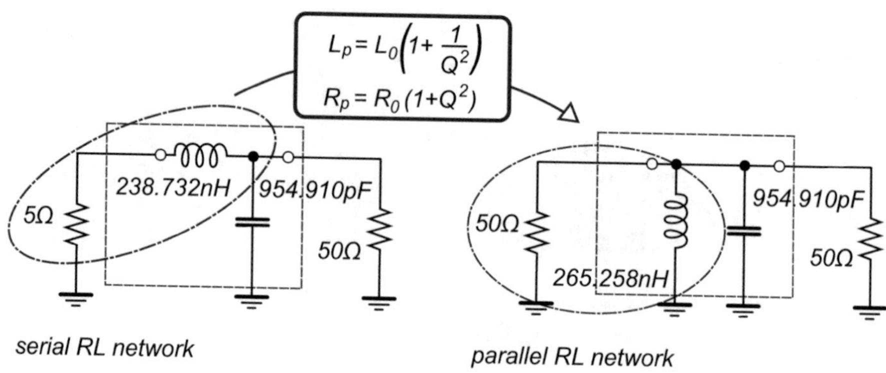

Fig. 11.21 Converting the serial RL subnetwork portion of the matching network (left) into its equivalent parallel RL subnetwork (right), for purposes of calculating the overall bandwidth

components is given by[3]

$$R_p = R_0\left(1 + Q^2\right) \tag{11.12}$$

$$L_p = L_0\left(1 + \frac{1}{Q^2}\right) \tag{11.13}$$

In this particular case, first we calculate Q_S factor of the equivalent input side non-ideal inductor consisting of serial (R_0, L_S) network as ratio of its complex and real impedances, i.e.

$$Q_S = \frac{X_S}{R_0} = \frac{|j\,\omega\,L_S|}{R_0} = \frac{2\pi \times 10\,\text{MHz} \times 238.732\,\text{nH}}{5\,\Omega} = \frac{15}{5} = 3$$

$$\therefore$$

$$R_p = 5\,\Omega\left(1 + 3^2\right) = 50\,\Omega$$

$$L_p = 238.732\,\text{nH}\left(1 + \frac{1}{3^2}\right) = 265.258\,\text{nH}$$

which is shown in Fig. 11.21 (right). Again, note that we calculated the Q_S factor for the stand-alone RL branch by itself. After this conversion, we see that a parallel LC resonant circuit is loaded by two 50 Ω resistors also in parallel, as in Fig. 11.21 (right), which makes the total resistance equal to $R_{\text{loaded}} = 50\,\Omega || 50\,\Omega = 25\,\Omega$ that completes the equivalent parallel RLC network.

In order to determine the 3 dB bandwidth of the matching circuit we need to calculate the Q factor of the "loaded" RLC network at resonance. By using either the capacitive or inductive reactance at resonance, we write as usual

$$Q_{\text{loaded}} = \frac{R_{\text{loaded}}}{X_P} = \frac{25\,\Omega}{16.667\,\Omega} = 1.5$$

$$\therefore$$

[3] See Sect. 10.5.

$$\Delta f = \frac{f_0}{Q_{\text{loaded}}} = \frac{10\,\text{MHz}}{1.5} = 6.67\,\text{MHz} \tag{11.14}$$

which although not perfectly accurate still provides a very good estimate of the bandwidth. Numerical simulations, see Fig. 11.9, show that result (11.14) underestimates the actual bandwidth by approximately 15% relative to the simulated 7.5 MHz, which is expected due to the assumptions and approximations being made.

11.3.2 Multisection Impedance Matching

Design of a single-stage matching network is constrained in a sense that after setting up LC network that is driven by source/load impedance ratio and the working frequency, the resulting bandwidth is already determined. There are circumstances, however, when we want to design a narrow bandwidth matching network. For example, input stages of RF amplifiers should be limited to only the minimum necessary bandwidth. Or, there are applications where it is necessary to increase the single-stage network's bandwidth.

In order to gain control over that parameter, we need to expand our single-stage, L-shaped matching network and add a second section. In principle, the two-section matching network is solved by repeating two times the methodology that we used for single-section networks. The additional step of freedom is achieved by introducing a temporary loading resistance R_{INT} (see Fig. 11.22). This allows us to split the two-section matching network problem into two single-section matching networks, where R_{INT} serves as the temporary load for the first section and as the source resistance for the second section.

11.3.2.1 Two-Stage Network: Bandwidth Increase

When the goal is to increase the existing single-stage network's bandwidth, we set the value of the temporary resistance R_{INT} to

$$\frac{R_0}{R_{\text{INT}}} = \frac{R_{\text{INT}}}{R_L} \quad \therefore \quad R_{\text{INT}} = \sqrt{R_0 R_L} \tag{11.15}$$

which is the geometrical mean between the source R_0 and load R_L resistances. The addition of the second section using condition (11.15) provides an optimal compromise in increasing the bandwidth.

While designing two or multi-stage matching networks, we have almost arbitrary freedom to pick the value of the temporary resistance R_{INT} parameter. However, in practice the decision which value of R_{INT} to use depends on the impedance levels of the terminations and practically realizable component values. For example, if the existing values of $[R_0, R_L]$ are already low, then it is more practical to

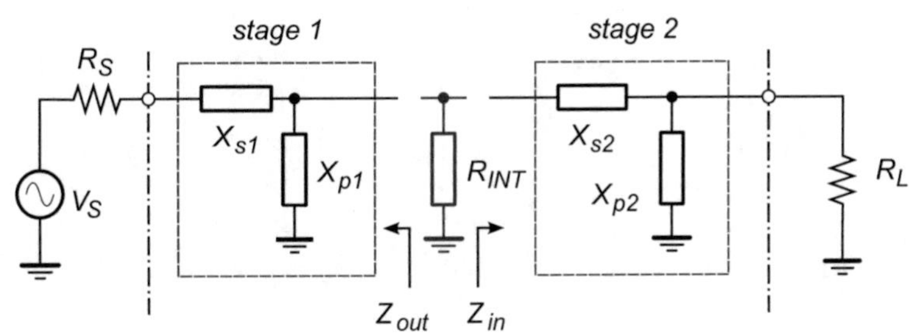

Fig. 11.22 Two-stage matching network with an intermediate temporary resistance R_{INT} for the purpose of intermediate calculations

select temporary resistance on the high side, otherwise we just pick a low value for the temporary resistance. Aside from these notes, there is nothing special about the two (or even multi) stage matching networks. It is good engineering practice however to design circuits with minimum number of components, hence, as the last step in the design of two-section matching networks multiple serial inductances should be replaced by a single component, as well as multiple parallel capacitances that should be replaced by a single capacitor.

Once again, the temporary resistance R_{INT} is a "ghost" value, not a real physical component; it is merely a number that would have been seen by looking into the matching network if it were split at the middle. In any case we follow the same idea: if branch resistance needs to be increased we add serial impedance, if branch resistance needs to be reduced we add parallel impedance. In the following two Examples 77 and 78 we illustrate design of two matching networks, one to increase and one to decrease the initial bandwidth.

Example 77: Q Matching Technique, Two-Stage Network, Increasing Bandwidth
Match 5 Ω source resistance to 50 Ω load resistance at 10 MHz. Design two-stage LC matching network while maintaining DC connection between source and load. One of the goals of this matching network design is to increase the bandwidth relative to the equivalent single LC matching network solution.

Solution 77: When one of the design goals is to increase bandwidth of the matching network then the intermediate resistance value is set using (11.15), i.e. geometrical mean of the two termination resistances. Then, the two-stage design problem is reduced to design of two single-stage LC matching networks, the first one designed to match source resistance R_S to R_{INT}, and the second one to match R_{INT} to the load R_L. Once again, resistance R_{INT} is *not a real resistor* component added in the network, it is only a fictitious number used to set impedance of the inner node in between the two matching network stages. For the given data given in this example, the matching network is designed as follows.

1. Calculation of the intermediate resistance:

$$R_{INT} = \sqrt{R_S\, R_L} = \sqrt{5\,\Omega \times 50\,\Omega} = 15.811\,\Omega \tag{11.16}$$

2. Calculation of Q factor at R_{INT} looking left into the first stage, and looking right into the second stage:

$$Q = \sqrt{\frac{R_{INT}}{R_S} - 1} = \sqrt{\frac{15.811\Omega}{5\Omega} - 1} = 1.470$$

$$Q = \sqrt{\frac{R_L}{R_{INT}} - 1} = \sqrt{\frac{50\Omega}{15.811\Omega} - 1} = 1.470$$

3. Calculation of first stage X_{S1} and X_{P1} components: matching 5–15.811 Ω results in

$$X_{S1} = Q\, R_S = 1.47 \times 5\,\Omega = 7.352\,\Omega \quad \therefore \quad L_{S1} = \frac{X_{S1}}{2\pi\, f} = 117\,\text{nH}$$

$$X_{P1} = \frac{R_{INT}}{Q} = \frac{15.811\,\Omega}{1.47} = 10.753\,\Omega \quad \therefore \quad C_{P1} = \frac{1}{2\pi\, f\, X_{P1}} = 1.480\,\text{nF}$$

(continued)

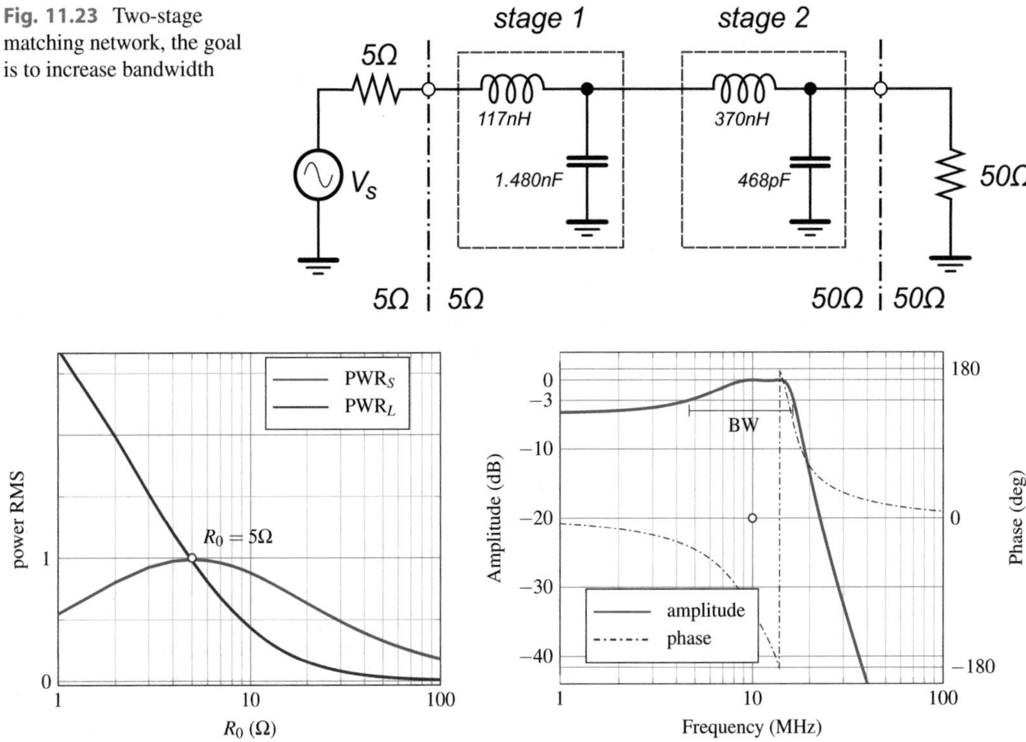

Fig. 11.23 Two-stage matching network, the goal is to increase bandwidth

Fig. 11.24 SPICE simulation of matching network in Example 77 shows maximal power level (normalized) delivered by source is when $R_0 = 5\,\Omega$ and $R_L = 50\,\Omega$ (left). AC simulation confirms that power delivered at $f = 10\,\text{MHz}$ is indeed maximal (right) while the bandwidths are enlarged to approximately 12 MHz

where series inductor and parallel capacitor maintain DC connection in the first stage.

4. Calculation of second stage X_{S2} and X_{P2} components: matching 15.811–50 Ω results in

$$X_{S2} = Q\,R_{INT} = 1.47 \times 15.811\,\Omega = 23.250\,\Omega \quad \therefore \quad L_{S2} = 370\,\text{nH}$$

$$X_{P2} = \frac{R_L}{Q} = \frac{50\,\Omega}{1.47} = 34\,\Omega \quad \therefore \quad C_{P1} = 468\,\text{pF}$$

The complete schema of this two-stage matching network is shown (Fig. 11.23).

Numerical simulation of matching circuit shown in Example 77 shows that the use of two-stage matching network provides a means to control bandwidth as well, Fig. 11.24. By sweeping the source resistance R_0 over two decades and measuring RMS power dissipated in the two real resistors, graph in Fig. 11.24 (left) confirms that the source delivers maximal power when $R_0 = 5\,\Omega$ that is at the same time equal to the power received by the 50 Ω load. At the same time AC simulation shows that maximal power is achieved at the desired $f = 10\,\text{MHz}$, Fig. 11.24 (right). In addition, comparison of the newly achieved bandwidth shows that relative to the initially obtained $BW \approx 7.5\,\text{MHz}$ in Example 71, this network allows $BW \approx 12\,\text{MHz}$, Fig. 11.24 (right).

We note that, as a rule, numerical values of matching network's LC components are rather precise and non-standard numbers. For that reason, in practice it is almost not possible to design antenna/RF amplifier matching network with the standard components. In addition, discrete components are manufactured within a certain tolerances, normally $\pm 1\%$, $\pm 5\%$, or $\pm 10\%$, which further complicates the matching network design that is to be centred at one RF carrier frequency. Practical approach, accordingly, is to use tuneable capacitors and custom-made inductors. As a result, there is an additional experimental effort required to tune LC values to their desired values.

11.3.2.2 Two-Stage Network: Bandwidth Reduction

When the goal is to decrease the already existing bandwidth of a single-stage matching network, instead of using relation (11.15) we need to select R_{INT} value outside of the interval $[R_S, R_L]$ of resistances. Thus, depending how R_{INT} is related to R_S, R_L resistances, there are two possibilities, either

$$R_{\text{INT}} > \max(R_S, R_L) \tag{11.17}$$

or

$$R_{\text{INT}} < \min(R_S, R_L) \tag{11.18}$$

which allows for practically arbitrary choice of R_{INT} that results in various bandwidths, Fig. 11.25. However, if the starting resistances R_S, R_L are already low it may not be practical to use the (11.18) relationship.

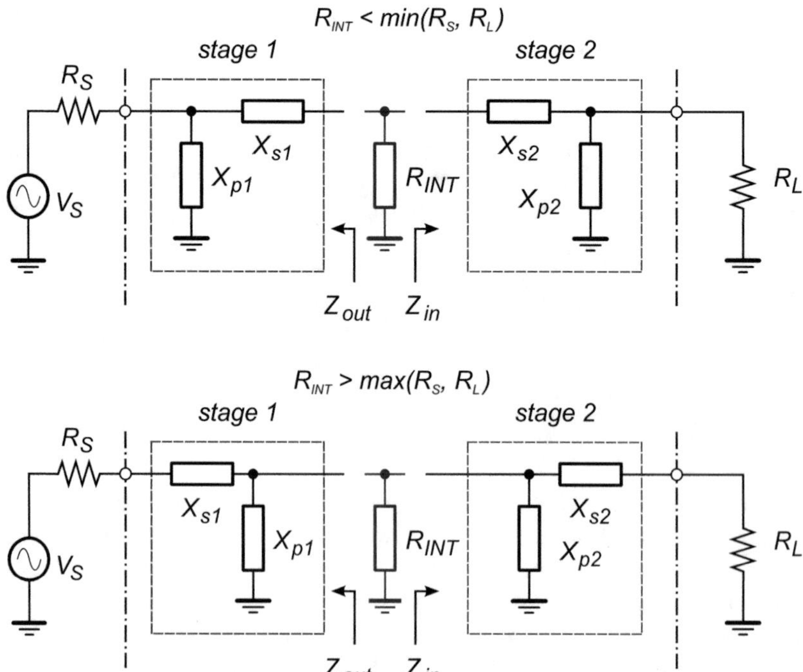

Fig. 11.25 Two-stage matching network design concept, $Z_{out} = R_{INT} = Z_{in}$. We note that if branch resistance needs to be increased we add serial impedance, if branch resistance needs to be reduced we add parallel impedance

In the following example, this technique based on "ghost" intermediate resistance R_{INT} is used to decrease the default bandwidth of matching network obtained in Example 71.

Example 78: Q Matching Technique, Two-Stage Network, Decreasing Bandwidth
Match 5 Ω source resistance to 50 Ω load resistance at 10 MHz by designing a general two–stage matching network. In this example, the goal is to decrease the bandwidth relative to the single LC matching network solution. Choose two ghost resistances R_{INT} and compare the resulting bandwidths.

Solution 78: To illustrate the bandwidth control, in this example we compare two possible bandwidths, first, when $R_{\mathrm{INT}} = 250\,\Omega$, then $R_{\mathrm{INT}} = 985\,\Omega$.

Case 1: $R_{\mathrm{INT}} > \max(R_S, R_L) = 250\,\Omega$, i.e. the ghost resistance is chosen as a convenient number greater than 50 Ω, for example 250 Ω.

1. Note that the following two calculated Q values, one looking left into the source side and one looking right into the load side, are not equal. However, the overall bandwidth is limited by the higher Q (i.e. narrower bandwidth), which is the one dominating the design.

$$Q_1 = \sqrt{\frac{R_{INT}}{R_S} - 1} = \sqrt{\frac{250}{5} - 1} = 7.0 \qquad Q_2 = \sqrt{\frac{R_{INT}}{R_L} - 1} = \sqrt{\frac{250}{50} - 1} = 2.0$$

   thus, $Q_1 = 7$ is the one limiting the total bandwidth.

2. First stage: Matching 5–250 Ω results in

$$X_{s1} = Q_1\, R_S = 7 \times 5\,\Omega = 35\,\Omega \qquad X_{p1} = \frac{R_{INT}}{Q_1} = \frac{250\,\Omega}{7} = 35.714\,\Omega$$

3. Second stage: Matching 250–50 Ω results in

$$X_{s2} = Q_2\, R_L = 2 \times 50\,\Omega = 100\,\Omega \qquad X_{p2} = \frac{R_{INT}}{Q_2} = \frac{250\,\Omega}{2} = 125\,\Omega$$

4. Considering solution, for example, that keeps DC connection between source and load, serial impedances are converted to inductances and parallel impedances into capacitances (after light rounding of numbers),

(continued)

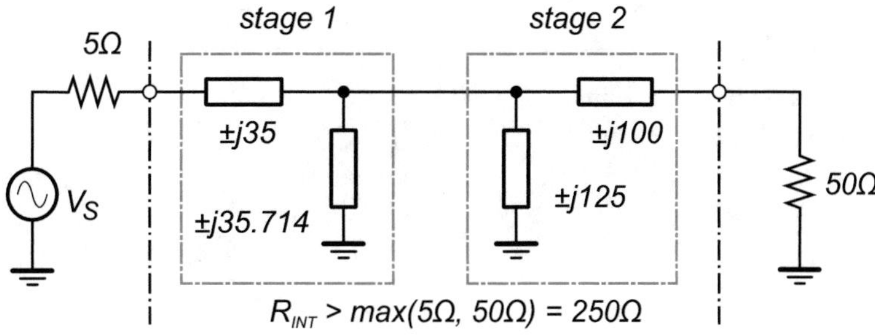

Fig. 11.26 Two-stage matching network, the goal is to decrease bandwidth

$$L_{s1} = \frac{X_{s1}}{2\pi f} = \frac{35\,\Omega}{2\pi\ 10\,\text{MHz}} = 557\,\text{nH} \quad C_{p1} = \frac{1}{2\pi\ f\ X_{p1}} = \frac{1}{2\pi\ 10\,\text{MHz} \times 35.714\,\Omega} = 445.6\,\text{pF}$$

$$L_{s2} = \frac{X_{s2}}{2\pi f} = \frac{100\,\Omega}{2\pi\ 10\,\text{MHz}} = 1.592\,\mu\text{H} \quad C_{p2} = \frac{1}{2\pi\ f\ X_{p2}} = \frac{1}{2\pi\ 10\,\text{MHz} \times 125\,\Omega} = 127.3\,\text{pF}$$

5. Thus, the complete two-stage matching network schematic may be reduced to T-type network after two parallel capacitors are combined into a single $C = 572.9\,\text{pF}$ component, (Fig. 11.26).

Case 2: $R_{\text{INT}} > \max(R_S, R_L) = 985\,\Omega$, i.e. the ghost resistance is chosen as a convenient number greater than $50\,\Omega$, for example $985\,\Omega$.

1. We repeat the same calculations, however using larger "ghost" resistor, therefore we find the new values as follows.

$$Q_1 = \sqrt{\frac{R_{INT}}{R_S} - 1} = \sqrt{\frac{985}{5} - 1} = 14.0 \qquad Q_2 = \sqrt{\frac{R_{INT}}{R_L} - 1} = \sqrt{\frac{985}{50} - 1} = 4.3$$

   thus, $Q_1 = 14$ is the one limiting the total bandwidth.
2. First stage: Matching 5–250 Ω results in

$$X_{s1} = Q_1\, R_S = 14 \times 5\,\Omega = 70\,\Omega \qquad X_{p1} = \frac{R_{INT}}{Q_1} = \frac{250\,\Omega}{14} = 70.4\,\Omega$$

3. Second stage: Matching 250–50 Ω results in

$$X_{s2} = Q_2\, R_L = 4.3 \times 50\,\Omega = 216.2\,\Omega \qquad X_{p2} = \frac{R_{INT}}{Q_2} = \frac{985\,\Omega}{4.3} = 227.8\,\Omega$$

4. Again, by keeping DC connection between source and load, serial impedances are converted to inductances and parallel impedances into capacitances (after light rounding of numbers),

$$L_{s1} = \frac{X_{s1}}{2\pi f} = \frac{70\,\Omega}{2\pi\ 10\text{MHz}} = 1.114\,\mu\text{H} \quad C_{p1} = \frac{1}{2\pi\ f\ X_{p1}} = \frac{1}{2\pi\ 10\,\text{MHz} \times 70.4\,\Omega} = 226.2\,\text{pF}$$

$$L_{s2} = \frac{X_{s2}}{2\pi f} = \frac{216.2\,\Omega}{2\pi\ 10\,\text{MHz}} = 3.441\,\mu\text{H} \quad C_{p2} = \frac{1}{2\pi\ f\ X_{p2}} = \frac{1}{2\pi\ 10\,\text{MHz} \times 227.8\,\Omega} = 69.87\,\text{pF}$$

   where the resulting circuit is as in Fig. 11.27, except for the new component values. Again, two parallel capacitors in the middle are combined into a single $C = 296\,\text{pF}$ component.

Numerical AC simulation of the two matching networks in this example clearly illustrates control of the overall bandwidth. Higher Q limits bandwidth, in this case, to 1.1 MHz, while lower Q permits 2.5 MHz, see figure below. In both cases, the network's bandwidth is reduced relative to 7.5 MHz in Example 71 (Fig. 11.28).

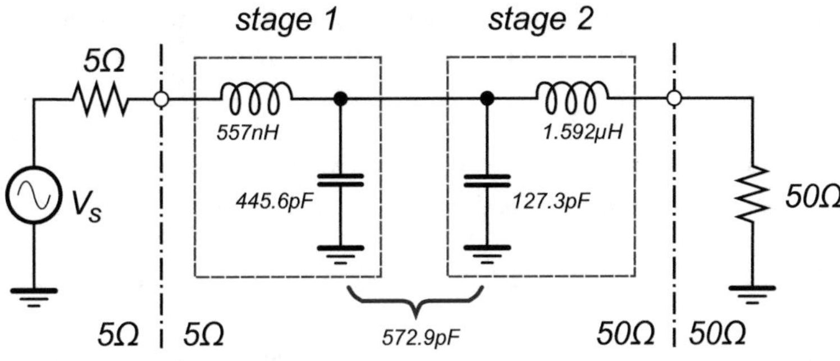

Fig. 11.27 Two-stage matching network, final version

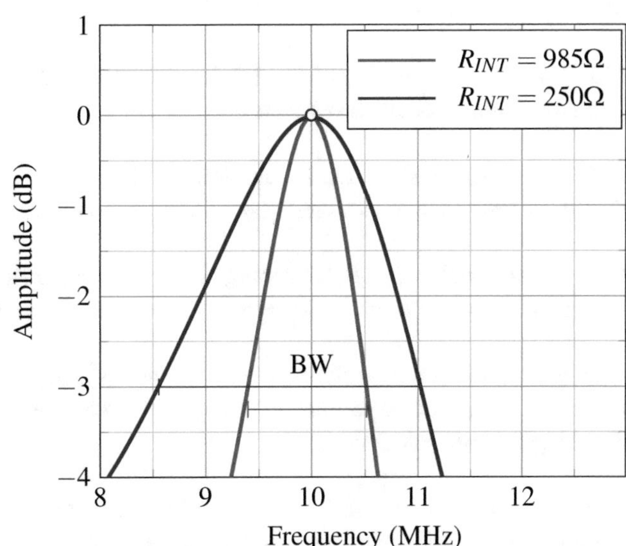

Fig. 11.28 Two-stage matching network: AC simulation comparison of Cases 1 and 2 networks (Example 78)

11.4 Summary

In this chapter, we familiarized ourselves with the basic concepts of power transfer between stages of a general multi-stage system. Having the main motivation of designing networks for use in wireless radio systems, it became our main priority to maximize power transfer of the RF signals. In order to achieve that goal, we introduced the concept of matching networks that serve as the gradual impedance converter between the source impedance and the load impedance. Their first application is between the antenna and the RF amplifier. We keep in mind that Q matching is just one of several practical techniques for matching network design and the solutions presented in this chapter are only some of many possibilities. The next logical step is to start using Smith Charts and other numerical tools for optimization, which offer a somewhat more elegant ways of designing RF matching networks, especially at higher frequencies.

Problems

11.1 Using Q matching technique, find equivalent parallel network to serial connection of $R_S = 5\,\Omega$ and $L_S = 2.8\,\text{nH}$ at $f = 100\,\text{MHz}$.

11.2 Design a single-stage LC matching network in between source with $R_S = 5\,\Omega$ and load $R_L = 50\,\Omega$ termination at $f = 100\,\text{MHz}$. Additional condition is to maintain DC connection between the source and load side.

11.3 Using results from Problem 11.2, find reflection coefficient Γ and mismatch loss ML at the interface between the serial and parallel parts of the matching network.

11.4 Using results from Problem 11.2, estimate the 3 dB bandwidth, assuming symmetrical network.

11.5 Using results from Problem 11.3, if the input signal changes to $f = 80\,\text{MHz}$ recalculate Γ and ML. Can you comment on the results?

11.6 Using results from Problem 11.3, if the input signal changes to 0.2 GHz recalculate Γ and ML. Can you comment on the last results?

11.7 Design single-stage LC matching network when parasitic inductance L_S exists in series with the source resistance R_S, where $R_S = 5\,\Omega$, $R_L = 50\,\Omega$, $L_S = 0.93\,\text{nH}$, $f = 0.85\,\text{GHz}$.

11.8 Design single-stage LC matching network when parasitic capacitance C_L exists in parallel with the load resistance R_L, where $R_S = 5\,\Omega$, $R_L = 50\,\Omega$, $C_L = 20\,\text{pF}$, $f = 0.1\,\text{GHz}$.

11.9 Match $5\,\Omega$ source resistance to $50\,\Omega$ load resistance at 200 MHz. Use two-stage LC matching network, LP-HP filter combination. The goal is to increase the bandwidth relative to the single LC matching network solution.

11.10 Match $5\,\Omega$ source resistance to $50\,\Omega$ load resistance at 200 MHz. Use two-stage LC matching network. The goal is to decrease the bandwidth relative to the single LC matching network solution. Make your own choice, and justify it, of the temporary resistance R_{tmp}.

11.11 Antenna impedance is assumed to be resistive $50\,\Omega$. Tuned RF amplifier is tuned at 665 kHz and has input impedance of $Z_{\text{in}} = 2\,\text{k}\Omega$. Design two possible matching networks using Q matching technique and comment on differences between the two solutions.

11.12 Using the parasitic absorption method match source impedance of $Z_S = (50 + j100)\,\Omega$ to a load impedance of $Z_L = (1000 - j750)\,\Omega$ (the capacitor is in parallel to R_L) at 100 MHz.

11.13 Using the parasitic resonance method, match source resistance of $R_S = 50\,\Omega$ to a load impedance that consists of a $C_L = 10\,\text{pF}$ in parallel to $R_L = 500\,\Omega$ at 100 MHz. The matching circuit should maintain DC connection from the input to the output.

RF and IF Amplifiers

12

After a weak radio frequency (RF) signal has arrived at the antenna, it is channelled to the input terminals of the RF amplifier through a passive matching network. The matching network is made to enable maximum power transfer of the receiving signal by equalizing the antenna impedance with the RF amplifier input impedance. After that, it is job of the RF amplifier to increase the power of the received signal and prepare it for further processing. In the first part of this chapter, we review the basic principles of linear baseband amplifiers and common circuit topologies. In the second part of the chapter, we introduce RF and IF amplifiers. Aside from their operating frequency, for all practical purposes, there is not much difference between the schematic diagrams of RF and IF amplifiers. In this book, unless we need to specifically separate the two functions, we refer to all tuned amplifiers as RF amplifiers.

12.1 Tuned Amplifiers

The class of "baseband amplifiers" assumes low-frequency operation, which is to say that all single-tone signals from DC to "not so high" frequencies are treated as equally interesting, desirable, and possible to amplify. As a result, aside from frequency limitations caused by LP filtering effects at the input stage of CE amplifiers due to the Miller effect or by high-pass/low-pass filtering effects due to the AC coupling at the RC input stage network, baseband amplifiers keep spending energy to amplify all possible tones that arrived at the amplifier's input terminal nodes.

This generous approach to signal amplification, aside from strictly technical difficulties, has at least two serious drawbacks if it is to be used for amplification of RF signals, see Fig. 12.1:

1. The frequency of the content of the message, for example the human voice, is limited to the range 20 Hz to 20 kHz. That is, high-fidelity (HiFi) sound reproduction does not require any of the tones outside of this frequency range and, strictly speaking, it is a waste of energy to amplify them. The amplification energy must come from somewhere and, by doing so, we unnecessarily drain the amplifier's batteries that must be charged and/or replaced more often. To make the things worse, we may need to provide an additional cooling mechanism to dissipate the excess heat generated by the components, not to mention the impact on the environment of disposing of the drained batteries.

R. Sobot, *Wireless Communication Electronics*,
https://doi.org/10.1007/978-3-030-48630-3_12

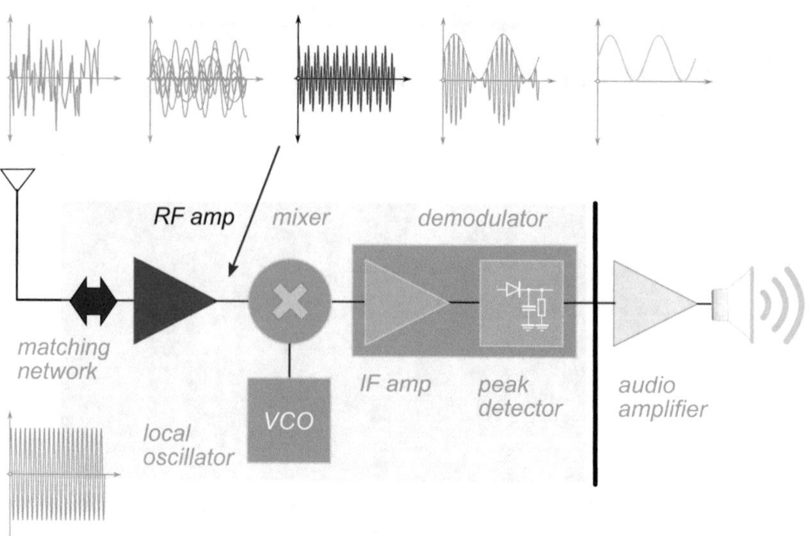

Fig. 12.1 Heterodyne AM radio receiver architecture—RF amplifier

2. Equally important drawback of wideband amplification is that all unwanted tones accepted into the amplifier contribute only to the increased noise level. After all, these tones are not needed and not desirable, hence they represent noise. The amplifier cannot possibly know what tones the user wants to hear, hence all tones are equally amplified. Because the overall noise energy is collected over a wider band than is necessary, it means the overall SNR is lowered.

There are other reasons why baseband amplifiers are not used in RF sections of radios. For the time being, the above two arguments should provide enough arguments to conclude that the overall result of using a baseband amplifier for radio signal amplification results in an expensive, bulky, power-hungry, and lesser quality RF amplifier. All that assumes we somehow managed to make the amplifier's bandwidth wide enough to start with. After recalling that modern RF carrier frequencies are in the order of MHz or GHz it becomes more evident that a wideband amplifier would have to be able to work from DC to the RF carrier frequency, which at least to say difficult and may not be even possible to do with the current technology.

In Chap. 11 we learned that in order to efficiently transport EM energy collected by an antenna to the input terminals of an amplifier, we need to design a matching network. At that time, we quietly accepted that the maximum power transfer was possible, in theory, at only one frequency, and in practice over a rather narrow range of frequencies determined by the Q factor of the matching network. In Sects. 10.2 and 10.8.1 we learned that any realistic RLC network behaves as a "bandpass filter".

The first three stages of a radio receiver, the antenna, the matching network, and the RF amplifier (see Fig. 12.1) are often referred to as the front end of the RF radio receiver. It is now time to ask how exactly an RF amplifier is different from a baseband amplifier. Before answering this question, let us first state that

(a) The frequency range of operation of an RF amplifier must be "aligned" (i.e. tuned) with the centre frequency of the matching network, which, in turn, is tuned with the antenna.
(b) The bandwidth of the RF amplifier should be similar to the bandwidth of the incoming message, approximately 20kHz in the case of music. That is, the RF amplifier bandwidth should be not too wide to cause a decrease of SNR, or too narrow to introduce signal distortions (recall Fourier).

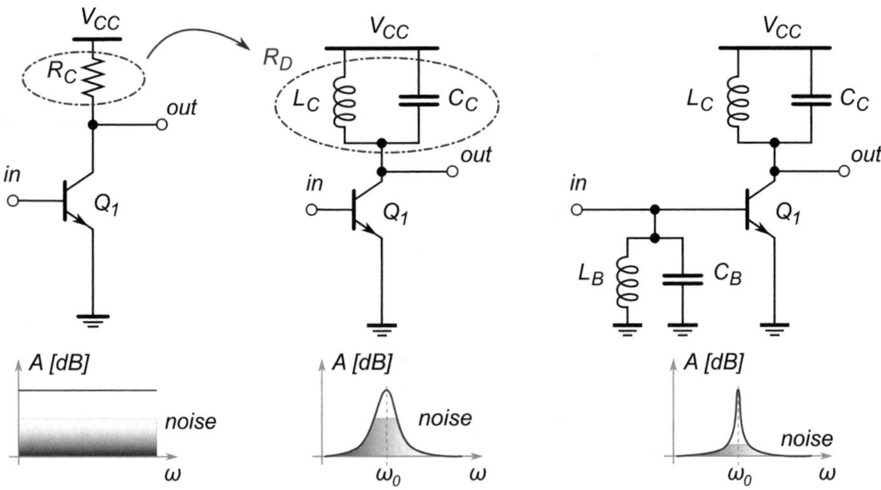

Fig. 12.2 CE amplifier (left) and its equivalent RF CE amplifier (centre and right)

Of the three single transistor amplifier types (i.e. CE, CB, and CC), the emitter follower is the only one that has voltage gain slightly less than unity, therefore we focus on the other two variants.

12.1.1 Single-Stage CE RF Amplifier

In principle, a single-stage wideband CE amplifier in Fig. 12.2 (left) is turned into a CE RF (i.e. narrowband) amplifier with two modifications. First, the collector's resistive load R_C is replaced by an $L_C\,C_C$ resonator as in Fig. 12.2 (centre). In Sect. 10.8.2 we learned how to estimate the equivalent dynamic resistance R_D of a resonator at the resonant frequency. It is to be noted that LC resonator is a BPF at the *output* node, thus all frequency components of the signal entering the amplifier are equally amplified, then filtered. Assuming, if $L_C\,C_C$ resonator is designed so that $R_D = R_C$, then the voltage gain $|A_v| = g_m R_C = g_m R_D$ (however, only at the resonant frequency ω_0).

Second, if $L_B\,C_B$ resonator is connected between the input node and the ground, Fig. 12.2 (right), then the input side signal bandwidth is limited *before* entering the amplifier and the overall effect is that both bandwidth and noise level are reduced. Both resonators are tuned to the same resonant frequency ω_0.

12.1.1.1 Intuitive View of CE RF Amplifier Operation

At the resonant frequency, $\omega_0 = 1/\sqrt{LC}$, both LC resonators are effectively equivalent to their respective dynamic resistances R_D. We already know that in the case of theoretically ideal LC components, a resonator's dynamic resistance R_D is infinite, while in the case of real LC components, the dynamic resistance RD is calculated[1] as $R_D = Q\,Z_L(\omega_0)$. Due to this (in the ideal case) infinite resistance at ω_0 connected to the input side node the amplifier does not "feel" any additional resistive load. In other words, there is no current splitting at the base node and 100% of the AC signal current is injected into the base. The only ramification of adding an ideal $L_B\,C_B$ resonator is that out of all possible frequencies only a single tone at ω_0 is able to pass through the $L_B\,C_B$ "entrance door" and

[1]We could have used Z_C instead; at the resonance, $Z_L = Z_C$, hence the dynamic resistance R_D is calculated as a product of the Q factor and either of the two impedances.

enter the transistor gate, all the other tones with frequencies $\omega \neq \omega_0$ are simply not aligned with the door and they "hit the wall", i.e. they are (ideally) attenuated down to zero amplitude. In the case of real $L_B C_B$ components, the entrance door is wider than a single frequency; hence not only ω_0 passes through but also the adjacent frequencies that are within the "width of the door". In technical terms, the input $L_B C_B$ resonator works as a narrowband bandpass filter, whose centre frequency is ω_0 and bandwidth $BW = \Delta\omega$, where $\Delta\omega = \omega_0/Q$ (note that if $Q = \infty \Rightarrow BW = 0$). This is how the amount of noise entering the amplifier through the input terminal is controlled.

At the same time, at the output side of the CE RF amplifier, the collector is experiencing very large resistive load R_D, which translates into large voltage gain $A_v = g_m R_D$. The transistor's transconductance g_m is set by the biasing network (not shown). Therefore, in the ideal case ($R_D = \infty$), the amplifier would achieve an infinite voltage gain, i.e. it would be able to amplify even an infinitely small single-tone signal at exactly ω_0 frequency and would "ignore" all other tones.[2] This ability to select and amplify only tones close to the resonant frequency while at the same time it "ignores" all other waveforms at frequencies not equal to ω_0 is the principal reason for using RF amplifiers.

In the real case, however, the gain is limited by finite R_D within the finite bandwidth BW but it is, nevertheless, still very high. Although the input side $L_B C_B$ resonator blocks all unwanted frequencies from entering the amplifier, we already know that there is internally generated noise that also needs to be filtered out by the $L_C C_C$ resonator. By means of these two LC resonators (effectively a double bandpass filter) along the signal path, the gain of the CE RF amplifier is optimized so that only the frequencies of interest within the bandwidth are amplified.

Although the intuitive interpretation of CE amplifier operation given so far ignores a few fine details, it is definitely useful in terms of understanding the overall functionality of a theoretical CE RF amplifier. With its high input resistance and high output current gain, the CE amplifier is considered one of the most important structures for voltage signal amplification. Let us now find out about the limitations of a simple CE RF amplifier and what needs to be done in order to make it truly practical at RF frequencies of interest.

12.1.1.2 Miler Effect

We already found out that the small bridging capacitance C_{CB} creates a feedback loop from the output terminals back to the input terminals, Fig. 12.3. As it turned out, the Miller capacitance is perceived by the input terminals as approximately A_v times greater than the real collector–base C_{CB} capacitance, with the consequence of a LP filter effect on the input side and drastic reduction of the signal bandwidth.

It would appear from the frequency analyses in Sect. 7.4.5 that a CE amplifier is hopelessly lost for all but low- to mid-frequency range RF applications. However, we can improve the frequency-dependent behaviour of a CE amplifier in RF applications by looking at the three conditions for the Miller effect one by one. Obviously, we cannot do anything about its inherent inverting signal nature and we do need to keep the high voltage gain. For all practical purposes, we cannot remove the Miller effect by modifying these two conditions. The only option left is to find out if we can do anything about the bridging I/O capacitance.

As a matter of fact, we learned at the end of Sect. 7.4.3 that not having the I/O bridging capacitance protects a CB amplifier configuration from the Miller effect. That gives us an idea of how to modify a simple CE stage and improve its bandwidth by turning it into a cascode amplifier, Sect. 7.4.7, which

[2]That is why we see voltages across LC resonator that are higher than the amplifier's power supply voltage level.

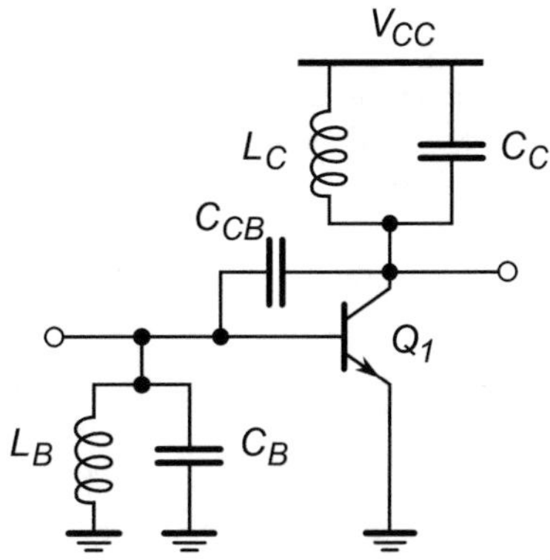

Fig. 12.3 CE RF amplifier with Miller capacitance C_{CB}

effectively removes the capacitive connection between the input and the output nodes of the cascode amplifier, see AC model of cascode amplifier in Fig. 7.28.

12.1.1.3 CE RF Amplifier Stability Issue

With a purely resistive load a CE amplifier is *inverting* the signal, i.e. the input and output signals have the phase difference of 180°. That being said, because the increase of the base current forces increase in the collector current, the input and the output current are in phase. By looking at the output node only, accordingly, the output voltage and current have the opposite phase.

At resonance, however, an LC resonator presents itself as the dynamic resistance R_D (Z_L and Z_C are same and have the opposite signs), hence from the amplifier's perspective it is the same as any other resistive load. Nevertheless, the statement is valid only at one frequency, $\omega = \omega_0$. At any other frequency than ω_0, the LC load becomes either capacitive (for $\omega > \omega_0$) or inductive (for $\omega < \omega_0$). This abrupt change of impedance causes the equally abrupt change of phase between $\pm 90°$. Depending upon the transistor gain, within a few signal cycles the feedback current becomes greater than the input current and the amplifier becomes unstable. Even the boundary condition, where the amplifier constantly switches between the stable and the unstable states, is itself an unstable condition. This instability of CE amplifier becomes more pronounced at higher frequencies and therefore smaller internal impedances.

12.1.1.4 Unilateralization of CE Amplifier

At least two techniques have been known that help improve the stability of CE RF amplifiers. In principle, the idea is very simple and general. Once we realized that main cause of CE amplifier instability is the feedback signal with the right phase relative to the input signal phase and recognized that this feedback signal is due to parasitic reactances inside the transistor that connect the output and input terminals of the amplifier, the solution to the problem came naturally. If another feedback path, external to the transistor, is created with a signal that is exactly the same as the parasitic feedback signal, but with the opposite phase, then the sum of the parasitic feedback signal and the external feedback signal can be made zero. In other words, the parasitic feedback signal is "neutralized" at the input terminal node. Cancellation of the feedback signal makes the transistor a truly unidirectional device (i.e. no feedback path), the process is sometimes referred to as "unilaterization".

In the first variant of signal neutralization, the idea is to "resonate out" the internal feedback capacitor C_{CB} by adding in parallel an external inductor L_n. However, that connection would create DC feedback path, thus the serial C_n capacitor must be added to remove the DC path from the output to the input terminal, Fig. 12.4 (left). The resonator path now includes two capacitors, C_{CB} and C_n, in series with the L_n inductor. A second variant of the same idea, shown in Fig. 12.4 (right), applies

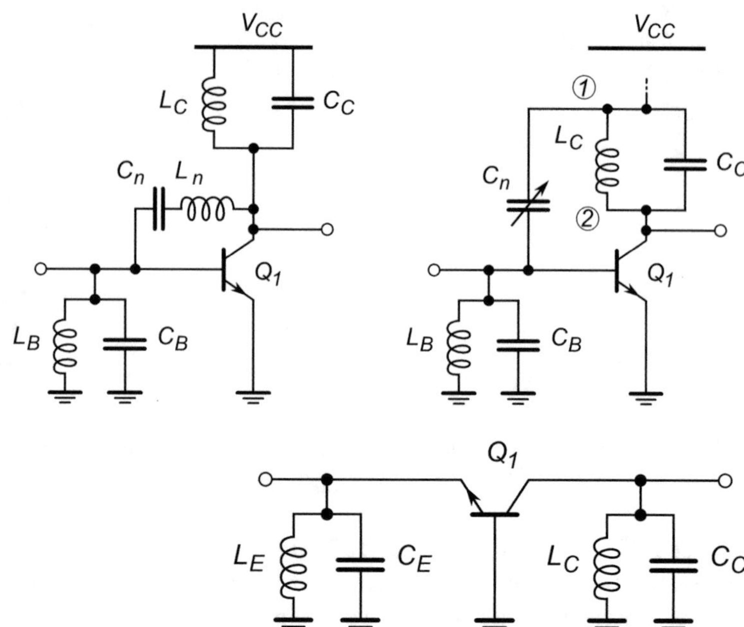

Fig. 12.4 Unilaterization by resonating out the internal capacitance (left), by tuneable capacitive feedback (right)

Fig. 12.5 A CB RF amplifier

an external capacitive feedback C_n in which the feedback signal is tapped from node 1 at the top of $L_C\,C_C$ tank, where the phase is opposite compared to node 2. The overall effect is, again, that the parasitic feedback signal is neutralized.

It should be noted that the neutralization techniques based on discrete passive components in the feedback path are limited by the component's self-resonant frequencies. Indeed, modern HF transceivers are designed mostly using HF IC technology.

12.1.2 Single-Stage CB as RF Amplifier

Frequency analysis in Sect. 7.4.4 shows that due to non-existing Miller effect a CB amplifier is inherently wideband, thus suitable for RF amplification. In addition, when the source RF signal is in the form of current (i.e. a very high-impedance signal source), it is beneficial to have an RF amplifier with low input resistance. We know that CB amplifiers satisfy the low input impedance requirement (Fig. 12.5) because $R_{in} \approx 1/g_m$ in the forward active mode. Both the input $L_E C_E$ and output $L_C\,C_C$ resonators are tuned to the same frequency. However, in this configuration there is no feedback path from the output to the input node, hence the CB amplifier is inherently stable.

12.1.3 Cascode RF and IF Amplifiers

Another amplifier configuration that is suitable for RF applications is cascode amplifier, in Sect. 7.4.7 it was shown that in CE-CB two-stage signal path Miller effect is removed. Following the same idea as for the other RF amplifiers, LC resonator load is used to create narrowband RF amplifier, Fig. 12.6. In order to minimize disturbance of the LC resonator[3] the output signal is inductively coupled by custom

[3]Very high-impedance low-capacitance RF probes are used to measure signals at LC resonators.

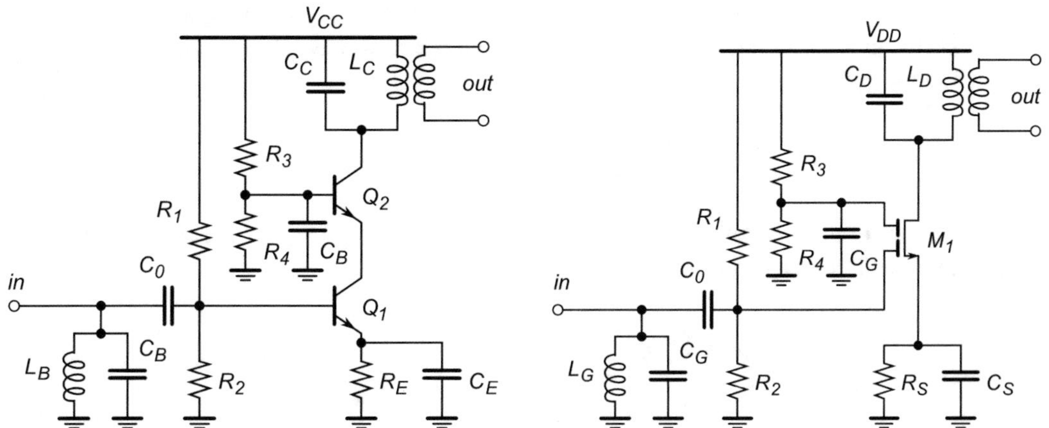

Fig. 12.6 Cascode BJT RF amplifier (left), and its equivalent dual-gate FET RF version (right)

designed transformer. The primary side serves as the inductor in LC resonator, while the secondary transmits the signal to the next stage.

In practice, a very common way to implement a cascode RF amplifier is by using a dual-gate MOSFET (often JFET) device, Fig. 12.6 (right). The two FET devices are manufactured on the same silicon substrate and packaged in the same package. This means that the manufactured dual device has exactly the same functionality as two cascode devices, with the advantage of reduced parasitic capacitances and greatly improved high-frequency (HF) performance compared to a configuration with two discrete devices.

From the previous discussion, for our practical purposes we conclude that, aside from the reasons related to Miller effect, using cascode amplifier is also recommended from the stability perspective because the CE feedback path is broken and cascode amplifier is inherently stable.

Schematic diagrams of two commonly used cascode RF amplifier structures are shown in Fig. 12.6. Advantage of using BJT devices is in higher g_m gains relative to MOSFET devices. On the other hand, advantage of MOSFET devices is in its very high input resistance, which makes FET input stage almost ideal load for a voltage source driver. To make advantage of both devices, in modern BiCMOS integrated technologies cascode amplifier is made as a combination of common-source (CS) and CB amplifiers.

12.2 Insertion Loss

Careful analysis of a parallel LC loading tank interaction with the active amplifying device in tuned RF amplifiers uncovers another interesting and important phenomena, which is actually the impedance matching problem in disguise. At low frequencies, i.e. below resonance ω_0, impedance of the inductor Z_L is very low. Consequently, there is *insertion loss* of voltage signal relative to the voltage level at the resonant frequency wo. On the other side of the resonance, i.e. at frequencies above the resonance, impedance of the capacitor Z_C is low, therefore causing the insertion loss again. Intuitively we conclude that in the case of ideal LC resonator, i.e. $R_D \to \infty$, there would be no insertion loss. However, the resonating tanks are real and their dynamic resonances are finite, and they appear in parallel with the collector resistance R_{ct}, therefore limiting R_D.

Being a property of a bandpass filter, insertion loss (IL) is important figure of merit of an LC resonator. In general, the overall RF amplifier output resistance Z_C consists of the parallel combination $R_{ct}||R_D$,

$$Z_C = \frac{R_{ct}\,R_D}{R_{ct} + R_D} = R_{ct}\,\frac{R_D}{R_{ct} + R_D} = R_{ct} \times IL \tag{12.1}$$

where insertion loss IL is defined by the resistive ratio $R_D/(R_{ct} + R_D)$. It is common practice to express the insertion loss in units of [dB], as

$$IL_{\text{dB}} = 20\log\frac{R_D}{R_{ct} + R_D} = 20\log\frac{1}{1 + \dfrac{R_{ct}}{R_D}} \quad [\text{dB}] \tag{12.2}$$

where the ideal case of ($R_D \to \infty$) translates into ($IL \to 0\,\text{dB}$) and in any other case the IL has negative number of decibels. For example, when $R_D = R_{ct}$ then $IL = -6\,\text{dB}$, and in the other extreme when ($R_D \to 0$) then ($IL \to -\infty$) to indicate that there is no power transfer through LC tank.

12.3 Case Study: An RF Amplifier

We now show a possible procedure of an RF amplifier design. The goal is to design an amplifier to be used in AM receiver that is intended to receive, for example, $f_0 = 10\,\text{MHz}$ AM modulated carrier signal. With this choice we can focus on designing voltage gain amplifier, as oppose to the power gain amplifier. The difference is that at the relatively low frequencies, such as $f_0 = 10\,\text{MHz}$, the parasitic LC effects of the connecting wires and the components are negligible. Therefore, the power reflection that is critical at HF and complex impedances is not as pronounced in this case because the involved impedances are very close to real resistances. That being said, we are free to focus on amplifier that is suitable for demonstrating the design principles learned so far in this book.

As the starting point, in the preliminary considerations we decide the following practical issues:

1. *Power supply:* let us assume the power supply voltage of the overall system is $V_{DD} = 15\,\text{V}$.
2. *Transistor:* we take the opportunity to use a JFET transistor, which is the type commonly used in RF circuits. The frequency range specification of this transistor should be at least one decade above the carrier's $f_0 = 10\,\text{MHz}$, i.e. its HF pole should be at 100 MHz or higher. In addition, the transistor's $V_{DS}(max)$ voltage specification should be greater than the chosen V_{DD}.
3. *Amplifier type:* we design CS cascoded JFET amplifier with degenerated source resistance to amplify a few mV signal generated by $50\,\Omega$ Thevenin source. The targeted gain specification is, for example, $A_V = 15\,\text{dB}$ or more. In accordance with (6.62), the cascoded transistor alone achieves gain of $g_m\,R_{Lt}$, where R_{Lt} is the *total* resistance at the drain node. These two numbers, i.e. R_{Lt} and g_m, are therefore among the first to be specified.
4. *Inductor:* in general, quality of capacitors is much better than quality of inductors. For that reason, it is critical to first choose an inductor that has acceptable specifications, most importantly in terms of its Q factor that is the consequence of its non-zero wire resistance. Depending upon its physical size, the inductor's wire resistance may be in the range from a few mΩ to tens of Ω. We recall (10.5) for the LC resonant frequency, which helps us choose the two components. For example, choosing the standard value $L = 3.3\,\mu\text{H}$ forces the choice of $C = 76.758\,\text{pF}$ so that in accordance to (10.5) the resonant frequency is set to 10 MHz. These two numbers illustrate

another practical issue when choosing LC components for setting up the resonator. It is not possible to use the standard component values for both L and C at the same time and to achieve the desired resonant frequency. Practical solution is that one of the two components must be "tuned" to a non-standard and very specific value. For all practical purposes inductors are not tuneable. On the other hand, "trim" capacitors either mechanical (see Fig. 2.16) or electronic (see Sect. 4.2.4) are commonly used exactly for this purpose. Therefore, $C = 76.758\,\text{pF}$ capacitance may be realized, for example, as the parallel combination of the standard 51 pF fixed value capacitor, and 0–50 pF trim capacitor.

Similar to the introductory example in Sect. 6.7, first, we choose and characterize JFET that is to be used in this design. The relevant specifications are found in transistor's datasheet, however, modern approach to circuit design is to use the transistor's SPICE model and derive its specifications by simulations.

1. **JFET DC characterization**: by replacing BJT in circuit shown in Fig. 6.28 with a generic JFET we can derive (I_D, V_{GS}) and (I_D, V_{DS}) transfer characteristics as in Fig. 12.7. From these two characteristics, by inspection we write the following specifications of this transistor:
   a. *JFET DC characteristics:* $(I_{DSS}, V_P) = (12\,\text{mA}, -3\,\text{V})$
   b. *Biasing point:* when using JFET transistors, it is a good practice to choose the biasing current I_D at 50% of I_{DSS} current, therefore by inspection of characteristics in Fig. 12.7 (left) we choose the biasing point at $(I_D, V_{GS}) = (6\,\text{mA}, -872\,\text{mV})$
   c. *Small signal parameters:* derivative of drain current at the biasing point gives $g_m = 5.62\,\text{mS}$, which is equivalent to say that source resistance is $r_s \approx 178\,\Omega$ (under condition $V_{DS} = 10\,\text{V}$). The drain resistance is found by inspection of characteristics at Fig. 12.7 (right) as $r_o \approx 75\,\text{k}\Omega$. And, we also find the minimal drain–source voltage $V_{DS}(\text{min}) \approx 2\,\text{V}$, which we use as the boundary condition that is necessary to maintain the constant current mode, before the transistor moves "down the slope" into the linear mode of operation. As a side note, in accordance to (6.63) derived in Sect. 6.6.2, we find that relative to the single transistor, indeed, cascoded transistor shows very high resistance at its drain, here the cascoded output resistance is $r_{oc} \approx 33\,\text{M}\Omega$.

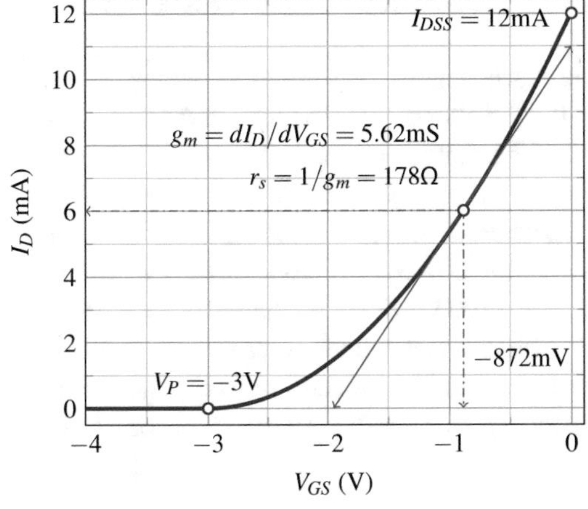

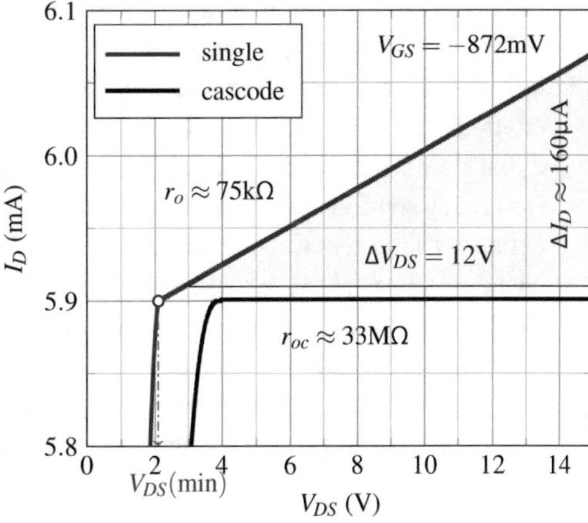

Fig. 12.7 JFET DC transfer characteristics

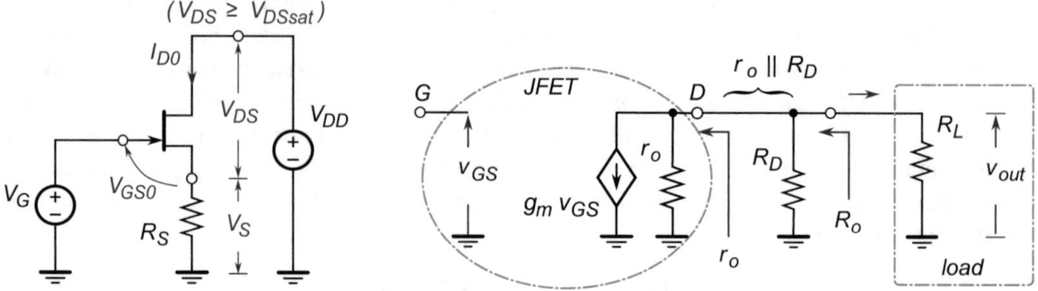

Fig. 12.8 JFET DC setup of the source resistor

2. **Estimate of the load resistance**: once the g_m is chosen, the next parameter is the desired voltage gain that controls the total drain load resistance. We start with an "envelope" gain estimate of unloaded amplifier so that $A_V = R_{Lt}/r_s = g_m\,R_{Lt} = 15\,\text{dB} \approx 5.623V/V$. By a direct calculation we find that, given g_m, the total load resistance should be approximately

$$R_{Lt} = \frac{A_V}{g_m} = \frac{5.623V/V}{5.62\,\text{mS}} = 1\,\text{k}\Omega$$

We already know how to evaluate the total resistance at a BJT collector.[4] For FET transistors, however, the gate DC current is zero (i.e. as if $\beta = \infty$ for MOS devices), thus by inspection of Fig. 12.8 (right) the total load resistance seen at the drain node is

$$R_{Lt} = r_o||R_D||R_L \approx r_o||R_D \quad \text{if} \quad R_L \gg r_o||R_D \tag{12.3}$$

where r_o is the output drain resistance, R_D the external resistor connected between the power supply and drain (or, the dynamic resistance R_D of LC resonator at the same place), and R_L is the input resistance of the following stage in the system, i.e. the load. For the moment, we assume voltage JFET amplifier, thus large $R_L \gg r_o||R_D$ resistance. For the given r_o and R_{Lt}, assuming a single transistor, from (12.3) we calculate $R_D \approx 1.1\text{k}\Omega$, which is in effect the specification for LC resonator. We know that r_o of cascoded transistor is much higher, therefore this value of R_D is "back of the envelope" number.

3. **LC resonator characteristics**: in Sect. 10.8 we learned how to estimate Q factor and RLC bandpass filter's bandwidth. In addition in Sect. 10.8.2 we learned a practical method to estimate dynamic resistance of an RLC network at resonant frequency. Given that in order to meet the frequency specification of this amplifier we decided to use $L = 3.3\,\mu\text{H}$, we must evaluate R_D and Q factor of the specific discrete component that is to be used. Simulation results in Sect. 10.8.2 are based on a typical average quality commercially available $L = 3.3\,\mu\text{H}$ inductor, thus we can reuse the same numbers: at $f_0 = 10\,\text{MHz}$ we found $R_D = 1100\,\Omega$, $Q = 5.4$, and $BW = 1.85\,\text{MHz}$. This non-ideal inductor is close to the intended design specification, otherwise if its R_D were lower we would have to either search for another higher quality inductor or to change the initial specifications. In the following steps, we use resistor $R_D = 1.1\,\text{k}\Omega$, until the LC resonator is put in its place.

[4]See Fig. 6.18.

4. **Degenerated source setup:** we know that emitter's (or, source) resistance $r_e = 1/g_m$ sets the upper voltage gain limit, see Sect. 4.3.8, however at the expense of DC biasing setup stability. Adding relatively larger $R_S \gg r_s$ resistor in series minimizes this instability at the cost of reduced voltage gain.[5] In order to restore the maximal gain, i.e. to remove R_S from the signal path only, while maintaining DC path, the bypass capacitor C_S is added in parallel to R_S to create virtual small signal ground at the source. For example, a standard resistor $R_S = 833\,\Omega \gg 178\,\Omega$, as in Fig. 12.8, sets DC voltage at the source terminal as $V_S = R_S\,I_{S0} = R_S\,I_{D0} = 833\,\Omega \times 6\,\text{mA} \approx 5\,\text{V}$. We cannot use too high value for R_S because of the overall headroom set by the power supply voltage. This reason is even more important in the case of cascoded amplifier. For the rest of the amplifier development, the power supply voltage is set to $V_{DD} = 15\,\text{V}$.
5. **Gate voltage:** if this JFET is to deliver $I_{D0} = 6\,\text{mA}$ at its drain terminal, its gate-to-source voltage must be set $V_{GS0} = -872\,\text{mV}$. In the degenerated source configuration the source voltage is set to $V_S = 5\text{V}$, therefore $V_G = V_S + V_{GS0} = 5 - 0.872 = 4.128\,\text{V}$, see Fig. 12.8 (left).
6. **Gain and bandwidth simulations:** in advance we know only approximately the internal capacitances of JFET (from its datasheet), thus we use AC simulation to determine BW of the amplifier's input stage (Fig. 12.9). In Sect. 7.3.2 we learned that the bypass capacitor C_E and R_E contribute to pole-zero pair in the transfer function, see (7.22) and (7.24). After limiting $\beta \to \infty$ for MOS transistors we estimate the range of C_S; for example, if $C_S = 1\,\text{nF}$, then

$$f_z = \frac{1}{2\pi\,R_S\,C_S} = \frac{1}{2\pi \times 833\,\Omega \times 1\,\text{nF}} = 191\,\text{kHz}$$

$$f_p = \frac{1}{2\pi C_S(R_S||r_s)} = \frac{1}{2\pi \times 1\,\text{nF} \times (833\,\Omega||178\,\Omega)} = 1.08\,\text{MHz}$$

AC simulation sweep of the C_S capacitor shows the influence of this pole-zero pair on the overall BW, see Fig. 12.9 (right). Although any of these capacitor values could be used, we want to minimize the input side BW so that the overall noise entering the amplifier is limited. On the other hand, we do not want to reduce the BW too much and therefore start limiting the signal

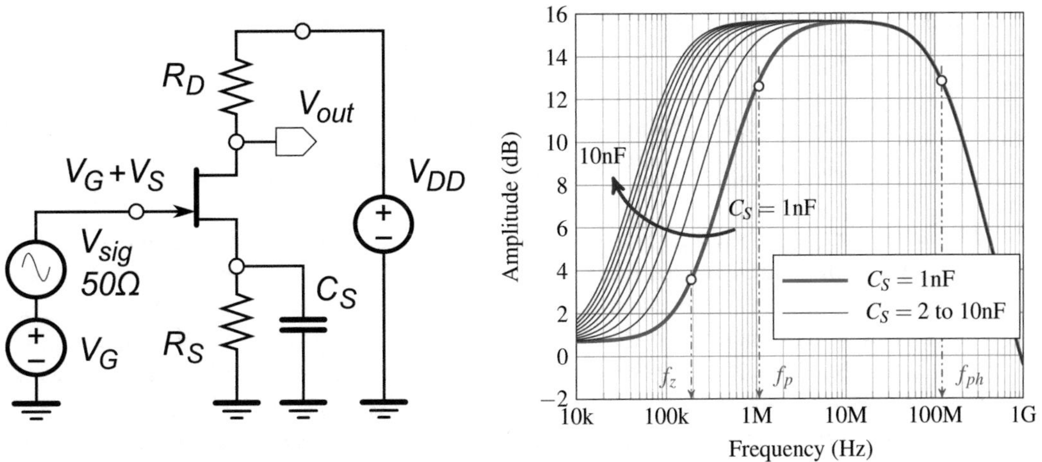

Fig. 12.9 JFET AC simulation setup (left) used to tune BW by sweeping C_S (right)

[5] We recall that voltage gain is the ratio of the total collector versus total emitter resistance.

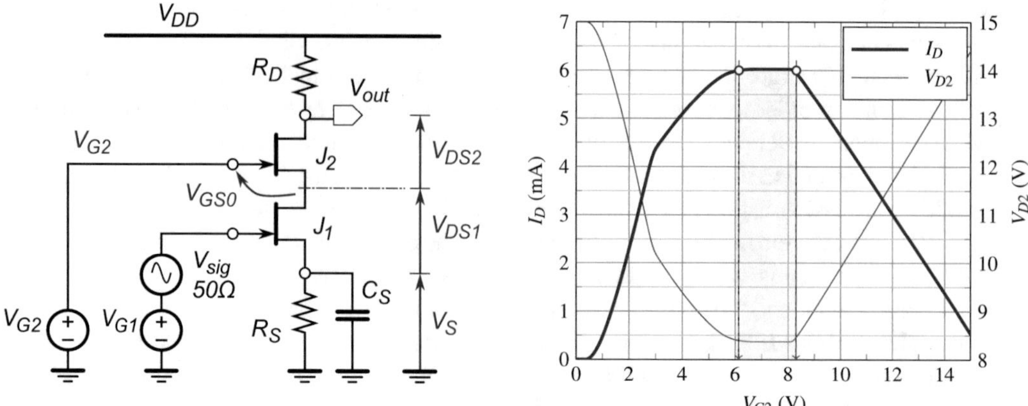

Fig. 12.10 JFET DC setup of the cascode transistor (left) used to determine V_{G2} voltage range (right)

itself. In addition, simulation shows high-frequency pole due to the parasitic capacitances of transistor at approximately $f_{ph} = 120\,\text{MHz}$. That being said, if $C_E = 1\,\text{nF}$, then the bandwidth of this amplifier is approximately cantered around the working $f_0 = 10\,\text{MHz}$ frequency, which is one possible justification to choose this capacitor value.

7. **Setting up the cascode transistor:** cascoded transistors are used extensively in RF circuit designs, so much that a "dual gate" JFET transistor is commonly available. That is, two identical cascoded transistors are manufactured on a single IC die, thus they are perfectly matched to each other, and then packaged in a single "dual gate" package. In Fig. 12.10 (left) the dual gate transistor is simulated as two separate transistors J_1 and J_2.

   In Fig. 12.7 (right) we see that for this particular JFET it is necessary to maintain $V_{DS} \geq 2\,\text{V}$. In order to estimate biasing voltage V_{G2} in Fig. 12.10 (left) we proceed as follows. Both transistor must permit $I_D = 6\text{mA}$, where J_1 is in the control (it is CS amplifier), while J_2 serves as the "current buffer" (it is CG amplifier). Starting at the ground level, $V_S = 5\,\text{V}$, then there is at least $V_{DS1} = 2\,\text{V}$ on top, and finally there is at least $V_{DS2} = 2\,\text{V}$ on top again. At the same time, by inspection of Fig. 12.10 (left), in order to calculate voltages V_{G2} and $V_{D2} = V_{\text{out}}$ we follow paths as

$$V_{G2} \geq V_S + V_{DS1} + V_{GS0} \geq 5\,\text{V} + 2\,\text{V} - 0.872\,\text{V} \geq 6.128\,\text{V}$$

$$V_{D2} = V_{\text{out}} = V_{DD} - V_{R_D} = V_{DD} - R_D\, I_D = 15\,\text{V} - 1100\,\Omega \times 6\,\text{mA} = 8.4\,\text{V}$$

   which are the output DC voltages that are maintained as long as both J_1 and J_2 permit $I_D = 6\,\text{mA}$ current. We keep in mind that for JFET transistors V_{GS} voltage is negative (i.e. $-V_P \leq V_{GS} = V_G - V_S \leq 0$) . That is to say that at its highest level the gate voltage should stay below the drain voltage as well. The conclusion is that J_2 gate voltage should be in the $6.128\,\text{V} \leq V_{G2} \leq 8.4\,\text{V}$ range. It is instructional to verify this conclusion by simulation. Indeed, in Fig. 12.10 (right) we clearly see that the drain current dropped I_D when either $V_{G2} \leq 6\,\text{V}$ or $V_{G2} \geq 8.4\,\text{V}$.

   Although any value of V_{G2} that is within the allowed interval is acceptable, we choose the minimum value that brings J_2 on the edge of constant current mode, in other words $V_{G2} \geq 6.128\,\text{V}$. The reason is that higher V_{G2} results in higher V_{DS} (min), which subsequently reduces the overall headroom available to the output signal. In addition, simulations confirm that frequency response of this cascode amplifier is, as expected, as same as the one for non-cascode

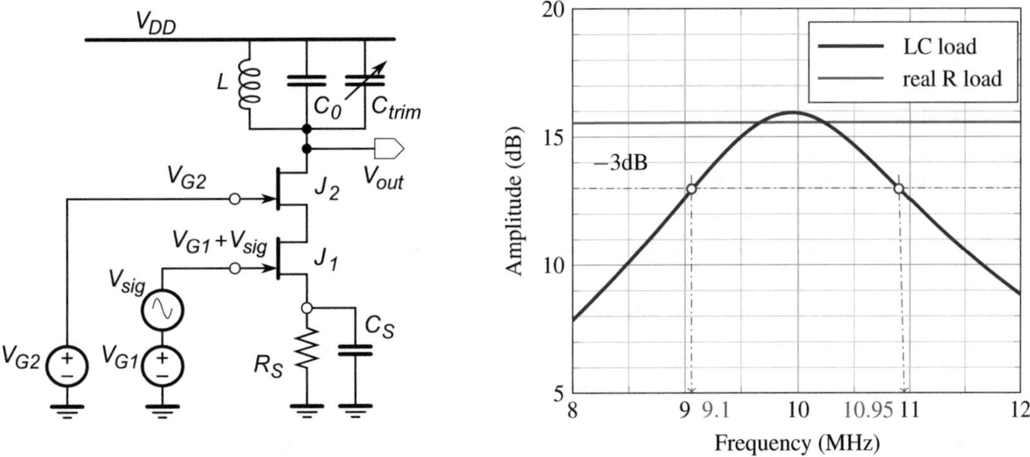

Fig. 12.11 JFET setup of LC resonator (left) and the comparison with the equivalent real resistive load (right)

version shown in Fig. 12.9 (right). Finally, the output resistance r_{oc} seen at the drain of J_2 is estimated, see (6.63), as

$$r_{oc} = 2r_o + g_m r_o^2 \approx 33\,\mathrm{M}\Omega \tag{12.4}$$

which is in agreement with the simulation results, as shown in Fig. 12.7.

8. **Setting up the LC resonator:** once the cascode transistor J_2 is set, it is now straightforward to replace the real resistor load R_D with the already characterized LC resonator, see Fig. 12.11 (left). Since the dynamic resistance of this resonator equals R_D in Fig. 12.10 (left), at its resonant frequency that is set to 10 MHz there is no difference in terms of the amplifier voltage gain. However, AC simulation clearly shows the narrow bandwidth $BW = 1.85\,\mathrm{MHz}$ that is now controlled by Q factor of this LC resonator. The difference between the two bandwidths is illustrated in Fig. 12.11 (right).
9. **Setting up the biasing references:** the required biasing voltage references V_{G_1} and V_{G_2} are implemented by using the voltage divider technique.[6] We recall that two resistors create voltage division as

$$V_{\text{ref}} = I_{(R_x,R_y)}\,R_y = \frac{V_{DD}}{R_x + R_y}R_y \quad \therefore \quad \frac{R_x}{R_y} = \frac{V_{DD}}{V_{\text{ref}}} - 1 \tag{12.5}$$

In order to set biasing voltage for $V_{G2} \geq 6.128\,\mathrm{V}$ we search the list of standard resistor values with, for example, 5% tolerance to find that $R_3 = 14\,\mathrm{k}\Omega$ and $R_4 = 11\,\mathrm{k}\Omega$ have the ratio of 1.36 that is close to (12.5). Similarly, we find that $R_1 = 39\,\mathrm{k}\Omega$ and $R_2 = 15\,\mathrm{k}\Omega$ are in $39/15 = 2.6$ ratio, which is close to the ideal 2.63 value. With these resistors we realized $V_{G1} = 4.167\,\mathrm{V}$ and $V_{G2} = 6.6\,\mathrm{V}$, both of which are very close to the required theoretical values. If necessary, these resistors could be fine-tuned after.

Although there is number of possible resistor values that could satisfy these ratios. For the modern mobile electronics we do not want to waste too much energy in these branches, thus we prefer higher resistor values. In addition, the input resistance of this RF amplifier is

[6]See Sect. 5.1.2.

$$R_{\text{in}} = R_1 || R_2$$

Being a voltage amplifier,[7] again we prefer higher R_{in}. Finally, time constant[8] due to C_1 is found by inspection as

$$\tau_1 = (R_S + R_{\text{in}})\, C_1 = (R_S + R_1||R_2)\, C_1 \approx (R_1||R_2)\, C_1 \quad \text{when} \quad R_S = 50\,\Omega \ll R_1||R_2$$

Therefore, it is not too difficult to achieve the required DC biasing setup and at the same time to control the position of ω_{τ_1} pole-zero frequency, see, for example (7.12). With that choice of the two resistors we calculate the input side resistance of this amplifier as

$$R_{\text{in}} = R_1||R_2 = 10.8\,\text{k}\Omega \gg R_S = 50\,\Omega$$

Last but not least, if we keep $C_1 = 1\text{nF}$ capacitor value, then its associated pole-zero frequency is

$$f_{\tau_1} = \frac{1}{2\pi\,(R_1||R_2)\, C_1} = \frac{1}{2\pi \times 10.8\,\text{k}\Omega \times 1\,\text{nF}} = 14.73\,\text{kHz}$$

which is still very far away from the 10 MHz working frequency, therefore it has negligible effect on the frequency response curve. Repeated AC simulation confirms that response in Fig. 12.11 (right) did not change. Transient analysis also confirms that the input voltage signal is indeed amplified by a factor of 5.6, which was the original gain target. We note that the output signal is inverted as expected, as well as almost 1 µs transition period before the output signal reached its maximum amplitude. This transition period is typical for all LC networks that need a certain amount of time to charge these two components (Fig. 12.12).

10. **The output resistance and the load range:** calculations so far assumed $R_L = \infty$, i.e. the "unloaded" amplifier. This calculation is useful to find theoretical gain limit, nevertheless,

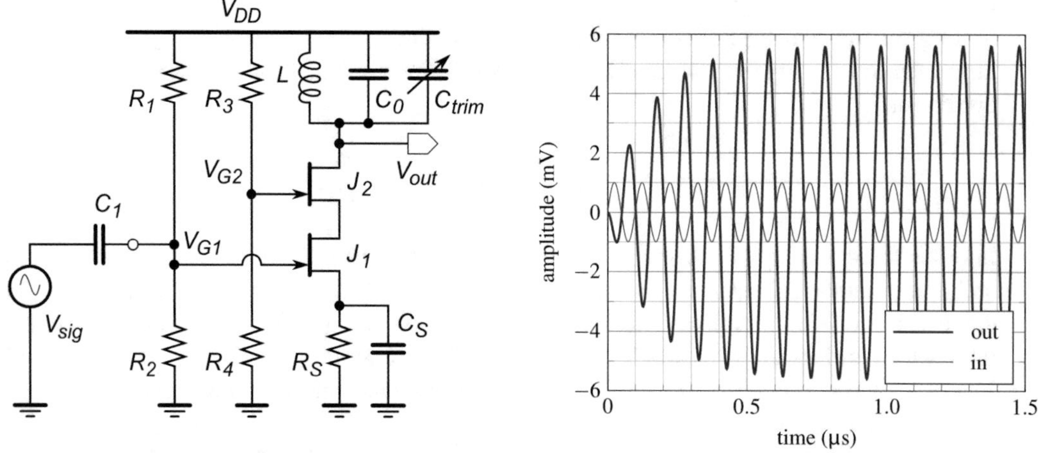

Fig. 12.12 JFET cascode RF amplifier schematic (left) and time domain I/O waveforms (right)

[7] See Sect. 6.1.2.

[8] See also Sect. 7.3.2 and Fig. 7.7.

as indicated in Fig. 12.8 (left) the load resistor directly affects the overall voltage gain. It is necessary, therefore, to specify some "useful" range of R_L (we keep in mind that r_{oc} is really large). In addition, there is decoupling capacitor C_0 that in connection with the drain's total resistance creates a pole, see Sect. 7.3.1 for CB amplifier, as

$$\omega_0 = \frac{1}{(R_D + R_L)C_0}$$

which, if we keep, for example $C_0 = 1\,\text{nF}$, in the worst case scenario when $R_D = R_L = 1.1\,\text{k}\Omega$ (i.e. the gain drops to 6 dB) set this pole around 70 kHz, again far enough not to influence the overall frequency response curve in Fig. 12.11 (right).

11. **Setting up the matching network:** let us assume that this amplifier is used to accept a $f_0 = 10\,\text{MHz}$ RF signal from a $R_0 = 50\,\Omega$ antenna, where the antenna is capacitively coupled to the amplifier. Therefore, we create matching network to between $R_0 = 50\,\Omega$ and $R_{in} = 10.8\,\text{k}\Omega$ resistances. This connection excludes the decoupling capacitor $C_1 = 1\,\text{nF}$, which is replaced with the matching network.
As per (11.2) we write

$$Q_S = Q_P = Q = \sqrt{\frac{R_L}{R_0} - 1} = \sqrt{\frac{10.8\,\text{k}\Omega}{50\,\Omega} - 1} = 14.66$$

and from (11.1) we write

$$X_S = Q_S R_0 = 14.66 \times 50\,\Omega = 733.14\,\Omega \quad \text{and} \quad X_P = \frac{R_L}{Q_P} = \frac{10.8\,\text{k}\Omega}{14.66} = 736.554\,\Omega$$

Therefore, the serial capacitor and parallel inductor are calculated as

$$C_S = \frac{1}{2\pi f_0 X_S} = \frac{1}{2\pi \times 10\,\text{MHz} \times 733.14\,\Omega} = 21.7\,\text{pF}$$

$$L_P = \frac{X_P}{2\pi f_0} = \frac{736.554\,\Omega}{2\pi \times 10\,\text{MHz}\times} = 11.7\,\mu\text{H}$$

In summary, this case study shows one possible procedure to design a practical RF amplifier for accepting 10 MHz carrier waveforms (even today similar frequencies are used by radio stations transmitting time reference signals).

12.4 Summary

In this section we reviewed fundamental concepts of RF amplifiers and developed intuitive views of their operation.

As the first step in amplifier design, gain of the active devices (either BJT or FET) is set by their DC operating point, and the subsequent signal analysis is simplified by omitting details of the biasing circuit, i.e. it is simply assumed that the active devices g_m gain is already set.

In the subsequent sections we demonstrated one possible design flow to transform a wideband amplifier into its RF version.

Both AC and transient simulation analyses are used illustrate the design steps and final operation of a cascode JFET RF amplifier.

Problems

12.1 For a single NPN BJT transistor, draw the schematic symbol and indicate potentials at the three terminals, i.e. the V_C, V_B, and V_E, and their relationship assuming the transistor is turned on, i.e. it is operating in the forward active region. Repeat the exercise using a PNP BJT transistor.

12.2 Based on the given set of numerical data, assuming the constant current mode of operation, estimate voltage gain A_v for circuit in Fig. 5.10 (right) if $R_C = 10\,\text{k}\Omega$ and $R_E = 100\,\Omega$. Express the result in [dB].

12.3 An ideal signal generator is coupled with the CE amplifier in Fig. 5.10 (right) through a serial capacitor $C = (10/2\pi)\,\mu\text{F}$. Estimate the range of frequencies where the CE amplifier should be used, if $R_C = 9.9\,\text{k}\Omega$, $R_E = 100\,\Omega$, $C_{CB} = (1/2\pi)\,\text{nF}$, $R_1 = 20\,\text{k}\Omega$, $R_2 = 20\,\text{k}\Omega$.

12.4 Estimate the range of frequencies where CE amplifier in Fig. 12.2 (right) should be used, if the base side inductor $L_B = 10/(2\pi)\,\mu\text{F}$, $C_B = 1/(2\pi)\,\text{nF}$.

12.5 Resistance seen by looking into a BJT emitter is $R_{\text{out}} = 100\,\Omega$. Resistance looking into the base is $R_{\text{in}} = 100\,\text{k}\Omega$. For $\beta = 99$, find reflected resistance at the base node, and R_E? (Note: ignore r_e.)

12.6 For grounded emitter amplifier powered from $V_{CC} = 10\,\text{V}$ with collector resistor $R_C = 5.1\,\text{k}\Omega$ estimate voltage gain for: (a) $V_{\text{out}} = 7.5\,\text{V}$, (b) $V_{\text{out}} = 5\,\text{V}$, (c) $V_{\text{out}} = 0.2\,\text{V}$.

12.7 If V_{BE} voltage of a BJT transistor changes by 18 mV what is change of I_C in expressed in dB? What if V_{BE} changes by 60 mV? Note: Use $kT/q = 25\,\text{mV}$.

12.8 For amplifier in Fig. 12.2, given that $R_C = 1\,\text{k}\Omega$ design $L_C C_C$ resonator if the amplifier is expected to receive RF carrier at $f_0 = 10\,\text{MHz}$.

12.9 For amplifier in Fig. 12.6 (left), given that $I_C(Q_1) = 1\,\text{mA}$ and that the amplifier is expected to receive RF carrier at $f_0 = 10\,\text{MHz}$ while providing voltage gain $A_V = 10$, design $L_C C_C$ resonator.

13 Sinusoidal Oscillators

Communication transceivers require oscillators that generate pure electrical sinusoidal signals (i.e. stable time reference signals) for further use in modulators, mixers, and other circuits. Although oscillators may be designed to deliver other waveforms as well, e.g. square, triangle, and sawtooth waveforms, if intended for applications in wireless radio communications, the sinusoidal and square waveforms are the most important ones. A good sinusoidal oscillator is expected to deliver either a voltage or a current signal that is stable both in amplitude and frequency. Because a variety of oscillator structures are available that are suitable for generation of periodic waveforms, circuit designers make the choice mostly based on their personal preference for one particular type of oscillator. In this chapter, we study several oscillator circuits, with emphasis on understanding the underlying principles, rather than very detailed analysis of any special oscillator type.

13.1 Closed-Loop Principle

Literally every modern communication system relies on the master timing reference, similar to a symphony orchestra musicians who must follow the rhythm given by the conductor. Synchronization of time among all blocks in a communication system is one of the main principles used in the communication theory.[1] The time synchronization is implemented with the help of a circuit known as a voltage-controlled "oscillator" (VCO) that produces a periodic waveforms whose frequencies are made as stable and precise as possible in the given technology. In addition, various precisely controlled periodic waveforms are needed to perform the mathematical multiplication operation, which is the key operation in RF systems. For example, an RF receiver in Fig. 13.1 contains one "local" oscillator (LO) whose output signal is used by the multiplying circuit (i.e. mixer) to shift frequency of the incoming RF carrier. Here, the term "local" simply means that the oscillator circuit itself is part of the receiver, not the external circuit.

Inherently the amplitude of the signal inside an oscillator circuit may (theoretically) infinitely increase, that is, in reality its output signal amplitude becomes large. Consequently, the conclusion is that small signal circuit analysis is not an applicable method anymore. Large signals imply *nonlinear operation*, which means that we have to apply numerical methods in order to estimate the circuit's internal states. The good news, however, is that all oscillator circuits belong to the general group of

[1] The alternative, i.e. asynchronous, systems are still the subject of research.

R. Sobot, *Wireless Communication Electronics*,
https://doi.org/10.1007/978-3-030-48630-3_13

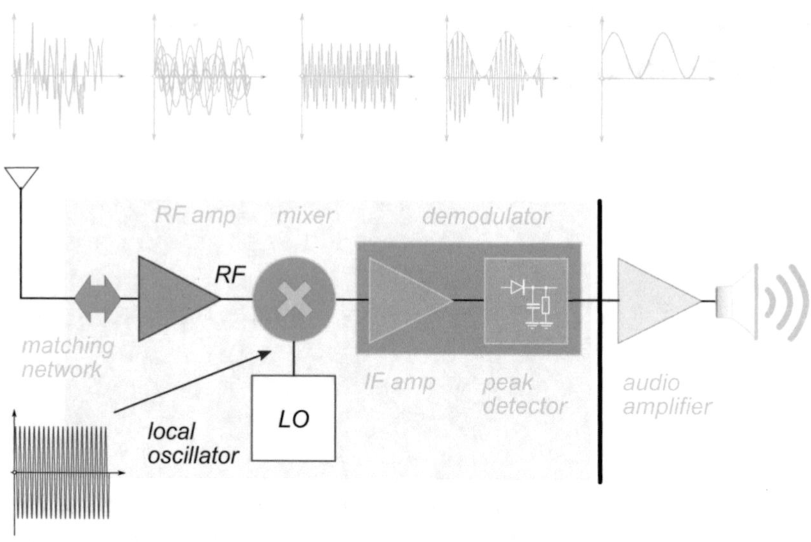

Fig. 13.1 Heterodyne AM radio receiver architecture—local oscillator (LO)

closed-loop feedback circuits (e.g. PLL, adaptive equalizer, $\Sigma\Delta$ modulator, etc.) whose fundamental theory is well understood, developed, and studied under the control theory subject.

As we observe RF receiver's block diagram in Fig. 13.1, first two questions that are usually asked are: if there is no "input" signal into LO circuit, how is it possible to create a stable, precisely controlled periodic waveform at the output terminal? Where does it come from exactly?

In order to answer these questions and to develop intuition about the loop operation, we start with the "open-loop" configuration, Fig. 13.2 (left), where the feedback signal v_f is disconnected from the signal v_i at the input terminal of the amplifier (as indicated by the jumper symbol). In the forward signal path, there is an amplifier with gain A, which may be either non-inverting or inverting.[2] The feedback path contains a passive network with gain of β (not to be confused with BJT transistor's current gain factor β). By inspection of Fig. 13.2 (left) we write

$$v_o = A\, v_i \tag{13.1}$$

$$v_f = \beta\, v_o = \beta\, A\, v_i \tag{13.2}$$

Therefore, a signal that follows the ($v_i \to v_o \to v_f \to v_i$) path is amplified by the "open-loop gain" A_o, which is found from (13.2) as

$$A_o = \frac{v_f}{v_i} = A\,\beta \tag{13.3}$$

Therefore, the design of an oscillator circuit consists of two steps: design of an amplifier A (as studied in Chaps. 6 and 7) to be used in the forward path, and design of a passive feedback network β that can take various forms. In this chapter we study four main RLC networks that are commonly used to implement the feedback path in oscillators.

[2]This is relevant for constructive or destructive signal addition.

oscillator: open loop

v_i v_o output

A

v_f

β

oscillator: clossed loop

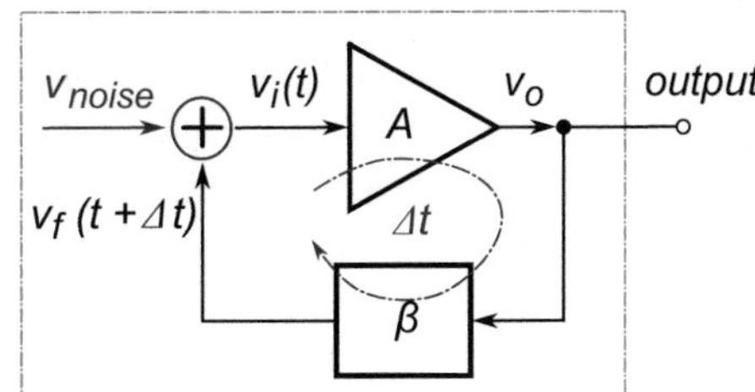

Fig. 13.2 Block diagram of a basic oscillator feedback loops: open (left) and closed (right)

When the jumper is closed, however, we create a closed-loop configuration, Fig. 13.2 (right) (for the moment we ignore the noise). There are three key mechanisms happening at the same time inside closed-loops that must be clearly understood.

1. *Addition of v_f and v_i signals at the amplifier's input terminal:* first key factor in understanding functionality of a closed-loop is to find how exactly this addition occurs. Let us follow the signal around the loop with respect to different moments in time. We start with v_i signal at the amplifier input terminal, see Fig. 13.2 (right). It becomes $v_o = A\, v_i$, however only after a certain delay that is caused by the *propagation time* through the amplifier. Going forward through the feedback network, before arriving at the amplifier's input terminal after the total loop propagation time Δt, the signal is amplified again to become $v_f = \beta\, v_o$. It is at this moment of arrival, when v_f is added to the *original* v_i signal, which is followed by yet another trip around the loop. Because this addition applies either to voltage or current signals, we use a generic $\oplus$ symbol at the input node of amplifier without showing the specific implementation details.

   Therefore, in reality it is the sum of signals $v_i(t)$ and $v_f(t + \Delta t)$ that is being amplified again and again. With this understanding, by inspection of Fig. 13.2 (right) we write expression for the "closed-loop gain" A_c as

$$v_o = A(v_i + v_f) = A\, v_i + A\, \beta v_o \quad \therefore \quad A_c = \frac{v_o}{v_i} = \frac{A}{1 - \beta A} \tag{13.4}$$

   Very important observation is that, as the consequence of closed-loop configuration, expression for the closed-loop gain (13.4) is a *rational* function that can limit to infinity if the denominator limits to zero. In other words, all feedback systems are inherently unstable.

2. *Creation of a periodic waveform:* so far we have not clarified how exactly the "original" v_i waveform was created in the first place? Where is it coming from?

   In the real circuits and systems, thermal noise is always present, see Chap. 8. What is more, its frequency spectrum is flat, Fig. 8.1, in other words it contains sine waveforms at all possible frequencies. We now follow the thermal noise voltage v_i around the closed loop. Assuming an ideal wide-band amplifier, all frequency components of the noise are equally amplified to produce v_o. The passive feedback network, however, we design as a RLC resonant network whose frequency transfer function is the one of bandpass filter[3] (BPF). As we already know,

[3] See Chap. 10.

the RLC resonant frequency f_0 is controlled by its LC components. With this setup, output v_f of the feedback network contains (in theory) only one sine waveform, while all other frequency components of the thermal noise are suppressed. In the beginning, being the noise component, this only surviving waveform is very weak. Nevertheless, it is the only one being amplified again and again by the loop gain and, eventually, its amplitude becomes very large. In conclusion, the internal thermal noise is responsible for the initial waveform seed, while RLC network removes all unwanted frequencies except the one at the resonant frequency of the oscillator. By tuning LC product, we control which frequency is selected out of the thermal noise spectrum.

3. *Constructive signal addition:* third fundamental factor deciding whether the closed-loop will actually create a steady waveform at its output or not is related to the constructive and destructive waveform additions.[4] Because there is inevitable time difference between arrivals of two waveforms that are about to be added, which can be expressed as the phase difference, it is very important to set up conditions favouring the constructive addition. That is to say, the two waveforms must be *in phase*, otherwise either partial or total destructive addition would destroy the only waveform that was selected for the amplification.

With this intuition of closed-loop functionality and the underlaying fundamental mechanism, an oscillator may be thought of as a positive feedback amplifier that is intentionally made unstable.

13.1.1 Criteria for Oscillations

Although control theory offers several commonly used methods for evaluating the stability of closed-loop systems (for instance, the Bode plot, the Routh–Hurwitz stability criterion, root-locus analysis, and the Nyquist stability criterion are applicable to linear, time-invariant (LTI) systems, while the Lyapunov stability criterion applies to nonlinear dynamic systems), none of these criteria is universal and self-sufficient. In practice, in addition to analytical and numerical methods, we usually use more than one of the criteria to reach a conclusion about the system's stability. For the purposes of determining under what conditions a linear electronic circuit oscillates, we introduce the intuitive (and also non-perfect) "Barkhausen Stability Criterion" which states that, if a feedback circuit is to maintain oscillations, then

1. The net gain around the feedback loop must be no less than one, i.e.

$$|A\beta| \geq 1 \tag{13.5}$$

2. The net phase shift $\Delta\phi$ around the loop must be a positive integer multiple of 2π radians, or

$$|\Delta\phi| = n \times 360^\circ \quad (n \text{ is an integer}) \tag{13.6}$$

The Barkhausen Criterion is a *necessary but not sufficient* condition for oscillation. Both $A = A(\omega)$ and $\beta = \beta(\omega)$ are frequency dependent, therefore the conditions listed in the Barkhausen Criterion are satisfied at the same time only at a single frequency. There are, therefore, two necessary conditions for sustaining the loop oscillations: one related to the loop gain and one to the phase shift. In practical designs, of course, the initial loop gain must be a little bit greater than unity in order to increase the

[4]See Sect. 1.4.3.

probability that the circuit can actually start oscillating, i.e. in a well-designed oscillator there should be no problem for the circuit in starting the oscillations on its own. In addition, it is necessary to build in some kind of mechanism to keep limiting the amplitude of the oscillation, so that the output signal does not become clipped or distorted.

Example 79: The Barkhausen Criterion
Given a CE amplifier biased at $I_C = 1\,\text{mA}$ and with $R_C = 1\,\text{k}\Omega$, estimate gain of the feedback network if the goal is to design an oscillator. Is there any additional specification that needs to be determined?

Solution 79: A CE amplifier biased at $I_C = 1\,\text{mA}$ at room temperature, i.e. $V_T \approx 25\,\text{mV}$, is therefore set to

$$g_m = \frac{1}{r_e} = \frac{1\,\text{mA}}{25\,\text{mV}} = 40\,\text{mS}$$

which means that the back of envelope estimate of voltage gain is

$$A_V = -g_m\,R_C = -40\,\text{mS} \times 1\,\text{k}\Omega = -40$$

In accordance to (13.5), it follows that the feedback network gain should be

$$\beta \geq \frac{1}{A} \geq \frac{1}{40}$$

In practice, the $\beta\,A$ gain is set only slightly larger than one. Additional condition for creating sustained oscillations is that the total phase shift must satisfy (13.6) as well. We know that CE amplifier by itself introduces 180° phase shift, therefore, it is necessary to design the feedback network with the additional 180° shift so that the total phase shift is set to zero.

13.2 Basic Oscillators

In a broad sense, oscillators are classified into two groups: digital oscillators that are based on closed-loop circuits that involve digital gates and produce square waveforms; and analog oscillators that are based on analog amplifiers and RLC networks that produce various periodic waveforms: sine, square, triangle, and sawtooth. In order to demonstrate several typical design techniques, in the following sections we review some of the most often used oscillator circuits.

13.2.1 Ring Oscillator

A simple example of a closed-loop circuit that can generate a square pulse waveform is based on the signal propagation delay through a chain of inverters by using the principle of an inverting amplifier that is driving its own input terminal. If we observe the input terminal of the first inverter at an arbitrary point in time t_0 and if, for the sake of argument, we observe a positive pulse, we could "join"

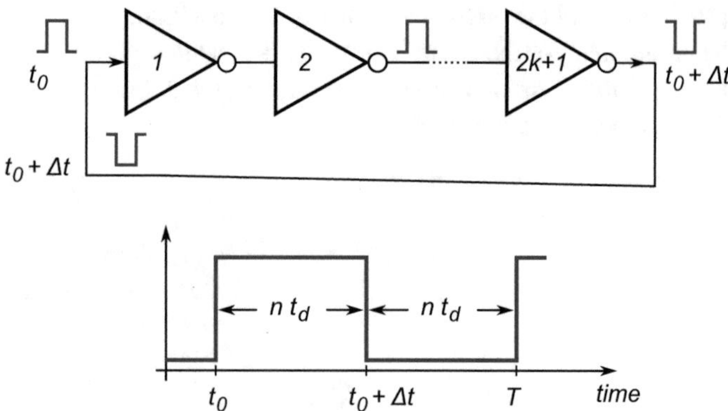

Fig. 13.3 Ring oscillator schematic diagram

the pulse on its trip around the loop, Fig. 13.3 (top). After propagating through the first inverter the pulse becomes inverted; after propagating through the second inverter the pulse is inverted again. It is straightforward to generalize and conclude that after every even inverter stage the pulse has as same polarity as the one at the first input, while after every odd inverter stage the pulse has the opposite polarity. Therefore, we conclude that for a chain of $n = (2k + 1), k = 1, 2, 3 \ldots$ inverters, a signal with opposite polarity takes $\Delta t = n\,t_d$ seconds to travel around the loop. That means, in this case, the output signal is always the opposite phase relative to the one at the input. Thus, the output node perpetually forces the input node to alternate its state.

The full period of a periodic signal is measured between two falling or two rising edges, hence a ring oscillator produces a square signal whose period is $T = 2\,t_d\,n\,t_d$ seconds, where t_d is the signal propagation time through each inverter stage, Fig. 13.3 (bottom). We should note that the "output" terminal is chosen arbitrary: it could be taken from any point around the loop.

Aside from their main application in the clock generator tree, ring oscillators are often used in IC technology as sensors for the process variations. The oscillator frequency depends upon the propagation delays of each inverting stage. Further, the stage propagation time depends upon the internal capacitances and resistances, which are very much process dependent. Therefore, by measuring the output frequency we are able to quantify the process variation.

Example 80: Ring Oscillator

An average propagation delay through a single inverter gate was estimated as $t_d = 0.998\,\text{ns}$. How many inverter gates are needed to design a ring oscillator working at $f = 1\,\text{MHz}$.

Solution 80: A 1 MHz signal is equivalent to the period of $T = 1\,\mu\text{s}$. Therefore, we calculate the number of required gates as

$$T = 2n\,t_d\,n\,t_d \quad \therefore \quad n = \left(\frac{T}{2\,t_d}\right) = \left(\frac{1\,\mu\text{s}}{2 \times 0.998\,\text{ns}}\right) = 501$$

where the "average" number comes from characterization of large number of manufactured digital gates in a given process.

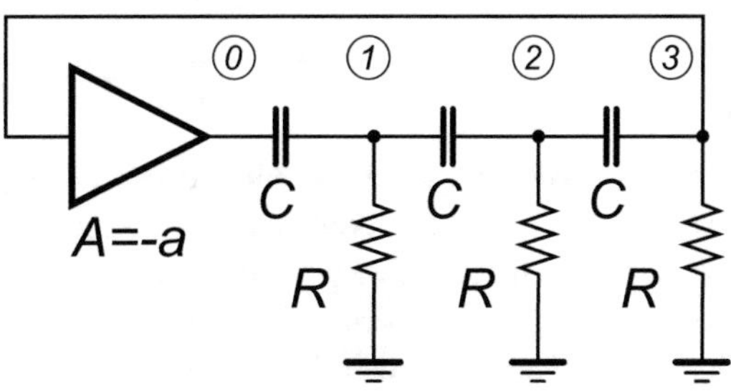

Fig. 13.4 Simplified schematic diagram of a phase shift oscillator

13.2.2 Phase Shift Oscillator

The oscillator architecture in Fig. 13.4 is known as a "phase-shift oscillator". An inverting amplifier with gain $A = -a$ that is used in the forward signal path is assumed to have infinite input resistance, i.e. there is no current flow into its input terminal. The feedback network with gain β consists of a classical RC ladder network of at least three RC sections. Although, this feedback network arrangement is also occasionally found with R and C interchanged, the arrangement shown here is more common. To satisfy the Barkhausen Criterion, a feedback path phase shift of exactly 180° is required in order to align the feedback signal with its initial phase, because the inverted amplifier gain introduces a first signal inversion of 180°. Therefore, the frequency of oscillation is equal to the frequency at which the phase shift introduced by the RC network is exactly 180°.

Systematic analysis of the ladder network starts at output node 3 of the network, which can be treated as the output node of the passive feedback path, and progresses back to input terminal 0 of the feedback path. Hence, by inspection of network in Fig. 13.4, we write

$$i_3 = \frac{v_3}{R} \qquad v_2 = v_3 + \frac{1}{j\omega C} i_3 = v_3 + \frac{v_3}{j\omega RC} \tag{13.7}$$

$$i_2 = \frac{v_2}{R} \qquad v_1 = v_2 + \frac{1}{j\omega C}(i_2 + i_3) = v_3 + \frac{3v_3}{j\omega RC} - \frac{v_3}{(\omega RC)^2} \tag{13.8}$$

$$i_1 = \frac{v_1}{R} \qquad v_0 = v_1 + \frac{1}{j\omega C}(i_1 + i_2 + i_3) = v_1 + \frac{(v_1 + v_2 + v_3)}{j\omega RC}$$

$$= v_3 + \frac{6v_3}{j\omega RC} - \frac{5v_3}{(\omega RC)^2} - \frac{v_3}{j(\omega RC)^3} \tag{13.9}$$

therefore,

$$v_0 = \left[v_3 - \frac{5v_3}{(\omega RC)^2}\right] + j\left[\frac{v_3}{(\omega RC)^3} - \frac{6v_3}{\omega RC}\right] = \Re(v_0) + j\Im(v_0) \tag{13.10}$$

The Barkhausen Criterion requires that the total phase shift around the loop should be exactly 2π. That is to say, the feedback network must introduce 180° by itself. If a complex number has 180° phase it means that its imaginary term must equal zero, i.e.

$$\Im(v_0) = 0 \quad \therefore \quad \frac{v_3}{(\omega RC)^3} - \frac{6v_3}{\omega RC} = 0 \quad \therefore \quad \omega_0 = \frac{1}{\sqrt{6}RC} \tag{13.11}$$

which defines the oscillation frequency. Substituting (13.11) into (13.10) gives

$$v_0 = v_3 - \frac{5\,v_3}{(1/\sqrt{6}RC)^2\,(RC)^2} \quad \therefore \quad \frac{v_3}{v_0} = -\frac{1}{29} = \beta \tag{13.12}$$

This is a very surprising result indeed: (13.12) states that the feedback path has a gain that is *independent* of the component values, i.e. $\beta = 1/29$. Following the Barkhausen Criterion $|\beta A| = 1$, we conclude that for this type of phase-shift oscillator, we must design the amplifier with inverting gain of at least $A = -29$. If the amplifier used inside the phase-shift oscillator has less than infinite input impedance, as would be the case for a real BJT amplifier, the derivations above would need to be modified. The modified equations are more difficult to solve and do not provide any further insight into this oscillator, thus they are omitted here.

Aside from being very good educational examples, phase-shift oscillators are used mostly at audio frequencies because the ladder RC network becomes impractical at higher radio frequencies.

Example 81: Phase Shift Oscillator
Estimate the minimum gain in dB of an inverting amplifier used in a phase-shift oscillator (Fig. 13.4).

Solution 81: To a first approximation, the oscillator loop gain must be at least one, hence the amplifier must compensate for the passive network attenuation of $\beta = 1/29 = -29.25\,\text{dB}$, by adding its own gain of $|A| = 29.25\,\text{dB}$. We keep in mind that the amplifier must be inverting to create the required phase.

13.3 RF Oscillators

Our introduction and treatment of RF oscillators follows a slightly different path from most textbooks. We first analyse four general RLC types of passive feedback network and then we use them inside some of the most common RF oscillator topologies. Although there is an infinite number of feedback network topologies that could be used in oscillators, if we impose the constraint that a minimal number of components is used, then only a small number of network topologies are suitable in the feedback path of an RF oscillator.

13.3.1 Tapped L, Centre-Grounded Feedback Network

Let us consider an RLC feedback network (see Fig. 13.5), while making the following assumptions:

1. The network operates near its resonant frequency ω_0.
2. The Q factor is high, i.e. ten or higher.
3. Inductors L_1 and L_2 are not coupled.
4. The network's Q factor is the effective Q of $L_1 + L_2$.
5. The inductors have equal values of Q factor.

In Sect. 13.1.1, we discussed a general model of a feedback loop (Fig. 13.2) and concluded that, in order to characterize the feedback network, the following three parameters are required: the loop's resonant frequency ω_0; the passive feedback path voltage gain β; and the effective input resistance R_{eff} of the feedback network.

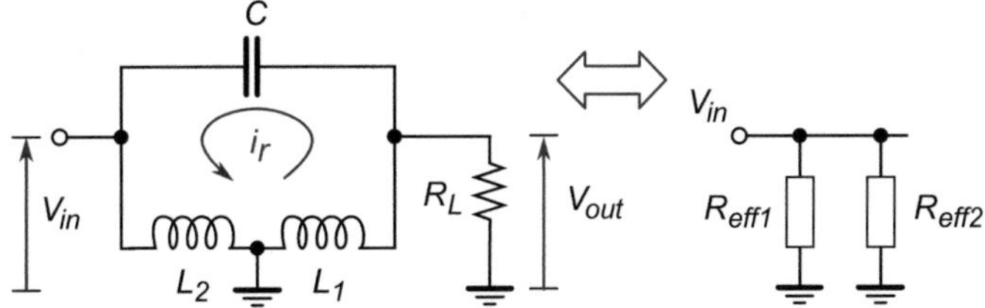

Fig. 13.5 Tapped L, centre-grounded network (left), and its equivalent representation for the energy dissipation calculations

In order to evaluate the resonant frequency ω_0, we need to recognize that the resonating current i_r stays within the L_1, L_2, C loop (see Fig. 13.5). The inductance of two inductors in series equals the sum of the two inductances, therefore we write

$$\omega_0^2 = \frac{1}{(L_1 + L_2)\,C} \tag{13.13}$$

The fact that the LC loop is tapped at two points (at the small signal ground between the two inductors and at the top of the loading resistor) does not influence the value of the resonant frequency—it is set by the total LC in the loop, as we concluded in Chap. 10.

The voltage gain $\beta = v_{out}/v_{in}$ of the feedback network can be evaluated as follows. At the resonant frequency, assuming a high Q factor, most of the power just circulates around the loop between the inductors and the capacitor (due to low thermal losses). The circulation of power around the resonant circuit may be represented by the continuous current i_r shown in Fig. 13.5 (left).

By inspection, we write the network equations as

$$v_{in} = i_r\; j\omega\, L_2 \qquad v_{out} = -i_r\; j\omega\, L_1 \quad \therefore \quad \beta \stackrel{\text{def}}{=} \frac{v_{out}}{v_{in}} = \frac{-i_r\; j\omega\, L_1}{i_r\; j\omega\, L_2} = -\frac{L_1}{L_2} \tag{13.14}$$

that is, the voltage gain β of a tapped L, centre-grounded feedback network is set by the inductive voltage divider.

Calculation of the effective resistance R_{eff} is a bit more complicated, due to its dependence upon the amplifier's input resistance value, which is modelled as the loading resistor R_L, Fig. 13.5 (left). Keep in mind that the input node of a feedback network is the one which is connected to the output node of the amplifier, while the output node of the feedback network is loaded by the input impedance of the amplifier, Fig. 13.5 (left). As already found, at resonance, the effective resistance R_{eff} of an LC resonator is purely resistive. Moreover, the resonator is loaded by impedance R_L (i.e. input impedance of the amplifier). One way of calculating the effective input impedance R_{eff} of the feedback network, which is a function of the load impedance, is by power calculation.

The total RMS power that is being put into the network is

$$P_{\text{rms}}(in) = \frac{v_{in}^2}{2\,R_{\text{eff}}} \tag{13.15}$$

Mathematically, we can imagine that the total input power is split between two effective resistances as follows.

One part of the input power is delivered to the external load impedance R_L at the output. After substituting (13.14), we write

$$P_{ext} = \frac{v_{out}^2}{2\ R_L} = \frac{\left[v_{in}\left(-\frac{L_1}{L_2}\right)\right]^2}{2\ R_L} = \frac{v_{in}^2\left(\frac{L_1}{L_2}\right)^2}{2\ R_L} = \frac{v_{in}^2}{2\ R_L\left(\frac{L_2}{L_1}\right)^2} = \frac{v_{in}^2}{2\ R_{\text{eff}}}$$

$$\therefore$$

$$R_{\text{eff}\,1} = R_L\left(\frac{L_2}{L_1}\right)^2 \tag{13.16}$$

Due to the finite Q factor, some of the total input power is dissipated in the internal LC circuit. At resonance, the LC resonator is equivalent to its dynamic resistance $R_D = Q\omega_0(L_1 + L_2)$ that is effectively connected between the top and the bottom of the LC circuit, i.e. between the input and output nodes in Fig. 13.5. Hence, the power P_{int} dissipated in this resistor is

$$\begin{aligned} P_{int} &= \frac{v_{R_D}^2}{2\ R_D} = \frac{(v_{in} - v_{out})^2}{2Q\omega_0(L_1 + L_2)} = \frac{\left[v_{in} + v_{in}\left(\frac{L_1}{L_2}\right)\right]^2}{2Q\omega_0(L_1 + L_2)} \\ &= \frac{v_{in}^2(L_1 + L_2)}{2\ Q\omega_0 L_2^2} = \frac{v_{in}^2}{2\ \frac{Q\omega_0 L_2^2}{L_1 + L_2}} \end{aligned}$$

$$\therefore$$

$$R_{\text{eff}\,2} = \frac{Q\omega_0 L_2^2}{L_1 + L_2} \tag{13.17}$$

Thus, from the input power perspective, because we referenced both powers P_{int} and P_{ext} relative to the input voltage v_{in}, it is being dissipated into two separate, parallel effective impedances, $R_{\text{eff}\,1}$ and $R_{\text{eff}\,2}$. Because both of these power dissipations occur simultaneously, the input power must be the sum of the two effective powers

$$P_{in} = P_{in1} + P_{in2} = \frac{v_{in}^2}{2R_{\text{eff}}} = \frac{v_{in}^2}{2(R_{\text{eff}\,1}||R_{\text{eff}\,2})} \tag{13.18}$$

Note that, from the power distribution perspective, the effective resistors are combined in parallel. Hence, the effective input impedance is estimated as

$$R_{\text{eff}} = R_{\text{eff}\,1}||R_{\text{eff}\,2} = R_L\left(\frac{L_2}{L_1}\right)^2\ ||\ \frac{Q\omega_0 L_2^2}{L_1 + L_2} \tag{13.19}$$

Equations (13.13), (13.14), and (13.19) define the three main parameters of a tapped L, centre-grounded feedback network. The oscillator design process now can be split into two parts: design of the amplifier for the forward signal path and design of the passive RLC network design. In order to acquire a complete set of commonly used feedback networks, we need to define three additional network configurations. The derivation process for the three main parameters of those network configurations is the same as the derivation process for the tapped L, centre-grounded feedback network. Therefore, the derivations of the formulas for ω_0, β, and R_{eff} for these RLC networks (see Fig. 13.5) are left as an exercise to the reader. Equations (8.17)–(8.25) complete the set of design parameters for passive RLC feedback networks that enable us to design commonly used RF oscillators.

We note that based on what type of the feedback network is used in oscillator, it is customary to refer to oscillators as Clapp (with tapped C, bottom-grounded network), Colpitts (with tapped C, bottom-grounded network), Hartley (with tapped L, bottom-grounded network).

13.3.2 Tapped C, Centre-Grounded Feedback Network

Figure 13.6 shows this type of feedback network. The equations for its main parameters are

$$\omega_0^2 = \frac{C_1 + C_2}{L C_1 C_2} \tag{13.20}$$

$$\beta = -\frac{C_2}{C_1} \tag{13.21}$$

$$R_{\text{eff}} = R_L \left(\frac{C_1}{C_2}\right)^2 \parallel Q\omega_0 L \left(\frac{C_1}{C_1 + C_2}\right)^2 \tag{13.22}$$

13.3.3 Tapped L, Bottom-Grounded Feedback Network

Figure 13.7 shows this type of feedback network. The equations for its main parameters are

$$\omega_0^2 = \frac{1}{C(L_1 + L_2)} \tag{13.23}$$

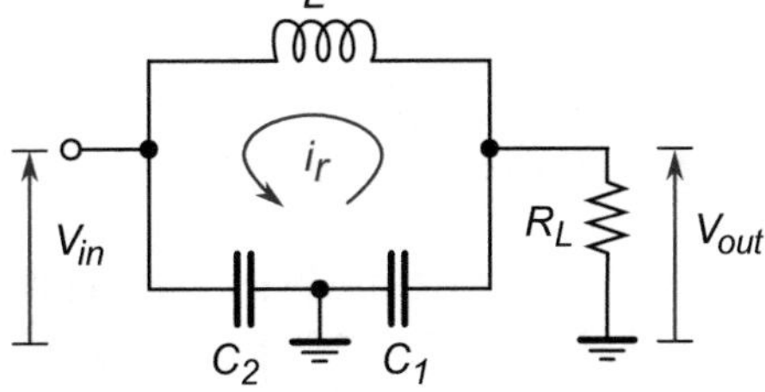

Fig. 13.6 Tapped C, bottom-grounded network

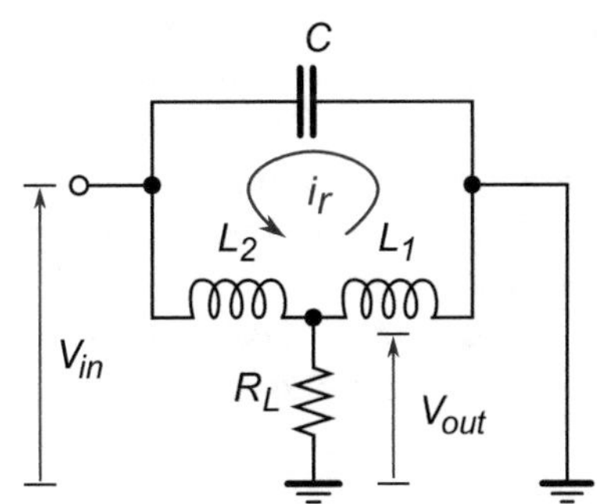

Fig. 13.7 Tapped L, bottom-grounded network, a.k.a. Hartley

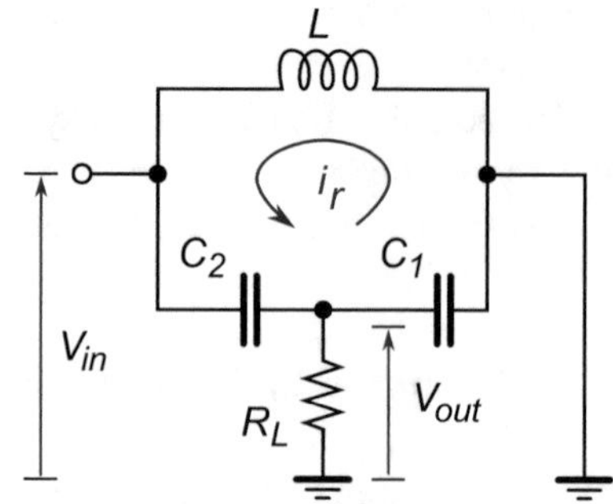

Fig. 13.8 Tapped C, bottom-grounded network, a.k.a. Clapp or Colpitts

$$\beta = \frac{L_1}{L_1 + L_2} \tag{13.24}$$

$$R_{\text{eff}} = R_L \left(\frac{L_1 + L_2}{L_1}\right)^2 \; || \; Q\omega_0(L_1 + L_2) \tag{13.25}$$

13.3.4 Tapped C, Bottom-Grounded Feedback Network

Figure 13.8 shows this type of feedback network. The equations for its main parameters are

$$\omega_0^2 = \frac{C_1 + C_2}{LC_1C_2} \tag{13.26}$$

$$\beta = \frac{C_2}{C_1 + C_2} \tag{13.27}$$

$$R_{\text{eff}} = R_L \left(\frac{C_1 + C_2}{C_2}\right)^2 \; || \; Q\omega_0 L \tag{13.28}$$

13.3.5 Tuned Transformer

An additional type of feedback network that is a member of the same family of RLC networks is the "tuned transformer", which is also very often used in RF sinusoidal oscillators. A tuned transformer is a type of passive feedback network that uses the primary, the secondary, or both transformer coils in parallel with their respective capacitors to create LC resonator tanks. A tuned transformer is said to be either inverting or non-inverting depending upon the relative orientation of the primary and secondary coils. These properties indicate that a transformer is very versatile device that allows for almost arbitrary ratios of the primary inductance L_P to the secondary inductance L_S, with the additional level of freedom to introduce a phase shift of either 0° or 2π between the primary and secondary sides.

In our brief analysis of non-inverting, primary tuned transformers, Fig. 13.9 (left), we make the following assumptions:

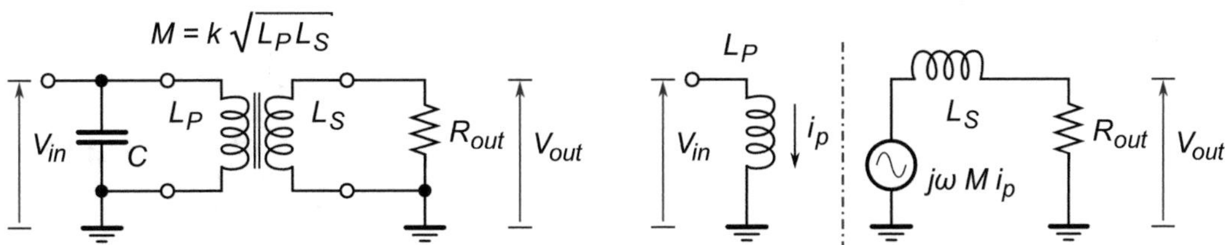

Fig. 13.9 Tuned primary transformer network (left), and its equivalent network (right)

1. The coupling factor k, $(0 \leqslant k \leqslant 1)$ between the primary and secondary is low, i.e. $k \ll 1$. Therefore, the loading effect of the primary on the secondary, and vice versa, may be ignored.
2. The output load resistance is much greater than the secondary impedance, i.e. $R_{out} \gg \omega L_S$. If this condition is not true, an additional phase shift is induced and the frequency of oscillation is different from ω_0.

In the equivalent circuit diagram, Fig. 13.9 (right), coupling between the primary and the secondary is represented by the AC voltage source in the secondary branch. The dashed vertical line indicates that the secondary and the primary are separate circuits. The three parameters of tuned transformer feedback network are determined as follows.

In order to determine the voltage gain factor β, we start by writing an expression for current in the primary network as

$$i_P = \frac{v_{in}}{Z_P} = \frac{v_{in}}{j\omega\, L_P} \tag{13.29}$$

which induces voltage in the secondary:

$$v_{ind} = \pm j\omega\, M\, i_P = \pm j\omega\, M\, \frac{v_{in}}{j\omega\, L_P} = \pm \frac{M}{L_P}\, v_{in} \tag{13.30}$$

where the $\pm$ sign indicates the phase difference between the primary and the secondary, which depends on the orientation of the transformer coils, and $M = k\sqrt{L_P L_S}$ is the mutual inductance. If the condition $R_{out} \gg j\omega\, L_s$ is satisfied, then it follows that

$$v_{out} = v_{ind} \tag{13.31}$$

which, after substituting (13.31) into (13.30) and rearranging, yields an expression for the voltage gain of the tuned amplifier as

$$\beta = \pm \frac{M}{L_P} \tag{13.32}$$

The effective resistance R_{eff} of this network is approximately just the impedance of the primary, because we assumed $k \ll 1$. Therefore, at resonance, we write

$$R_{\text{eff}} \approx Q_P \omega_0\, L_P \tag{13.33}$$

where Q_P is the Q factor of the primary transformer.

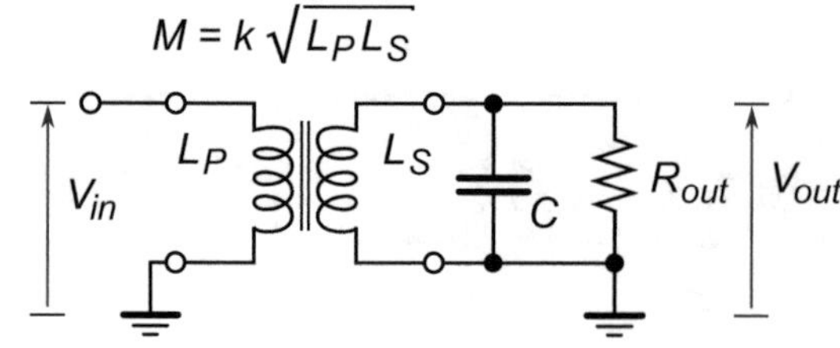

Fig. 13.10 Tuned secondary transformer network

The resonant frequency ω_0, we simply write as

$$\omega_0^2 = \frac{1}{L_P\, C} \tag{13.34}$$

Let us take a brief look at the case of a tuned secondary transformer network (Fig. 13.10). The analysis is similar to the previous case and it leads to the following expressions for the three parameters of the network:

$$\omega_0^2 = \frac{1}{L_S\, C} \tag{13.35}$$

$$\beta = \pm \frac{j M Q_{S\,\text{eff}}}{L_P} \tag{13.36}$$

$$R_{\text{eff}} = j\omega\, L_P \tag{13.37}$$

The assumptions made in this section are often not used in textbooks, which are usually for a specific type of oscillator network.

13.4 Amplitude-Limiting Methods

At this moment it is logical to ask the following questions:

1. If noise, with its infinite frequency spectrum, is responsible for starting the oscillation process, how is it that at the output terminal we see only a single tone?
2. If the output signal is amplified with each pass through the loop, what keeps its amplitude stable and finite in real circuits?

To answer the first question, we recall that although the internal thermal noise is responsible for providing the initial stimulus to the input terminals of the forward path amplifier, the feedback RLC path is designed to be a very selective bandpass network (by means of the high Q factor). Hence, of all possible tones from the noise frequency spectrum only the one with frequency equal to ω_0 is actually amplified, while all other tones are suppressed. That frequency-selection behaviour of oscillator feedback is the key property of any oscillator circuit.

Now that we have determined intuitively how an oscillator locks on a single tone, the second question needs to be answered. We have already concluded that, in order to start oscillations, it is necessary to design the loop gain larger than unity. Consequently, immediately after powering up the oscillator, the output signal amplitude increases with each passing cycle. Eventually, if the gain stays constant, the output signal becomes a non-sinusoidal (i.e. square) waveform with amplitude that theoretically increases indefinitely. Therefore, in order to generate a non-distorted sinusoidal waveform, some form of amplitude-limiting mechanism is required that prevents the signal amplitude from becoming too large. A few amplitude-limiting schemes are listed here.

1. Automatic Gain Control (AGC)
2. Clamp Biasing
3. Gain Reduction by Use of Temperature Dependent Resistors
4. Device Saturation Combined with a Tuned Output

We leave studies of these techniques to the more advanced courses. For the time being we accept that there are multiple limiting mechanisms that help oscillating circuits produce practically perfect sinusoidal waveforms.

13.5 Crystal-Controlled Oscillators

Piezoelectric crystalline materials, quartz being one of the best known, exhibit reciprocal properties relative to their mechanical and electrical behaviour. That is, if an electric potential is applied across a thin sheet of piezoelectric crystal, it physically bends. In return, if a piezoelectric crystal is physically deformed, then the internal electrical charges are separated and a voltage is produced across its plates. Consequently, if a sinusoidal electrical signal with frequency that is equal to the crystal's mechanical resonant frequency is applied across its plates, then a sheet of piezoelectric crystal exhibits both electrical and mechanical resonance. Moreover, the mechanical resonant frequency is very stable and can be controlled over several orders of magnitude by precisely cutting the quartz sheet into specific shapes and dimensions. Typical values of the fundamental tone resonant frequency are in the range of kHz to about 50 MHz. For higher frequencies, the physical dimensions of the crystal become too small and higher-order resonant tones are used instead. In other words, a crystal with a fundamental resonance at 30 MHz can also be used at 60, 90, 120 MHz, and sometimes even at 150 MHz.

The resonant frequency stability of ordinary crystals at room temperature is in the order of about one part in a million (1 ppm); an order of magnitude improvement in stability can be achieved if the crystal is mounted inside a temperature-controlled oven. By using special technologies, the upper achievable limit of frequency stability for crystals is about 0.1–1 ppb, i.e. less than one part in a billion. To put it in perspective, 0.1ppb is equivalent to a ratio of 10^{10}, which is equivalent to approximately 1 s in 300 years. Not surprisingly, the main application of piezoelectric crystals in electronics is to serve as timing references for clock signals. Although the manufacturing process of crystals is not difficult by modern standards, in practice it is common that the crystals are manufactured to precisely match several "standard" reference frequencies that are commonly used in wired and wireless communications. Of course, modern circuits operate at frequencies far higher than the above-mentioned 150 MHz overtone. In addition, the crystal's resonant frequency is fixed by its physical dimensions (i.e. the resonant frequency is precise but is not tunable); by itself, crystal produces very tiny currents meaning that it always needs some active buffering circuit that improves its driving capability. Various frequencies are derived from the crystal reference frequency by means of a closed-loop circuit known as a "phase-locked loop" (PLL).

Accurate behavioural modelling of piezoelectric crystals requires that a set of differential equations describing both mechanical and electrical properties is solved, which is usually done numerically using modern multiphysics simulators. In our work, however, we are concerned only about the electrical properties of the crystals, which may be described in terms of a passive RLC electrical network (Fig. 13.11). A typical simple electrical model is based on a serial RLC branch in parallel with a small capacitor, Fig. 13.11 (right). For instance, values used to model a $f_0 = 9.545\,\text{MHz}$

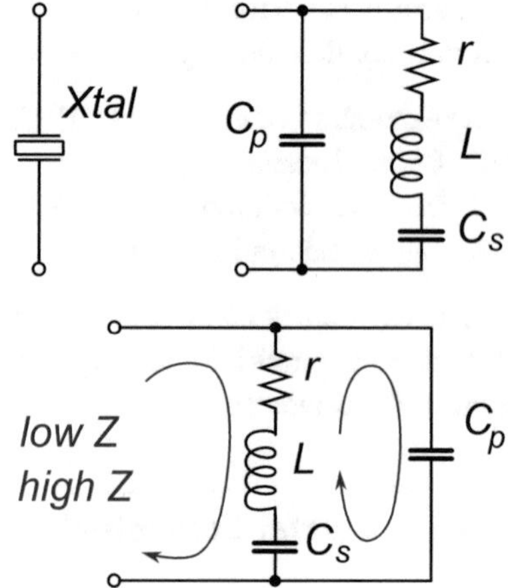

Fig. 13.11 The symbol for a quartz crystal (left); a basic electrical model (right)

Fig. 13.12 Serial and parallel resonant modes of quartz crystal

crystal (that is used in AM radio receivers) are as follows: $L = 2.54647909\,\text{mH}$, $r = 6.4\,\Omega$, $C_s = 109.1813\,\text{fF}$, and $C_p = 24.8844947\,\text{pF}$, therefore (see Sect. 10.3)

$$Q = \frac{\omega_0 L}{r} = \frac{1}{\omega_0\, r\, C_s} = \frac{1}{2\pi \times 9.545\,\text{MHz} \times 6.4\,\Omega \times 109.1813\,\text{fF}} \approx 24 \times 10^3 \tag{13.38}$$

Note that, for the frequency of operation, this is a very high value of the inductance, which is combined with a small internal resistance r to yield a high Q factor. However, it is the mechanical property of the crystal, rather than the electrical property, that gives this equivalent high inductance value. In addition, we note that in this model the numbers are expressed with large number of decimal places.

Because of the two parallel branches, there are two possible resonant frequencies: a series resonance that is determined only by the serial branch of the equivalent circuit in Fig. 13.12 and a parallel resonance determined by both the series branch and the parallel capacitance C_p (Fig. 13.12). In order to demonstrate these two important resonant modes of a crystal, let us evaluate our example model. The series resonance (i.e. in low impedance mode) is approximately

$$\omega_s = \frac{1}{\sqrt{LC_s}} = \frac{1}{\sqrt{2.54647909\,\text{mH} \times 109.1813\,\text{fF}}} = 59.973 \times 10^6 \ \text{[rad/s]}$$

$$\therefore$$

$$f_s = \frac{\omega_s}{2\pi} = 9.5450\,\text{MHz} \tag{13.39}$$

In the parallel resonance (i.e. high impedance) mode, the capacitances C_s and C_p are perceived as being in series around the resonant loop (thus, the equivalent capacitance is

$$\frac{1}{C_{eq}} = \frac{1}{C_s} + \frac{1}{C_p} = \frac{1}{109.1813\,\text{fF}} + \frac{1}{24.8844947\,\text{pF}} = \frac{1}{108.704357\,\text{fF}} \tag{13.40}$$

which, of course, is just a little bit smaller than C_s, therefore it sets just a little bit higher resonant frequency of

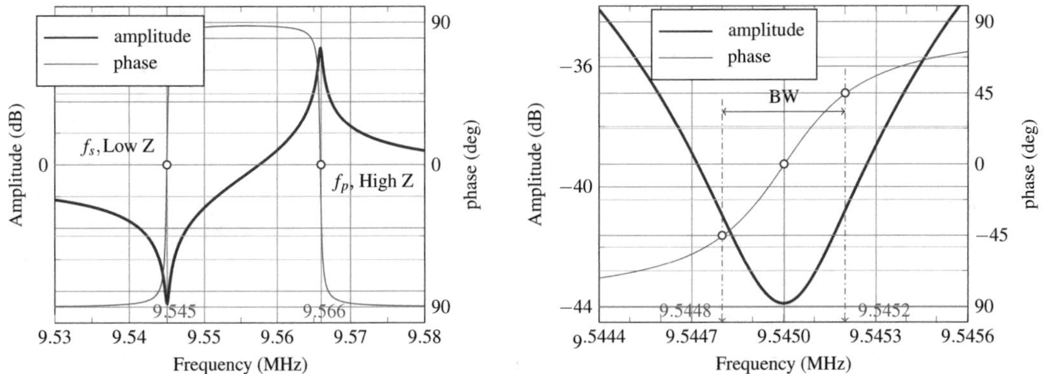

Fig. 13.13 XTAL AC simulations: serial and parallel resonance (left) and zoom-in at the serial resonant frequency (right)

$$\omega_p = \frac{1}{\sqrt{L\,C_{eq}}} = \frac{1}{\sqrt{2.54647909\,\text{mH} \times 108.704357\,\text{fF}}} = 60.1044278 \times 10^6 \;\; [\text{rad/s}]$$

$$\therefore$$

$$f_p = 9.5659\,\text{MHz} \tag{13.41}$$

Note that these frequencies are very close to one another, about 0.2% apart in this example. Simulation of the model in Fig. 13.12 with the component values as in (13.38) to (13.41) confirms the two resonant modes, Fig. 13.13 (left). In addition, note very rapid phase change of 180° at each of the two frequencies, which is the consequence of very high Q factor. A zoom at, for example, serial frequency $f_s = 9.5450\,\text{MHz}$ region shows that phase change from capacitive (i.e. $-45°$) to inductive (i.e. $+45°$)[5] happens over very narrow bandwidth BW, which can be used to confirm the Q factor by simulations as

$$BW \approx (9.545196 - 9.544803)\,\text{MHz} = 393\,\text{Hz}$$

$$\therefore$$

$$Q \overset{\text{def}}{=} \frac{f_0}{BW} = \frac{9.545\,\text{MHz}}{393\,\text{Hz}} = 24.3 \times 10^3 \approx 24 \times 10^3 \tag{13.42}$$

where a longer simulation with more points would result in better approximation. This rapid change of phase requires about 40 ppm change of frequency (i.e. 392 Hz versus 9.545 MHz). Aside from this important property of dual resonant frequency, we also need to examine how the impedance of a piezoelectric element behaves at frequencies in proximity to these two resonant frequencies, Fig. 13.13. At frequencies below the serial resonance ω_s the crystal reactance is dominated by the large serial capacitance, i.e. it is negative. At the serial resonant frequency ω_s the reactance is zero (i.e. $Z_L = Z_C$) and the overall impedance is at its minimum close to zero, i.e. $Z = r$. Between the two resonant frequencies, the reactance is inductive and tends to infinity (in reality, a very high value), while the overall impedance follows the trend. At the parallel resonant frequency, ω_p the overall impedance is at its maximum. Above the parallel resonant frequency, the reactance is again negative and the overall impedance decreases. Due to the proximity of these two modes of operation,

[5]See Sects. 3.2.2 and 3.2.3.

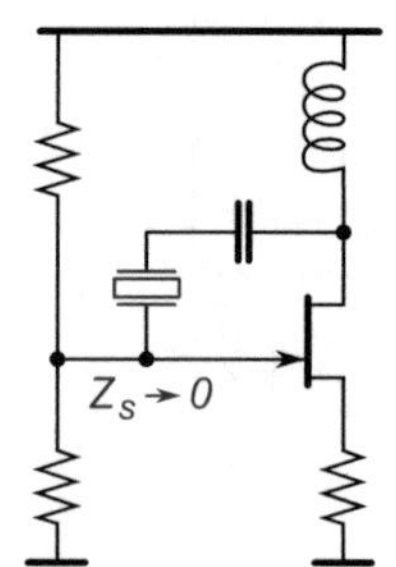

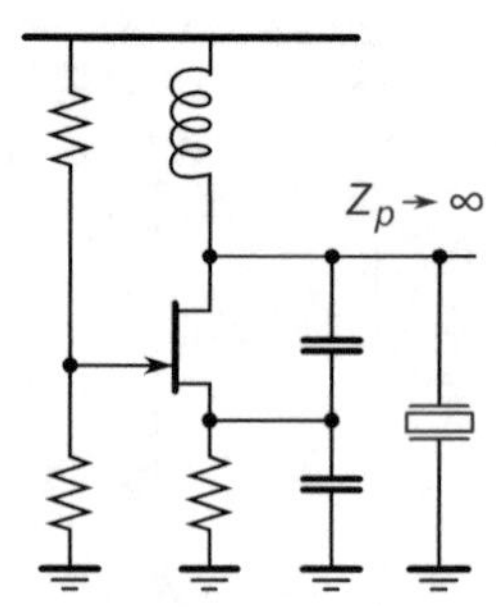

Fig. 13.14 Series mode crystal controlled oscillators, i.e. low impedance mode (left), and parallel (i.e. high impedance, a.k.a Pierce oscillator) mode crystal controlled oscillators (right)

it is very important to recognize that serial resonance is associated with minimum overall impedance and parallel resonance is associated with very high impedance; this determines how crystals are used in oscillator circuits.

Many different oscillator circuit arrangements use crystals. However, the general rule is that the low impedance mode (i.e. at series resonance) is used when the crystal is connected in series with the other elements (the two diagrams on the left of Fig. 13.14) and the high impedance mode is used in parallel with other circuit elements (the two diagrams on the right of Fig. 13.14). In other words, aside from controlling the oscillator's resonant frequency, insertion of a crystal into an oscillator should cause minimum interference with flow of the internal currents.

13.6 Voltage-Controlled Oscillators

The ability to generate a single-tone, sinusoidal waveform with precisely controlled frequency is of vital importance for wireless communication systems and much engineering effort has gone into designing various forms of oscillator. However, communication systems require more than just one specific value of the frequency. For instance, every radio and TV receiver is capable of receiving signals from more than one transmitting station. As we already know from our daily lives, in order to select the desired station, we must tune the receiver to the particular frequency associated with the station. If we were only able to design an oscillator circuit capable of delivering a single frequency, we would either need to carry one radio receiver unit for each station to which we would like to listen or our receivers would be very bulky and complicated indeed. Obviously, that is not the case; we invented frequency-tunable oscillators whose output frequency depends on a control variable, either voltage or current, i.e. $\omega_0 = f(V_{\mathtt{ctrl}}, I_{\mathtt{ctrl}})$. The resonant frequency of an LC resonator is determined by component values of the inductor L and the capacitor C that are used in the resonator tank. Therefore, frequency tunability (or simply, "tunability") is achieved by varying the value of the inductance, the value of the capacitance, or both. In principle, there are two possible ways of implementing a variable capacitor or inductor, "discrete" and "continuous".

The "discrete" method simply means that a bank of serial or parallel components is connected and each component is independently switched in or out of the network. Obviously, this method is feasible only with a finite number of components and switches, hence it can deliver only a discrete set of component values. If a relatively fine change of the capacitive or inductive value is achieved with each switching step, then it is sometimes referred to as a "quasi-continuous" method.

A truly "continuous" method means that a component is capable of physically changing its value smoothly in response to the control variable. For instance, a rotary variable capacitor (Fig. 13.15) is created by mechanically controlling the overlapping area between two plates of a capacitor. According to (2.26) the capacitance of a plate capacitor is a linear function of the capacitor's surface area S, which in this case means the *overlapping* area between the two plates. The overlapping capacitor area

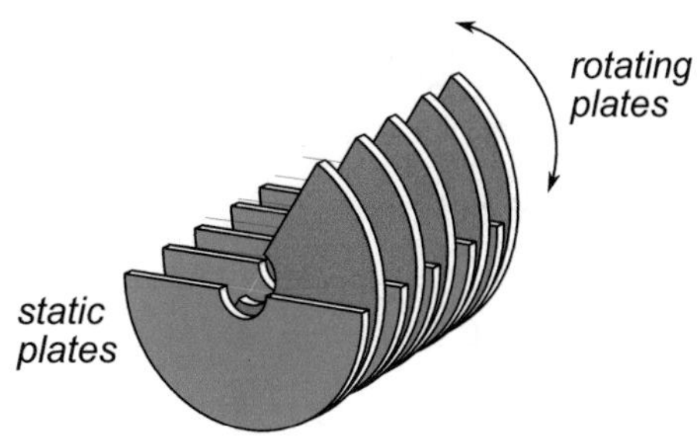

Fig. 13.15 Rotary air-gap variable capacitor used on radios for tuning the RF stage and the local oscillator

varies with the angle of rotation. Each pair of static and rotating plates makes one capacitor—there are five pairs in Fig. 13.15 that make, in total, nine tunable capacitors in parallel.

Both methods of creating variable components (discrete and continuous) are used in practice. From a practical perspective, it is much easier to design and manufacture tunable capacitors than tunable inductors. Indeed, for over 100 years, the rotary variable capacitor was used almost exclusively for continuous tuning of LC resonators in commercial radio receivers. As you have already noticed, this kind of capacitor is very bulky and long ago it became the largest component by far inside a radio receiver. This implies that the mechanical rotary capacitor is not suitable for miniaturization and higher frequencies. This is further limited because tunability is achieved by manual control of a knob. Although miniature versions of fundamentally the same design, the trimmer capacitor, are still in use for semi-permanent tuning of radio receiver sections, modern high-frequency (HF) oscillators are based on a semiconductor device known as a "varicap diode" (varactor), see Sect. 4.2.4. At the same time, design of miniature variable inductors is still in the research domain, with some progress being made mostly due to advances in micro-electro-mechanical system (MEMS) technologies.

A simplified schematic diagram (Fig. 13.16 (left)) of RF amplifier with electronically tunable LC resonator shows a varicap capacitor C_D and its DC biasing voltage V_D connected to an LC resonator tank through a large capacitor. The role of the C_∞ capacitor is simply to block the DC voltage V_D that keeps varicap C_D reverse biased without interfering with the biasing of the BJT current source, it is assumed a short at all non-zero frequencies. From an AC perspective, the tunable capacitor $C_D = f(V_D)$ is connected in parallel with the resonator capacitor C (see Fig. 13.16 (right)), the inductor L, and BJT collector resistance r_o, hence the resonating frequency is calculated as

$$\omega_0 = \frac{1}{\sqrt{L\,(C + C_D(V_D))}} \quad \therefore \quad \omega_0 = f(V_D) \tag{13.43}$$

Now that we understand how a varicap diode is used in an LC resonator tank to control its resonant frequency, let us take a look at a simplified schematic diagram (Fig. 13.17) of a VCO circuit that incorporates, for instance, a tapped C bottom-grounded passive feedback network (see Fig. 13.8) with a CB amplifier (see Fig. 6.7). In order to implement electronic control of its resonant frequency, a varicap diode is added[6] in the LC resonating loop, which is perceived by the loop as being in series with the C_1 and C_2 resonator capacitors. The CB amplifier is biased through an RFC (a.k.a. a "choke") inductor that provides DC connection to the power supply line and, at the same time, blocks AC current in the RF range. In textbooks, an oscillator configuration that uses a tapped C, bottom-grounded feedback network is usually referred to as a "Clapp oscillator".

In the first approximation, i.e. ignoring parasitic elements, the resonant frequency of the oscillating current is approximately set by the passive feedback network only. Capacitors C_1 and C_2 are in series, hence their equivalent capacitance C_S is

[6]For simplicity, varicap DC biasing is not shown.

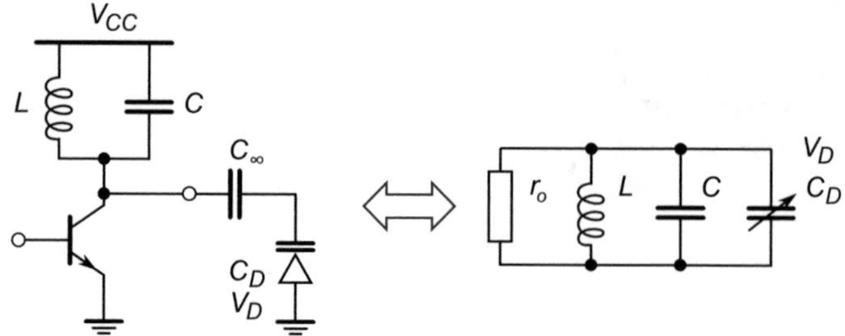

Fig. 13.16 Schematic diagram of RF amplifier showing principle of an LC resonator tuning by the means of varicap voltage control voltage V_D. AC equivalent resonator is shown on the left

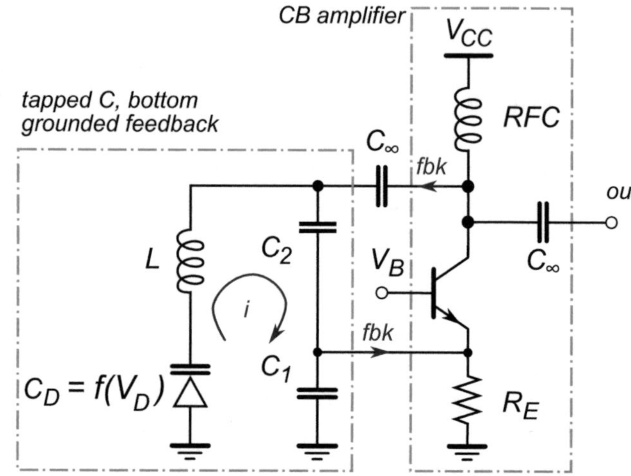

Fig. 13.17 Simplified schematic diagram of VCO that uses CB amplifier in the forward path, and tapped C bottom-grounded network in the feedback path

$$C_S = \frac{C_1 C_2}{C_1 + C_2} \tag{13.44}$$

It, in turn, is in series with the varicap, hence the total equivalent capacitance C in the LC loop is written as

$$C = \frac{C_D C_S}{C_D + C_S} \quad \therefore \quad \omega_0 = \frac{1}{\sqrt{LC}} \tag{13.45}$$

For all practical purposes, we can look at a VCO as a *voltage-to-frequency converter*.

Example 82: Clapp VCO

For a given LC Clapp oscillator the resonant frequency is set to $\omega_0 = 2\pi\ 10\,\text{MHz}$ by varicap diode voltage by $V_0 = 6\,\text{V}$. The ratio of varicap capacitance to the serial capacitance in the resonator tank is $n = 0.1$. Estimate the frequency deviation constant k of this oscillator.

Solution 82:

A straight implementation of (4.26) results in

$$k == -\frac{5}{4}\frac{f_0}{(1-2V_D)(1+n)} = -\frac{5}{4}\frac{10\,\text{MHz}}{2\,(1+0.1)\,(1+2\times 6\,\text{V})} = -437\frac{\text{kHz}}{\text{V}}$$

(continued)

We conclude that, for each volt of change in varicap bias, the oscillator resonant frequency moves about 437 kHz around f_0 frequency. Depending upon characteristics of the specific diode used in this design, the tuneable frequency range is found in accordance with graphs similar to Fig. 4.10.

13.7 Time and Amplitude Jitter

A realistic periodic waveform produced by an oscillator suffers from a short-term frequency fluctuation that is referred to as "phase noise". At the same time, amplitude variations of the waveform are always present to a certain extent. For instance, an oscillator's sinusoidal output with phase variations $\varphi(t)$ and amplitude variations $A(t)$ may be expressed as

$$v_s(t) = V_s[1 + A(t)]\ \sin[\omega_c t + \varphi(t)] \tag{13.46}$$

where V_s is the average peak voltage of the output signal. The phase and amplitude variations may be random or discrete or both. Individual spectral components at the oscillator output are referred to as "spurious responses"; noise, in this context, refers to the random variations of both frequency and phase. The engineering term for these random variations is "jitter". A very useful and practical method for estimating the amplitude and time jitter of a periodic signal with period T is to create an "eye plot" (Fig. 13.18). Instead of plotting a waveform from time zero to the last data point, the full data vector is split into sections so that each section contains a set of data only half a period long. Then, all sections are overlapped (similar to a deck of playing cards). The newly created plot looks similar to an open eye if the amount of jitter is not too excessive. Amplitude jitter becomes clearly visible and easily measurable on the vertical axis, while timing jitter is easily measured on the horizontal axis around the cross-over point between rising and falling edges of the waveform (see Fig. 8.16). Commonly, timing jitter is expressed relative to the waveform period T, e.g. T, e.g.

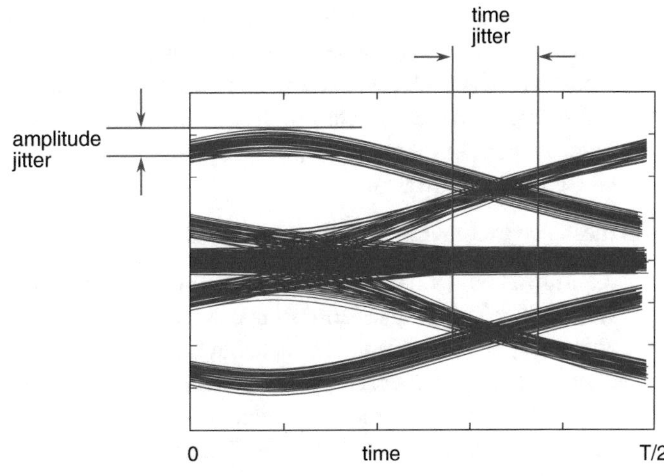

Fig. 13.18 Eye diagram of a long waveform showing time and amplitude jitter. The periodic waveform is split in sections each half a period long, i.e. T/2, and overlapped

$t_{\texttt{jitter}} = T/8$. Almost all modern oscilloscopes have a built-in eye diagram function, which makes it extremely easy for the user to create the plot in real time.

Detailed statistical analysis of phase noise is the subject of more advanced courses in communication theory, hence it is omitted in this text.

13.8 Case Study: RF Oscillator

Here, we learn one possible procedure to design an RF oscillator intended for AM receiver application. The AM radio carrier frequencies are in the frequency range of 540–1600 kHz that are assigned at 10 kHz intervals. The only purpose of this oscillator is to serve as the local reference that is needed to facilitate the shift of AM carrier frequency down to the intermediated frequency (IF), which is standardized to $f_{\text{IF}} = 455\,\text{kHz}$. In other words, if we are to receive, for example, $f_0 = 10\,\text{MHz}$ AM modulated carrier signal, then the local oscillator (LO) frequency must be $f_{\text{LO}} = f_{\text{IF}} - f_0 = (10000 - 455)\,\text{kHz} = 9.545\,\text{MHz}$.

An oscillator is built by an amplifier and the feedback LC network, see Sect. 13.1.1. In regard to the choice of amplifier and feedback network, as the starting point, in the preliminary considerations we decide the following practical issues:

1. *Power supply:* similarly to the RF amplifier studied in Sect. 12.3, we assume $V_{DD} = 15\,\text{V}$.
2. *Inductor:* There is no specific requirement for the inductor except that $Q \geq 10$, thus we reuse $L = 3.3\,\mu\text{H}$ inductor that was characterized in the RF amplifier example. Its $Q = 10.2$ and when connected in parallel with $C = 84.2508824\,\text{pF}$ the LC resonator bandwidth is $BW = 938\,\text{kHz}$ at $f_0 = 9.545\,\text{MHz}$. In addition, as per (10.52) its dynamic resistance is calculated as

$$R_D = Q\,2\pi\,f_0\,L = 2019\,\Omega \approx 2\,\text{k}\Omega$$

   which is also confirmed by simulation.
3. *Transistor:* again, we reuse JFET transistor from Sect. 12.3 whose DC characteristics and biasing point are: $(I_{DSS}, V_P) = (12\,\text{mA}, -3\,\text{V})$, $(I_D, V_{GS}) = (6\,\text{mA}, -884\,\text{mV})$, $g_m = 1/r_s = 5.67\,\text{mS}$, $r_s = 176.3\,\Omega$, and $r_o = 77\,\text{k}\Omega$.
4. *Amplifier:* although it is not mandatory and any amplifier type could be used to make an oscillator, here for the oscillator design we use CG amplifier due to its simplicity and wide frequency range. In addition, its current mode of operation is suitable to drive LC resonator that is used in the feedback path (which is also the method used in the simulations of LC resonators).
5. *LC Resonator:* in principle, any of the LC networks studied in Sect. 13.3.1 to 13.3.5 could be used. Let us choose, for example, tapped C, bottom-grounded feedback network shown in Sect. 13.3.4 and (see Fig. 13.8 (right)).

With the above preliminary decisions, we develop CG oscillator for $f_0 = 9.545\,\text{MHz}$ as follows.

1. **CG amplifier**: DC setup of CG/CB amplifier is simpler than for CS/CE amplifiers, see Fig. 13.19 (left). The gate terminal is connected to DC ground by a large resistor and, in order to further suppress possible noise, capacitor C_S is added in parallel. Values of these two components are not critical, however their pole frequency $\omega = 1/\tau = R_G C_G$ should be low.

   Due to the fact that the gate potential is set to zero, it is sufficient to elevate potential at the source terminal to $V_S = +884\,\text{mV}$. Therefore, the gate to source voltage is set at $VGS0) = -884\,\text{mV})$. Given that $I_S = I_D = 6\,\text{mA}$, as a result $R_S = 884\,\text{mV}/6\,\text{mA} = 147\,\Omega$.

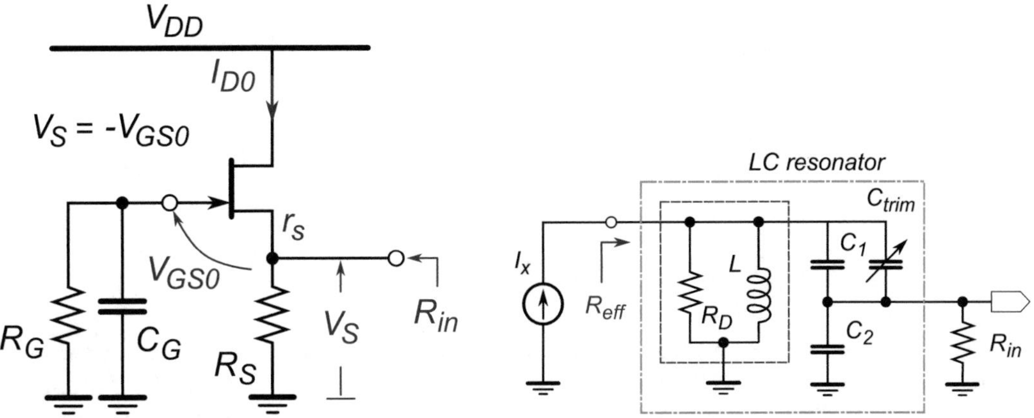

Fig. 13.19 CG amplifier DC setup

With the I_D current set, we now calculate the input resistance of CB amplifier. Looking into the source, by inspection[7] we find

$$R_{\text{in}} = R_S || r_s = 147\,\Omega || 176\,\Omega = 80\,\Omega$$

2. **LC feedback resonator**: we recall parameters β, ω_0, and R_{eff} of tapped C, bottom-grounded LC network, as already shown in (13.26) to (13.28), by repeating the three formulas:

$$\beta = \frac{C_2}{C_1 + C_2} \quad ; \quad \omega_0^2 = \frac{C_1 + C_2}{LC_1C_2} = \frac{1}{L\,C_1\,\beta} \quad ; \quad R_{\text{eff}} = \frac{R_{\text{in}}}{\beta^2} \;||\; Q\omega_0 L$$

where we note that loading R_{in} is the loading resistance for the feedback network. We should note that design of LC resonating network is an iterative process that requires very precise tuning of circuit parameters. In practice, the exact values of capacitors are achieved by using trimmer capacitors in parallel, Fig. 13.19 (right). What is more, depending on the component models and the simulator, often one has to patiently tune the simulator itself and sometimes run rather long simulations until satisfactory results are achieved. That being said, the final tuning and fine adjustments are always in the actual prototype of the circuit.

The goal is to determine capacitors C_1, C_2 so that the Barkhausen gain criterion for oscillations[8] $A = A_V\,\beta \geq 1$ is satisfied. In terms of the Barkhausen phase criterion, CG does not invert phase of its signal. Further around the feedback loop, when LC network is in resonance, its dynamic resistance is real, therefore the input and output signals are in phase. In total, the loop phase is zero, thus both conditions for the oscillations are satisfied.

We can start, for example, by assuming a modest voltage gain of CG amplifier. This assumption is valid, because the LC feedback resonator is loaded by the rather low $R_{\text{in}} = 80\,\Omega$, thus its effective resistance R_{eff} is reduced, which in effect limits the overall voltage gain. Let us assume $A_V = 5$ of the transistor and the load resistance only, i.e. excluding the voltage divider at the input side that unavoidably reduces the theoretical gain limit. Necessarily, in order to satisfy the Barkhausen gain criterion it follows that $\beta \geq 1/A_V = 0.2$, which gives us the opportunity to do the calculations as follows.

[7] See Sect. 6.3.2 and assume that $\beta = \infty$ for FET transistors.

[8] See Sect. 13.1.1.

$$\omega_0^2 = \frac{C_1 + C_2}{LC_1C_2} = \frac{1}{L\,C_1\,\beta}$$

$$\therefore$$

$$C_1 = \frac{1}{L\,\omega_0^2\,\beta} = 421.25441\,\text{pF} \quad \therefore \quad C_2 = \frac{1}{\dfrac{1}{\beta} - 1} = 105.31360\,\text{pF}$$

$$R_{\text{eff}} = \frac{R_{\text{in}}}{\beta^2} \;||\; Q\omega_0 L = 2\,\text{k}\Omega||2\,\text{k}\Omega = 1\,\text{k}\Omega$$

3. **Open-loop simulations**: we need to verify that the amplifier indeed perceives dynamic resistance $R_D = 1\,\text{k}\Omega$, and provides the gain as expected. This test is done in the "open-loop" configurations, that is to say, as if this was a CG RF amplifier with the LC resonator serving as the load and the output is taken at the drain node. By inspection[9] of Fig. 13.20 (left), similar to (6.30), we calculate voltage gain as

$$A_v = A_1 \times A_2 = \frac{R_{\text{in}}}{R_{\text{x}} + R_{\text{in}}} \times g_m\,(R_D||r_o) = \frac{80\,\Omega}{80\,\Omega + 80\,\Omega} \times 5.67\,\text{mS}\,(2\,\text{k}\Omega||77\,\text{k}\Omega)$$
$$\approx 0.5 \times 11 = 5.5$$

where, in order to illustrate the idea, in this numerical example the signal generator's resistance is $R_{\text{x}} = 80\,\Omega$ and $V_x = 100\,\text{mV}$. As originally assumed, this voltage gain is not too high (it is also function of the source resistance), and it is easily confirmed by simulations, Fig. 13.20 (right). We note that due to LC elements, there is about 1.5 µs transition period before the full signal amplitude at the output is reached. Of course, being an RF amplifier, average of the output voltage signal is at V_{DD} and the maximum voltage is above the power supply rail.

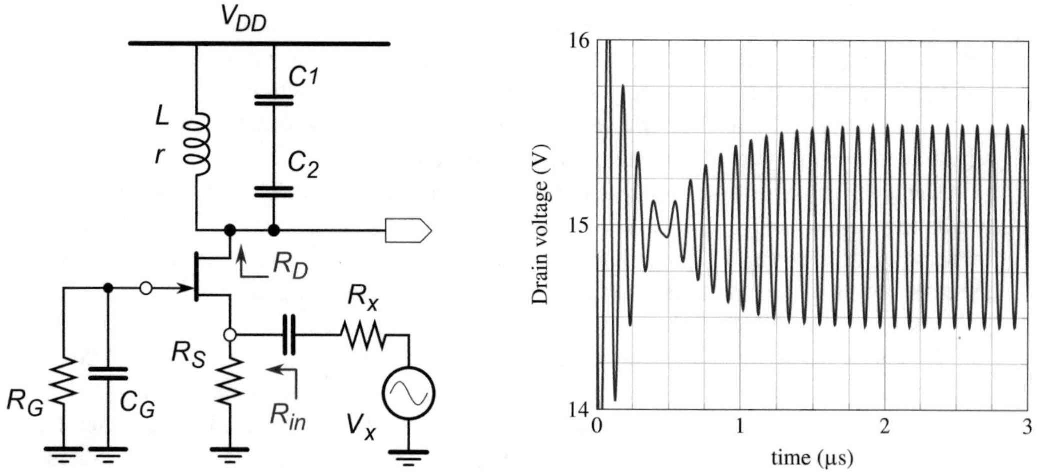

Fig. 13.20 CG oscillator—open loop simulation, schematic diagram (left) and time domain waveform (right)

[9] See Sect. 6.3.4.

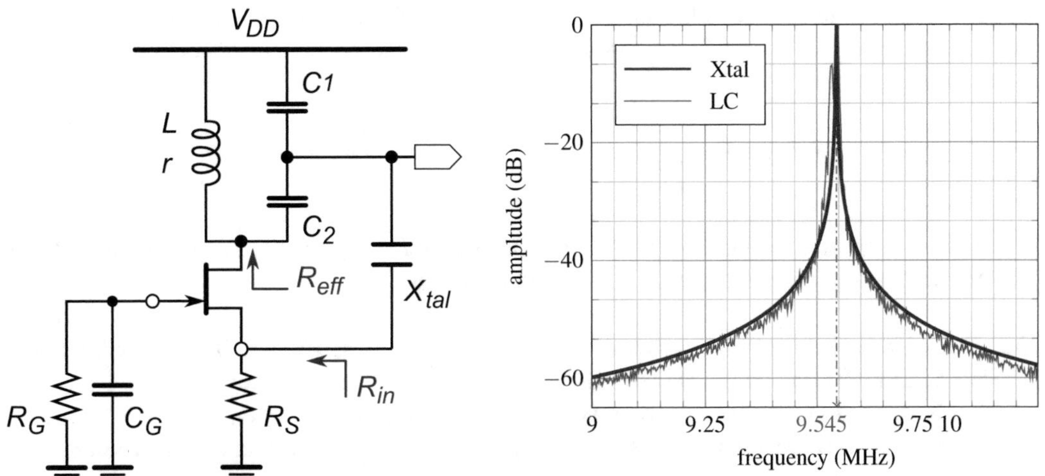

Fig. 13.21 CG oscillator—closed loop simulation, complete schematic diagram including Xtal in series feedback connection (left) and frequency domain response (right). Note, due to very high Q factor that requires very long simulations, Xtal response is approximated

4. **Closed loop simulations**: once the feedback network and amplifier are designed and tested by the open-loop simulations, it is time to close the loop and to verify that the oscillator really can start and sustain its oscillations, Fig. 13.21 (right). In the first experiment, the loop is closed without the X_{tal} whose serial resonant frequency $f_s = 9.545$ MHz, see Sect. 13.5. Clearly, this oscillator is capable to maintain the resonant frequency $f_0 = 9.545$ MHz that is set by LC_{eq}. In order to reduce phase jitter and improve the stability of the resonant frequency, a crystal component is inserted in the feedback path. We note the model in Fig. 13.12 and its frequency response in Fig. 13.13 are associated with very high $Q \approx 24\,000$. As a direct consequence of this high Q factor, the oscillator needs about 25 ms to finish its transition period and reach full signal amplitude.
   In the practical experiment, adding this component and observing the response on spectrum analyser is not difficult at all. However, to produce a good frequency response curve by Fourier analysis of transient data, it is necessary that the simulation is run for a long time, thus the resulting data file is accordingly large. The problem is that the resolution of the produced data file must be sufficient so that SPICE can separate the serial and parallel frequency modes of the crystal. Comparative plot of the oscillator transient domain simulations with and without crystal are shown in Fig. 13.21 (right).

13.9 Summary

Basic techniques for the analysis of general oscillator circuits are presented in this chapter in which we have learned about closed-loop feedback networks that are used to generate sinusoidal waveforms. Circuit conditions that lead into steady oscillations are specified in terms of the loop gain and the phase shift. Because oscillators are, in general, large-signal systems, small signal techniques are used only to estimate the initial conditions that are required to establish steady oscillations. Nonlinear numerical analysis techniques are required for detailed circuit design. Because four types of passive RLC network are used in most typical oscillators, we accepted the open-loop design methodology where the forward active path, consisting of an amplifier, is designed to compensate for the gain of the passive RLC feedback network. At the same time, the feedback network is responsible for setting up the correct phase shift around the loop. VCOs were introduced as very important for practical radio communication systems.

Problems

13.1 Derive expression for the loop gain for the general case of a loop consisting of a forward path amplifier with gain A and the feedback circuit path with gain.

13.2 One of several phase oscillator versions, Fig. 13.22 (left), is based on a CE amplifier and three RC stages in the feedback loop. Derive expression for the minimal transistor gain factor $\beta_{\min}$ (not to be confused with the feedback loop parameter), and resonant frequency ω_0, under the following assumptions: (a) the transistor's output resistance r_o is infinite; (b) all capacitors have the same value; (c) all resistors have the same values, while the transistor's base resistance r_b is absorbed in the left most resistor R; (d) for simplicity, details of biasing network are not shown; and (e) all elements are ideal, ignore the small emitter resistance r_e and base collector capacitance C_{BC}. Then, calculate values for resistors R and capacitors C, if $R_C = 10\,\text{k}\Omega$ assuming $f = 10\,\text{MHz}$.

13.3 Estimate resonant frequency ω_0 of an oscillator whose feedback network is shown in Fig. 13.22 (centre), if $L_1 = 0.5\,\mu\text{H}$, $L_2 = 1.5\,\mu\text{H}$, and $C = 126.65\,\text{pF}$.

13.4 Using the same data as in Problem 13.3 estimate the feedback network's gain factor β.

13.5 Using the same data as in Problems 13.3 and 13.4 estimate the effective resistance R_{eff} that this feedback network presents to the output of the oscillator's amplifier whose input impedance is $R_{\text{in}} = 10\,\text{k}\Omega$. Also, the effective inductor's L_{eff} Q factor is $Q = 50$.

13.6 Repeat problems 13.3 to 13.5 for the other three types of feedback network shown in the textbook.

13.7 For the circuit shown Fig. 13.22 (right) derive: (a) expression for the resonant frequency ω_0; and (b) expression for g_m of BJT transistor. As an example, use these two formulas to calculate the

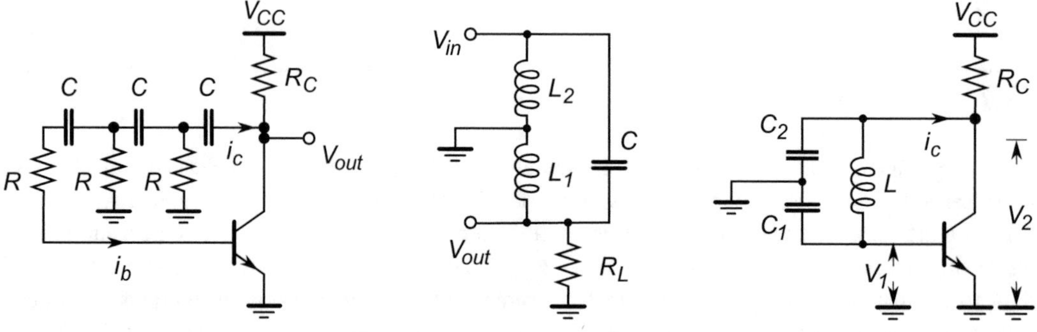

Fig. 13.22 Simplified schematic of a phase oscillator, Problems 13.2, 13.3, and 13.7

resonant frequency and g_m using the following data: $R_C = 10\,\text{k}\Omega$, BJT output resistance $r_c = 10\,\text{k}\Omega$, $L = 2\,\mu\text{H}$, $C_1 = C_2 = 253.3\,\text{pF}$, and $Q_L \to \infty$. Details of biasing network are omitted for simplicity.

Repeat the same problem assuming finite Q_L, derive the new equation(s) for the resonant frequency ω_0 and g_m. Again using, for example, $Q_L = 50$ recalculate these two values and compare with the ideal case solutions.

13.8 For the Clapp oscillator shown Fig. 13.17 calculate the oscillating frequency at: (a) zero bias of the varicap diode, and (b) at $V_D = -7\,\text{V}$.

Data: $L = 100\,\mu\text{H}$, $C_1 = C_2 = 300\,\text{pF}$, and $C_0 = 20\,\text{pF}$.

14 Frequency Shifting

In this chapter, we focus on the mathematical operation of "frequency shifting" that is fundamental to wireless communication systems. Frequency shifting (or "frequency translation") is complementary to the frequency tuning mechanism used in VCOs. However, as will be shown, it is a much broader concept with a much wider range of applications. As it turns out, mathematical multiplication of two sinusoidal waveforms with given frequencies results in waveforms that contain both higher and lower frequencies. This phenomenon is known as "frequency shifting", where the term "upconversion" refers to the process of shifting a lower frequency tone to the upper frequency range (used in RF transmitters), while "downconversion" refers to the frequency shifting from higher to lower frequency ranges (used in RF receivers). Hence, in a complete wireless communication system, the information-carrying signal is shifted in both directions.

14.1 Frequency Shifting

Even if it were practical to build wireless systems that operate in the audio frequency band for communications over longer distances than those achieved by natural speech, there is a practical problem. We would create a world crammed with very loud sources all transmitting at the same time, all the time. Moreover, everyone would be able to hear them without the help of any artificial equipment. Instead, in practice we use mixer circuit to perform the frequency shifting, see Fig. 14.1.

Human speech, including music, requires a very narrow frequency band (only about 20 kHz wide, known as the "audio band"). In comparison, the EM spectrum is immense. Splitting that enormous frequency space into abutting "strips" 20 kHz wide would create many parallel "pipes" (i.e. audio communication channels) each of them wide enough to conduct the full audio spectrum. These communication channels are strictly separated only in the frequency domain; in real space, they co-exist at all times and everywhere (Fig. 14.2). Having the ability to visualize the same signal in all three of these domains, i.e. frequency, time, and space, is essential to understanding wireless communication systems. Only then it is possible to understand how each of these communication channels could be made to connect a receiver with a specific transmitter that is located somewhere in space while, at the same time, an arbitrary number of other transmitter–receiver pairs also maintain their connections. With that in mind, we conceptualize a wireless communication system where each transmitter–receiver pair is assigned its own frequency channel for the duration of the communication.

R. Sobot, *Wireless Communication Electronics*,
https://doi.org/10.1007/978-3-030-48630-3_14

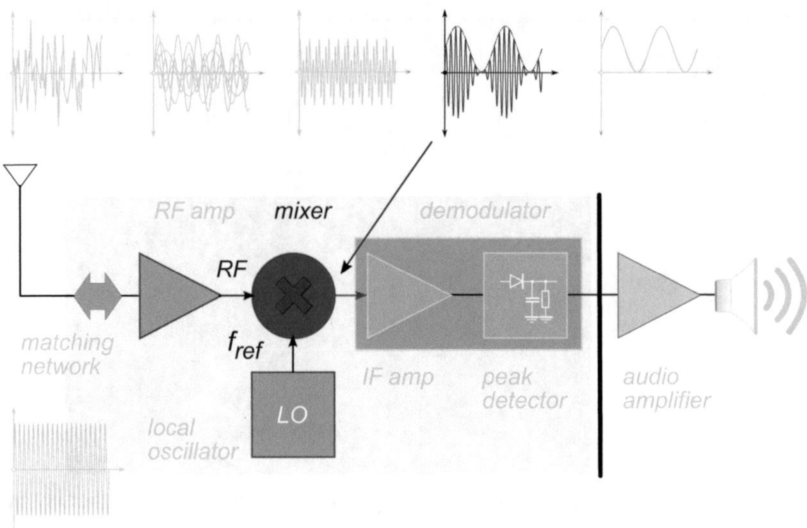

Fig. 14.1 Heterodyne AM radio receiver architecture—mixer

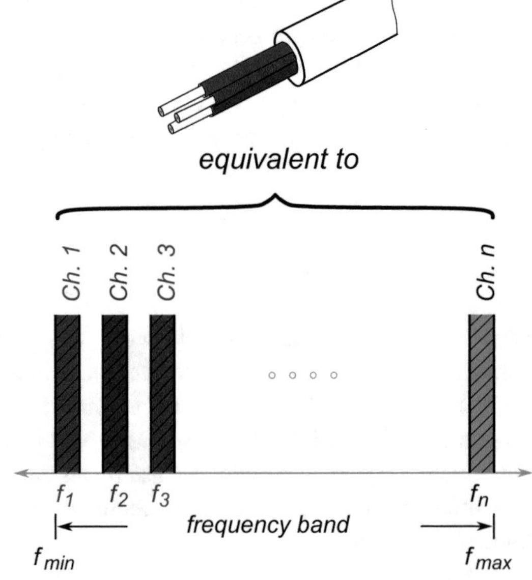

Fig. 14.2 Frequency band with multiple channels compared to a multi-wire communication cable

However, in order for this multiple frequency band approach to be practical and for us to make use of it, we need to resolve the following design issues:

1. how to shift the frequency of each individual audio signal up and align it exactly with its assigned channel;
2. how to force the receiving equipment to listen only to that particular channel while ignoring communications in all other parallel channels; and
3. how to shift the frequency of the received signal back to the audio range and decode the original message.

Solutions to these three fundamental steps are required for virtually all communication systems invented so far.

Practical implementation of the required frequency shifting is based on well-known trigonometric identities for product of two sinusoidal forms, i.e.

$$\cos\alpha \cos\beta = \frac{1}{2}\left(\cos(\alpha-\beta)+\cos(\alpha+\beta)\right) \tag{14.1}$$

$$\sin\alpha \sin\beta = \frac{1}{2}\left(\cos(\alpha-\beta)-\cos(\alpha+\beta)\right) \tag{14.2}$$

$$\sin\alpha \cos\beta = \frac{1}{2}\left(\sin(\alpha+\beta)+\sin(\alpha-\beta)\right) \tag{14.3}$$

$$\cos\alpha \sin\beta = \frac{1}{2}\left(\sin(\alpha+\beta)-\sin(\alpha-\beta)\right) \tag{14.4}$$

We note that in all four cases, i.e. regardless of what (sin $\leftrightarrow$ cos) product is calculated, the result always consists of another two sine terms whose arguments are *sum* and *difference* of the original two arguments. It should also be noted that both input sine functions have amplitudes equal to one, while the two functions at the output have amplitudes multiplied by "1/2" factor.

If the two abstract mathematical arguments α and β are given physical meaning, for example, frequency f_1 and frequency f_2, the term "frequency shifting" becomes apparent.

Example 83: Frequency Shifting
Given two sine waveforms, with $f_1 = 10\,\text{MHz}$ and $f_2 = 9.545\,\text{MHz}$, what sine waveforms are found at the output of a frequency multiplier?

Solution 83: Assuming "true" sinusoids, which is not critical for the final result, we use (14.2) to find

$$\begin{aligned}
\sin(\omega_1\,t)\,\sin(\omega_2\,t) &= \frac{1}{2}\,(\cos(\omega_1-\omega_2)\,t-\cos(\omega_1+\omega_2)\,t)\\
&= \frac{1}{2}\,(\cos 2\pi\,(10\,\text{MHz}-9.545\,\text{MHz})\,t-\cos 2\pi\,(10\,\text{MHz}+9.545\,\text{MHz})\,t)\\
&= \frac{1}{2}\,(\cos 2\pi\,(455\,\text{kHz})\,t-\cos 2\pi\,(19.545\,\text{MHz})\,t)
\end{aligned}$$

This result illustrates how multiplication of f_1 and f_2 frequencies produces the output spectrum that contains "shifted" frequencies at 455 kHz and 19.545 MHz. We note "the sum" and "the difference" effect in the result.

14.2 Signal Mixing Mechanism

An electronic circuit that can multiply two AC signals is called a mixer. A mixer in RF systems always refers to a circuit with a nonlinear transfer characteristics that, for two input single-tone signals ω_1 and ω_2, produces single-tone output signals that have frequencies found at the sum (i.e. $\omega_1+\omega_2$) and

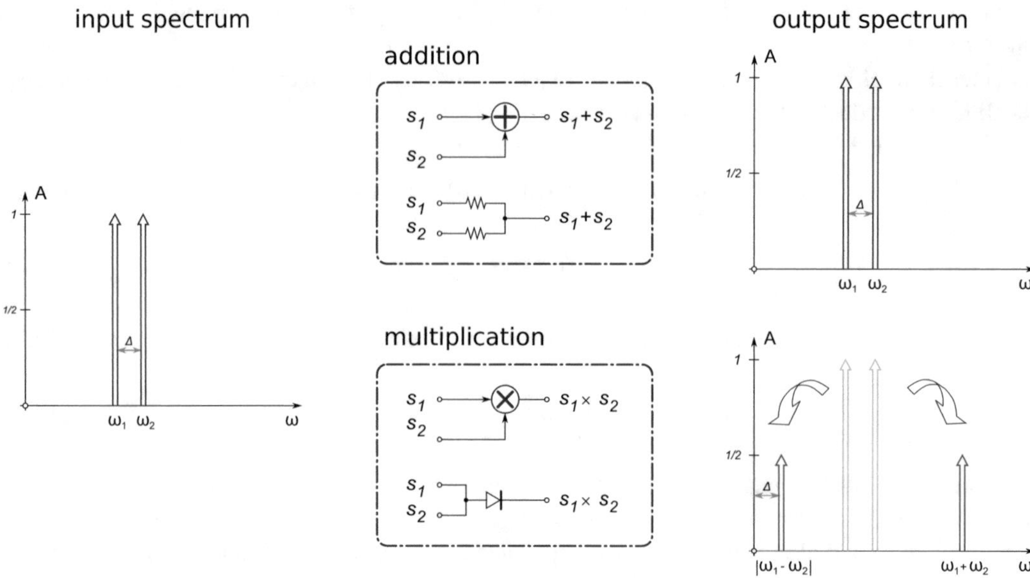

Fig. 14.3 Linear summing (top) and nonlinear mixing (bottom) functions

the difference (i.e. $|\omega_1 - \omega_2|$[1]) of the input frequencies. Simultaneously, the input side frequencies are not found in the output side spectrum. Note that, in audio systems, operators refer to "mixing" two soundtracks, which is not mixing in the RF sense: it is a linear operation that simply adds two tones as opposed to a nonlinear operation of RF multiplication. The symbolic representation of these two operations (see Fig. 14.3) illustrates the difference between linear addition and the nonlinear mixing of two AC signals.

Because ideal LTI systems cannot possibly produce output signals with spectral components not present at the input, mixers must be either nonlinear or time-varying elements in order to provide the frequency translation. Historically, many devices (e.g. electrolytic cells, magnetic ribbons, brain tissue, and rusty scissors, in addition to more traditional devices such as vacuum tubes and transistors) have been used as the nonlinear elements, demonstrating that virtually any nonlinear device can be used as a mixer. Of course, some nonlinearities work better than others, so we focus only on practical RF mixer types.

At the core of all modern mixers is the multiplication of two sinusoidal signals in the time domain. The fundamental usefulness of the multiplication may be understood from the basic trigonometric identities[2]

$$\cos(\omega_1 t) \times \cos(\omega_2 t) = \frac{1}{2} \; [\cos(|\omega_1 - \omega_2| t) + \cos(\omega_1 + \omega_2) t] \tag{14.5}$$

$$\sin(\omega_1 t) \times \sin(\omega_2 t) = \frac{1}{2} \; [\cos(|\omega_1 - \omega_2| t) - \cos(\omega_1 + \omega_2) t] \tag{14.6}$$

---

[1]The absolute value is equivalent to a geometrical distance on the horizontal axis, i.e. $|\omega_1 - \omega_2| = |\omega_2 - \omega_1|$.

[2]In strict mathematical syntax:
$\sin x \cdot \sin y = 1/2\,[\cos(|x - y|) - \cos(x + y)]$ and $\cos x \cdot \cos y = 1/2\,[\cos(|x - y|) + \cos(x + y)]$.

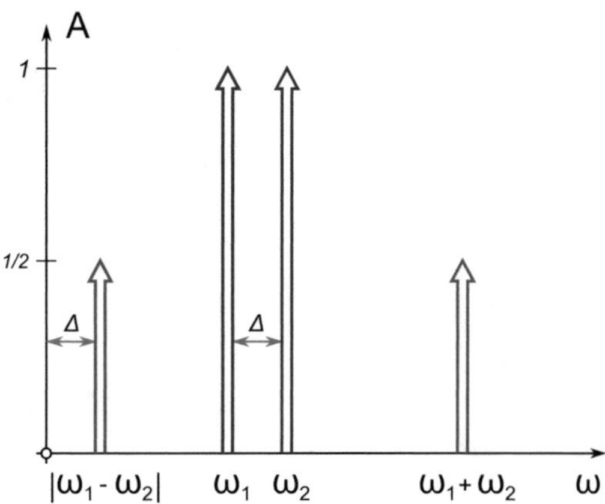

Fig. 14.4 Multiplication of ω_1 and ω_2 tones with amplitudes "1" results in two new tones $\omega_1 + \omega_2$ and $|\omega_1 - \omega_2|$ with amplitudes "½"

which shows that the multiplication of two sinusoidal functions[3] results in two *new* sinusoids (Figs. 14.3 and 14.4). In other words, (14.5) and (14.6) indicate that the multiplication of two sinusoidal waveforms with their respective arguments $(\omega_1 t)$ and $(\omega_2 t)$ results in one "upconverted" tone, i.e. whose frequency is $(\omega_1 + \omega_2)$, and one "downconverted" tone, i.e. whose frequency is $|\omega_1 - \omega_2|$ relative to the two input tones. Because a time-domain signal shape does not reveal information about its frequency content, it is much more practical to observe signals in the frequency domain (Fig. 14.4). Depending on the final objective any of the two newly produced tones can be used, while the other is removed by additional band-filtering.

An ideal theoretical model that delivers only one shifted tone (e.g. upconverted) is realized, for instance, by subtracting (14.6) from (14.1), which after substituting the arguments for frequency results (after normalizing the product of the two amplitudes to one) in

$$\cos[(\omega_1 + \omega_2)t] = \cos(\omega_1 t)\cos(\omega_2 t) - \sin(\omega_1 t)\sin(\omega_2 t) \tag{14.7}$$

which could be directly synthesized by a circuit whose block diagram is shown in Fig. 14.5. Assuming the existence of $\cos(\omega_1 t)$ and $\cos(\omega_2 t)$ that need to be frequency shifted, a step-by-step literal implementation of (14.7) is as follows:

1. PS_1: $\sin(\omega_1 t) = \cos(\omega_1 t + 90°)$, we need a 90° phase-shift circuit
2. PS_2: $\sin(\omega_2 t) = \cos(\omega_2 t + 90°)$, we need a 90° phase-shift circuit
3. M_1: $\cos(\omega_1 t) \times \cos(\omega_2 t) = 1/2\ [\cos(|\omega_1 - \omega_2|t) + \cos(\omega_1 + \omega_2)t]$, a multiplier to implement (14.1)
4. M_2: $\sin(\omega_1 t) \times \sin(\omega_2 t) = 1/2\ [\cos(|\omega_1 - \omega_2|t) - \cos(\omega_1 + \omega_2)t]$, a multiplier to implement (14.6)
5. Adder Σ: the last step is to add two waveforms generated by M_1 (to positive input) and M_2 (to inverting input), as

$$\frac{1}{2}\left[\cancel{\cos(|\omega_1 - \omega_2|t)} + \cos(\omega_1 + \omega_2)t\right] - \frac{1}{2}\left[\cancel{\cos(|\omega_1 - \omega_2|t)} - \cos(\omega_1 + \omega_2)t\right]$$
$$= \cos(\omega_1 + \omega_2)t$$

[3]Keep in mind that sin and cos functions have the same shape, they are just phase-shifted versions of each other, i.e. they have different starting points.

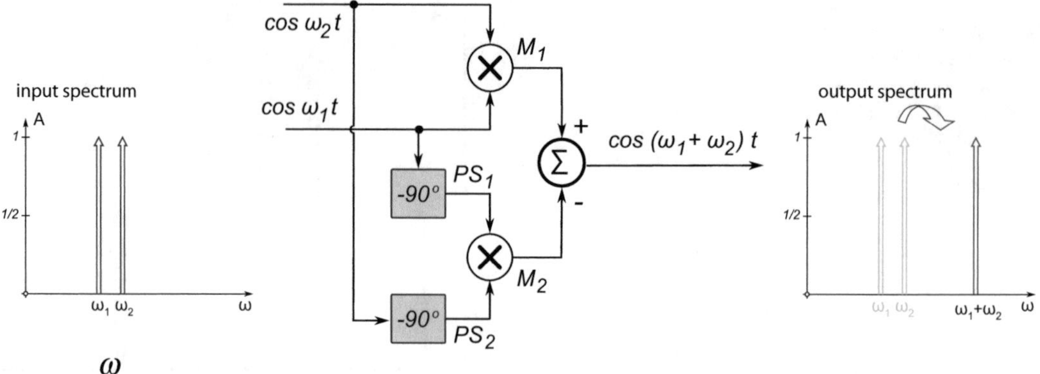

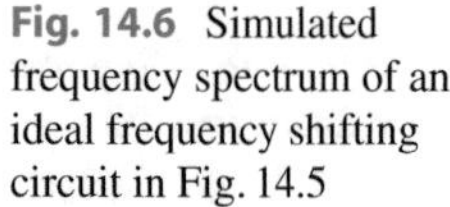

Fig. 14.5 Ideal frequency shifting circuit based on literal implementation of (14.7)

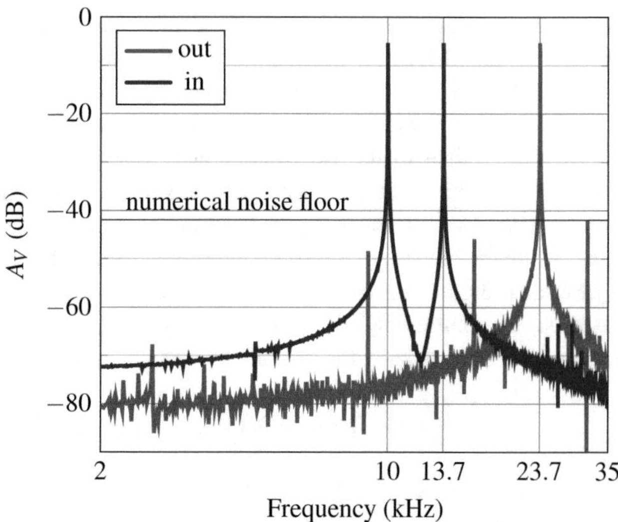

Fig. 14.6 Simulated frequency spectrum of an ideal frequency shifting circuit in Fig. 14.5

Thus, the ideal frequency shifter circuit in Fig. 14.5 produces a single-tone output at the frequency that is the sum of the input frequencies.

Simulated frequency spectrum of an ideal frequency shifting model in Fig. 14.5 is shown in Fig. 14.6. (We note "numerical noise" of Fourier calculations.) As an example, the two input tone frequencies are 10 and 13.7 kHz. Clearly, the output spectrum contains only a single tone whose frequency is 23.7 kHz. As a practical detail, in order to aide Fourier's numerical algorithm it is a good practice to use non-common factor frequencies so that the artificial numerical harmonics are reduced. In addition, the observed noise level is "numerical noise".

However, a practical implementation of circuit based on the block diagram in Fig. 14.5 is not trivial. It would be relatively straightforward to build the adder and multiplier blocks as IC devices. However, the wideband 90° phase-shift circuit is where the problem arises. It is relatively easy to design a narrowband 90° phase shift at a given fixed frequency, but over a wide range of frequencies there is no convenient way of producing an exact phase shift over the whole range. For this reason, the scheme in Fig. 14.5, which in theory works over any range of frequencies, is seldom attempted. Most frequency shifting is done using a single multiplier, as already indicated, in less than perfect relationship between (14.1) and (14.6).

The unwanted component which could be either $(\omega_1 + \omega_2)$ or $(\omega_1 - \omega_2)$ is removed by filtering. Most practical frequency translation circuits combine the processes of multiplying and filtering in order to achieve frequency translation. Hence, this methodology performs a quasi-multiplication because, besides the wanted product, one or more other frequency components are generated and must be removed by some form of filtering. In the following sections we take a look at some of the most commonly used mixer circuits.

14.3 Diode Mixers

A diode mixer is a very simple circuit (Fig. 14.7) that is useful up to very high frequencies. Because it works at almost any frequency, it is commonly used in measuring equipment which is expected to work over a range of frequencies. Two voltage single-tone signals

$$v_1 = V_1 \cos(\omega_1 t) \tag{14.8}$$

$$v_2 = V_2 \cos(\omega_2 t) \tag{14.9}$$

are first added and then passed through an ideal diode whose voltage–current function is given as

$$i_D = I_S \left[\exp\left(\frac{v_D}{V_t}\right) - 1 \right] \tag{14.10}$$

which is the nonlinear element that is required for frequency shifting. In the following analysis, for the simplicity, we assume a small diode current and ignore the voltage drop across the loading resistor R_L. That is, the diode voltage V_D is approximately equal to the voltage at node①, i.e. $V_D \approx V(1)$.

Two equal resistors R serve as a linear voltage adder. Because of their voltage-dividing property the voltage at node① is at half the sum of the two inputs, i.e.

$$v_D = v_1 = \frac{1}{2}(v_1 + v_2) = \frac{1}{2}[V_1 \cdot \cos(\omega_1 t) + V_2 \cdot \cos(\omega_2 t)] \tag{14.11}$$

Following the signal path after node①, the diode voltage v_D is converted into current i_D. The diode voltage is assumed to be small (implying that $v_D < V_t$ so that the higher-order terms in (14.13) are approximately zero), which allows the exponential term in (14.10) to be expanded into the Taylor series around the diode's biasing point, where series expansion for an exponential function is well known as

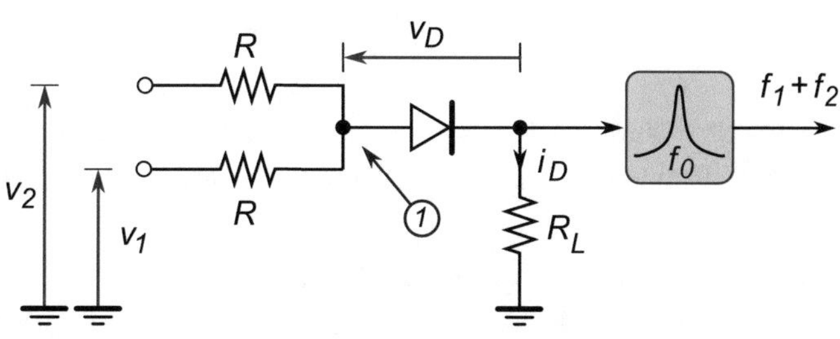

Fig. 14.7 Simplified schematic diagram of diode mixer

$$e^x = \sum_{n=0}^{\infty} \frac{x^n}{n!} = 1 + x + \frac{x^2}{2} + \frac{x^3}{6} + \frac{x^4}{24} + \cdots \quad (14.12)$$

hence, after substitution of $x = v_D/V_t$ into (14.12) and application of the exponential term in (14.10), we write

$$i_D = I_S \left\{ \left[1 + \frac{v_D}{V_t} + \frac{1}{2}\left(\frac{v_D}{V_t}\right)^2 + \frac{1}{6}\left(\frac{v_D}{V_t}\right)^3 + \frac{1}{24}\left(\frac{v_D}{V_t}\right)^4 + \cdots \right] - 1 \right\} \quad (14.13)$$

We now examine each of the terms on the right of (14.13) separately and find out about the signal's total spectrum (note that 1 and -1 cancel). Obviously, the exact series includes an infinite number of terms. In the first approximation, because the assumption is that the signal is small, the third- and higher-order terms can be ignored (they are smaller and smaller numbers divided by larger and larger numbers). After substituting(14.11) into (14.13), we focus on the first two terms:

1. The linear term:

$$\frac{v_D}{V_t} = \frac{1}{2V_t}[V_1 \cdot \cos(\omega_1 t) + V_2 \cdot \cos(\omega_2 t)] = f(\omega_1, \omega_2) \quad (14.14)$$

   We conclude that the linear term of the series expansion has a frequency spectrum that is equal to the original spectrum of the signal v_D, i.e. ω_1 and ω_2. We already had that tone at the input side; hence, this term is not much use.
2. The square term:

$$\begin{aligned}
\frac{1}{2}\left(\frac{v_D}{V_t}\right)^2 &= \frac{1}{2V_t^2}\left[\frac{1}{2}(V_1 \cdot \cos(\omega_1 t) + V_2 \cdot \cos(\omega_2 t))\right]^2 \\
&= \frac{1}{8V_t^2}\left[V_1^2 \cos^2(\omega_1 t) + 2V_1 V_2 \cos(\omega_1 t)\cos(\omega_2 t) + V_2^2 \cos^2(\omega_2 t)\right] \\
&= \frac{1}{8V_t^2}\left[V_1^2 \frac{1}{2}(1 + \cos(2\omega_1 t)) + V_1 V_2(\cos(|\omega_1 - \omega_2| t) + \cos((\omega_1 + \omega_2)t)\right. \\
&\quad \left. + V_2^2 \frac{1}{2}(1 + \cos(2\omega_2 t))\right]
\end{aligned} \quad (14.15)$$

   which states that the output frequency spectrum due to the second (nonlinear) term contains

$$\frac{1}{2}\left(\frac{v_D}{V_t}\right)^2 = f\left[(\omega_1 - \omega_2), 2\omega_1, 2\omega_2, (\omega_1 + \omega_2)\right]$$

   In other words, aside from the up- and down-shifted tones $(\omega_1 + \omega_2)$ and $(|\omega_1 - \omega_2|)$, there are additional tones ($2\omega_1$ and $2\omega_2$) present that are not present in the result of the ideal multiplication operation.

Simulated frequency spectrum of a diode mixer, see Fig. 14.8, using same two-tone input as for the ideal mixer, clearly shows non-ideal characteristics of practical mixer circuits. While the up- and down-shifted $(f_1 + f_2)$ and $(|f_1 - f_2|)$ tones are visible as in the spectrum (see Fig. 14.8 (left)) as predicted by (14.6) and (14.1), we also find the input two tones f_1 and f_2 as well as all the other tones

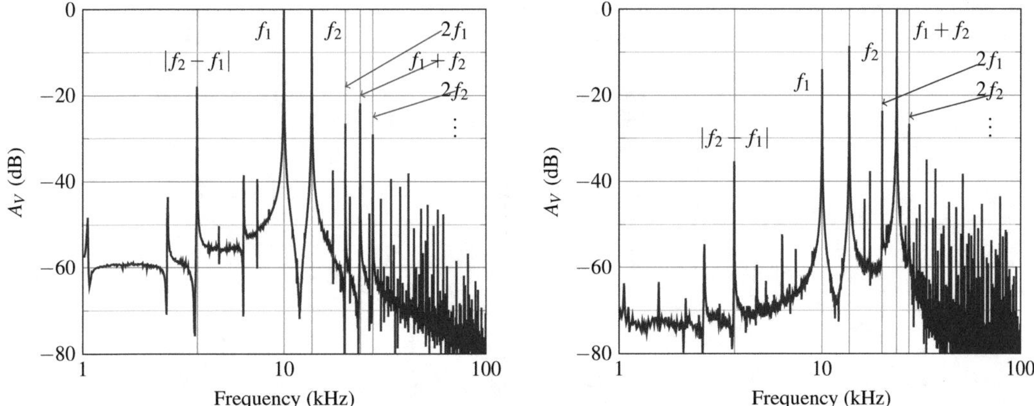

Fig. 14.8 Simulated frequency spectrum of diode mixer, see Fig. 14.7, before BPF (left) and after a high-Q BPF (right) centred at $f_1 + f_2$. Note: amplitudes are normalized to 0 dB

indicated in (14.14) and (14.15). For that reason, in order to isolate the desired f_1 and f_2 tone it is necessary to use a high-Q BPF (see Fig. 14.8 (right)). In addition, the power levels are reduced (e.g. to note the $1/8$ factor in (14.15)), which necessitate immediate use of an amplification to restore the lost signal levels.

Therefore, a diode is the simplest active device that serves as a practical mixer and works over a wide range of frequencies. In addition, a more specific small signal condition should be stated as $V_1 V_2 < V_t^2$. The additional, unwanted tones are usually filtered out with an LC resonator.

In conclusion, using a diode as a nonlinear element for the purpose of multiplying two single-tone signals does produce the desired theoretical tones the $(\omega_1 - \omega_2)$ and $(\omega_1 + \omega_2)$. However, it also produces tones that are not part of the ideal solution (i.e. ω_1, ω_2, $2\omega_1$, $2\omega_2$, ...). In addition, if the higher-order harmonics in (14.13) are not neglected, many more tones are observed in the output frequency spectrum. Therefore, for good performance this mixer is restricted to rather low input signal levels. Because all tones that are not needed must be filtered out afterwards, a diode is a simple but very inefficient multiplying element. Nevertheless, at very high frequencies it may be the only practical choice.

14.4 Transistor Mixer

Active mixers are also based on nonlinear exponential functions of BJTs and metal-oxide semiconductor field-effect transistors (MOSFETs). If two voltage signals

$$v_1 = V_1 \cos(\omega_1 t) \tag{14.16}$$

$$v_2 = V_2 \cos(\omega_2 t) \tag{14.17}$$

are added and then applied to the base of an ideal BJT Q_1 (Fig. 14.9 (left)) or v_1 is applied to the base of Q_2 and v_2 is applied to the source node by means of a 1:1 ratio transformer (Fig. 14.9 (right)), then the two signals are mixed (both methods are applicable to both BJT and MOS devices). Assuming

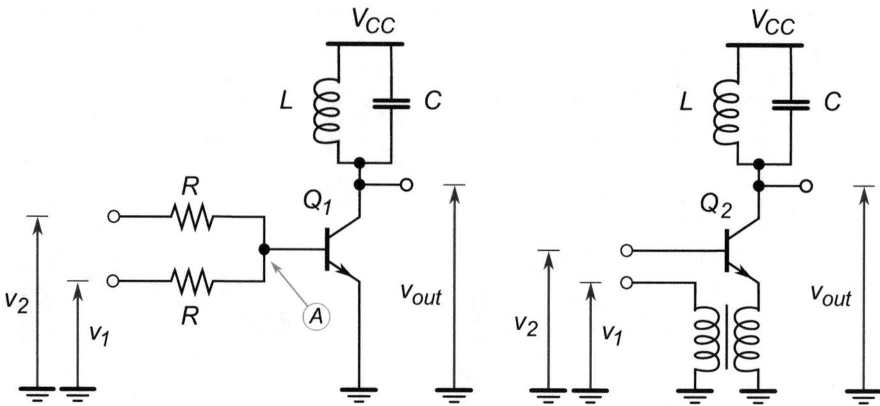

Fig. 14.9 Simplified schematic of two versions of BJT mixers

ideal transistors with a current gain of β, the two variants of BJT mixer are very similar; therefore, the following two equations are written by inspection:[4]

$$V_{BE}(Q_1) = \frac{1}{2}(v_1 + v_2) \tag{14.18}$$

$$V_{BE}(Q_2) = v_1 - v_2 \tag{14.19}$$

The relationship between the collector current IC of a BJT versus the base–emitter voltage v_{BE} is the same as for a forward-biased diode

$$i_C = I_S \left[\exp\left(\frac{v_{BE}}{V_t}\right) - 1 \right] \tag{14.20}$$

which is to say that the expression for the square term of interest in the output frequency spectrum of the circuit in Fig. 14.9 (left) is similar to the one for a diode, with the addition of the β factor:

$$I_C = \beta\, I_S \frac{V_1 V_2}{8V_t^2} \left[\cos(|\omega_1 - \omega_2| t) + \cos((\omega_1 + \omega_2) t)\right]$$

The corresponding expression for the circuit in Fig. 14.9 (right) is only slightly different. It is important to note that, because of the β factor, a BJT mixer has much better efficiency than a simple diode mixer and it is possible even to have a "conversion gain". That means that it is possible for the output tone (either the downconverted $|\omega_1 - \omega_2|$ or upconverted $\omega_1 + \omega_2$ tone) to have more power than the input signal. On the other hand, a BJT mixer has the same limitation as the diode in terms of the input signal amplitude relative to the V_t voltage. Both circuits in Fig. 14.9 use an LC resonator in the collector branch that is tuned to either of the two tones of interest, i.e. either to $|\omega_1 - \omega_2|$ or $\omega_1 + \omega_2$, and they filter out all unwanted tones in the frequency spectrum.

[4]Voltage addition is done by means of a voltage divider with two equal R.

14.5 JFET Mixer

In RF amplifiers, it is common practice to replace BJTs with JFETs. JFET gate current is much lower than the base current and has higher transconductance than a MOSFET transistor. Therefore it is often used in the front end of low-noise, high-input-impedance RF amplifiers.

Two input voltage signals

$$v_1 = V_1 \cos(\omega_1 t) \tag{14.21}$$

$$v_2 = V_2 \cos(\omega_2 t) \tag{14.22}$$

are applied to the gate of JFET that is used instead of Q_1 in Fig. 14.9. We use a procedure similar to that in the previous sections; the main difference is that the current–voltage characteristics between the drain current I_D and the gate-source voltage v_{GS} of a JFET obey the following relationship:

$$I_D = I_{DSS} \left(1 - \frac{v_{GS}}{V_p}\right)^2 \tag{14.23}$$

where I_{DSS} is the JFET saturation drain current, V_{GS} is the gate-source voltage, and V_p is the pinch-off voltage. In the JFET case, there is no exponential term, which makes the derivation a bit simpler. Therefore, a straightforward expansion of (14.23) leads to

$$I_D = I_{DSS} \left[1 - 2\,\frac{v_{GS}}{V_p} + \frac{v_{GS}^2}{V_p^2}\right] \tag{14.24}$$

By focusing only on the nonlinear terms in (14.24), the square term is

$$\begin{aligned} I_D &\sim -I_{DSS}\,\frac{1}{4}\,\frac{[V_1 \cdot \cos(\omega_1 t) + V_2 \cdot \cos(\omega_2 t)]^2}{V_p^2} \\ &\sim -I_{DSS}\,\frac{V_1 V_2}{2\,V_p^2}\,[\cos(|\omega_1 - \omega_2| t) + \cos((\omega_1 + \omega_2) t)] \end{aligned} \tag{14.25}$$

where (14.25) focuses only at the cos product term from the previous step. It is interesting to note that, because there was only a second-order term in (14.24) and no higher-order terms, there was no need to apply power series expansion as in the cases of the diode and BJT. That is, there is no strict limitation to the amplitudes of V_1 and V_2, as long as the JFET is not cut off or becomes forward biased. Again, similar to the BJT circuits from Sect. 14.4, the LC resonator simultaneously filters out all harmonics except the desired one. JFETs are commonly used in RF mixer applications because of their tolerance for high signal levels and good conversion efficiency.

14.6 Dual-Gate MOSFET Mixer

We have already learned (Sect. 6.6 and Fig. 6.25) that a cascode amplifier configuration is very useful in RF applications because of its high output impedance and its resilience to the Miller effect. Putting the two transistors into a single package and creating a dual-gate device was a natural development. In this section, we learn about the application of dual-gate transistors to the design of RF mixers. The additional important property of a dual-gate transistor is that the two devices are almost perfectly "matched"—they are "twins" in respect of their electrical properties. Consequently, two independent input signals applied to the two gates control the drain current at the same time and equally well.

Two input voltage signals

$$v_1 = V_{DC1} + V_1 \sin(\omega_1 t) \tag{14.26}$$

$$v_2 = V_{DC2} + V_2 \sin(\omega_2 t) \tag{14.27}$$

are applied to a dual-gate FET transistor that is used as a mixer (Fig. 14.10). In a standard cascode configuration, transistor M_1 is set as the CS amplifier for the v_1 signal, while M_2 serves as the CG current buffer. Therefore, assuming $v_2 = \text{const} = V_{DC2}$, we write equations for the M_1 transistor in saturation (ignoring its nonlinear effects) as

$$I_D = k(V_{GS} - V_{th})^2 = k(v_1 - V_{th})^2$$

$$\therefore$$

$$g_m{}' \stackrel{\text{def}}{=} \frac{dI_D}{dV_{GS}} = 2k(v_1 - V_{th}) = 2k[V_1 \sin(\omega_1 t) - V_{th}] \tag{14.28}$$

where $V_{DC1,2}$ are constant biasing DC voltages, $k = (\mu_n C_{ox} W)/(2L)$, and V_{th} is the MOS threshold voltage, $g_m{}'$ is the circuit's overall g_m under the condition that the gate of M_2 is at its small signal ground. Drain current I_D passes through the current buffer M_2 with no loss (i.e. the two transistors

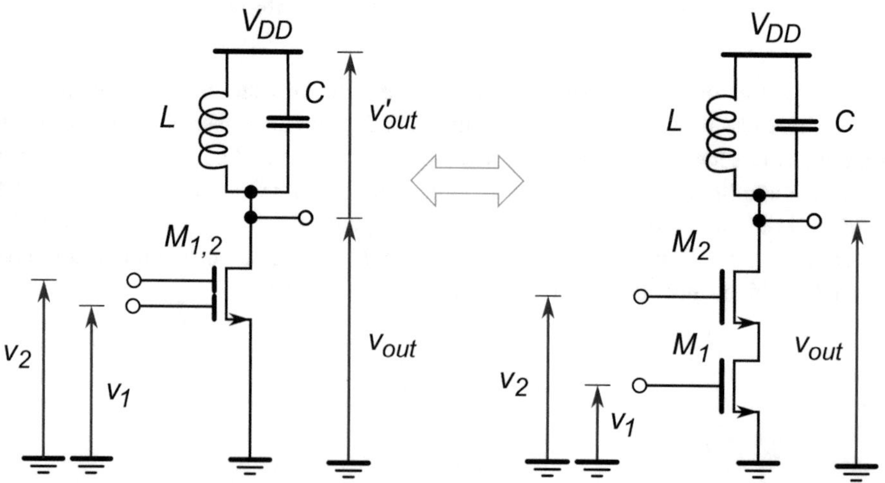

Fig. 14.10 Simplified schematic of a dual-gate FET mixer and its equivalent circuit diagram, where M_1 and M_2 are assumed identical

have the same drain current), which is followed by LC load. Therefore, the voltage across the LC load at resonance is approximately bounded by its dynamic resistance R_D multiplied by the drain current, which is the same as the output voltage V'_{out} relative to V_{DD}. In other words,

$$V'_{out} = I_D\, R_D = g_m{}'\, v_1\, R_D = g_m{}' R_D V_1 \sin(\omega_1 t) \tag{14.29}$$

That is to say, when the gate of M_2 is at the small signal ground, in respect of signal v_1 the circuit works as a CS cascoded amplifier. However, when the gate of M_2 is used as the input terminal for the second signal, for example, when the dual-gate MOS mixer signal v_2 comes from a local oscillator (LO), the common drain current is additionally controlled by v_2. Variation of the drain current because of variation of the v_2 voltage is manifested as a change in the circuit's overall g_m, as

$$\begin{aligned}
I_D &= k[(V_{DC2} + V_{GS2}) - V_{th}]^2 \\
&\therefore \\
g_m &\equiv \frac{dI_D}{d(V_{DC2} + V_{GS2})} = 2k[(V_{DC2} + V_{GS2}) - V_{th}] \\
&= 2kV_{DC2} + 2k(V_2 \sin(\omega_2 t) - V_{th}) \\
&\sim g_m{}' + g_{m\Delta} \sin(\omega_2 t)
\end{aligned} \tag{14.30}$$

where $g_m{}'$ is part of the circuit's g_m due to v_1 (14.28), while $g_{m\Delta}$ is variation of the circuit's g_m due to v_2 whose common mode is (i.e. it is centred around) at V_{DC2}. It is important to note that V_{DC2} is *not* constant anymore and that this arrangement works because the two transistors are identical with the same drain current.

After replacing $g_m{}'$ in (14.29) with g_m from (14.30) it follows that

$$\begin{aligned}
V_{out} &= [g_m{}' + g_{m\Delta} \sin(\omega_2 t)]\, R_D V_1 \sin(\omega_1 t) \\
&= g_m{}'\, R_D V_1 \sin(\omega_1 t) + g_{m\Delta}\, R_D V_1 \sin(\omega_2 t) \sin(\omega_1 t) \\
&\sim g_{m\Delta}\, R_D V_1\, [\cos(|\omega_1 - \omega_2| t) + \cos((\omega_1 + \omega_2) t)]
\end{aligned} \tag{14.31}$$

where (14.31) focuses only on the cos product term and the LC resonator is tuned to either of the two desired tones and filters out all the other harmonics.

In conclusion, a dual-gate FET mixer is commonly used in the design of an RF mixer to multiply the incoming RF signal with the LO. Setting the appropriate LO frequency, the RF signal is then precisely shifted in the frequency domain, i.e. either "downconverted" or "upconverted". In addition, from (14.31), it becomes obvious that the v_2 signal amplitude should be as large as possible so that the $g_{m\Delta}$ term is maximized, which is one of the advantages of this circuit.

14.7 Image Frequency

A less obvious, but very important, consequence of signal multiplication (14.1) and (14.6) is that for any given frequency ω_1 there are two separate single tones ω_2 and ω_3 that produce exactly the same

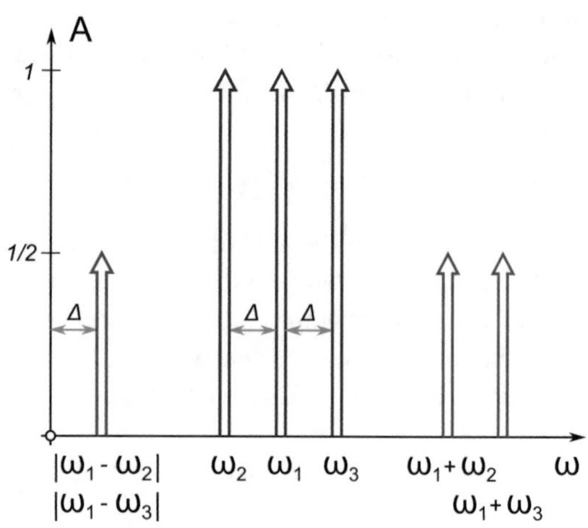

Fig. 14.11 Frequency domain diagram of the relative positions of the main and image frequencies

$|\omega_1 - \omega_2| = |\omega_1 - \omega_3|$ tones (Fig. 14.11[5]). At the same time, the higher frequency tones $(\omega_1 + \omega_2)$ and $(\omega_1 + \omega_3)$ are easily distinguished. This phenomenon of dual frequencies entering the mixer and producing the same output tone is so important in wireless communication systems that it is commonly referred to as "image frequency" or a "ghost image". For instance, if the original intent was to multiply frequencies ω_1 and ω_2, then signal ω_3 would be declared a ghost image. Similarly, ω_2 would be declared a ghost image if the original intent was to do frequency shifting of the ω_1 and ω_3 tones.

14.7.1 Image Rejection

The problem of ghost images is very real and is dealt with by using the following two methods. First, the transmitting frequencies are licensed and assigned at a country level—some frequencies are forbidden for communications because they would represent ghost images to their dual frequencies, which are already in use. Second, radio receiver front-end electronics are required to be able to suppress image frequencies relative to the desired tones by a specified amount.

The front end of a receiver consists of one or more parallel tuned resonant circuits that act as a high-Q bandpass filter centred around the desired frequency. To really appreciate the need for high Q resonators, let us try to find out what happens when the incoming signal frequency is not exactly the same as the LC tank resonant frequency. In other words, at what distance $\Delta\omega$ is the image frequency found from the resonant frequency suppressed by the Q factor of the front-end LC tank?

14.7.1.1 LC Tank Admittance

In order to estimate the amount of signal suppression for a tone that is not centred at the resonant LC frequency we need to evaluate the frequency dependence of a realistic LC tank model (Fig. 10.10). A realistic inductor is modelled as a serial combination of an ideal inductor L and an ideal resistor R that embodies the total wire resistance in the resonant loop (including the inductor's DC resistance), while the capacitor is still assumed to be ideal. We already derived an expression, (10.60), that is repeated here for convenience

[5]Although all three signals ω_1, ω_2, and ω_3 are shown as having the same amplitude, in general it does not have to be the case.

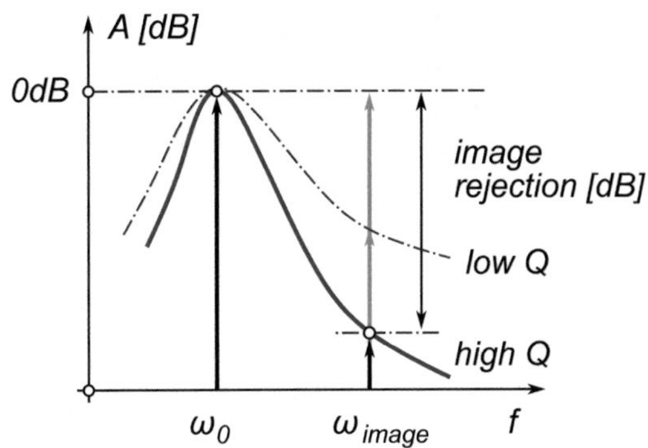

Fig. 14.12 Graphical representation of image rejection measure

$$|Y| = Y_0\sqrt{1 + (\delta Q)^2} \tag{14.32}$$

where ω is close to ω_0 (i.e. it is less than one decade away), so that $\omega/\omega_0 \approx 1$ and

$$\delta = \frac{\omega}{\omega_0} - \frac{\omega_0}{\omega} \tag{14.33}$$

The graph in Fig. 14.12 demonstrates the relationship between the resonant tone ω_0 and the nearby tone $\omega_{\texttt{image}}$ for cases of high and low Q resonators. Lower Q means a wider bandwidth, which means that signal ($\omega_{\texttt{image}}$) is more attenuated than the resonant tone normalized to 0dB level. Higher Q provides a narrower bandwidth, which means that the same image signal is even more attenuated; hence, it is more suppressed relative to the desired tone at wo.

A straightforward implementation of (14.32) to calculate the output voltage at the image frequency, assuming the signal current I_S, yields

$$|V_0| = \frac{|I_S|}{|Y|} = \frac{I_S}{Y_0\sqrt{1 + (\delta\, Q)^2}} \tag{14.34}$$

which, at resonance, reduces to $V_0(\omega_0) = I_S/Y_0$. Hence, the relative voltage amplitude between the non-resonant and resonant voltages is given by

$$A_r \triangleq \frac{|V_0|}{V_0(\omega_0)} = \frac{1}{\sqrt{1 + (\delta Q)^2}} \tag{14.35}$$

for a single tuned circuit. If several tuned circuits are included and isolated by amplifiers, then the overall response is given by the product $A_r(tot) = A_{r1}A_{r2}\cdots A_{rn}$. Further improvement of the image rejection is usually achieved by using a double-conversion radio receiver architecture.

Example 84: Image Suppression
An AM broadcast receiver is tuned to 500 kHz with LC resonator whose $Q = 50$. Calculate signal rejection in dB of unwanted signal being transmitted at 1430 kHz.

Solution 84: Because the front-end LC tank is tuned at $f_0 = 500$ kHz, radio transmitter emitting at that frequency is the desired signal. It is then straightforward to implement (14.32) as follows:

(continued)

$$\delta\, Q = \left(\frac{\omega}{\omega_0} - \frac{\omega_0}{\omega}\right) Q = \left(\frac{1430}{500} - \frac{500}{1430}\right) 50 = 126$$

which, after substituting in (14.35) and approximating $\sqrt{1 + 126^2} \approx 126$, yields

$$A_r = 20 \log \frac{1}{126} = -42\,\text{dB}$$

Therefore, if a second radio station is transmitting at 1430 kHz, its signal is received as 126 times weaker than the signal from the desired radio station. Using two tuned amplifiers would double the selectivity and further suppress the image signal down to −84 dB.

14.8 Case Study: A Dual-Gate JFET RF Mixer

Once we have developed RF amplifier in Sect. 13.8, and because we used JFET cascode topology, it requires only a minor modification to create RF mixer for the same AM receiver. As we have seen in Fig. 14.1, this mixer is intended to downconvert the incoming 10 MHz RF carrier to the 455 kHz IF frequency.

Dual-gate FET transistor is a natural device to perform the frequency multiplication, as already shown in Sect. 14.6; therefore, in this case study we simply reuse the already existing RF amplifier where the cascode transistor J_2 is now used in the same way as J_1: the gate of J_1 serves as the input for RF signal, while the gate of J_2 accepts LO signal, see Fig. 14.13.

The modification that is required is that the LC resonator must be centred at $f_{IF} = 455$ kHz, which is implemented with $L = 15\,\mu$H and $C = 8.156920$ nF. As usual, a capacitance such as this one is implemented by a parallel connection of fixed and trim capacitors. In addition, in order to satisfy $BW = 10$ kHz specification we must choose the inductor with $Q = 46$ or equivalently $R_D = 1950\,\Omega$.

As a test, we run simulation with two sinusoidal non-modulated waveforms whose amplitudes have been already amplified by RF amplifier and the local oscillator circuits. That is to say, the input signals into mixer are already in order of 1 to 2 V. Transient analysis shows that this circuit

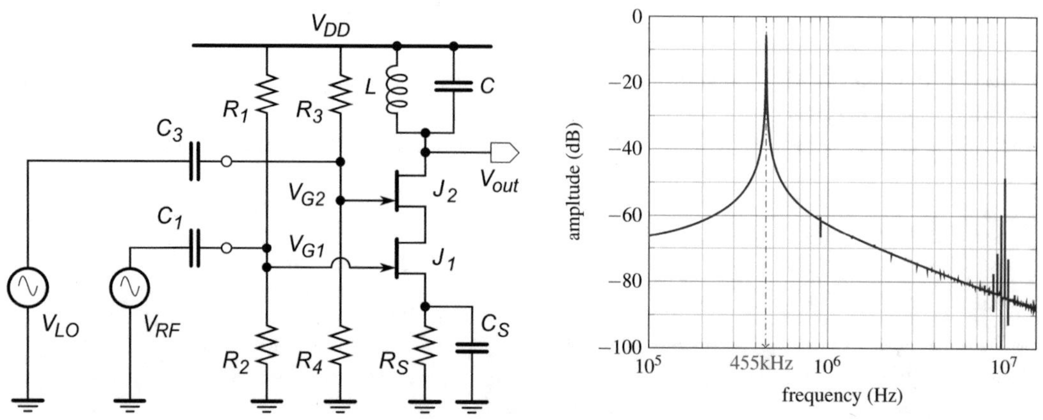

Fig. 14.13 Dual gate JFET mixer schematic diagram (left) and frequency response (right)

needs around 100 μs for the transition period before the full amplitude is produced at its output. Subsequently, Fourier analysis clearly shows that IF signal is almost 50 dB stronger than the original RF and LO signals and their products found the proximity of 10 MHz and the other multiples, see Fig. 14.13 (right).

14.9 Summary

In this section, we have learned about the frequency-shifting mechanism that is fundamental to radio communication systems. The underlying mathematics is based on multiplying two sinusoidal forms, while the practical realization is based on passing the two single tones through a nonlinear element. Because of imperfect multiplication in realistic systems based on diodes, BJT or FET devices and additional filtering are required to remove unwanted tones. As a side product of the multiplication operation, we learned about the existence of the ghost image and its influence on the wanted signals. Image suppression is an important requirement for the front end of radio receivers; hence, we worked out a formula for estimating the image (or any other side signal, for that matter) suppression relative to the desired tone that is aligned with the LC resonant frequency.

Problems

14.1 For this problem use these four single-tone signals:
$S_1 = V_1 \sin(\omega_1 t)$, $S_2 = V_2 \sin(\omega_2 t)$, $S_3 = \cos(|\omega_1 - \omega_2| t)$, and $S_4 = \cos((\omega_1 + \omega_2) t)$.
Assuming $f_1 = 1\,\text{MHz}$, $f_2 = 20\,\text{MHz}$, $V_1 = 2\,\text{V}$, and $V_2 = 3\,\text{V}$ do the following:

(a) Find expression for $S = S_1 S_2$. Using a graphing software of your choice, plot S, $(V_1 V_2) S_1$, and $-(V_1 V_2) S_1$ in the same window. Observe relative relationship between these signals.
(b) Now plot $S_o = 1/2 \cdot (V_1 V_2) \cdot (S_3 - S_4)$. What can you conclude?

14.2 Starting from $S_1 = \sin(2\pi \times 10\,\text{MHz} \times t)$, find *two* other single-tones that could be used to generate a single tone at $f = 1\,\text{kHz}$. Explain the process and the result.

14.3 A radio receiver is tuned to receive AM modulated wave transmitted at carrier frequency of $f_{RF} = 980\,\text{kHz}$. Local oscillator inside the receiver is set at $f_{LO} = 1435\,\text{kHz}$. Find:

(a) frequencies coming out of the receiver's mixer,
(b) which one is IF frequency,
(c) frequency of a radio station which would represent image frequency to the radio station in this problem,
(d) frequency graph of the frequencies involved.

14.4 An RF amplifier has LC tank with $Q = 20$ and it is tuned at RF frequency f_0. Estimate attenuation of the image signal, if the image frequency is 10% higher than the RF signal.

Modulation 15

In a broad sense, the term "modulation" implies a change in time of a certain parameter, where the "change" itself is the message being transmitted. For instance, while listening to a steady single-tone signal with constant amplitude and frequency coming out of a speaker, we merely receive the simplest message that conveys information only about the existence of the signal source and nothing else. If the source is turned off, then we cannot even say if there is a signal source out there or not. For the purpose of transmitting a more sophisticated message, the communication system must use at least the simplest modulation scheme, based on time divisions, i.e. turning on and off the signal source. By listening to short and long beeps, we can decode complicated messages letter by letter. As slow and inefficient as it is, Morse code does work and is used even today in special situations, for example, in a very low SNR environment.

In this chapter, we study the main modulation techniques for wireless communications, which are based on the time variation of periodic electrical signals.

15.1 Amplitude Modulation

Conceptually, amplitude modulation (AM) is the simplest form of carrier modulation. In this technique, the amplitude of the carrier signal is made to replicate the shape of the information-carrying signal (for instance, the voice or pulse signal), Fig. 15.1. We keep in mind that, in accordance with Fourier theorem, a complicated signal consists of multiple single tones that occupy a certain bandwidth determined by the lowest f_{min} and highest f_{max} tones in the frequency spectrum. Fourier's theorem enables us to do analytical signal processing of complicated signals by decomposing it into its fundamental single-tone harmonics. The theorem also enables us to synthesize a complicated signal waveform that takes practically any time domain shape, e.g. basic mathematical forms of square, by adding a series of single tones.

We demonstrate analytically AM technique using two single-tone time-varying signals (for simplicity, assume zero initial phase). That is to say, instead of analysing a really complicated message in this hand-calculations we use only a single-tone waveform as the representative of a complicated signal. We keep in mind though that, in accordance with Fourier's theorem, the same analysis applies to all other harmonics of a complicated message.

R. Sobot, *Wireless Communication Electronics*,
https://doi.org/10.1007/978-3-030-48630-3_15

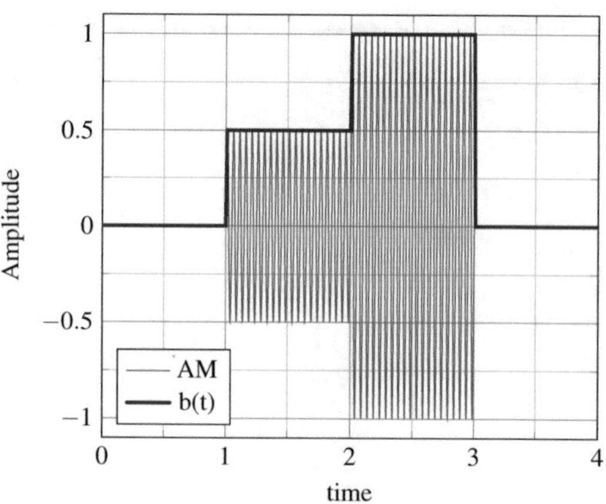

Fig. 15.1 A time domain plot of superimposed AM (red) and $b(t)$ (blue) waveforms. The modulating signal information $b(t)$ is clearly embedded by the control of the carrier's amplitude

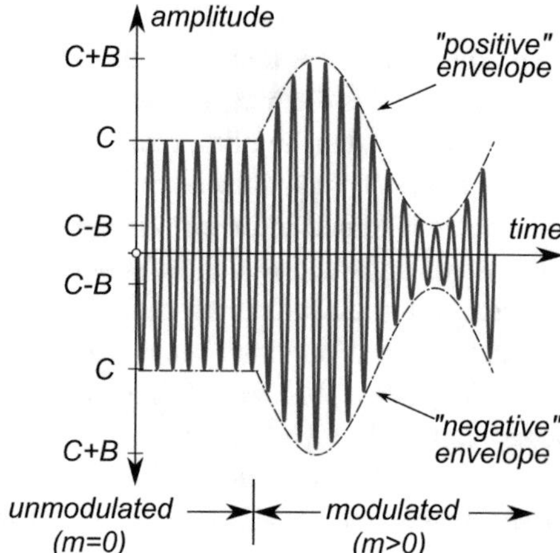

Fig. 15.2 Amplitude-modulated signal, the initially unmodulated carrier, and the information signal in the form of an envelope

$$b(t) = B \sin \omega_b t \tag{15.1}$$

$$c(t) = C \sin \omega_c t \tag{15.2}$$

where $b(t)$ is the LF information (i.e. the "baseband" modulating signal), B is its maximum amplitude (here, simply amplitude of $b(t)$ sinusoid), ω_b is its angular frequency (here a single frequency instead of the frequency band), and $c(t)$ is the high-frequency single-tone carrier, C is its amplitude, ω_c is its angular frequency.

The sum of the modulating signal $b(t)$ and the carrier's maximum amplitude C is called the "envelope wave" $e(t)$ which is described as

$$e(t) = C + b(t) \tag{15.3}$$

Note that for the unmodulated AM signal, i.e. $b(t) = 0$, the envelope $e(t)$ equals the carrier's amplitude C and it is in that case, therefore, constant. Otherwise, the envelope of AM signal replicates the message, regardless if it is a single tone (as in this analysis) or a voice (see Fig. 15.2).

An analytical expression for the modulated carrier amplitude $c_{AM}(t)$ is derived by replacing the carrier's amplitude C in (15.2) the with expression for the envelope (15.3), which is the equivalent of saying that the carrier amplitude is modulated by the baseband signal, i.e.

$$
\begin{aligned}
c_{AM}(t) &= e(t)\sin\omega_c t \\
&= (C + B\sin\omega_b t)\sin\omega_c t \\
&= C\left(1 + \frac{B}{C}\sin\omega_b t\right)\sin\omega_c t \\
&= C\,(1 + m\sin\omega_b t)\sin\omega_c t \qquad (15.4)\\
&= \sin\omega_c t + m\sin\omega_b t\sin\omega_c t \qquad (15.5)\\
&= \sin\omega_c t + \frac{m}{2}\left[\cos|\omega_c - \omega_b|\,t - \cos(\omega_c + \omega_b)\,t\right] \qquad (15.6)
\end{aligned}
$$

where the modulation index is defined as

$$
m \overset{\text{def}}{=} \frac{\text{the modulating signal's maximum amplitude}}{\text{the carrier's maximum amplitude}} = \frac{B}{C} \qquad (15.7)
$$

Note that, without losing in generality, after setting $C = 1$ in (15.4) everything afterwards is consequently normalized to the carrier's amplitude.

The AM index m is an important communication parameter that shows the ratio of the baseband and the carrier maximum amplitudes (see Fig. 15.3). In the interests of efficient power transfer and high SNR, it is desirable to have the amplitude of the modulating signal as high as possible relative to the carrier's amplitude. In that respect, there are three possible relations:

1. $m < 1$: If the carrier's maximal amplitude is greater than the modulating signal's amplitude, then the embedded envelope is a faithful representation of the information (in this case, a clean sinusoidal shape), however, SNR is not at its maximum, Fig. 15.3 (left). In other words, there is still some space for increasing the signal's amplitude.
2. $m = 1$: If the carrier's maximal amplitude is equal to the modulating signal's amplitude, then the embedded envelope is still a faithful copy of the information (i.e. $(b(t))$ and SNR is at its maximum, Fig. 15.3 (centre). In other words, there is no more space for increasing the signal's amplitude.
3. $m > 1$: If the carrier's maximal amplitude is less than the modulating signal's amplitude, then the embedded envelope is not a faithful copy of the transmitted information, Fig. 15.3 (right), because the signal's amplitude is greater than the carrier's amplitude. Consequently, the "signal clipping" occurs and the envelope is not anymore a faithful copy of the information being transmitted; here, the envelope is not perfect replica of $b(t)$.

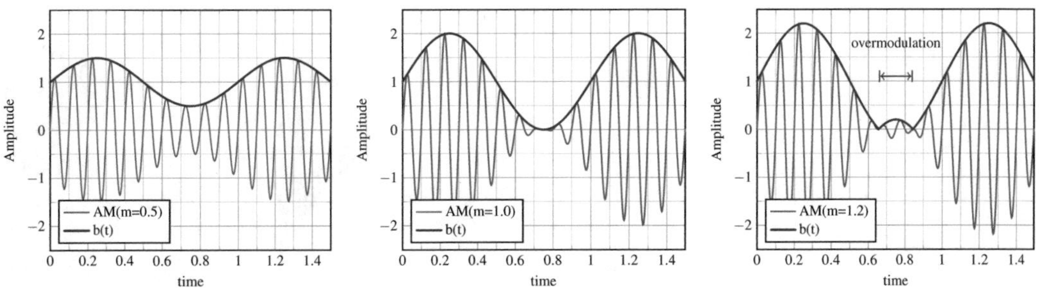

Fig. 15.3 Time domain plot of sinusoidal amplitude modulation (15.5) for three values of the AM modulating index

Note that the amplitude-modulated signal has two symmetrical envelopes, one positive and one negative, that carry the same information. As long as the two envelopes are kept separate and do not overlap (i.e. the positive envelope stays positive and the negative envelope stays negative), it is possible to recover the information from either of the two envelopes, Fig. 15.3 (left and centre). However, once the two envelopes overlap, Fig. 15.3 (right), the information is distorted because sections of the positive envelope cross over and become part of the negative envelope, and vice versa, which causes signal clipping. This is referred to as "over-modulation": neither the positive nor the negative envelope looks like the original information (it looks like a clipped sinusoidal). Note that in case of over-modulation, (15.5) is not valid within the clipping region. In practical systems, the modulation index is held close to one for most of the time and relative to the strongest harmonics in the spectrum.

15.1.1 Trapezoidal Patterns and the Modulation Index

Except for the trivial case of a single-tone modulating signal (Fig. 15.3), observing amplitude-modulated signals in the time domain using an oscilloscope is cumbersome because it is difficult to synchronize the time sweep for all harmonics in the given bandwidth. Instead, non-periodic signals, for instance voice, are observed using the "trapezoidal method", which is usually used to plot Lissajous curves. In this method, the AM signal $c_{AM}(t)$ is fed into channel A and the modulating signal $b(t)$ is fed into channel B of the oscilloscope. By setting the plotting mode so that channel A is on the vertical axis and channel B is on the horizontal axis, the amplitude-modulated signal plots trapezoidal patterns similar to the ones in Fig. 15.4.

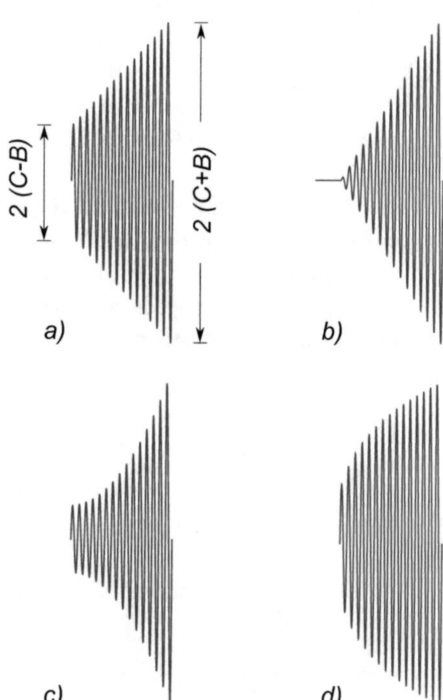

Fig. 15.4 Trapezoidal patterns of an AM signal: (**a**) for $m < 1$; (**b**) for $m > 1$, with the clipping section easily visible as the straight line tail; (**c**) for $m < 1$ and weak RF driver, i.e. the carrier signal is too strong; and (**d**) for $m < 1$ and nonlinear modulator, with visible nonlinear gain for high amplitudes

The expanded form of the AM modulation index is written as

$$m \stackrel{\text{def}}{=} \frac{B}{C} = \frac{(C+B)-(C-B)}{(C+B)+(C-B)} = \frac{2(C+B)-2(C-B)}{2(C+B)+2(C-B)} \tag{15.8}$$

which is easily correlated with the geometrical sizes of the plots in Fig. 15.4. Hence, by measuring lengths of the long and short trapezoidal sides directly on the oscilloscope and applying (15.8), we calculate the modulation index.

15.1.2 Frequency Spectrum of AM Modulated Signal

By inspection of (15.4) and (15.6) we realize that amplitude modulation is based on the multiplication operation discussed in Sect. 14.2. Therefore, the frequency content of the amplitude-modulated signal contains the two side tones, the upper-side $(\omega_c + \omega_b)$ and the lower side $|\omega_c - \omega_b|$ harmonics. In addition, the AM signal also contains harmonic ω_c at the carrier frequency. It is important to notice that the amplitude of the side tones is multiplied by $m/2$, which (in the best case of $m = 1$) means that amplitude of the side tones is half of the carrier amplitude tone (see Fig. 15.5). We also observe that, for a baseband signal whose highest harmonic is ω_b and $\omega_c > \omega_b$, the amplitude-modulated signal occupies the bandwidth

$$BW \stackrel{\text{def}}{=} \omega_{max} - \omega_{min} = (\omega_c + \omega_b) - (\omega_c - \omega_b) = 2\,\omega_b \tag{15.9}$$

that is centred around the carrier frequency. Because $\omega_b \ll \omega_c$ it follows that all three tones, i.e. $|\omega_c - \omega_b|$, ω_c, and $(\omega_c + \omega_b)$, in the AM spectrum are very close to each other, in other words they are HF signals that can be wirelessly transmitted by an antenna.

15.1.2.1 Average Power

We now quantify the amount of energy contained in each of the three harmonics of an AM signal (15.6) as

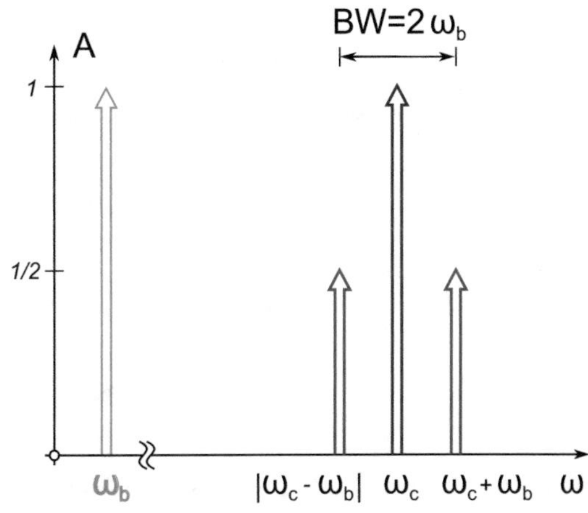

Fig. 15.5 Frequency spectrum of an AM signal containing a tone at the carrier frequency ω_c in addition to the two side tones $(\omega_c + \omega_b)$ and $|\omega_c - \omega_b|$ (wb « wo)

$$\begin{aligned} c_{AM}(t) &= \sin\omega_c t + \frac{m}{2}\cos\omega_L t - \frac{m}{2}\cos\omega_U t \\ &= c_C + c_L - c_U \end{aligned} \tag{15.10}$$

where c_C is the instantaneous carrier voltage, c_L is the instantaneous voltage of the lower-side harmonic $\omega_L = |\omega_c - \omega_b|$, and c_U is the instantaneous voltage of the upper-side harmonic $\omega_U = (\omega_c + \omega_b)$. Hence, the instantaneous power of the AM wave across a resistor R is

$$\begin{aligned} p_{AM} &\stackrel{\text{def}}{=} \frac{c_{AM}^2}{R} \\ &= \frac{(c_C + c_L - c_U)^2}{R} = \frac{c_C^2}{R} + \frac{c_L^2}{R} + \frac{c_U^2}{R} + \frac{2}{R}(c_C c_L - c_L c_U - c_C c_U) \end{aligned} \tag{15.11}$$

where the three squared terms denote the instantaneous power of each of the wave components: the carrier, the lower-side harmonic, and the upper-side harmonic.

Let us first evaluate the cross-product term $(c_C c_L - c_L c_U - c_C c_U)$. As we discussed in Sect. 14.2, the product of two sinusoidal terms is another sinusoidal term (it is irrelevant for our discussion that it is frequency shifted). Furthermore, the average value of a sinusoidal waveform is zero, therefore all three cross products have zero average values and do not contribute to the total average power calculations.

With reference to (15.2) and (15.10), for each of the three squared terms in (15.11), starting with the carrier voltage, we write expressions for their average power as

$$\langle P_C \rangle = \frac{c_{C\mathrm{rms}}^2}{R} = \frac{\left(\frac{C}{\sqrt{2}}\right)^2}{R} = \frac{C^2}{2R} \tag{15.12}$$

$$\langle P_L \rangle = \frac{c_{L\mathrm{rms}}^2}{R} = \frac{\left(\frac{mC/2}{\sqrt{2}}\right)^2}{R} = \frac{m^2}{4}\frac{C^2}{2R} = \frac{m^2}{4}\langle P_C \rangle \tag{15.13}$$

$$\langle P_U \rangle = \frac{c_{U\mathrm{rms}}^2}{R} = \frac{\left(\frac{mC/2}{\sqrt{2}}\right)^2}{R} = \frac{m^2}{4}\frac{C^2}{2R} = \frac{m^2}{4}\langle P_C \rangle = \langle P_L \rangle \tag{15.14}$$

Hence, the total average power P_T of an AM modulated waveform is therefore

$$\langle P_T \rangle = \langle P_C \rangle + \frac{m^2}{4}\langle P_C \rangle + \frac{m^2}{4}\langle P_C \rangle = \langle P_C \rangle \left(1 + \frac{m^2}{2}\right) \tag{15.15}$$

In order to simplify the syntax, we recall that (15.15) refers to the average power simply write

$$P_T = P_C\left(1 + \frac{m^2}{2}\right) \tag{15.16}$$

Again, the value of the AM factor m is important for the overall power transfer efficiency, with (15.16) showing that the total average power in the case of $m = 1$ is $P_T = 1.5P_C$, while each of the sidebands transfers only $1/4\,P_C$. We conclude that even for 100% AM scheme, i.e. $m = 1$,

$$P_T = P_C\left(1+\frac{m^2}{2}\right) = \frac{3}{2}P_C \quad ; \quad P_U = P_C\frac{m^2}{4} = \frac{1}{4}P_C \quad ; \quad P_L = P_C\frac{m^2}{4} = \frac{1}{4}P_C$$

which is to say that the ratios of the total power to the sidebands powers are

$$\frac{P_U}{P_T} = \frac{\frac{1}{4}\cancel{P_C}}{\frac{3}{2}\cancel{P_C}} = \frac{1}{6} \quad \text{and} \quad \frac{P_L}{P_T} = \frac{\frac{1}{4}\cancel{P_C}}{\frac{3}{2}\cancel{P_C}} = \frac{1}{6}$$

i.e. only $1/6$ of the total power is present in each of the sidebands (each contains its own copy of the useful information), while $2/3$ of the total power is in the carrier (which contains no information whatsoever).

Although the above analysis focused on a single-tone signal, we keep in mind that a non-sinusoidal modulating signal consists of a number of sine waves, not necessarily harmonically related. The overall average power is then the sum of the individual single-tone average powers:

$$P_T = P_C\left(1+\frac{m_1^2}{2}+\frac{m_2^2}{2}+\cdots\right) \tag{15.17}$$

where m_i $(i = 1, 2, \ldots, n)$ is the modulation index of tone i. A detailed analysis of random signals similar to speech involves statistical mathematical models and is the subject of advanced courses in signal processing, hence it is omitted in this book.

15.1.2.2 Double-Sideband and Single-Sideband Modulation

The amplitude modulation scheme described so far is the most straightforward form of signal modulation. Its full name is "double-sideband-full carrier" (DSB–FC) modulation. In summary, it has the advantage of being very simple to modulate and demodulate, but has the following disadvantages:

1. It can be over-modulated, which causes signal distortion (generation of new tones that can end up being shifted outside the assigned bandwidth).
2. It is inefficient in the use of power (most of the transmitted power is in the carrier, which contains no information).
3. Its required bandwidth is twice the modulation signal's bandwidth, that is, it does not use the frequency bandwidth efficiently, which is important because the overall available bandwidth is limited and directly controls the maximum number of the users.

The power inefficiency and bandwidth requirement of the DSB–FC modulating scheme present serious concerns for modern battery-powered RF equipment. By inspection of (15.10) and (15.16), we conclude that major power savings could be achieved by removing the carrier tone from the AM signal frequency spectrum before it reaches the transmitting antenna. Such a modulation scheme is known as "double-sideband–suppressed carrier" (DSB–SC) modulation.

Indeed, the whole class of symmetrical modulating circuits, commonly known as "balanced modulators", was developed to perform carrier tone removal. In the mathematical terms, a balanced modulator truly implements the trigonometric identities (14.1) and (14.6), where there are only USB and LSB frequencies but not the carrier frequency ω_c itself.

In general, transmission of a single tone is rarely used, instead complicated signals such as speech or music are most often transmitted. From the conceptual perspective, our main concern is to determine the frequency bandwidth that is required by the signal. For instance, an audio signal occupies an approximately 20 kHz-wide bandwidth, which means that we must allocate a 20 kHz frequency bandwidth for its transmission. Hence, we need to use the terms "upper sideband" (USB) and "lower sideband" (LSB) to indicate that the transmitted signal consists of more than a single tone, where the sideband is bounded by the frequencies of its lowest and highest tones. After the carrier tone is removed by the use of balanced modulators, the modulated signal still consists of both upper and

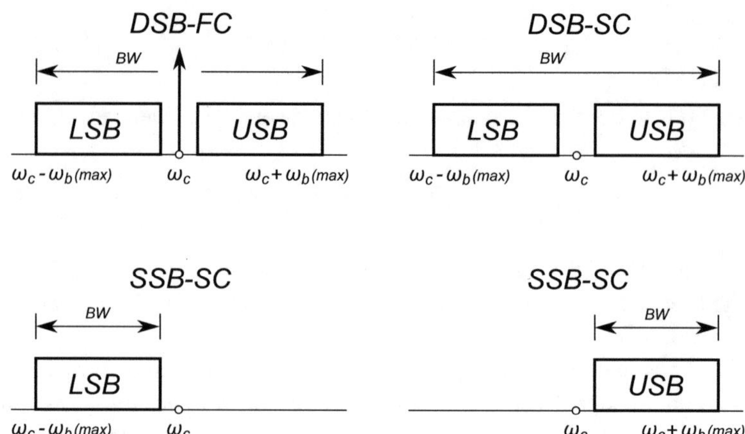

Fig. 15.6 Amplitude-modulated signal spectrum, (**a**) double-sideband full carrier (DSB–FC); (**b**) double-sideband suppressed carrier (DSB–SC); (**c**) single-sideband suppressed carrier (SSB–SC) using lower sideband (LSB); (**d**) single-sideband suppressed carrier (SSB–SC) using upper sideband (USB)

lower sidebands and occupies. By using bandpass filters in the last stages of the AM transmitter, we can remove either LSB or USB. This AM scheme is known as "single-sideband–suppressed carrier" (SSB–SC) modulation using either the USB or the LSB frequency range. The frequency spectrums of these four major AM schemes are summarized in Fig. 15.6.

Example 85: USB, LSB, the Carrier
The full 20 kHz audio band is transmitted by a 10 MHz RF carrier. If the USB and LSB are separated by $\Delta f = 100\,\text{kHz}$, what are frequency ranges of the two sidebands?

Solution 85: Because USB and LSB are symmetrically centred around the carrier frequency, it follows that USB starts at $f_{min}(USB) = 10\,\text{MHz} + (100/2)\,\text{kHz} = 10.050\,\text{MHz}$. Therefore, the full audio band finishes at $f_{max}(USB) = f_{min}(USB) + 20\,\text{kHz} = 10.070\,\text{MHz}$.

Similarly, LSB occupies space as $f_{max}(LSB) = 10\,\text{MHz} - (100/2)\,\text{kHz} = 9.950\,\text{MHz}$ to $f_{min}(USB) = f_{max}(LSB) - 20\,\text{kHz} = 9.030\,\text{MHz}$.

15.1.2.3 Bandpass Filters for SSB Modulation

Although functionally very simple, the design of bandpass filters suitable for SSB suppression is limited by the available technology. The main problem becomes more obvious after we take a look at the commonly cited formula for the required Q factor of the bandpass filter, which is expressed in terms of the amount of required suppression A_{dB} as

$$Q = \frac{f_C}{\Delta f} \frac{\sqrt{10^{\left(\frac{A_{\text{dB}}}{20}\right)}}}{4} \tag{15.18}$$

where f_C is the carrier frequency, Δf is the separation between the USB and the LSB, and A_{dB} is the required attenuation expressed in units of dB (see Fig. 15.7). A practical implementation of SSB suppression bandpass filters is based on the following options:

1. Surface acoustic wave (SAW) filters are hybrid filters based on electromechanical signal conversion using piezoelectric materials. In the AM frequency range, this kind of filter may be able to achieve a Q factor as high as 35 000 (the literature data is not always conclusive).

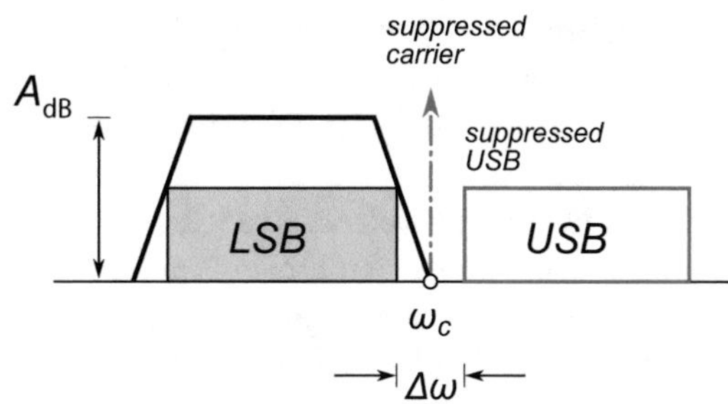

Fig. 15.7 Example of bandpass filter definition for LSB suppression, (15.18), of a DSB–SC signal

2. Crystal filters are another form of quartz crystal that can achieve a Q factor in the order of 20 000 (again, we take this number as indicative rather than definitive).
3. Mechanical filters are based on the mechanical resonance of various metallic materials. We assume that their Q factor is of the order of 10 000, however, some nickel–iron alloys apparently achieve a Q factor as high as 25 000.
4. Ceramic filters are made of ceramic alloys and may be able to achieve Q factors in the order of 2000.
5. RLC filters are based on discrete components. With careful design, they may provide Q factors in the order of 500.

The Q factor numbers here are only for illustrative purposes. Because the Q factor is defined for a specific centre frequency and bandwidth, which means that it may not always be possible to compare fairly the various types of SSB bandpass filter. Nevertheless, for the sake of the exercise, we assume that all these filters are comparable and on an equal footing.

Example 86: SSB Filter
In a typical AM radio system, the signal bandwidth is $\Delta f = \pm 100\,\text{Hz}$. Estimate the type of SSB filter that is needed to suppress the LSB by $A_{\text{dB}} = 80\,\text{dB}$, if the centre frequency is: (a) $f_C = 100\,\text{kHz}$ and (b) $f_C = 1\,\text{MHz}$

Solution 86: By direct implementation of (15.18) we write
(a)

$$Q = \frac{f_C}{\Delta f}\frac{\sqrt{10^{\left(\frac{A_{\text{dB}}}{20}\right)}}}{4} = \frac{100\,\text{kHz}}{200\,\text{Hz}}\frac{\sqrt{10^{\left(\frac{80}{20}\right)}}}{4}$$

$$= 12\,500 \quad \text{i.e. we need a crystal filter or better.}$$

(b)

$$Q = \frac{1\,\text{MHz}}{200\,\text{Hz}}\frac{\sqrt{10^{\left(\frac{80}{20}\right)}}}{4}$$

$$= 125\,000 \quad \text{i.e. we need several SAW filters in cascade.}$$

Aside from the "brute force approach" of using a high Q bandpass SSB suppression filter to directly remove one of the sidebands, there are number of more sophisticated techniques in use. To illustrate the possibilities, let us take a look at a typical representative method known as the "phase shift method".

In the phase shift method, two identical balanced modulators are used in parallel (Fig. 15.8). The message signal $b(t)$ is fed directly into modulator A and with a phase shift of 90° into modulator B. The carrier frequency is provided by the crystal oscillator and it is fed into modulator A with the 90° phase shift. The two output waves $b_A(t)$ and $b_B(t)$ are first added in the summing block and the output wave is an SSB with suppressed lower sideband. Main property of balanced modulators is that they cancel the carrier frequency. In order to see how the LSB cancellation happens, we need to take a look at the mathematics of this system.

Assuming that the message signal is $b(t) = \sin(\omega_b t))$ and the carrier signal is $c(t) = \sin(\omega_c t)$, where $\omega_c > \omega_b$ then a balanced modulator output is

$$\sin(\omega_c t) \times \sin(\omega_b t) = \frac{1}{2} \left[\cos(\omega_c t - \omega_b t) - \cos(\omega_c t - \omega_b t\right] \tag{15.19}$$

and if, for instance, the carrier frequency ω_c is shifted by 90°, then (15.19) takes the form

$$\sin((\omega_c t + 90°)) \times \sin(\omega_b t) = \frac{1}{2} \left[\cos((\omega_c t + 90°) - \omega_b t) - \cos((\omega_c t + 90°) - \omega_b t\right] \tag{15.20}$$

A phase shift modulator is based on two balanced modulators whose output signals are (see Fig. 15.8)

$$\begin{aligned} b_A(t) &= \cos[(\omega_c t + 90°) - \omega_b t] - \cos[(\omega_c t + 90°) + \omega_b t] \\ &= \cos[\omega_c t - \omega_b t + 90°] - \cos[\omega_c t + \omega_b t + 90°] \end{aligned} \tag{15.21}$$

$$\begin{aligned} b_B(t) &= \cos[\omega_c t - (\omega_b t + 90°)] - \cos[\omega_c t + (\omega_b t + 90°)] \\ &= \cos[\omega_c t - \omega_b t - 90°] - \cos[\omega_c t + \omega_b t + 90°] \end{aligned} \tag{15.22}$$

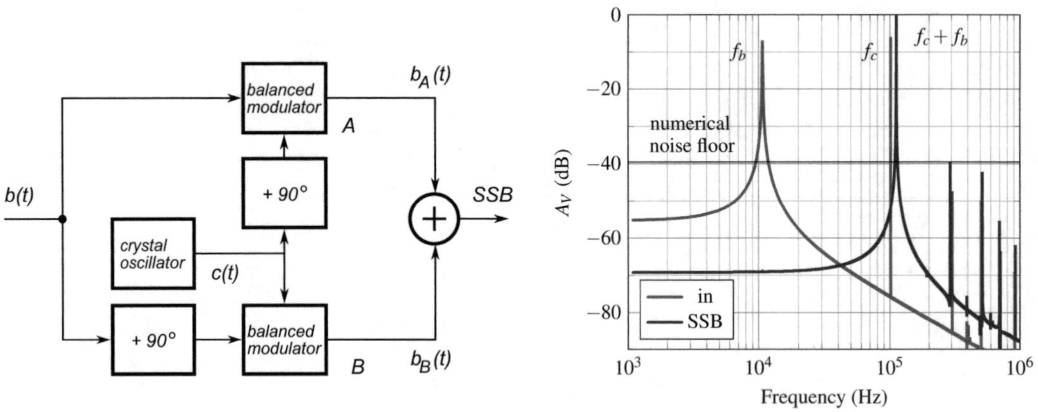

Fig. 15.8 Block diagram of a phase shift of generating SSB AM signal (left), and its frequency spectrum (right)

The first of the two cosine terms in (15.21) and (15.22) are LSB terms that are opposite in phase, the one in (15.21) is leading by 90° while the one in (15.22) is lagging by 90°. Therefore, they cancel when added in the summing block[1] and the output of the summing block is

$$SSB = b_A(t) + b_B(t) = -2\,\cos[\omega_c\,t + \omega_b t + 90^\circ] = 2\,\sin[\omega_c\,t + \omega_b t] \tag{15.23}$$

which is the USB signal at $\omega_c\,t + \omega_b t$ frequency whose amplitude is two. We note that this circuit produces true SSB AM output spectrum, that by controlling the phase of the LO we choose to cancel either USB or LSB, and that there is no more need to have the SSB filter. On the negative side, the need for a wide band phase shifting circuit limits the practical range of this scheme. In addition, the demodulator on the receiving side needs to be synchronized with the incoming SSB wave; any mismatch in the transmitter and the receiver local waveforms will cause unwanted tones. Considering the number of SSB schemes that have been developed over time, we limit our discussion to the two basic techniques presented in this section.

15.1.3 The Need for Frequency and Phase Synchronization

In a transmitter's AM circuit, the information-carrying signal $b(t)$ is upconverted using the local carrier tone $c(t)$ that is generated by the transmitter's VCO. Once the amplitude-modulated signal departs from the antenna into space, the receiving circuit must tune in the appropriate frequency band and down-convert the incoming RF signal to the baseband. This frequency shifting is done by the receiver's mixer, which uses the receiver's VCO as the source of the high-frequency reference for its multiplication operation. Aside from technical details of the tuning process itself, there is another important issue that at first is not obvious.

A common assumption in analysis and simulations is that the receiver's local VCO generates exactly the same frequency ω_c as the one generated by the transmitter's VCO (which is located faraway from the receiver). Considering the vast distances between the transmitter and the receiver, it is natural to ask how these two frequencies are synchronized and if there is any consequence if they are not equal, either in frequency or phase. To answer these questions, let us take a look what happens when the receiver's local VCO generates a tone that is only slightly off both in frequency and in phase relative to the tone generated by the transmitter's VCO, i.e. instead of the correct AM carrier wave $c(t) = f(t)\cos\omega_c\,t$ generated by the transmitter, the receiver's VCO generates a slightly incorrect $c'(t) = \cos\left[(\omega_c + \Delta\omega_c)\,t + \theta\right]$, which has an error both in frequency ($\Delta\omega_c \neq 0$) and in phase ($\theta \neq 0$). For simplicity, the modulating signal is $f(t)$, so that the multiplication operation is performed by the receiver to generate $R(t)$ received signal as[2]

$$\begin{aligned}
R(t) &= f(t)\,\cos\omega_c t \times \{\cos\left[(\omega_c + \Delta\omega_c)t + \theta\right]\}\\
&= f(t)\,\left[\cos\omega_c\,t\,\cos\omega_c\,t\,\cos(\Delta\omega_c\,t + \theta) - \cos\omega_c\,t\,\sin\omega_c\,t\,\sin(\Delta\omega_c\,t + \theta)\right]\\
&= \frac{1}{2}\,f(t)\cos(\Delta\omega_c\,t + \theta)
\end{aligned}$$

[1]See Sect. 15.1.2.2. constructive adding.

[2]Trigonometric identity: $\cos(\alpha \pm \beta) = \cos\alpha\cos\beta \mp \sin\alpha\sin\beta$.

$$
\begin{aligned}
&+ \underbrace{\frac{1}{2} f(t) \cos 2\omega_c t \, \cos(\Delta\omega_c t + \theta)}_{\text{filtered out}} \\
&- \underbrace{\frac{1}{2} f(t) \sin 2\omega_c t \, \sin(\Delta\omega_c t + \theta)}_{\text{filtered out}}
\end{aligned} \tag{15.24}
$$

where the last three terms in (15.24) denote the frequency spectrum due to the frequency and phase mismatch. To verify, setting $\Delta\omega_c = 0$ and $\theta = 0$ shows that the (15.24) identity is correct.[3]

The last two terms in (15.24) are at a high frequency, close to $(2\,\omega_c\,t)$, i.e. $2\,\omega_c \pm \Delta\omega_c \approx 2\,\omega_c$, and are removed by a bandpass filter centred at $\Delta\omega_c$. The first of the terms, however, shows that instead of correctly recovering the information signal f(t), the resulting waveform $R(t)$ is a function of $\cos(\Delta\omega_c t + \theta)$. This is a rather serious issue since each time the cosine argument equals an odd number multiple of $\pi/2$, i.e. (after filtering out the HF harmonics)

$$
\begin{aligned}
R(t) &= \frac{1}{2} f(t) \cos(\Delta\omega_c t + \theta) = 0 \\
&\text{when} \quad (\Delta\omega_c t + \theta) = (2n+1)\,\pi/2, \qquad (n = 1, 2, 3, \ldots)
\end{aligned} \tag{15.25}
$$

the entire $f(t)$ signal disappears. The user perceives this as a strong audio effect known as "beating" of the output signal.

Therefore, in order to correctly demodulate the incoming DSB signal, it is necessary to multiply it by using the local tone with exactly correct phase and frequency. For that reason, in practice, PLL circuits are fundamental for wireless communication because they are capable of synthesizing periodic waveforms (either sinusoidal or square) that are both phase- and frequency-locked to the waveform of the RF carrier. Hence, they can (ideally) eliminate problems related to the phase and frequency offsets between the local VCO and the RF carrier waveforms. However, if there is no frequency and phase mismatch, it is possible to simultaneously transmit at the same frequency on a second DSB channel but with carrier phase $\theta = 90°$ with respect to the first channel. These two signals can be received independently of one another. This transmitting scheme is known as "quadrature multiplexing".

In summary, DSB and SSB transmitting schemes have the following main advantages:

1. They are efficient in respect to the signal power, there is no waste due to the carrier tone.
2. DSB can transmit two channels simultaneously by using the quadrature multiplexing.
3. SSB modulation is efficient in terms of its frequency bandwidth requirements.

The main disadvantages of these AM schemes are that a much more complicated transceiver is required and that SSB is not suitable for pulse (digital) communication or music. Today, a number of various schemes are used in either DSB or SSB modulation receivers.

15.1.4 Amplitude Modulator Circuits

From the mathematical point of view, amplitude modulation operation is equivalent to the frequency-shifting operation described in Sect. 14.2. In other words, AM modulated signal $c_{AM}(t)$ is produced by

[3] $\cos 2x = \cos^2 x - \sin^2 x$, and $\cos^2 x + \sin^2 x = 1$.

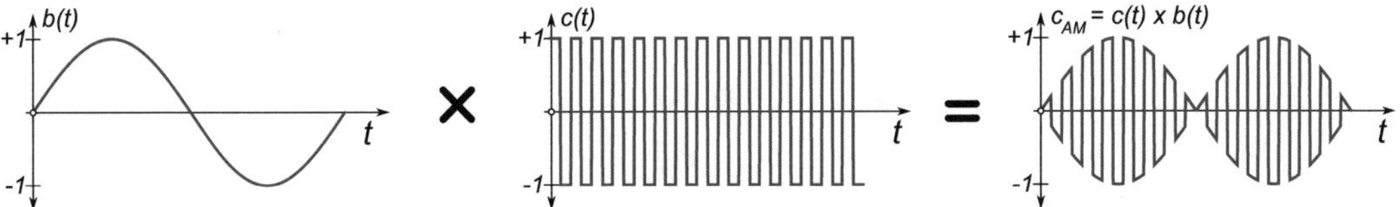

Fig. 15.9 Multiplication of signal $b(t)$ with ± 1 pulse stream $c(t)$ results in amplitude-modulated carrier signal $c_{AM}(t) = b(t) \times c(t)$

Fig. 15.10 AM modulator principles

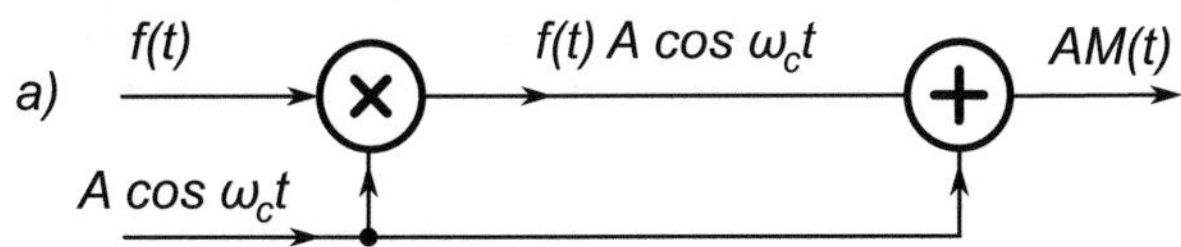

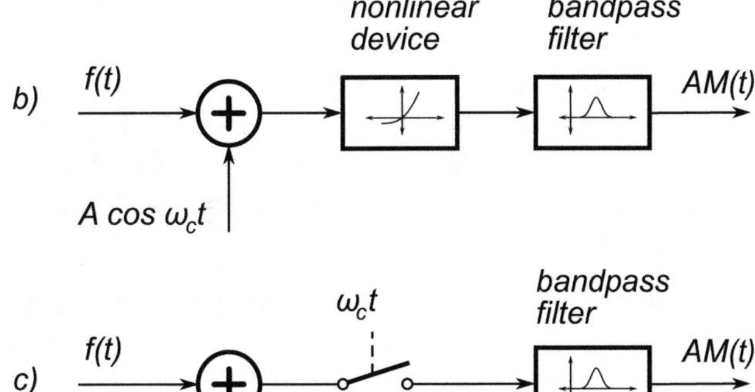

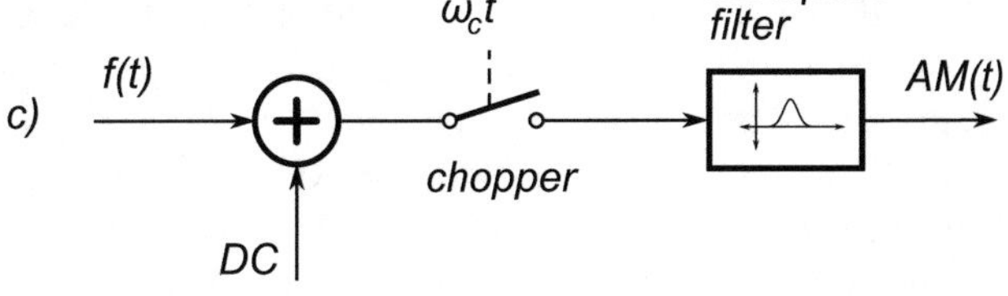

multiplying the modulation signal $b(t)$ with ± 1 pulse stream, Fig. 15.9. Therefore, the mixer circuits described in Sects. 14.4–14.6 are perfectly valid as amplitude modulator circuits. The diode mixer introduced in Sect. 14.3 does not have gain; hence, it is used only at very high frequencies where the BJT/FET active devices also lose their gain but much faster.

Modulation and upconversion are done inside the transmitter circuit and a wide range of circuits are used for AM. A detailed study of transmitter circuits is left for another occasion; in this book, we focus only on the general principles and a few typical modulation circuits.

Amplitude modulation may be done either at a low level in the transmitter, i.e. where the signal power is still relatively low, for example, at the base of an input BJT or at a high level in the transmitter hierarchy, i.e. both the carrier and the modulating signals are combined close to the antenna.

There are at least three possible low-level schemes used to generate AM waves, as shown by the block diagrams in Fig. 15.10:

1. A literal implementation of the AM waveform model (15.5), where $f(t) = m \cos \omega_b t$, Fig. 15.10a.
2. A nonlinear device that produces an AM waveform, which is mathematically approximated by a polynomial as

$$v_o = a_0 + a_1 v_i + a_2 v_i^2 + \cdots \tag{15.26}$$

After substituting $v_i = f(t) + A \cos \omega_c t$ it follows that

$$\begin{aligned} v_o = & a_0 \\ & + a_1 f(t) + a_1 \cos \omega_c t \\ & + a_2 f^2(t) + 2a_2 A f(t) \cos \omega_c t + a_2 A^2 \cos^2 \omega_c t \\ & + \cdots \end{aligned} \tag{15.27}$$

which, after expanding the squared cosine terms,[4] shows that a nonlinear device generates spectrum components that do not exist in the original signal. Hence, a bandpass filter is needed to remove the frequency components that are not close to the carrier frequency, so that

$$v(t) \approx a_1 \cos \omega_c t + 2 a_2 A f(t) \cos \omega_c t \tag{15.28}$$

which is the desired AM waveform, Fig. 15.10b.

3. Nonlinearity does not have to be provided by an active component. Inherently, any switching function is also nonlinear, in fact it is very nonlinear. A switching device that is controlled by a periodic signal ω_c is usually referred to as a "chopper" (see Fig. 15.10c) and may be used to generate an AM waveform. Effectively, the chopper works as a multiplier for its input signal $b(t)$ and the switching square wave (i.e. the switch is a binary device) at ω_c frequency. Again, the chopper is followed by a bandpass filter that is needed to filter out harmonics from the pulse spectrum.

One of the main disadvantages of low-level AM modulators is that the modulation must be followed by a linear amplifier. Linear amplifiers are relatively inefficient for power transfer applications, hence these modulating schemes are not used for high-power RF transmitters in commercial broadcasting radio stations or for modern battery-powered wireless devices. Instead, some topology based on a high-level scheme that employs a class C amplifier is used more often. In the following pages, we briefly introduce several commonly used circuits for amplitude modulation.

15.1.4.1 BJT AM Circuit

At least in principle, one of the simplest ways to do amplitude modulation is to feed the carrier tone and the modulation signal to a single active device as shown in Fig. 15.11. The base serves as the input terminal for the RF signal $c(t)$, while the emitter serves as the input terminal to the modulation signal $b(t)$. The nonlinear characteristics of the active device provides frequency shifting, while the LC resonant circuit at the output node is tuned to either of the two sidebands. As we already know, the resonator tank presents high impedance at the chosen sideband frequency, while effectively shorting to the ground all other tones in the signal spectrum.

15.1.4.2 Balanced AM Circuits

The main property of a balanced modulator is that it outputs the product of the two input signals $b(t)$ and $c(t)$ while suppressing one or both of them. Based on whether only one or both of the input tones is removed from the output spectrum, the balanced modulator is said to be either single or double balanced. That is, in the ideal case of a double-balanced modulator, the output spectrum contains $(\omega_c \pm \omega_b)$ tones but neither ω_c nor ω_b themselves. If signal-suppressing input is used by the carrier $c(t)$, then a balanced modulator produces the DSB–SC spectrum. A large number of balanced

[4]Use trigonometric identities for $\cos^2 \theta = \dfrac{1 + \cos 2\theta}{2}$ and $\sin^2 \theta = \dfrac{1 - \cos 2\theta}{2}$.

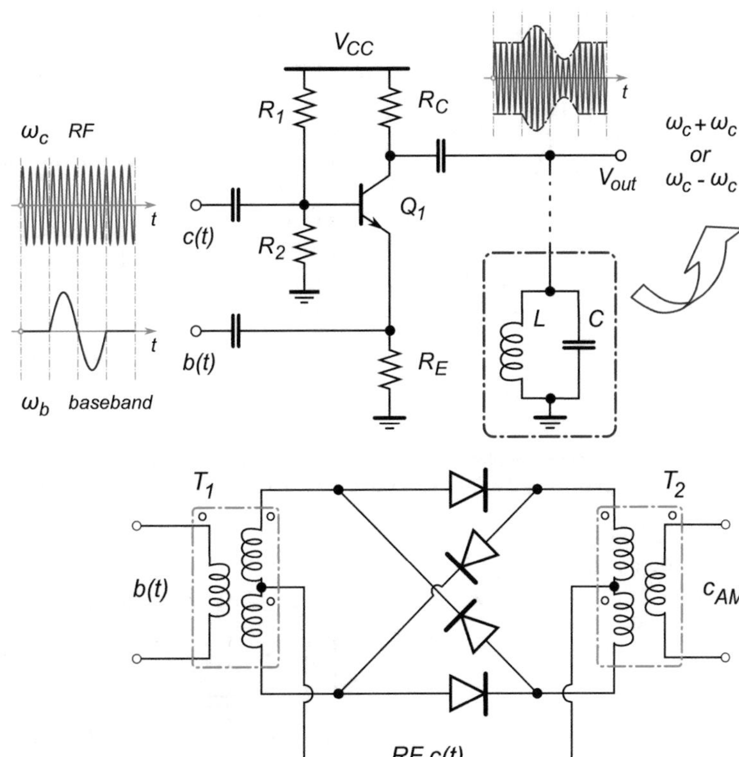

Fig. 15.11 Simplified schematic a typical BJT amplitude modulator circuit

Fig. 15.12 Schematic of double balanced diode ring modulator

modulator designs are in use; we review the operation of three typical circuits: diode ring modulators, balanced FET modulators, and IC balanced modulators.

15.1.4.3 Double-Balanced Diode Ring Modulator

A simple AM circuit commonly used for low-frequency applications in telephone networks is based on four diodes and two transformers (see Fig. 15.12). The bulkiness of the two transformers is probably the main reason this modulator is not used in more applications. The amplitude-modulated output produced by the diode ring modulator is "double balanced with suppressed carrier". Diode pairs D_1–D_2 and D_3–D_4 are alternatively switched on and off by the HF carrier $c(t)$, where the carrier signal could be either a sinusoidal or a square wave at frequency ω_c whose amplitude (after accounting for the real diode voltage drop) is greater than amplitude of the modulation signal $b(t)$ (Fig. 15.13).

In the ideal circuit, all four diodes are perfectly matched and the two transformers are perfectly symmetrical, hence, when HF signal is zero, $c(t) = 0$, then the output signal is also zero, $c_{AM}(t) = 0$. By inspection, we follow the HF current entering the centre tap of transformer T_1, which then splits and passes in parallel through diodes D_1 and D_2, only to converge again at the centre tap of T_2 and return to the HF source. The two HF currents that enter the primary side of transformer T_2 in opposite directions induce voltages of equal magnitude and opposite polarity in the T_2 secondary, which therefore cancel each other and produce zero voltage output.

The square wave switching function $c(t)$ can be written using its Fourier transformation with amplitude $A = \pi/2$ as

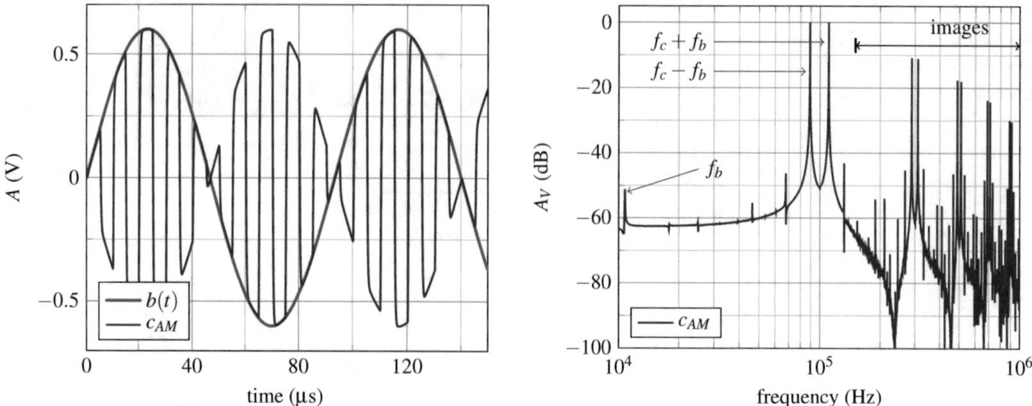

Fig. 15.13 Simulated waveforms of a double balanced diode ring modulator: time domain AM signal (left), and frequency spectrum (right). It is clearly visible that the carrier tone is suppressed

Fig. 15.14 Schematic of singe balanced FET modulator

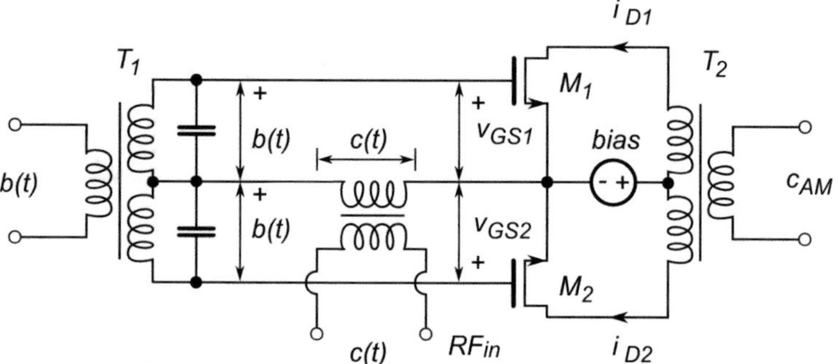

$$c(t) = \sin \omega_c t + \frac{1}{3} \sin 3\,\omega_c t + \cdots + \frac{1}{n} \sin n\,\omega_c t + \cdots \tag{15.29}$$

where $n = 2k + 1$ is an odd number. Further mathematical analysis shows that multiplication of the time-varying signal c(t) and the sinusoidal modulation $b(t) = \sin \omega_b t$, after substituting (15.29), results in

$$\begin{aligned} c_{AM} &= b(t) \times c(t) \\ &= \sin \omega_b t \, \sin \omega_c t + \sin \omega_b t \, \frac{1}{3} \sin 3\,\omega_c t + \cdots + \sin \omega_b t \, \frac{1}{n} \sin n\,\omega_c t + \cdots \end{aligned} \tag{15.30}$$

which, after expanding all products, shows that the output spectrum contains only $(n\omega_c \pm \omega_b)$ terms with neither ω_c nor ω_b terms by themselves, hence this is a double-balanced modulator.

15.1.4.4 Single Balanced FET Modulator

Instead of using diodes, two FET transistors are connected as in Fig. 15.14 to produce a DSB–SC amplitude-modulated waveform. The modulation signal $b(t)$ is split into two copies at the symmetrical secondary of T_1, while the carrier signal $c(t)$ is injected into the circuit through its own 1:1 transformer. The two FET transistors are perfectly matched, which is usually achieved by using IC components.

By inspection, we write

$$v_{GS1} = c(t) + b(t) = C \cos\omega_c t + B \cos\omega_b t \tag{15.31}$$

$$v_{GS2} = c(t) - b(t) = C \cos\omega_c t - B \cos\omega_b t \tag{15.32}$$

hence, the drain currents of the two FET transistors are approximated by the second-order polynomials as

$$i_{D1} \approx I_0 + a_1\, v_{GS1} + a_2\, v_{GS1}^2 \tag{15.33}$$

$$i_{D2} \approx I_0 + a_1\, v_{GS2} + a_2\, v_{GS2}^2 \tag{15.34}$$

hence, after substituting (15.31) and (15.32), we write

$$i_{D1} = I_0 + a_1\,(C \cos\omega_c t + B \cos\omega_b t) + a_2\,(C \cos\omega_c t + B \cos\omega_b t)^2 \tag{15.35}$$

$$i_{D2} = I_0 + a_1\,(C \cos\omega_c t - B \cos\omega_b t) + a_2\,(C \cos\omega_c t - B \cos\omega_b t)^2 \tag{15.36}$$

The total current in the primary of transformer T_2 is the current difference $(i_{D1} - i_{D2})$, which subsequently induces the output voltage AM waveform as

$$c_{AM} = M\,\frac{d(i_{D1} - i_{D2})}{dt} \tag{15.37}$$

where M is the mutual inductance between the primary and secondary inductances of transformer T_2 (in an ideal case $M = 1$). Therefore,

$$\begin{aligned}(i_{D1} - i_{D2}) &= 2a_1\, B \cos(\omega_b t) + 4a_2\, BC \cos(\omega_b t) \cos(\omega_c t)\\ &= 2a_1\, B \cos(\omega_b t) + 2a_2\, BC \cos(\omega_c - \omega_b)t + 2a_2\, BC \cos(\omega_c + \omega_b)t\end{aligned} \tag{15.38}$$

By inspection of (15.38) we conclude that its first term contains frequency component ω_b of the modulating signal $b(t)$, the second term is at frequency $(\omega_c - \omega_b)$, and the third term is at $(\omega_c + \omega_b)$. The linear addition of these three terms and derivation (15.37) do not change the frequency spectrum, hence the output voltage c_{AM} waveform contains only the two side tones $(\omega_c \pm \omega_b)$ and the modulation signal $b(t)$, Fig. 15.15. It should be noted that if waveforms $b(t)$ and $c(t)$ exchanged input terminals, then the carrier tone would have survived and $b(t)$ would have been suppressed from the output frequency spectrum.

15.1.4.5 Double-Balanced IC Modulator

In order to reduce the size of the electronic equipment, the use of transformers in modern mobile devices is avoided if possible. Ideally, the goal is to design all critical functions that are needed for radio equipment communication using IC technology. As the technology developed so did balanced modulators in IC form (Fig. 15.16). A typical example of an AM modulator circuit in IC technology is a Gilbert-cell-based differential multiplier. A Gilbert cell itself is a very versatile circuit that has many applications, e.g. if used in switching mode, it works as a balanced AM multiplier circuit.

The circuit works as a multiplier of the two differential input signals. When it is used as a balanced modulator, an elementary analysis of the circuit is as follows. First, we find the output signal when

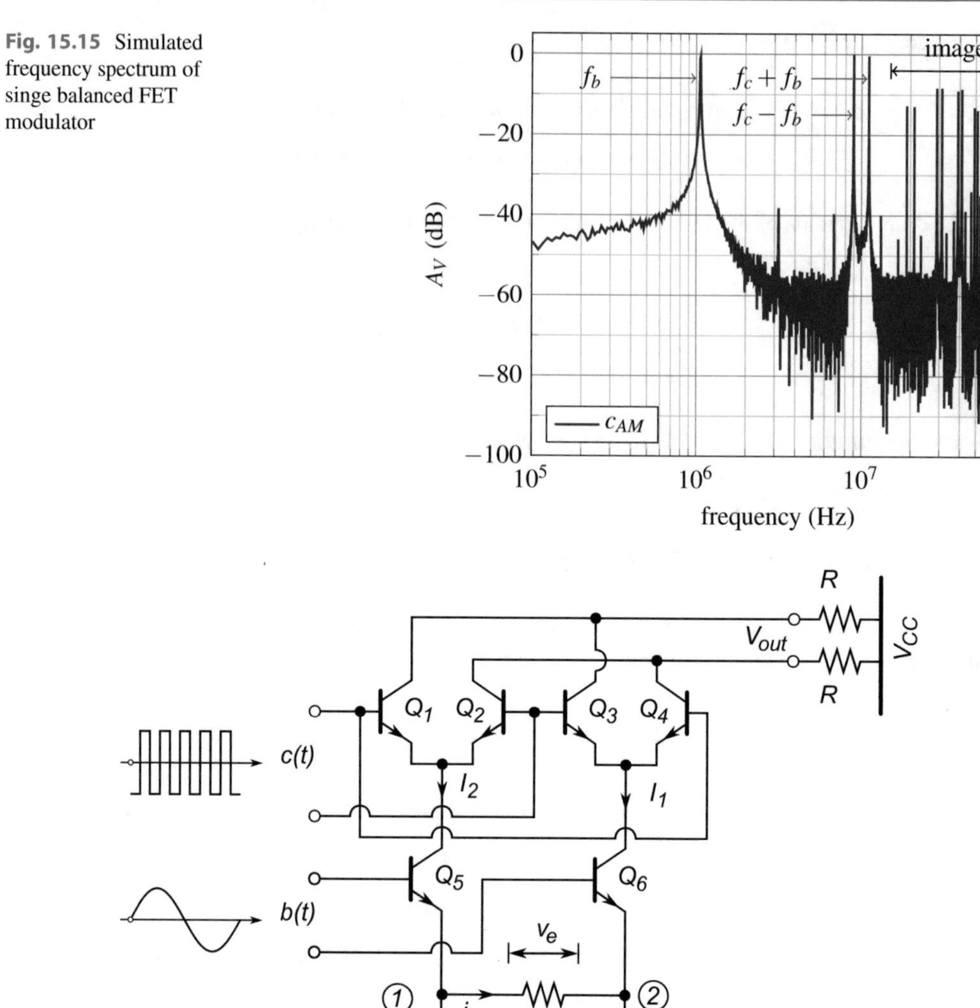

Fig. 15.15 Simulated frequency spectrum of singe balanced FET modulator

Fig. 15.16 Double balanced IC modulator based on Gilbert cell

the carrier signal is absent. Second, we add the carrier signal as the product. The carrier is considered to be a high-level switching voltage that alternately switches transistors pairs Q_1, Q_4 and Q_2, Q_3 on and off.

With no carrier signal applied, and assuming that the base currents are negligible, after summing the currents at junctions node① and node②, we write

$$I_2 = I + i_e$$
$$I_1 = I - i_e \tag{15.39}$$

Hence, the differential output voltage v_o' is

$$v_o' = v_2 - v_1 = R(I_2 - I_1) = R(2i_e) \tag{15.40}$$

Application of KVL to the loop that contains the modulating voltage signal $b(t)$ and resistance R_e yields

$$b(t) = V_{be5} + v_e - V_{be6} \tag{15.41}$$

$$\therefore$$

$$b(t) \approx v_e \tag{15.42}$$

because the circuit operates with small signal current $I \gg i_e$ and keeps $V_{be5} \approx V_{be6}$. Therefore

$$i_e = \frac{v_e}{R_e} = \frac{b(t)}{R_e} \tag{15.43}$$

$$\therefore$$

$$v'_{out} = \frac{2R}{R_e}\, b(t) = \frac{2R}{R_e}\, \sin \omega_b\, t \tag{15.44}$$

after substituting (15.43) back into (15.40). After the carrier signal $c(t)$ is added the modulator output delivers their product. In the case of a square carrier signal, which contains an infinite number of odd harmonics, we approximate its function as

$$c(t) = \sin(\omega_c\, t) + \frac{1}{3}\, \sin(3\omega_c\, t) + \frac{1}{5}\, \sin(5\omega_c\, t) + \cdots \approx \sin(\omega_c\, t) \tag{15.45}$$

where all higher harmonics $(3\omega_c, 5\omega_c, \ldots)$ are filtered out, which leads to an expression for the output voltage v_o when both the carrier $c(t)$ and the modulation $b(t)$ signals are present,

$$\begin{aligned} v_{out} &\approx v'_{out} \times c(t) = \frac{2R}{R_e}\, \sin(\omega_b\, t) \times \sin(\omega_c\, t) \\ &= \frac{R}{R_e}\, [\cos(\omega_c - \omega_b)t - \cos(\omega_c + \omega_b)t] \end{aligned} \tag{15.46}$$

that is, the output contains only the upper and lower side tones, while the carrier itself does not appear in the output. This modulator circuit is a typical example of how, by taking advantage of IC technology where the components are manufactured as perfect copies of each other, almost perfectly balanced voltages and currents are possible without the external bulky components.

15.2 Angle Modulation

Following the discussion in Sect. 9.2 in regard to (9.4), we now proceed to find out how the two angular parameters of the carrier signal $c(t)$, ω and phase ϕ can be modulated by the modulating signal $b(t)$. Although, frequency and PM are similar and are often studied together under the more inclusive name "angle modulation", the two are different in very important details. Frequency modulation (FM) is commonly used for HiFi broadcasting of music and speech, because of its lower sensitivity to noise, while PM requires a more complicated receiver and is used in some wireless LAN standards, and military and space applications.

15.2.1 Frequency Modulation

The modulating signal $b(t)$ in (15.1) is used to vary the frequency ω_c of the carrier waveform $c(t)$ in the time domain. Let the change in carrier frequency be

$$\Delta\omega_c = k\,b(t) \tag{15.47}$$

where k is a constant known as the "frequency deviation constant"; then the instantaneous carrier frequency is

$$\omega(t) = \omega_c + \Delta\omega_c = \omega_c + k\,b(t) \tag{15.48}$$

where ω_c is the unmodulated carrier frequency. After substituting $b(t) = B\cos\omega_b t$ in (15.48) the instantaneous frequency $f(t)$ of FM waveform becomes

$$\omega(t) = \omega_c + k\,B\,\cos\omega_b t \tag{15.49}$$

$$\therefore$$

$$f(t) = \frac{\omega(t)}{2\pi} = f_c + \frac{k\,B}{2\pi}\cos\omega_b t \tag{15.50}$$

that is, the maximum and minimum values of the instantaneous frequency are

$$f_{max} = f_c + \frac{k\,B}{2\pi} \tag{15.51}$$

$$f_{min} = f_c - \frac{k\,B}{2\pi} \tag{15.52}$$

where the maximum swing of the instantaneous frequency from the unmodulated carrier frequency f_c is called "peak frequency deviation" Δf and is defined as

$$\Delta f \equiv f_{max} - f_c = \frac{k\,B}{2\pi} \tag{15.53}$$

which enables us to define frequency modulation index m_f, and deviation ratio δ as

$$m_f \overset{\text{def}}{=} \frac{\Delta f}{f_m} = \frac{k\,B}{\omega_b}\qquad \delta \overset{\text{def}}{=} \frac{\Delta f}{f_c} = \frac{k\,B}{\omega_c} \tag{15.54}$$

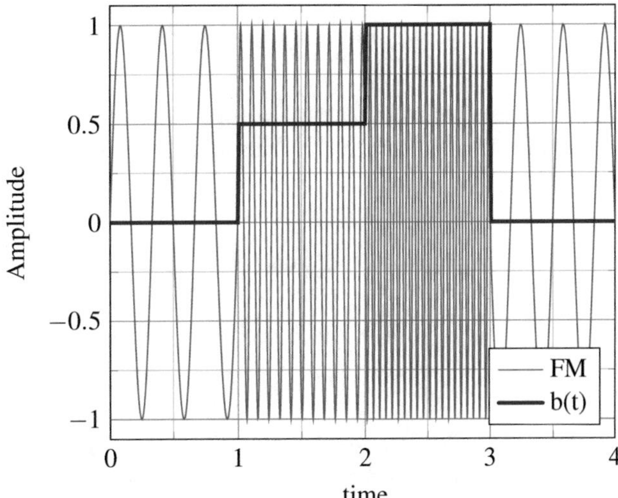

Fig. 15.17 An example of time domain plot of FM (red) and $b(t)$ (blue) waveforms. The modulating signal information $b(t)$ is clearly embedded by the control of the carrier's frequency

An example sketch diagram of the instantaneous frequency $\omega_c(t)$ variation over time is shown in Fig. 15.17. It is important to understand that this graph illustrates the frequency–time curve, Fig. 15.17 (red). The amplitude–time function is shown in Fig. 15.17 (blue).

An analytical expression for the FM waveform may be derived as follows. The unmodulated carrier is a sine wave, therefore

$$c(t) = C \sin(\omega_c t + \phi) \tag{15.55}$$

Equation (15.55) is only a special case of the more general case

$$c(t) = C \sin[\theta(t)] \tag{15.56}$$

where $\theta(t)$ is an arbitrary time-dependent function. By definition, the angular frequency $\omega_c(t)$ is the rate of change in time of $\theta(t)$. Only when the frequency is constant is the particular form of (15.55) valid. When the frequency is time dependent, as in FM, an instantaneous angular frequency may be defined as

$$\omega_c(t) = 2\pi\, f_c(t) = \frac{d\theta(t)}{dt} \tag{15.57}$$

$$\therefore$$

$$\theta(t) = \int \omega_c(t)\, dt \tag{15.58}$$

The instantaneous frequency $\omega_c(t)$ is related to the modulated frequency through relation (15.50). For instance, in the case of constant (unmodulated) angular frequency ω_c, we write

$$\theta(t) = \int \omega_c\, dt = \omega_c t + \phi \tag{15.59}$$

where ϕ is the integration constant. In the case of sinusoidal modulation, after substituting (15.55) into (15.58) we have

$$\theta(t) = \int 2\pi\,(f_c + \Delta f\,\cos\omega_b\,t)\,dt = \omega_c\,t + \frac{\Delta f}{f_m}\,\sin\omega_b\,t + \phi \tag{15.60}$$

The integration constant ϕ may be made equal to zero by an appropriate choice of time reference axis, while the expression for the sinusoidally modulated FM wave is obtained by substituting (15.60) into (15.56) as[5]

$$\begin{aligned} c_{FM} &= C\,\sin\left(\omega_c\,t + \frac{\Delta f}{f_m}\,\sin\omega_b\,t\right) \\ &= C\,\sin\left(\omega_c\,t + m_f\,\sin\omega_b\,t\right) \\ &= C\,\left[\sin(\omega_c\,t)\,\cos(m_f\,\sin\omega_b\,t) + \cos(\omega_c\,t)\,\sin(m_f\,\sin\omega_b\,t)\right] \end{aligned} \tag{15.61}$$

We note that, unlike the AM index m, the frequency modulation index m_f can be greater than unity.

It turns out that mathematicians have already found suitable expansions for functions of type "$\cos(x\,\sin y)$" using Bessel functions, i.e.

$$\cos(m_f\,\sin\omega_b\,t) = J_0(m_f) + 2\sum_{n=1}^{\infty} J_{2n}(m_f)\,\cos(2n\omega_b\,t) \tag{15.62}$$

$$\sin(m_f\,\sin\omega_b\,t) = 2\sum_{n=0}^{\infty} J_{2n+1}(m_f)\,\sin[(2n+1)\omega_b\,t] \tag{15.63}$$

where $J_n(m_f)$ is the Bessel function of the first kind and of n-th order. After substituting (15.62) and (15.63) into (15.61), and after expending the sinusoidal products, the analytical expression of FM waveform for the case of sinusoidal modulation (15.61) is written as

$$\begin{aligned} c_{FM} = {} & J_0(m_f)\,C\,\sin\omega_c\,t \\ & + J_1(m_f)\,C\,[\sin(\omega_c + \omega_b)\,t - \sin(\omega_c - \omega_b)\,t] \\ & + J_2(m_f)\,C\,[\sin(\omega_c + 2\omega_b)\,t + \sin(\omega_c - 2\omega_b)\,t] \\ & + J_3(m_f)\,C\,[\sin(\omega_c + 3\omega_b)\,t - \sin(\omega_c - 3\omega_b)\,t] \\ & + \cdots \end{aligned} \tag{15.64}$$

where Bessel function $J_n(m_f)$ is defined by the series

$$\begin{aligned} J_n(m_f) = \frac{m_f^n}{2^n\,n!}\Bigg[1 &- \frac{m_f^2}{2(2n+2)} + \frac{m_f^4}{2(4)(2n+2)(2n+4)} \\ &- \frac{m_f^6}{2(4)(6)(2n+2)(2n+4)(2n+6)} + \cdots\Bigg] \end{aligned} \tag{15.65}$$

[5]Use the trigonometric identity: $\sin(\alpha \pm \beta) = \sin\alpha\cos\beta \pm \cos\alpha\sin\beta$.

It is handy to have Bessel functions (15.65) both in graphical form (Fig. 15.18) and in tabular form (Table 15.1). For instance, by reading values from the table for FM index $m_f = 1.0$, we find that first five significant spectral component amplitudes are:

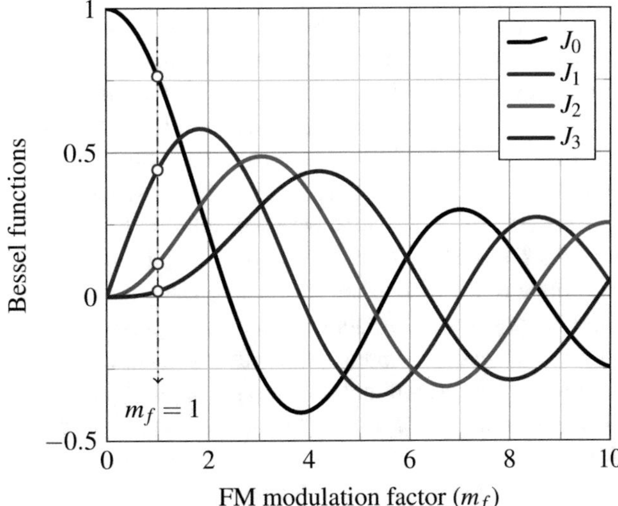

Fig. 15.18 Bessel functions, $J_n(m_f)$ for $n = 0, 1, 2, 3$. Example points shown for $m_f = 1$

Table 15.1 Bessel functions of order 1–10, for modulation factors 0–10

m_f	J_0	J_1	J_2	J_3	J_4	J_5	J_6	J_7	J_8	J_9	J_{10}
0.0	1.000	–	–	–	–	–	–	–	–	–	–
0.2	0.990	0.099	0.005	–	–	–	–	–	–	–	–
0.4	0.960	0.196	0.020	0.001	–	–	–	–	–	–	–
0.5	0.938	0.242	0.030	0.002	–	–	–	–	–	–	–
0.6	0.912	0.287	0.044	0.004	–	–	–	–	–	–	–
0.8	0.846	0.369	0.076	0.010	0.001	–	–	–	–	–	–
1.0	0.765	0.440	0.115	0.020	0.002	–	–	–	–	–	–
1.2	0.671	0.498	0.159	0.033	0.005	0.001	–	–	–	–	–
1.4	0.567	0.542	0.207	0.050	0.009	0.001	–	–	–	–	–
1.5	0.512	0.558	0.232	0.061	0.012	0.002	–	–	–	–	–
1.6	0.455	0.570	0.257	0.072	0.015	0.002	–	–	–	–	–
1.8	0.340	0.582	0.306	0.099	0.023	0.004	0.001	–	–	–	–
2.0	0.224	0.577	0.353	0.129	0.034	0.007	0.001	–	–	–	–
2.5	−0.048	0.497	0.446	0.217	0.074	0.019	0.004	0.001	–	–	–
3.0	−0.260	0.339	0.486	0.309	0.132	0.043	0.011	0.002	–	–	–
3.5	−0.380	0.137	0.459	0.387	0.204	0.080	0.025	0.008	0.001	–	–
4.0	−0.397	−0.066	0.364	0.430	0.281	0.132	0.049	0.015	0.004	0.001	–
4.5	−0.321	−0.231	0.218	0.425	0.348	0.195	0.084	0.030	0.009	0.002	0.001
5.0	−0.178	−0.328	0.467	0.365	0.391	0.261	0.131	0.053	0.018	0.005	0.001
6.0	0.151	−0.277	−0.243	0.115	0.358	0.362	0.246	0.130	0.056	0.021	0.007
7.0	0.300	−0.005	−0.301	−0.168	0.158	0.348	0.339	0.234	0.128	0.059	0.023
8.0	0.172	0.235	−0.113	−0.291	−0.105	0.186	0.338	0.321	0.223	0.126	0.061
9.0	−0.090	0.245	0.145	−0.181	−0.265	−0.055	0.204	0.327	0.305	0.215	0.125
10.0	−0.246	0.043	0.255	0.058	−0.220	−0.234	−0.014	0.217	0.318	0.292	0.207

Carrier frequency	(f_c)	$J_0(1.0) = 0.765$
First-order side frequencies	$(f_c \pm f_b)$,	$J_1(1.0) = 0.440$
Second-order side frequencies	$(f_c \pm 2f_b)$,	$J_2(1.0) = 0.115$
Third-order side frequencies	$(f_c \pm 3f_b)$,	$J_3(1.0) = 0.020$
Fourth-order side frequencies	$(f_c \pm 4f_b)$,	$J_4(1.0) = 0.002$

The fact that the spectrum components around the carrier frequency decrease in amplitude does not mean that the carrier wave is amplitude modulated. The carrier wave is the sum of all harmonics in its spectrum, and the harmonics add up to produce the constant amplitude FM waveform, Fig. 15.17 (red). The main distinction to note is that the FM modulated carrier is not a sine wave, whereas each of the spectrum components around the carrier frequency is. In addition, negative amplitudes in Table 15.1 only indicate the phase inversion.

Existence of the negative harmonic amplitudes implies that there are values of the FM index m_f for which the corresponding harmonic amplitude is zero, for instance if $m_f = 2.4, 5.5, 8.65, \ldots$ amplitude of the tone at the carrier frequency becomes zero. It is important to distinguish this case from its matching AM-balanced case of suppressed carrier. For the FM waveform, if the tone at the carrier frequency is suppressed it only means that its energy is redistributed to the side tones, while the FM waveform amplitude is always constant. This statement emphasizes the point that it is only the sinusoidal component in the FM spectrum that is at the carrier frequency, not the FM carrier itself. Amplitude of the carrier tone may become zero, therefore it varies from positive to negative peak values as the modulation index changes.

The ideal FM waveform frequency spectrum consists of the infinite number of harmonic tones uniformly spaced by the modulation frequency f_b (see Fig. 15.19). Therefore, the procedure for establishing the required FM waveform bandwidth involves approximate methods. One commonly used method for estimating FM waveform bandwidth is based on approximation that sets the bandwidth limits by inclusion of the highest relevant harmonics on both sides as

$$B_{FM} = 2n\, f_b \tag{15.66}$$

where n is the highest order of the side frequency harmonic tone whose amplitude is significant (i.e. not negligible). By careful observation, using values from Table 15.1, it was found that if the order

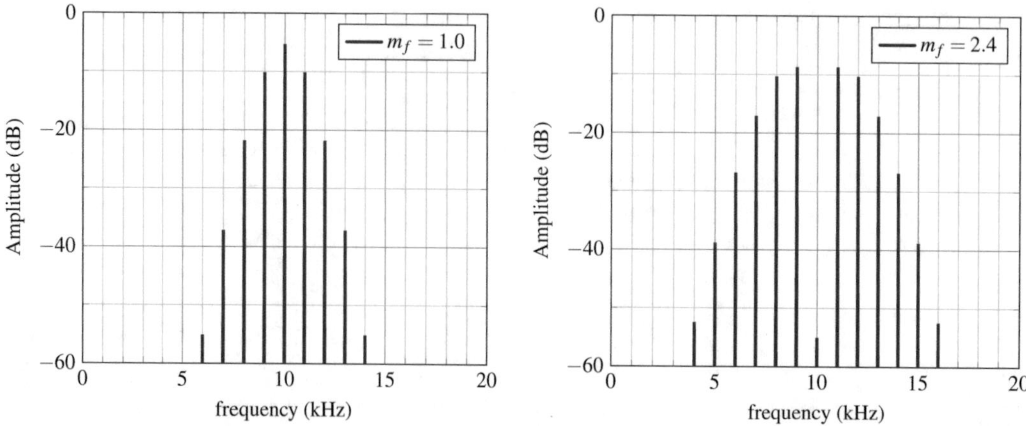

Fig. 15.19 Diagram of two frequency spectrums of an FM waveform, for $m_f = 1$ and $m_f = 2.4$, showing only the first few relevant side tones. In this example, $f_c = 10\,\text{kHz}$ and $f_b = 1\,\text{kHz}$

of the side frequency is greater than $(m_f + 1)$, the FM waveform amplitude is within 5% of the unmodulated carrier amplitude. Using this approximation as the guide for estimating the bandwidth requirement, (15.66) is rewritten as

$$B_{FM} = 2(m_f + 1) f_b = 2(\Delta f + f_b) \tag{15.67}$$

Relation (15.67) is known as "Carson's rule" . In order to illustrate the application of this rule, let us take a look at the following numerical examples:

1. If $\Delta f = 75\,\text{kHz}$, $f_b = 0.1\,\text{Hz}$ than $B_{FM} = 150\,\text{kHz}$;
2. If $\Delta f = 75\,\text{kHz}$, $f_b = 1.0\,\text{kHz}$ than $B_{FM} = 152\,\text{kHz}$;
3. If $\Delta f = 75\,\text{kHz}$, $f_b = 10\,\text{kHz}$ than $B_{FM} = 170\,\text{kHz}$.

Thus, although the modulation frequency changes by a factor of 100, the bandwidth occupied by the spectrum is almost constant.

Bessel functions relate the voltage amplitude of each of the sinusoidal side frequency components to the unmodulated carrier amplitude. That is

$$E_n = J_n\, E_c \tag{15.68}$$

where E_n is the amplitude of the n-th harmonic, J_n is Bessel's function of n-th order, and E_c is the amplitude of the carrier tone. Assuming that the amplitudes E_n and E_c are their RMS values, the power contained in the n-th sinusoidal component is given as

$$P_n = \frac{E_n^2}{R} \tag{15.69}$$

After noticing that there is only one component at the carrier frequency and pairs of components for each frequency n, the total power in the FM waveform is simply the sum of all harmonics, i.e.

$$\begin{aligned} P_T &= P_0 + 2P_1 + 2P_2 + \cdots \\ &= \frac{E_0^2}{R} + \frac{2E_1^2}{R} + \frac{2E_2^2}{R} + \cdots \\ &= \frac{J_0 E_c^2}{R} + \frac{2J_1 E_c^2}{R} + \frac{2J_2 E_c^2}{R} + \cdots \\ &= P_c \left(J_0^2 + 2\left(J_1^2 + J_2^2 + \cdots\right)\right) \end{aligned} \tag{15.70}$$

where P_c is the power of the unmodulated carrier and J_n are Bessel's functions for the given value of modulation index m_f. Again, the total power in the modulated waveform remains constant for all values of the modulation index. This is illustrated by the fact that the sum of the squares of the Bessel function coefficients in (15.70) for a given value of m_f is always unity. For instance, if $m_f = 1.5$, the total power P_T relative to the unmodulated carrier power P_c is found using values from Table 15.1 as

$$\frac{P_T}{P_c} = 0.512^2 + 2(0.558^2 + 0.232^2 + 0.061^2 + 0.012^2 + 0.002^2) = 1.000258$$

That is, if only the first five side harmonics are used then the rounding error is 0.026%.

15.2.2 Phase Modulation

The third method of RF carrier modulation is PM, which is somewhat similar to the FM technique. In today's communication systems, it is often used for satellite and deep-space missions because, like FM, its noise properties are superior to AM but, unlike FM, it can be produced in a simple circuit driven from a frequency stable, crystal-controlled carrier oscillator. A VCO is intentionally made very variable with respect to frequency to produce high deviations and a high modulation index.

The derivation process of an analytical expression for the PM waveform (Fig. 15.21) is similar to the one used for deriving the FM waveform expression. Start with the unmodulated carrier that is given by

$$c(t) = \sin(\omega_c + \phi_c) \tag{15.71}$$

When phase modulated, the carrier phase ϕ_c is replaced with the instantaneous phase $\phi(t)$), where

$$\phi(t) = \phi_c + K\, b(t) \tag{15.72}$$

where K is the phase deviation constant (analogous to k for FM) and $b(t)$ is the modulating signal. Because ϕ_c is constant, we can set its value to zero by choosing the appropriate reference point. After substituting $b(t) = B\, m(t)$, where $m(t)$ is a general time-dependent function, (15.72) becomes

$$\phi(t) = \Delta\phi\, m(t) \tag{15.73}$$

where $\Delta\phi = K\, B$ is the maximum phase deviation. After substituting (15.73) back into (15.71), the expression for the phase-modulated waveform is written as

$$c_{PM}(t) = \sin[\omega_c t + \Delta\phi\, m(t)] = \sin[\omega_c t + m_p \sin\omega_m t] \tag{15.74}$$

for sinusoidal modulation. After renaming the phase deviation $\Delta\phi$ to PM index m_p, a comparison of (15.74) and (15.61) shows a similarity between FM and PM schemes.

A simplified block diagram of a PM waveform generator is derived after expanding (15.74) and using the narrowband approximation $\Delta\phi < 0.25$, i.e. $\cos\Delta\phi \approx 1$ and $\sin\Delta\phi \approx \Delta\phi$. After applying these approximations, the following result is obtained (remember that cosine and sinusoidal functions are the same, except for a 90° phase difference):

$$v(t) = A\cos\omega_c t - A\,(m_p \cos\omega_b)\,\sin\omega_c t \tag{15.75}$$

Equation (15.75) is presented in a graphical form in Fig. 15.20, which suggests one possible block-level diagram of a simple PM transmitter implementation scheme. To emphasize the phase shifting,

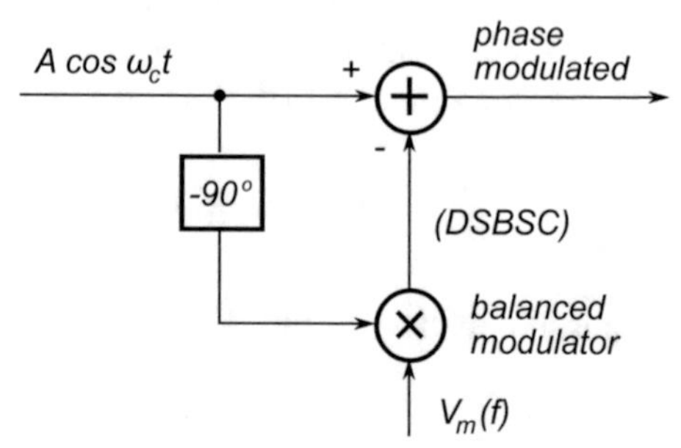

Fig. 15.20 Phase modulator block diagram

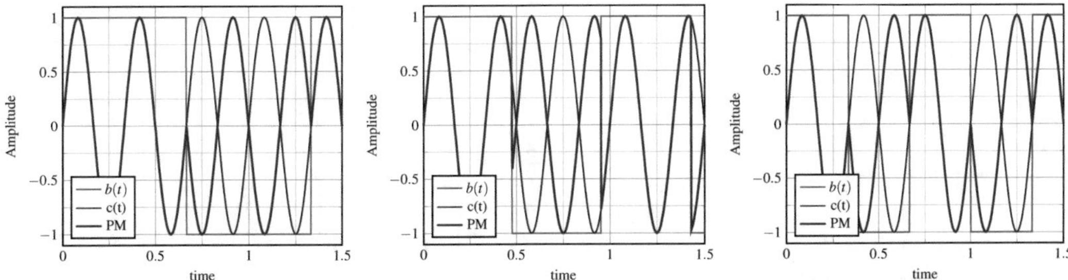

Fig. 15.21 Examples of phase-modulated waveforms PM (blue) for three relative ratios of carrier and modulation signal frequencies. The pulse modulation signal $b(t)$ (red) is clearly embedded and controls the carrier's phase

the carrier's phase is controlled by hard-switching pulse stream. The multiplication of carrier and pulse waveforms, in mathematical sense, translates in a simple multiplication by ± 1 which hard-flips the carrier's phase $\pm 90°$. As an illustration, simulated waveforms are shown in Fig. 15.21 for three relative ratios of the carrier's and pulse waveforms. As a reference, unmodulated carrier $c(t)$ is also shown in dotted line.

15.2.3 Angle Modulator Circuits

Frequency modulation is done by means of a VCO circuit whose varicap biasing control terminal serves as the input terminal for the modulation voltage $V_D = b(t)$. Therefore, any circuit described in Sect. 13.6 is perfectly suited for the role of an angle modulator circuit. In this section, we study two additional principles that are used in VCO circuits for modulating the frequency of the output waveform: the reactance modulator and the varicap-based phase modulator.

15.2.3.1 Reactance Modulator

The main disadvantage of varicap-based FM circuits is their narrow tuning range, which is due to the diode's very narrow small signal operational zone. Instead, at frequencies that are not ultrahigh, wider tunable impedance variation can be achieved with a circuit known as "reactance modulator". Its operation is based on the intentional increase and exploitation of Miller's effect by a circuit that effectively behaves as voltage-controlled capacitive impedance (Fig. 15.22) between the output node and the ground (Fig. 15.23).

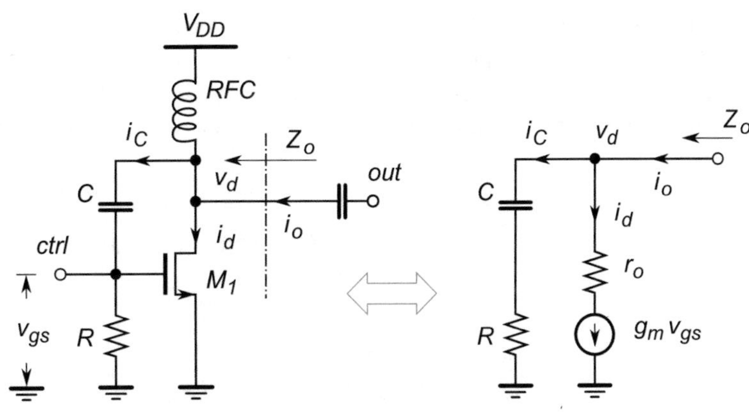

Fig. 15.22 Simplified schematic diagram of reactance modulator circuit and its equivalent small signal network

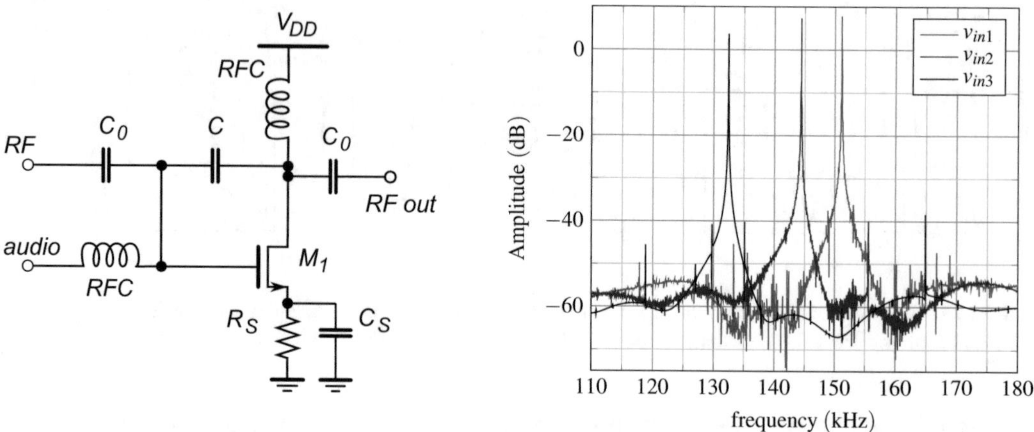

Fig. 15.23 A practical reactance FM modulator circuit and its frequency response for three input voltage amplitudes

Use of a FET device indicates an assumption that the gate current $i_g = 0$, which is the first approximation used in the analysis of a reactance modulator circuit. In addition, the use of an RFC, i.e. "RF choke", inductor enables DC biasing of the FET while blocking AC currents. The equivalent schematic circuit diagram for such an arrangement is shown in Fig. 15.22 (right).

Effective output impedance Z_o, as seen by looking into the output node, is found by definition. That is, by the ratio of the output voltage v_o to the output current i_o, as

$$i_C = \frac{v_o}{R + Z_C} = \frac{v_o}{R - j\dfrac{1}{\omega C}} \tag{15.76}$$

$$\therefore$$

$$v_{gs} = R\, i_C = \frac{R\, v_o}{R - j\dfrac{1}{\omega C}} \tag{15.77}$$

which leads into expression for M_1 drain current i_d as

$$i_d = g_m v_{gs} = g_m \frac{R\, v_o}{R - j\dfrac{1}{\omega C}} \tag{15.78}$$

It is time to introduce the following assumptions: current i_C through the capacitor C branch needs to be much smaller than the M_1 drain current id, i.e. $i_d \gg i_C$ or $i_d + i_C \approx i_d$; and the capacitor C impedance X_C is much greater than resistance R, that is $R - X_C \approx -X_C$.

Application of these approximation enables us to write an expression for the output current i_o, as

$$i_o = i_C + i_d \approx i_d = g_m \frac{R\, v_o}{R - j\dfrac{1}{\omega C}} \approx g_m \frac{R\, v_o}{-j\dfrac{1}{\omega C}} = \frac{v_o}{-j\dfrac{1}{\omega\, g_m\, R C}}$$

$$\therefore$$

$$Z_o \equiv \frac{v_o}{i_o} = -j\frac{1}{\omega\,(g_m\, R C)} = -j\frac{1}{\omega\, C_{RM}} \tag{15.79}$$

where $C_{RM} = (g_m\, \omega\, R\, C)$ depicts the effective capacitance as seen at the output node of the reactance modulator. Through (15.79), we have shown that disposition of the output impedance Z_o of the reactance modulator is very well approximated by a tunable capacitance $C_{RM} = f(g_m)$, which is controlled by the M_1 gate-source voltage, i.e. $C_{RM} = f(v_{gs})$. Hence, the reactance modulator behaves as a voltage-controlled capacitor, which can be connected to an LC resonator tank for purposes of controlling its resonant frequency and, therefore, enabling FM. As a closing note, if the positions of the resistor R and capacitor C are swapped inside the network, the output impedance would effectively become inductive.

15.2.3.2 Varicap Diode-Based Phase Modulator

A steady RF waveform $c(t)$ generated by a crystal oscillator with constant amplitude and phase is injected into the PM circuit whose straightforward implementation is shown in Fig. 15.24. For simplicity, the varicap biasing control voltage $b(t)$ is not shown. It is implicitly assumed that the varicap capacitance is a function of the modulation voltage, i.e. $C_D(t) = f(b(t))$. The variation of diode capacitance C_D alters the phase angle of the phase modulator's tuned-circuit admittance and then the phase angle of its RF voltage.

Time dependence of the RF waveform phase is implemented by adding voltage-controlled phase variation on the phase of the constant RF waveform, as already seen in (15.72), which is repeated here for convenience

$$\phi(t) = \phi_c + K b(t) = K\, b(t) \tag{15.80}$$

where K is the phase deviation constant, $b(t)$ is the modulating signal, and ϕ_c is the phase of the RF waveform $c(t)$. By setting $\phi_c = 0$ in (15.80) the phase variation $\phi(t) = K\, b(t)$ is expressed relative to ϕ_c.

Using a procedure similar to that used in Sect. 4.2.4, the phase deviation constant K of the simple phase modulator (Fig. 15.24), after the derivative term is expanded into three terms, is derived as follows

$$K = \frac{d\phi}{dV_D} = \frac{d\phi}{dC}\,\frac{dC}{dC_D}\,\frac{dC_D}{dV_D} \tag{15.81}$$

where the total tuning capacitance C consists of the varicap capacitance C_D and capacitance C_1 in serial connection, i.e.

$$C = \frac{C_D\, C_1}{C_D + C_1}$$

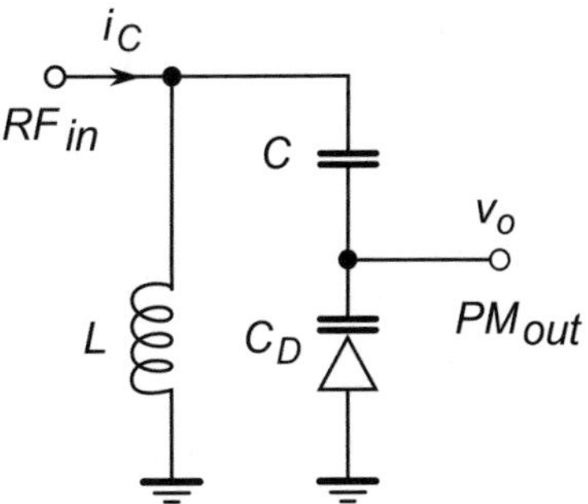

Fig. 15.24 Simplified schematic diagram of phase modulator circuit based on a varicap diode

$$\therefore$$

$$\frac{dC}{dC_D} = \left(\frac{C_1}{C_1 + C_{D0}}\right)^2 = \frac{1}{(1+n)^2} \tag{15.82}$$

where n is the ratio of varicap capacitance C_{D0} to fixed capacitance C_1 at varicap biasing voltage V_0. Practical implementation of FM modulator circuit and its frequency response is shown in Fig. 15.25.

As we know, admittance Y and phase[6] of a tuned LC circuit with dynamic resistance R_D are

$$Y = \frac{1}{R_D} + j\left(\omega_0 C - \frac{1}{\omega_0 L}\right) \tag{15.83}$$

$$\therefore$$

$$\tan\phi = \left[R_D\left(\omega_0 C - \frac{1}{\omega_0 L}\right)\right] \approx \phi \tag{15.84}$$

where for small angles we applied approximation $\tan\phi \approx \phi$, hence

$$\frac{d\phi}{dC} = \omega_0 R_D = \frac{Q}{C} = \frac{Q\,(C_1 + C_D)}{C_1 C_D} = \frac{Q(1+n)}{C_{D0}} \tag{15.85}$$

after substituting $R_D = Q/\omega_0 C$ into (15.85) and after applying biasing voltage V_0. We already found the sensitivity of varicap capacitance versus diode voltage, hence, after substituting (4.26), (15.85) and (15.82) into (15.81), we write

$$K = \left[\frac{Q(1+n)}{C_{D0}}\right]\left[\frac{1}{n+1}\right]^2\left[-\frac{C_{D0}}{1+2V_0}\right]$$

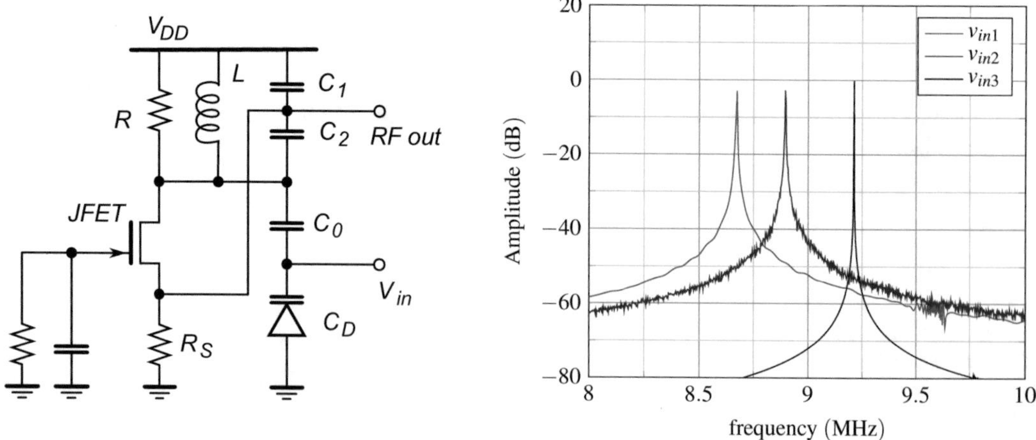

Fig. 15.25 One possible implementation of a practical FM modulator circuit with a varicap as the tunable element, and its frequency response for three input voltage amplitudes

[6]Use the Pythagorean theorem on complex numbers.

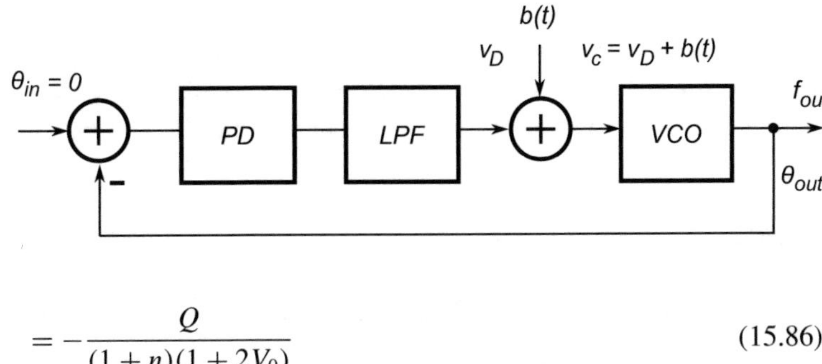

Fig. 15.26 Simplified block diagram of PLL phase modulator circuit

$$= -\frac{Q}{(1+n)(1+2V_0)} \tag{15.86}$$

Example 87: Phase Modulation Constant
Estimate PM deviation constant K for phase modulator in Fig. 15.24.

Data: $V_0 = 15\,\text{V}$, $C = 10\,C_{D0}$, and $Q = 70$.

Solution 87: Straightforward implementation of (15.86) yields

$$K = -\frac{Q}{(1+n)(1+2V_0)} = -\frac{70}{(1+10)(1+2\times 15\,\text{V})} = -0.2\,\text{rad/v} \tag{15.87}$$

which means that if biasing voltage across the varicap diode changes by 1 V phase of the output waveform changes by 11.46°.

15.2.3.3 PLL Modulator

Main function of a PLL circuit, namely to produce a phase and frequency tunable waveform, is also used to generate PM or FM waveforms. By careful inspection of the PLL circuit in Fig. 15.26, we note that by adding a modulation signal b(t) to the original VCO control signal v_D, we effectively push the VCO away from the reference point. If the loop bandwidth is wide enough, the loop quickly responds and generates the cancelling signal because the two phases at the input terminals of PD are forced by the loop to be equal. Hence, the total control signal is then $v_c = v_D + b(t)$ and must be constant because the input reference phase θ_{in} is constant. Therefore, the input phase $-\theta_{in}$ is proportional to v_D and θ_{out} is proportional to the modulation signal $b(t)$. Modern RF IC transceivers use PLL circuits for multiple applications, as a modulator, as a clock generator, or as a PM/FM modulator.

15.3 Summary

The three main modulation schemes, AM, FM, and PM, are briefly introduced in this chapter. Modulation is a process by which the useful baseband signal $b(t)$ (i.e. the information) is embedded into the HF single-tone signal $c(t)$ by altering its amplitude, frequency, or phase and is frequency-shifted upward and centred around the HF carrier frequency ω_0. Therefore, very loosely, the mixer circuits discussed in Sect. 14.2 may be considered as a special case of modulators because mixers are analysed only from the perspective of the frequency-shifting operation (i.e. signal multiplication). Frequency upshifting is primarily done in the transmitting part of the wireless radio system. Because they are similar, frequency and PM schemes are often studied together under the common designation of "angle modulation techniques". Although modern wireless systems employ mostly FM and PM schemes, because of their lower sensitivity to amplitude noise, AM is the simplest and oldest of the three schemes and is still widely used in long-distance Earth-bound broadcasting systems

Problems

15.1 An audio signal $A_a = 0.5\,\text{V}\sin(2\pi \times 1500\,t)$ modulates a carrier $A_c = 1\,\text{V}\sin(2\pi \times 100\,000\,t)$.

(a) Sketch the audio signal; and sketch the carrier waveform
(b) Construct the modulated wave
(c) Determine the modulation factor and percent modulation
(d) What are the frequencies of the audio signal and the carrier?
(e) What frequencies show are found in the frequency spectrum of this AM wave?

15.2 How many AM broadcasting stations can be accommodated in a 100 kHz bandwidth if the highest frequency modulating the carriers is 5 kHz?

15.3 A radio receiver is tuned to receive AM modulated wave transmitted at carrier frequency of $f_{RF} = 980\,\text{kHz}$. The local oscillator inside the receiver is set at $f_{LO} = 1435\,\text{kHz}$. Estimate,

(a) frequencies generated by the receiver's mixer,
(b) which frequency is used is IF frequency,
(c) frequency of a radio station which would present image frequency to the radio station in this problem,
(d) frequency spectrum graph showing the involved frequencies.

15.4 Tuned RF amplifier has LC tank with $Q = 100$ tuned at RF frequency f_0. Estimate attenuation of the image signal, if the image frequency is 5% lower than the RF signal.

15.5 For frequency modulation index of $m_f = 1.5$ and modulation signal $f_b = 10\,\text{kHz}$, find:

(a) estimated required bandwidth B_{FM} (using Carson's rule)
(b) ratio of the total power relative P_T to the power in the FM unmodulated waveform
(c) which harmonics has the highest amplitude

15.6 Determine the power content of each of the sidebands and of the carrier of an AM signal that has a percent modulation of 85% and contains 1200 W of total power.

15.7 Using plots in Fig. 15.3 sketch the corresponding trapezoidal forms.

15.8 An AM signal whose carrier waveform is modulated 70% contains 1500 W at the carrier frequency. Determine the power content of the upper and lower sidebands for this percent modulation. Calculate the power at the carrier and the power of each of the sidebands when the percent modulation drops to 50%.

15.9 A standard AM broadcast receiver has an intermediate frequency (IF) of 455 kHz.

(a) Calculate the required frequency of the local oscillator f_{LO} when the receiver is tuned to $f_c = 540\,\text{kHz}$, if the local oscillator tracks above the frequency of the received signal.
(b) Repeat (a) if the local oscillator tracks below the frequency of the received signal.

15.10 A $f_c = 107.6\,\text{MHz}$ carrier is frequency modulated by a $f_m = 7\,\text{kHz}$ sine wave. The resultant FM signal has frequency deviation of $\Delta f = 50\,\text{kHz}$.

(a) calculate the carrier swing of the FM signal,
(b) determine the highest and the lowest frequencies attained by the modulated signal,
(c) what is the modulation index of the FM wave?

15.11 FM transmitter has total power of $P_T = 100\,\text{W}$ and modulation index of $m_f = 2.0$.

(a) find power levels contained in the first six frequency components,
(b) estimate the bandwidth requirement if modulation signal is $f_m = 1.0\,\text{kHz}$.

15.12 For circuit Fig. 15.22 find value of capacitor C.

Data: $f_{out} = 3.5\,\text{MHz}$, $C_T = 83.4\,\text{nF}$, $L_T = 20\,\text{nH}$, $R = 10\,\Omega$, $g_m(M_1) = 10\,\text{mS}$.

AM and FM Signal Demodulation

16

When a modulated signal arrives at the receiving antenna, the embedded information must somehow be extracted by the receiver and separated from the HF carrier signal. This information recovery process is known as "demodulation" or "detection". It is based on an underlying mechanism similar to the one used in mixers, where a nonlinear element is used to multiply two waves and accomplish the frequency shifting. However, the demodulation process is centred around the carrier frequency ω_0 and the signal spectrum is shifted downward to the baseband and returned to its original position in the frequency domain. Both modulation and demodulation involve a frequency-shifting process; both processes shift the frequency spectrum by a distance ω_0 on the frequency axis; and both processes require a nonlinear circuit to accomplish the task. Although very similar, the two processes are different in very subtle but important details. In the modulating process the carrier wave is generated by the LO circuit, and then combined with the baseband signal inside the mixer. In the demodulating process, however, the carrier signal is already contained in the incoming modulated signal and it can be recovered at the receiving point.

16.1 AM Demodulation Principle

Signal delivered by the mixer stage is in the form of downconverted AM waveform whose IF (in the case of AM radio receiver) is $f_{IF} = 455\,\text{kHz}$. It is now necessary to de-embed the transmitted message, see Fig. 16.1, which is the function of demodulator circuit. In its principal implementation, the demodulator continuously "measures" the IF waveform's amplitude, which by itself is the representation of the transmitted message $b(t)$.

In order to introduce the AM demodulation process analytically, let us consider a simple square law device, with one input and one output terminal, whose voltage–current characteristic is

$$i(t) = a_2\, c_{AM}^2(t) \tag{16.1}$$

where a_2 is a constant, $c_{AM}(t)$ is an AM modulated wave of the following form

$$c_{AM}(t) = C\,(1 + m\cos\omega_b\, t)\,\cos\omega_c\, t \tag{16.2}$$

R. Sobot, *Wireless Communication Electronics*,
https://doi.org/10.1007/978-3-030-48630-3_16

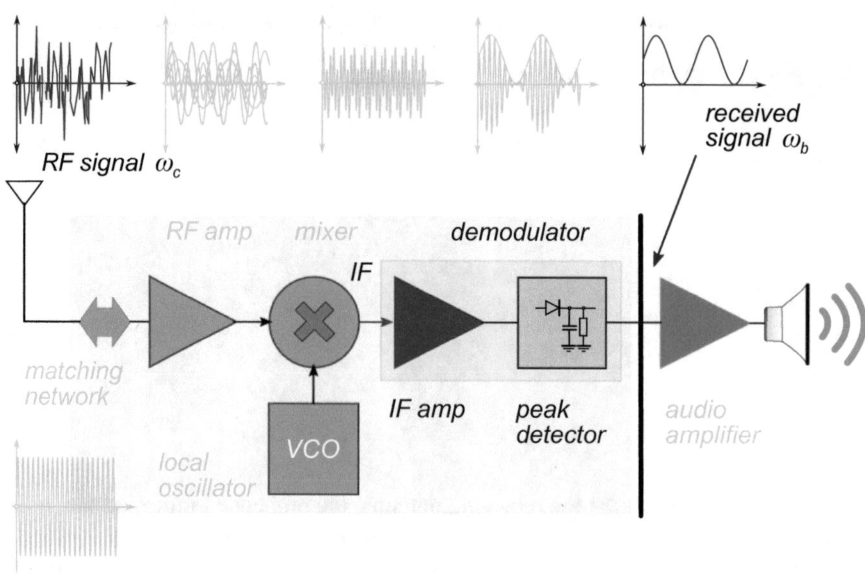

Fig. 16.1 Heterodyne AM radio receiver architecture—demodulator

where $b(t)$ is the information baseband signal and m is the amplitude modulation index that is presented at its input terminal. Then, the output signal contains the following terms

$$
\begin{aligned}
i(t) &= a_2\, C^2\, [1 + m\cos\omega_b\, t]^2\, \cos^2\omega_c\, t \\
&= a_2\, C^2\, [1 + 2m\cos\omega_b\, t + m^2\cos^2\omega_b\, t] \left[\frac{1}{2} + \frac{1}{2}\cos 2\omega_c\, t\right] \\
&= a_2\, C^2 \left[\left(\frac{1}{2} + \frac{m^2}{4}\right)_{DC} + m\,\cos\omega_b\, t + \frac{m^2}{4}\cos 2\omega_b\, t\right. \\
&\quad + \left(\frac{1}{2} + \frac{m^2}{2}\right)\underbrace{\cos 2\omega_c\, t}_{LPF} + \frac{m}{2}\underbrace{\cos(2\omega_c + \omega_b)\, t}_{LPF} + \frac{m}{2}\underbrace{\cos(2\omega_c - \omega_b)\, t}_{LPF} \\
&\quad \left. + \frac{m^2}{8}\underbrace{\cos(2\omega_c + 2\omega_b)\, t}_{LPF} + \frac{m^2}{8}\underbrace{\cos(2\omega_c - 2\omega_b)\, t}_{LPF}\right]
\end{aligned}
\tag{16.3}
$$

That is, the output spectrum contains tones at ω_b, $2\omega_b$, $2\omega_c$, $(2\omega_c + \omega_b)$, $(2\omega_c - \omega_b)$, $2(\omega_c + \omega_b)$, and $2(\omega_c + \omega_b)$, with the carrier frequency ω_c being absent. There is a wide separation between the baseband frequency ω_b and the HF carrier ω_c (i.e. $\omega_c \gg \omega_b$), and even larger separation between ω_b and higher harmonics $n\,\omega_c$, or between ω_b and any other tone $(n\omega_c \pm \omega_b)$ for that matter. The point is that, even with a relatively simple LP filter, we are able to suppress all higher-frequency tones and approximate the output current signal $i_(t)$ with

$$
i(t) \approx a_2\, m\, C^2 \left[\cos\omega_b\, t + \frac{m}{4}\underbrace{\cos 2\omega_b\, t}_{LPF}\right]
\tag{16.4}
$$

which consists only of the desired information signal ω_b and its second harmonic $2\omega_b$, with DC term (i.e. $1/2 + m^2/4$) is removed. It is now matter of designing an LP filter with steep frequency transfer curve so that the attenuation of the second harmonic is "good enough" relative to ω_b and $2\omega_b$.

16.2 Diode AM Envelope Detector

There is a debate in the literature about the distinction between the terms "detector" and "demodulator" in respect to whether the diode AM envelope detector is a real demodulator or not. The argument is mostly semantic, with claims that a "true" demodulator must involve two input signals, the local carrier signal and the AM signal, not just the AM signal. Having acknowledged the argument, we proceed into analysis of the simplest possible AM envelope detector (also known as the "peak detector") for extracting information from the envelope of the AM signal. It also happens to be one of the most versatile little circuits in electronics and is used in a wide range of applications.

The diode peak detector circuit (see Fig. 16.2) has a built-in time constant $\tau = RC$ that is fundamental to its operation. The diode serves as a switch that controls the flow of the AM signal. On the positive swing of the sinusoidal input voltage V_{AM}, the diode is forward biased, the charge flows into the capacitor, and therefore capacitor voltage v_C follows as $v_C = v_{AM}$ because the AM signal source is directly connected to the capacitor. When the input AM voltage reaches its maximum value V_m, the capacitor is fully charged. From that moment on, the diode becomes reverse biased and it turns off. That is, the capacitor voltage is at $V_C = V_m$ and the capacitor is disconnected from the AC source (because the diode is turned off). Hence, the capacitor starts to discharge linearly through resistor R, which is the only available path to ground, with time constant $\tau = RC$,[1] Fig. 16.3. The discharging process lasts as long as $v_C + V_t \geq v_{AM}$, i.e. until the next upswing of the input AM voltage when the diode turns on again and the cycle repeats (see Fig. 16.3), where V_t is the minimal diode voltage ("diode drop"), see Fig. 4.1.

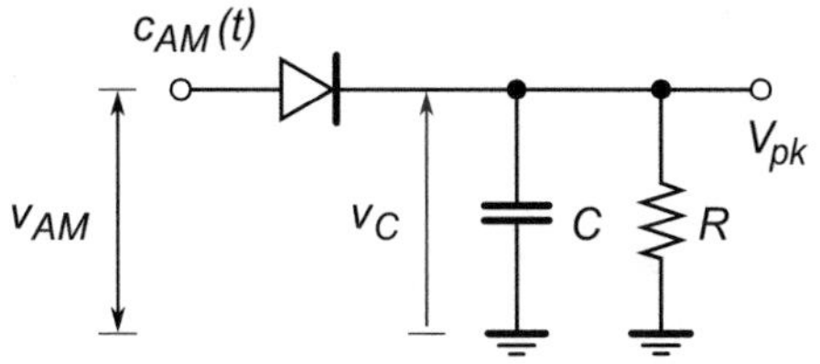

Fig. 16.2 Diode AM envelope detector

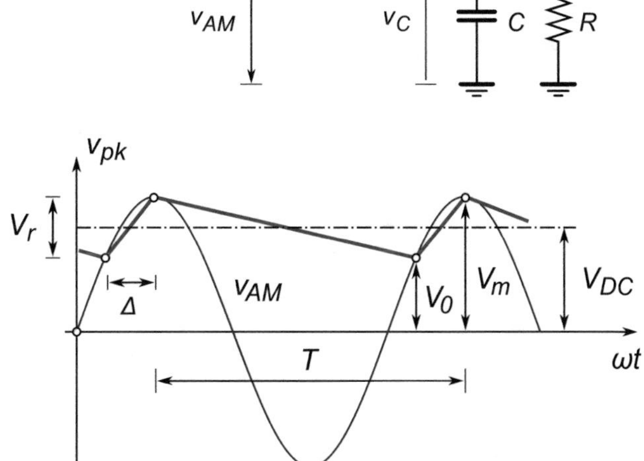

Fig. 16.3 The piece-wise approximate shape of the envelope wave as decoded by the diode AM envelope detector; the value of voltage drop V_r is exaggerated relative to the maximum amplitude V_m; in reality $V_m \gg V_r$ and $\Delta \to 0$

[1]See Sect. 3.2.2.

An exact analysis of the recovered signal wave $V_{pk}(t)$ is a bit more involved, both numerically and in using calculus. For the purposes of our analysis, we use the approximate engineering approach, which yields reasonably accurate results.

16.2.1 Ripple Factor

With reference to Fig. 16.3, it is assumed that the output voltage v_{pk} is approximated with a linear function within the time window that is labelled as Δ in each cycle. In reality, during that time period $V_{pk} = v_{AM}$ (because the diode is turned on). Indeed, the approximation is valid because the maximum amplitude V_m of the two signals (v_{AM} and V_{pk}) is $V_m \gg V_r$, where V_r is the amount of voltage by which the capacitor is discharged before the diode turns again on the subsequent upswing of the input wave. Consequently, the value of the time window $\Delta \to 0$ is very small, which also means that period T of the sawtooth v_{pk} function is approximately equal to the period of the sinusoidal v_{AM} function. In addition, it is assumed that the time constant $\tau = RC \gg T$, hence the exponential capacitor discharge[2] is approximated with the linear function within one period T time window. For the sake of clarity, in Fig. 16.3 the value of amplitude V_r is exaggerated relative to the amplitude of V_m.

With these approximations in mind, we write an expression for the average value VDC of the extracted sawtooth voltage as

$$V_{DC} = V_m - \frac{V_r}{2} = I_{DC}\, R \tag{16.5}$$

where I_{DC} is the average discharge capacitor current. The value of I_{DC} is approximated as follows. Starting with a fully charged capacitor C whose voltage is $V_C = V_m$, the diode is turned off and the capacitor discharge current I_{DC} is controlled by resistor R; the discharge rate is assumed constant because of the $\tau \gg T$ approximation. The value of the constant discharging current is calculated as the initial current at the beginning of the discharging cycle when the capacitor voltage is $V_C = V_m$, hence

$$I_{DC} = \frac{V_m}{R} \tag{16.6}$$

The RMS amplitude value of the sawtooth voltage wave is

$$V_{r\,\mathtt{rms}} = \frac{V_r}{2\sqrt{3}} \tag{16.7}$$

After one full time period T, the capacitor voltage has dropped by V_r (Fig. 16.3), which is controlled by the time constant $\tau = RC$ as

$$V_r = V_m - V_0 = V_m - V_m\, e^{-\dfrac{T}{RC}} = V_m \left[1 - e^{-\dfrac{T}{RC}} \right]$$

$$\therefore$$

[2]See Sect. 2.3.3.5.

$$\approx V_m \left[1 - \left(1 - \frac{T}{RC}\right)\right] = \frac{V_m}{R}\frac{T}{C} = I_{DC}\frac{T}{C} = \frac{I_{DC}}{f\,C} \tag{16.8}$$

where $T = 1/f$ is the period of the carrier wave and the exponential function was approximated by (14.12), using only the linear terms. After substituting (16.8) into (16.7), we write

$$V_{r\,\mathtt{rms}} = \frac{I_{DC}}{f\,C\,2\sqrt{3}} \tag{16.9}$$

The ripple factor r_F of the extracted V_{pk} signal is defined as

$$r_F \stackrel{\text{def}}{=} \frac{V_{r\,\mathtt{rms}}}{V_{DC}} = \frac{\dfrac{I_{DC}}{f\,C\,2\sqrt{3}}}{I_{DC}\,R} = \frac{1}{f\,RC\,2\sqrt{3}} \tag{16.10}$$

Naturally, the ripple factor reduces if:

1. The frequency of the input signal is reduced—in the limiting case for DC input (of course), there is no ripple.
2. A larger capacitor, which stores more charges and increases the time constant τ, is used—in the limiting case of $C \to \infty$ the capacitor voltage never changes.
3. A larger resistor is used—in the limiting case $R \to \infty$ no current flows, therefore $\tau \to \infty$ and the voltage across the capacitor never changes.

The ripple factor increases at higher frequencies or if smaller RC components are used. In most cases, (16.10) is adequately accurate, especially in the case of small ripple values.

16.2.2 Detection Efficiency

Internal diode resistance r_D causes, effectively, a voltage divider formed by the diode resistance and the load resistor R. Hence, the amplitude of the incoming AM wave v_{AM} is proportionately reduced, which is quantified by the "detection efficiency factor" envelope. A reasonably accurate analytical expression for detection accuracy may be derived after making the following assumptions. First, let us approximate the current–voltage characteristic of the diode with a linear function, Fig. 16.4 (right), which is reasonable around the diode's biasing point. Second, let us assume that the voltage V_{DC} across the RC load is constant over the AM wave period, i.e. ripple factor $r_F = 0$, Figs. 16.4 (left) and 16.5). Third, the diode current i_D is assumed to be

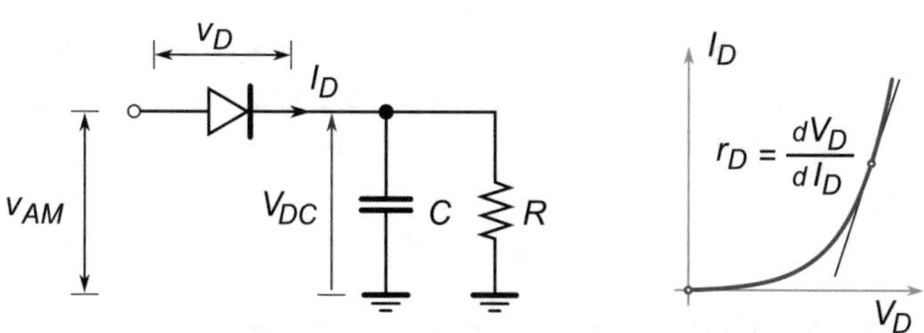

Fig. 16.4 Diode resistance r_D and variables used in an approximate analysis of diode detector efficiency

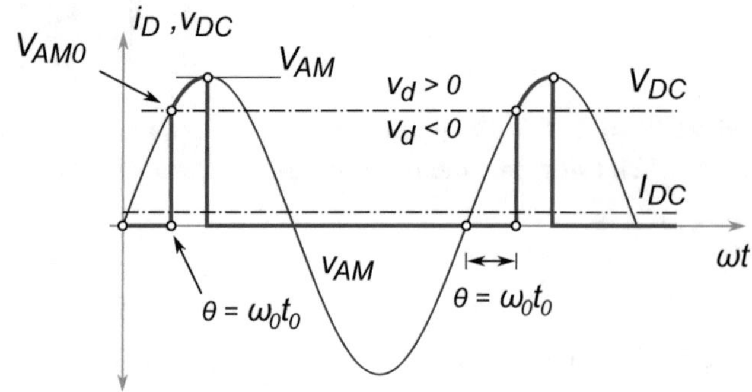

Fig. 16.5 Definitions used in the approximate analysis of diode current i_D

$$i_D = \begin{cases} \dfrac{v_D}{r_D}, & \text{for } v_D > 0 \\ 0, & \text{for } v_D < 0 \end{cases} \tag{16.11}$$

where the last two assumptions are valid for $\tau \gg RC$. At the same time, the value of DC through the diode is also calculated from the resistor load side as

$$I_D = \frac{V_{DC}}{R} \tag{16.12}$$

The AM wave v_{AM} that enters the diode AM detector is described as

$$v_{AM} = C\,(1 + m\, \cos\omega_b\, t)\, \cos\omega_c\, t = V_{AM}\, \cos\omega_c\, t \tag{16.13}$$

where $V_{AM} = C\,(1 + m\, \cos\omega_b\, t) = f(C, m, \omega_b, t)$ is the time-varying amplitude of the carrier wave $\cos\omega_c\, t$, which is introduced simply for convenience of writing the following equations, m is the amplitude modulation index, and $\cos\omega_b\, t$ is the modulation baseband wave.

With these assumptions, in reference to Fig. 16.5, we write an expression for voltage across the diode while in conducting mode, i.e. $v_D > 0$, as

$$v_D = v_{AM} - V_{DC} = V_{AM}\, \cos\omega_c\, t - V_{DC} \tag{16.14}$$

$$\therefore$$

$$i_D = \begin{cases} \dfrac{V_{AM}}{r_D}\, \cos\omega_c\, t - \dfrac{V_{DC}}{r_D}, & \text{for } v_D > 0 \\ 0, & \text{for } v_D < 0 \end{cases} \tag{16.15}$$

where r_D is the diode resistance for the given biasing point. The point in the AM signal cycle $\theta = \omega_c\, t_0$ corresponds to the crossover point when $v_{AM} > V_{DC}$ and the diode turns on, therefore the diode current $i_D > 0$ becomes larger than zero. That particular amplitude value when $v_{AM}(\theta) = V_{DC} = V_{AM0}$ is important for our analysis (Fig. 16.5) and we note that the following relation holds

$$V_{AM}\, \cos\omega_c\, t_0 = V_{AM}\, \cos\theta = V_{DC} \tag{16.16}$$

By inspection of Fig. 16.5, we note that frequency spectrum of the instantaneous diode current i_D must contain a very large number of harmonics because of its sharp switching characteristic, however, we will focus only on its DC and the first harmonic term at frequency ω_c. Therefore, we find the average value of the diode current I_{DC} by integrating i_D over one period, i.e. by definition

$$I_D = \frac{1}{2\pi}\int_0^{2\pi} i_D \, d\omega_c t$$

$$\therefore$$

$$I_D = \frac{1}{\pi}\int_0^{\theta} i_D \, d\theta \tag{16.17}$$

because θ changes only within the $(0, \pi)$ window. Therefore, we continue the integration as

$$\begin{aligned} I_D &= \frac{1}{\pi r_D}\int_0^{\theta} (V_{AM} \cos\theta - V_{DC}) \, d\theta \\ &= \frac{1}{\pi r_D} (V_{AM} \sin\theta - V_{DC}\theta) \end{aligned} \tag{16.18}$$

which, after substituting (16.16), becomes

$$I_D = \frac{1}{\pi r_D} V_{AM} \, (\sin\theta - \theta\cos\theta) \tag{16.19}$$

From (16.12), (16.16), and (16.19), we write

$$I_D = \frac{V_{DC}}{R} \tag{16.20}$$

$$= \frac{V_{AM}\cos\theta}{R} \tag{16.21}$$

$$= \frac{V_{AM}}{\pi r_D} (\sin\theta - \theta\cos\theta) \tag{16.22}$$

$$\therefore$$

$$\frac{r_D}{R} = \frac{1}{\pi} \frac{\sin\theta - \theta\cos\theta}{\cos\theta} \tag{16.23}$$

$$= \frac{1}{\pi} (\tan\theta - \theta) \tag{16.24}$$

which gives the ratio of the diode resistance and the load resistor $((r_D/R))$ as a function of θ but not the two resistances by themselves.

Now we have all elements needed to define the detection efficiency η of the diode detector as the ratio of the average value of the load voltage V_{DC} relative to the peak AM input wave V_{AM} by using (16.16) as

$$\eta = \frac{V_{DC}}{V_{AM}} = \cos\theta \tag{16.25}$$

From (16.20) and (16.22), we also write

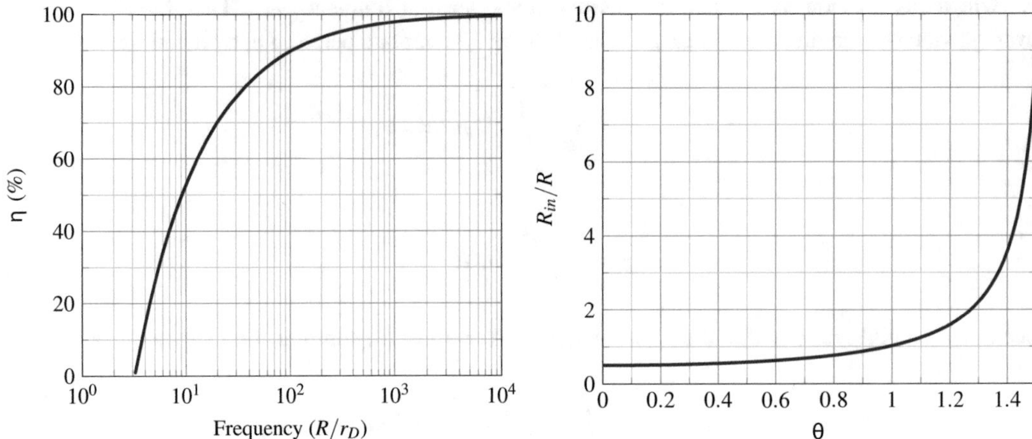

Fig. 16.6 Ratio of effective input resistance to resistor R_{eff}/R—(16.31) (left); Detection efficiency η against the R/r_D plot of (16.25) and (16.26) (right)

$$\eta = \frac{R}{\pi r_D} (\sin\theta - \theta\cos\theta) \tag{16.26}$$

which provides the detection efficiency η as a function of $(R/r_D, \theta)$. We note that it does not depend on V_{AM} by itself, which implies that the detection efficiency is also not a function of the amplitude modulation index m. These two equations, (16.25) and (16.26), provide the designer with a tool to determine the required type of diode (i.e. its resistance) and loading resistor for the desired detection efficiency. It is not easy to write explicit analytical expressions for $\eta = f(R/r_D)$; instead we use the two equations to produce a graphical relationship of$\eta = f(R/r_D)$, (see Fig. 16.6).

Example 88: Peak Detector Efficiency

For a given diode, whose resistance is $r_D = 100\,\Omega$, determine the value of the loading resistor R if the desired detection efficiency for the diode AM detector is $\eta = 80\%$.

Solution 88: From (16.25), we find

$$\eta = \cos\theta \quad \therefore \quad \theta = \arccos(0.8) = 0.6435 \tag{16.27}$$

then we write

$$\frac{R}{r_D} = \frac{\pi\,\eta}{(\sin\theta - \theta\cos\theta)} = 29.5 \tag{16.28}$$

which, for the given $r_D = 100\,\Omega$ yields $R = 29.5 \times 100\,\Omega = 2.95\,k\Omega$.

16.2.3 Input Resistance

Similarly to any other electronic circuit, it is important to find an expression for the effective input resistance R_{eff} of the diode AM detector for given resistance R. Due to the nonlinear nature of the circuit, the most often used analytical method is based on analysis of power absorbed by the detector. In this case, we make an approximation by taking into account only the fundamental harmonic of the diode current i_D, whose maximum value is found by direct implementation of the Fourier coefficient as

$$\begin{aligned} I_{D\,max} &= \frac{1}{\pi}\int_0^{2\pi} i_D \cos\omega_c t \; d(\omega_c t) \\ &= \frac{2}{\pi}\int_0^{\theta} \frac{1}{r_D}\,(V_{AM}\cos\alpha - V_{DC})\,\cos\alpha\, d\alpha \\ &= \frac{2}{\pi r_D}\int_0^{\theta} V_{AM}\cos^2\alpha d\alpha - \int_0^{\theta} V_{DC}\;\cos\alpha\; d\alpha \\ &= \frac{2}{\pi r_D}\left(\frac{1}{4}\sin 2\alpha + \frac{\alpha}{2}\right)\Bigg|_0^{\theta} - V_{DC}\;\sin\alpha|_0^{\theta} \\ &= \frac{2V_{AM}}{\pi r_D}\left(\theta + \frac{1}{2}\sin 2\theta - \sin\theta\cos\theta\right) \\ &= \frac{V_{AM}}{\pi r_D}\,(\theta - \sin\theta\cos\theta) \end{aligned} \tag{16.29}$$

The power P dissipated in the diode and the resistor is, by definition

$$P = \frac{1}{T}\int_0^{T} v_{AM}\, i_D\; dt = \frac{V_{AM} I_{D\,max}}{2} = \frac{V_{AM}^2}{2\pi r_D}(\theta - \sin\theta\cos\theta)$$

$$\therefore$$

$$\frac{R_{\text{eff}}}{r_D} \overset{\text{def}}{=} \frac{V_{AM}^2}{2P} = \frac{\pi}{\theta - \sin\theta\cos\theta} \tag{16.30}$$

which only yields the ratio of the input resistance R_{eff} and the diode resistance r_D as a function of θ. In order to find out how resistor R influences the input resistance, we substitute (16.24) into (16.30) and write

$$\frac{R_{\text{eff}}}{R} = \frac{R_{\text{eff}}}{r_D}\frac{r_D}{R} = \frac{\tan\theta - \theta}{\theta - \sin\theta\cos\theta} \tag{16.31}$$

which only gives the ratio of the input effective resistance R_{eff} and resistor R as a function of θ. In the ideal case, detection efficiency is high, i.e. $\eta \to 1$, which implies very low $\theta \to 0$. We expand the sinusoidal terms into their respective power series

$$\sin\theta = \sum_{n=0}^{\infty} \frac{(-1)^n}{(2n+1)!}\,x^{2n+1} = \theta - \frac{\theta^3}{3!} + \frac{\theta^5}{5!} - \cdots \approx \theta - \frac{\theta^3}{6}$$

$$\cos\theta = \sum_{n=0}^{\infty} \frac{(-1)^n}{(2n)!} x^{2n} = 1 - \frac{\theta^2}{2!} + \frac{\theta^4}{4!} - \cdots \approx 1 - \frac{\theta^2}{2}$$

and we take only the first two terms of the series. After substituting (16.25) into (16.31), we derive

$$\begin{aligned}\frac{R_{\text{eff}}}{R} &= \frac{1}{\cos\theta} \frac{\sin\theta - \theta\cos\theta}{\theta - \sin\theta\cos\theta} \\ &= \frac{1}{\eta}\left[\frac{\left(\theta - \frac{\theta^3}{6}\right) - \theta\left(1 - \frac{\theta^2}{2}\right)}{\theta - \left(\theta - \frac{\theta^3}{6}\right)\left(1 - \frac{\theta^2}{2}\right)}\right] \\ &= \frac{1}{\eta}\left[\frac{\frac{\theta^3}{3}}{\frac{2\theta^3}{3} + \frac{\theta^5}{2}}\right] \approx \frac{1}{\eta}\left[\frac{\frac{\theta^3}{3}}{\frac{2\theta^3}{3}}\right] \\ &= \frac{1}{2\eta} \approx \frac{1}{2}\end{aligned} \tag{16.32}$$

for the case of high detection efficiency $\eta \to 1$. Therefore, we approximate the effective input resistance as $R_{\text{eff}} \approx 1/2R$. The detailed functional relationship of $R_{\text{eff}}/R = f(\theta)$, as derived in (16.31), is shown in Fig. 16.6.

16.2.4 Distortion Factor

So far we have used linearized approximations to estimate the parameters of a diode AM amplitude detector, which worked quite well and produced reasonably accurate expressions. One of the approximations that we made was linearization of the diode $I_D V_D$ characteristic, which is one source of distortion in the output wave due to the nonlinearity of the exponential function. For large decoding efficiency circuits and strong modulation signals, this distortion source contributes a few percentage points; for weak modulation signals, however, this source of distortion may be as high as about 25%.

The second, and less obvious, source of distortion is caused by the capacitor discharge current being constant and set by the time constant $\tau = RC$. The problem is that the choice of time constant is always a compromise between opposing requirements. In an AM wave, peaks and troughs of the envelope signal may come almost randomly, hence it is realistic that "clipping" may occur (see Fig. 16.7) when the time constant is too long and the slope of the discharging current is not steep enough to follow the envelope downslope accurately. Consequently, the recovered waveform does not accurately follow the embedded AM envelope and the recovered signal is distorted.

In order to reduce the ripple factor, the time constant needs to be long relative to the period T of the carrier signal. However, if it is made too long, clipping occurs. We need to estimate the maximum allowable value of the time constant in order to prevent clipping and yet to make the response of the diode detector fast enough to follow the slope of the envelope signal, where the clipping factor is determined by making this compromise.

The most critical condition for the peak detector occurs when the modulation frequency ω_b is highest. The envelope wave equation is given by

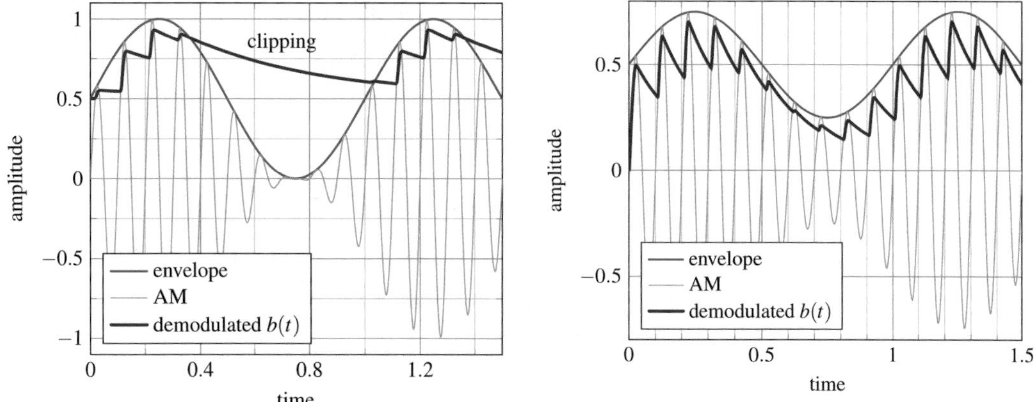

Fig. 16.7 AM wave: the wrong time constant in the diode detector causes clipping (left) and correct timing (right). For illustration, in this example, the ratio of carrier relative to the modulating frequency is 10

$$b(t) = C_0\,(1 + m\ \cos\omega_b\,t) \tag{16.33}$$

where C_0 is the maximum amplitude of the carrier signal and m is the amplitude modulation index. At any moment in time $t = t_0$, the value and slope of the modulation envelope of the modulation signal are

$$b(t_0) = C_0\,(1 + m\ \cos\omega_b\,t) \tag{16.34}$$

$$\left(\frac{db(t)}{dt}\right)\bigg|_{t_0} = -\omega_b\ m\,C_0 \sin\omega_b\,t \tag{16.35}$$

By setting the potential across the capacitor equal to the modulation voltage at $t = t_0$, we write

$$v_C = C_0\,(1 + m\ \cos\omega_b\,t) \tag{16.36}$$

After considering $t > t_0$, the capacitor signal decays at the following rate

$$v_C = V_{C0}\,e^{-\dfrac{t-t_0}{RC}} \tag{16.37}$$

$$\therefore$$

$$\left(\frac{dv_C}{dt}\right)\bigg|_{t_0} = -\frac{1}{RC}\,v_C = -\frac{C_0}{RC}\,(1 + m\ \cos\omega_b\,t) \tag{16.38}$$

In order to avoid diagonal clipping, the capacitor voltage v_C must be equal to or less than the envelope voltage v_b for time $t > t_0$ and the slope must be equal to or less than the envelope slope at $t = t_0$ (which is clearly not the case in Fig. 16.7). These conditions are written as

$$-\frac{C_0}{RC}\,(1 + m\ \cos\omega_b\,t) \leqslant -\omega_b\ m\,C_0 \sin\omega_b\,t \tag{16.39}$$

$$\therefore$$

$$\frac{1}{RC} \geqslant \omega_b \frac{m \sin \omega_b t}{1 + m \cos \omega_b t} \tag{16.40}$$

The fastest RC constant is at the point when

$$\frac{d}{dt} \frac{m \sin \omega_b t}{1 + m \cos \omega_b t} = 0$$

$$\therefore$$

$$m \, \omega_b \frac{[\cos \omega_b t (1 + m \cos \omega_b t) + m \sin^2 \omega_b t]}{(1 + m \cos \omega_b t)^2} = 0 \tag{16.41}$$

which is equivalent to the following conditions

$$\cos \omega_b t (1 + m \cos \omega_b t) + m \sin^2 \omega_b t = 0 \tag{16.42}$$

$$\cos \omega_b t + m (\cos^2 \omega_b t + \sin^2 \omega_b t) = 0 \quad \therefore \quad \cos \omega_b t = -m \tag{16.43}$$

After substituting $\cos \omega_b t = -m$ into (16.42), we give the second solution as

$$-m (1 - m^2) + m \sin^2 \omega_b t = 0 \quad \therefore \quad \sin \omega_b t = \sqrt{1 - m^2} \tag{16.44}$$

Values of the two sinusoidal functions, (16.43) and (16.44), at this particular instance in time are substituted back into (16.40), which yields the boundary condition for the time constant RC where the capacitor voltage has the greatest difficulty in following the modulation signal as

$$\frac{1}{RC} \geqslant \omega_b \frac{m \sqrt{1 - m^2}}{1 - m^2}$$

$$\therefore$$

$$\frac{1}{RC} \geqslant \omega_b \frac{m}{\sqrt{1 - m^2}} \tag{16.45}$$

which is the commonly cited condition that needs to be satisfied if the output voltage v_C is to follow the AM waveform envelope even under the worst conditions. The formula is very approximate in the sense that, for instance, it implies that for the maximum modulation index $m = 1$, the RC time constant would have to be zero, which further implies that the output waveform is equal to the input AM carrier waveform, i.e. no envelope detection is possible. A more conservative condition, which was found experimentally, modifies (16.45) to

$$\frac{1}{RC} \geqslant m \, \omega_b \tag{16.46}$$

which gives a guide to the designer in how to select the passive component values for the design of a diode AM envelope decoder (Fig. 16.8).

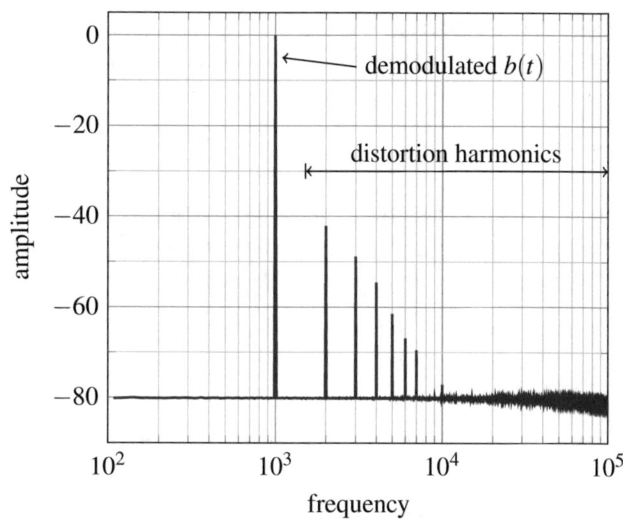

Fig. 16.8 Frequency spectrum of demodulated AM wave. For illustration, in this example, the ratio of carrier relative to the modulating frequency is 455

16.3 FM Wave Demodulation

The recovery process for information embedded into an FM wave carrier is based on a two-step procedure where the frequency variation of the carrier is first converted into an amplitude variation, which is then converted back into the baseband modulation signal by conventional AM demodulators.

In principle, an FM demodulation system includes a chain of processing sub-blocks (see Fig. 16.9): a frequency-to-amplitude converter, an AM envelope detector, and an LP filter. The transfer function of a frequency-to-amplitude converter is

$$H(j\omega) = \frac{V_{AM}(j\omega)}{V_{FM}(j\omega)} \tag{16.47}$$

$$\therefore$$

$$v_{AM}(t) = \frac{dv_{FM}}{dt} \tag{16.48}$$

which serves as a time domain differentiator of the FM wave. For a frequency-modulated signal at carrier frequency ω_0, we write

$$v_{FM}(t) = A\,\cos(\omega_0 t + \theta(t)) \tag{16.49}$$

where A is the FM wave's fixed amplitude and $\theta(t)$ is the time-varying phase angle. Therefore, using (16.48), the output of the frequency-to-amplitude converter is

$$v_{AM}(t) = -A\left[\omega_0 + \frac{d\theta}{dt}\right]\,\sin(\omega_0 t + \theta(t)) \tag{16.50}$$

whose amplitude portion is first approximately detected by the envelope detector as

$$v_{pk}(t) = A\left[\omega_0 + \frac{d\theta}{dt}\right] \tag{16.51}$$

where the first term $\omega_0 A$ is a DC component and is to be removed by the LP filter. The second term contains the embedded information signal $b(t)$ through (15.60) that may be written as

$$\theta(t) = m_f \int b(t)\, dt \tag{16.52}$$

where m_f is the FM index. Therefore, output of the envelope detector (16.51) contains the information $b(t)$ that is subsequently "cleaned up" and fully recovered by the LP filter

There are three main types of FM demodulator circuit that are reviewed in the following sections:

1. Slope detectors and FM discriminators
2. Quadrature detectors
3. PLL demodulators

They are used to implement the general system shown in Fig. 16.9 and described by (16.47)–(16.52).

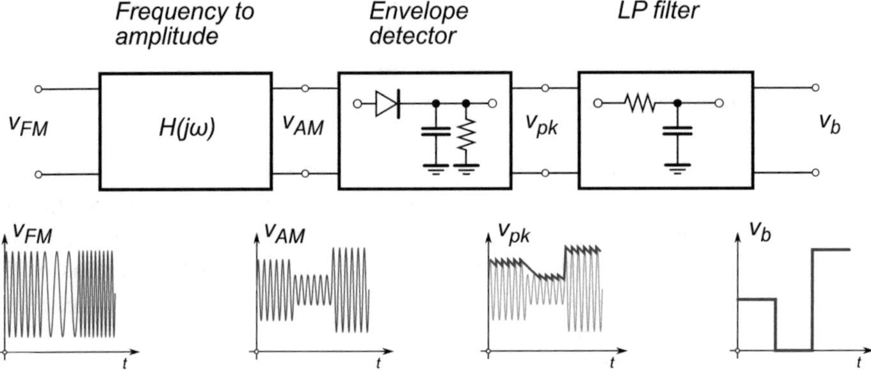

Fig. 16.9 FM wave demodulation chain

16.3.1 Slope Detectors and FM Discriminators

Although slope detectors are not much used any more, their simplicity and obvious operation allows us to easily understand the basic principle of frequency-to-amplitude conversion, therefore they serve the educational purpose well. At its core, a slope detector employs a simple LC resonator tank and a diode AM detector in series (see Fig. 16.10). Although it is a very simple network, the exact analysis of a slope detector circuit is very complicated because the input signal v_{FM} is frequency modulated and, therefore, a simple steady-state analysis does not apply; instead, transient analysis is required.

Fig. 16.10 A slope detector circuit using a simple LC resonator (circled) and an AM slope detector

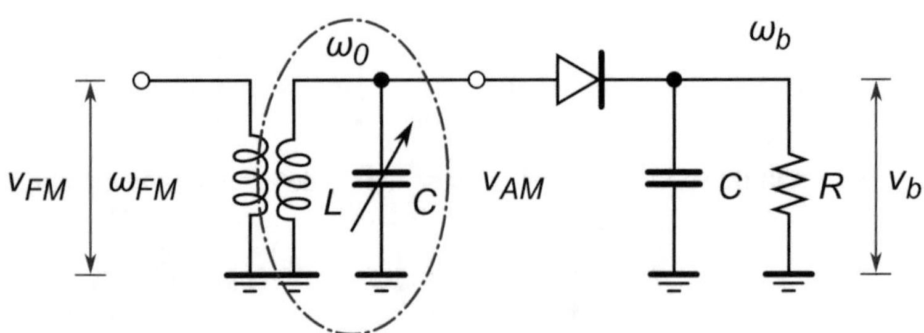

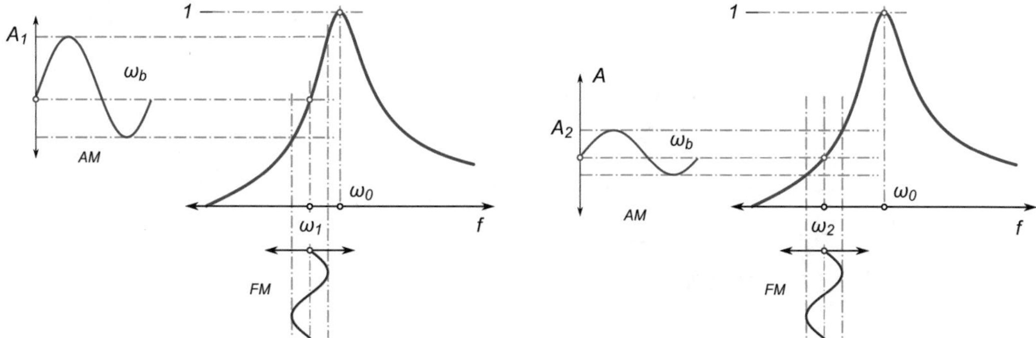

Fig. 16.11 Slope detection using a simple LC resonator

Nevertheless, we illustrate its operation graphically in Fig. 16.11. The resonant LC tank is tuned at ω_0 frequency and the carrier frequency of the incoming FM wave v_{FM} is ω_{FM} not equal to ω_0, i.e. $\omega_{FM} \neq \omega_0$. Because of that arrangement, the non-modulated tone of the incoming FM wave falls on the slope of the LC resonator's frequency characteristic, Fig. 16.11. As we discussed in Sect. 10.2.2, the vertical axis of the LC resonator frequency characteristic shows the amplitude of the incoming tone relative to the amplitude of the resonant tone at ω_0 that is normalized to one. Thus, the amplitude of the incoming tone at ω_{FM} is reduced to A_0. As the frequency of the incoming FM wave changes to $\omega_{FM} \pm \Delta\omega$, the amplitude of the recovered FM wave also changes in accordance with the slope of the LC tank characteristic. Once the FM wave passes through the resonator tank it becomes amplitude modulated in accordance with its embedded modulation signal $b(t)$, which is to say that the conventional diode AM detector is now able to extract the modulation signal $b(t)$ as usual.

We note that either side of the frequency characteristic may be used for frequency-to-amplitude conversion. On the right side of the characteristic, the increase in frequency corresponds to a decrease in amplitude; on both sides of the characteristic, the tuning range is very narrow. In addition, we also note that the recovered signal is distorted by the nonlinear characteristic. For example, the amplitude of the recovered sinusoid is not symmetrical around the A_0 point—the positive side is slightly larger (Fig. 16.11).

16.3.1.1 Dual Slope Detector

A simple evolutionary step for improving the slope detector's performance is to design a "dual slope detector" by creating a symmetrical circuit that literally contains two mirrored versions of the slope detector (see Fig. 16.12). The resonant frequencies of the two resonators are tuned to two separate frequencies that are slightly off on each side relative to the ω_{FM} value. Due to the symmetrical topology of the circuit, i.e. signals flowing through the two sides of the circuit are opposite in phase, the newly created frequency characteristic has a wider linear region centred around the FM carrier frequency ω_{FM} (Fig. 16.12). The main strength of the dual slope FM decoder, namely the linearity that is created by the two offset resonators, is also the source of its main weakness. The circuit depends on three key frequencies (the FM carrier frequency and the two side frequencies), which means that it enables the reception of radio signals at each of these three frequencies instead of only one. In addition, it requires two variable capacitors, which further increases its complexity.

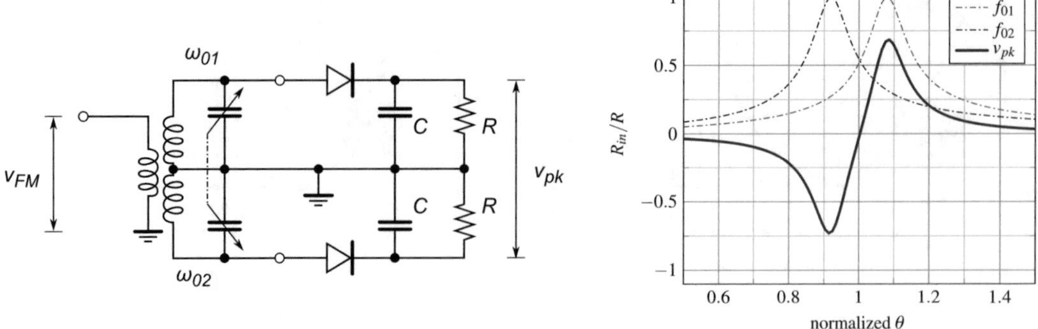

Fig. 16.12 Dual slope FM detection using two slightly offset LC resonators (left), and frequency transfer function (right)

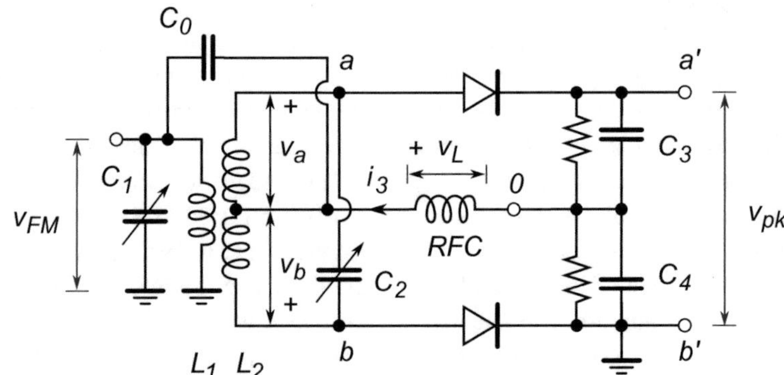

Fig. 16.13 A Foster–Seeley discriminator

16.3.1.2 Foster–Seeley Dual Slope Detector

A modified version of a dual slope FM detector, known as "Foster–Seeley" (see Fig. 16.13), includes a shunting capacitor C_0 between the primary L_1 and the centre tap of the secondary inductance L_2 of the input transformer, a shunting capacitor C_2 across the secondary inductance L_2, and an RF choke RFC. It would be possible to use a resistor instead of RFC, however, that would reduce detection efficiency of the peak detectors (the RFC blocks the RF signals and provides a DC path). The input transformer is dual side tuned, i.e. both the L_1, C_1 and L_2, C_2 resonators are tuned to the non-modulated carrier frequency ω_0 of the incoming FM wave.

A limited analysis of a Foster–Seeley decoder is possible and is educational. We start by assuming that reactances X_{C0} and X_{C4} are very small, while reactance $X_L = X_{RFC}$ are very high. With these two assumptions, we approximate the voltage across RFC by following the (C_0, RFC, C_4) path, as

$$\mathbf{v}_L = jX_L \, \mathbf{i}_3 = jX_L \frac{\mathbf{v}_{FM}}{-jX_{C_0} + jX_L - jX_{C_4}} \approx jX_L \frac{\mathbf{v}_{FM}}{+jX_L} = \mathbf{v}_{FM} \tag{16.53}$$

where bold letters indicate vector variables. By assuming a small mutual inductance between primary and secondary transformer inductances, we can approximate the resonant current $\mathbf{i}_1$ inside the L_1C_1 resonant tank as

$$\mathbf{i}_1 \approx \frac{\mathbf{v}_{FM}}{r_1 + jX_{L1}} \approx \frac{\mathbf{v}_{FM}}{jX_{L1}} \tag{16.54}$$

for a high Q primary inductor, where r_1 is the internal resistance of the L_1 inductor. This current induces EM potential $\mathbf{v}_2$ inside the secondary coil. After substituting (16.54), we write

$$\mathbf{v}_2 = \pm j\omega M\,\mathbf{i}_1 = \pm\frac{M}{L_1}\,\mathbf{v}_{FM} \tag{16.55}$$

The secondary coil is loaded by capacitor C_2, hence, after ignoring the loading effect of the diodes (and keeping the low mutual coupling approximation, i.e. the term $\omega^2 M^2/r_p \approx 0$), we write an expression for the secondary resonant current $\mathbf{i}_{C2} = \mathbf{i}_{L2} = \mathbf{i}_2$ as

$$\mathbf{i}_2 \approx \frac{\mathbf{v}_2}{r_2 + jX_{L2} - jX_{C2}} = \pm\frac{\mathbf{v}_{FM}}{r_2 + j(X_{L2} - X_{C2})}\,\frac{M}{L_1} \tag{16.56}$$

after substituting (16.55), where r_2 is the internal resistance of the L_2 inductor.

The same secondary current causes voltages $\mathbf{v}_{a0}$ and $\mathbf{v}_{b0}$ across their respective half L_2 inductance, after arbitrarily picking the positive sign of the current, as

$$\mathbf{v}_{a0} = \mathbf{i}_2\,\frac{jX_{L2}}{2} = \frac{\mathbf{v}_{FM}}{r_2 + j(X_{L2} - X_{C2})}\,\frac{jX_{L2}\,M}{2L_1} \tag{16.57}$$

$$\mathbf{v}_{b0} = -\mathbf{i}_2\,\frac{jX_{L2}}{2} = \frac{\mathbf{v}_{FM}}{r_2 + j(X_{L2} - X_{C2})}\,\frac{jX_{L2}\,M}{2L_1} \tag{16.58}$$

$$\therefore$$

$$\mathbf{v}_{ab} = \mathbf{i}_2\,X_{L2} = \frac{\mathbf{v}_{FM}}{r_2 + j(X_{L2} - X_{C2})}\,\frac{jX_{L2}\,M}{L_1} \tag{16.59}$$

which means that $\mathbf{v}_{a0}$ and $\mathbf{v}_{b0}$ have 180° phase difference and $|\mathbf{v}_{ab}|$ has the same shape as in Fig. 16.12 (right). We note that the DC potential at the output terminals $V_{a'}$ is proportional to the peak voltage V_a, while potential $V_{b'}$ is proportional to V_b, relative to the ground, therefore the total output DC voltage is

$$V_{a'b'} = V_{a'0} + V_{0b'} = V_{a'0} - V_{b'0} \tag{16.60}$$

Now, we consider the following three cases:

1. *A non-modulated FM wave, i.e. the instantaneous frequency $\omega(t)$ is equal to the FM carrier ω_0 frequency*: The secondary circuit is at resonance and the two reactances are equal ($X_{L2} = X_{C2}$), therefore, from (16.57) and (16.58), it follows that $|\mathbf{v}_{a0}| = |\mathbf{v}_{b0}|$, which after taking into account (16.60) leads to the conclusion that the output DC voltage is $V_{a'b'} = 0$.
   It is instructive to present the internal voltages on a vector diagram. At resonance, both the primary and the secondary resonators have real impedances and there is, therefore, 90° difference between voltages $\mathbf{v}_{RF}$ and $\mathbf{v}_{a0}$. At the same time, voltage $\mathbf{v}_L$ across RFC is in phase with $\mathbf{v}_{RF}$ (16.53), as shown in Fig. 16.14 (left).
2. *A modulated FM wave for which the instantaneous frequency $\omega(t)$ is lower than the FM carrier ω_0 frequency*: The secondary circuit is at resonance and reactance ($X_{L2} < X_{C2}$), therefore, from (16.57) and (16.58) follows that $|\mathbf{v}_{a0}| < |\mathbf{v}_{b0}|$ (Fig. 16.14 (centre)).
3. *A modulated FM wave for which the instantaneous frequency $\omega(t)$ is higher than the FM carrier ω_0 frequency*: The secondary circuit is at resonance and reactance ($X_{L2} > X_{C2}$), therefore, from (16.57) and (16.58) it follows that $|\mathbf{v}_{a0}| > |\mathbf{v}_{b0}|$ (Fig. 16.14 (right)).

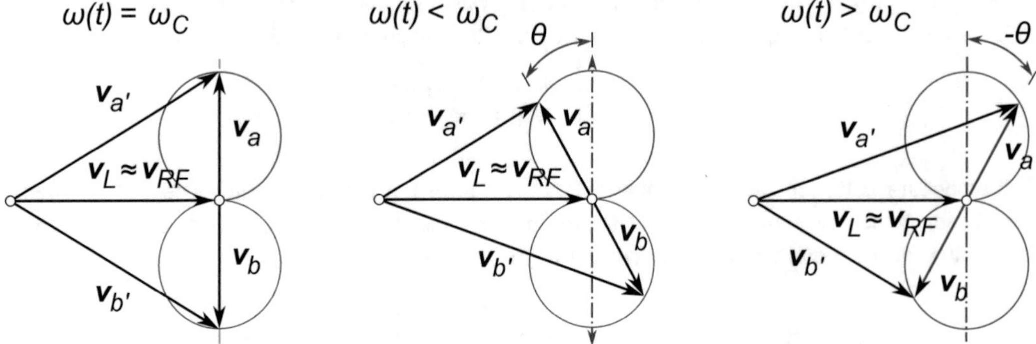

Fig. 16.14 Vector diagram for Foster–Seeley discriminator internal voltages

In order to find phase angle θ between vectors $\mathbf{v}_{\mathbf{FM}}$ and $\mathbf{v}_{\mathbf{ab}}$, i.e. between the input FM voltage and the inducted voltage across L_2, we rearrange and approximate (16.59) assuming high Q value, as

$$
\begin{aligned}
\frac{\mathbf{v}_{ab}}{\mathbf{v}_{FM}} &= \frac{1}{r_2 + j(X_{L2} - X_{C2})}\,\frac{M}{L_1} = \frac{1}{1 + j\left[\dfrac{1}{r_2}(X_{L2} - X_{C2})\right]}\,\frac{jX_{L2}\,M}{r_2\,L_1} \\
&= \frac{1}{1 + j\left[\dfrac{1}{r_2}(X_{L2} - X_{C2})\right]}\,\frac{jX_{L2}\,M}{r_2\,L_1} = \frac{1}{1 + j\,\dfrac{1}{r_2}\left(\omega L_2 - \dfrac{1}{\omega C_2}\right)}\,\frac{jX_{L2}\,M}{r_2\,L_1} \\
&= \frac{1}{1 + j\dfrac{\omega L_2}{r_2}\left(1 - \dfrac{1}{\omega^2 L_2 C_2}\right)}\,\frac{jX_{L2}\,M}{r_2\,L_1} = \frac{1}{1 + j\dfrac{\omega L_2}{r_2}\left(1 - \dfrac{\omega_0^2}{\omega^2}\right)}\,\frac{jX_{L2}\,M}{r_2\,L_1} \\
&= \frac{1}{1 + j\dfrac{\omega_0 L_2}{r_2}\left(\dfrac{\omega}{\omega_0} - \dfrac{\omega_0}{\omega}\right)}\,\frac{jX_{L2}\,M}{r_2\,L_1} = \frac{1}{1 + j\dfrac{\omega_0 L_2}{r_2}\,\delta}\,\frac{jX_{L2}\,M}{r_2\,L_1} \\
&= \frac{1}{1 + jQ\,\delta}\,\frac{jX_{L2}\,M}{r_2\,L_1} = \frac{1}{1 + jQ\,\delta}\,\frac{j\omega\,L_2\,M}{r_2\,L_1} \\
&= \frac{1}{1 + jQ\,\delta}\,\frac{jQ\,M}{L_1} = \frac{1}{1 + jQ\,\delta}\,jQ\,k\sqrt{\frac{L_2}{L_1}}
\end{aligned}
\tag{16.61}
$$

where Q is the Q factor of the secondary coil, $M = k\sqrt{L_1 L_2}$, and δ is sometimes referred to as the "detuning factor". At resonance when $\omega = \omega_0$ then $\delta = 0$, leading to

$$
\frac{\mathbf{v}_{ab}}{\mathbf{v}_{FM}} = \frac{j\omega\,L_2\,M}{r_2\,L_1} = jQ\,k\sqrt{\frac{L_2}{L_1}}
\tag{16.62}
$$

that is, at resonance when the primary and the secondary resonators are left only with their real impedances, there is a 90° phase difference between the $\mathbf{v}_{ab}$ and $\mathbf{v}_{FM}$ vectors due to the C_0 capacitor. For any other case, when the instantaneous frequency is off from the resonant frequency, there will be a positive or negative phase angle θ added to the 90° average phase angle (see Fig. 16.14). Phase shift θ is small and is caused by the first term of (16.61), which is approximated[3] as

$$\theta = \arg \frac{1}{1 + jQ\,\delta} \approx \arg (1 - j\,Q\,\delta) = -\arctan (Q\,\delta) \approx -\arctan \frac{2\,Q\,\Delta\omega}{\omega_0} \tag{16.63}$$

after detuning factor δ was approximated by using substitution $\omega = \omega_0 + \Delta\omega$, as

$$\delta = \frac{\omega_0 + \Delta\omega}{\omega_0} - \frac{\omega_0}{\omega_0 + \Delta\omega} = \frac{2\,\Delta\omega}{\omega_0 + \Delta\omega} \approx \frac{2\,\Delta\omega}{\omega_0} \tag{16.64}$$

The last parameter that we need to define for the discriminator is its sensitivity factor k_d. In other words, we are interested in finding out how much change in the output DC voltage V_{ab} is generated for a unit change in frequency, i.e.

$$k_d \stackrel{\text{def}}{=} \frac{d V_{ab}}{df} \quad \left[\frac{\text{V}}{\text{Hz}}\right] \tag{16.65}$$

and we determine Foster–Seeley parameters using both analytical results in this section and the vector diagram.

Therefore, the overall function of the Foster–Seeley discriminator is to produce a changing DC voltage at the output terminals whose amplitude is proportional to the amplitude of the FM embedded signal. Very often in the literature, we find another version of the Foster–Seeley discriminator called a "ratio detector". With a minor tweak, the ratio detector achieves better AM rejection than the discriminator with about 6dB (theoretically) lower sensitivity. Subsequently, a number of modified versions of the ratio detector have been designed and used.

Example 89: Foster–Feeley Discriminator
One of the most common IF in FM receivers is $f_0 = 10.7\,\text{MHz}$ while the maximum allowed frequency deviation from the carrier frequency is $\Delta f = 75\,\text{kHz}_{pk}$, i.e. $\Delta f = 150\,\text{kHz}_{pp}$. The internal components of a Foster–Seeley discriminator are scaled so that

$$K = \frac{Q\,M}{2\,L_1} = 0.5$$

and it has $Q = 23.259$. The output voltage is measured as $V_a = 1\,V_{\mathtt{rms}}$. Determine the peak output voltage V_{ab} and discriminator sensitivity.

Solution 89: Phase shift θ is calculated from (16.64) as

$$\delta = -\arctan \frac{2\,Q\,\Delta\omega}{\omega_0} = -\arctan \frac{2 \times 23.259 \times 75\,\text{kHz}_{pk}}{10.7\,\text{MHz}} = -18^\circ \tag{16.66}$$

Knowing θ, we construct a vector diagram that contains a right triangle with 90°, 18°, and 72°, as shown in Fig. 16.15. From the $K = 0.5$ data point and (16.61), we conclude that $|\mathbf{v}_L| =$

(continued)

[3]For small ($b \ll 1$) it follows that ($b^2 \approx 0$), hence: $\dfrac{1}{1+jb} = \dfrac{1}{1+jb}\,\dfrac{1-jb}{1-jb} = \dfrac{1-jb}{1+b^2} \approx 1 - jb.$

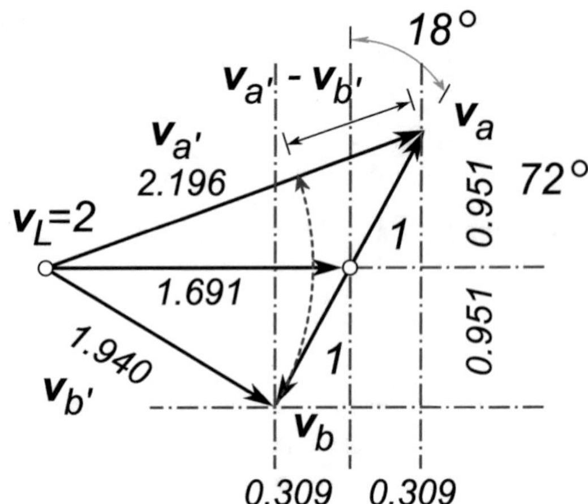

Fig. 16.15 Dual slope vector Example 89

$|\mathbf{v}_a/K = 2$ (keep in mind that $\mathbf{v}_L = \mathbf{v}_{RF}$). With that information, it is straightforward to apply Pythagoras' theorem on right triangles to Fig. 16.15 and conclude that $V_{ab} = V_{a'} - V_{b'} = 2.196\,V_{\text{rms}} = \sqrt{2} \times 2.196 V_{\text{rms}} = 3.106\,\text{V}DC$.

The calculated voltage $V_{ab} = 3.106\,\text{V}DC$ is generated by the circuit due to the maximum frequency deviation therefore $k_d = 3.106\,\text{V}DC/75\,\text{kHz} \approx 41.5\,\mu\text{V/Hz}$.

16.3.2 Quadrature Detector

A quadrature detector is based on a similar principle to a slope detector. However, instead of converting the incoming FM wave into an AM wave, a quadrature detector first converts an FM wave into a PM wave. This conversion is done by means of a phase-shifting network whose phase versus frequency characteristic is linear, hence variation of the carrier frequency $\Delta\omega_c$ creates a frequency-dependent phase shift. In principle, the function of a quadrature decoder is relatively simple. As its name implies, it is based on two signals whose phase difference is 90° (hence, they are "in quadrature"), where the incoming FM wave is split and transmitted through two separate paths. The first path is a simple short connection, while the second path leads through a phase-shifting network (see Fig. 16.16), which adds first a fixed 90° phase shift due to capacitor C_0 and then an additional $\theta = k_\omega\,\Delta\omega_0$ shift due to the $R_pL_pC_p$ resonator network. Consequently, the original FM wave arrives at node ① with its original $\theta = 0$ phase, while its copy arrives at node ② with a new phase of $\theta = {}^{\pi}/_{2} + k_\omega\,\Delta\omega_0$, where k_ω is the proportionality constant.

In the following simplified analysis, we extract the term causing the frequency-dependent phase shift and show that it is indeed linear for small frequency variations. Along the way, we are not concerned about exact expressions for the amplitude of a sinusoidal wave, because it is always set by passive component values of an RLC resonant circuit and it may contain a large number of polynomial terms if calculated exactly. Instead, we focus only on the phase-altering terms. In addition, we assume high Q values, which simplifies the serial to parallel transformation of RLC resonating networks. We note that the resonant frequency of the phase-shifting network is

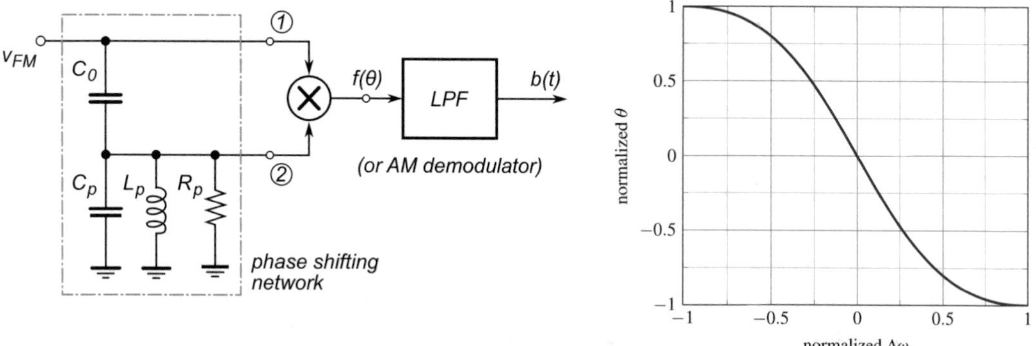

Fig. 16.16 A quadrature decoder circuit (left), quadrature decoder phase characteristic showing linear conversion from frequency variations $\Delta\omega$ of an FM wave to phase $\theta(\omega)$ (right)

$$\omega_0 = \frac{1}{\sqrt{(C_0 + C_p)\,L_p}} = \omega_c \tag{16.67}$$

because, looking into node ②, the two capacitors appear in parallel (C_0 is connected to the AC ground through the FM signal source).

To show how the linear $\theta = f(\Delta\omega)$ phase shift characteristic is implemented, let us take a look at a simplified schematic diagram of a quadrature decoder and its phase-shifting network (Fig. 16.16). First, we note that reactance X_{C0} of capacitor C_0 and impedance Z_{RLC} of parallel $R_p L_p C_p$ resonator are in series, and they effectively create a voltage divider at node ②, hence we write

$$v_2 = \frac{Z_{RLC}}{X_{C0} + Z_{RLC}}\, v_1 = A_0\, v_1 \tag{16.68}$$

where term A_0 is determined by the exact set of values (C_0, R_p, L_p, C_p) of what is just another RLC resonating circuit. Exact derivation of A_0 involves a set of serial–parallel transformations and contains a number of polynomial terms. However, we already showed details of a similar derivation in (16.61), where we concluded that the transfer function A_0 of any resonant RLC network may be simplified into the following general form

$$\begin{aligned} A_0 &= \frac{j\,K}{1 + j\,\alpha} = \frac{j\,K}{\sqrt{1+\alpha^2}}; \quad \angle(\arctan\alpha) \\ &= \frac{K}{\sqrt{1+\alpha^2}}; \quad \angle\left(\frac{\pi}{2} + \arctan\alpha\right) \end{aligned} \tag{16.69}$$

where $K = f(Q)$ is a real constant that controls the amplitude. However, $\alpha = f(\omega, \omega_0, Q)$ is a function of the passive component values[4] and the instantaneous frequency ω. The general form (16.69) is very handy because we can determine its amplitude and phase[5] simply by inspection. After substituting (16.69) into (16.68), we write

[4]Keep in mind that $\omega_0 = f(RLC)$ and $Q = f(\omega_0 L, R)$.

[5]Keep in mind that $\phi = \arctan \Im/\Re$, which reduces to $\phi = \arctan \Im$ when $\Re = 1$.

$$
\begin{aligned}
v_2 &= \frac{K}{\sqrt{1+\alpha^2}}\, v_1; \quad \angle\Big(\frac{\pi}{2} + \arctan\alpha\Big) \\
&= \frac{K}{\sqrt{1+\alpha^2}}\, V_1 \cos\omega_c t; \quad \angle\Big(\frac{\pi}{2} + \arctan\alpha\Big) \\
&= \frac{K_1}{\sqrt{1+\alpha^2}} \cos\Big(\omega_c t + \frac{\pi}{2} + \arctan\alpha\Big)
\end{aligned} \tag{16.70}
$$

where $K_1 = K\,V_1$ is the new amplitude proportionality constant. The time domain term affects the phase $\theta(t)$ and we need to find its average value relative to the sinusoidal variation. Therefore, we evaluate the time-dependent term by integrating it over carrier time $T/2$, as

$$
\begin{aligned}
I &= \frac{\omega}{\pi}\int_0^{\frac{\pi}{\omega}} \cos\Big(\omega_c t + \frac{\pi}{2} + \arctan\alpha\Big)\, dt \\
&= -\frac{2}{\pi}\frac{\alpha}{\sqrt{1+\alpha^2}}
\end{aligned}
$$

$\therefore$

$$
phase(v_2) \propto -\frac{2}{\pi}\frac{K_1}{\sqrt{1+\alpha^2}}\frac{\alpha}{\sqrt{1+\alpha^2}} = -K_2\frac{\alpha}{1+\alpha^2} \tag{16.71}
$$

where $K_2 = (K_1 \times 2/\pi)$ is the new amplitude proportionality constant. A plot of (16.71) shows the liner phase dependence versus frequency variations (see Fig. 16.16 (right)).

Now that we have found out how the $\Delta\omega$ to phase conversion is implemented, we need to find out how the embedded information signal $b(t)$ is extracted. That extraction is done by the multiplier circuit in the frequency domain after signals v_1 and v_2 have reached its input terminals. Hence, output of the multiplier circuit, after substituting $k_\omega\Delta\omega = \arctan\alpha$ and assuming an ideal multiplier, is

$$
\begin{aligned}
f(\theta) = v_1 \times v_2 &= K_0\Big[\cos\omega_c t \times \cos\Big(\omega_c t + \frac{\pi}{2} + k_\omega\Delta\omega\Big)\Big] \\
&= -K_0\left[\cos\omega_c t \times \sin\left(\omega_c t + k_\omega\Delta\omega\right)\right] \\
&= -\frac{K_0}{2}\left[\sin\left(2\omega_c t + k_\omega\Delta\omega\right) - \sin\left(k_\omega\Delta\omega\right)\right] \\
&\approx \frac{K_0}{2}\sin\left(k_\omega\Delta\omega\right)
\end{aligned} \tag{16.72}
$$

where the approximation was introduced after the signal passed through the LP filter and the high-frequency tone close to $2\omega_c$ was removed from the signal spectrum (see Fig. 16.17). As the last step of signal recovery, we note that

$$
b(t) \propto \sin\left(k_\omega\Delta\omega\right) \approx k_\omega\Delta\omega \tag{16.73}
$$

for small variations of the sine argument. Therefore, for small frequency shifts, a quadrature decoder has a reasonably linear characteristic. We note that implementation of the multiplier and the LP filter is very important for operation of an analog quadrature decoder. However, if v_1 wave is digital, i.e. a square pulse stream, then a simple digital multiplier (in form of an AND gate) is employed.

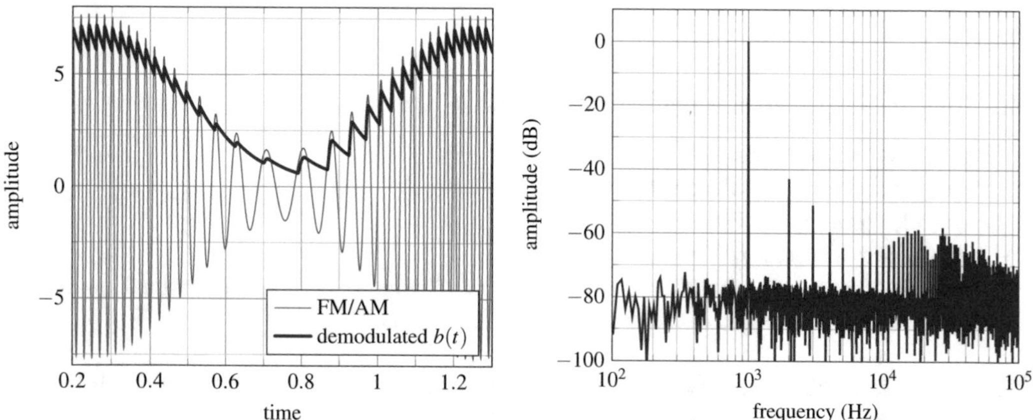

Fig. 16.17 Simulated quadrature demodulator in Fig. 16.16, time domain (left) and frequency domain (right) response

16.3.3 PLL Demodulator

By inspection of the PLL circuit (Fig. 15.26), we note that if, instead of looking into the VCO's output node that generates either a sinusoidal or a square wave at frequency ω_0, we probe the VCO's input node ② where the (quasi) DC voltage level is generated, then without any additional circuitry we have realized a phase or a frequency demodulator. Note that the voltage at node ② is directly proportional to the change of frequency $\Delta\omega$ of the wave entering the input of the PD. If the input wave is an FM wave, then the VCO control voltage accurately tracks the FM, in other words the envelope of the modulation signal $b(t)$ that is embedded into the FM wave.

There are two slightly different cases of PLLs for FM demodulation. In the first case, when the loop bandwidth is wide enough to match the bandwidth of the modulation signal, then the PLL works as a frequency demodulator. On the other hand, if the loop bandwidth is very narrow, then the PLL is locked to the unmodulated carrier signal ω_0 so that the reference phase is averaged, that is, the phase detector holds almost constant phase that serves as a reference for comparison with the VCO phase.

16.4 Summary

Basic techniques for recovery of the received information $b(t)$ are based on a very simple diode-rectifying circuit that is a fundamental component of both AM and FM demodulators. Accurate reproduction of the envelope wave is obviously important and the amount of imperfection of the recovered information signal $b(t)$ is referred to as "distortion". When the recovered signal carries audio information, the distortion from a low-quality demodulator is perceived by our hearing system as "bad sound". Similarly, if the recovered signal carries digital information, the distortion may cause "bit errors" if the binary signal levels shift too far from their acceptable levels, as defined by digital noise margins. Modern, more sophisticated, integrated versions of radio transceivers heavily employ PLL circuits both for modulation and demodulation of digital waves.

Problems

16.1 Using the simplified schematic in Figs. 16.2 and 16.3, assuming $R = 2\,\text{k}\Omega$, estimate:

(a) Detector input impedance
(b) Total power delivered to the detector
(c) $v_0(max)$, $v_0(min)$, and $V_0(DC)$
(d) Average output current $I_0(DC)$
(e) Appropriate capacitor value C to prevent diagonal clipping distortion for maximal modulation frequency $f_m(max) = 5\,\text{kHz}$ and maximal modulation index $m_a = 0.9$

16.2 AM diode detector whose diode I_D vs. V_D characteristics is also shown in Fig. 16.18, is intended to recover a signal V_s that is embedded in IF waveform found at the node one.

(a) Sketch frequency spectrum of the signals that can be found in this circuit;
(b) Sketch AM waveform shapes at nodes 1–5;
(c) Sketch the equivalent circuit at $f = 5\,\text{kHz}$. Calculate amplitude ratio of the input signal and signal at node 3;
(d) Sketch the equivalent circuit at $f = 665\,\text{kHz}$. Calculate amplitude ratio of the input signal and signal at node 3;
(e) Comment on the results.

Data: $f_{\text{IF}} = 665\,\text{kHz}$, $f_s = 5\,\text{kHz}$, $C_1 = 220\,\text{pF}$, $C_2 = 5\,\text{nF}$, $R_1 = 470\,\Omega$, $R_2 = 4.7\,\text{k}\Omega$, $R_L = 50\,\text{k}\Omega$

16.3 RF signal as received by a 50Ω antenna has RMS amplitude of 8 μV. Subsequently, the signal is processed by the receiver in Fig. 16.19. Estimate:

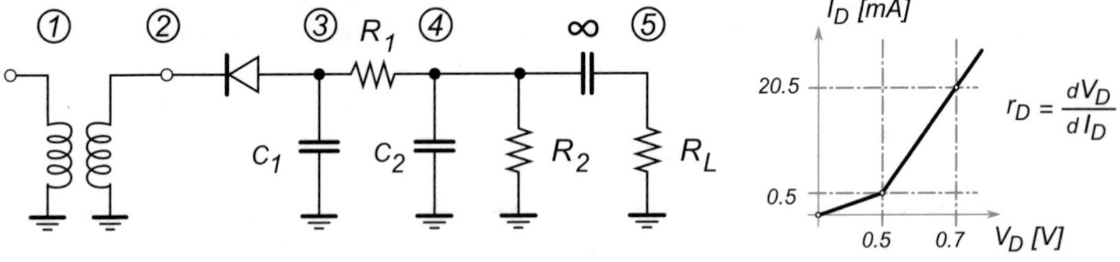

Fig. 16.18 Simplified schematic and voltage–current characteristics for Problem 16.2

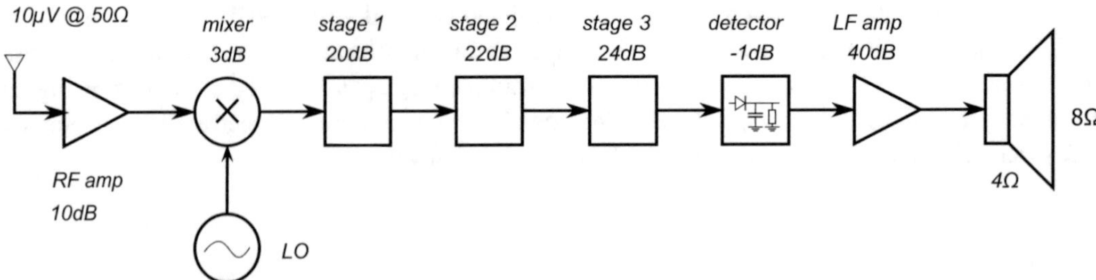

Fig. 16.19 Block diagram for Problem 16.3

(a) input signal power in W and dBm units,
(b) power delivered to the speaker in dBm and W.

16.4 A modulation wave signal has symmetrical triangular shape with zero DC component and amplitude of $V_b = 2\,\mathrm{V_{pp}}$ while the carrier wave has amplitude of $V_C = 2\,\mathrm{V_p}$. Calculate modulation index and find the ratio of the side lengths in the corresponding trapezoidal pattern.

16.5 For unmodulated signal, AM current in the antenna is $I_0 = 1\,\mathrm{A}$, while sinusoidal modulation wave causes the antenna current to be $I_m = 1.1\,\mathrm{A}$. Calculate the modulation index.

17 RF Receivers

In a general sense, a radio receiver is an electronic system that is expected to detect the existence of a single, very specific EM wave in the overcrowded air space, separate it from the rest of the frequency spectrum, and extract a message. Hence, the literal implementation of the receiver function, which is known as a TRF receiver, consists only of a receiving antenna, an RF amplifier, and an audio amplifier. On the other hand, advanced radio receiver versions include one or more mixers and VCO blocks, which are meant to perform either a single-step frequency downconversion (also known as a "heterodyne receiver") or multiple step frequency downconversions (also known as a "super-heterodyne receiver") in order to shift the HF wave down to the baseband.

In this chapter, we review basic radio receiver topologies, the nonlinear effects caused by less than ideal electronic circuitry used to implement the receiver, and receiver specification parameters.

17.1 Basic Radio Receiver Topologies

In its simplest form, a radio receiver is just an LC resonator with an envelope detector. The simplest possible implementation is known as "crystal radio" (Fig. 17.1) and it consists of the antenna (a long wire), an inductor with several taps (i.e. a quasi-tunable inductor with, for example, three discrete "taps" that permit the choice of three different inductances), a diode, and high-impedance headphones. The resonance is achieved by the antenna-inductor connection. To understand how this works, we keep in mind that, for instance, the commercial AM radio band is in the 530–1710 kHz range, that is, the associated wavelengths are from 566 to 174 m, or equivalently 141 to 44 m quarter wavelength. Using an antenna of quarter of the wavelength ($\lambda/4$ is common practice), which means that even for the upper AM band we would need a wire at least 44 m long. Usually, we settle for a wire about 20 m

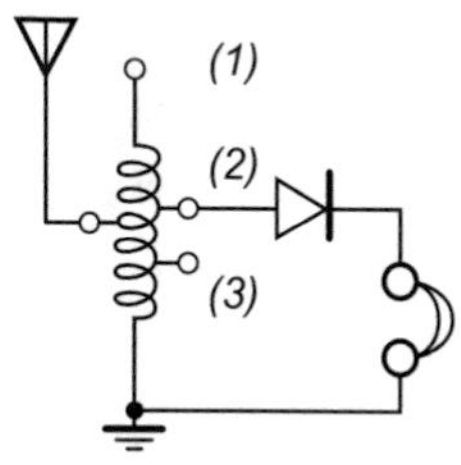

Fig. 17.1 "Crystal" radio receiver topology schematic diagram

R. Sobot, *Wireless Communication Electronics*,
https://doi.org/10.1007/978-3-030-48630-3_17

long (we have to be *very careful* with the trees and houses in the neighbourhood), which means that at these relatively low frequencies (thus, $Z_L \rightarrow 0$) the antenna is mostly capacitive. Indeed, a 20 m long antenna behaves very much like a 250–300 pF capacitor. Thus, knowing the resonance equation, it is straightforward to calculate the required inductive size and create LC resonator.

Even with such a simple LC resonator, the desired RF AM signal can be sufficiently isolated and it takes a simple envelope detector to de-embed the message. In this case, the envelope detector is built of a diode and high-impedance headphones that serve as the resistive load in combination with the antenna and parasitic capacitances. Because of the high impedances within the circuit, i.e. small currents, the amount of energy collected in the antenna is sufficient to generate an audio signal in the headphones. Hence, there is no need for an external power supply, which was the reason why this kind of radio receiver was used very much by soldiers during World War I (when the radio was named a "foxhole radio"). Obviously, this simple structure is much too simple for commercial use and had to be improved in many ways.

The most direct and oldest implementation of a commercial radio receiver was based on the tuned radio frequency receiver (TRF) topology (see Fig. 17.2). Although a TRF receiver may contain more than one RF tuned amplifier, each RF amplifier must be directly tuned to its carrier frequency ω_c, with subsequent stages tuned appropriately. Tuning the resonator stage in a TRF receiver to the carrier implies that the envelope detector must decode the message from the HF carrier, which means that the carrier frequency must be relatively low. In addition, a relatively wide bandwidth (i.e. low Q factor) of a single front-end LC resonator allows a number of tones to pass through and enter the envelope detector, which directly affects the overall SNR of the receiver as well as its selectivity. To make things worse, the RF amplifier gain is a function of the signal frequency, hence different carrier frequencies were received with different gains. At times when only a handful of radio stations were broadcasting, it was relatively easy to separate their carrier frequencies so that they did not interfere with each other, even with the low Q resonator tanks being used. As the radio broadcasting industry grew, the air space became more crowded and the only way to increase Q, and therefore selectivity,

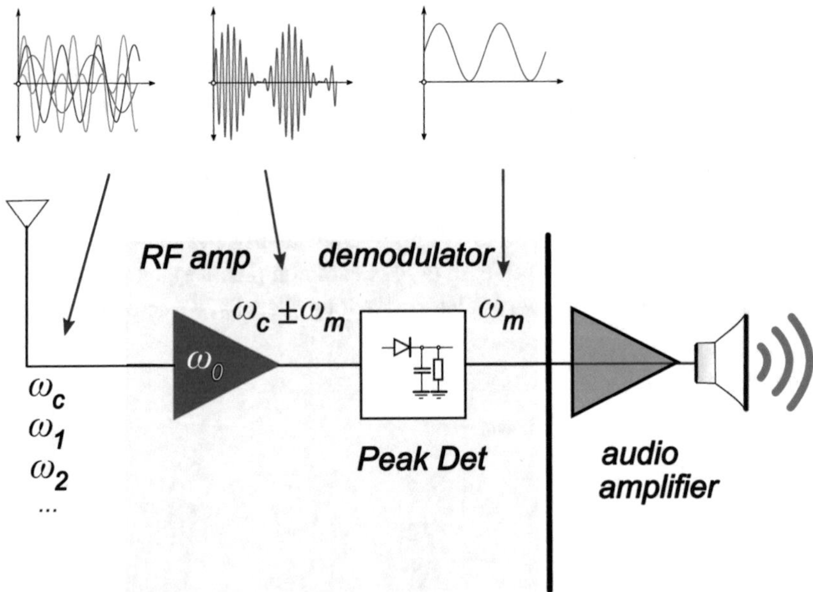

Fig. 17.2 TRF receiver topology block diagram.

was to add a cascade of LC tanks (see Sect. 10.9). However, this was at a cost of increased complexity and increased effort to keep all the resonators properly tuned. To its credit, this topology was the main technology of the day until it was replaced by the heterodyne receiver.

The solution to improve TRF receiver was to introduce, first, the heterodyne receiver topology (with one mixer/VCO stage), as shown in Fig. 17.3, and then the super-heterodyne receiver topology (with two or more mixer/VCO stages, also known as "double conversion"). Similarly to a TRF receiver, heterodyne receivers first tune to the HF carrier frequency. However, after the RF amplifier stage separates the desired carrier frequency from the crowded frequency spectrum, the carrier-centred signal is down-shifted in frequency to some IF that is fixed for the given receiver. This makes it much easier to design the downstream stages: they always work at the same frequency regardless to what carrier frequency the RF stage is tuned to. The amount of shifting is determined by the current frequency of the local VCO, which is tuned in tandem with the RF stage.

At the system level, a radio receiver is analysed and characterized using common metrics, so that we can compare the performance of various designs. Some of the most common parameters that are compared are:

1. *Selectivity*: the minimum separation between the desired carrier frequency and its first neighbouring frequency, under the condition that the receiver can safely receive the intended signal.
2. *Sensitivity*: the minimum amplitude of the incoming RF signal that the receiver can decode, under the condition of the required SNR.
3. *Dynamic range*: the amplitude ratio of the strongest and weakest signals that the receiver can decode.

We establish detailed metrics for each of these parameters in the following sections; however, in the meantime, we need to familiarize ourselves with terminology and several key consequences of the fact that RF circuits are *nonlinear systems*.

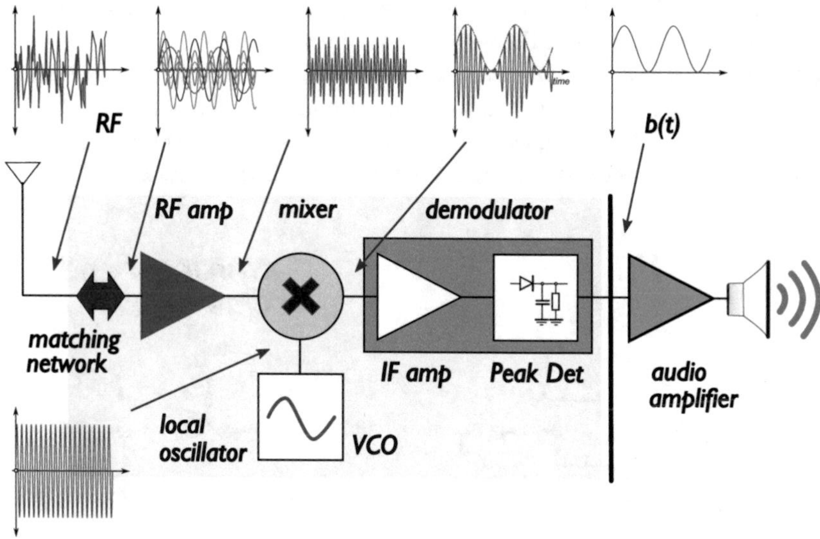

Fig. 17.3 Heterodyne receiver topology block diagram. Super-heterodyne topology contains two or more cascaded mixer/VCO stages that perform frequency down shifting in more than one step

17.2 Nonlinear Effects

Understanding the characterization of general systems is important for understanding the behaviour of radio systems. Let us review basic terminology from systems theory. We loosely define a *linear system* as one in which the output signal consists of the sum of proportionally scaled input signals. In mathematical terminology, it satisfies the superposition law, i.e.

$$F[a_1 x_1(t) + a_2 x_2(t)] = a_1 F(x_1(t)) + a_2 F(x_2(t)) \tag{17.1}$$

where, a_1 and a_2 are constants independent of time. If the system does not satisfy the superposition law, then it is *nonlinear*.

A system is *time invariant* if a time shift in the input results in the same time shift at the output, i.e. in mathematical terminology, if $x(t) \rightarrow (t)$, then $x(t-\tau) \rightarrow (t-\tau)$ for all values of τ. Systems that are both linear and time invariant are known as "LTI systems".

We define a *memoryless system* as one whose output does not depend on the past values of its input. For instance, a memoryless linear system obeys the relation $y(t) = a\,x(t)$, where a is a constant. If $a(t)$ is a function of time, then relation $y(t) = a(t)\,x(t)$ describes a memoryless time-variant system. We can define a memoryless nonlinear system by using the general polynomial relation

$$y(t) = a_0 + a_1 x(t) + a_2 x^2(t) + a_3 x^3(t) + \cdots \tag{17.2}$$

where, a_i is constant in time, otherwise we define time-variant memoryless nonlinear system. Clearly, if in practice all terms in (17.2) disappear (or are negligibly small) except the first two, then the linear approximation of $y(t)$ would be valid. Example of two networks, one that is LTI, and one that is both nonlinear and time variant, is shown in Fig. 17.4. Note that, in Fig. 17.4 (right), the switch by itself is a nonlinear element, and due to dependence of the output variable $y(t)$ on the switching frequency ω_c time invariance is broken. General radio systems are analysed using (17.2) where $a_i = const$ because they are approximated as memoryless time-invariant nonlinear systems.

We define *amplitude distortion* in the case when the output amplitude is not a linear function of the input amplitude. In general *distortion* is any difference between the original and the output form of the signal, Fig. 17.5. The following are most commonly studied effects due to nonlinearity of the transfer characteristic: harmonic distortion, gain compression, desensitization, and intermodulation.

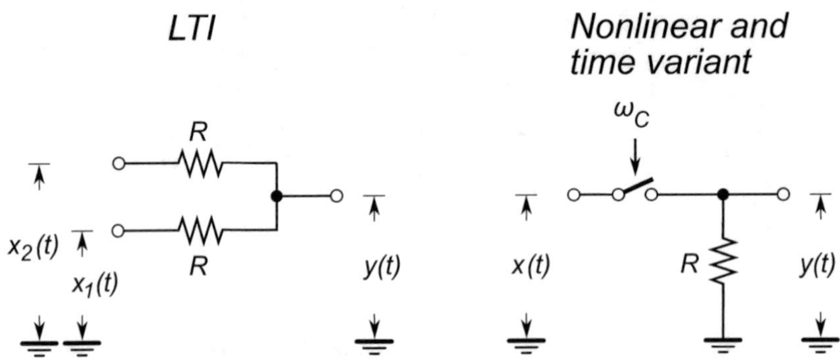

Fig. 17.4 Examples of LTI, and nonlinear and time-variant systems

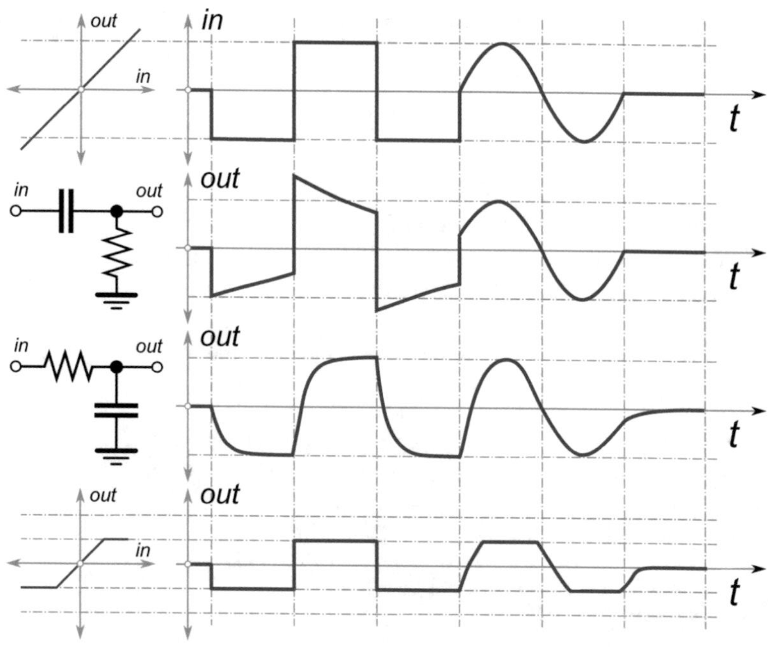

Fig. 17.5 Examples of distorted waveforms due to the nonlinear transfer functions

17.2.1 Harmonic Distortion

After injecting a single-tone signal $x(t) = B \cos \omega t$ into a nonlinear system whose transfer function is described as (17.2) with the DC term removed (i.e. $a_0 = 0$), the output shows at the output node as

$$y(t) = a_1 B \cos \omega t + a_2 B^2 \cos^2 \omega t + a_3 B^3 \cos^3 \omega t + \cdots \tag{17.3}$$

$$= a_1 B \cos \omega t + \frac{a_2 B^2}{2} (1 + \cos 2\omega t) + \frac{a_3 B^3}{4} (3 \cos \omega t + \cos 3\omega t) + \cdots$$

$$= \frac{a_2 B^2}{2} + \left(a_1 B + \frac{3\, a_3 B^3}{4} \right) \cos \omega t + \frac{a_2 B^2}{2} \cos 2\omega t + \frac{a_3 B^3}{4} \cos 3\omega t + \cdots$$

$$= b_0 + b_1 \cos \omega t + b_2 \cos 2\omega t + b_3 \cos 3\omega t + \cdots \tag{17.4}$$

where b_0 is the output signal's DC term. In addition, we make the following observations: the input signal spectrum $x(\omega)$ contains only one tone ω, while the output signal spectrum $y(\omega)$ contains higher-order harmonics 2ω, 3ω, etc. that did not exist in the input spectrum. They have been created by the nonlinear transfer function. Even-order harmonics (i.e. terms with $2\omega t$, $4\omega t$, ...) are associated with the even-order constants a_i, $i = 2k$, therefore these terms disappear if the system transfer function has odd symmetry, for instance, transfer functions of differential circuits. For large amplitude, $B \gg 1$ the nth harmonics is approximately proportional to B^n. These are very important observations that give clues about the frequency spectrum of a nonlinear system.

One of the commonly used quantitative measures of nonlinearity is "total harmonic distortion" (THD). The individual percentage distortions are calculated as

$$D_2 = \frac{b_2}{b_1} \quad D_3 = \frac{b_3}{b_1} \quad D_4 = \frac{b_4}{b_1} \quad \cdots \tag{17.5}$$

relative to the first harmonic coefficient. Then, by definition, we calculate THD for voltage or current as

$$THD = \sqrt{D_2^2 + D_3^2 + D_4^2 \cdots} \tag{17.6}$$

Example 90: THD Calculation

A cosine current was measured at the output of a non-inverting amplifier. The three experimentally determined pairs of the input voltage V_{in} and the matching output current I_{out} are: $V_{max} \Rightarrow I_{max} = 1\,\text{mA}$, $V_b \Rightarrow I_b = 0.01\,\text{mA}$, $V_{min} \Rightarrow I_{max} = -0.95\,\text{mA}$, where V_b is the biasing voltage at the midpoint between the maximum and minimum input voltage amplitudes. Based on the available data, estimate the THD of the system.

Solution 90: The collected experimental data correspond to the cosine wave input voltage function (the non-inverting amplifier), therefore we know the associated ωt angles.[1] After substitution back into (17.4), this results in:

$$V_{in} = V_{max} \quad \therefore \quad \omega t = 0 \quad \therefore \quad I_{max} = b_0 + b_1 + b_2$$

$$V_{in} = V_b \quad \therefore \quad \omega t = \frac{\pi}{2} \quad \therefore \quad I_b = b_0 - b_2$$

$$V_{in} = V_{min} \quad \therefore \quad \omega t = \pi \quad \therefore \quad I_{max} = b_0 - b_1 + b_2$$

which is solved as

$$b_0 = \frac{I_b}{2} + \frac{I_{max}}{4} + \frac{I_{min}}{4} = 17.5\,\mu A$$

$$b_1 = \frac{I_{max}}{2} - \frac{I_{min}}{2} = 975\,\mu A$$

$$b_2 = -\frac{I_b}{2} + \frac{I_{max}}{4} + \frac{I_{min}}{4} = 7.5\,\mu A$$

By definition, we write

$$D_2 = \frac{b_2}{b_1} \times 100\% = 0.77$$

$$THD = \sqrt{D_2^2 + D_3^2 + D_4^2 \cdots} = \sqrt{D_2^2} = D_2 = 0.77\%$$

because with only three measurements we can solve for up to the second-order term in (17.4). If more detailed measurement were done, for instance with five measured points that would add amplitudes at $V_{in}(\pm 1/2)$, then we would have $\omega t = \pi/3$ and $\omega t = 2\pi/3$ corresponding angles as well, which would enable us to calculate b_0, b_1, b_2, b_3, and b_4 constants.

[1] Simply plot a cosine function and find arguments for its maximum, minimum, middle, and $\pm 1/2$ amplitude points.

17.2.1.1 Gain Compression

A common property of most amplifier circuits is that as the input signal power level increases, at first the output signal level increases proportionally. That is, for low-power signals, the output–input relationship is linear $P_{out} = A\,P_{in}$, where A is the gain that is calculated as the $A = d\,P_{out}/d\,P_{in}$ derivative. However, eventually, the output signal level is limited by the circuit's power supply level or the reduced biasing current of its active devices. In other words, the small signal linearity relationship does not hold for large input signal levels.

We define the 1 dB *compression point* as the input signal power level $S_{in}(-1\,\text{dB})$ which corresponds to the gain $A_{(-1\,\text{dB})}$ for which the output signal level is 1 dB lower relative to the linear model (see Fig. 17.6, noting that the plot is in log-log scale).

The 1 dB compression point is determined both analytically and experimentally. Let us take a nonlinear system described by (17.4) and try to find the 1 dB compression point. The first term in (17.4) is the DC term, hence its derivative is zero and it is not part of the gain equation. The second term describes the output signal of the input $x(t) = A\cos\omega t$, hence we write the equations for the linear gain function (i.e. if all nonlinear terms in (17.3) are ignored) and the nonlinear gain function as

$$|V_{out}| \approx a_1 B\,\cos\omega t$$

$$|V'_{out}| \approx \left(a_1 B + \frac{3\,a_3 B^3}{4}\right)\cos\omega t$$

$$\therefore$$

$$\left|\frac{V'_{out}}{V_{out}}\right| = \left(1 + \frac{3\,a_3\,B^2}{4\,a_1}\right)$$

where, the ratio of V'_{out}/V_{out} is apparent gain between the linear and the nonlinear functions. Clearly, if $a_3/a_1 < 0$ and $\left|\frac{3\,a_3\,B^2}{4\,a_1}\right| < 1$, then there is compression in the gain. After conversion into dB scale[2] we write expression for the apparent gain as

$$20\log V'_{out} - 20\log V_{out} = 20\log\left(1 + \frac{3\,a_3\,B^2}{4\,a_1}\right)$$

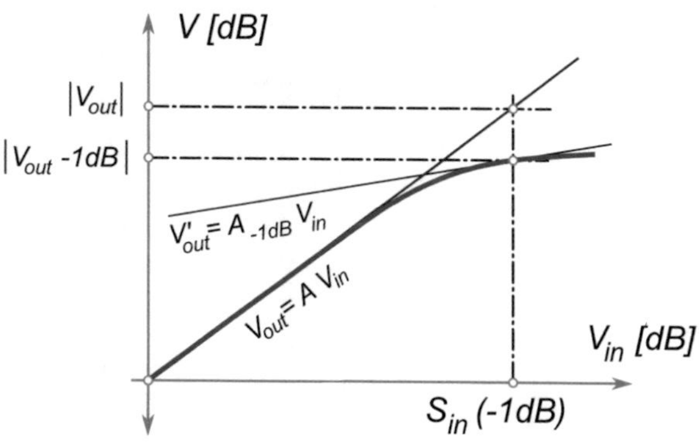

Fig. 17.6 Output signal level vs. input signal level, and the 1 dB compression point

[2]Keep in mind that $\log a/b = \log a - \log b$.

$$-1\,[\text{dB}] = 20\,\log\left(1 + \frac{3\,a_3\,B^2}{4\,a_1}\right)$$

$$10^{-1/20} - 1 = \frac{3\,a_3\,B^2}{4\,a_1}$$

$$\therefore$$

$$B(-1\,\text{dB}) = \sqrt{0.145\,\left|\frac{a_1}{a_3}\right|} \tag{17.7}$$

where, the input signal level $S_{in}(-1\,\text{dB})$ was introduced in [dB], therefore

$$S_{in}(-1\,\text{dB}) = 20\log\left[B(-1\,\text{dB})\right]\;\;[\text{dB}] \tag{17.8}$$

Interestingly enough, (17.7) shows that the 1 dB compression point of the first harmonic is, through a_3, connected to the third harmonic of the input signal. We formalize this connection in the following sections.

17.2.2 Intermodulation

As opposed to harmonic distortion, which is caused by self-mixing of one input signal and where the higher-order harmonics in (13.4) are relatively easy to suppress by LP filtering, *intermodulation* involves two input tones with close frequencies ω_a and ω_b. Consequently, in case of any nonlinearity the output spectrum must contain various harmonics of the fundamental tones; however, it also contains tones that are not harmonics of the input frequencies.

Let us assume the input signal is the sum of $x(t) = B_1 \cos\omega_a t + B_2 \cos\omega_b t$ (see Fig. 17.7), then (17.3) becomes

$$y(t) = a_1\,(B_1 \cos\omega_a t + B_2 \cos\omega_b t)$$

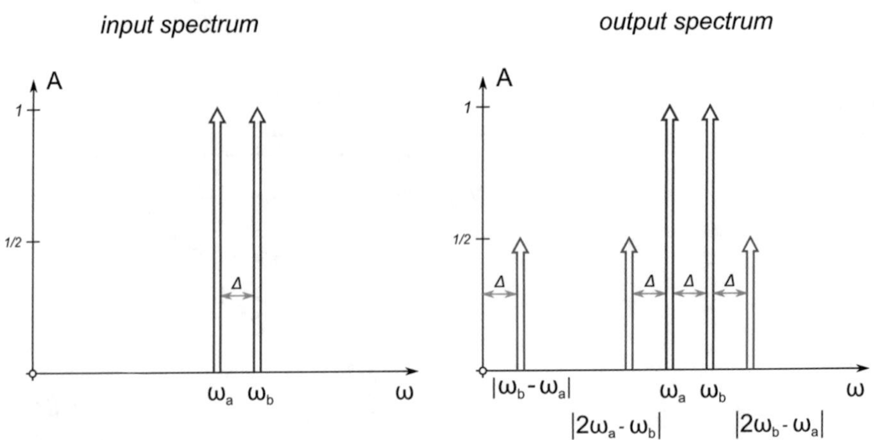

Fig. 17.7 Part of the intermodulation frequency spectrum showing the third-order terms $2\omega_a \pm \omega_b$ close to the fundamental tones

$$+ a_2 (B_1 \cos \omega_a t + B_2 \cos \omega_b t)^2$$
$$+ a_3 (B_1 \cos \omega_a t + B_2 \cos \omega_b t)^3 + \cdots \tag{17.9}$$

which, after expanding and collecting the frequency terms, yields the following terms:

$$\begin{aligned}
y(t) =& \frac{a_2(B_1^2 + B_2^2)}{2} && (DC \text{ term}) \\
&+ \left(a_1 B_1 + \frac{3}{4} a_3 B_1^3 + \frac{3}{2} a_3 B_1 B_2^2\right) \cos \omega_a t && (fundamental\ \text{terms}) \\
&+ \left(a_1 B_2 + \frac{3}{4} a_3 B_2^3 + \frac{3}{2} a_3 B_2 B_1^2\right) \cos \omega_b t \\
&+ \frac{a_2}{2} \left(B_1^2 \cos 2\omega_a t + B_2^2 \cos 2\omega_b t\right) && (second\text{-order terms}) \\
&+ a_2 B_1 B_2 \left[\cos(\omega_a + \omega_b)t + \cos|\omega_a - \omega_b| t\right] \\
&+ \frac{a_3}{4} \left(B_1^3 \cos 3\omega_a t + B_2^3 \cos 3\omega_b t\right) && (third\text{-order terms}) \\
&+ \frac{3a_3}{4} \left\{B_1^2 B_2 \left[\cos(2\omega_a + \omega_b)t + \cos(2\omega_a - \omega_b)t\right]\right. \\
&\quad \left. + B_1 B_2^2 \left[\cos(2\omega_b + \omega_a)t + \cos(2\omega_b - \omega_a)t\right]\right\}
\end{aligned} \tag{17.10}$$

Frequency spectrum analysis (17.10) comes in handy for the "two-tone test" that uses two slightly different tones with the same small amplitude $B_1 = B_2 = B$, which means that the higher harmonics are negligible and (17.10) simplifies to

$$\begin{aligned}
y(t) =& a_2 B \\
&+ B \left(a_1 + \frac{9}{4} a_3 B^2\right) \cos \omega_a t + B \left(a_1 + \frac{9}{4} a_3 B^2\right) \cos \omega_b t \\
&+ \frac{B^2 a_2}{2} (cos 2\omega_a t + \cos 2\omega_b t) + a_2 B^2 \left[\cos(\omega_a + \omega_b)t + \cos|\omega_a - \omega_b| t\right] \\
&+ \frac{B^3 a_3}{4} (\cos 3\omega_a t + \cos 3\omega_b t) \\
&+ \frac{3B^3 a_3}{4} \{[\cos(2\omega_a + \omega_b)t + \cos(2\omega_a - \omega_b)t] \\
&\quad + [\cos(2\omega_b + \omega_a)t + \cos(2\omega_b - \omega_a)t]\}
\end{aligned} \tag{17.11}$$

With an assumption of a small amplitude B , i.e. $B^2 \to 0$, the amplitudes of the fundamental terms are approximated as

$$\left(a_1 + \frac{9}{4} a_3 B^2\right) \approx a_1 \tag{17.12}$$

We are especially interested in the power of tones at $(2\omega_b \pm \omega_a)$ relative to the power of the fundamental tones. In a similar fashion to the derivation of the 1dB compression point, let us take a look at the input signal level that causes the power of the third-order term to be equal to the power of the fundamental, i.e.

$$a_1 B = \frac{3B^3 a_3}{4}$$
$$\therefore$$
$$B(\text{IIP3}) = \sqrt{\frac{4}{3}\left|\frac{a_1}{a_3}\right|} \tag{17.13}$$

where the amplitude of the fundamental was approximated as (17.12) and $B(IIP3)$ refers to the input signal level known as the *third-order intercept point* (IIP3). By comparing (17.13) with (17.7), we write

$$B(-1\,\text{dB}) = \sqrt{\frac{4}{3}\left|\frac{a_1}{a_3}\right|\,0.11} = IIP3 - 9.6\,\text{dB} \tag{17.14}$$

We note that IIP3 gives nonlinearity because of the third-order terms and that the initial assumption was that the two input tones had small amplitude. That is, expression (17.13) is not valid for strong signals. Because of that, IIP3 is the theoretical point that is extrapolated from the linear portions of the gain plot, shown in Fig. 17.8 (right). It is interesting to note that the slope of the third-order term is three times the slope of the fundamental. This observation leads to a graphical solution for the third-order IIP3 from the experimental data (Fig. 17.9). The input power of the fundamental tones is measured and compared with the power of the third-order term on a spectrum analyser, where the power difference ΔP is in [dB], Fig. 17.9 (left), which is translated into the I/O power plot, Fig. 17.9 (right). Using similar triangles, we conclude that the IIP3 point must be at

$$IIP3 = P_{in} + \frac{\Delta P}{2} \quad [\text{dB}] \tag{17.15}$$

which is a practical way to estimate the IIP3 by measurement. In this analysis, we have ignored any effects of the second-order terms. They have less influence in narrowband systems relative to the third-order terms; however, in the case of low IF or direct conversion systems, the second-order terms

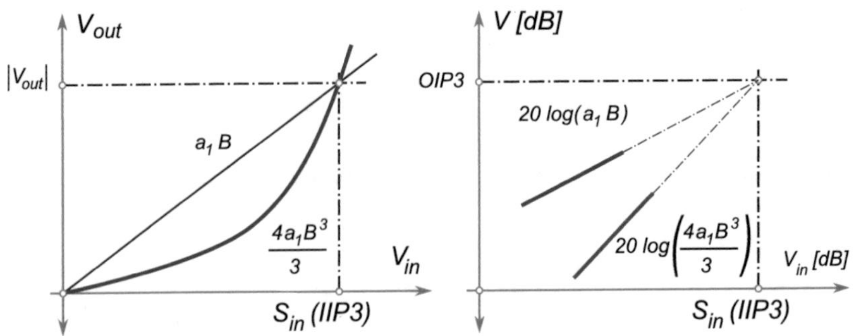

Fig. 17.8 Third-order intercept point extrapolation

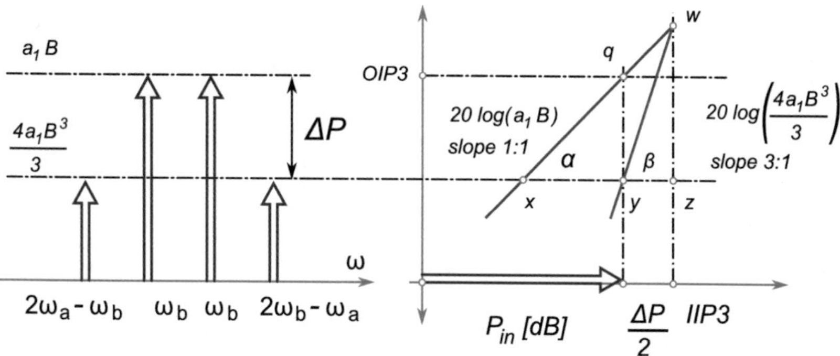

Fig. 17.9 Graphical solution for third-order intercept point

come very close to the baseband signal. If not taken care of, they may even "overwrite" the desired tone. More detailed study of intermodulation terms is beyond the scope of this book.

17.2.3 Cross-modulation

There are two important cases of cross-modulation that we need to become familiar with. In the first scenario, two signals arrive at the antenna, one much stronger than the other. The problem is that the desired signal is the "weak" one. As an illustration, imagine using a cell phone in a crowded bus with another cell phone user very close by. The signal leaving the neighbouring cell phone is very strong, but unfortunately it is not for you. The one that you are trying to hear is already at the end of its journey and is very weak, barely dumping its leftover energy into the antenna. Unfortunately for you, the other user is doing the same and your signal may be "blocked" or "jammed".

Let us take a closer look at the case from the mathematical perspective. The incoming signal

$$x(t) = B_1 \cos \omega_a t + B_2 \cos \omega_b t; \qquad B_2 \gg B_1 \tag{17.16}$$

is processed by a nonlinear circuit whose gain equation is given by (17.2), which, after substitution of (17.16) becomes

$$\begin{aligned} y(t) &\approx \left(a_1 B_1 + \frac{3}{4} a_3 B_1^3 + \frac{3}{2} a_3 B_1 B_2^2 \right) \cos \omega_a t + \cdots \\ (B_2 \gg B_1) \\ &\approx \left(1 + \frac{3}{2} \frac{a_3}{a_1} B_2^2 \right) a_1 B_1 \cos \omega_a t + \cdots \end{aligned} \tag{17.17}$$

where, we focus only on the first fundamental term of the desired tone at ω_a. Most circuits are compressive, therefore $a_3/a_1 < 0$, which leads to the conclusion that, under the right circumstances and large amplitude B_2 of the blocking signal, the amplitude of the desired signal ω_a may be reduced to zero, i.e.

$$0 = \left(1 - \frac{3}{2} \left| \frac{a_3}{a_1} \right| B_2^2 \right) \quad \therefore \quad B_2 = \sqrt{\frac{2}{3} \left| \frac{a_1}{a_3} \right|} \tag{17.18}$$

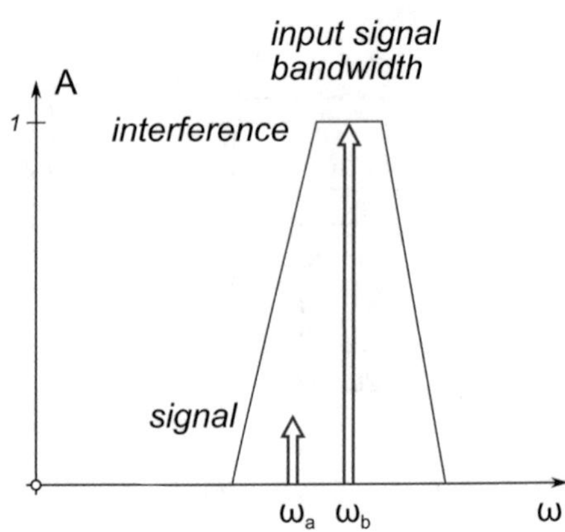

Fig. 17.10 Strong interference and weak signal in the same band

Modern RF equipment is expected to correctly decode the desired signal in presence of an interfering signal that may be 60–70 dB stronger.

In the second scenario (see Fig. 17.10), the receiving antenna is exposed to two signals, the desired one at frequency ω_a and a strong AM signal, i.e.

$$x(t) = B_1 \cos \omega_a t + B_2(1 + m \cos \omega_b t) \cos \omega_c t \tag{17.19}$$

Using the same approach again, we focus only on the main harmonic of the desired signal, i.e.

$$\begin{aligned} y(t) &\approx \left[a_1 B_1 + \frac{3}{2} a_3 B_1 B_2^2 \left(1 + \frac{m^2}{2} + \frac{m^2}{2} \cos 2\omega_b t + 2m \cos \omega_b t \right) \right] \cos \omega_a t + \cdots \\ &= f(\omega_b, 2\omega_b) \cos \omega_a t \end{aligned} \tag{17.20}$$

in other words, the receiving signal is modulated by the AM signal, which is superimposed on the original message. Depending on the exact circumstances, the desired signal may be completely blocked by the strong AM signal.

17.2.4 Image Frequency

The main limitation of a TRF receiver, its limited selectivity over a wide range of receiving frequencies, was a strong motivation for development of heterodyne receiver topology. Even though it is much more complicated than the simple TRF receiver structure, advances in IC technology enable very sophisticated heterodyne and super-heterodyne receivers to be manufactured as a subcircuit of even more complex communication integrated systems. Indeed, it is a standard expectation for modern equipment to have one or more integrated RF transceivers included for a fractional increase in the overall cost.

However, the solution to the selectivity problem, which was enabled by the addition of a VCO-mixer combination, comes with its own issue, known as the “image frequency”, which is sometimes referred to as a “ghost frequency”. This inherent issue comes from the fact that a mixer generates two tones, $\omega_a \pm \omega_b$, at its output terminal (see Fig. 17.11). In order to see how the ghost frequency issue arises, let us take a look at the following scenario. Let us say that an audio signal with $f_m = 1\,\text{kHz}$ is embedded into a carrier signal, $f_C = 10\,\text{MHz}$. At the receiving side, the LO is tuned to $f_{VCO} =$

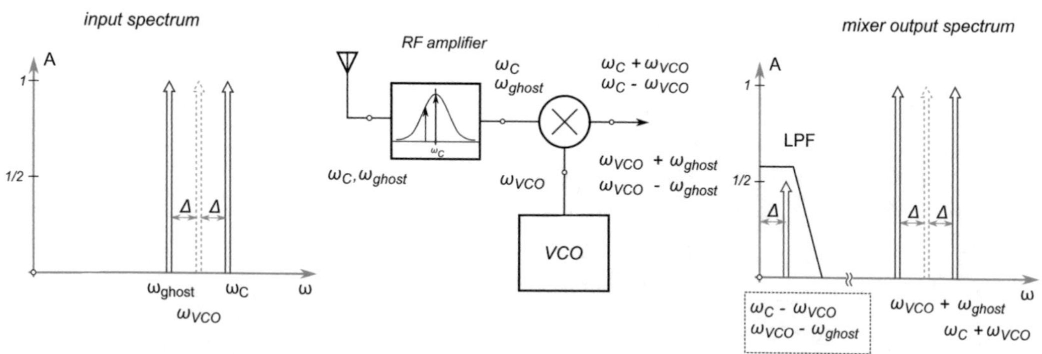

Fig. 17.11 Frequency domain relation among the carrier frequency ω_c, the image frequency ω_{ghost}, the local oscillator frequency ω_{VCO}, and the sum and difference tones generated by mixer. This illustration assumes the $\omega_{VCO} < \omega_c$ case. If $\omega_{VCO} > \omega_c$, then the roles of the carrier and the image frequency are swapped

9.999 MHz. Routinely, we state that the frequency spectrum at output of the ideal mixer must contain

$$\begin{aligned} f_1 &= f_C + f_{VCO} = 10\,\text{MHz} + 9.999\,\text{MHz} = 19.999\,\text{MHz} \\ f_2 &= f_C - f_{VCO} = 10\,\text{MHz} - 9.999\,\text{MHz} = 1\,\text{kHz} \end{aligned} \tag{17.21}$$

where $f_2 = 1\,\text{kHz}$ is the desired signal, and $f_1 = 19.999\,\text{MHz}$ is the high-frequency tone that is removed by an LP filter. However, a more careful analysis reveals that, in the case of another signal arriving at the receiving antenna, we may have the following scenario. Let us take a look at the frequency that is located, in this case, at two times the modulation frequency f_m below the carrier frequency, i.e.

$$f_{ghost} = f_C - 2\,f_m = 10\,\text{MHz} + 2 \times 1\,\text{kHz} = 9.998\,\text{MHz} \tag{17.22}$$

which is close enough to the carrier frequency and, therefore, passes through the RF amplifier's resonator and enters the mixer. Consequently, output of the mixer must contain the following tones

$$\begin{aligned} f_3 &= f_{VCO} + f_{ghost} = 9.999\,\text{MHz} + 9.998\,\text{MHz} = 19.997\,\text{MHz} \\ f_4 &= f_{VCO} - f_{ghost} = 9.999\,\text{MHz} - 9.998\,\text{MHz} = 1\,\text{kHz} \end{aligned} \tag{17.23}$$

To our surprise, we find out that we have received not the desired message but another message carried by another carrier at the image frequency. Indeed, the second message was generated by a second (real) transmitter working at $f_{ghost} = 9.998\,\text{MHz}$ frequency, and it is irreversibly mixed with the desired message.

The issue of image frequency must be dealt with before the first mixer stage. The following methods are most often used to deal with it:

1. Increasing the Q factor of the input front-end resonator and further rejecting the image (see Sect. 14.7.1).
2. Keeping a minimum distance between any two neighbouring radio-transmitting frequencies.
3. Declaring "forbidden" frequencies within the frequency spectrum.
4. Introducing super-heterodyne receiver topology with a second VCO-mixer pair that further separates the troubling tones from the desired one.

In reality, the radio system design process involves a number of specifications and standards that provide guidelines and working boundaries to the designer.

Example 91: Image Frequency
For a standard AM receiver that is tuned to a carrier signal of $f_C = 620\,\text{kHz}$ and IF frequency of $f_{IF} = 455\,\text{kHz}$, determine the image frequency f_{image} if the receiver is designed to have $f_{VCO} > f_C$.

Solution 91: With reference to Fig. 17.11, we write an expression for the frequency of the LO f_{VCO} as the difference between

$$f_{IF} = f_{VCO} - f_C \quad \therefore \quad f_{VCO} = f_{IF} + f_C = 1075\,\text{kHz}$$

$$\therefore$$

$$f_{IF} = f_{image} - f_{VCO} \quad \therefore \quad f_{image} = f_{IF} + f_{VCO} = 1530\,\text{kHz}$$

17.3 Radio Receiver Specifications

System-level radio designers aim to improve the selectivity of the systems by designing architectures that are better equipped to deal with the intermodulation and image frequency issues. It is common for modern radio receiver architectures that are implemented using IC technologies to be able to select a signal from a wide range of carrier frequencies that span over several "standard" frequency bands. For example, the latest cell phones are capable of covering up to three GSM frequency bands, such as the 2100–1900–850 MHz combination. The rule is that each user must conform to its assigned channel boundaries, i.e. just as it is not desirable to have signal cross-talk within a multi-wire bundle, it is not desirable to have "spilling over" of frequency spectrum among wireless channels.

17.3.1 Dynamic Range

The term *dynamic range* refers to the ratio of the largest and smallest values that the system is capable of processing. For instance, if the lowest signal amplitude that an amplifier can detect and amplify is 1 mV and the largest amplitude is 1 V, then its dynamic range is 1 : 1000. It is common practice in technical and science literature to describe 1V relative to 1 : 1000 as a 60 dB dynamic range; that is, dynamic range is a *dimensionless* number.

State-of-the-art electronic equipment often exhibits a dynamic range of more than 100 dB. In order to put a perspective on these numbers, 100 dB is a ratio of 100 000 : 1 (the equivalent of 1 mV relative to 100 V). That is equivalent to a ratio of, for instance, the height of the CN Tower in Toronto relative to an ant.

17.3.1.1 Noise Floor

The upper limit of the dynamic range is set by circuit nonlinearities. The most commonly used metric for quantifying the dynamic range of a circuit is by specifying its 1dB compression point or, equivalently, its IIP3. Therefore, control of the upper signal limit is, to a large extent, under the control of the designer. For instance, a straightforward way of increasing the upper signal limit is to design the circuit to operate with increased power supply voltage level.

Determining the minimum signal level that can be detected against the background noise starts by establishing the total amount of noise in the system. We introduced thermal noise in (8.30), which is repeated here

$$P_n = k\,T\,\Delta f \quad [\mathrm{W}] \tag{17.24}$$

which can be normalized for $\Delta f = 1\mathrm{Hz}$, as

$$P_n = k\,T \quad [\mathrm{W/Hz}] \tag{17.25}$$

Unless specifically stated, we assume "room temperature" for the environment, i.e. $T = 290\,\mathrm{K}$, and write

$$P_n = k\,T = 1.38 \times 10^{-23}\,\mathrm{J/K} \times 290\,\mathrm{K} = -174\,\mathrm{dBm} \tag{17.26}$$

This number is commonly used to set the "noise floor" at room temperature. Reducing the environment temperature, of course, reduces the noise floor. That approach is used in high-end receivers for radio astronomy where the incoming signal is very low. Indeed, the approximate power of the radio signal that was transmitted by the Galileo space probe and arrived at Earth was in the order of 10×10^{-21} W or $-170\,\mathrm{dBm}$ and requires a 70-meter-long DSN antenna. However, the cooling system for temperatures close to 0 K is not suitable for general use. Circuit designers reduce the system noise by controlling the frequency bandwidth Δf .

17.3.1.2 Sensitivity

Defining the sensitivity of a receiver requires that we put together all the knowledge that we have collected in this book and apply the following reasoning. The receiver input signal is referenced relative to the noise floor. Depending upon the circuit bandwidth, there is additional $10\log\Delta f$ noise added into the system. Narrowband systems are the obvious conclusion; however, this opportunity for noise reduction can be exploited only so much. Therefore, for any bandwidth above 1Hz (17.24) is extended as

$$P_n = -174\,\mathrm{dBm} + 10\log\Delta f \quad [\mathrm{dBm}] \tag{17.27}$$

Progressing through the receiver circuit, the internally generated noise is quantified by the noise figure NF, which needs to be added into the noise budget, hence

$$P_n = -174\,\mathrm{dBm} + 10\log\Delta f + NF \quad [\mathrm{dBm}] \tag{17.28}$$

which sets the "real" noise floor for the receiver. In order to be useful, the receiver must be able to process signals above the real noise floor; in other words, it has to be designed for a certain desired signal-to-noise ratio, $SNR_{desired}$.

We now define the receiver sensitivity (S) as the signal level

$$S_n = P_n + SNR_{desired} \quad [\mathrm{dBm}] \tag{17.29}$$

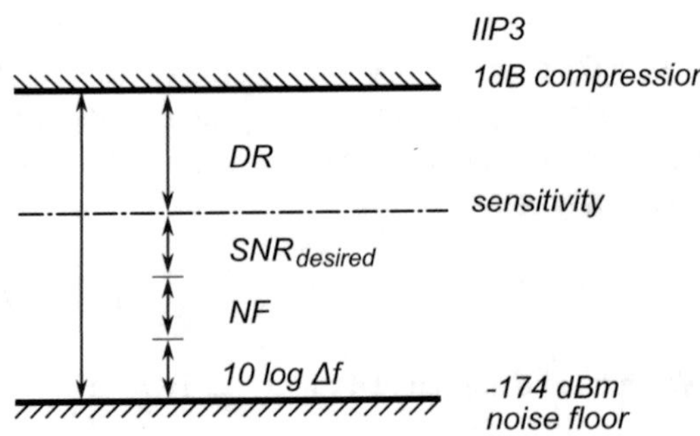

Fig. 17.12 Elements of dynamic range at room temperature

where P_n represents the level, for the given bandwidth Δf, at which the signal power is equal to the noise power. That is, the level is equivalent to the case when the SNR of the receiver is 0 dB (see Fig. 17.12). With this discussion in mind, we define the ideal dynamic range as the difference between the 1dB compression point and the receiver's sensitivity, i.e.

$$DR = 1\,\text{dB}_{point} - S_n \quad [\text{dB}] \tag{17.30}$$

which is a somewhat optimistic result. In practice, it is often adjusted by about 30% down to 2/3 DR. Clearly, it is a goal to design a receiver with as wide a dynamic range as possible. The current state of the art is about 100 dB.

Example 92: Receiver sensitivity
Determine the sensitivity of a receiver at room temperature whose $NF = 5\,\text{dB}$, $BW = 1\,\text{MHz}$, and desired $SNR = 10\,\text{dB}$

Solution 92: A straightforward implementation of (17.28) yields

$$S = -174\,\text{dBm} + 10\log 1\,\text{MHz} + 5\,\text{dB} + 10\,\text{dB} = -99\,\text{dBm}$$

which is a relatively typical number for state-of-the-art receivers.

17.4 Summary

Figures of merit serve the purpose of comparing various design solutions and looking for ways to improve them. Radio receivers deal with very low signal powers, cell phone for instance receives signals as low as −110 dBm. Thermal noise presents the lower power limit under which the desired signal becomes irreparably drowned in the background noise. On the upper side limit, nonlinear effects in the receiver circuit and the signal distortion become determining factors for establishing the dynamic range.

Problems

17.1 An AM receiver is designed to receive RF signals in the 500–1600 kHz frequency range with the required bandwidth of $BW = 10\,\text{kHz}$ at $f_0 = 1050\,\text{kHz}$. The RF amplifier uses inductor $L = 1\,\mu\text{H}$.

1. calculate bandwidth at $f = 1600\,\text{kHz}$ and capacitance C
2. calculate bandwidth at $f = 500\,\text{kHz}$ and capacitance C
3. comment on the results

17.2 An AM receiver is designed to receive RF signals in the 500–1600 kHz frequency range. All incoming RF signals are shifted to intermediate frequency $IF = 465\,\text{kHz}$. AM receiver tuning is commonly done by a knob that simultaneously tunes resonating capacitors in the RF and LO oscillator sections. For the receiver architecture in Fig. 17.13 (left) (matching network not shown),

(a) calculate tuning ratio $C_{RF}(max)/C_{RF}(min)$ of the resonator capacitor in the RF amplifier,
(b) calculate tuning ratio $C_{LO}(max)/C_{LO}(min)$ of the resonator capacitor in the local oscillator LO,
(c) recommend the resonating frequency for the local oscillator.

17.3 The LO oscillator's frequency is 11 MHz, and RF signal frequency is 10 MHz. What is the image frequency?

17.4 Input–output power characteristics of an amplifier is given in Fig. 17.13 (right). Estimate the gain, 1 dB compression point, and the third-order intercept point.

17.5 A receiver operates in the 3–30 MHz range while using 10.7 MHz IF frequency. Estimate the range of oscillator frequencies and the range of image frequencies. Can you suggest filters to be used with this receiver?

17.6 A double conversion receiver architecture is based on two IF frequencies, $IF_1 = 10.7\,\text{MHz}$ and $IF_2 = 455\,\text{kHz}$. If the receiver is tuned to a 20 MHz signal, find frequencies of the local oscillators and the image frequencies.

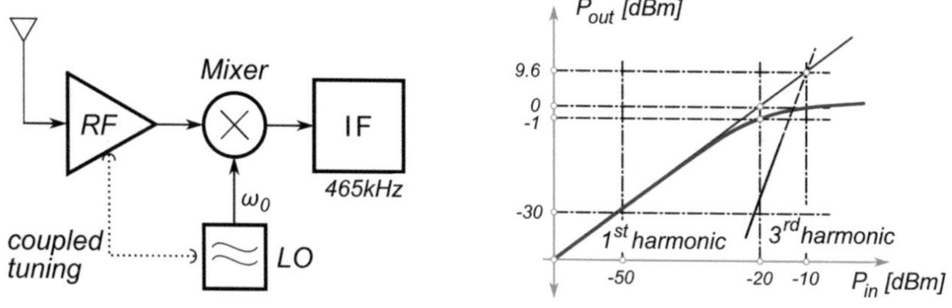

Fig. 17.13 AM receiver block diagram for Problem 17.2, and diagram for Problem 17.4

Physical Constants and Engineering Prefixes

A

Tables A.1, A.2, and A.3

Table A.1 Basic physical constants

Physical constant	Symbol	Value
Speed of light in vacuum	c	299 792 458 m/s
Magnetic constant (vacuum permeability)	μ_0	$4\pi \times 10^{-7}$ N/A^2
Electric constant (vacuum permittivity)	$\varepsilon_0 = 1/(\mu_0 c^2)$	$8.854\ 187\ 817 \times 10^{-12}$ F/m
Characteristic impedance of vacuum	$Z_0 = \mu_0 c$	376.730 313 461 Ω
Coulomb's constant	$k_e = 1/4\pi\varepsilon_0$	$8.987\ 551\ 787 \times 10^{9}$ Nm^2/C^2
Elementary charge	e	$1.602\ 176\ 565 \times 10^{-19}$ C
Boltzmann constant	k	$1.380\ 6488 \times 10^{-23}$ J/K

Table A.2 Basic engineering prefix system

Tera	Giga	Mega	Kilo	Hecto	Deca	Deci	Centi	Milli	Micro	Nano	Pico	Femto	Atto
T	G	M	k	h	da	d	c	m	μ	n	p	f	a
10^{12}	10^{9}	10^{6}	10^{3}	10^{2}	10^{1}	10^{-1}	10^{-2}	10^{-3}	10^{-6}	10^{-9}	10^{-12}	10^{-15}	10^{-18}

Table A.3 SI system of fundamental units

Name	Symbol	Quantity	Symbol
Metre	m	Length	l
Kilogram	kg	Mass	m
Second	s	Time	t
Ampere	A	Electric current	I
Kelvin	K	Thermodynamic temperature (−273.16 °C)	T
Candela	cd	Luminous intensity	Iv
Mole	mol	Amount of substance	n

R. Sobot, *Wireless Communication Electronics*,
https://doi.org/10.1007/978-3-030-48630-3

Second-Order Differential Equation

B

The three basic elements have voltages at their respective terminals as:

$$v_R = i R \quad ; \quad v_L = L \frac{di}{dt} \quad ; \quad v_C = \frac{q}{C} \tag{B.1}$$

If they are put together in a series circuit that includes a voltage source $v(t)$, after applying KVL, the circuit equation is

$$v(t) = v_L + v_R + v_C$$

$$\therefore$$

$$v(t) = L \frac{di}{dt} + i\ R + \frac{q}{C} \tag{B.2}$$

However, we know that a current is a derivative of charge with respect to time, hence we have the second-order differential equation

$$v(t) = L \frac{d^2q}{dt^2} + R \frac{dq}{dt} + \frac{1}{C} q$$

$$\therefore$$

$$v(t) = \frac{d^2q}{dt^2} + \frac{R}{L} \frac{dq}{dt} + \frac{1}{LC} q \tag{B.3}$$

This is solved, starting with its auxiliary quadratic equation

$$0 = x^2 + \frac{R}{L} x + \frac{1}{LC} \tag{B.4}$$

and its general solution with complex roots are

$$r_{1\,2} = \frac{1}{2} \left(-\frac{R}{L} \pm \sqrt{\left(\frac{R}{L}\right)^2 - \frac{4}{LC}} \right) \tag{B.5}$$

R. Sobot, *Wireless Communication Electronics*,
https://doi.org/10.1007/978-3-030-48630-3

Complex Numbers

C

A complex number is a neat way of presenting a point in (mathematical) *space* with two coordinates or, equivalently, it is a neat way to write two equations in the form of one. A general complex number is $Z = a + jb$, where a and b are real numbers referred to as real and imaginary parts, i.e. $\Re(Z) = a$ and $\Im(Z) = b$. Here is a reminder of the basic operations with complex numbers. Keep in mind that $j^2 = -1$.

$$(a + jb) + (c + jd) = (a + c) + j(b + d) \tag{C.1}$$

$$(a + jb) - (c + jd) = (a - c) + j(b - d) \tag{C.2}$$

$$(a + jb)\,(c + jd) = (ac - bd) + j(bc + ad) \tag{C.3}$$

$$\frac{(a + jb)}{(c + jd)} = \frac{(a + jb)}{(c + jd)}\,\frac{(c - jd)}{(c - jd)} = \frac{ac + bd}{c^2 + d^2} + j\,\frac{bc - ad}{c^2 + d^2} \tag{C.4}$$

$$(a + jb)^* = (a - jb) \tag{C.5}$$

$$|(a + jb)| = \sqrt{(a + jb)(a - jb)} = \sqrt{(a^2 + b^2)} \tag{C.6}$$

It is much easier to visualize complex numbers and operations if we use vectors and the trigonometry of a right triangle, i.e. Pythagoras' theorem. The imaginary part always takes its value from the y axis and the real part is always on the x axis (see Fig. C.1).

Therefore, an alternative view of complex numbers is based on geometry, i.e.

$$(a + jb) \equiv (|Z|\;\theta) \tag{C.7}$$

where, of course, the absolute value of Z is the length of the hypotenuse and the real and imaginary parts are the two legs of the right-angled triangle, i.e.

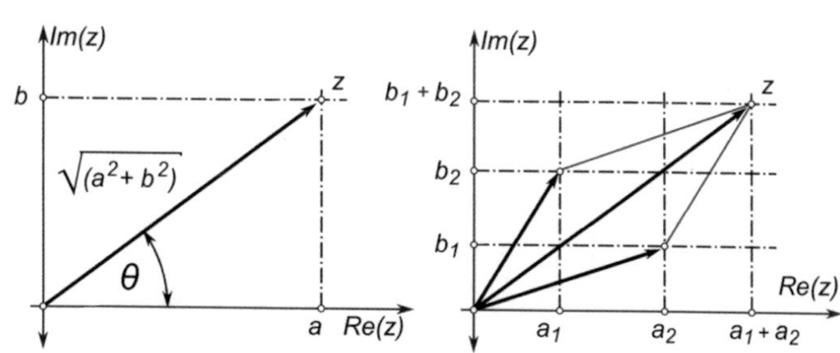

Fig. C.1 Complex numbers in $[\Re(Z)\;\Im(Z)]$ space, their equivalence to Pythagoras' theorem and vector arithmetic

R. Sobot, *Wireless Communication Electronics*,
https://doi.org/10.1007/978-3-030-48630-3

$$|Z| = \sqrt{Z\,Z^*} = \sqrt{(a^2 + b^2)} \qquad \theta = \arctan\left(\frac{b}{a}\right) \tag{C.8}$$

where θ is the phase angle. After using Euler's formula, this becomes

$$e^{jx} \equiv \cos x + j \sin x \tag{C.9}$$

and enables us to write a really compact form of complex numbers

$$Z = a + jb = |Z|\, e^{j\theta} \tag{C.10}$$

which leads into another simple way of doing complex arithmetic, by using the absolute values and the arguments in combination with the algebraic rules of exponential numbers, for example

$$\left(A\, e^{j\theta_A}\right)\, \left(B\, e^{j\theta_B}\right) = AB\, e^{j(\theta_A + \theta_B)} \tag{C.11}$$

and we have the final link,

$$A\, e^{j\theta} \equiv A\, (\cos\theta + j \sin\theta) \tag{C.12}$$

where

$$\Re\left(A\, e^{j\theta}\right) = A\, \cos\theta \qquad \Im\left(A\, e^{j\theta}\right) = A\, \sin\theta \tag{C.13}$$

Basic Trigonometric Identities

D

$$\sin(\alpha + {}^{\pi}/_{2}) = +\cos\alpha \tag{D.1}$$

$$\cos(\alpha + {}^{\pi}/_{2}) = -\sin\alpha \tag{D.2}$$

$$\sin(\alpha + \pi) = -\sin\alpha \tag{D.3}$$

$$\cos(\alpha + \pi) = -\cos\alpha \tag{D.4}$$

$$\sin(\alpha \pm \beta) = \sin\alpha\cos\beta \pm \cos\alpha\sin\beta \tag{D.5}$$

$$\cos(\alpha \pm \beta) = \cos\alpha\cos\beta \mp \sin\alpha\sin\beta \tag{D.6}$$

$$\cos\alpha\cos\beta = {}^{1}/_{2}\,(\cos(\alpha - \beta) + \cos(\alpha + \beta)) \tag{D.7}$$

$$\sin\alpha\sin\beta = {}^{1}/_{2}\,(\cos(\alpha - \beta) - \cos(\alpha + \beta)) \tag{D.8}$$

$$\sin\alpha\cos\beta = {}^{1}/_{2}\,(\sin(\alpha + \beta) + \sin(\alpha - \beta)) \tag{D.9}$$

$$\cos\alpha\sin\beta = {}^{1}/_{2}\,(\sin(\alpha + \beta) - \sin(\alpha - \beta)) \tag{D.10}$$

$$\sin^2\alpha = {}^{1}/_{2}\,(1 - \cos 2\alpha) \tag{D.11}$$

$$\cos^2\alpha = {}^{1}/_{2}\,(1 + \cos 2\alpha) \tag{D.12}$$

$$\sin^3\alpha = {}^{1}/_{4}\,(3\sin\alpha - \sin 3\alpha) \tag{D.13}$$

$$\cos^3\alpha = {}^{1}/_{4}\,(3\cos\alpha + \cos 3\alpha) \tag{D.14}$$

$$\sin^2\alpha\cos^2\alpha = {}^{1}/_{8}\,(1 - \cos 4\alpha) \tag{D.15}$$

$$\sin^3\alpha\cos^3\alpha = {}^{1}/_{32}\,(3\sin 2\alpha - \sin 6\alpha) \tag{D.16}$$

$$\sin\alpha \pm \sin\beta = 2\sin\left(\frac{\alpha \pm \beta}{2}\right)\cos\left(\frac{\alpha \mp \beta}{2}\right) \tag{D.17}$$

$$\cos\alpha + \cos\beta = 2\cos\left(\frac{\alpha + \beta}{2}\right)\cos\left(\frac{\alpha - \beta}{2}\right) \tag{D.18}$$

$$\cos\alpha - \cos\beta = -2\sin\left(\frac{\alpha + \beta}{2}\right)\sin\left(\frac{\alpha - \beta}{2}\right) \tag{D.19}$$

R. Sobot, *Wireless Communication Electronics*,
https://doi.org/10.1007/978-3-030-48630-3

E Useful Algebraic Equations

1. *Binomial formula*

$$(x \pm y)^2 = x^2 \pm 2xy + y^2 \tag{E.1}$$

$$(x \pm y)^3 = x^3 \pm 3x^2y + 3xy^2 \pm y^3 \tag{E.2}$$

$$(x \pm y)^4 = x^4 \pm 4x^3y + 6x^2y^2 \pm 4xy^3 + y^4 \tag{E.3}$$

$$(x \pm y)^n = x^n + nx^{n-1} + \frac{n(n-1)}{2!}x^{n-2}y^2 + \frac{n(n-1)(n-2)}{3!}x^{n-3}y^{3\circ} \cdots + y^n \tag{E.4}$$

where $n! = 1, 2, 3, \ldots n$ and $0! \overset{\text{def}}{=} 1$.

2. *Special cases*

$$x^2 - y^2 = (x - y)(x + y) \tag{E.5}$$

$$x^3 - y^3 = (x - y)(x^2 + xy + y^2) \tag{E.6}$$

$$x^3 + y^3 = (x + y)(x^2 - xy + y^2) \tag{E.7}$$

$$x^4 - y^4 = (x^2 - y^2)(x^2 + y^2) = (x - y)(x + y)(x^2 + y^2) \tag{E.8}$$

3. *Useful Taylor series*

$$e^x = \sum_{n=0}^{\infty} \frac{x^n}{n!} = 1 + x + \frac{x^2}{2!} + \frac{x^3}{3!} + \cdots \tag{E.9}$$

$$\sin x = \sum_{n=0}^{\infty} \frac{(-1)^n}{(2n+1)!}x^{2n+1} = x - \frac{x^3}{3!} + \frac{x^5}{5!} - \cdots \quad \text{odd, for all } x \tag{E.10}$$

$$\cos x = \sum_{n=0}^{\infty} \frac{(-1)^n}{(2n)!}x^{2n} = 1 - \frac{x^2}{2!} + \frac{x^4}{4!} - \cdots \quad \text{even, for all } x \tag{E.11}$$

$$\tan x = \sum_{n=1}^{\infty} \frac{B_{2n}(-4)^n(1-4^n)}{(2n)!}x^{2n-1} = x + \frac{x^3}{3} + \frac{2x^5}{15} + \cdots \quad \text{for } |x| < \frac{\pi}{2} \tag{E.12}$$

R. Sobot, *Wireless Communication Electronics*,
https://doi.org/10.1007/978-3-030-48630-3

Bessel Polynomials

F

1. *Bessel differential equation*

$$x^2\frac{d^2y}{dx^2} + x\frac{dy}{dx} + \left(x^2 - \alpha^2\right) y = 0 \tag{F.1}$$

2. *Relation with trigonometric functions*

$$\cos(x \sin\alpha) = J_0(x) + 2\ [J_2(x)\cos 2\alpha + J_4(x)\cos 4\alpha + \cdots] \tag{F.2}$$

$$\sin(x \sin\alpha) = 2\ [J_1(x)\sin\alpha + J_3(x)\sin 3\alpha + J_5(x)\sin 5\alpha + \cdots] \tag{F.3}$$

$$\cos(x \cos\alpha) = J_0(x) - 2\ [J_2(x)\cos 2\alpha - J_4(x)\cos 4\alpha + J_6(x)\cos 6 - J_8(x)\cos 8\alpha \cdots] \tag{F.4}$$

$$\sin(x \cos\alpha) = 2\ [J_1(x)\cos\alpha - J_3(x)\sin 3\alpha + J_5(x)\sin 5\alpha + \cdots] \tag{F.5}$$

3. *Bessel series*

$$J_0(x) = 1 - \frac{x^2}{2^2} + \frac{x^4}{2^2 \ldots 4^2} - \frac{x^6}{2^2 \ldots 4^2 \ldots 6^2} + \cdots \tag{F.6}$$

$$J_1(x) = \frac{x}{2}\left[1 - \frac{x^2}{2^2 \cdots 2} + \frac{x^4}{2 \cdots 2^4 \cdots 2 \cdots 3} + \cdots\right] \tag{F.7}$$

$$J_n(x) = \frac{x^n}{2^n n!}\left[1 - \frac{x^2}{2^2 \cdots (n+1)} + \frac{x^4}{2 \cdots 2^4 \cdots (n+1) \cdots (n+2)} \right.$$
$$\left. + \frac{(-1)^p x^{2p}}{p!\ 2^{2p}\ (n+1)(n+2)\cdots(n+p)} + \cdots\right] \tag{F.8}$$

4. *Bessel approximations*
   For very large x Bessel function reduces to

$$J_n(x) = \sqrt{\frac{2}{\pi\ x}}\cos\left(x - \frac{n\ \pi}{2} - \frac{\pi}{4}\right) \tag{F.9}$$

R. Sobot, *Wireless Communication Electronics*,
https://doi.org/10.1007/978-3-030-48630-3

Glossary

This glossary of technical terms is provided for reference only. The reader is advised to further study the terms in appropriate books, for example, a technical dictionary.

1 dB gain compression point The point at which the power gain at the output of a nonlinear device or circuits is reduced by 1 dB relative to its small signal linear model predicted value.

Absolute zero The theoretical temperature at which entropy would reach its minimum value. By international agreement, absolute zero is defined as 0 K on the Kelvin scale and as $-273.15\,°C$ on the Celsius scale.

Active device An electronic component that has signal gain larger than one, for example, a transistor. Compare to *passive device*.

Active mode Condition for a BJT transistor where emitter-base junction is forward biased, while the collector-base junction is reverse biased.

Admittance The measure of how easily AC current flows in a circuit (in Siemens [S]). The reciprocal of *impedance*.

Ampere [A] The unit of electric current defined as the flow of one coulomb of charge per second.

Ampère's law A current flowing into a wire generates a magnetic flux that encircles the wire following the "right hand rule" (the thumb points the direction of the current flow and the other curled fingers show direction of the magnetic field). Study *Maxwell's equations* for more details.

Amplifier A liner device that implements mathematical equation $y = A\,x$, where y is the amplified output signal, A is the gain coefficient, and x is the input signal.

Analogue The general class of devices and circuits meant to process a continuous signal. Compare with *digital* and sampled signals.

Attenuation Gain lower than one.

Attenuator A device that reduces a gain without introducing phase or frequency distortion.

R. Sobot, *Wireless Communication Electronics*,
https://doi.org/10.1007/978-3-030-48630-3

Automatic gain control A closed loop feedback system designed to hold the overall gain as constant as possible.

Average power The power averaged over one time period.

Bandwidth The difference between upper and lower frequencies at which the amplitude response is 3 dB below the maximum level. It is equivalent to *half-power* bandwidth.

Base The region of BJT transistor between the emitter and the collector.

Bel [B] is a dimensionless unit used to express the *ratio* of two powers. A more practical unit is the [dB].

Beta β The current gain of a BJT. It is the ratio of the change in collector current to the change in the base current, $\beta = dI_C/dI_B$.

Bias A steady current or voltage used to set the operating conditions of a device.

Breakdown voltage The voltage at which the reverse current of a reverse biased p-n junction suddenly rises. If the current is not limited, the device is destroyed.

Capacitance The ratio of the electric charge and voltage between two conductors.

Capacitor A device made of two conductors separated by an insulating material for the purpose of storing electric charge, i.e., energy.

Celsius [°C] A unit increment of temperature unit defined as 1/100 between the freezing point (0 °C) and boiling point (100 °C) of water. Compare with Kelvin and Fahrenheit.

Characteristic curve A family of I–V plots shown for several parameter values.

Characteristic impedance The entry point impedance of an infinitely long transmission line.

Charge A basic property of elementary particles of matter (electrons, protons, etc.) responsible for creating a force field.

Circuit The interconnection of devices, both passive and active, for the purpose of synthesizing a mathematical function.

Common-base A single BJT amplifier configuration in which the base potential is fixed, the emitter serves as the input and the collector as the output terminal. Also known as "current buffer". Equivalent to a "common-gate" configuration for MOS amplifiers.

Common-collector A single BJT transistor configuration in which the collector potential is fixed, the base serves as the input and the emitter as the output terminal. Also known as "voltage buffer" or voltage follower. Equivalent to a "common-drain" configuration for MOS amplifiers.

Common-emitter A single BJT amplifier configuration in which the emitter potential is fixed, while the base serves as the input and the collector as the output terminal. Also known as the g_m stage. Equivalent to a "common-source" configuration for MOS amplifiers.

Common mode The average value of a sinusoidal waveform.

Conductivity The ability of a matter to conduct electricity.

Conductor A material that easily conducts electricity.

Coulomb [C] The unit of electric charge defined as the charge transported through a unity area in 1 s by an electric current of one ampere. An electron has a charge of 1.602×10^{-19} C.

Coulomb's law A definition of the force between two electric charges in space.

Current A transfer of electrical charge through a unity size area per unit of time.

Current gain The ratio of current at the output terminals to the current at the input terminals of a device or circuit.

Current source A device capable of provided constant current value regardless of the voltage at its terminals.

DC See *Direct current.*

DC analysis A mathematical procedure to calculate the stable operating point.

DC biasing The process of setting the stable operating point of a device.

DC load line A straight line across a family of I–V curves that shows movement of the operating point as the output voltage changes for a given load.

Decibel [dB] A dimensionless unit used to express the ratio of two powers. A decibel is ten times smaller than a *bel* [B].

Device A single discrete device, for instance, a resistor, a transistor, or a capacitor.

Dielectric A material that is not good in conducting electricity, i.e. the opposite of a conductor. Characterized by the dielectric constant.

Differential amplifier An amplifier that operates on differential signals.

Differential signal A difference between two sinusoidal signals of the same frequency, same amplitude, same common mode, and with phase difference of 180°.

Digital The general class of devices and circuits meant to process a sampled signal. Compare with *analogue* and continuous signals.

Diode A nonlinear, two-terminal device that obeys the exponential transfer function. Used as unidirectional switch.

Direct current (DC) Current that flows in one direction only.

Discrete device An individual electrical component that exhibits behaviour associated with a resistor, a transistor, a capacitor, an inductor, etc. Compare with *distributed* components.

Dynamic range The difference between the maximum acceptable signal level and the minimum acceptable signal level.

Electric field An energy field generated by an electric charge, detected by the existence of the electric force within a space surrounding the charge.

Electrical noise Any unwanted electrical signal.

Electromagnetic (EM) wave A phenomenon exhibited by a flow of electromagnetic energy through space. In the special case of a *standing* wave this definition may need more explanation.

Electron A fundamental particle that carries negative charge.

Electronics The branch of science and technology that makes use of the controlled motion of electrons through different media and a vacuum.

Electrostatics The branch of science that deals with the phenomena arising from stationary or slow-moving electric charges.

Emitter A region of a BJT from which charges are injected into the base. One of the three-terminal points of a BJT device.

Energy A concept that can be loosely defined as the ability of a body to perform work.

Equivalent circuit A simplified version of a circuit that performs the same function as the original.

Equivalent noise temperature The absolute temperature at which a perfect resistor would generate the same noise as its equivalent real component at room temperature.

Fall time The time during which a pulse decreases from 90% to 10% of its maximum value (sometimes defined between the 80% and 20% points).

Farad [F] The unit of capacitance of a capacitor. One farad is very large; the capacitance of the Earth's ionosphere with respect to the ground is around 50 mF.

Faraday cage An enclosure that blocks out external static electric fields.

Faraday's law The law of electromagnetic induction. See also *Faraday's cage*.

Feedback The process of coupling output and input terminals through an external path. Negative feedback increases the stability of an amplifier for the cost of reduced gain, positive feedback boosts gain and is needed to create oscillating circuits.

Field A concept that describes a flow of energy through space.

Field-effect transistor (FET) A transistor controlled by two perpendicular electrical fields used to change resistivity of the semiconductor material underneath the gate terminal and to force current between the source and drain terminals.

Flicker noise A random noise in semiconductors whose power spectral density is, to the first approximation, inverse to frequency (1/f noise.).

Frequency The number of complete cycles per second.

Frequency response A curve showing the gain and phase change of a device as a function of frequency.

Gain The ratio of signal values measured at output and input terminals.

Gauss's law A law relating the distribution of electric charge to the resulting electric field.

Ground An arbitrary potential reference point that all other potentials in a circuit are compared against. The difference between the ground potential and the node potential is expressed as voltage at that node. The ground node may or may not be the lowest potential in the circuit.

Henry [H] The unit of measurement for self and mutual inductance.

Hertz [Hz] The unit of measurement for frequency, equal to one cycle per second.

Impedance Resistance of a two-terminal device at any frequency.

Inductance A property whereby a change in the electric current through a circuit induces an electromotive force (EMF) that opposes the change in current.

Inductor A passive electrical component that can store energy in a magnetic field created by an electric current passing through it.

Input Current, voltage, power, or other driving force applied to a circuit or device.

Insertion loss The attenuation resulting from inserting a circuit between source and load.

Insulator A material with very low conductivity.

Intermediate frequency (IF) A frequency to which a carrier frequency is shifted as an intermediate step in transmission or reception.

Intermodulation products Additional harmonics created by a nonlinear device processing two or more single tone signals.

Junction A joining of two semiconductor materials.

Junction capacitance Capacitance associated with p-n junction region.

Kelvin [K] The unit increment of temperature on the absolute temperature scale.

Kirchhoff's current law (KCL) The law of conservation of charge: at any instant, the total current entering any point in a network is equal to the total current leaving the same point.

Kirchhoff's voltage law (KVL) The law of conservation of energy given or taken by a potential field (not including energy taken by dissipation): at any instant, the algebraic sum of all electromotive forces and potential differences around a closed loop is zero.

Large signal A signal with an amplitude large enough to move the operating point of a device far away from its original biasing point. Hence, nonlinear model of the device must be used.

Large signal analysis A method used to describe the behaviour of devices stimulated by large signals. It describes nonlinear devices in terms of the underlying nonlinear equations.

Law of conservation of energy The fundamental law of nature. It states that energy can be neither created nor destroyed, it can only be transformed from one state to another.

Linear network A network in which the parameters of resistance, inductance, and capacitance are constant with respect to current or voltage, and in which the voltage or current of sources is independent of or directly proportional to other voltages and currents, or their derivatives, in the network.

Load A device that absorbs energy and converts it into another form.

Local oscillator (LO) An oscillator used to generate a single-tone signal that is needed for upconversion and downconversion operations.

Lossless A theoretical device that does not dissipate energy.

Low noise amplifier (LNA) An electronic amplifier used to amplify very weak signals captured by an antenna.

Lumped element A self-contained and localized element that offers one particular property, for example, resistance over a range of frequencies.

Magnetic field A field generated by magnetic energy, detected by the existence of a magnetic force within space surrounding a magnet.

Matching A concept of connecting two networks for purpose to enable maximum energy transfer between them.

Matching circuit A passive circuit designed to interface two networks to enable maximum energy transfer between the two networks.

Maxwell's equations A set of four partial differential equations that relate electric and magnetic fields to their sources, charge density, and current density. These equations can be combined to show that light is an electromagnetic wave. Individually, the equations are known as Gauss's law, Gauss's law for magnetism, Faraday's law of induction, and Ampère's law with Maxwell's correction. These four equations and the Lorentz force law make up the complete set of laws of classical electromagnetism.

Metal-oxide-semiconductor field-effect transistor (MOSFET) Originally, a sandwich of aluminium–silicone dioxide–silicon was used to manufacture FET transistors. Although metal is no longer used to create gates for FET transistors, the name has stuck.

Microwaves Waves in the frequency range of 1–300 GHz, i.e. with a wavelength of 300–1 mm.

Mixer A nonlinear three-port device used for frequency-shifting operation.

Negative resistance The resistance of a device or circuit where an increase in the current entering a port results in a decrease in voltage across the same port.

Noise Any unwanted signal that interferes with a wanted signal.

Noise figure (NF) A measure of degradation of the signal-to-noise ratio (SNR), caused by the internal noise generated by components in a radio frequency (RF) signal chain.

Nonlinear circuit A system that does not satisfy the superposition principle or whose output is not directly proportional to its input.

Norton's theorem Any collection of voltage sources, current sources, and resistors with two terminals is electrically equivalent to an ideal current source in parallel with a single resistor. This is the twin of *Th'evenin's theorem.*

NPN transistor A transistor with p-type base and n-type collector and emitter.

Octave The interval between any two frequencies having a ratio of 2:1

Ohm [Ω] Unit of resistance, as defined by Ohm's law.

Ohm's law The change of current through a conductor between two points is directly proportional to the change of voltage across the two points and inversely proportional to the resistance between them.

Open-loop gain The ratio of the output signal and the input signals of an amplifier with no feedback path present.

Oscillator An electronic device that generates a single tone (or some other regular shape) signal at predetermined frequency.

Output Current, voltage, power, or driving force delivered at the output terminals.

Passive device A component that does not have gain larger than one. Compare to *active device*.

Phase The angular property of a wave.

Phase shifter A two-port network that provides a controllable phase shift of the RF signals.

Phasor A mathematical representation of a sine wave by a rotating vector.

Power The rate at which work is performed.

Quality factor (Q factor) A dimensionless parameter that characterizes a resonator's bandwidth relative to its centre frequency.

Radio frequency (RF) Any frequency at which coherent radiation of energy is possible.

Reactance The opposition of a circuit element to a change of current, caused by the build-up of electric or magnetic fields in the element.

Reactive element An inductor and capacitor.

Reflected waves The waves reflected from a discontinuity in the medium in which they are traveling.

Resistance A measure of an object's opposition to the passage of a steady electric current.

Resistor A lumped element designed to have a certain resistance.

Resonant frequency The frequency at which a given system or circuit responds with maximum amplitude when driven by an external single tone.

Root mean square (RMS) The square root of the arithmetic mean (average) of the squares of the original values.

Saturation A condition in which an increase of the input signal to a circuit does not produce an expected linearly-proportional change at the output.

Self-resonant frequency The frequency at which all real devices or circuits start to oscillate due to the internal parasitic inductances and capacitances.

Signal An electrical quantity containing information that is carried by a voltage or current.

Single-ended circuit A circuit operating on single-ended (as opposed to differential) signals.

Skin effect The tendency of an alternating current (AC) to distribute itself within a conductor so that the current density near the surface of the conductor is greater than at its core. That is, the electric current tends to flow at the "skin" of the conductor, at an average depth called the "skin depth".

Small signal A low-amplitude signal that occupies a very narrow region that is centred at the biasing point. Hence, linear model always applies.

Small-signal amplifier An amplifier that operates only in the linear region.

Space The boundless, three-dimensional extent in which objects and events occur and have relative position and direction.

Stability The ability of a circuit to stay away from the self-resonating frequency.

Standing wave A wave whose maximum and minimum points remain at a constant position. It can arise in a stationary medium as a result of interference between two waves traveling in opposite directions. For waves of equal amplitude traveling in opposite directions, there is on average no net propagation of energy.

Standing wave ratio (SWR) The ratio of the maximum to the minimum value of current or voltage in a standing wave.

Thévenin's theorem Any combination of voltage sources, current sources, and resistors with two terminals is electrically equivalent to a single voltage source and a single series resistor. This is a twin of *Norton's theorem*.

Third-order intercept point (IP3) A measure of weakly nonlinear systems and devices, for example, receivers, linear amplifiers, and mixers.

Time A concept used to order a sequence of events.

Transmission line Any system of conductors capable of efficiently conducting electromagnetic energy.

Tuned circuit A circuit consisting of inductance and capacitance that can be adjusted for resonance at the desired frequency.

Tuning The process of adjusting resonant frequency of a *tuned circuit*.

Varactor A two-terminal p-n junction used as a voltage-controlled capacitor.

Volt [V] A unit of measurement for potential difference.

Voltage-controlled oscillator (VCO) An oscillator whose output frequency is controlled by a voltage.

Voltage divider A simple linear circuit that produces an output voltage that is a fraction of its input voltage.

Voltage follower amplifier An amplifier that provides electrical impedance transformation from one circuit to another. Also known as a "voltage buffer amplifier".

Voltage source A device capable of providing a constant voltage value regardless of the current at its terminals.

Wave A disturbance that progresses from one point in space to another.

Wavefront A cross-sectional surface having constant phase.

Wavelength A distance in space between two consecutive points having the same phase.

Wave propagation The journey of a wave through space.

White noise A random signal that consists of all possible frequencies from zero to infinity.

Work The advancement in space of a point under application of a force.

References

G

[Amo90] S.W. Amos, *Principles of Transistor Circuits* (Butterworths, London, 1990). Number 0-408-04851-4

[BMV05] J.S. Beasley, G.M. Miller, J.K. Vasek, *Modern Electronic Communication* (Prentice Hall, Pearson, 2005). Number 0-13-113037-4

[BG03a] L. Besser, R. Gilmore, *Practical RF Circuit Design for Modern Wireless Systems I* (Artech House, Boston, 2003). Number 1-58053-521-6

[BG03b] L. Besser, R. Gilmore, *Practical RF Circuit Design for Modern Wireless Systems II* (Artech House, Boston, 2003). Number 1-58053-522-4

[Bro90] J.J. Brophy, *Basic Electronics for Scientist* (McGraw-Hill, New York, 1990). Number 0-07-008147-6

[Bub84] P. Bubb, *Understanding Radio Waves* (Lutterworth Press, Glasgow, 1984). Number 0-7188-2581-0

[CC03a] D. Comer, D. Comer, *Advanced Electronic Circuit Design* (Wiley, London, 2003). Number 0-471-22828-1

[CC03b] D. Comer, D. Comer, *Fundamentals of Electronic Circuit Design* (Wiley, London, 2003). Number 0-471-41016-0

[CL62] D.R. Corson, P. Lorrain, *Introduction to Electromagnetic Fields and Waves* (Freeman, New York, 1962). Number 62-14193

[DA01] W.A. Davis, K.K. Agarwal, *Radio Frequency Circuit Design* (Wiley Interscience, New York, 2001). Number 0-471-35052-4

[DA07] W.A. Davis, K.K. Agarwal, *Analysis of Bipolar and CMOS Amplifiers* (CRC Press, West Palm Beach, 2007). Number 1-4200-4644-6

[Ell66] R.S. Elliott, *Electromagnetics* (McGraw Hill, New York, 1966). Number 66-14804

[FLS05] R.P. Feynman, R.B. Leighton, M. Sands, *The Feynman Lectures on Physics* (Pearson Addison Wesley, Reading, 2005). Number 0-8053-9047-2

[Fle08] D. Fleisch, *A Student's Guide to Maxwell's Equations* (Cambridge University, Cambridge, 2008). Number 978-0-521-87761-9

[Gol48] S. Goldman, *Frequency Analysis, Modulation and Noise* (McGraw-Hill, New York, 1948). Number TK6553.G58 1948

[JN71] R.H. Good Jr., T.H. Nelson, *Classical Theory of Electric and Magnetic Fields* (Academic Press, New York, 1971). Number 78-137-628

[GM93] P.G. Gray, R.G. Meyer, *Analysis and Design of Analog Integrated Circuits* (Wiley, New York, 1993). Number 0-471-57495-3

[Gre04] B. Green, *The Fabric of Cosmos* (Vintage Books, New York, 2004). Number 0-375-72720-5

[Gri84] J. Gribbin. *In Search of Schrödinger's Cat, Quantum Physics and Reality* (Bantam Books, New York, 1984). Number 0-553-34253-3

[HH89a] T.C. Hayes, P. Horowitz, *Student Manual for the Art of Electronics* (Cambridge University, Cambridge, 1989). Number 0-521-37709-9

[Jr.89] W.H Hayt Jr., *Engineering Electromagnetics* (McGraw Hill, New York, 1989). Number 0-07-024706-1

[JK93] W.H. Hayt Jr., J.E. Kemmerly, *Engineering Circuit Analysis* (McGraw Hill, New York, 1993). Number 0-07-027410-X

[HH89b] P. Horowitz, W. Hill, *The Art of Electronics* (Cambridge University, Cambridge, 1989). Number 0-521-37095-7

[Hur10] P.G. Huray, *Maxwell's Equations* (Wiley, New York, 2010). Number 978-0-470-54276-7

[II99] U.S. Inan, A.S. Inan, *Electromagnetic Waves* (Prentice Hall, Englewood, 1999). Number 0-201-36179-5

[Kin09] G.C. King, *Vibrations and Waves* (Wiley, New York, 2009). Number 978-0-470-01189-8

[Kon75] J.A. Kong, *Theory of Electromagnetic Waves* (Wiley, New York, 1975). Number 0-471-50190-5

R. Sobot, *Wireless Communication Electronics*,
https://doi.org/10.1007/978-3-030-48630-3

[KB80] H.L. Krauss, C.W. Bostian, *Solid State Radio Engineering* (Wiley, New York, 1980). Number 0-471-03018-X

[Lee05] T.H. Lee, *The Design of CMOS Radio-Frequency Integrated Circuits* (Cambridge University Press, Cambridge, 2005). Number 0-521-63922-0

[Lov66] W.F. Lovering, *Radio Communication* (Longmans, London, 1966). Number TK6550.L546 1966

[LB00] R. Ludwig, P. Bretchko, *RF Circuit Design, Theory and Applications* (Prentice Hall, Englewood, 2000). Number 0-13-095323-7

[PP99] Z. Popovic, D. Popovic, *Electromagnetic Waves* (Prentice Hall, New York, 1999). Number 0-201-36179-5

[Pur85] E.M. Purcell, *Electricity and Magnetism* (McGraw Hill, New York, 1985). Number 0-07-004908-4

[Rad01] M.M. Radmanesh, *Radio Frequency and Microwave Electronics* (Prentice Hall, New York, 2001). Number 0-13-027958-7

[Raz98] B. Razavi, *RF Microelectronics* (Prentice Hall, New York, 1998). Number 0-13-887571-5

[RR67] J.H. Reyner, P.J. Reyner, *Radio Communication* (Pitman, London, 1967)

[RC84] D. Roddy, J. Coolen, *Electronic Communications* (Reston Publishing Company, Reston, 1984). Number 0-8359-1598-0

[Rut99] D.B. Rutledge, *The Electronics of Radio* (Cambridge University Press, Cambridge, 1999). Number 0-521-64136-5

[Sch92] R.J. Schoenbeck, *Electronic Communications Modulation and Transmission* (Prentice Hill, New York, 1992). Number 0-675-21311-8

[Scr84] M.G. Scroggie, *Foundations of Wireless and Electronics*, 10th edn. (Newnes Technical Books, London, 1984). Number 0-408-01202-1

[See56] S. Seely, *Radio Electronics* (McGraw Hill, New York, 1956). Number 55-5696

[Sim87] R.E. Simpson, *Introductory Electronics for Scientist and Engineers* (Allyn and Bacon, Boston, 1987). Number 0-205-08377-3

[SB00] B. Streetman, S. Banerjee, *Solid State Electronic Devices* (Prentice Hall, New York, 2000). Number 0-13-025538-6

[Sze81] S.M. Sze, *Physics of Semiconductor Devices* (Wiley, New York, 1981). Number 0-471-05661-8

[Ter03] D. Terrell, *Electronics for Computer Technology* (Thompson Delmar Learning, Clinton Park, 2003). Number 0-7668-3872-2

[Tho06] M.T. Thompson, *Intuitive Analog Circuit Design* (Newnes, London, 2006). Number 0-7506-7786-4

[Wik10a] Wikipedia.org., *Electromagnetic Wave Equation* (2010). http://en.wikipedia.org/wiki/Electromagnetic_wave_equation

[Wik10b] Wikipedia.org., *Waves, Wavelength* (2010). http://en.wikipedia.org/wiki/Wave

[Wol91] D.H. Wolaver, *Phase-Locked Loop Circuit Design* (Prentice Hall, New York, 1991). Number 0-13-662743-9

[You04] P.H. Young, *Electronic Communication Techniques* (Prentice Hill, Englewood, 2004). Number 0-13-048285-4

Solutions to Selected Problems

Solutions to Selected Problems in Chap. 1

1.1

$$f_1 = 10\,\text{MHz},\ T_1 = 100\,\text{ns},\ \lambda \approx 30\,\text{m},\ \beta = 0.21\,\text{m}^{-1},\ v_p \approx 3\times 10^8\,\text{m/s}$$

$$f_2 = 100\,\text{MHz},\ T_1 = 10\,\text{ns},\ \lambda \approx 3\,\text{m},\ \beta = 2.1\,\text{m}^{-1},\ v_p \approx 3\times 10^8\,\text{m/s}$$

$$f_3 = 10\,\text{GHz},\ T_1 = 0.1\,\text{ns},\ \lambda \approx 30\,\text{mm},\ \beta = 210\,\text{m}^{-1},\ v_p \approx 3\times 10^8\,\text{m/s}$$

1.2

$$\begin{aligned}
\text{sine}: &\quad \text{AVG:}\quad 0, \quad \text{RMS:}\quad 1/\sqrt{2}\\
\text{square}: &\quad \text{AVG:}\quad 0, \quad \text{RMS:}\quad 1\\
\text{triangle}: &\quad \text{AVG:}\quad 0, \quad \text{RMS:}\quad 1/\sqrt{3}\\
\text{sawtooth}: &\quad \text{AVG:}\quad 0, \quad \text{RMS:}\quad 1/\sqrt{3}
\end{aligned}$$

1.3 $f = 1\,\text{kHz}$, at $t = 125\,\mu\text{s}$ the instantaneous phase is $\phi(125\,\mu\text{s}) = \pi/2$.

1.4 $i(t_0) = v(t_0)/R = 1\,\text{V}/1\,\text{k}\Omega = 1\,\text{mA}$, $\Delta\phi = \pi/2 \stackrel{\text{def}}{=} T/4$, therefore

$$f_1 = 10\,\text{kHz} : \Delta t = T/4 = 25\,\mu\text{s},\ \Delta x \approx 7.5\,\text{km}$$

$$f_2 = 10\,\text{MHz} : \Delta t = T/4 = 25\,\text{ns},\ \Delta x \approx 7.5\,\text{m}$$

$$f_3 = 10\,\text{GHz} : \Delta t = T/4 = 25\,\text{ps},\ \Delta x \approx 7.5\,\text{mm}$$

1.5 First three terms in Fourier polynom of sawtooth waveform.

1.8 $E = P \times t$, therefore $P_1 = 1.728 \times 10^6\,\text{J}$, and $P_1 = 60 \times 10^3\,\text{J}$

1.9 $p(t) = v(t)\,i(t) = 1\,100\,\sin(2\omega t)$, and $\Delta\phi = 0$ (purely reactive circuit), therefore $\langle P\rangle = 0$.

1.10 $G = 5 \times 8 = 40$, and $G_{\text{dB}} = 6.99 + 9.03 = 16\,\text{dB}$.

R. Sobot, *Wireless Communication Electronics*,
https://doi.org/10.1007/978-3-030-48630-3

Solutions to Selected Problems in Chap. 2

2.1 As the consequence of the first derivative, each fast edge creates a pulse, thus "edge detector" (Fig. G.1).

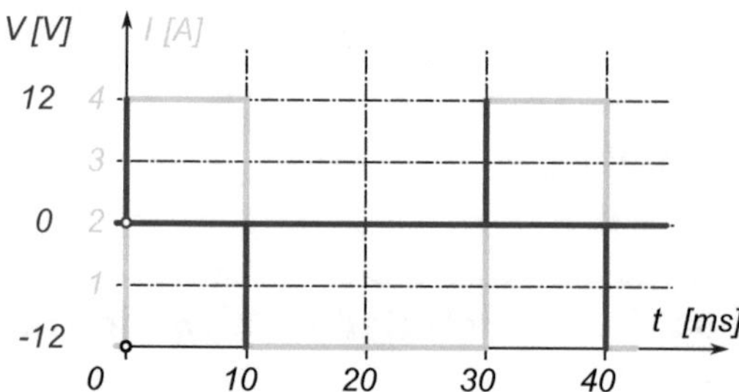

Fig. G.1 Illustration for Problem 2.1

2.2 Charging/discharging time constant is $\tau = RC = 1\,\text{ms}$. It takes approximately $5\,\tau$ to reach 99% level (and theoretically, never 100%) (Fig. G.2).

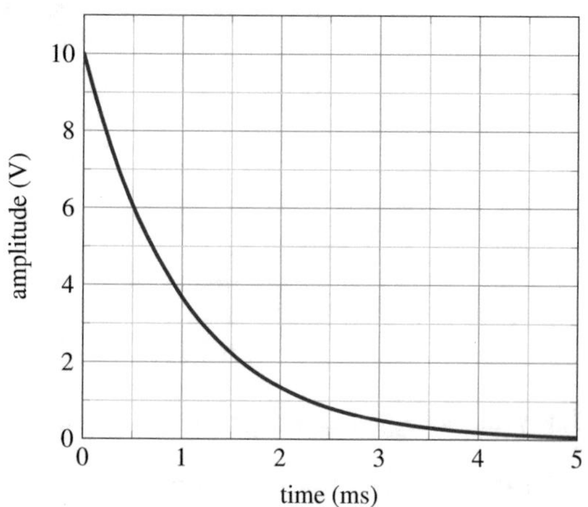

Fig. G.2 Illustration for Problem 2.2

2.3

(a) $R_{AB} = R + 10R = 11R \approx 10R$, $R_{AB} \neq f(\omega)$
(b) $R_{AB} = R - j/\omega C$, $R_{AB}(DC) = \infty$, $R_{AB}(\infty) = R$
(c) $R_{AB} = R||Z_C = R/(1 + j\omega RC)$, $R_{AB}(DC) = R$, $R_{AB}(\infty) = 0$
(d) $R_{AB} = R||10R \approx R$, $R_{AB} \neq f(\omega)$

2.4

(a) $A_V = Z_2/(Z_1 + Z_2)$, $(\neq f(\omega))$ (1) $A_V \approx 0$, (2) $A_V \approx 1$
(b) $A_V = R||Z_C/(R + R||Z_C)$, $A_V(DC) = 1/2$, $A_V(\infty) = 0$

2.5 (a) 267.75 mm, (b) 166.71 mm, (c) 45.54 mm

2.6 (a) 6.954 pF, (b) 5.216 pF, (c) 3.477 pF

2.7 $C \approx 10.6\,\text{pF}$, $L \approx 320\,\text{pH}$. Therefore,

(a) $Z_R = (500 + j\,0)\,\Omega$
(b) $Z_R = (500 - j\,1/(2\pi \times 10.6\,\text{pF} \times f))\,\Omega$
(c) $Z_R = (500 + j\,2\pi \times 320\,\text{pH} \times f)\,\Omega$

2.8 $C = 100\,\text{pF}$, $L \approx 1\,\mu\text{H}$. Therefore,

(a) $Z_C = (-j\,1/(2\pi \times 100\,\text{pF} \times f))\,\Omega$
(b) $Z_C = (+j\,2\pi \times 1\,\mu\text{H} \times f)\,\Omega$

2.9 $L \approx 3.3\,\mu\text{H}$, $C = 0.1\,\text{nF}$. Therefore,

(a) $Z_L = (+j\,2\pi \times 3.3\,\mu\text{H} \times f)\,\Omega$
(b) $Z_L = (-j\,1/(2\pi \times 0.1\,\text{nF} \times f))\,\Omega$

2.10 It takes at least $5\,\tau$ to charge/discharge a capacitor, that is to say that absolute minimum timing must be $5\,\tau = T/2$, therefore

(a) $\tau = RC = 1\,\text{ns}$, $f_{\text{min}} = 100\,\text{MHz}$,
(b) $\tau = RC = 1\,\mu\text{s}$, $f_{\text{min}} = 100\,\text{kHz}$,
(c) $\tau = RC = 1\,\text{ms}$, $f_{\text{min}} = 100\,\text{Hz}$.

Solutions to Selected Problems in Chap. 3

3.1 $A_V = 1 \times 10 \times 0.9756 \times 1 \times 0.99 \times 2 \times 0.9524 = 18.397\,\text{V/v} = 25.3\,\text{dB}$

3.2 It takes two design conditions to uniquely calculate both resistors, not only the I/O voltage ratio.

$$\left.\begin{aligned}\frac{R_1}{R_2} &= 1.5\\ \frac{R_1 R_2}{R_1 + R_2} &= 6\,\text{k}\Omega\end{aligned}\right\}\quad \text{therefore,}\qquad R_1 = 15\,\text{k}\Omega,\; R_2 = 10\,\text{k}\Omega$$

3.3 It is the ratio $k = R_1/R_2$ that determines the output voltage. That is to say, given one resistor the other is determined by the k factor. Therefore,

(a) $V_{out} = 1\,\text{V} \Rightarrow R_1 = 9\,R_2$, current source (large output resistance relative to the load, i.e. $R_1 \gg R_2$),
(b) $V_{out} = 5\,\text{V} \Rightarrow R_2 = R_1$, no preference,
(c) $V_{out} = 9\,\text{V} \Rightarrow R_2 = 9\,R_1$, voltage source (small output resistance relative to the load, i.e. $R_1 \ll R_2$).

3.4 Note that the impedance ratio k is now a complex number, therefore we calculate the ratio as

$$\left|\frac{\frac{1}{j\omega C}}{R + \frac{1}{j\omega C}}\right| = \left|\frac{1}{1 + j\omega RC}\right| = \frac{1}{\sqrt{1 + (\omega RC)^2}} = k \quad \text{therefore,} \quad C = \frac{\sqrt{(1/k)^2 - 1}}{2\pi\, f\, R}$$

Given $R = 1\,\text{k}\Omega$ and $f_1 = 10\,\text{kHz}$:

(a) $V_{out} = 1\,\text{V} \Rightarrow R_1 = 9\,Z_C$, $C = 142.352\,\text{nF}$
(b) $V_{out} = 5\,\text{V} \Rightarrow R_1 = Z_C$, $C = 27.566\,\text{nF}$
(c) $V_{out} = 9\,\text{V} \Rightarrow Z_C = 9\,R_1$, $C = 7.708\,\text{nF}$.

Given $R = 1\,\text{k}\Omega$ and $f_1 = 1\,\text{MHz}$:

(a) $V_{out} = 1\,\text{V} \Rightarrow R_1 = 9\,Z_C\,,\, C = 1.423\,\text{nF}$
(b) $V_{out} = 5\,\text{V} \Rightarrow R_1 = Z_C\,,\, C = 275.664\,\text{pF}$
(c) $V_{out} = 9\,\text{V} \Rightarrow Z_C = 9\,R_1\,,\, C = 77.042\,\text{pF}$.

Given $R = 1\,\text{k}\Omega$ and $f_1 = 1\,\text{GHz}$:

(a) $V_{out} = 1\,\text{V} \Rightarrow R_1 = 9\,Z_C\,,\, C = 1.423\,\text{pF}$
(b) $V_{out} = 5\,\text{V} \Rightarrow R_1 = Z_C\,,\, C = 275.664\,\text{fF}$
(c) $V_{out} = 9\,\text{V} \Rightarrow Z_C = 9\,R_1\,,\, C = 77.04\,\text{fF}$.

In each case, -3 dB frequency is calculated as $\omega = RC$.

Solutions to Selected Problems in Chap. 4

4.1 Very quickly, the exponential term becomes much larger than one. To illustrate the point

(a) $V_D = 200\,\text{mV} \Rightarrow I_D = 160.9\,\mu\text{A}$, and $I_D = 161.0\,\mu\text{A}$; the error is close to zero percent (about 1 nA relative to 161.0 μA).
(b) $V_D = 70\,\text{mV} \Rightarrow I_D = 1.000\,\mu\text{A}$, and $I_D = 1.073\,\mu\text{A}$; the error is close to 7% (about 70 nA relative to 1 μA).

When the diode voltage V_D is a few times greater than kT/q, the approximation error is negligible.

4.2 (a) OFF; (b) OFF; (c) ON; (d) ON; (e) OFF; In order for a diode to conduct, it is necessary to have the anode V_A voltage superior to the cathode V_C voltage.

4.4 Equation for collector current in function of V_D is rewritten as

$$V_D = \frac{kT}{q}\ln\left(\frac{I_D}{I_S}\right) \Rightarrow V_D = 247.4 \pm 2.7\,\text{mV}, \quad \text{or} \quad 247.4\,\text{mV} \pm 1\% \tag{G.10}$$

Note that 10% in current change translates only to 1% in voltage change, thus a diode is often used as a voltage reference.

4.7 For a transistor to hold $I_C =$ const. there are two conditions to be met: (1) base–emitter diode is ON, and (2) base-collector diode is OFF. Therefore,

(a) The BE diode is ideal so any positive voltage V_{BE} will turn it on. There is assumed 1 mA current through R_1, R_2, therefore any $R_2 > 0$ provides positive forward bias voltage for the base–emitter diode. At the same time it must be satisfied that base-collector diode is reverse biased, that is $V_B - V_C \leq 0$, in other words $V_C \geq V_B$.
(b) The base–emitter diode is $V_{BE} > 0.7\,\text{V}$, therefore voltage across R_2 must be greater than 0.7 V, which is achieved if $R_2 > 0.7\,\text{V}/1\,\text{mA} = 700\,\Omega$. At the same time, base-collector diode will be OFF as long as being reverse biased, that is to say $V_B - V_C \leq 0.7\,\text{V}$ i.e. $V_C \geq V_B - 0.7\,\text{V} = 0$. More accurate estimate is to account for the minimum $V_{CE}(\text{min}) \approx 0.1$–$0.2\,\text{V}$, so that $V_C \geq V_E + V_{CE}(\text{min}) \approx 0.2\,\text{V}$.

4.8 For a transistor to hold $I_C =$ const. there are two conditions to be met: (1) base–emitter diode is ON, and (2) base-collector diode is OFF. Therefore,

(a) Starting with the emitter voltage,

$$V_E = R_E\, I_E = 1\,\text{k}\Omega \times 1\,\text{mA} = 1\,\text{V}$$

The BE diode is ideal so any positive voltage V_{BE} will turn it on, that is to say $V_B > V_E = 1\,\text{V}$, therefore, $R_1/R_2 = 9$. At the same time it must be satisfied that base-collector diode is reverse biased, that is $V_B - V_C \leq 0$, in other words $V_C \geq V_B$.

(b) The base–emitter diode is $V_{BE} > 0.7\,\text{V}$, therefore the required $V_B > V_E + 0.7\,\text{V} = 1.7\,\text{V}$, which is generated by voltage divider ratio $R_1/R_2 = 4.882$. At the same time, base-collector diode is reverse biased if $V_B - V_C \leq 0.7\,\text{V}$, i.e. $V_C \geq V_B - 0.7\,\text{V} = 1\,\text{V}$. More accurate estimate is to account for the minimum $V_{CE}(\text{min}) \approx 0.1$–$0.2\,\text{V}$, so that $V_C \geq V_E + V_{CE}(\text{min}) \approx 1.2\,\text{V}$.

4.9 In the first approximation, without knowing anything else, by inspection we write: (a) $R_{\text{out}} = R_1||R_2$, (b) $R_{\text{in}} = (\beta + 1)R_E$, (c) $R_{\text{out}} = R_E||[(R_1||R_2)/(\beta + 1)]$

4.10 Thermal voltages are: $V_T = kT/q = 18.8\,\text{mV}, 25.7\,\text{mV}, 34.3\,\text{mV}$. Consequently, the diode current at room temperature is $I_D = 174\,\mu\text{A}$.

4.12 This problem must be solved iteratively. By reusing voltage V_T calculated at room temperature in Problem 4.10, i.e. $V_T = kT/q = 25.7\,\text{mV}$, we create the iterative equation as:

$$V_D = V_{CC} - V_R = V_{CC} - I_D\, R$$

$$\therefore$$

$$V_D = V_T\, \ln\left(\frac{V_{CC} - V_D}{I_S\, R} + 1\right) \quad \therefore \quad V_D = f(V_D) \tag{G.11}$$

Given the initial guess (as $n = 0$ iteration) that argument in (G.11) is $V_D = V_{CC} = 5\,\text{V}$ (which is not all-important), then we work out the following $V_D(n)$ iterations:

1. $V_D(0) = 5\,\text{V}$, then from (G.11): $V_D(25\,°\text{C}) = 0\,\text{V}$;
2. $V_D(1) = 0\,\text{V}$, then from (G.11): $V_D(25\,°\text{C}) = 751\,\text{mV}$;
3. $V_D(2) = 751\,\text{mV}$, then from (G.11): $V_D(25\,°\text{C}) = 747\,\text{mV}$;
4. $V_D(3) = 747\,\text{mV}$, then from (G.11): $V_D(25\,°\text{C}) = 747\,\text{mV}$;

therefore, the diode voltage converges to $V_D(25\,°\text{C}) = 747\,\text{mV}$.

Solutions to Selected Problems in Chap. 5

5.2 Given data, where $V_{BE2} = V_{BE1} + 18\,\text{mV}$, the ratio of two collector currents is

$$\frac{I_{C2}}{I_{C1}} = \exp\left(\frac{18\,\text{mV}}{25\,\text{mV}}\right) = 2 = 6\,\text{dB}$$

5.3 Given data,

$$100\,\text{k}\Omega = (\beta + 1)R_E \Rightarrow R_E = 1\,\text{k}\Omega$$

$$100\,\Omega = \frac{R_B}{\beta + 1} \Rightarrow R_B = 10\,\text{k}\Omega$$

5.4 Given data, back of the envelope calculations gives

$$I_C = \frac{V_{CC} - V_{\text{out}}}{R_C} \Rightarrow I_C = 2.5\,\text{mA}, 5\,\text{mA}, 9.8\,\text{mA}$$

$$\therefore$$

$$r_e = \frac{V_T}{I_C} \Rightarrow r_e = 10\,\Omega, 5\,\Omega, 2.5\,\Omega$$

$$\therefore$$

$$|A_V| = \frac{R_C}{r_e} \Rightarrow A_V = 100, 200, 400\,\text{V/v}$$

This example illustrates large g_m variations in a simple CE amplifier, therefore its voltage gain variations, which is function of the instantaneous signal amplitude at the output node. For this reason the external $R_E \gg r_e$ is used in series so that the overall gain is function of $R_E + r_e$ instead of r_e alone.

5.5 Impedance $Z_C = \infty$, therefore back of the envelope estimate is $|A_V| = R_C/R_E = 10\text{V/v} = 20\,\text{dB}$.

5.6 Given frequency and three capacitance values we calculate their respective impedances as $Z_C = 9.95\,\text{k}\Omega, 100\,\text{m}\Omega, 0\,\Omega$. After calculating that collector current as $I_C = 10\,\text{mA}$ we can estimate the total resistance at the emitter node as $R'_E = r_e + R_E||Z_C$. Finally, we calculate back of the envelope voltage gain as the ratio of total collector and emitter resistances. That is, $|A_V| = R_C/R'_E = 20.8\,\text{dB}, 71.5\,\text{dB}, 71.8\,\text{dB}$. This example illustrates how realistic C_E values are used to achieve voltage gain that is close to the theoretical back of the envelope value (i.e. when $C_E \to \infty$), while at the same time taking advantage of $R_E \gg r_e$ resistor.

5.7 Given data, the input side network creates a HPF, where $\tau = R_1||R_2\,C = 1\,\text{k}\Omega \times 1\,\mu\text{F} = 1\,\text{ms}$. Therefore, signals above $f > 160\,\text{Hz}$ are amplified.

5.8 Given data, and voltage gains calculated in Problem 5.6 as $A_V[\text{V/V}] = 113{,}760$, and 3906, then by definition $C_M = (A_V + 1)C_{CB} = 12\,\text{pF}, 3.76\,\text{nF}$, and $3.90\,\text{nF}$.

5.9 Given data, by inspections and definitions we calculate:

$$V_B = \frac{V_{CC}}{3} = 3\,\text{V} \Rightarrow V_E = V_B - V_{BE} = 2.3\,\text{V} \Rightarrow R_E = \frac{V_E}{I_E} = 1.15\,\text{k}\Omega$$

$$V_B = \frac{V_{CC}}{3} \Rightarrow \left.\begin{array}{r} \dfrac{R_1}{R_2} = 2 \\ R_1 + R_2 = 45\,\text{k}\Omega \end{array}\right\} \text{ therefore,} \quad R_1 = 30\,\text{k}\Omega,\ R_2 = 15\,\text{k}\Omega$$

also,

$$I_C = \frac{\beta}{\beta + 1} I_E = 1.98\,\text{mA} \Rightarrow r_e = \frac{V_T}{I_C} = \frac{25\,\text{mV}}{1.98\,\text{mA}} = 12.6\,\Omega \Rightarrow g_m = \frac{1}{r_e} = 79.2\,\text{mS}$$

therefore, due to β, $C_E \to \infty$ (i.e. $R_{\text{in}}/(\beta + 1) \to 0$ and R_E is shorted), we have $|A_V| = g_m\,R_C$, which leads into

$$R_C = |A_V|\,r_e = 8 \times 12.6\,\Omega = 101\,\Omega$$

Finally, the input side resistance consists of $R_{\text{in}} = R_1||R_2||R_{\text{sig}} = 5\,\text{k}\Omega$.

5.10 Given data and $\beta \to \infty$, inductor L forms and ideal LC band-pass filter with C_M. Therefore, there is only one frequency that is amplified,

$$f_0 = \frac{1}{2\pi\sqrt{LC_M}} = \frac{1}{2\pi\sqrt{L(A_V+1)C_{CB}}} = \frac{1}{2\pi\sqrt{2.533\,\mu\text{H} \times (99+1) \times 1\,\text{pF}}} = 10\,\text{MHz}$$

If $\beta \ll \infty$, the consequence would be that $(\beta + 1)R_E$ resistance is projected in parallel with the LC resonator, thus the associated bandwidth is not zero, i.e. there would be a range of frequencies centred around f_0 that would also be amplified to a certain degree. (In Chap. 10 we study LC resonators and their bandwidth in more details.)

5.12 Given data, by inspection we write:

$$V_C = V_{CC} - R_C I_C = 1.25\,\text{V} \quad \text{and} \quad g_m = \frac{I_C}{V_T} = 20\,\text{mS}$$

also, $v_{BE} = -v_i$, thus

$$v_C = R_C\, i_C = -R_C g_m v_{BE} = R_C g_m v_i \quad \therefore \quad A_V = \frac{v_C}{v_i} = g_m R_C = 150\,\text{V/v} = 43.5\,\text{dB}$$

Solutions to Selected Problems in Chap. 6

6.1 Given data, we write:

(a) Ideal voltage amplifier, see Fig. G.3 (top), must have gain as

$$v_{out} = A_R\, i_{in} = A_R\, G_m v_{in} \quad \therefore \quad A_V \overset{\text{def}}{=} \frac{v_{out}}{v_S} = G_m\, A_R = 100 \tag{G.12}$$

where $i_{in} = i_{out}$ and $v_S = v_{in}$. One possible choice would be, for example, if $G_m = 100\,\text{mS}$, then $A_R = 1\,\text{k}\Omega$.

(b) In addition to A_V gain, real voltage amplifier has one current and two voltage dividers: one voltage divider at the input interface, one current divider in the middle, and one voltage divider at the output interface, see Fig. G.3 (bottom). Thus, from the two voltage dividers we write:

$$v_{in} = \frac{v_S}{R_S + R_{i1}} R_{i1} \quad \text{and} \quad v_{out} = \frac{A_R\, i_{in}}{R_L + R_{o2}} R_L \tag{G.13}$$

The current divider equation is found from the expression for voltage across $R_{o1}||R_{i2}$ and R_{i2} as

$$G_m\, v_{in} \times R_{o1}||R_{i2} = i_{in} \times R_{i2} \tag{G.14}$$

By elimination of i_{in} and v_{in} found in (G.13), and substitution in (G.14), we write

$$A_V \overset{\text{def}}{=} \frac{v_{out}}{v_S} = \underbrace{\frac{R_{i1}}{R_S + R_{i1}}}_{<1} \times \underbrace{\frac{R_{o1}\, \cancel{R_{i2}}}{R_{o1} + R_{i2}}}_{<1} \times \frac{1}{\cancel{R_{i2}}} \times G_m A_R \times \underbrace{\frac{R_L}{R_L + R_{o2}}}_{<1} \tag{G.15}$$

Equation (G.15) helps us conclude the following: the maximal theoretical gain is $A_V = G_m A_R$, which can be achieved only if all three dividers are made equal to one. That is to say if $R_S = 0$

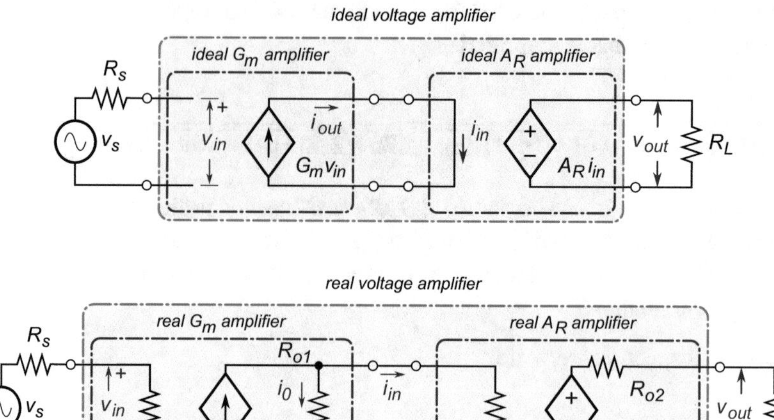

Fig. G.3 Ideal voltage amplifier for Problem 6.1

or $R_{i1} \to \infty$, and if $R_{i2} = 0$ or $R_{i1} \to \infty$, and if $R_{02} = 0$ or $R_L \to \infty$. In other words, if we use the ideal model in Fig. G.3 (left) that serves as our limiting case that we try to approach as close as possible.

6.2 Given data and design flow that is summarized in Fig. 6.31, from characteristics in Fig. 6.13 we could choose, for example, biasing current $I_{C0} = 4\,\text{mA}$. With that choice, the following scenario develops.

1. From I_C vs. V_{BE} characteristics we read $V_{BE0} = 690\,\text{mV}$
2. From I_C vs. V_{BE} characteristics, at (4 mA, 690 mV) biasing point by using the graphical method we estimate $g_m \approx 145\,\text{mS}$. Alternatively, we calculate g_m as,
   - At room temperature ($T = 300\,\text{K}$):

$$V_T = \frac{kT}{q} = 25.8\,\text{mV} \quad \therefore \quad g_m = \frac{I_{C0}}{V_T} = 155\,\text{mS} \tag{G.16}$$

   - and, by definition we have

$$r_e = \frac{1}{g_m} = 6.45\,\Omega \tag{G.17}$$

3. Therefore, the total resistance found at collector node must be

$$A_V = g_m R \quad \therefore \quad R = \frac{A_v}{g_m} = 690\,\Omega \tag{G.18}$$

4. With this collector resistance, from I_C vs. V_{CE} characteristics we take $V_{CC} = 15\,\text{V}$, which forces DC voltage at the collector to be

$$V_C = V_{CC} - R I_{C0} = 12.24\,\text{V} \tag{G.19}$$

This voltage is rather close to V_{CC} because the upper headroom is $15\,\text{V} - 12.24\,\text{V} = 2.76\,\text{V}$. However, because voltage gain equals $A_V = 40\,\text{dB} = 100$, this headroom sets the limit to the input signal amplitude to be smaller than $(2.76/100)\,\text{V} = 27.6\,\text{mV}$. If this is the case, then this amplifier would be acceptable.

5. In an attempt to set $V_C = V_{CC}/2$, we could try to first calculate the required R so that

$$V_C = \frac{V_{CC}}{2} = 7.5\,\text{V} \quad \therefore \quad R = \frac{V_C}{I_{C0}} = \frac{7.5\,\text{V}}{4\,\text{mA}} = 1875\,\Omega \tag{G.20}$$

As a consequence, if we are to keep the voltage gain $A_V = 100$, then

$$A_V = g_m R \quad \therefore \quad g_m = \frac{A_V}{R} = 53\,\text{mS} \quad \therefore \quad I_{C0} = g_m V_T = 1.376\,\text{mA} \tag{G.21}$$

which is contradictory to $I_{C0} = 4\,\text{mA}$ that we started from. We note that, accepting $I_{C0} = 1.376\,\text{mA}$ again sets the collector voltage at

$$V_C = V_{CC} - R I_{C0} = 15\,\text{V} - 1875\,\Omega \times 1.376\,\text{mA} = 12.42\,\text{V} \tag{G.22}$$

The conclusion is that, aside from pure luck, the initial set of specifications, here, the choice of transistor and the exact voltage gain, may not be possible.

Given that the exact voltage gain specification is very difficult to achieve under all variations, one possible strategy could be to relax the gain specification from $A_V = 40\,\text{dB}$ to $A_V \geq 40\,\text{dB}$. In this case, we have one more step of freedom, and one possible design scenario could be as follows:

1. Choose $I_{D0} = 4\,\text{mA}$, therefore $g_m = 155\,\text{mS}$,
2. Set V_C voltage with $R = 1875\,\Omega$
3. Calculate voltage gain as $A_V = g_m R = 290 = 49\,\text{dB}$, and accept it.

With this gain margin of 9 dB at the beginning of the design, we give ourselves a chance to make compromises in the later phases of the design as long as $A_V \geq 40\,\text{dB}$. Simulations confirm that our hand analysis and conclusion are indeed very close to the reality.

6.3 Given data and characteristics in Fig. 6.13, we can estimate from I_C vs. $V_C E$ graph that $V_{CE}(\text{min}) \approx 200\,\text{mV}$. In addition, assuming $I_{C0} = 4\,\text{mA}$ and $V_{CC} = 15\,\text{V}$ we write

$$V_{CE}(\text{min}) + R_{\text{max}} I_{C0} = V_{CC} \quad \therefore \quad R_{\text{max}} = \frac{V_{CC} - V_{CE}(\text{min})}{I_{C0}} = 3.7\,\text{k}\Omega \tag{G.23}$$

which forces $V_C = V_{CE}$ voltage to its minimum value, beyond that point the transistor is forced to enter the linear region (i.e. $V_{CE} < V_{CE}(\text{min})$) and reduce the biasing current. When that happens, the transistor does not follow exponential equation (4.31) and the original assumption of constant biasing current I_{C0}.

Solutions to Selected Problems in Chap. 7

7.2 The equivalent impedances are

$$Z_{AB} = R_1 || \left(R_2 + \frac{1}{j\omega C}\right) \quad \text{and} \quad Z_{AB} = R_1 || \left(R_2 + j\omega L\right)$$

then, we need to find their real and imaginary parts.

7.3 By inspection of voltage dividers, we write

$$H(jw) = \frac{R_2||j\omega L}{R_1 + R_2||j\omega L}$$

$$H(j\omega) = \frac{R_2 + R_3||j\omega L}{R_1 + R_2 + R_3||j\omega L}$$

$$H(j\omega) = \frac{R_3}{R_3 + R_2 + R_1||j\omega L}$$

then, we need to find their real and imaginary parts.

7.4 Emitter node is associated with low impedance, thus its pole is dominant (see (7.24)), i.e. $f_L = f_p$, so we find

$$C_E = \frac{1}{\left[\left(\frac{R_S||R_B}{\beta + 1} + r_e\right)||R_E\right] 2\pi\, f_L} = 24.6\,\text{nF}$$

To make it highest dominant pole, the other two poles should be at least one decade lower, similarly we find $C_1 \geq 155\,\text{pF}$ and $C_2 \geq 80\,\text{pF}$.

7.8 By the following definitions we find

$$r_e = \frac{V_T}{I_C} = 25\,\Omega \quad \therefore \quad r_{\text{in}} = r_\pi = (\beta + 1)\, r_e = 2.525\,\text{k}\Omega$$

$$A_0 = g_m\, r_o = \frac{r_o}{r_e} = 2000\,\text{V/v} = 66\,\text{dB}$$

$$R_{o1} = r_o = 50\,\text{k}\Omega \quad \therefore \quad R_{o2} \approx \beta \times r_o = 5\,\text{M}\Omega$$

Solutions to Selected Problems in Chap. 8

8.1 Given data, by definitions we calculate

$$S_n = 4.14 \times 10^{-21}\,\text{W/Hz},\ P_n = 4.14 \times 10^{-15}\,\text{W},\ P_s = 5 \times 10^{-15}\,\text{W},\ SNR = 1.208 = 0.82\,\text{dB}$$

8.2 Given data, by definitions we calculate

$$R = 50\,\Omega \quad \therefore \quad E_n^2 = 1.601 \times 10^{-14}\,\text{V}^2 \quad \therefore \quad E_n = 126.5\,\text{nV}$$

$$R = 5\,\text{k}\Omega \quad \therefore \quad E_n^2 = 3.202 \times 10^{-13}\,\text{V}^2 \quad \therefore \quad E_n = 565.8\,\text{nV}$$

$$R = 5\,\text{M}\Omega \quad \therefore \quad E_n^2 = 3.202 \times 10^{-10}\,\text{V}^2 \quad \therefore \quad E_n = 17.9\,\mu\text{V}$$

8.5 Effective bandwidth is $\Delta f_{\text{eff}} = 1/(4RC) = 12.5\,\text{kHz}$, and the total noise voltage generated is $e_n = \sqrt{kT/C} = 14\,\mu\text{V}$.

8.6 Using Friss formula and definition for noise temperature we find $NF = 10\log(2.27) = 3.56\,\text{dB}$ and $T_n = 380\,\text{K}$.

8.7 By definitions, $F = 4$ and $NF = 6\,\text{dB}$.

8.8 Shot noise current i_{sn} generated in a diode is $i_{\text{sn}}^2 = 2qI_DB$, therefore $i_{\text{sn}} = 18\,\text{nA}$, with $V_T \approx 26\,\text{mV}$ it follows that $r_D \approx 26\,\Omega$, which gives $e_n = i_{\text{sn}}\,r_D = 463\,\text{nV}$.

8.9 Shot noise current is found as $I_{ns} = \sqrt{2qI_{DCn}BW} = 1.79\,\text{nA}$, which generates noise voltage in the source resistance as $e_{ns}(R_S) = I_{ns}\,R_S = 85.9\,\text{nV}$. At the same time, noise voltages generated by the two resistors are $e_n(R_S) = 2.88\,\mu\text{V}$ and $e_n(R_{in}) = 12.87\,\mu\text{V}$. Therefore, the total noise voltage is found as $v_n = \sqrt{e_{ns}(R_S)^2 + e_n(R_S)^2 + e_n(R_{in})^2 +} = 13.19\,\mu\text{V}$. Relative to v_S the total noise voltage causes $SNR = 17.6\,\text{dB}$.

8.10 By definitions, $12\,\text{dB} = 15.85$ and $50\,\text{dB} = 1 \times 10^5$. Therefore, $T\text{rec} = (15.85 - 1) \times 300\,\text{K} = 4455\,\text{K}$ and $T\text{sys} = 90\,\text{K} + 4455\,\text{K}/1 \times 10^5 \approx 90\,\text{K}$.

Solutions to Selected Problems in Chap. 10

10.1 Given data, by definitions we find: $R_s = 628\,\text{m}\Omega$, $R_p = 25\,\text{k}\Omega$, $C = 126\,\text{pF}$, $R = 2.778\,\text{k}\Omega$.

10.9 Given data and frequency response curve in Fig. 10.15, by definitions we find: $BW = f_2 - f_1 = 10\,\text{kHz}$, $Q = f_0/BW = 45.5$, $L = 122.35\,\text{nH}$, $R = \omega L/Q = 76.877\,\text{m}\Omega$.

10.11 Given data and by definitions, we find: $R_p = R_s(1 + Q^2) = 50\,\Omega$, $X_p = X_s(1 + 1/Q^2) = 25.024\,\Omega$, therefore $C_p = 6.36\,\text{pF}$.

Solutions to Selected Problems in Chap. 11

11.3 We need to convert parallel networks into their equivalent versions, so that $R_s(R_p) = 5\,\Omega$, $X_s(X_p) = 15\,\Omega$, which allows us to write the equivalent series network impedance on the load side as $X_p = (5 - j15)\,\Omega$. Therefore, $|Z_s| = |Z_p| = 15.811\,\Omega$.

By definition,

$$\Gamma = \frac{Z_s - Z_p}{Z_s + Z + p} = 0 = \infty\ \text{dB} \quad \therefore \quad ML = \frac{1}{1 - \Gamma^2} = 1 = 0\,\text{dB}$$

Solutions to Selected Problems in Chap. 12

12.2 By inspection, given data, back of the envelope voltage gain is $A_V = R_C/R_E = 100\,\text{V/v} = 40\,\text{dB}$.

12.3 By inspection, given data, back of the envelope voltage gain is $A_V = R_C/R_E = 99\,\text{V/v}$. Therefore, Miller capacitance is $C_M \approx 160\,\text{pF}$. Time constant of C and $R_1||R_2$ defines pole of HPF at $f = 10\,\text{Hz}$. At that frequency the impedance of Miller capacitance is approximately $Z_{C_M}(10\,\text{Hz}) \approx 100\,\text{M}\Omega$, that is to say it has no visible impact.

However, time constant of Miller capacitance and $R_1||R_2$ defines pole of LPF at $f = 100\,\text{kHz}$. At that frequency $Z_C(100\,\text{kHz}) \approx 1\,\Omega$, that is to say, this impedance is negligible if current source is used.

12.4 By definition, assuming the output side resonator is tuned at the same frequency, $f_0 = 10\,\text{MHz}$.

Solutions to Selected Problems in Chap. 13

13.2 Analysis of this circuit gives equations

$$\omega(x) = \frac{1}{RC\sqrt{4x+6}} \quad \text{and}$$

$$\beta(x) = \frac{29}{x} + 4x + 23 \quad \therefore \quad x \approx \pm 2.6926 \quad \therefore \quad \beta \approx 44.5 \quad \therefore \quad R = 3.714\,\text{k}\Omega$$

where $x = R_C/R$. Therefore, $C = 1\,\text{pF}$.

13.3 By inspection $f_0 = 10\,\text{MHz}$.

13.4 For this tapped L network we find $\beta = -1/3$.

13.5 For this network, given data, we find $R_{\text{eff}} = 3.4\,\text{k}\Omega$

13.7 A general analysis shows same result as what we find by inspection $\omega_0^2 = (C_1 + C_2)/L(C_1 C_2)$, therefore $f_0 = 10\,\text{MHz}$. Under condition of oscillation, $g_m = C_1/(R_C \| r_c\, C_2) = 200\,\mu\text{S}$

13.8 At zero bias voltage $C = 17.65\,\text{pF}$, therefore $f_0 = 3.789\,\text{MHz}$. With the biasing $V_D = -7\,\text{V}$ it follows that $C = 5\,\text{pF}$, therefore $f_0 = 7.126\,\text{MHz}$.

Solutions to Selected Problems in Chap. 14

14.2 Given 10 MHz waveform, a mixer can produce 1 kHz waveform after multiplying it with 9.999 MHz and 10.001 MHz.

14.4 Given $Q = 20$, a wave form close to the resonant frequency is attenuated as

$$A_r = \frac{1}{\sqrt{1 + (\delta Q)^2}} = 0.455 = -6.8\,\text{dB}$$

where $\delta = \omega/\omega_0 - \omega_0/\omega$.

Solutions to Selected Problems in Chap. 15

15.1 Literal implementation of AM modulated waveform in (15.5) gives

$$\begin{aligned} c_{AM}(t) &= \sin(2\pi \times 100\,000\,t) + 0.5 \sin(2\pi \times 1\,500\,t) \sin(2\pi \times 100\,000\,t) \\ &= \sin(2\pi \times 100\,000\,t)\,[1 + 0.5 \sin(2\pi \times 1\,500\,t)] \end{aligned}$$

where $m = 0.5\,\text{V}/1\,\text{V} = 0.5 = 50\%$, thus produces waveform in Fig. 15.3 (left). The modulated waveform amplitude is $1\,\text{V} \pm 0.5\,\text{V}$. Carrier frequency is $f_c = 100\,\text{kHz}$ and signal frequency is $f_s = 1.5\,\text{kHz}$. Therefore, the output frequency spectrum shows three frequencies, $(f_c - f_s,\ f_c \cdot f_c + f_s)$.

15.2 AM waveform occupies two times the signal frequency, i.e. $2 \times 5\,\text{kHz} = 10\,\text{kHz}$. Therefore, in the given bandwidth it is possible to create maximum of $100/10 = 10$ channels. However, in reality every communication standard requires some "guard-band" between the channels, thus smaller number of channels is available.

15.3 Receiver's mixer generates 2.415 MHz and 455 kHz, where the downconverted waveform is used as IF, i.e. 455 kHz. If there is another radio transmitter operating in the same region, and its carrier frequency is 1.890 MHz, there would be second 455 kHz waveform generated by receiver's mixer that would destroy the wanted signal.

15.5 Carson's rule gives $B_{FM} = 2(m_f + 1) f_b = 50$ kHz. The sum of Bessel's functions results in

$$\frac{P_T}{P_c} = J_0^2 + 2\left(J_1^2 + J_2^2 + \cdots\right) = 1.000258$$

for the first five functions. That is to say, the total power is constant, just redistributed among the harmonics. When $m_f = 1.5$ the first sideband harmonic has highest amplitude $J_1 = 0.558$ relative to the unmodulated signal.

15.6 By definition,

$$P_T = P_c\left(1 + \frac{m^2}{2}\right) \;\therefore\; P_c = 881.5\,\text{W} \;\therefore\; P_{\text{USB}} = P_T - P_c = 318.5\,\text{W} \;\therefore \tag{G.24}$$

$$P_{\text{USB}} = P_{\text{LSB}} = 159.25\,\text{W}$$

15.8 By definition, for $m = 0.7$ we find

$$P_{\text{USB}} = P_{\text{LSB}} = \frac{m^2 P_c}{4} = 183.75\,\text{W}\;(m = 0.7), \quad \text{and} \tag{G.25}$$

$$P_{\text{USB}} = P_{\text{LSB}} = \frac{m^2 P_c}{4} = 93.75\,\text{W}\;(m = 0.5)$$

15.9 The IF frequency is found as difference between the carrier and LO frequencies, thus

$$f_{LO} = 995\,\text{kHz}\;(f_{LO} > f_c), \quad \text{and} \quad f_{LO} = 85\,\text{kHz}\;(f_{LO} < f_c)$$

15.10 Given $\Delta f = 50$ kHz, the carrier swing is ± 50 kHz $= 100$ kHz. Therefore, the carrier frequency is between $f_c = 107.55$ MHz and $f_c = 107.65$ MHz. By definition, $m_f = \Delta f / f_m = 7.143$.

15.11 In accordance with Bessel functions $m_f = 2.0$, for we write: $P_0 = 100\,\text{W} \times 0.224^2 = 5.0176\,\text{W}$, $P_1 = 100\,\text{W} \times 2 \times 0.577^2 = 66.5858\,\text{W}$, $P_2 = 100\,\text{W} \times 2 \times 0.353^2 = 24.9218\,\text{W}$, $P_3 = 100\,\text{W} \times 2 \times 0.129^2 = 3.3282\,\text{W}$, $P_4 = 100\,\text{W} \times 2 \times 0.034^2 = 0.2312\,\text{W}$, $P_5 = 100\,\text{W} \times 2 \times 0.007^2 = 0.0098\,\text{W}$, $P_6 = 100\,\text{W} \times 2 \times 0.001^2 = 0.0002\,\text{W}$, which sums up to 100 W (when neglecting all remaining harmonics). Using Carson's rule we estimate $B_{\text{FM}} = 6$ kHz.

15.12 This circuit is known as reactance modulator, where $C_{eq} = g_m RC$ and

$$f_{out} = \frac{1}{2\pi\sqrt{L_T(C_T + C_{eq})}} \quad \therefore \quad C_{eq} = 20\,\text{nF} \quad \therefore \quad C = 200\,\text{nF}$$

Solutions to Selected Problems in Chap. 16

16.2 The incoming IF signal contains 660 kHz, 665 kHz, 670 kHz. After the nonlinear diode multiplier, the frequency spectrum contains the sums and differences: 1325 kHz, 1330 kHz, 1335 kHz, 5 kHz, and 10 kHz. (With the help of an LPF, 5 kHz signal is separated from the other frequencies.)

Due to the orientation of the diode, the negative side envelope is recovered. At 5 kHz, the circuit components present impedances: $Z_{C_1} = 145\,\text{k}\Omega$, $Z_{C_2} = 6.3\,\text{k}\Omega$, $R_D = 10\,\Omega$. Therefore, the equivalent voltage divider consists of R_D and $R = Z_{C_1}||(R_1 + Z_{C_2}||R_2||R_L) = 3\,\text{k}\Omega$, which is to say that with $R_D = 10\,\Omega$ voltage divider the 5 kHz signal is almost not attenuated.

At 665 kHz, however, the equivalent time constant of this circuit is dominated by $C_2 = 5\,\text{nF}$ capacitor. Consequently, the 665 kHz frequency is suppressed while the 5 kHz signal is not significantly affected. Compare the simulation results if, for example, $C_2 = 22\,\text{pF}$.

16.3 By definition, the input side power is $P_{in} = v^2/R = 1.28\,\text{pW} = -88.9\,\text{dBm}$. The output power is found by adding gains of each stage to the power of the input signal, i.e. $P_{out} = 30\,\text{dBm} = 1\,\text{W}$.

16.4 From the corresponding trapezoidal pattern, by inspection we find

$$V_{\max} = V_C + \frac{V_b}{2} = 3\,\text{V}, \;\therefore\; V_{\min} = 1\,\text{V} \quad \therefore \quad m = 0.5$$

Therefore, ratio of the side lengths in the corresponding trapezoidal pattern is found as $3\,\text{V}/1\,\text{V} = 3$.

16.5 It can be deduced that

$$m = \sqrt{2\left[\left(\frac{1.1}{1}\right)^2 - 1\right]} = 0.648$$

Solutions to Selected Problems in Chap. 17

17.1 By definition, Q factor of an LC resonator is $Q = f_0/BW = 105$. At the higher end of frequency range we find $BW(f_{\max}) = 1.6\,\text{MHz}/105 = 15.328\,\text{kHz}$, therefore $C = 9.895\,\text{nF}$. At the lower end of frequency range we calculate $BW(f_{\min}) = 500\,\text{kHz}/105 = 4.762\,\text{kHz}$, which requires $C = 101.321\,AE$. Assuming constant Q factor, bandwidth is not constant. Consequently, this receiver should be used only to process signals whose bandwidth is $B \leq 4.762\,\text{kHz}$, however if the channel spacing Δf between various AM radio stations is constant, then it must be set to $\Delta f \geq BW(f_{\max}) = 15.238\,\text{kHz}$ to avoid the inter-channel interference.

17.2 Ratio of maximum and minimum RF frequency leads into

$$\frac{f_{RF}(\max)}{f_{RF}(\min)} = 3.2 = \sqrt{\frac{C_{\max}}{C_{\min}}} \quad \therefore \quad \frac{C_{\max}}{C_{\min}} = 10.24$$

At the same time, there are two possibilities at the mixer's output:

1. Case $f_{LO} > f_{RF}$: in this case we find $f_{LO}(\min) = 965\,\text{kHz}$ and $f_{LO}(\max) = 2065\,\text{kHz}$. Again,

$$\frac{f_{LO}(\max)}{f_{LO}(\min)} = 2.14 = \sqrt{\frac{C_{\max}}{C_{\min}}} \quad \therefore \quad \frac{C_{\max}}{C_{\min}} = 4.6$$

2. Case $f_{LO} < f_{RF}$: in this case we find $f_{LO}(\min) = 35\,\text{kHz}$ and $f_{LO}(\max) = 1135\,\text{kHz}$. Then,

$$\frac{f_{LO}(\max)}{f_{LO}(\min)} = 32.43 = \sqrt{\frac{C_{\max}}{C_{\min}}} \quad \therefore \quad \frac{C_{\max}}{C_{\min}} = 1052$$

These results provide guidelines for choosing the receiver architecture and tuning range of capacitors. Obviously, it is not reasonable to have a practical tuneable capacitor whose tuning ratio is greater than one thousand.

17.3 Frequency of the image signal is at 12 MHz.

17.4 By inspection we find −1 dB compression point at $P_{in} = -20\,\text{dBm}$ and the third-order intercept point (IIP3) is at $P_{in} = -10\,\text{dBm}$.

17.6 Summary of this superheterodyne architecture is shown in Fig. G.4.

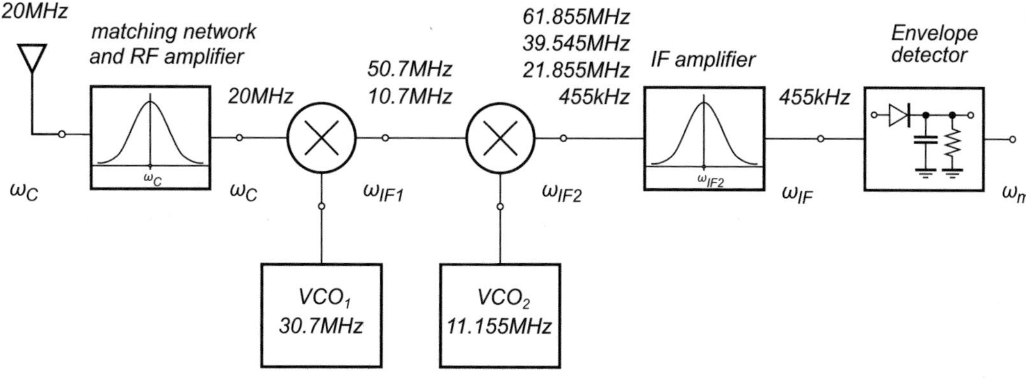

Fig. G.4 Superheterodyne AM receiver block diagram for Problem 17.6

Index

R. Sobot, *Wireless Communication Electronics*,
https://doi.org/10.1007/978-3-030-48630-3

D

S

T

Printed in the United States
by Baker & Taylor Publisher Services